QUANTITATIVE SEISMOLOGY

A Series of Books in Geology

EDITOR: Allan Cox

QUANTITATIVE SEISMOLOGY
Theory and Methods

VOLUME I

Keiiti Aki
MASSACHUSETTS INSTITUTE OF TECHNOLOGY

Paul G. Richards
COLUMBIA UNIVERSITY

W. H. FREEMAN AND COMPANY
SAN FRANCISCO

Sponsoring Editor: John H. Staples
Manuscript Editor: Dick Johnson
Designer: Marie Carluccio
Production Coordinator: Linda Jupiter
Illustration Coordinator: Cheryl Nufer
Artist: Catherine Brandel
Compositor: Syntax International
Printer: The Maple-Vail Book Manufacturing Group

Library of Congress Cataloging in Publication Data

Aki, Keiiti, 1930–
Quantitative seismology.

(A Series of books in geology)
Includes bibliographies and index.
1. Seismology—Mathematics. I. Richards, Paul G., 1943– joint author. II. Title.
QE539.A64 551.2′2 79-17434
ISBN 0-7167-1058-7 (v. 1)

Printed in the United States of America

1 2 3 4 5 6 7 8 9

Contents

VOLUME I

VOLUME II

Preface

In the past decade, seismology has matured as a quantitative science through an extensive interplay between theoretical and experimental workers. Several specialized journals have recorded this progress in thousands of pages of research papers, yet such a forum does not bring out key concepts systematically. Because many graduate students have expressed their need for a textbook on this subject and because many methods of seismogram analysis now used almost routinely by small groups of seismologists have never been adequately explained to the wider audience of scientists and engineers who work in the peripheral areas of seismology, we have here attempted to give a unified treatment of those methods of seismology that are currently used in interpreting actual data.

We develop the theory of seismic-wave propagation in realistic Earth models. We study specialized theories of fracture and rupture propagation as models of an earthquake, and we supplement these theoretical subjects with practical descriptions of how seismographs work and how data are analyzed and inverted.

Our text is arranged in two volumes. Volume I gives a systematic development of the theory of seismic-wave propagation in classical Earth models, in which material properties vary only with depth. It concludes with a chapter on seismometry. This volume is intended to be used as a textbook in basic courses for advanced students of seismology. Volume II summarizes progress made in the major frontiers of seismology during the past decade. It covers a range of special subjects, including chapters on data analysis and inversion, on successful methods for quantifying wave propagation in media varying laterally (as well as with depth), and on the kinematic and dynamic aspects of motions near a fault plane undergoing rupture. The second volume may be used as a textbook in graduate courses on tectonophysics, earthquake mechanics, inverse problems in geophysics, and geophysical data processing.

Many people have helped us. Armando Cisternas worked on the original plan for the book, suggesting part of the sequence of subjects we eventually adopted. Frank Press's encouragement was a major factor in getting this project started. Chapter 12, on inverse problems, grew out of a course given at M.I.T. by one of the authors and Theodore R. Madden, to whom we are grateful for many helpful discussions. Our students' Ph.D. theses have taught us much of what we know, and have been freely raided. We drew upon the explicit ideas and results of several hundred people, many of them colleagues, and hope their contributions are correctly acknowledged in the text. Here, we express our sincere thanks.

Critical readings of all or part of the manuscript were undertaken by Roger Bilham, Jack Boatwright, David Boore, Roger Borcherdt, Michel Bouchon, Arthur Cheng, Tom Chen, Wang-Ping Chen, Bernard Chouet, George Choy, Vernon Cormier, Allan Cox, Shamita Das, Bill Ellsworth, Mike Fehler, Neil Frazer, Freeman Gilbert, Neal Goins, Anton Hales, David Harkrider, Lane Johnson, Bruce Julian, Gerry LaTorraca, Wook Lee, Dale Morgan, Bill Menke, Gerhard Müller, Albert Ng, Howard Patton, Steve Roecker, Tony Shakal, Euan Smith, Teng-Fong Wong, Mai Yang, and George Zandt. We appreciate their attention, their advice, and their encouragement.

About fifteen different secretaries typed for us over the four years during which we prepared this text. Linda Murphy at Lamont-Doherty carried the major burden, helping us to salvage some self-respect in the way we handled deadlines. We also thank our manuscript editor, Dick Johnson, for his sustained efforts and skill in clarifying the original typescript.

We acknowledge support from the Alfred P. Sloan Foundation and the John Simon Guggenheim Memorial Foundation (P.G.R.). This book could not have been written without the support given to our research projects over the years by several funding agencies: the U.S. Geological Survey and the Department of Energy (K.A.); the Advanced Research Projects Agency, monitored through the Air Force Office of Scientific Research (K.A. and P. G. R.); and the National Science Foundation (K.A. and P.G.R.).

Keiiti Aki
Paul G. Richards

June 1979

QUANTITATIVE SEISMOLOGY

CHAPTER 1

Introduction

Seismology is a science based on data called seismograms, which are records of mechanical vibrations of the Earth. These vibrations may be caused artificially by man-made explosions, or they may be caused naturally by earthquakes and volcanic eruptions. Both natural phenomena have strongly attracted the attention of mankind for centuries, even today arousing feelings of fear and mystery as well as our intellectual curiosity.

The great progress made in seismology in the past hundred years has been stimulated principally by the availability of steadily improving data. The major steps in this progress have been initiated by scientists well grounded in the methods of mathematical physics. Each generation of seismologists has worked toward quantitative results, with barriers to computation being pushed back first by mechanical hand-calculators and most recently by advances in digital microprocessing. In the past two decades, computers have become effective enough to handle a large fraction of the information contained in seismograms, so that the quantitative picture of seismology today involves a massive interplay between high-quality data, detailed models of seismic source mechanisms, and models of the Earth's internal structure.

Today seismology is used in structural engineering to aid in the design of earthquake-resistant buildings and in mineral prospecting and exploration for oil and natural gas. Other uses arise in far-ranging political, economic, and social problems associated with the detection of nuclear explosions (and distinguishing them from natural earthquakes) and in the reduction of hazards by helping to eliminate unsafe sites for large power plants and dams. More recently, there have been developments in seismology that make accurate earthquake prediction a foreseeable goal. There is great pressure on seismologists to pursue this aspect of their subject, as can be seen by quoting some facts: over 630,000 people were killed in an earthquake in 1976 July 28 in the People's Republic of China; the cost of damage done by the Guatemala earthquake of 1976 February 4 was comparable with that country's annual gross national product; even the relatively small San Fernando, California, earthquake of 1971 February 9 caused damage estimated at well over $550,000,000. Figures such as these make accurate earthquake prediction so important that the subject of seismology itself is likely to change and grow considerably in the next decade, just as it grew in the 1960's in response to the need to monitor any nuclear test-ban treaty that might be contemplated between the U.S. and the U.S.S.R. (The first global network of calibrated seismographs as well as several large-aperture arrays were set up initially to improve the capability of seismology to detect and identify underground nuclear tests.) The reading list at the end of this chapter includes some books and papers that cover this wide range of applications of modern seismological techniques.

Seismology is at an extreme of the whole spectrum of Earth sciences. First, it is concerned only with mechanical properties and dynamics of the Earth. Second, it offers a means by which investigation of the Earth's interior can be carried out to the greatest depths, and with resolution and accuracy higher than are attainable in any other branch of geophysics. The high resolution and accuracy are attainable because seismic waves have the shortest wavelength of any wave that can be observed after modulation by passing through structures inside the Earth. Seismic waves undergo the least distortion in waveform and the least attenuation in amplitude, as compared with other geophysical observables, such as heat flow, static displacement, strain, gravity, or electromagnetic phenomena.

A third unique characteristic of seismology is that it contributes to our knowledge of only the *present* state of the Earth's interior. Because of its emphasis on *current* tectonic activity, seismology attracts a rather direct public interest.

The methods of seismology, like other geophysical methods, are applicable to tremendous ranges of scale. These ranges may be classified according to the size of the seismic source (both man-made and natural) and according to the size of the seismograph network. The explosive charges used in seismological investigations range in size from less than a gram to more than a megaton (a factor of 10^{12}). From the smallest detectable microearthquake to such great events as the Chilean earthquake of 1960 May 22, the range of natural earthquakes is even greater, amounting to a factor of about 10^{18} in terms of the equivalent point-source strength (seismic moment). The linear dimensions of seismograph networks range from tens of meters for an engineering foundation survey to 10,000 km for the global array of seismological observatories, or a factor of 10^6.

The interpretation of seismograms has progressed in the usual scientific manner, starting with an initial guess that is later supported or corrected after testing its consequence against new data. We simplify the problem of interpreting seismograms by artificially separating the effect of the source from the effect of the medium. Historically, our knowledge of the seismic source and our knowledge of the Earth medium have advanced in a see-saw fashion. For example, at one stage the source may be better known than the medium, in which case new data are used to improve the knowledge of the medium, assuming that the source is known. In the next stage, new data are combined with the improved knowledge of the medium to revise our knowledge of the source.

As in all other branches of geophysics, the effects of source and medium are strongly coupled in seismology. Double errors, one in the source and another in the medium, can produce a result consistent with observation. A deep understanding of physical principles is required to avoid being lured by an apparent consistency. A fascinating story of such double errors concerns the identification of *P*- and *S*-waves. In the early days of seismology, it was controversial whether the main motion of a local earthquake is due to compressional waves or shear waves. The main motion was called the *S*-phase because it was the secondary arrival, preceded by the smaller *P*-phase, so called because it was the primary, i.e., first, arrival. In 1906, F. Omori, the founder of seismology in Japan, investigated this problem using the seismograms of an earthquake observed at what was then the world's best local station network. Using his own formula relating the time between *S* and *P* arrivals to the distance between seismometer and earthquake epicenter, and using also the relative arrival times at several stations, he located the epicenter at about 500 km south of the coast of Honshu. Then he found that the particle motion of the *S*-phase is mainly in the north-south direction—that is, the *S*-phase is apparently longitudinally polarized. If, at this point, he had insisted that the *S*-phase should be shear waves, having a particle motion perpendicular to the direction of wave propagation, then he could have correctly put the focal depth of the earthquake about 500 km *beneath* Honshu to resolve the inconsistency. Instead, he erroneously concluded that the *S*-phase does not consist of shear waves. This double error was actually in harmony with then dominating ideas about earthquake foci and seismic waves. At that time, the concept of isostasy was already well known to explain gravity observations, and nobody dreamed of earthquake foci deep in what was then thought to be a ductile part of the Earth. The conclusion about the *S*-phase was also in harmony with the so-called Mallet's doctrine, which held that the main motion in the epicentral area is due to longitudinal waves. Robert Mallet, who was also the first person to measure seismic velocity in the field using explosives, arrived at this doctrine from the first scientific field study of earthquake-damaged structures, which he examined in the epicentral area of the Neapolitan earthquake of 1857.

In 1906, the existence of compressional waves and shear waves in solids was well known. Since the discovery of Hooke's law in 1660, major advances in elasticity theory were made by Navier's study in 1821, on the general equation of equilibrium and vibration, as well as by Fresnel's experiments indicating that light consists of transversely polarized waves propagating through ether. Before these experiments, it was generally considered that only longitudinal waves could propagate through an unbounded continuum. Progress in the theory of elastic wave propagation continued with Cauchy (who by 1822 had developed the concept of six independent components of stress, and six of strain) and with Poisson (who used a Newtonian concept of intermolecular forces within a solid, so that the force between any pair of molecules is assumed to be proportional to the distance away from their equilibrium separation). Poisson found the two types of waves we now know as *P* and *S*, and concluded for his restricted model that the *P*-wave speed is $\sqrt{3}$ times the *S*-wave speed. A firmer foundation for the theory was given by Green, who invoked the existence of a strain energy function with 21 independent coefficients for an arbitrary anisotropic body. The number of coefficients reduces to two for an isotropic body.

Love (1892; reprinted 1944) gave, in the introduction to his classic textbook, an excellent historical sketch of the development of elasticity theory. The early history of observational seismology, on the other hand, is well described by Dewey and Byerly (1969).

The explanation of Rayleigh waves (Rayleigh, 1887), which can propagate over the free surface of an elastic body, postdated the first recording of earthquake waves. The first theoretical seismogram was constructed by Lamb (1904) for a point impulsive source buried in a homogeneous half-space. The resultant seismogram at the surface consists of a sequence of three pulses corresponding to *P*-, *S*-, and Rayleigh waves, much too simple as compared with observed records.

When the first earthquake seismogram was written in the early 1880's, seismologists were puzzled why the oscillations lasted so long. We shall find that Rayleigh waves can be dispersed (so that waves having different frequency travel at different speeds), and this is one reason for long-lasting oscillations. But there are also oscillations after the arrival of *P*- and *S*-waves and before the arrival of surface (e.g., Rayleigh) waves. Jeffreys (1931) examined and rejected a host of explanations, concluding that "the only suggestion which survives is that the oscillations are due to reflexions of the original pulse within the surface layers." When the first seismogram from the Moon was obtained in 1969, seismologists were again puzzled by the great length of time for which oscillations continued. Again, the explanation appears to lie in the scattering of waves by heterogeneities.

The application of Lamb's methods to actual earthquakes and explosions in the Earth had to be postponed to about 1960, when high-quality data on long-period seismic waves became available through the efforts of Hugo Benioff, Maurice Ewing, Frank Press, and others. Long-period waves average out the short-wavelength heterogeneity of the Earth, and the Earth then behaves as if it were an equivalent homogeneous body, The process at the earthquake source is also simpler at long periods. For this reason, the extremely simple model of Lamb's problem can be of practical use in the interpretation of long-period seismograms.

The Earth models considered in this book are very simple. In most cases, the medium is homogeneous or heterogeneous only in one direction, such as the layered half-space or sphere, in which material properties change only vertically or radially.

Models in seismology are mathematical frameworks within which observed seismograms are related to the Earth's interior via model parameters. For example, if a homogeneous, unbounded, isotropic elastic body is used as the model of the Earth in interpreting seismograms, the parameters obtainable from such interpretations are, at best, Lamé's constants, λ and μ, and a constant density, ρ. On the other hand, when the model is vertically heterogeneous, we can determine $\lambda(z)$, $\mu(z)$, and $\rho(z)$ as functions of depth. Of course, a three-dimensionally heterogeneous and arbitrarily anisotropic medium is the most desirable model, but the numerical effort to deal with it becomes too great to be practical. Furthermore, it has more parameters than we can expect to elucidate from data presently available. So far, the most productive model has been a vertically heterogeneous half-space or sphere. The heart of this book is devoted to surface waves (Chapter 7), free oscillations (Chapter 8), and body waves (Chapter 9) in such models.

To prepare the reader for these chapters, we start with basic and practically useful theorems applicable to general problems of elastodynamics, such as the reciprocity theorem and a representation theorem (Chapter 2). In Chapter 3 we formulate the representation of localized internal seismic sources as the starting point for developing the theory of seismic motions in the Earth. More specific aspects of seismic source mechanisms are postponed to Chapters 14 and 15.

The most productive source representation for an earthquake has been the displacement discontinuity across an internal surface, called the dislocation model. We also consider a volume source in which transformational strain is prescribed within a volume.

A complete description of seismic motion from a point dislocation source in a homogeneous medium is given in Chapter 4. The analysis is extended to a smoothly varying medium, using curvilinear coordinates fixed by the geometrical ray paths. This chapter, among other things, offers the basis for determining the fault plane solution of an earthquake from body waves.

The properties of plane waves, such as reflections and transmissions at a plane interface, phase shifts, inhomogeneous (evanescent) waves, attenuation, and physical dispersion, are exhaustively studied in Chapter 5. In Chapter 6 we solve Lamb's problem, in which a spherical wave from a point source interacts with a plane surface. Three major types of waves emerge from this interaction: waves that are directly reflected from, or transmitted through, the boundary; waves that travel from source to receiver along the boundary (head waves); and waves of the Rayleigh, or Stoneley type, with amplitude decaying exponentially with distance from the interface. We study these waves using the Cagniard method, as well as Fourier transform methods, to prepare the ground for Chapters 7 through 9, giving practical methods for calculating seismograms in vertically heterogeneous structures.

The ordering of the three chapters on vertically heterogeneous media (surface waves, free oscillations, and body waves) reflects the historical development of wave-theoretical analysis of seismograms, as well as the degree of difficulty of the analysis. The fundamental modes of Love and Rayleigh waves are the first waves whose entire records were understood quantitatively in terms of the parameters of realistic models of the Earth and earthquakes. The analysis of body waves is more difficult, partly because we cannot set up the seismograph station at any desired position along the wave path, but only at its endpoint. A complete analysis of free oscillations is also more difficult than that of surface waves, but in this case the reason is the work involved in manipulating the long records, which contain many hundreds of modes. The methods of calculating seismograms for one-dimensionally heterogeneous Earth models described in these three chapters are now well established.

Our final subject in Volume I is the problem of how seismic data may be acquired. Thus, in Chapter 10, we describe principles of seismometry, together with a survey of seismic signals and noises for a wide range of frequencies, sources, and source-receiver distances to help in designing instrumentation for a given experiment. This concluding chapter is accessible to anyone with some knowledge of classical physics (properties of pendulums and elementary electronic circuit theory).

Volume I is intended to be a self-contained description of the basic elements of modern seismology. Five further chapters constitute Volume II, which is concerned with a variety of special subjects. These build upon the material of Volume I and cover the principal methods now used in data analysis, inverse theory, wave propagation in structures with three-dimensional inhomogeneity, and kinematics and dynamics of seismic sources.

SUGGESTIONS FOR FURTHER READING

EARTHQUAKE ENGINEERING

Lomnitz, C., and E. Rosenblueth (editors). *Seismic Risk and Engineering Decisions*. Amsterdam: Elsevier, 1976.

Newmark, N. M., and E. Rosenblueth. *Fundamentals of Earthquake Engineering*. Englewood Cliffs, New Jersey: Prentice-Hall, 1971.

SEISMIC PROSPECTING

Dix, C. H. *Seismic Prospecting for Oil.* New York: Harper & Row, 1952.

Grant, F. S., and G. F. West. *Interpretation Theory in Applied Geophysics.* New York: McGraw-Hill, 1965.

Telford, W. M., L. P. Geldart, R. E. Sheriff, and D. A. Keys. *Applied Geophysics.* Cambridge Univ. Press, 1976.

SEISMIC DETECTION AND DISCRIMINATION OF NUCLEAR EXPLOSIONS

Bolt, B. A. *Nuclear Explosions and Earthquakes: The Parted Veil.* San Francisco: W. H. Freeman and Company, 1976.

Dahlman, O., and H. Israelson. *Monitoring Underground Nuclear Explosions.* Amsterdam: Elsevier Scientific Publishing Co., 1977.

Rodean, H. C. *Nuclear-Explosion Seismology.* Washington, D.C.: U.S. Atomic Energy Comm., 1971.

Thirlaway, H. I. S. Forensic seismology, *Quarterly Journal of the Royal Astronomical Society,* 14, 297–310, 1973.

EARTHQUAKE PREDICTIONS

Panel on Earthquake Prediction of the Committee of Seismology. *Predicting Earthquakes: A Scientific and Technical Evaluation—With Implications for Society.* Washington, D.C.: National Academy of Sciences, 1976.

Panel on the Public Policy Implications of Earthquake Prediction. *Earthquake Prediction and Public Policy.* Washington, D.C.: National Academy of Sciences, 1975.

Press, F. Earthquake prediction. *Scientific American,* May 1975.

Rikitake, T. *Earthquake Prediction.* Amsterdam: Elsevier Scientific Publishing Co., 1976.

White, G. F., and J. E. Haas. *Assessment of Research on Natural Hazards.* Cambridge: MIT Press, 1975.

CHAPTER 2

Basic Theorems in Dynamic Elasticity

An analytical framework for studying seismic motions in the Earth must incorporate, at the very least, the following three components: a description of seismic sources, equations for the motions that can propagate once motion has somewhere been initiated, and a theory coupling the source description into the particular solution sought for the equations of motion. It will be useful if the theory can be simplified by taking full advantage of our conjectures about seismic motion (though such a theory may mislead the user if the conjectures are invalid). For example, there is the conjecture that two sets of small motions may be superimposed without interfering with each other in a nonlinear fashion. Another conjecture is that the seismic motions set up by some physical source should be uniquely determined by the combined properties of that source and of the medium of wave propagation. These conjectures, and many others that are generally assumed by seismologists to be true, are properties of infinitesimal motion in classical continuum mechancis for an elastic medium with a linear stress-strain relation; such a theory will provide the mathematical framework for almost all of this text.

Seismology is largely an observational science, so that the ability to interpret seismograms is fundamental to progress. For this reason, there is a need to know what information about the motion in one part of a medium is enough to determine uniquely the motion that may be observed in another part. As a practical example, we often need to know how to characterize a seismic source (an explosion or a spontaneous fault motion) and how to allow for boundary conditions at the Earth's free surface in order to determine the resulting motion at an array of receivers. Fortunately, for an elastic medium, this problem has a definite solution, in that prescribed source conditions (in terms of body forces) and boundary conditions can readily be stated in forms that do enforce uniqueness for the resulting motions. After giving a formulation of the problem (i.e., establishing notation; defining displacement, strain, traction, body force, and stress; and stating constraints on the motion), we prove the two fundamental theorems of uniqueness and reciprocity. Reciprocity is used together with a Green function to obtain a representation of motion at a general point in the medium in terms of body forces and information on boundaries. This method of representation is fairly recent in elastodynamics, being due to Knopoff (1956) and de Hoop (1958), but it has many familiar parallels in complex number theory, in potential theory, and in the theory of the scalar wave equation for a homogeneous medium.

BOX **2.1**

Examples of representation theorems

1 If $f(z)$ is an analytic function of the complex variable z, then

$$f(z) = \frac{1}{2\pi i}\oint \frac{f(\zeta)\,d\zeta}{\zeta - z},$$

where the circuit integral is taken counterclockwise on any path C around the point z. (No singularities of f are allowed inside C.) This formula is then a *representation* of the function f, which allows f to be evaluated everywhere inside C merely by knowing the values of f on C itself.

2 If $\phi(x, y, z)$ satisfies the Poisson equation $\nabla^2\phi = -4\pi\rho$, then

$$\phi(\mathbf{x}) = \iiint_V \frac{\rho(\boldsymbol{\xi})\,dV(\boldsymbol{\xi})}{|\mathbf{x} - \boldsymbol{\xi}|},$$

where V is a volume including all of the density distribution ρ that contributes to ϕ. This too is a *representation* of ϕ, but one that does not involve values of ϕ itself.

The elastodynamic representation theorem involves both the above types of representation, and also incorporates time dependence.

It is often useful to have the equations of elastic motion referred to general orthogonal curvilinear coordinate systems, since, in many instances, the (curved) coordinate surfaces are just those on which it is natural to state some boundary condition. We give derivations of the displacement-stress equations and of the strain-displacement equations, using the physical components of displacement, stress, and strain in a general orthogonal system.

This chapter may seem at first sight to consist mainly of formal results—of proofs that must be established once, by one person, to legitimize the specific problem-solving methods expounded in later chapters. However, the reader who wishes to develop the ability to solve problems in applied seismology on his or her own will soon face the question of how a problem is "set up." That is, how does one translate the physical description of a seismic source—and the general problem of calculating the ensuing motions at nearby and/or distant receivers—into a specific mathematical problem? In large part, the ability to set up such problems will stem from mastery of the representation theorem, given in various forms by equations (2.41)–(2.43) and (3.1)–(3.3). We shall frequently refer to these equations in later chapters.

2.1 Formulation

Two different methods are widely used to describe the motions and the mechanics of motion in a continuum. These are the Lagrangian description, which emphasizes the study of a particular particle that is specified by its original position at some reference time, and the Eulerian description, which emphasizes the study of whatever particle happens to occupy a particular spatial location. For most applications in seismology, the linear theory of elasticity is conceptually simpler to develop with the Lagrangian description, and this is the framework we shall almost always adopt. After all, a seismogram is the record of motion of a particular part of the Earth (namely, the particles to which the seismometer was attached during installation), so it is directly a record of Lagrangian motion.

We shall work in this chapter with a Cartesian coordinate system (x_1, x_2, x_3), and all tensors here are Cartesian tensors. We use the term *displacement*, regarded as a function of space and time, and written as $\mathbf{u} = \mathbf{u}(\mathbf{x}, t)$, to denote the vector distance of a particle at time t from the position $\mathbf{x}$ that it occupies at some reference time t_0. Since $\mathbf{x}$ does not change with time, it follows that the *particle velocity* is $\partial\mathbf{u}/\partial t$ and that the *particle acceleration* is $\partial^2\mathbf{u}/\partial t^2$.

To analyze the distortion of a medium, whether it be solid or fluid, elastic or inelastic, we use the *strain tensor*. If a particle initially at position $\mathbf{x}$ is moved to position $\mathbf{x} + \mathbf{u}$, then the relation $\mathbf{u} = \mathbf{u}(\mathbf{x})$ is used to describe the displacement field. To examine the distortion of the part of the medium that was initially in the vicinity of $\mathbf{x}$, we need to know the new position of the particle that was initially at $\mathbf{x} + \delta\mathbf{x}$. This new position is $\mathbf{x} + \delta\mathbf{x} + \mathbf{u}(\mathbf{x} + \delta\mathbf{x})$. Any distortion is liable to change the relative position of the ends of the line-element $\delta\mathbf{x}$. If this change is $\delta\mathbf{u}$, then $\delta\mathbf{x} + \delta\mathbf{u}$ is the new vector line-element, and by writing down the difference between its end points we obtain

$$\delta\mathbf{x} + \delta\mathbf{u} = \mathbf{x} + \delta\mathbf{x} + \mathbf{u}(\mathbf{x} + \delta\mathbf{x}) - (\mathbf{x} + \mathbf{u}).$$

BOX **2.2**

Notation

We shall use boldface symbols (e.g., $\mathbf{u}$, $\boldsymbol{\tau}$) for vector and tensor fields, and subscripts (e.g., u_i, τ_{kl}) to designate vector and tensor components in a Cartesian coordinate system. Useful references for the properties of Cartesian tensors are Jeffreys (1965) and Chapter 3 of Jeffreys and Jeffreys (1972).

For unit vectors (other than $\mathbf{v}$, $\mathbf{l}$, $\mathbf{n}$, $\mathbf{b}$), the circumflex is used (e.g., $\hat{\mathbf{x}}$). Scalar products are written as $\mathbf{a} \cdot \mathbf{b}$, and vector products are written as $\mathbf{a} \times \mathbf{b}$.

Overdots are used to indicate time derivatives (e.g., $\dot{\mathbf{u}} = \partial \mathbf{u}/\partial t$, $\ddot{\mathbf{u}} = \partial^2 \mathbf{u}/\partial t^2$), and a comma between subscripts is used in spatial derivatives (e.g., $u_{i,j} = \partial u_i/\partial x_j$).

The summation convention for repeated subscripts is followed throughout (e.g., $a_i b_i = a_1 b_1 + a_2 b_2 + a_3 b_3 = \mathbf{a} \cdot \mathbf{b}$), and frequent use is made of the Kronecker symbol δ_{ij} and the alternating tensor with components ε_{ijk}:

$$\delta_{ij} = 0 \quad \text{for} \quad i \neq j, \quad \text{and} \quad \delta_{ij} = 1 \quad \text{for } i = j;$$

$$\varepsilon_{ijk} = 0 \quad \text{if any of } i, j, k \text{ are equal,}$$

otherwise

$$\varepsilon_{123} = \varepsilon_{312} = \varepsilon_{231} = -\varepsilon_{213} = -\varepsilon_{321} = -\varepsilon_{132} = 1.$$

The most important properties of these symbols are then

$$a_i = \delta_{ij} a_j, \qquad \varepsilon_{ijk} a_j b_k = (\mathbf{a} \times \mathbf{b})_i,$$

and they are linked by the property $\varepsilon_{ijk}\varepsilon_{ilm} = \delta_{jl}\delta_{km} - \delta_{jm}\delta_{kl}$.

The second-order tensor $\mathbf{t}$ is symmetric if and only if $\varepsilon_{ijk} t_{jk} = 0$.

Since $|\delta\mathbf{x}|$ is arbitrarily small, we can expand $\mathbf{u}(\mathbf{x} + \delta\mathbf{x})$ as $\mathbf{u} + (\delta\mathbf{x} \cdot \nabla)\mathbf{u}$ plus negligible terms of order $|\delta\mathbf{x}|^2$. It follows that $\delta\mathbf{u}$ is related to gradients of $\mathbf{u}$ and to the original line-element $\delta\mathbf{x}$ via

$$\delta\mathbf{u} = (\delta\mathbf{x} \cdot \nabla)\mathbf{u}, \qquad \text{or} \qquad \delta u_i = \frac{\partial u_i}{\partial x_j} \delta x_j. \tag{2.1}$$

Not all of the nine independent components of the tensor $u_{i,j}$ are needed, however, to specify true distortion in the vicinity of $\mathbf{x}$, since part of the motion is due merely to an infinitesimal rigid-body rotation of the neighborhood of $\mathbf{x}$.

This can be seen from the identity $(u_{i,j} - u_{j,i})\,\delta x_j = \varepsilon_{ijk}\varepsilon_{jlm}u_{m,l}\,\delta x_k$, so that equation (2.1) can be rewritten as

$$\delta u_i = \tfrac{1}{2}(u_{i,j} + u_{j,i})\,\delta x_j + \tfrac{1}{2}(\text{curl }\mathbf{u} \times \delta\mathbf{x})_i, \tag{2.2}$$

and the rigid-body rotation is of amount $\frac{1}{2}$ curl $\mathbf{u}$. The interpretation of the last term in (2.2) as a rigid-body rotation is made possible if $|u_{i,j}| \ll 1$. If displacement gradients were not "infinitesimal" in the sense of this inequality, then we should instead have to analyze the contribution to $\delta\mathbf{u}$ from a *finite* rotation—a much more difficult matter, since finite rotations do not commute and cannot be expressed as vectors.

In terms of the infinitesimal strain tensor, defined to have components

$$e_{ij} \equiv \tfrac{1}{2}(u_{i,j} + u_{j,i}), \tag{2.3}$$

the effect of true distortion on any line-element δx_i is to change the relative position of its end points by $e_{ij}\,\delta x_j$. Rotation does not affect the length of the element, and the new length is

$$
\begin{aligned}
|\delta\mathbf{x} + \delta\mathbf{u}| &= (\delta\mathbf{x}\cdot\delta\mathbf{x} + 2\,\delta\mathbf{u}\cdot\delta\mathbf{x})^{1/2} && (\text{neglecting } \delta\mathbf{u}\cdot\delta\mathbf{u})\\
&= (\delta x_i\,\delta x_i + 2e_{ij}\,\delta x_i\,\delta x_j)^{1/2} && (\text{from (2.2), and using } (\text{curl }\mathbf{u}\times\delta\mathbf{x})\cdot\delta\mathbf{x} = 0)\\
&= |\delta\mathbf{x}|(1 + e_{ij}\nu_i\nu_j),
\end{aligned}
$$

where $\boldsymbol{\nu}$ is the unit vector $\delta\mathbf{x}/|\delta\mathbf{x}|$. It follows that the extensional strain of a line-element originally in the $\boldsymbol{\nu}$ direction is $e_{ij}\nu_i\nu_j$.

To analyze the internal forces acting mutually between adjacent particles within a continuum, we use the concepts of *traction* and *stress tensor*. Traction is a vector, being the force acting per unit area across an internal surface within the continuum, and quantifies the contact force (per unit area) with which particles on one side of the surface act upon particles on the other side. For a given point of the internal surface, traction is defined (see Fig. 2.1) by considering the infinitesimal force $\delta\mathbf{F}$ acting across an infinitesimal area δS of the surface, and taking the limit of $\delta\mathbf{F}/\delta S$ as $\delta S \to 0$. With a unit normal $\mathbf{n}$ to the surface S, the convention is adopted that $\delta\mathbf{F}$ has the direction of force due to material on the side to which $\mathbf{n}$ points and acting upon material on the side from which $\mathbf{n}$ is pointing; the resulting traction is denoted as $\mathbf{T}(\mathbf{n})$. Thus, in a fluid, the pressure is $-\mathbf{n}\cdot\mathbf{T}(\mathbf{n})$. For a solid, shearing forces can act across internal surfaces, and so $\mathbf{T}$ need not be parallel to $\mathbf{n}$. Furthermore, the magnitude and direction of traction depend on the orientation of the surface element δS across which contact forces are taken (whereas pressure at a point in a fluid is the same in all directions). To appreciate this orientation-dependence of traction at a point,

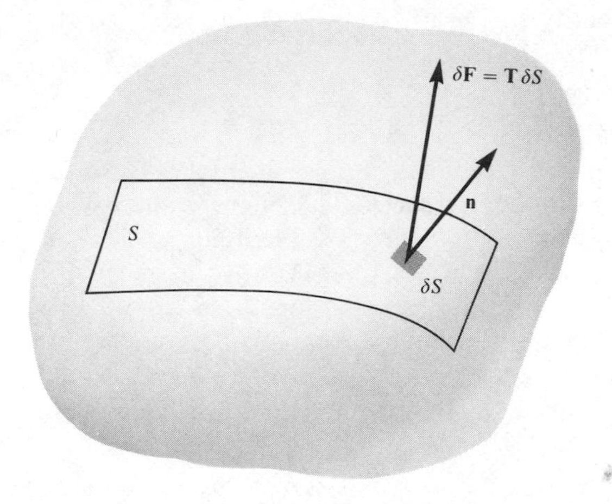

FIGURE 2.1
The definition of traction $\mathbf{T}$ acting at a point across the internal surface S with normal $\mathbf{n}$. For a fluid medium, the pressure would be $-\mathbf{n} \cdot \mathbf{T}$.

consider a point P, as shown in Figure 2.2, on the exterior surface of a house. For an element of area on the surface of the wall at P, the traction $\mathbf{T}(\mathbf{n}_1)$ is zero (neglecting atmospheric pressure and winds); but for a horizontal element of area within the wall at P, the traction $\mathbf{T}(\mathbf{n}_2)$ may be large.

The forces acting upon particles in a solid or fluid medium consist not only of the contact forces between adjacent particles, but also of (i) forces between particles that are not adjacent, and (ii) forces due to the application of physical processes external to the medium itself. An example of type (i) would be the mutual gravitational forces acting between particles of the Earth. Type (ii) is illustrated by the forces on buried particles of iron when a magnet is moved around outside the medium in which the iron is contained. To these noncontact forces, we give the name *body forces*, and use the notation $\mathbf{f}(\mathbf{x}, \mathbf{t})$ to denote the body force acting per unit volume on the particle originally at position $\mathbf{x}$ at some reference time. It will often be useful to consider the special case of a force applied impulsively to one particular particle at $\mathbf{x} = \boldsymbol{\xi}$ and time $t = \tau$. If this force is in the direction of the x_n-axis, it follows that $f_i(\mathbf{x}, t)$ is proportional to the three-dimensional Dirac delta function $\delta(\mathbf{x} - \boldsymbol{\xi})$, specifying the spatial location; to the one-dimensional Dirac delta function $\delta(t - \tau)$, specifying the timing of the impulse; and to the Kronecker delta function δ_{in}, signifying the directional

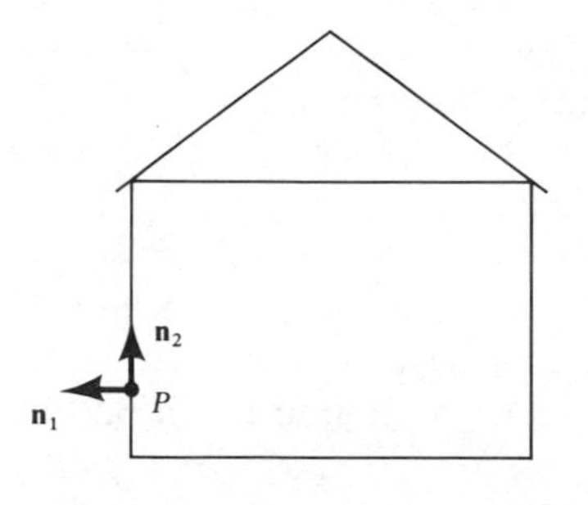

FIGURE 2.2
$\mathbf{T}(\mathbf{n}_1) \neq \mathbf{T}(\mathbf{n}_2)$.

property that $f_i = 0$ for $i \neq n$. Thus the body-force distribution is given by

$$f_i(\mathbf{x}, t) = A\,\delta(\mathbf{x} - \boldsymbol{\xi})\delta(t - \tau)\delta_{in}, \tag{2.4}$$

where A is a constant giving the strength of the impulse. Note that the dimensions of f_i, $\delta(\mathbf{x} - \boldsymbol{\xi})$, and $\delta(t - \tau)$ are, respectively, force per unit volume, 1/unit volume, and 1/unit time. The Kronecker delta is dimensionless, so A does have the correct physical dimension for an impulse (force × time).

We are now in a position to state a constraint on the accelerations, body forces, and tractions acting throughout a volume V with surface S (see Fig. 2.3). By equating the rate of change of momentum of particles constituting V to the forces acting on these particles, we find

$$\frac{\partial}{\partial t}\iiint_V \rho \frac{\partial \mathbf{u}}{\partial t}\,dV = \iiint_V \mathbf{f}\,dV + \iint_S \mathbf{T}(\mathbf{n})\,dS. \tag{2.5}$$

This relation is based on a Lagrangian description, and V and S move with the particles. The left-hand side can thus be written as $\iiint_V \rho(\partial^2\mathbf{u}/\partial t^2)\,dV$, since the particle mass $\rho\,dV$ is constant in time.

Our first use of (2.5) is to obtain an explicit form for the functional relationship $\mathbf{T} = \mathbf{T}(\mathbf{n})$ and to introduce the stress tensor. Consider a particle P within the medium for which the acceleration, the body force, and the tractions are all nonsingular. Surround this particle by a small volume ΔV, and consider the relative magnitude of the three terms in (2.5) as ΔV shrinks down onto P. The volume integrals will be of order ΔV, but the surface integral is of order $\iint_S dS$ taken over the surface of ΔV. In general, such integrals are of order $(\Delta V)^{2/3}$, tending to zero more slowly than ΔV. Dividing (2.5) through by $\iint_S dS$, it follows that

$$\frac{\left|\iint \mathbf{T}\,dS\right|}{\iint dS} = O(\Delta V^{1/3}) \to 0 \quad \text{as} \quad \Delta V \to 0. \tag{2.6}$$

Now suppose that ΔV is a disc, with opposite faces having outward normals $\mathbf{n}$

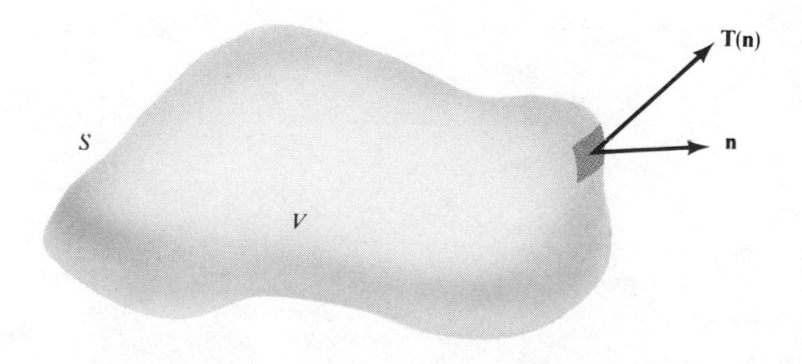

FIGURE **2.3**
A material volume V of the continuum, with surface S.

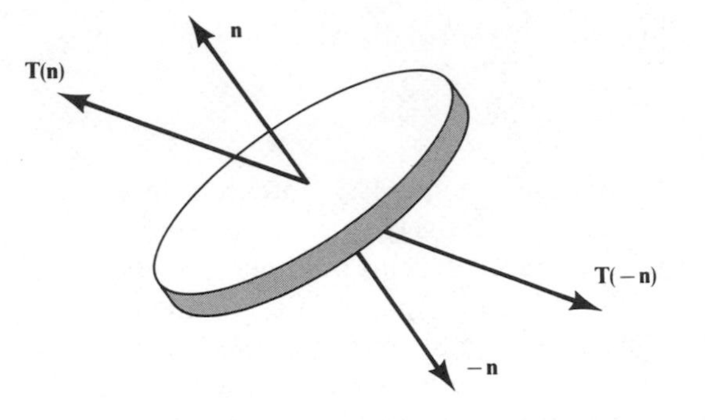

FIGURE **2.4**
A small disc within a stressed medium.

and $-\mathbf{n}$ (see Fig. 2.4) and the edge having insignificant area. Equation (2.6) then implies the result

$$\mathbf{T}(-\mathbf{n}) = -\mathbf{T}(\mathbf{n}). \tag{2.7}$$

Next, suppose ΔV to be a small tetrahedron, with three of its faces in the coordinate planes (see Fig. 2.5) and the fourth having $\mathbf{n}$ as its outward normal.

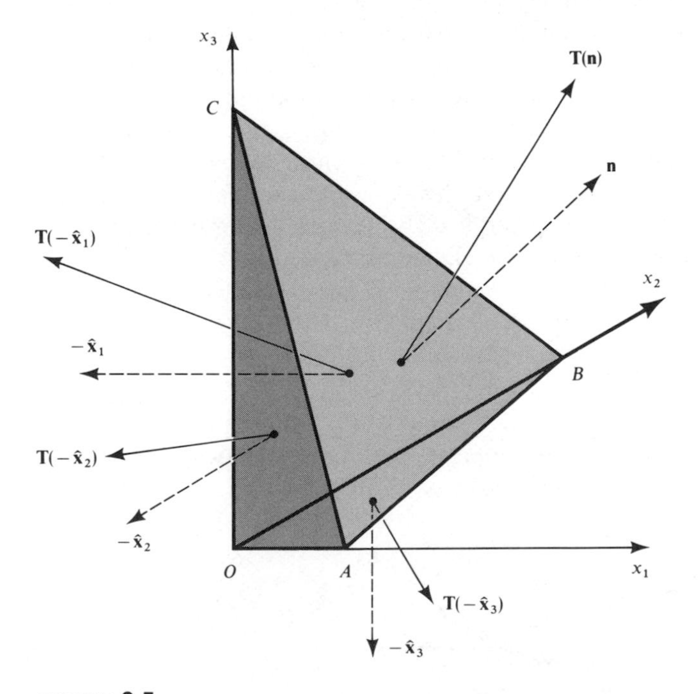

FIGURE **2.5**
The small tetrahedron OABC has three of its faces in the coordinate planes, with outward normals $-\hat{\mathbf{x}}_j (j = 1, 2, 3)$, and the fourth face has normal $\mathbf{n}$.

Equation (2.6) then implies

$$\frac{\mathbf{T}(\mathbf{n})ABC + \mathbf{T}(-\hat{\mathbf{x}}_1)OBC + \mathbf{T}(-\hat{\mathbf{x}}_2)OCA + \mathbf{T}(-\hat{\mathbf{x}}_3)OAB}{ABC + OBC + OCA + OAB} \to \mathbf{0} \tag{2.8}$$

as $\Delta V \to 0$. Here, the symbols ABC etc. denote areas of triangles, and one can show geometrically that the components of $\mathbf{n}$ are given by $(n_1, n_2, n_3) = (OBC, OCA, OAB)/ABC$. Then (2.8) and (2.7) yield

$$\mathbf{T}(\mathbf{n}) = \mathbf{T}(\hat{\mathbf{x}}_j)n_j, \tag{2.9}$$

which is a specific and important relationship between traction $\mathbf{T}(\mathbf{n})$ and $\mathbf{n}$ in terms of three tractions acting across coordinate planes. The properties (2.7) and (2.9) are trivial for a static medium, but we have shown them to be true even during acceleration.

The stress tensor is introduced by defining the nine quantities

$$\tau_{kl} = T_l(\hat{\mathbf{x}}_k),$$

so that τ_{kl} is the lth component of the traction acting across the plane normal to the kth axis due to material with greater x_k acting upon material with lesser x_k. Thus

$$T_i = \tau_{ji}n_j. \tag{2.10}$$

Our second use of (2.5) is to obtain the equation of motion of a general particle. Applying (2.10) and Gauss's divergence theorem to give

$$\iint_S T_i \, dS = \iint_S \tau_{ji}n_j \, dS = \iiint_V \tau_{ji,j} \, dV, \tag{2.11}$$

we find for a general volume V that

$$\iiint_V (\rho\ddot{u}_i - f_i - \tau_{ji,j}) \, dV = 0. \tag{2.12}$$

This integrand must be zero wherever it is continuous, otherwise a volume V could be found that violates (2.12), hence

$$\rho\ddot{u}_i = f_i + \tau_{ji,j}, \tag{2.13}$$

which is our first form for the equation of motion.

BOX **2.3**

Euler or Lagrange?

A closer look at the application of Gauss's theorem in (2.11) shows that our Lagrangian approach is inappropriate for the spatial differentiations in (2.11)–(2.13). The particles constituting S at time t have, in general, moved from their position at the reference time t_0, so that

$$\iint_S \tau_{ji} n_j \, dS = \iiint_V \frac{\partial}{\partial X_j} \tau_{ji} \, dV,$$

where $\mathbf{X} = \mathbf{x} + \mathbf{u}$, and the spatial differentiation that must be conducted on points throughout V at time t is $\partial/\partial X_j$. For finite motions, the exact equation for motion in the continuum is therefore, in our notation,

$$\rho \frac{\partial^2}{\partial t^2} u_i = f_i + \frac{\partial}{\partial X_j} \tau_{ji}. \qquad \text{(2.13, strict form)}$$

The Eulerian approach instead discusses field variables directly as a function of $\mathbf{X}$ and t (taking $\mathbf{u}$ to be the displacement of the particle at $\mathbf{X}$ and time t from its position $\mathbf{x}$ at time t_0), and τ_{ji} would be a stress component at $(\mathbf{X}, t)$. This offers the advantage of allowing one to work with independent variables that are natural for interpreting the right-hand side of the equation of motion, but has the disadvantage of cumbersome expressions for the rate of change of properties carried by the particles. For example, particle velocity $\mathbf{v}$ at $(\mathbf{X}, t)$ is difficult to express in terms of the displacement field $\mathbf{u} = \mathbf{u}(\mathbf{X}, t)$: consideration of the distance travelled in time δt yields

$$\mathbf{v}\,\delta t = \mathbf{u}(\mathbf{x} + \mathbf{v}\,\delta t, t + \delta t) - \mathbf{u}(\mathbf{X}, t)$$

and hence

$$v_i = \left(\frac{\partial u_i}{\partial t}\right)_{\mathbf{X}\text{ fixed}} + v_j \left(\frac{\partial u_i}{\partial X_j}\right)_{t\text{ fixed}}$$

is the implicit equation to be solved for $\mathbf{v}$ (components of $\mathbf{v}$ appear on both sides of the equation). Once the particle velocity is found, the acceleration at $(\mathbf{X}, t)$ is given by the material derivative $\partial \mathbf{v}/\partial t + (\mathbf{v} \cdot \nabla)\mathbf{v}$.

In seismology, the distinction between Lagrangian and Eulerian approaches rarely needs to be made, since spatial fluctuations in the displacements, strains, accelerations, and stresses have wavelengths much greater than the amplitude of particle displacements. In this case, it makes no practical difference whether a spatial gradient is evaluated at a fixed position (Euler) or for a particular particle (Lagrange). We have chosen to emphasize the Lagrangian approach, in part because there is then a simple relationship between particle velocity and particle displacement, $\mathbf{v} = \partial \mathbf{u}(\mathbf{x}, t)/\partial t$. In fluid mechanics, where there is little interest in particle displacement as a field variable, the Eulerian approach is more useful.

A final acknowledgement: the "Eulerian" and "Lagrangian" approaches were both developed by Leonhard Euler.

Another constraint upon the mechanics of motion is given by equating the rate of change of angular momentum about the origin of coordinates to the moment of forces acting on the particles in V. Thus

$$\frac{\partial}{\partial t}\iiint_V \mathbf{X} \times \rho\dot{\mathbf{u}}\, dV = \iiint_V \mathbf{X} \times \mathbf{f}\, dV + \iint_S \mathbf{X} \times \mathbf{T}\, dS, \tag{2.14}$$

where $\mathbf{X} = \mathbf{x} + \mathbf{u}$. Since $\partial\mathbf{x}/\partial t$, $\dot{\mathbf{u}} \times \dot{\mathbf{u}}$, and $\partial(\rho\, dV)/\partial t$ are all zero, the left-hand side here is $\iiint_V \mathbf{X} \times \rho\ddot{\mathbf{u}}\, dV$. Using the strict interpretation of (2.13) developed in Box 2.3, it follows that

$$\begin{aligned}\iiint_V \varepsilon_{ijk}X_j \frac{\partial}{\partial X_l}\tau_{lk}\, dV &= \iiint_V \varepsilon_{ijk}X_j(\rho\ddot{u}_k - f_k)\, dV \\ &= \iint_S \varepsilon_{ijk}X_jT_k\, dS \qquad \text{(from (2.14))} \\ &= \iint_S \varepsilon_{ijk}X_j\tau_{lk}n_l\, dS. \qquad \text{(from (2.10))}\end{aligned}$$

Applying the divergence theorem to this surface integral and using $\partial X_j/\partial X_l = \delta_{jl}$, one obtains

$$\iiint_V \varepsilon_{ijk}\tau_{jk}\, dV = 0 \qquad \text{for any volume } V,$$

implying $\varepsilon_{ijk}\tau_{jk} = 0$ everywhere, and hence that the stress tensor is symmetric:

$$\tau_{kj} = \tau_{jk}. \tag{2.15}$$

With this fundamental result, we can finally state the formulas for traction components as

$$T_i = \tau_{ij}n_j \tag{2.16}$$

and the equation of motion as

$$\rho\ddot{u}_i = f_i + \tau_{ij,j}. \tag{2.17}$$

The spatial derivative here should be carried out with respect to X_j, but (as discussed in Box 2.3) differentiation with respect to x_j is usually adequate in seismology, and will henceforth be assumed.

2.2 Stress-strain Relations and the Strain-energy Function

A medium is said to be *elastic* if it possesses a natural state (in which strains and stresses are zero) to which it will revert when applied forces are removed. Under the influence of applied loads, stress and strain will change together, and the relation between them, called the constitutive relation, is an important characteristic of the medium. That there is such a relation we prove below by thermodynamic arguments. The relation itself is a proper subject for experimental determination, and Robert Hooke's measurements of "springy bodies" led him, over 300 years ago, to the conclusion that stress is proportional to strain. His statements on this matter were somewhat enigmatic, as today's concepts of traction and tensor were then unavailable. Augustin Cauchy, in the early nineteenth century, was the first to develop many of our modern ideas of traction, and it is clear that he understood many results that today are more easily communicated in terms of tensors, which did not come into general use until the present century. The modern generalization of Hooke's law is that each component of the stress tensor is a linear combination of all components of the strain tensor, i.e., that there exist constants c_{ijkl} such that

$$\tau_{ij} = c_{ijpq}e_{pq}. \tag{2.18}$$

A body that obeys the constitutive relation (2.18) is said to be *linearly elastic*. The quantities c_{ijkl} are components of a fourth-order tensor, and have the symmetries

$$c_{jipq} = c_{ijpq} \qquad (\text{due to } \tau_{ji} = \tau_{ij}) \tag{2.19}$$

and

$$c_{ijqp} = c_{ijpq} \qquad (\text{due to } e_{qp} = e_{pq}). \tag{2.20}$$

It is also true from a thermodynamic argument that $c_{pqij} = c_{ijpq}$, as we now shall show.

Suppose that an elastic body occupies the volume V with surface S. The first law of thermodynamics states that the body possesses an internal (or, intrinsic) energy, which may change with deformations of the body, and the energy balance is:

Rate of doing mechanical work + Rate of heating

= Rate of increase of (kinetic + internal energies). (2.21)

Let us analyze each of these terms separately.

(1) The rate of mechanical work is given by

$$\begin{aligned}
&\iiint_V \mathbf{f}\cdot\dot{\mathbf{u}}\,dV + \iint_S \mathbf{T}\cdot\dot{\mathbf{u}}\,dS \\
&\quad = \iiint_V [f_i\dot{u}_i + (\tau_{ij}\dot{u}_i)_{,j}]\,dV \qquad \text{(from (2.16) and Gauss's divergence theorem)} \\
&\quad = \iiint_V (\rho\ddot{u}_i\dot{u}_i + \tau_{ij}\dot{u}_{i,j})\,dV \qquad \text{(from (2.17))} \\
&\quad = \frac{\partial}{\partial t}\iiint_V \frac{1}{2}\rho\dot{u}_i\dot{u}_i\,dV + \iiint_V \tau_{ij}\dot{e}_{ij}\,dV \qquad \text{(from symmetries in } \tau_{ij} \text{ and } e_{ij}). \qquad (2.22)
\end{aligned}$$

(2) Let $\mathbf{h}(\mathbf{x}, t)$ be the heat flux, such that $\mathbf{h}\cdot\mathbf{n}$ is the rate at which heat is transmitted (per unit area) in the $\mathbf{n}$ direction across area elements normal to $\mathbf{n}$. Let the heat input per unit volume be $\mathscr{Q} = \mathscr{Q}(\mathbf{x}, t)$. Then the rate of heating is given by

$$-\iint_S \mathbf{h}\cdot\mathbf{n}\,dS = \frac{\partial}{\partial t}\iiint_V \mathscr{Q}\,dV, \qquad (2.23)$$

and clearly $\dot{\mathscr{Q}} = -\nabla\cdot\mathbf{h}$.

(3) The rate of increase of kinetic energy is given by

$$\frac{\partial}{\partial t}\iiint_V \frac{1}{2}\rho\dot{u}_i\dot{u}_i\,dV. \qquad (2.24)$$

(4) Let $\mathscr{U}$ be the internal energy per unit volume. Then from (2.21)–(2.24), we conclude that

$$\dot{\mathscr{U}} = -h_{i,i} + \tau_{ij}\dot{e}_{ij}, \quad \text{or} \quad \dot{\mathscr{U}} = \dot{\mathscr{Q}} + \tau_{ij}\dot{e}_{ij}. \qquad (2.25)$$

If $\mathscr{U}$, $\mathscr{Q}$, and e_{ij} are measured as small perturbations away from a state of thermodynamic equilibrium, then (2.25) is equivalent to

$$\begin{aligned}
d\mathscr{U} &= d\mathscr{Q} + \tau_{ij}\,de_{ij} \\
&= \mathscr{T}\,d\mathscr{S} + \tau_{ij}\,de_{ij} \qquad \text{(for reversible processes)}, \qquad (2.26)
\end{aligned}$$

in which $\mathscr{S}$ is the entropy per unit volume and $\mathscr{T}$ is the absolute temperature.

Equation (2.26) implies that entropy and strain components are the state variables in terms of which internal energy is completely and uniquely specified. In particular, internal energy does not depend on the time history of strain.

It is occasionally useful to work with a function $\mathscr{W}$ of the strain components, in terms of which the stresses are

$$\tau_{ij} = \frac{\partial \mathscr{W}}{\partial e_{ij}}. \tag{2.27}$$

A function with this property is called a *strain-energy function.* Note from (2.26) the formal result

$$\tau_{ij} = \left(\frac{\partial \mathscr{U}}{\partial e_{ij}}\right)_{\mathscr{S}}. \tag{2.28}$$

It is also true that $\tau_{ij} = (\partial \mathscr{F} / \partial e_{ij})_{\mathscr{T}}$, where $\mathscr{F} = \mathscr{U} - \mathscr{T}\mathscr{S}$ is the *free energy* per unit volume (for which $d\mathscr{F} = -\mathscr{S}\, d\mathscr{T} + \tau_{ij}\, de_{ij}$). If the deformation process takes place so slowly as to be isothermal, as in some tectonic processes, it is natural to form τ_{ij} from changes in the free energy, and one would choose $\mathscr{W} = \mathscr{F}$.

However, if the processes of deformation are adiabatic, so that $\mathbf{h} = \mathbf{0}$ and $\dot{\mathscr{Q}} = 0$, then the actual changes in $\mathscr{U}$ associated with changes in strain do occur at constant entropy, and we can choose $\mathscr{W} = \mathscr{U}$ and use (2.28). This is usually the situation in seismology for all wavelengths greater than a few millimeters, since the time constant of thermal diffusion in rock ((distance)^2/diffusivity) is very much longer than the period of seismic waves (wavelength/velocity). Unless stated otherwise, we always assume adiabatic conditions so that the existence of a strain-energy function is assured.

Combining the properties of the strain-energy function with Hooke's law, we find

$$\frac{\partial \mathscr{W}}{\partial e_{ij}} = \tau_{ij} = c_{ijpq} e_{pq}, \tag{2.29}$$

which implies

$$c_{pqij} = c_{ijpq} \qquad \left(\text{from } \frac{\partial^2 \mathscr{W}}{\partial e_{ij}\, \partial e_{pq}} = \frac{\partial^2 \mathscr{W}}{\partial e_{pq}\, \partial e_{ij}}\right). \tag{2.30}$$

Since all the first derivatives of $\mathscr{W}$ are homogeneous (of order one) in strain components, and $\mathscr{W}$ can be taken as zero in the natural state, $\mathscr{W}$ itself must be homogeneous (of order two) in the form

$$\mathscr{W} = d_{ijpq} e_{ij} e_{pq}. \tag{2.31}$$

This quadratic is the same as $\frac{1}{2}(d_{ijpq} + d_{pqij})e_{ij}e_{pq}$, but differentiation of (2.31) to give τ_{ij} shows that $(d_{ijpq} + d_{pqij}) = c_{ijpq}$, hence the strain-energy function is, explicitly,

$$\mathscr{W} = \tfrac{1}{2}c_{ijkl}e_{ij}e_{kl} = \tfrac{1}{2}\tau_{ij}e_{ij}. \tag{2.32}$$

In adiabatic and isothermal conditions, the strain-energy function is positive except for the natural state (where $\mathscr{W} = 0$), so that $\frac{1}{2}c_{ijkl}e_{ij}e_{kl}$ is a positive definite quadratic form. ($\mathscr{W} \geq 0$, because we assume the natural state is stable.)

The c_{ijkl} are independent of strain, which is why they are called "elastic constants," but in fact they are varying functions of position in the Earth. The elasticity theory used in seismology is to a large extent characterized by a preoccupation with inhomogeneous media, particularly with a spherically symmetric medium that is everywhere isotropic. In general, the symmetries (2.19), (2.20), and (2.30) reduce the number of independent components in c_{ijkl} from 81 to 21. There is considerable simplification in the case of an isotropic medium, since $\mathbf{c}$ must be isotropic. It can be shown (Jeffreys and Jeffreys, 1972) that the most general isotropic fourth-order tensor, having the symmetries of $\mathbf{c}$, has the form

$$c_{ijkl} = \lambda\delta_{ij}\delta_{kl} + \mu(\delta_{ik}\delta_{jl} + \delta_{il}\delta_{jk}). \tag{2.33}$$

This involves only two independent constants, λ and μ, known as the Lamé constants.

Note that the results we have obtained in the present section are specialized to the case of small perturbations away from a reference state in which strain and stress are both zero. In the Earth's interior, self-gravitation is responsible for pressures of up to around 1 megabar. Even if one postulates a state of zero stress and strain for Earth materials, it is clear that the results of this section cannot directly be applied in seismology, since strains due to such pressures are not small. Using such a reference state, one must work with a theory of finite strain, in which the stress-strain relation is nonlinear. Alternatively, one can choose the static equilibrium configuration of the Earth, prior to an earthquake, as a reference state. This is the usual procedure in seismology. By definition, the reference state is one of zero strain, but now the initial stress is nonzero, and seismic motions are studied in terms of a linear relationship between strains and *incremental stresses*. Thus the stress is $\boldsymbol{\sigma}^0$ at zero strain, and is $\boldsymbol{\sigma}^0 + \boldsymbol{\tau}$ at nonzero strain, where $\tau_{ij} = c_{ijkl}e_{kl}$, and components σ_{ij}^0 can be of the same order as components c_{ijkl} ($\sim$1 megabar).

For the present, we shall continue to neglect the effects of initial stress $\boldsymbol{\sigma}^0$. This simplification is justified in Chapter 8, where initial stresses are correctly taken into account and where a brief review is given of those aspects of the theory that need revision (Box 8.5). To quantify the effects of self-gravitation, we shall in Chapter 8 adopt an Eulerian approach.

2.3 Theorems of Uniqueness and Reciprocity

It is natural to introduce the discussion of uniqueness (for the displacement field $\mathbf{u}$ throughout a body with volume V and surface S) with some general remarks concerning the ways in which motion can be set up. Because the displacement is constrained to satisfy (2.17) throughout V, the application of body forces will generate a displacement field, as will the application of tractions on the surface S. We shall show that specification of the body forces throughout V, and tractions over all of S, is enough to determine uniquely the displacement field that will develop throughout V from given initial conditions. An alternative way to specify the influence of S on the displacement field is to give a boundary condition for displacement itself (on S) instead of for the traction. For example, S might be rigid. It might seem at first that the traction on S and the displacement on S are independent properties of the displacement field throughout V. This is not so, however, and it is important for an intuitive understanding of Sections 2.3–2.5 to appreciate that traction over S determines the displacement over S, and vice versa.

UNIQUENESS THEOREM

The displacement $\mathbf{u} = \mathbf{u}(\mathbf{x}, t)$ throughout the volume V with surface S is uniquely determined after time t_0 by the initial values of displacement and particle velocity at t_0, throughout V; and by values at all times $t \geq t_0$ of (i) the body forces $\mathbf{f}$ and the heat $\mathcal{Q}$ supplied throughout V; (ii) the tractions $\mathbf{T}$ over any part S_1 of S; and (iii) the displacement over the remainder S_2 of S, with $S_1 + S_2 = S$. (Either of S_1 or S_2 can be the whole of S.)

PROOF

Suppose $\mathbf{u}_1$ and $\mathbf{u}_2$ are any solutions for $\mathbf{u}$ that satisfy the same initial conditions and are set up by the same values for (i)–(iii). Then the difference $\mathbf{U} \equiv \mathbf{u}_1 - \mathbf{u}_2$ is a displacement field having zero initial conditions, and is set up by zero body forces, zero heating, zero traction on S_1, and $\mathbf{U} = \mathbf{0}$ on S_2. It remains to prove that $\mathbf{U} = \mathbf{0}$ throughout V for $t > t_0$.

The rate of doing mechanical work in the displacement field $\mathbf{U}$ is clearly zero throughout V and S_1 and S_2 (see (2.22)) for $t \geq t_0$. The third equality in (2.22) can be integrated from t_0 to t, and, together with the zero initial conditions and the use of a strain-energy function ($\mathbf{U}$ involves adiabatic changes), it follows that

$$\iiint_V \frac{1}{2}\rho \dot{U}_i \dot{U}_i \, dV + \iiint_V \frac{1}{2} c_{ijkl} U_{i,j} U_{k,l} \, dV = 0.$$

Both the kinetic and strain energies are positive definite, so that $\dot{U}_i = 0$ for $t \geq t_0$. But $U_i = 0$ at $t = t_0$, and hence $\mathbf{U} = \mathbf{0}$ throughout V for $t \geq t_0$.

BOX **2.4**

Use of the term "homogeneous," as applied to equations and boundary conditions

The equation for elastic displacement is $\mathbf{L}(\mathbf{u}) = \mathbf{f}$, where $\mathbf{L}$ is the vector differential operator defined on the components of $\mathbf{u}$ by

$$(\mathbf{L}(\mathbf{u}))_i \equiv \rho \ddot{u}_i - (c_{ijkl} u_{k,l})_{,j}.$$

If body forces are absent, then the equation $\mathbf{L}(\mathbf{u}) = \mathbf{0}$ for $\mathbf{u}$ is said to be *homogeneous*. A *homogeneous boundary condition* on the surface S is one for which *either* the displacement *or* the traction vanishes at every point of the surface.

This terminology is reminiscent of linear algebra, for which a system of n equations in n unknowns, in the form $\mathbf{A}\mathbf{x} = \mathbf{0}$, is also said to be homogeneous. Here, $\mathbf{x}$ is a column vector and $\mathbf{A}$ is some $n \times n$ matrix. It is well known that nontrivial solutions ($\mathbf{x} \neq \mathbf{0}$) can exist, but only if $\mathbf{A}$ has a special property (namely, a zero determinant). The corresponding result in dynamic elasticity is that motions can occur throughout a finite elastic volume V without any body forces and with a homogeneous boundary condition over the surface of V. These are the *free oscillations* or *normal modes* of the body, which can occur only at certain frequencies. See Chapter 8.

RECIPROCITY THEOREMS

We shall state and prove several general relationships between a pair of solutions for the displacement through an elastic body V.

Suppose that $\mathbf{u} = \mathbf{u}(\mathbf{x}, t)$ is one of these displacement fields, and that $\mathbf{u}$ is due to body forces $\mathbf{f}$ and boundary conditions on S and initial conditions at time $t = 0$. Let $\mathbf{v} = \mathbf{v}(\mathbf{x}, t)$ be another displacement field due to body forces $\mathbf{g}$ and to boundary conditions and initial conditions (at $t = 0$) which in general are different from the conditions for $\mathbf{u}$. To distinguish the tractions on surfaces normal to $\mathbf{n}$ in these two cases, we shall use the notation $\mathbf{T}(\mathbf{u}, \mathbf{n})$ for the traction due to the displacement $\mathbf{u}$ and, similarly, $\mathbf{T}(\mathbf{v}, \mathbf{n})$ for the traction due to $\mathbf{v}$.

The first reciprocal relation to note between $\mathbf{u}$ and $\mathbf{v}$ is

$$\iiint_V (\mathbf{f} - \rho \ddot{\mathbf{u}}) \cdot \mathbf{v} \, dV + \iint_S \mathbf{T}(\mathbf{u}, \mathbf{n}) \cdot \mathbf{v} \, dS = \iiint_V (\mathbf{g} - \rho \ddot{\mathbf{v}}) \cdot \mathbf{u} \, dV + \iint_S \mathbf{T}(\mathbf{v}, \mathbf{n}) \cdot \mathbf{u} \, dS. \quad (2.34)$$

This result is due to Betti. It can easily be proved by subsitution from (2.17) and (2.16) and then applying the divergence theorem to reduce the left side to

BOX **2.5**

Parallels

A rearrangement of Betti's relation (2.34) gives

$$\iiint_V \{v_i(c_{ijkl}u_{k,l})_{,j} - u_i(c_{ijkl}v_{k,l})_{,j}\} \, dV = \iint_S \{v_i T_i(\mathbf{u}, \mathbf{n}) - u_i T_i(\mathbf{v}, \mathbf{n})\} \, dS.$$

This is a vector theorem for the second-order spatial derivatives occurring in the wave equation of elasticity, which is analogous to Green's theorem

$$\iiint_V (\psi \nabla^2 \phi - \phi \nabla^2 \psi) \, dV = \iint_S \left(\psi \frac{\partial \phi}{\partial n} - \phi \frac{\partial \psi}{\partial n} \right) dS$$

for scalars and the Laplacian operator. Green's theorem is a working tool for studying inhomogeneous equations, such as $\nabla^2 \phi = -4\pi\rho$, and we shall use Betti's theorem for the elastic wave equation, in which the inhomogeneity is the body-force term.

There are many further analogies between Dirichlet problems (for potentials that are zero on S) and elasticity problems with rigid boundaries; and between Neumann problems ($\partial\phi/\partial n = 0$ on S) and traction-free boundaries.

$\iiint_V c_{ijkl} v_{i,j} u_{k,l} \, dV$. Similarly, the right-hand side reduces to $\iiint_V c_{ijkl} u_{i,j} v_{k,l} \, dV$, and (2.34) follows from the symmetry $c_{ijkl} = c_{klij}$.

Note that Betti's theorem does not involve initial conditions for $\mathbf{u}$ or $\mathbf{v}$. Furthermore, it remains true even if the quantities $\mathbf{u}$, $\ddot{\mathbf{u}}$, $\mathbf{T}(\mathbf{u}, \mathbf{n})$, and $\mathbf{f}$ are evaluated at time t_1 but $\mathbf{v}$, $\ddot{\mathbf{v}}$, $T(\mathbf{v}, \mathbf{n})$, $\mathbf{g}$ are evaluated at a different time t_2. If we choose $t_1 = t$ and $t_2 = \tau - t$ and integrate (2.34) over the temporal range 0 to τ, then the acceleration terms reduce to terms that depend only on the initial and final values, since

$$\begin{aligned} \int_0^\tau \rho\{\ddot{\mathbf{u}}(t) \cdot \mathbf{v}(\tau - t) - \mathbf{u}(t) \cdot \ddot{\mathbf{v}}(\tau - t)\} \, dt \\ = \rho \int_0^\tau \frac{\partial}{\partial t} \{\dot{\mathbf{u}}(t) \cdot \mathbf{v}(\tau - t) + \mathbf{u}(t) \cdot \dot{\mathbf{v}}(\tau - t)\} \, dt \\ = \rho\{\dot{\mathbf{u}}(\tau) \cdot \mathbf{v}(0) - \dot{\mathbf{u}}(0) \cdot \mathbf{v}(\tau) + \mathbf{u}(\tau) \cdot \dot{\mathbf{v}}(0) - \mathbf{u}(0) \cdot \dot{\mathbf{v}}(\tau) \end{aligned}$$

If there is some time τ_0 before which $\mathbf{u}$ and $\mathbf{v}$ are everywhere zero throughout V (and hence $\dot{\mathbf{u}} = \dot{\mathbf{v}} = \mathbf{0}$ for $\tau \leq \tau_0$), then the convolution

$$\int_{-\infty}^{\infty} \rho\{\ddot{\mathbf{u}}(t) \cdot \mathbf{v}(\tau - t) - \mathbf{u}(t) \cdot \ddot{\mathbf{v}}(\tau - t)\} \, dt$$

is zero. We deduce from Betti's theorem the important result, for displacement fields with a quiescent past, that

$$\int_{-\infty}^{\infty} dt \iiint_V \{\mathbf{u}(\mathbf{x}, t) \cdot \mathbf{g}(\mathbf{x}, \tau - t) - \mathbf{v}(\mathbf{x}, \tau - t) \cdot \mathbf{f}(\mathbf{x}, t)\}\, dV$$

$$= \int_{-\infty}^{\infty} dt \iint_S \{\mathbf{v}(\mathbf{x}, \tau - t) \cdot \mathbf{T}(\mathbf{u}(\mathbf{x}, t), \mathbf{n}) - \mathbf{u}(\mathbf{x}, t) \cdot \mathbf{T}(\mathbf{v}(\mathbf{x}, \tau - t), \mathbf{n})\}\, dS. \quad (2.35)$$

2.4 Introducing Green's Function for Elastodynamics

A major aim of this chapter and the next is the development of a representation for the displacements that typically occur in seismology. The representation will be a formula for the displacement (at a general point in space and time) in terms of the quantities that originated the motion, and we have seen (in the uniqueness theorem) that these are body forces and applied tractions or displacements over the surface of the elastic body under discussion. For earthquake faulting, the seismic source is complicated in that it extends over a finite fault plane (or a finite volume) and over a finite amount of time, and in general involves motions (at the source) that have varying direction and magnitude. We shall find that the representation theorem is really nothing but a bookkeeping device by which the displacement from realistic source models is synthesized from the displacement produced by the simplest of sources—namely, the unidirectional unit impulse, which is localized precisely in both space and time.

The displacement field from such a simple source is the elastodynamic Green function. If the unit impulse is applied at $\mathbf{x} = \boldsymbol{\xi}$ and $t = \tau$ and in the n-direction (see (2.4), taking A = unit constant with dimensions of impulse), then we denote the ith component of displacement at general $(\mathbf{x}, t)$ by $G_{in}(\mathbf{x}, t; \boldsymbol{\xi}, \tau)$. Clearly, this Green function is a tensor (we shall work throughout with Cartesian tensors, and therefore do not distinguish between tensors and dyadics). It depends on both receiver and source coordinates, and satisfies the equation

$$\rho \frac{\partial^2}{\partial t^2} G_{in} = \delta_{in}\delta(\mathbf{x} - \boldsymbol{\xi})\delta(t - \tau) + \frac{\partial}{\partial x_j}\left(c_{ijkl} \frac{\partial}{\partial x_l} G_{kn}\right) \quad (2.36)$$

throughout V. We shall invariably use the initial conditions that $\mathbf{G}(\mathbf{x}, t; \boldsymbol{\xi}, \tau)$ and $\partial\{\mathbf{G}(\mathbf{x}, t; \boldsymbol{\xi}, \tau)\}/\partial t$ are zero for $t \leq \tau$ and $\mathbf{x} \neq \boldsymbol{\xi}$. To specify $\mathbf{G}$ uniquely, it remains to state the boundary conditions on S, and we shall use a variety of different boundary conditions in different applications.

If the boundary conditions are independent of time (e.g., S always rigid), then the time origin can be shifted at will, and we see from (2.36) that $\mathbf{G}$ depends on t

and τ only via the combination $t - \tau$. Hence

$$\mathbf{G}(\mathbf{x}, t; \boldsymbol{\xi}, \tau) = \mathbf{G}(\mathbf{x}, t - \tau; \boldsymbol{\xi}, 0) = \mathbf{G}(\mathbf{x}, -\tau; \boldsymbol{\xi}, -t), \tag{2.37}$$

which is a reciprocal relation for source and receiver times.

If $\mathbf{G}$ satisfies homogeneous boundary conditions on S, then (2.35) can be used to obtain an important reciprocal relation for source and receiver positions. One takes $\mathbf{f}$ to be a unit impulse applied in the m-direction at $\mathbf{x} = \boldsymbol{\xi}_1$ and time $t = \tau_1$, and $\mathbf{g}$ to be a unit impulse applied in the n-direction at $\mathbf{x} = \boldsymbol{\xi}_2$ and time $t = -\tau_2$. Then $u_i = G_{im}(\mathbf{x}, t; \boldsymbol{\xi}_1, \tau_1)$ and $v_i = G_{in}(\mathbf{x}, t; \boldsymbol{\xi}_2, -\tau_2)$, so that (2.35) directly yields

$$G_{nm}(\boldsymbol{\xi}_2, \tau + \tau_2; \boldsymbol{\xi}_1, \tau_1) = G_{mn}(\boldsymbol{\xi}_1, \tau - \tau_1; \boldsymbol{\xi}_2, -\tau_2). \tag{2.38}$$

Choosing $\tau_1 = \tau_2 = 0$, this becomes

$$G_{nm}(\boldsymbol{\xi}_2, \tau; \boldsymbol{\xi}_1, 0) = G_{mn}(\boldsymbol{\xi}_1, \tau; \boldsymbol{\xi}_2, 0), \tag{2.39}$$

which specifies a purely spatial reciprocity. Choosing $\tau = 0$ in (2.38) gives

$$G_{nm}(\boldsymbol{\xi}_2, \tau_2; \boldsymbol{\xi}_1, \tau_1) = G_{mn}(\boldsymbol{\xi}_1, -\tau_1; \boldsymbol{\xi}_2, -\tau_2), \tag{2.40}$$

which specifies a space-time reciprocity.

The actual computation of an elastodynamic Green function can itself be a complicated problem. We shall take up this subject in Chapter 4 for the simplest of elastic solids (homogeneous, isotropic, infinite) and also for the case of large separation between source and receiver in inhomogeneous media.

2.5 Representation Theorems

If the integrated form of Betti's theorem, our equation (2.35), is used with a Green function for one of the displacement fields, then a representation for the other displacement field becomes available.

Specifically, suppose we are interested in finding an expression for the displacement $\mathbf{u}$ due both to body forces $\mathbf{f}$ throughout V and to boundary conditions on S. We substitute into (2.35) the body force $g_i(\mathbf{x}, t) = \delta_{in}\delta(\mathbf{x} - \boldsymbol{\xi})\delta(t)$, for which the corresponding solution is $v_i(\mathbf{x}, t) = G_{in}(\mathbf{x}, t; \boldsymbol{\xi}, 0)$, and find

$$\begin{aligned} u_n(\boldsymbol{\xi}, \tau) = {} & \int_{-\infty}^{\infty} dt \iiint_V f_i(\mathbf{x}, t) G_{in}(\mathbf{x}, \tau - t; \boldsymbol{\xi}, 0)\, dV \\ & + \int_{-\infty}^{\infty} dt \iint_S \{G_{in}(\mathbf{x}, \tau - t; \boldsymbol{\xi}, 0) T_i(\mathbf{u}(\mathbf{x}, t), \mathbf{n}) \\ & - u_i(\mathbf{x}, t) c_{ijkl} n_j G_{kn,l}(\mathbf{x}, \tau - t; \boldsymbol{\xi}, 0)\}\, dS. \end{aligned}$$

Before giving a physical interpretation of this equation, it is helpful to interchange the symbols $\mathbf{x}$ and $\boldsymbol{\xi}$ and the symbols t and τ. This permits $(\mathbf{x}, t)$ to be the general position and time at which a displacement is to be evaluated, regarded as an integral over volume and surface elements at varying $\boldsymbol{\xi}$ with a temporal convolution. The result is

$$\begin{aligned} u_n(\mathbf{x}, t) = &\int_{-\infty}^{\infty} d\tau \iiint_V f_i(\boldsymbol{\xi}, \tau) G_{in}(\boldsymbol{\xi}, t - \tau; \mathbf{x}, 0)\, dV(\boldsymbol{\xi}) \\ &+ \int_{-\infty}^{\infty} d\tau \iint_S \{G_{in}(\boldsymbol{\xi}, t - \tau; \mathbf{x}, 0) T_i(\mathbf{u}(\boldsymbol{\xi}, \tau), \mathbf{n}) \\ &- u_i(\boldsymbol{\xi}, \tau) c_{ijkl}(\boldsymbol{\xi}) n_j G_{kn,l}(\boldsymbol{\xi}, t - \tau; \mathbf{x}, 0)\}\, dS(\boldsymbol{\xi}). \end{aligned} \tag{2.41}$$

This is our first representation theorem. It states a way in which displacement $\mathbf{u}$ at a certain point is made up from contributions due to the force $\mathbf{f}$ throughout V, plus contributions due to the traction $\mathbf{T}(\mathbf{u}, \mathbf{n})$ and the displacement $\mathbf{u}$ itself on S. However, the way in which each of these three contributions is weighted is unsatisfactory, since each involves a Green function with source at $\mathbf{x}$ and observation point at $\boldsymbol{\xi}$. (Note that the last term in (2.41) involves differentiation with respect to ξ_l.) We want $\mathbf{x}$ to be the observation point, so that the total displacement obtained there can be regarded as the sum (integral) of contributing displacements at $\mathbf{x}$ due to each volume element and surface element. The reciprocal theorem for $\mathbf{G}$ must be invoked, but this will require extra conditions on Green's function itself, since the equation $G_{in}(\boldsymbol{\xi}, t - \tau; \mathbf{x}, 0) = G_{ni}(\mathbf{x}, t - \tau; \boldsymbol{\xi}, 0)$ (see (2.39)) was proved only if $\mathbf{G}$ satisfies homogeneous boundary conditions on S, whereas (2.41) is valid for *any* Green function set up by an impulsive force in the n-direction at $\boldsymbol{\xi} = \mathbf{x}$ and $\tau = t$.

We shall examine two different cases. Suppose, first, that Green's function is determined with S as a rigid boundary. We write $\mathbf{G}^{\text{rigid}}$ for this function and $G_{in}^{\text{rigid}}(\boldsymbol{\xi}, t - \tau; \mathbf{x}, 0) = 0$ for $\boldsymbol{\xi}$ in S. Then (2.41) becomes

$$\begin{aligned} u_n(\mathbf{x}, t) = &\int_{-\infty}^{\infty} d\tau \iiint_V f_i(\boldsymbol{\xi}, \tau) G_{ni}^{\text{rigid}}(\mathbf{x}, t - \tau; \boldsymbol{\xi}, 0)\, dV \\ &- \int_{-\infty}^{\infty} d\tau \iint_S u_i(\boldsymbol{\xi}, \tau) c_{ijkl} n_j \frac{\partial}{\partial \xi_l} G_{nk}^{\text{rigid}}(\mathbf{x}, t - \tau; \boldsymbol{\xi}, 0)\, dS. \end{aligned} \tag{2.42}$$

Alternatively, we can use $\mathbf{G}^{\text{free}}$ as Green's function, so that the traction $c_{ijkl} n_j (\partial/\partial \xi_l)\, G_{kn}^{\text{free}}(\boldsymbol{\xi}, t - \tau; \mathbf{x}, 0)$ is zero for $\boldsymbol{\xi}$ in S, finding

$$\begin{aligned} u_n(\mathbf{x}, t) = &\int_{-\infty}^{\infty} d\tau \iiint_V f_i(\boldsymbol{\xi}, \tau) G_{ni}^{\text{free}}(\mathbf{x}, t - \tau; \boldsymbol{\xi}, 0)\, dV \\ &+ \int_{-\infty}^{\infty} d\tau \iint_S G_{ni}^{\text{free}}(\mathbf{x}, t - \tau; \boldsymbol{\xi}, 0) T_i(\mathbf{u}(\boldsymbol{\xi}, \tau), \mathbf{n})\, dS. \end{aligned} \tag{2.43}$$

Equations (2.41)–(2.43) are all different forms of the representation theorem, and each has its special uses. Taken together, they seem to imply a contradiction to the question of whether $\mathbf{u}(\mathbf{x}, t)$ depends upon displacement on S (see (2.42)) or traction (see (2.43)) or both (see (2.41)). But since traction and displacement cannot be specified independently on the surface of an elastic medium, there is no contradiction.

The surface on which values of traction (or displacement) are explicitly required has been taken, in this chapter, as external to the volume V. It is often useful to take this surface to include two adjacent internal surfaces, being the opposite faces of a buried fault. Specialized forms of the representation theorem can then be developed, which enable one to analyze the earthquakes set up by activity on a buried fault. This subject is central to earthquake source theory, and is taken up in the following chapter.

So far, we have considered only Cartesian coordinate systems. In practice, the seismologist is often required to use different coordinates that allow the physical relationship between components of displacement, stress, and strain to be simplified for the geometry of a particular problem. In particular, it is often found that a boundary condition must be applied on a surface that can be chosen as the surface on which a general curvilinear coordinate is constant. Vector operations grad, div, curl, and ∇^2 are derived for general orthogonal coordinates in many texts, but rather more is needed to analyze the vector operations required in elasticity.

BOX **2.6**

General properties of orthogonal curvilinear coordinates

Consider a point at the vector position $\mathbf{x}$ to be specified by three parameters, c^1, c^2, c^3. That is, each of the three components of $\mathbf{x}$ (in some Cartesian coordinate system) is a scalar function of the c^p:

$$x_i = x_i(c^1, c^2, c^3) \qquad (i = 1, 2, 3).$$

We suppose that these functions x_i have continuous derivatives and that there are inverse functions

$$c^p = c^p(x_1, x_2, x_3) \qquad (p = 1, 2, 3) \quad \text{or} \quad c^p = c^p(\mathbf{x}),$$

so that the equation $c^p = $ constant can be thought of as a coordinate surface for each p, and these three surfaces intersect in lines along which only one of the c^1, c^2, c^3 is varying. We use superscripts for quantities identified with the general curvilinear system.

Let $\mathbf{n}^p$ be the unit normal to the coordinate surface $c^p = $ constant, and suppose $\mathbf{x}$ and $\mathbf{x} + d\mathbf{x}$ both lie in this surface. Then $c^p(\mathbf{x}) = c^p(\mathbf{x} + d\mathbf{x})$, and hence $d\mathbf{x} \cdot \nabla c^p = 0$, using the Taylor expansion of $c^p(\mathbf{x} + d\mathbf{x})$. Since $d\mathbf{x}$ is *any* line element within the surface, it follows that ∇c^p is normal to $c^p = $ constant, and ∇c^p must be parallel to $\mathbf{n}^p$.

Let the length of vector ∇c^p be $1/h^p$ (a scaling factor). Then

$$\mathbf{n}^p = h^p \nabla c^p. \tag{1}$$

(We drop the summation convention for superscripts, but retain it for subscripts, these being related to the original Cartesian system.)

We shall assume that c^1, c^2, c^3 form a *right-handed orthogonal system*, i.e., that

$$\mathbf{n}^p \cdot \mathbf{n}^q = \delta^{pq} \qquad \text{(the Kronecker delta)}, \tag{2}$$

and that $\mathbf{n}^3 = \mathbf{n}^1 \times \mathbf{n}^2$.

Using n_i^p for the ith Cartesian component of $\mathbf{n}^p$, we can now obtain an important relation relation between $\mathbf{n}^p$ and $\partial \mathbf{x}/\partial c^p$, as follows:

$$\mathbf{n}^p = n_i^p \hat{\mathbf{x}}_i = n_i^p \frac{\partial \mathbf{x}}{\partial x_i} = \sum_q n_i^p \frac{\partial \mathbf{x}}{\partial c^q} \frac{\partial c^q}{\partial x_i} \qquad \text{(the chain rule)}$$

$$= \sum_q n_i^p \frac{n_i^q}{h^q} \frac{\partial \mathbf{x}}{\partial c^q} \text{(from (1))} = \sum_q \frac{\delta^{pq}}{h^q} \frac{\partial \mathbf{x}}{\partial c^q} \qquad \text{(from (2))}$$

and hence

$$\mathbf{n}^p = \frac{1}{h^p} \frac{\partial \mathbf{x}}{\partial c^p}. \tag{3}$$

A small change $d\mathbf{x}$ in position is associated with a small change in each of coordinates c^1, c^2, c^3 by $d\mathbf{x} = \sum_p (\partial \mathbf{x}/\partial c^p)\, dc^p$, and the magnitude of this change is given by

$$(ds)^2 = d\mathbf{x} \cdot d\mathbf{x} = \sum_p \frac{\partial \mathbf{x}}{\partial c^p} dc^p \cdot \sum_q \frac{\partial \mathbf{x}}{\partial c^q} dc^q$$

$$= (h^1\, dc^1)^2 + (h^2\, dc^2)^2 + (h^3\, dc^3)^2 \qquad \text{(from (3) and (2))}. \tag{4}$$

This result leads to one of the quickest ways of actually finding the scaling functions: the Euclidean distance associated with increment dc^1 along $\mathbf{n}^1$ is $h^1 dc^1$; similarly for h^2 and h^3.

In Section 2.6, we need formulas for derivatives of the type $\partial \mathbf{n}^p/\partial c^q$ in terms of the undifferentiated normals. From (2) and (3), the equations to be satisfied are

$$\left.\begin{aligned} &\mathbf{n}^p \cdot \frac{\partial \mathbf{n}^q}{\partial c^r} + \mathbf{n}^q \cdot \frac{\partial \mathbf{n}^p}{\partial c^r} = 0 \qquad \text{(18 different scalar equations)} \\ &\frac{\partial}{\partial c^q}(h^p \mathbf{n}^p) = \frac{\partial}{\partial c^p}(h^q \mathbf{n}^q) \qquad \text{(3 nontrivial vector equations)} \end{aligned}\right\} \tag{5}$$

The above are 27 different scalar equations for the 27 scalar unknowns in $\partial \mathbf{n}^p/\partial c^q$, and hence are exactly enough to determine the solution. In vector form, this solution is

$$\frac{\partial \mathbf{n}^p}{\partial c^q} = \frac{\mathbf{n}^q}{h^p} \frac{\partial h^q}{\partial c^p} - \delta^{pq} \left[\frac{\mathbf{n}^1}{h^1} \frac{\partial h^p}{\partial c^1} + \frac{\mathbf{n}^2}{h^2} \frac{\partial h^p}{\partial c^2} + \frac{\mathbf{n}^3}{h^2} \frac{\partial h^p}{\partial c^3} \right], \tag{6}$$

as may be verified by direct substitution back into (5).

2.6 Strain-displacement relations and Displacement-stress Relations in General Orthogonal Curvilinear Coordinates

Continuing with the notation developed in Box 2.6, we shall first obtain relations between strain components e^{pq} and displacement components u^r that generalize the usual Cartesian result $e_{ij} = \frac{1}{2}(\partial u_i/\partial x_j + \partial u_j/\partial x_i)$. By e^{pq}, we merely mean the components of the Cartesian second-order tensor $\mathbf{e}$, referred to rotated *Cartesian axes*, which are defined (at the point of interest) to lie along the directions $\mathbf{n}^1$, $\mathbf{n}^2$, $\mathbf{n}^3$. Thus we emphasize the physical components of strain, rather than the general tensor components (which may not even have the dimensions of strain). Our problem is to express e^{pq} in terms of derivatives (with respect to c^1, c^2, c^3) of the physical components of displacement also resolved along $\mathbf{n}^1$, $\mathbf{n}^2$, $\mathbf{n}^3$: the difficulties that arise are due (a) to spatial changes in the scaling functions h^1, h^2, h^3, and (b) to spatial changes in the directions $\mathbf{n}^1$, $\mathbf{n}^2$, $\mathbf{n}^3$.

Direction cosines of the rotated Cartesian axis along $\mathbf{n}^p$ are $(n_1{}^p, n_2{}^p, n_3{}^p)$, referred to the Cartesian axes $\hat{\mathbf{x}}_1$, $\hat{\mathbf{x}}_2$, $\hat{\mathbf{x}}_3$ (which are in the same fixed direction at every point). Therefore, from the fundamental transformation property of Cartesian vector and tensor components,

$$u^p = n_i^p u_i \qquad \text{(summation is retained for repeated subscripts)} \tag{2.44}$$

$$\begin{aligned}
e^{pq} &= n_i^p n_j^q e_{ij} \\
&= \frac{1}{h^p h^q}\frac{\partial x_i}{\partial c^p}\frac{\partial x_j}{\partial c^q}\frac{1}{2}\left(\frac{\partial u_i}{\partial x_j} + \frac{\partial u_j}{\partial x_i}\right) \qquad &&\text{(from (3) in Box 2.6; no summation over superscripts)} \\
&= \frac{1}{2h^p h^q}\left(\frac{\partial x_i}{\partial c^p}\frac{\partial u_i}{\partial c^q} + \frac{\partial x_i}{\partial c^q}\frac{\partial u_i}{\partial c^p}\right) &&\text{(reversing the chain rule in the previous line)} \\
&= \frac{1}{2h^q}\left[\frac{\partial}{\partial c^q}\left(\frac{u_i}{h^p}\frac{\partial x_i}{\partial c^p}\right) - u_i\frac{\partial}{\partial c^q}\left(\frac{1}{h^p}\frac{\partial x_i}{\partial c^p}\right)\right] \\
&\quad + \frac{1}{2h^p}\left[\frac{\partial}{\partial c^p}\left(\frac{u_i}{h^q}\frac{\partial x_i}{\partial c^q}\right) - u_i\frac{\partial}{\partial c^p}\left(\frac{1}{h^q}\frac{\partial x_i}{\partial c^q}\right)\right] \\
&= \frac{1}{2h^q}\frac{\partial u^p}{\partial c^q} + \frac{1}{2h^p}\frac{\partial u^q}{\partial c^p} - \frac{1}{2}u_i\left[\frac{1}{h^q}\frac{\partial}{\partial c^q}n_i^p + \frac{1}{h^p}\frac{\partial}{\partial c^p}n_i^q\right] \\
& &&\text{(repeated use of (3) in Box 2.6 and (2.44))} \\
&= \frac{1}{2h^q}\frac{\partial u^p}{\partial c^q} + \frac{1}{2h^p}\frac{\partial u^q}{\partial c^p} - \frac{1}{2}\mathbf{u}\cdot\left[\frac{1}{h^q}\frac{\partial \mathbf{n}^p}{\partial c^q} + \frac{1}{h^p}\frac{\partial \mathbf{n}^q}{\partial n^p}\right].
\end{aligned}$$

In this form, we can use the final equation of Box 2.6 to obtain

$$e^{pq} = \frac{1}{2}\left[\frac{h^p}{h^q}\frac{\partial}{\partial c^q}\left(\frac{u^p}{h^p}\right) + \frac{h^q}{h^p}\frac{\partial}{\partial c^p}\left(\frac{u^q}{h^q}\right)\right] + \frac{\delta^{pq}}{h^q}\left[\frac{u^1}{h^1}\frac{\partial h^p}{\partial c^1} + \frac{u^2}{h^2}\frac{\partial h^p}{\partial c^2} + \frac{u^3}{h^3}\frac{\partial h^p}{\partial c^3}\right], \tag{2.45}$$

in which all reference to the Cartesian system (x_1, x_2, x_3) has at last been eliminated. Only the first square bracket is required for the off-diagonal components $(p \neq q)$, but for a typical diagonal component (2.45) reduces to, e.g.,

$$e^{11} = \frac{1}{h^1}\frac{\partial u^1}{\partial c^1} + \frac{u^2}{h^1 h^2}\frac{\partial h^1}{\partial c^2} + \frac{u^3}{h^3 h^1}\frac{\partial h^1}{\partial c^3}. \tag{2.46}$$

To obtain the displacement-stress relations for general orthogonal components of $\mathbf{u}$ and $\boldsymbol{\tau}$, we follow steps similar to the derivation of $\rho\ddot{u}_i = \tau_{ij,j}$ given in Section 2.1 for fixed Cartesian directions. The principal difficulty lies in interpreting $\iint_S \mathbf{T}\, dS$, the integral of traction acting across the surface S with volume V. With $\boldsymbol{\nu}$ as the outward normal on dS,

$$\begin{aligned} T_i(\boldsymbol{\nu})\, dS &= \tau_{ij}\nu_j\, dS \\ &= \sum_{p,q} \tau^{pq} n_i^p n_j^q \nu_j\, dS \qquad \text{(transformation to components in rotated Cartesians)} \\ &= \sum_{p,q} \tau^{pq} n_i^p \nu^q\, dS, \end{aligned}$$

where ν^q is the component of the normal to dS, resolved along $\mathbf{n}^q$.

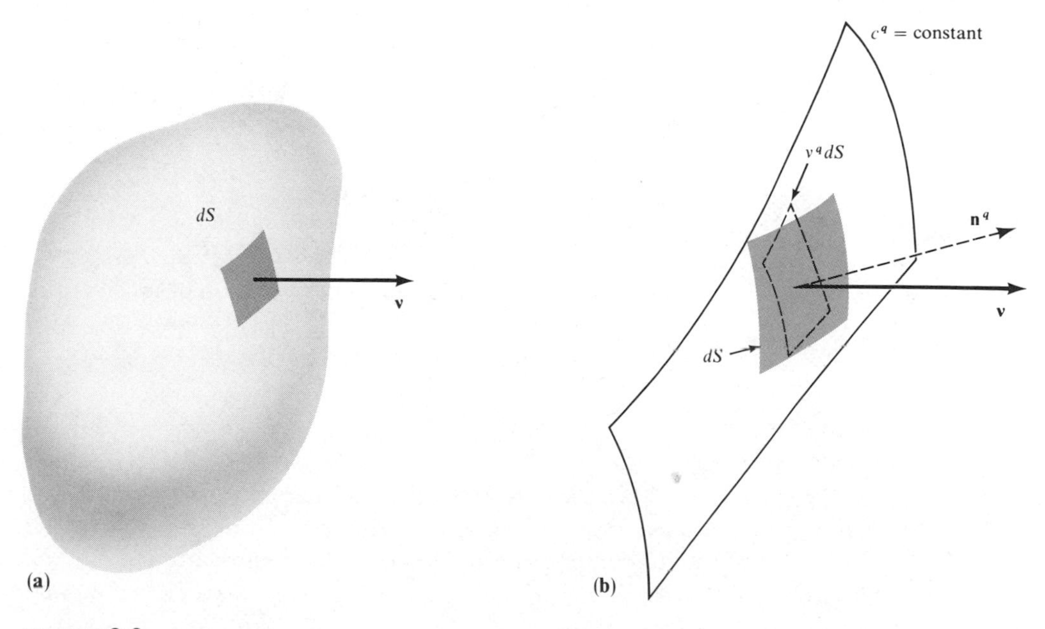

FIGURE **2.6**
The projection of dS onto the surface c^q = constant. The resulting area on the coordinate surface is $\nu^q\, dS$. (**a**) Shown here is dS as part of the surface of V. (**b**) The projection of dS, in broken outline, onto the coordinate surface c^q = constant.

Now $\nu^q\, dS$ is the projection of dS onto the surface c^q = constant (see Fig. 2.6), so that $\nu^1\, dS = h^2h^3\, dc^2\, dc^3$; similarly for $\nu^2\, dS$ and $\nu^3\, dS$. It follows that

$$\iint_S T_i\, dS = \sum_p \iint [\tau^{p1}n_i^p h^2h^3\, dc^2\, dc^3 + \tau^{p2}n_i^p h^3h^1\, dc^3\, dc^1 + \tau^{p3}n_i^p h^1h^2\, dc^1\, dc^2]$$

$$= \sum_p \iiint_V \left[\frac{\partial}{\partial c^1}(\tau^{p1}n_i^p h^2h^3) + \frac{\partial}{\partial c^2}(\tau^{p2}n_i^p h^3h^1) + \frac{\partial}{\partial c^3}(\tau^{p3}n_i^p h^1h^2)\right] dc^1\, dc^2\, dc^3.$$

But the physical volume element dV is $h^1h^2h^3\, dc^1\, dc^2\, dc^3$, so from steps parallel to the proof given in Section 2.1, we find here that

$$\rho\frac{\partial^2\mathbf{u}}{\partial t^2} = \mathbf{f} + \frac{1}{h^1h^2h^3}\sum_{p,q}\frac{\partial}{\partial c^q}\left(\tau^{pq}\mathbf{n}^p\frac{h^1h^2h^3}{h^q}\right). \tag{2.47}$$

Again, the derivative $\partial\mathbf{n}^p/\partial c^q$ is needed (see (6) in Box 2.6), and by resolving (2.47) along direction $\mathbf{n}^1$ we find

$$\rho\frac{\partial^2 u^1}{\partial t^2} = f^1 + \frac{1}{h^1h^2h^3}\left[\frac{\partial}{\partial c^1}(\tau^{11}h^2h^3) + \frac{\partial}{\partial c^2}(\tau^{12}h^3h^1) + \frac{\partial}{\partial c^3}(\tau^{31}h^1h^2)\right] + \frac{\tau^{12}}{h^1h^2}\frac{\partial h^1}{\partial c^2} + \frac{\tau^{31}}{h^3h^1}\frac{\partial h^1}{\partial c^3} - \frac{\tau^{22}}{h^1h^2}\frac{\partial h^2}{\partial c^1} - \frac{\tau^{33}}{h^3h^1}\frac{\partial h^3}{\partial c^1}. \tag{2.48}$$

Similar results for $\rho\ddot{u}^2$ and $\rho\ddot{u}^3$ can be found from a permutation of superscripts in (2.48).

The stress-strain relation, $\tau_{ij} = c_{ijkl}e_{kl}$, becomes

$$\tau_{ij} = \lambda\delta_{ij}e_{kk} + 2\mu e_{ij} \tag{2.49}$$

in isotropic media. We have used (2.33) here: λ and μ are (in general) functions of position, and $e_{kk} = e_{11} + e_{22} + e_{33}$ is the volumetric strain. Equation (2.49) is expressed in terms of components in the fixed-direction Cartesian system, but the corresponding result for physical components in the general orthogonal system has the same form. It is

$$\tau^{pq} = \lambda\delta^{pq}\sum_r e^{rr} + 2\mu e^{pq}, \tag{2.50}$$

since isotropy of the medium implies $c^{pqrs} = c_{pqrs,}$ and we can again use (2.33). The only difference in the form of (2.49) and (2.50) is due to our using a summation convention for subscripts but not for supercripts.

Applications of (2.46), (2.48), and (2.50) are common in spherical polars (r, θ, ϕ), for which the scaling functions h^1, h^2, h^3 become, respectively, $1, r, r \sin \theta$; and, in cylindrical polars, (r, ϕ, z) with scaling functions $1, r, 1$. In Chapter 4, we shall use orthogonal curvilinear coordinates associated with the wavefronts and rays that emanate from a point source in an inhomogeneous isotropic medium. Our convention of superscripts is convenient for the derivation of (2.45)–(2.50), but in applications the superscripts are usually replaced by subscripts that directly indicate the coordinate of interest. Thus, if (c^1, c^2, c^3) are the spherical polars (r, θ, ϕ), one refers to e^{12} as $e_{r\theta}$, to u^3 as u_ϕ, and to $\mathbf{n}^2$ as $\hat{\boldsymbol{\theta}}$.

SUGGESTIONS FOR FURTHER READING

Achenbach, J. D. *Wave Propagation in Elastic Solids*. Amsterdam: North-Holland, 1973.

Fung, Y. C. *Foundations of Solid Mechanics*. Englewood Cliffs, New Jersey: Prentice-Hall, 1965.

Jeffreys, H. *Cartesian Tensors*. Cambridge University Press, 1965.

Love, A. E. H. *A Treatise on the Mathematical Theory of Elasticity*. New York: Dover Publications, 1944.

Malvern, L. E. *Introduction to the Mechanics of a Continuous Medium*. Englewood Cliffs, New Jersey: Prentice-Hall, 1969.

PROBLEMS

2.1 What happens to the stress in a body if temperature is raised at fixed strain? Does the stress obey Hooke's law (2.18) or must this be modified in some way? (Recall that seismological applications of (2.18) are usually for adiabatic loading.)

2.2 We have shown how the displacement field $\mathbf{u}(\mathbf{x}, t)$ for an elastic body is given uniquely (e.g., by applied body forces and tractions). Show that body forces and tractions are given uniquely once $\mathbf{u}(\mathbf{x}, t)$ is known everywhere. (A proof "by construction" is very quick and simple.)

2.3 Do the relations (2.21)–(2.25) change if stress depends on strain rate (e.g., for a viscous medium)?

2.4 Obtain the traction due to displacement field **u** acting on area elements normal to **n**, in the form

$$\mathbf{T}(\mathbf{u}, \mathbf{n}) = \lambda(\nabla \cdot \mathbf{u})\mathbf{n} + \mu\left(2\frac{\partial \mathbf{u}}{\partial n} + \mathbf{n} \times (\nabla \times \mathbf{u})\right).$$

Here

$$\frac{\partial \mathbf{u}}{\partial n} = (\mathbf{n} \cdot \nabla)\mathbf{n}.$$

2.5 The traction **T** in the previous question is a function of position **x**, in the sense that $\mathbf{T} = \mathbf{T}(\mathbf{u}(\mathbf{x}), \mathbf{n})$.

a) Modify our derivation of (2.7) to show that traction is a continuous function of position, in the sense that

$$\mathbf{T}(\mathbf{x} + \delta\mathbf{x}) - \mathbf{T}(\mathbf{x}) \to \mathbf{0} \quad \text{as} \quad \delta\mathbf{x} \to \mathbf{0},$$

provided $\delta\mathbf{x}$ is taken parallel to the direction **n** that defines the orientation of area elements on which traction is evaluated.

b) Consider a book resting on a flat table. Is it true that traction is a continuous function of position on the surface of the table?

c) Check that your answers to a) and b) are not in conflict.

d) Show that τ_{yz}, τ_{zx}, τ_{zz} are continuous functions of z in any medium, but that τ_{zz} need not be continuous in the x- or y-directions; and that τ_{xx}, τ_{yy} and τ_{xy} need not be continuous in the z-direction.

2.6 For a point at pressure P in a fluid, the stress tensor is isotropic and has components $\tau_{ij} = -P\delta_{ij}$. To emphasize the differences between stresses that are possible in a solid and those that are present in a fluid, it is convenient to define *deviatoric stresses* τ'_{ij} by $\tau_{ij} = \frac{1}{3}\tau_{kk}\delta_{ij} + \tau'_{ij}$ and *deviatoric strains* by $e_{ij} = \frac{1}{3}e_{kk}\delta_{ij} + e'_{ij}$. Show then that the strain energy $\mathscr{U}$ in an isotropic elastic medium is given by

$$\mathscr{U} = \tfrac{1}{2}\left[(\lambda + \tfrac{2}{3}\mu)e_{ii}e_{kk} + 2\mu e'_{ij}\,e'_{ij}\right].$$

Show that e_{ii} is the change in volume per unit volume (i.e., the volumetric strain). Hence $\mathscr{U}$ can be regarded as a sum of dilatational energy, $\frac{1}{2}(\lambda + \frac{2}{3}\mu)e_{ii}e_{kk}$, and strain energy $\mu e'_{ij}e'_{ij}$. Why must $\lambda + \frac{2}{3}\mu$ (often called the *bulk modulus*, denoted by κ) and μ be positive? Is it natural to call κ the *compressibility* or the *incompressibility*?

CHAPTER 3

Representation of Seismic Sources

Seismic waves are set up by winds, ocean waves, meteorite impacts, rocket launchings, and atmospheric explosions—even by people walking around in the vicinity of seismometers. These, however, are examples of sources external to the solid Earth, and they can usually be analyzed within the simple framework of time-varying tractions applied to the Earth's surface. Other sources that, for many practical purposes, are also external, include volcanic eruptions, vented explosions, and spalling (free fall of a surface layer thrown upward by an underground explosion). For internal sources, such as earthquakes and underground explosions, the analytical framework is more difficult to develop, because the equations governing elastic motion (2.17)–(2.18) do not hold throughout the solid Earth. This chapter is about internal sources, and we shall distinguish two different categories: faulting sources and volume sources.

A faulting source is an event associated with an internal surface, such as slip across a fracture plane. A volume source is an event associated with an internal volume, such as a sudden (explosive) expansion throughout a volumetric source region. We shall find that a unified treatment of both source types is possible, the common link being the concept of an internal surface across which discontinuities can occur in displacement (for the faulting source) or in strain (for the volume source).

The mathematical description of internal seismic sources has classically been pursued along two different lines: first, in terms of a body force applied to certain elements of the medium containing the source; and second, by some discontinuity in displacement or strain (e.g., across a rupturing fault surface or across the surface of a volume source). The second approach can usefully be incorporated into the first, and we begin our analysis by developing body-force equivalents in some detail for simple shearing across a fault surface, showing that radically different systems of forces can be equivalent to exactly the same displacement discontinuity. We then develop the general theory for faulting sources, following Burridge and Knopoff (1964), and finally we outline the theory for a volume source.

The motions recorded in a seismogram are a result both of propagation effects and of source effects. Thus a major reason for seeking a better understanding of the source mechanism has been to isolate the propagation effects, since these bear information on the Earth's internal structure. More recently, earthquake source mechanisms have been studied to chart the motions of tectonic plates and to learn how the plates are driven. Source theory is now widely being developed with a view to predicting earthquake hazards at engineering sites on the basis of geological and geophysical data on the properties of nearby faults and the distribution of regional stresses.

3.1 Representation Theorems for an Internal Surface; Body-force Equivalents for Discontinuities in Traction and Displacement

The representation theorems obtained in Chapter 2 can be a powerful aid in seismic source theory if the surface S is chosen to include two adjacent surfaces internal to the volume V. The motivation here comes from the work of H. F. Reid, whose study of the San Andreas fault before and after the 1906 San Francisco earthquake has led to general recognition that earthquake motion is due to waves radiated from spontaneous slippage on active geological faults. We shall discuss this source mechanism in more detail in Sections 3.2 and 3.3, and the dynamical processes involved (and other source mechanisms) in Chapter 15. Our present concern is simply to show how the process of slip on a buried fault, and the waves radiated from it, can naturally be analyzed by our representation theorems.

For applications of (2.41)–(2.43), we shall take the surface of V to consist of an external surface labeled S (see Fig. 3.1) and two adjacent internal surfaces, labeled Σ^+ and Σ^-, which are opposite faces of the fault. If slip occurs across Σ, then the displacement field is discontinuous there and the equation of motion is no longer satisfied throughout the interior of S. However, it *is* satisfied throughout the "interior" of the surface $S + \Sigma^+ + \Sigma^-$, and to this we can apply our previous representation results.

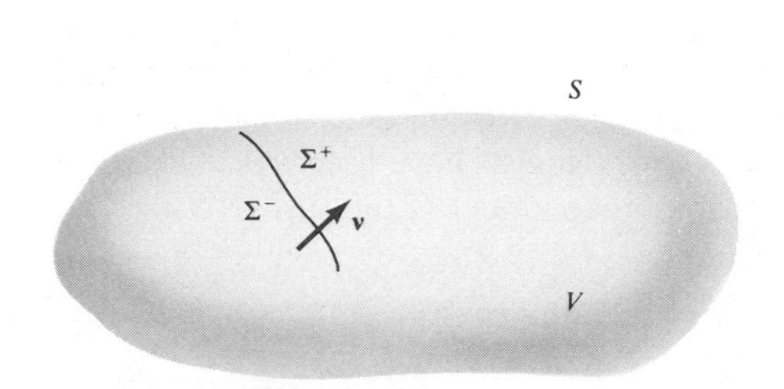

FIGURE **3.1**

A finite elastic body, with volume V and external surface S, and an internal surface Σ (modeling a buried fault) across which discontinuities may arise. That is, displacements on the Σ^- side of Σ may differ from displacements on the Σ^+ side of Σ. The normal to Σ is $\boldsymbol{\nu}$ (pointing from Σ^- to Σ^+), and the displacement discontinuity is denoted by $[\mathbf{u}(\boldsymbol{\xi}, \tau)]$ for $\boldsymbol{\xi}$ on Σ, with square brackets referring to the difference $\mathbf{u}(\boldsymbol{\xi}, \tau)|_{\Sigma^+} - \mathbf{u}(\boldsymbol{\xi}, \tau)|_{\Sigma^-}$. In general, a similar difference may be formed for the tractions (due to external applied forces on Σ), but for spontaneous rupture the tractions must be continuous, and $[\mathbf{T}(\mathbf{u}, \boldsymbol{\nu})] = \mathbf{0}$.

The surface S is no longer of direct interest (it may be the surface of the Earth), and we shall assume that both $\mathbf{u}$ and $\mathbf{G}$ satisfy homogeneous boundary conditions on S. Then

$$
\begin{aligned}
u_n(\mathbf{x}, t) = &\int_{-\infty}^{\infty} d\tau \iiint_V f_p(\boldsymbol{\eta}, \tau) G_{np}(\mathbf{x}, t - \tau; \boldsymbol{\eta}, 0)\, dV(\boldsymbol{\eta}) \\
&+ \int_{-\infty}^{\infty} d\tau \iint_\Sigma \{[u_i(\boldsymbol{\xi}, \tau) c_{ijpq} \nu_j \partial G_{np}(\mathbf{x}, t - \tau; \boldsymbol{\xi}, 0)/\partial \xi_q] \\
&- [G_{np}(\mathbf{x}, t - \tau; \boldsymbol{\xi}, 0) T_p(\mathbf{u}(\boldsymbol{\xi}, \tau), \boldsymbol{\nu})]\}\, d\Sigma.
\end{aligned}
\qquad (3.1)
$$

This formula uses $\boldsymbol{\eta}$ as the general position within V, and $\boldsymbol{\xi}$ as the general position on Σ. Square brackets are used for the difference between values on Σ^+ and Σ^- (see legend of Fig. 3.1).

As yet, nothing has been assumed for the boundary conditions on Σ. Although the choice for $\mathbf{u}$ must conform to actual properties of displacement and traction across a rupturing fault surface, the choice for $\mathbf{G}$ can be made in any fashion that turns out to be useful. Thus, for $\mathbf{u}$, the slip on a fault leads to a nonzero value for $[\mathbf{u}]$, but the continuity of traction (see the proof of (2.7), and Problem 2.5) implies $[\mathbf{T}(\mathbf{u}, \boldsymbol{\nu})] = \mathbf{0}$. The simplest and most commonly used way of establishing a defining property of $\mathbf{G}$ on Σ is to take Σ as an artificial surface across which $\mathbf{G}$ and its derivatives are continuous, so that $\mathbf{G}$ satisfies the equation of motion (2.36) even on Σ. This is the easiest Green function to compute for the volume V, and (in the absence of body forces for $\mathbf{u}$) it gives the representation

$$
u_n(\mathbf{x}, t) = \int_{-\infty}^{\infty} d\tau \iint_\Sigma [u_i(\boldsymbol{\xi}, \tau)] c_{ijpq} \nu_j\, \partial G_{np}(\mathbf{x}, t - \tau; \boldsymbol{\xi}, 0)/\partial \xi_q\, d\Sigma. \qquad (3.2)
$$

It is not surprising that displacement on the fault is enough to determine displacement everywhere: this feature of (3.2) might have been expected from the uniqueness theorem. Nevertheless, it *is* at first sight surprising that no boundary conditions on Σ are needed for the Green function that describes waves propagating from the source. One might expect that motions occurring on the fault would set up waves that are themselves diffracted in some fashion by the fault surface. Although this interaction complicates the determination of the slip function $[\mathbf{u}(\xi, \tau)]$, it does not enter into the determination of the Green function used in (3.2), and many seismologists have used this formula to compute the motions set up by some assumed model of the slip function. We shall describe such integrations in Chapter 14.

BODY-FORCE EQUIVALENTS

The earthquake model we have just described does not directly involve any body forces, though the representation (3.2) does give displacement at $(\mathbf{x}, t)$ as an integral over contributing Green functions, each of which is set up by a body force. Thus there must be some sense in which an active fault surface can be regarded as a surface distribution of body forces.

To determine this body-force equivalent, we start with (3.1) and assume still that Σ is transparent to $\mathbf{G}$. Making no assumptions about $[\mathbf{u}]$ and $[\mathbf{T}(\mathbf{u}, \mathbf{n})]$ across Σ (so that sources of traction are also allowed), we find

$$\begin{aligned} u_n(x, t) = & \int_{-\infty}^{\infty} d\tau \iiint_V f_p(\boldsymbol{\eta}, \tau) G_{np}(\mathbf{x}, t - \tau; \boldsymbol{\eta}, 0)\, dV(\boldsymbol{\eta}) \\ & + \int_{-\infty}^{\infty} d\tau \iint_\Sigma \{[u_i(\xi, \tau)] c_{ijpq} \nu_j G_{np,q}(\mathbf{x}, t - \tau; \xi, 0) \\ & - [T_p(\mathbf{u}(\xi, \tau), \boldsymbol{\nu})] G_{np}(\mathbf{x}, t - \tau; \xi, 0)\}\, d\Sigma(\xi). \end{aligned} \tag{3.3}$$

The discontinuities on Σ can be localized within V by using the delta function $\delta(\boldsymbol{\eta} - \xi)$. For example, $[\mathbf{T}]\, d\Sigma(\xi)$ has the dimensions of force, and its body-force distribution (i.e., force/unit volume) is $[\mathbf{T}]\delta(\boldsymbol{\eta} - \xi)\, d\Sigma$ as $\boldsymbol{\eta}$ varies throughout V. The traction discontinuity therefore contributes the displacement

$$\int_{-\infty}^{\infty} d\tau \iiint_V \left\{ -\iint_\Sigma [T_p(\mathbf{u}(\xi, \tau), \boldsymbol{\nu})] \delta(\boldsymbol{\eta} - \xi)\, d\Sigma \right\} G_{np}(\mathbf{x}, t - \tau; \boldsymbol{\eta}, 0)\, dV.$$

Since this expression has precisely the form of a body-force contribution (see the first term in the right-hand side of (3.3)), the body-force equivalent of a traction discontinuity on Σ is given by $\mathbf{f}^{[\mathbf{T}]}$, where

$$\mathbf{f}^{[\mathbf{T}]}(\boldsymbol{\eta}, \tau) = -\iint_\Sigma [\mathbf{T}(\mathbf{u}(\xi, \tau), \boldsymbol{\nu})] \delta(\boldsymbol{\eta} - \xi)\, d\Sigma(\xi). \tag{3.4}$$

The displacement discontinuity is harder to interpret, displacement being less simply related to force than is traction. We use the delta function derivative $\partial\delta(\boldsymbol{\eta} - \boldsymbol{\xi})/\partial\eta_q$ to localize points of Σ within V. This function has the property

$$\frac{\partial}{\partial\xi_q} G_{np}(\mathbf{x}, t - \tau; \boldsymbol{\xi}, 0) = -\iiint_V \frac{\partial}{\partial\eta_q} \delta(\boldsymbol{\eta} - \boldsymbol{\xi}) G_{np}(\mathbf{x}, t - \tau; \boldsymbol{\eta}, 0)\, dV(\boldsymbol{\eta}),$$

so that the displacement discontinuity in (3.3) contributes the displacement

$$\int_{-\infty}^{\infty} d\tau \iiint_V \left\{ - \iint_\Sigma [u_i(\boldsymbol{\xi}, \tau)] c_{ijpq} \nu_j \frac{\partial}{\partial\eta_q} \delta(\boldsymbol{\eta} - \boldsymbol{\xi})\, d\Sigma \right\} G_{np}(\mathbf{x}, t - \tau; \boldsymbol{\eta}, 0)\, dV$$

at position $\mathbf{x}$ and time t. The body-force equivalent $\mathbf{f}^{[\mathbf{u}]}$, of a displacement discontinuity on Σ can now be recognized from this expression as

$$f_p^{[\mathbf{u}]}(\boldsymbol{\eta}, \tau) = - \iint_\Sigma [u_i(\boldsymbol{\xi}, \tau)] c_{ijpq} \nu_j \frac{\partial}{\partial\eta_q} \delta(\boldsymbol{\eta} - \boldsymbol{\xi})\, d\Sigma. \tag{3.5}$$

Although the integrand here involves 27 terms (summation over i, j, q), which are different for each p, we shall find important examples in which only two or three terms are nonzero. The body-force equivalents (3.4) and (3.5) hold for a general inhomogeneous anistropic medium, and they are remarkable in their dependence on properties of the elastic medium only at the fault surface itself.

BOX **3.1**

On the use of effective slip and effective elastic constants in the source region

We are using the words "fault plane" and "fault surface," symbolized by Σ, as mathematical entities that have no thickness. Yet there are many places in the world where Earth scientists have direct access to fault regions, and one often finds there a zone of crushed and deformed rock, perhaps several meters thick, so that geologists often speak of "fault gouge" and a "fault zone." What, then, is meant by our claim that body-force equivalents depend only on elastic constants at the fault surface?

The thickness of the fault zone itself is almost always far less than the wavelengths of detectable seismic radiation, hence it is the displacement change across the whole fault zone that is the apparent displacement discontinuity, initiating waves which propagate out of the source region. Therefore, in almost all practical cases, the elastic constants for equations (3.2), (3.3), and (3.5) are the constants appropriate for the competent (unaltered) rock adjoining the fault zone.

Since faulting within the volume V is an internal process, the total momentum and total angular momentum must be conserved. It follows that the total force due to $\mathbf{f}^{[\mathbf{u}]}$, and the total moment of $\mathbf{f}^{[\mathbf{u}]}$ about any fixed point, must be zero. Thus

$$\iiint_V \mathbf{f}^{[\mathbf{u}]}(\boldsymbol{\eta}, \tau)\, dV(\boldsymbol{\eta}) = \mathbf{0} \qquad \text{for all } \tau, \tag{3.6}$$

and

$$\iiint_V (\boldsymbol{\eta} - \boldsymbol{\eta}_0) \times \mathbf{f}^{[\mathbf{u}]}(\boldsymbol{\eta}, \tau)\, dV(\boldsymbol{\eta}) = \mathbf{0} \qquad \text{for all } \tau \text{ and any fixed } \boldsymbol{\eta}_0. \tag{3.7}$$

To verify (3.6), note that the p-component of the left-hand side is $-\iint_\Sigma [u_i] c_{ijpq} \nu_j \{\iiint_V \partial\delta(\boldsymbol{\eta} - \boldsymbol{\xi})/\partial\eta_q dV\}\, d\Sigma(\boldsymbol{\xi})$. The volume integral here is $\iint_S \delta(\boldsymbol{\eta} - \boldsymbol{\xi}) n_q\, dS(\boldsymbol{\eta})$, which vanishes because $\boldsymbol{\eta}$ on S can never equal $\boldsymbol{\xi}$ (S and Σ having no common point).

To verify (3.7), write the m-component of the left-hand side as

$$\iiint_V \varepsilon_{mnp}(\eta_n - \eta_{0n}) f_p^{[\mathbf{u}]}\, dV$$

$$= -\iint_\Sigma c_{ijpq}\nu_j[u_i] \left\{ \iiint_V \varepsilon_{mnp}(\eta_n - \eta_{0n}) \frac{\partial}{\partial \eta_q} \delta(\boldsymbol{\eta} - \boldsymbol{\xi})\, dV \right\} d\Sigma \qquad \text{(from (3.5))}$$

$$= +\iint_\Sigma \varepsilon_{mqp} c_{ijpq} \nu_j [u_i]\, d\Sigma \qquad \left(\text{using } \frac{\partial}{\partial \eta_q}(\eta_n - \eta_{0n}) = \delta_{nq}\right)$$

$$= 0 \qquad (\text{using the symmetry } c_{ijpq} = c_{ijqp}).$$

As a simple example of a body force that is equivalent to a field discontinuity, consider the case of a body force applied at just one point, and in a particular direction (e.g., the body force for a Green function, given by (2.4)). This can instead be regarded as a discontinuity in a component of stress. To obtain the equivalence, take x_3 as the depth direction and consider a vertical point force, with magnitude F, applied at $(0, 0, h)$ and time $\tau = 0$ and held steady. Then

$$\mathbf{f}(\boldsymbol{\eta}, \tau) = (0, 0, F)\delta(\eta_1)\delta(\eta_2)\delta(\eta_3 - h)H(\tau). \tag{3.8}$$

This can instead be regarded as a discontinuity in traction across one point of the plane $\xi_3 = h$, with

$$[\mathbf{T}(\boldsymbol{\xi}, \tau)]_{\boldsymbol{\xi}=(\xi_1, \xi_2, h^-)}^{\boldsymbol{\xi}=(\xi_1, \xi_2, h^+)} = -(0, 0, F)\delta(\xi_1)\delta(\xi_2)H(\tau), \tag{3.9}$$

i.e., τ_{13}, τ_{23} are continuous, and the jump is in τ_{33}. The equivalence of (3.8) and (3.9) can be shown by a straightforward application of (3.4).

The most important example of a body-force equivalent in seismology is found in shear faulting, and we next take up this subject in some detail.

3.2 A Simple Example of Slip on a Buried Fault

The seismic waves set up by fault slip are the same as those set up by a distribution on the fault of certain forces with canceling moment. The distribution (for given fault slip) is not unique, but in an isotropic medium it can always be chosen as a surface distribution of double couples. This conclusion is ironical, in view of arguments used in a long-lasting debate on the question of whether earthquakes are modeled by a single couple or a double couple. Those who advocated the single-couple theory did believe that earthquakes were due to slip on a fault, but they intuitively thought this was equivalent to a single couple (composed of two forces corresponding to the motions on opposite sides of the fault). An intuitive approach is often dangerous in elastodynamics. On the other hand, some of those who advocated the double-couple theory thought that an earthquake must be voluminal collapse under pre-existing shear stress. In recent years, the fault theory (now recognized as the equivalent of a double couple) has been gaining strong support from increasing amounts of data obtained very close to earthquake sources as well as support from the radiation patterns observed at great distances.

As Figure 3.2 shows, we shall take the fault Σ to lie in the plane $\xi_3 = 0$, so that $[\mathbf{u}]$ has no component in the ξ_3-direction. (This is the case we have called

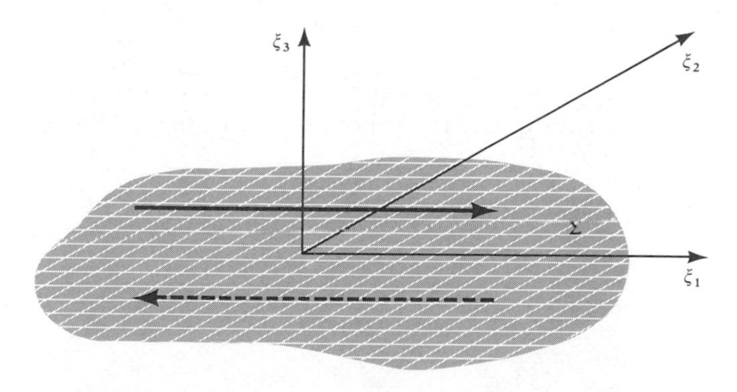

FIGURE **3.2**

A fault surface Σ within an isotropic medium is shown lying in the $\xi_3 = 0$ plane. Slip is presumed to take place in the ξ_1-direction across Σ, as shown by the heavy arrows. Motion on the side Σ^+ (i.e., $\xi_3 = 0^+$) is along the direction of ξ_1 increasing, and on the side Σ^- is along ξ_1 decreasing.

"fault slip"; i.e., $[\mathbf{u}]$ is parallel to Σ.) Let ξ_1 be the direction of slip, so that $[u_2] = [u_3] = 0$ and $\nu_1 = \nu_2 = 0$. Then the body-force equivalent, from (3.5), reduces to

$$f_p(\boldsymbol{\eta}, \tau) = -\iint_\Sigma [u_1(\boldsymbol{\xi}, \tau)] c_{13pq} \frac{\partial}{\partial \eta_q} \delta(\boldsymbol{\eta} - \boldsymbol{\xi})\, d\xi_1\, d\xi_2.$$

In isotropic (though still possibly inhomogeneous) media, we can find from (2.33) that all c_{13pq} vanish, except $c_{1313} = c_{1331} = \mu$. Hence

$$\begin{aligned} f_1(\boldsymbol{\eta}, \tau) &= -\iint_\Sigma \mu(\boldsymbol{\xi})[u_1(\boldsymbol{\xi}, \tau)]\delta(\eta_1 - \xi_1)\delta(\eta_2 - \xi_2) \frac{\partial}{\partial \eta_3} \delta(\eta_3)\, d\xi_1\, d\xi_2, \\ f_2(\boldsymbol{\eta}, \tau) &= 0, \\ f_3(\boldsymbol{\eta}, \tau) &= -\iint_\Sigma \mu[u_1] \frac{\partial}{\partial \eta_1} \delta(\eta_1 - \xi_1)\delta(\eta_2 - \xi_2)\delta(\eta_3)\, d\xi_1\, d\xi_2. \end{aligned} \tag{3.10}$$

First, let us look at f_1, which we shall find represents a system of single couples (forces in $\pm\eta_1$-direction, arm along η_3-direction, moment along η_2-direction) distributed over Σ. The integral above yields

$$f_1(\boldsymbol{\eta}, \tau) = -\mu(\boldsymbol{\eta})[u_1(\boldsymbol{\eta}, \tau)] \frac{\partial}{\partial \eta_3} \delta(\eta_3). \tag{3.11}$$

As shown in Figure 3.3, this component may be thought of as point forces distributed over the plane $\eta_3 = 0^+$ and opposed forces distributed over the plane $\eta_3 = 0^-$.

The total force due to f_1 vanishes (see discussion of (3.6)), but the moment of this force component alone does not. The total moment about the η_2-axis is

$$\iiint_V \eta_3 f_1\, dV = -\iiint_V \eta_3 \mu[u_1] \frac{\partial}{\partial \eta_3} \delta(\eta_3)\, d\eta_1\, d\eta_2\, d\eta_3 = \iint_\Sigma \mu[u_1(\boldsymbol{\xi}, \tau)]\, d\Sigma.$$

If slip is averaged over Σ to obtain the quantity

$$\bar{u}(\tau) \equiv \frac{\iint_\Sigma [u_1(\boldsymbol{\xi}, \tau)]\, d\Sigma}{A},$$

where $A \equiv \iint_\Sigma d\Sigma$ is the fault area, and if the fault region is homogeneous (so that μ is constant), then the total moment about the η_2-axis due to $f_1(\boldsymbol{\xi}, \tau)$ is simply $\mu \bar{u} A$ along the direction of η_2 increasing.

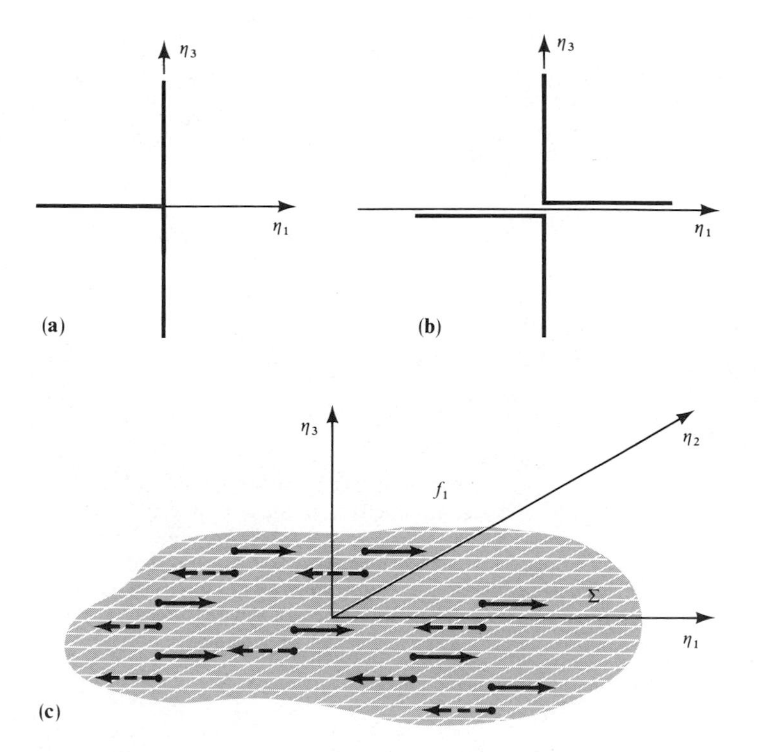

FIGURE **3.3**
Interpretive diagrams for the first component, f_1, of the body-force equivalent to fault slip of the type shown in Figure 3.2. (**a**) The spike $(-\delta(\eta_3), 0, 0)$ is plotted against η_3. (That is, a spike in the $-\eta_1$-direction, acting at $\eta_3 = 0$.) (**b**) The derivative $((-\partial/\partial\eta_3)\delta(\eta_3), 0, 0)$ is plotted against η_3. The body force $(f_1, 0, 0)$ is proportional to this quantity (see equation (3.11)). (**c**) Heavy arrows show the distribution of f_1 over the Σ^+ side of Σ and over the Σ^- side (broken arrows). This is the body-force component that would obviously be expected in any body-force model of the motions shown in Figure 3.2.

The body-force equivalent, given in (3.10), also involves f_3, and we shall find that this represents a system of single forces. Taking the η_1-derivative outside the integration, we find

$$f_3(\boldsymbol{\eta}, \tau) = -\frac{\partial}{\partial\eta_1}\{\mu[u_1(\boldsymbol{\eta}, \tau)]\}\delta(\eta_3). \tag{3.12}$$

Although this component is not itself a couple at each point on Σ, in the sense that we have shown f_1 to be a couple, the whole distribution of f_3 across Σ does have a net moment. Figure 3.4 shows how f_3 can reverse direction at different

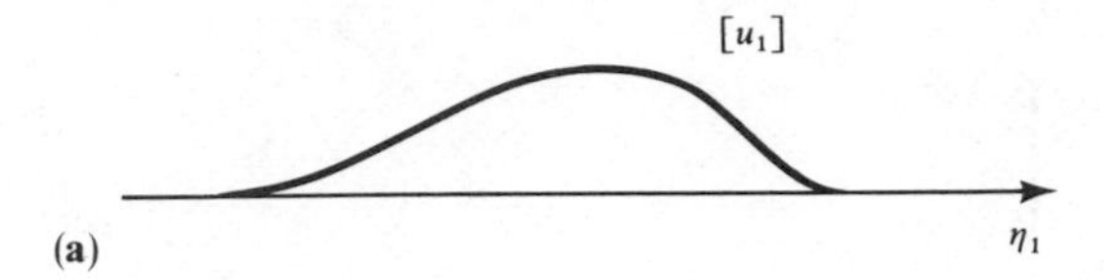

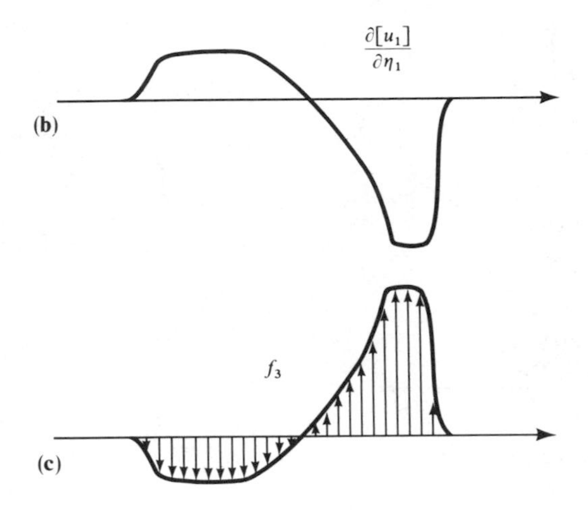

FIGURE **3.4**
Interpretive diagrams for the third component, f_3, of the body-force equivalent to fault slip $[u_1]$. (**a**) An assumed variation of slip $[u_1]$ with η_1, at fixed η_2 and τ. (**b**) The corresponding derivative $\partial[u_1]/\partial\eta_1$. (**c**) The distribution of single forces f_3 with varying η_1 (see equation (3.12)). This distribution will clearly yield a net couple, with moment in the $-\eta_2$-direction.

points of Σ. The total moment about the η_2-axis is

$$\iiint_V \varepsilon_{213}\eta_1 f_3 \, dV = \iiint_V \eta_1 \frac{\partial}{\partial \eta_1}\{\mu[u_1]\}\delta(\eta_3)\, d\eta_1\, d\eta_2\, d\eta_3$$

$$= \iint_\Sigma \xi_1 \frac{\partial}{\partial \xi_1}\{\mu[u_1]\}\, d\xi_1\, d\xi_2 = -\iint_\Sigma \mu[u_1]\, d\xi_1\, d\xi_2.$$

(This last equality follows from an integration by parts, using a fault surface Σ defined to have $[\mathbf{u}] = \mathbf{0}$ around its perimeter.) In a homogeneous source region, it follows that the total moment due to f_3 is $-\mu\bar{u}A$, which is equal in magnitude to the total moment of f_1, but acts in the opposite direction. We obtained this result in more general form in (3.7), but have found here the two canceling contributions that arise.

We have now shown that fault slip is equivalent to a distribution of single couples (f_1), plus a distribution of single forces (f_3) that have the net effect of an opposing couple. Yet the classical force equivalent for fault slip is a double-couple distribution over Σ, as was first shown for a finite fault by Maruyama in 1963. The fact is that force equivalents to a given fault slip are not unique. A direct way to see this, and to obtain the double-couple density as well as the single-couple/single-force density, is to write out representation (3.2) for the fault slip described in Figure 3.2. The result is

$$u_n(\mathbf{x}, t) = \int_{-\infty}^{\infty} d\tau \iint_{\Sigma} \mu[u_1]\left\{\frac{\partial G_{n1}}{\partial \xi_3} + \frac{\partial G_{n3}}{\partial \xi_1}\right\} d\Sigma. \tag{3.13}$$

The first term here in curly brackets, $\partial G_{n1}(\mathbf{x}, t - \tau; \boldsymbol{\xi}, 0)/\partial \xi_3$, is the limit of

$$\frac{(G_{n1}(\mathbf{x}, t - \tau; \boldsymbol{\xi} + \varepsilon\hat{\boldsymbol{\xi}}_3, 0) - G_{n1}(\mathbf{x}, t - \tau; \boldsymbol{\xi} - \varepsilon\hat{\boldsymbol{\xi}}_3, 0))}{2\varepsilon}$$

as $\varepsilon \to 0$. (We take $\hat{\boldsymbol{\xi}}_i$ as a unit vector in the ξ_i-direction.) This is the single-couple distribution shown in Figure 3.5a. The second term in (3.13) involves the limit of

$$\frac{G_{n3}(\mathbf{x}, t - \tau; \boldsymbol{\xi} + \varepsilon\hat{\boldsymbol{\xi}}_1, 0) - G_{n3}(\mathbf{x}, t - \tau; \boldsymbol{\xi} - \varepsilon\hat{\boldsymbol{\xi}}_1, 0)}{2\varepsilon},$$

and this single-couple distribution is shown in Figure 3.5b. These two systems form a double-couple distribution, and we must ask why the earlier set of body-force equivalents we derived, (3.10), made up a single couple plus a single force.

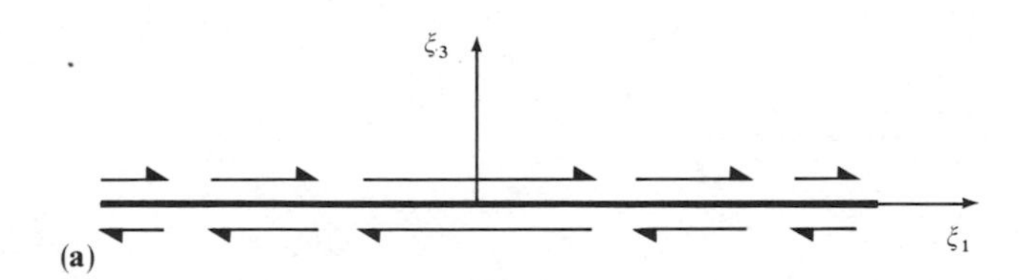

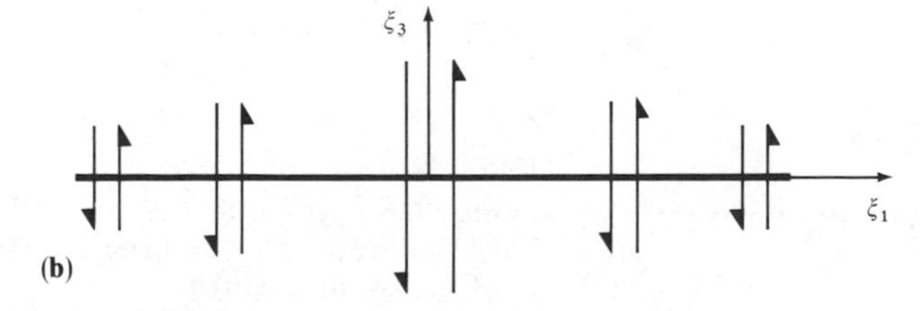

FIGURE **3.5**

The two single-couple distributions that, taken together, are equivalent (in the sense of radiating the same waves) to fault slip. Note that there is no net couple, and no net force, acting on any element of area in the fault plane ($\xi_3 = 0$).

The answer can be seen if (3.13) is integrated by parts, giving

$$u_n(\mathbf{x}, t) = \int_{-\infty}^{\infty} d\tau \iint_{\Sigma} \mu \left([u_1] \frac{\partial G_{n1}}{\partial \xi_3} - \left[\frac{\partial u_1}{\partial \xi_1}\right] G_{n3} \right) d\Sigma. \tag{3.14}$$

This force system is illustrated in Figure 3.6; clearly it is the same as the system we found first of all, shown in Figures 3.3 and 3.4. There is always a single couple (f_1, Figs. 3.3, 3.5a, and 3.6a) made up of forces in the same direction as fault-surface displacements (Fig. 3.2). But a complete equivalent to fault slip has another part, which may be a single force (f_3, Figs. 3.4 and 3.6b), a single couple (Fig. 3.5b), or an appropriate linear combination of these alternative extremes. For a given element of area $d\Sigma$ on the fault, these force systems are physically quite different: from the integrand in representation (3.13), there appears to be no force or moment acting on $d\Sigma$; but from (3.14), there does appear to be both force and moment acting on $d\Sigma$, although we showed earlier that f_1 and f_3 integrate to give zero net force and zero net moment on the whole of Σ.

We have brought out these results in some detail, because they show the limited utility of force equivalents for studying the dynamics of fault slip. It is the whole fault surface that is radiating seismic waves, and we cannot assess from (3.13) or (3.14) the actual contribution made to the radiation by individual elements of fault area. This makes sense in physical terms, because individual

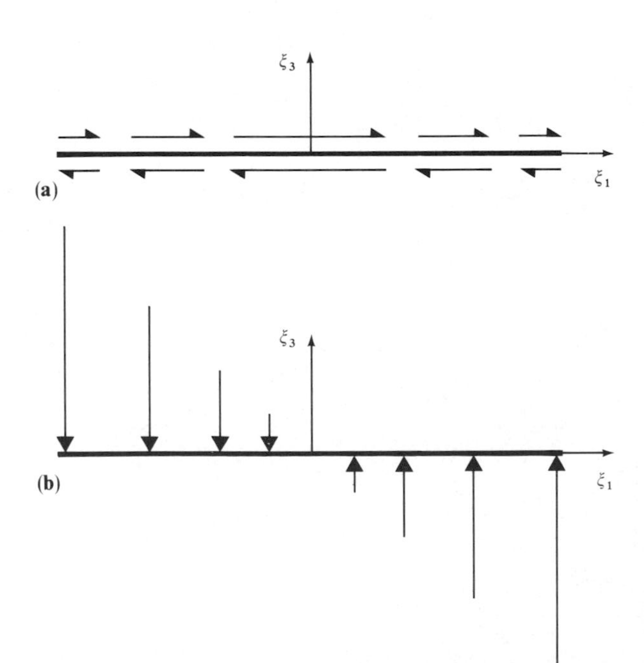

FIGURE **3.6**
Another force system that is equivalent to fault slip (compare with Fig. 3.5). (**a**) and (**b**) here constitute a single-couple plus single-force system, which has zero total couple and zero total force for the whole fault surface. But individual elements of area are acted on by a couple and a force.

elements of fault area do not move dynamically in isolation from other parts of the fault. Force equivalents (usually chosen as the double-couple distribution) find their main use only when the slip function $[\mathbf{u}(\xi, \tau)]$ has been determined (or guessed), and then they are important because they enable one to compute the radiation by weighting Green's functions.

At great distance from a rupturing fault, it often occurs that the only waves observed are those with wavelength much greater than linear dimensions of Σ, the causative fault. (Higher-frequency components are relatively weak even at the source—see Chapter 15—and are more effectively attenuated during propagation.) In such cases, Σ acts as a point source, and the slip is localized by replacing $[\mathbf{u}(\xi_1, \xi_2, \tau)]$ by the concentrated distribution $\bar{\mathbf{u}}A\delta(\xi_1)\delta(\xi_2)H(\tau)$. Then f_3, as well as f_1, becomes a single couple, and the double-couple equivalent to fault slip is

$$\begin{aligned} f_1(\boldsymbol{\eta}, \tau) &= -M_0\delta(\eta_1)\delta(\eta_2)\frac{\partial}{\partial\eta_3}\delta(\eta_3)H(\tau) \\ f_2 &= 0 \\ f_3(\boldsymbol{\eta}, \tau) &= -M_0\frac{\partial}{\partial\eta_1}\delta(\eta_1)\delta(\eta_2)\delta(\eta_3)H(\tau), \end{aligned} \tag{3.15}$$

where

$$M_0 = \mu\bar{u}A = \mu \times \text{average slip} \times \text{fault area.} \tag{3.16}$$

We call M_0 *seismic moment.* It is perhaps the most fundamental parameter we can use to measure the strength of an earthquake caused by fault slip. Measured values of M_0 range from about 10^{30} dyne-cm (1960 Chilean earthquake, 1964 Alaskan earthquake) down to around 10^{12} dyne-cm for microearthquakes, and 10^5 dyne-cm for microfractures in laboratory experiments on loaded rock samples. Even for geophysics, twenty-five orders of magnitude is an exceptionally large range to be spanned by a single physical variable. The first person to obtain the double-couple equivalence for an effective point source of slip was Vvedenskaya (1956). The first estimate of seismic moment was made by Aki (1966) for the Niigata earthquake of 1964, using long-period Love waves observed by the World-Wide Standard Seismograph Network.

We have defined M_0 as a constant, but for some purposes it is useful to recognize the seismic moment as a function of time, given by $\mu\bar{u}(t)A$, in which $\bar{u}$ is averaged at time t. In these cases, $M_0H(\tau)$ in (3.15) is replaced by $M_0(\tau)$, and (in the terminology of Chapter 14) we speak of the "rise time" being different from zero.

Note that there is a fundamental ambiguity in identifying the fault plane associated with the point-source double couple (3.15). We have worked with a fault surface normal to the x_3-direction and slip parallel to the x_1-direction. If the fault surface instead is taken normal to the x_1-direction and slip parallel to the x_3-direction with the same moment, then the equivalent body force is again (3.15). It follows that there can be no information in the seismic radiation from an effective point-source of slip that will enable one to distinguish between the fault plane and its *auxiliary plane* (i.e., the plane perpendicular to both the fault and the slip).

3.3 General Analysis of Displacement Discontinuities Across an Internal Surface Σ

In this section we shall introduce the seismic *moment tensor*, **M**. This is a quantity that depends on source strength and fault orientation, and it characterizes all the information about the source that can be learned from observing waves whose wavelengths are much longer than the linear dimensions of Σ. In this case, the source is effectively a point source with an associated radiation pattern, and the moment tensor can often be estimated in practice for a given earthquake by using long-period teleseismic data. In practice, seismologists use moment tensors that are confined to sources having a body-force equivalent given by couples alone. Such sources include geologic faults (shearing) and explosions (expansion), with **M** as a second-order tensor. For forces differentiated more than once, sources can be characterized by higher-order moment tensors.

For sources of finite extent, we shall introduce the seismic *moment density tensor*, **m**, which can often be thought of as $d\mathbf{M}/d\Sigma$, or as $d\mathbf{M}/dV$ for a volume source.

There are two ways in which this section generalizes Section 3.2. First, the coordinate axes are not taken in directions related to directionalities of the source. (This generality is important, because the direction of slip and the orientation of the fault plane are not usually known *a priori*, but must be deduced from the radiated seismic waves.) Second, discontinuities are to be allowed in the displacement component normal to the fault plane, so that apparent expansions or compressions will be allowed to occur.

Our starting point for the general analysis of displacement discontinuities is the representation (3.2), but using now the convolution symbol $*$ so that

$$u_n(\mathbf{x}, t) = \iint_{\Sigma} [u_i] \nu_j c_{ijpq} * \frac{\partial}{\partial \xi_q} G_{np} \, d\Sigma. \tag{3.17}$$

If X_0 is the amplitude of a force applied in the p-direction at $\boldsymbol{\xi}$ with general time variation, then the convolution $X_0 * G_{np}$ gives the n-component of displacement at $(\mathbf{x}, t)$ due to the varying point force at $\boldsymbol{\xi}$. More generally, if the force applied at $\boldsymbol{\xi}$ is $\mathbf{F}(\boldsymbol{\xi}, \tau)$, then we can sum over p and write $F_p * G_{np}$ for the n-component of displacement at ($\mathbf{x}$, t). As in (3.17), there is also a derivative of G_{np} with respect to the source coordinate ξ_q. Such a derivative, we saw in Section 3.2, can be thought of physically as the equivalent of having a single couple (with arm in the ξ_q-direction) on Σ at $\boldsymbol{\xi}$. The sum over q in (3.17) is then telling us that each displacement component at $\mathbf{x}$ is equivalent to the effect of a sum of couples distributed over Σ.

For three components of force and three possible arm directions, there are nine generalized couples, as shown in Figure 3.7 Thus the equivalent surface force corresponding to an infinitesimal surface element $d\Sigma(\boldsymbol{\xi})$ can be represented as a combination of nine couples. In general, we need "couples" with force and arm in the same direction (cases (1,1), (2,2), (3,3) of Fig. 3.7), and these are sometimes called *vector dipoles*.

Since $[u_i]v_j c_{ijpq} * \partial G_{np}/\partial \xi_q$ in (3.17) is the field at $\mathbf{x}$ due to couples at $\boldsymbol{\xi}$, it follows that $[u_i]v_j c_{ijpq}$ is the strength of the (p, q) couple. The dimensions of

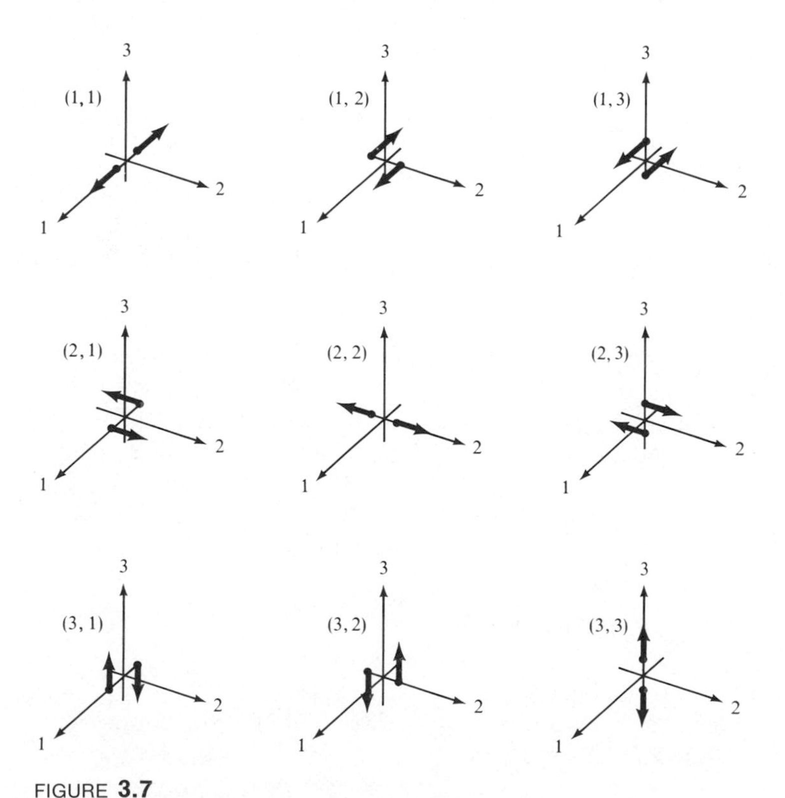

FIGURE **3.7**
The nine possible couples that are required to obtain equivalent forces for a generally oriented displacement discontinuity in anisotropic media.

$[u_i]\nu_j c_{ijpq}$ are moment per unit area, and this makes sense because the contribution from $\boldsymbol{\xi}$ has to be a surface density, weighted by the infinitesimal area element $d\Sigma$ to give a moment contribution. We define

$$m_{pq} \equiv [u_i]\nu_j c_{ijpq} \tag{3.18}$$

to be the components of the *moment density tensor*, **m**. In terms of this symmetric tensor, which is time dependent, the representation theorem for displacement at **x** due to general displacement discontinuity $[\mathbf{u}(\boldsymbol{\xi}, \tau)]$ across Σ is

$$u_n(\mathbf{x}, t) = \iint_{\Sigma} m_{pq} * G_{np,q}\, d\Sigma. \tag{3.19}$$

When we have learned more about the Green function (in Chapter 4), we shall find that the time dependence of the integrand in (3.19) is quite simple, because if **x** is many wavelengths away from $\boldsymbol{\xi}$, then convolution with **G** gives a field at $(\mathbf{x}, t)$ that depends on what occurs at $\boldsymbol{\xi}$ only at "retarded time," i.e., t minus some propagation time between $\boldsymbol{\xi}$ and **x**.

For an isotropic body, it follows from (2.33) and (3.18) that

$$m_{pq} = \lambda \nu_k [u_k(\boldsymbol{\xi}, \tau)]\delta_{pq} + \mu(\nu_p[u_q(\boldsymbol{\xi}, \tau)] + \nu_q[u_p(\boldsymbol{\xi}, \tau)]). \tag{3.20}$$

Further, if the displacement discontinuity (or slip) is parallel to Σ at $\boldsymbol{\xi}$, the scalar product $\boldsymbol{\nu} \cdot [\mathbf{u}]$ is zero and

$$m_{pq} = \mu(\nu_p[u_q] + \nu_q[u_p]). \tag{3.21}$$

In the case of Σ lying in the plane $\xi_3 = 0$, with slip only in the ξ_1-direction, we have the source model considered in Section 3.2, and for this the moment density tensor is

$$\mathbf{m} = \begin{pmatrix} 0 & 0 & \mu[u_1(\boldsymbol{\xi}, \tau)] \\ 0 & 0 & 0 \\ \mu[u_1(\boldsymbol{\xi}, \tau)] & 0 & 0 \end{pmatrix},$$

which is the now familiar double couple.

In the case of a tension crack in the $\xi_3 = 0$ plane, only the slip component $[u_3]$ is nonzero, and from (3.20) we find

$$\mathbf{m} = \begin{pmatrix} \lambda[u_3(\boldsymbol{\xi}, \tau)] & 0 & 0 \\ 0 & \lambda[u_3(\boldsymbol{\xi}, \tau)] & 0 \\ 0 & 0 & (\lambda + 2\mu)[u_3(\boldsymbol{\xi}, \tau)] \end{pmatrix}.$$

Thus a tension crack is equivalent to a superposition of three vector dipoles with magnitudes in the ratio $1:1:(\lambda + 2\mu)/\lambda$ (see Fig. 3.8).

The above results have been developed for a fault plane Σ of finite extent, but in practice the seismologist often has data that are good only at periods for which the whole of Σ is effectively a point source. For these waves, the contributions from different surface elements $d\Sigma$ are all approximately in phase, and the whole surface Σ can be considered as a system of couples located at a point, say the center of Σ, with *moment tensor* equal to the integral of moment density over Σ. Thus, for an effective point source,

$$u_n(\mathbf{x}, t) = M_{pq} * G_{np,q}, \tag{3.22}$$

where the moment tensor components are

$$M_{pq} = \iint_{\Sigma} m_{pq}\, d\Sigma = \iint_{\Sigma} [u_i] v_j c_{ijpq}\, d\Sigma, \qquad \text{i.e, } m_{pq} = \frac{dM_{pq}}{d\Sigma}. \tag{3.23}$$

In (3.22) we have one of the most important equations of this chapter. Later in this book, we shall evaluate the Green function and the different waves it contains. Thus in Chapter 4 we shall use ray theory for $\mathbf{G}$ and interpret (3.22) in terms of body waves excited by given $\mathbf{M}$ (equation (4.91)). In Chapter 7, we shall find the surface waves excited by $\mathbf{M}$ (equations (7.147)–(7.150)), and in Chapter 8 the normal modes of the whole Earth (8.37).

In terms of seismic moment M_0, and with the choice of coordinate axes made in Section 3.2, the moment tensor for an effective point source of slip is

$$\mathbf{M} = \begin{pmatrix} 0 & 0 & M_0 \\ 0 & 0 & 0 \\ M_0 & 0 & 0 \end{pmatrix}. \tag{3.24}$$

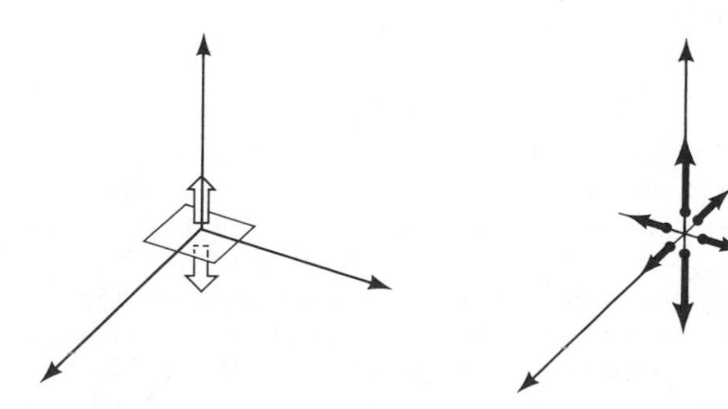

FIGURE **3.8**
The body-force equivalent for a tension crack in an isotropic medium.

Equations (3.23) justify the name "moment tensor density" for **m**. In the case of a finite source, we can now interpret the representation (3.19) as an areal distribution of point sources, each point having the moment tensor **m** $d\Sigma$.

BOX **3.2**

On uses of the word "moment" in seismic source theory

In rotational mechanics, it is often enough to speak of a couple possessing the qualities of magnitude and a single direction. The magnitude of a couple is then a scalar quantity called the moment. In our study of displacement discontinuities, however, and body-force equivalents, we imply more directional qualities behind the word "couple" than is the case in rigid-body rotational mechanics. For us, "couple" involves the directions of both force and lever arm. A result of this is that the quantity "moment" jumps up from scalar to tensor.

Second-order Cartesian tensors in mathematical physics are usually quantities that relate one physical vector to another (e.g., equation (2.16), in which the stress tensor is a device for obtaining traction from the vector orientation of an area element; and the inertia tensor I, which gives angular momentum **h** from angular velocity $\boldsymbol{\omega}$ via $h_i = I_{ij}\omega_j$). In seismic source theory, however, the moment tensor is an input rather than a filter, and it is operated on by a third-order tensor to yield vector displacement (see (3.19) and (3.22)).

BOX **3.3**

Body-force equivalents and the seismic moment tensor

For a general displacement discontinuity across Σ, it follows from (3.5) that

$$f_p = -\frac{\partial}{\partial \eta_q}\{[u_i]\nu_j c_{ijpq}\delta(\Sigma)\},$$

where, by $\delta(\Sigma)$, we mean a one-dimensional spatial Dirac delta function that is zero off Σ. Thus, if Σ lies in the plane $\eta_3 = 0$, $\delta(\Sigma) = \delta(\eta_3)$ for points (η_1, η_2) on Σ.

It must be emphasized that **f** is a force *per unit volume*, and it is unique. (Once $[u_i]$ is given on Σ, then **u** is determined everywhere, and $\mathbf{f} = \mathbf{L}(\mathbf{u})$, where **L** is given in Box 2.4.) The ambiguities mentioned in Section 3.2 arise only when equivalent *surface* forces are sought. Thus the above formula for f_p does not give a distribution of couples and dipoles. Such a distribution arises only after the displacement representation $\iiint_V G_{np}\{f_p\}\, dV$ has been integrated by parts and the η_3 integration completed to give (3.17), which may then be interpreted in terms of equivalent surface forces. These are nonunique (see (3.13) and (3.14)), but a surface distribution of couples and vector dipoles is always possible.

We have introduced the seismic moment tensor in the form $M_{pq} = \iint_\Sigma [u_i] v_j c_{ijpq}\, d\Sigma$, but from the above formula for body force it is easy to show that

$$M_{pq} = \iiint_V f_p \eta_q \, dV(\boldsymbol{\eta}).$$

This result can be used to extend the definition of **M**, since it can be used for any body-force distribution, and not just for the body-force equivalent to a displacement discontinuity. With this definition, the moment (in the ordinary sense of rotational mechanics) of body forces **f** about the ith axis is $\iiint_V \varepsilon_{ijk} \eta_j f_k \, dV = \varepsilon_{ijk} M_{kj}$, which is zero whenever the moment tensor is symmetric (e.g., in (3.23)).

We conclude this section with an interesting use of "seismic moment," suggested by Brune (1968), involving the kinematic motions of tectonic plates. Such motions lead frequently to a type of regional seismicity in which many different earthquakes share the same fault plane (although any one event will involve slip over only a part of the whole fault area). If M_0^i is the seismic moment of the ith earthquake in a series of N earthquakes, it follows from the definition of M_0^i that the total slip due to the whole series is

$$\Delta U = \frac{\sum_{i=1}^{N} M_0^i}{\mu S}, \tag{3.25}$$

where S is the total area broken in the series. ΔU is averaged over all of S, and all the terms in the right-hand side of (3.25) can be estimated. If the earthquake series includes all the significant events in a given period of time ΔT, and if the seismicity during ΔT is representative of the activity on that plate margin for longer time scales, then $\Delta U / \Delta T$ is an estimate of the relative velocity of the plates, regarded as slow-moving rigid bodies, which can be obtained from seismic data alone.

BOX **3.4**

The strain energy released by earthquake faulting

Within a medium that initially has a static stress field $\boldsymbol{\sigma}^0$, we suppose that a displacement discontinuity develops across an internal surface Σ. This leads to a displacement field $\mathbf{u}(\mathbf{x}, t)$, measured with reference to the initial configuration, and from $\mathbf{u}$ we can determine the additional time-dependent strain and the additional stress $\boldsymbol{\tau}$. Then the total stress is $\boldsymbol{\sigma} = \boldsymbol{\sigma}^0 + \boldsymbol{\tau}$, and after all motions have died down the new static stress field is $\boldsymbol{\sigma}^1$. If ΔE is defined as the change in strain energy throughout the medium, from its initial static configuration to its final static configuration, it can be shown that

$$\Delta E = -\tfrac{1}{2}\int_\Sigma [u_i](\sigma_{ij}^0 + \sigma_{ij}^1)\nu_j \, d\Sigma, \tag{1}$$

where $[\mathbf{u}]$ is the final offset. (See Fig. 3.1 for definitions of $[\quad]$ and $\boldsymbol{\nu}$.) Equation (1) is known as the Volterra relation (Steketee, 1958; Savage, 1969a).

This result (which we derive below) can be simply restated in terms of work apparently done by tractions on the fault surface. We can say from (1) that the drop in strain energy throughout the medium, $-\Delta E$, is the positive quantity obtained by imagining a quasistatic growth of traction that is linear with offset:

$$T = T^0 + (T^1 - T^0)\frac{U}{[u]} \qquad \text{for } 0 \le U \le [u] \tag{2}$$

(for each component of traction T and displacement U). Integrating from 0 to $[u]$ to get the total work done on Σ then gives (1).

Several points now need to be made about this relation between ΔE and the average stress.

The liberated energy, $-\Delta E$, supplies the work actually done on the two faces Σ^+ and Σ^- as they grind past each other, plus the work done in initiating the process of fracture. We discuss these two types of work in Chapter 15. Moreover, $-\Delta E$ supplies the seismic energy E_s that is radiated away from the source region. It is natural to introduce the *seismic efficiency*, η, as the ratio $E_s/(-\Delta E)$. Then

$$E_s = -\eta\,\Delta E = \tfrac{1}{2}\eta\int_\Sigma [u_i](\sigma_{ij}^0 + \sigma_{ij}^1)\nu_j \, d\Sigma. \tag{3}$$

If the average of the two static tractions does not vary strongly over Σ, then for the type of tangential slip shown in Figure 3.2 we see that (3) can be expressed in terms of the moment $M_0 = \mu \int_\Sigma [u_1]\, d\Sigma$. This gives

$$E_s = \eta M_0 \bar{\sigma}/\mu, \tag{4}$$

where $\bar{\sigma} = \tfrac{1}{2}(\sigma_{13}^0 + \sigma_{13}^1)$.

From estimates that can be made of E_s, M_0 and μ, it thus becomes possible from (4) to estimate the product $\eta\bar{\sigma}$, called the *apparent stress* by Wyss and Brune (1968, 1971).

Since the slip function $[\mathbf{u}(\boldsymbol{\xi}, t)]$ in (3.17) determines all displacements (and hence strain and stress increments) throughout the medium, it also determines the stress drop, $\boldsymbol{\sigma}^0 - \boldsymbol{\sigma}^1$. But there is no way one can work purely from observations of the radiated field $\mathbf{u}(\mathbf{x}, t)$ and learn anything about the absolute level of stress in the source region. Putting this another way, and using (1), one can make the following statement. If the same slip function $[\mathbf{u}(\boldsymbol{\xi}, t)]$ occurs on Σ in two different faulting events with different initial stresses, then all the seismic displacements will be the same for the two events: but the strain energies liberated for the two events will be different.

It remains, then, to prove our main result (1). This is a formula of great generality, and a correct derivation can in fact be given by considering the quasistatic deformation we described in (2). We shall give an explicit proof, however, for the special case in which the internal strain energy $\mathscr{U}$ is given by a strain-energy function $\mathscr{W}$ (see Section 2.2). Further, we assume there is an accessible reference state of zero stress and zero strain. The initial stresses and strains just prior to faulting are σ_{ij}^0 and e_{ij}^0, and $\mathbf{u}$ is measured from this state.

From (2.32) applied to the total stresses and strains, we get

$$\begin{aligned}\mathscr{W} &= \tfrac{1}{2}(\sigma_{ij}^0 + \tau_{ij})(e_{ij}^0 + u_{i,j}) \qquad \text{(using symmetry of } \sigma_{ij})\\ &= \mathscr{W}^0 + \tfrac{1}{2}\sigma_{ij}u_{i,j} + \tfrac{1}{2}c_{ijkl}u_{k,l}e_{ij}^0\\ &= \mathscr{W}^0 + \tfrac{1}{2}\sigma_{ij}u_{i,j} + \tfrac{1}{2}\sigma_{kl}^0 u_{k,l} \qquad \text{(using (2.30))}.\end{aligned}$$

Thus the increase in internal energy in the new static configuration is

$$\Delta E = \int_V (\mathscr{W}^1 - \mathscr{W}^0)\, dV = \tfrac{1}{2}\int_V (\sigma_{ij}^1 + \sigma_{ij}^0)u_{i,j}\, dV, \tag{5}$$

where V is the whole elastic volume containing Σ (see Fig. 3.1). Since σ_{ij}^0 and σ_{ij}^1 are static stress fields, (2.17) implies $\sigma_{ij,j}^0 = \sigma_{ij,j}^1 = 0$ (we assume there are no body forces). From (5), we obtain

$$\Delta E = \tfrac{1}{2}\int_V \{(\sigma_{ij}^0 + \sigma_{ij}^1)u_i\}_{,j}\, dV,$$

to which we can apply Gauss's divergence theorem, regarding V as the interior of $S + \Sigma^+ + \Sigma^-$. This does give (1) if S is either a rigid surface or if, like the surface of the Earth, it is free.

3.4 Volume Sources: Outline of the Theory and Some Simple Examples

In order to develop equations for seismic waves from buried explosions or from rapid phase transformations, it is necessary to introduce the concept of a volume source. We shall describe such a source in terms of a transformational (or stress-free) strain introduced in the source volume, and shall develop properties of an associated seismic moment tensor.

Let us illustrate this concept by a set of imaginary cutting, straining, and welding operations described by Eshelby (1957). First, we separate the source material by cutting along the surface Σ enclosing it and removing the source volume (the "inclusion") from its surroundings (the "matrix"). At this stage, we suppose that the material removed is held in its original shape by tractions having the same value over Σ as the tractions imposed across Σ by the matrix before the cutting operation. Second, we let the source material undergo transformational strain Δe_{rs}. By this, we mean that Δe_{rs} occurs without changing the stress within the inclusion, hence the name "stress-free strain." It is this strain that characterizes the seismic source. Processes that can be described by stress-free strain include phase transformation, thermal expansion, and some plastic deformations. Stress-free strain is a static concept. Third, we apply extra surface tractions that will restore the source volume to its original shape: this will result in an additional stress field $-c_{pqrs}\,\Delta e_{rs} = -\Delta\tau_{pq}$ throughout the inclusion, and the additional tractions applied on its surface Σ are $-c_{pqrs}\,\Delta e_{rs}\nu_q$, where ν_q is the outward normal on Σ. Since $\Delta\tau_{pq}$ is a static field, $\Delta\tau_{pq,q} = 0$. The stress in the matrix is still unchanged, being held at its original value by tractions imposed across the internal surface Σ, and having the same value as tractions imposed on the matrix by the inclusion before it was cut out. Fourth, we put the inclusion back in its hole (which is exactly the correct shape) and weld the material across the cut. The traction on Σ^- is now an amount $-c_{pqrs}\,\Delta e_{rs}\nu_q$ greater than that on Σ^+, leading to a traction discontinuity (in the $\boldsymbol{\nu}$-direction) of amount $+c_{pqrs}\,\Delta e_{rs}\nu_q$. This traction is due to applied surface forces that are external to the source and which act on the inclusion to maintain its correct shape. Fifth, we release the applied surface forces over Σ^-. Since traction is actually continuous across Σ, this amounts to imposing an apparent traction discontinuity of $-(c_{pqrs}\,\Delta e_{rs})\nu_q$. The elastic field produced in the matrix by the whole process is that due to the apparent traction discontinuity across Σ.

The above procedure can be extended to a dynamic case of seismic wave generation, since, at any given time, a transformational strain Δe_{rs} can be defined for the unrestrained material. For each instant, it is still true that $\Delta\tau_{pq,q} = 0$. The seismic displacement generated by the discontinuity in traction was given by (3.3). Putting $[T_p] = -(c_{pqrs}\,\Delta e_{rs})\nu_q$ in (3.3), we get

$$u_n(\mathbf{x}, t) = \int_{-\infty}^{\infty} d\tau \iint_{\Sigma} c_{pqrs}\,\Delta e_{rs}\nu_q G_{np}(\mathbf{x}, t - \tau; \boldsymbol{\xi}, 0)\, d\Sigma(\boldsymbol{\xi}). \tag{3.26}$$

If the integrand and its derivatives with respect to $\boldsymbol{\xi}$ are continuous, we can apply the Gauss theorem to obtain

$$u_n(\mathbf{x}, t) = \int_{-\infty}^{\infty} d\tau \iiint_{V} \frac{\partial}{\partial \xi_q}\{c_{pqrs}\,\Delta e_{rs} G_{np}(\mathbf{x}, t - \tau; \boldsymbol{\xi}, 0)\}\, dV(\boldsymbol{\xi}) \tag{3.27}$$

(V here refers only to the volume of the inclusion, i.e., the source volume). Using $\partial(c_{pqrs}\,\Delta e_{rs})/\partial\xi_q = \Delta\tau_{pq,q} = 0$, we can rewrite (3.27) and obtain

$$u_n(\mathbf{x}, t) = \iiint_{V} c_{pqrs}\,\Delta e_{rs} * \frac{\partial G_{np}}{\partial \xi_q}\, dV. \tag{3.28}$$

Comparing this volume integral with the surface integral in (3.17), one sees that it is natural to introduce a moment-density tensor

$$\frac{dM_{pq}}{dV} = c_{pqrs}\,\Delta e_{rs} \tag{3.29}$$

with the dimensions of moment per unit volume (compare also with (3.23)). Then

$$u_n(\mathbf{x}, t) = \iiint_V \frac{dM_{pq}}{dV} * \frac{\partial G_{np}}{\partial \xi_q}\,dV. \tag{3.30}$$

Note that $\Delta\tau_{pq} = dM_{pq}/dV$ is not the stress drop (the difference between the initial equilibrium stress and the final equilibrium stress in the source region), as is clear from its definition. The stress drop is not limited to the source volume, but $\Delta\tau_{pq}$ vanishes outside the source volume. $\Delta\tau_{pq}$ is called the "stress glut" by Backus and Mulcahy (1976).

For long waves, for which the whole of V is effectively a point source, the whole volume V can be considered a system of couples located at a point, say the center of V, with moment tensor equal to the integral of moment density over V. Thus, for an effective point source, (3.22) applies, with the moment tensor components

$$M_{pq} = \iiint_V c_{pqrs}\,\Delta e_{rs}\,dV. \tag{3.31}$$

For example, if a shear collapse occurs in a homogeneous isotropic body of volume V with the nonzero transformational strain components $\Delta e_{13} = \Delta e_{31}$, say, the moment tensor is

$$\mathbf{M} = 2\mu V \begin{pmatrix} 0 & 0 & \Delta e_{13} \\ 0 & 0 & 0 \\ \Delta e_{13} & 0 & 0 \end{pmatrix}. \tag{3.32}$$

The seismic radiation is identical to the point source equivalent to a fault-slip, except that the seismic moment M_0 is given by $2\mu\,\Delta e_{13} V$. For a group of earthquakes in an intraplate seismic zone, a cumulative strain may be more meaningful than a cumulative slip given by (3.25). Kostrov (1974) suggested summing moments for a group of earthquakes sharing the same source mechanism in a given volume to find the total strain in the volume. From (3.32), the total strain ΔE_{13} may be estimated as

$$\Delta E_{13} = \frac{\sum_{i=1}^{N} M_0^i}{2\mu V}, \tag{3.33}$$

where M_0^i is the moment of the ith earthquake.

Finally, let us consider a spherical volume with radius a undergoing a transformational expansion. The stress-free strain components in this case are $\Delta e_{12} = \Delta e_{13} = \Delta e_{23} = 0$ and $\Delta e_{11} = \Delta e_{22} = \Delta e_{33} = \frac{1}{3}\,\Delta\theta$, where $\Delta\theta$ is the fractional change in volume. For a homogeneous isotropic body, we have from (3.31)

$$\mathbf{M} = \frac{4\pi}{3} a^3 \begin{pmatrix} \Delta p & 0 & 0 \\ 0 & \Delta p & 0 \\ 0 & 0 & \Delta p \end{pmatrix} \tag{3.34}$$

where $\Delta p = (\lambda + \frac{2}{3}\mu)\,\Delta\theta$. Thus a spherical source with transformational volume expansion is equivalent to three mutually perpendicular dipoles, as shown in Figure 3.7. In the above equation, Δp should not be confused with the pressure jump at the spherical surface at radius a (see Problem 3.5).

SUGGESTIONS FOR FURTHER READING

Backus, G., and M. Mulcahy. Moment tensors and other phenomenological descriptions of seismic sources—I. Continuous displacements. *Geophysical Journal of the Royal Astronomical Society*, **46**, 341–361, 1976.

Backus, G., and M. Mulcahy. Moment tensors and other phenomenological descriptions of seismic sources—II. Discontinuous displacements. *Geophysical Journal of the Royal Astronomical Society*, **47**, 301–329, 1976.

Burridge, R., and L. Knopoff. Body force equivalents for seismic dislocations. *Bulletin of the Seismological Society of America*, **54**, 1875–1888, 1964.

Eshelby, J. D. The determination of the elastic field of an ellipsoidal inclusion and related problems. *Proceedings of the Royal Society*, **A241**, 376–396, 1957.

Kostrov, B. V. The theory of the focus for tectonic earthquakes. *Izvestia, Physics of the Solid Earth*, 258–267, April 1970.

Kostrov, B. V., Seismic moment and energy of earthquakes and seismic flow of rock *Izvestia, Physics of the Solid Earth*, 13–21, January 1974.

Maruyama, T. On force equivalents of dynamic elastic dislocations with reference to the earthquake mechanism. *Bulletin of the Earthquake Research Institute, Tokyo University*, **41**, 467–486, 1963.

Nyland, E. Body force equivalents as sources of anelastic processes. *Canadian J. Earth Sci.*, **8**, 1184–1189, 1971.

Press, F., and C. Archambeau. Release of tectonic strain by underground nuclear explosions. *Journal of Geophysical Research*, **67**, 337–342, 1962.

Savage, J. C. Steketee's paradox. *Bulletin of the Seismological Society of America*, **59**, 381, 1969.

Steketee, J. A. Some geophysical applications of the theory of dislocations. *Canadian Journal of Physics*, **36**, 1168–1198, 1958.

PROBLEMS

3.1 Equations (3.25) and (3.33) are written as scalar equations, because in their derivation it has been assumed that earthquakes in a given region (on S, or within V) all have moment tensors with the same orientations.

Generalize (3.25) to a vector equation and (3.33) to a tensor equation in cases where earthquakes in the series (on S or in V) have moment tensors of arbitrary orientation. (For (3.25), however, continue to assume that the displacement discontinuity for each event is a shear and that S is planar.)

3.2 In our derivation of (3.2), we have assumed that the elastic constants are continuous across Σ and that G_{np} and $\partial G_{np}/\partial \xi_q$ are continuous. If the elastic constants are *not* continuous across Σ, interpret part of the integrand in (3.2) as a traction, and show that this representation is still valid, although $\partial G_{np}/\partial \xi_q$ may not be continuous across the surface. (*Note*: For purposes of computing $\mathbf{G}$, assume Σ to be a welded surface.)

3.3 We have stated that the time-dependent seismic moment is given by $M_0(t) = \mu \bar{u}(t) A$. Is $\bar{u}(t)$ here averaged over the area $A(t)$ that has ruptured at time t or is it averaged over $A(\infty)$, the area that ultimately is ruptured during the seismic event under consideration?

3.4 Show that the body-force equivalent to a point source at ξ with moment tensor M_{pq} is given by

$$f_p(\mathbf{x}, t) = -M_{pq}(t) \frac{\partial}{\partial x_q} \delta(\mathbf{x} - \boldsymbol{\xi}).$$

3.5 Consider a spherical cavity with radius a inside a homogeneous isotropic body. When a uniform pressure jump ΔP is applied at the surface of the cavity, spherically symmetric waves with displacement component only in the radial direction will be generated. Show that the moment tensor of the point source equivalent to this seismic source has a form similar to (3.34) except that the factor $\frac{4}{3}$ is replaced by $(\lambda + 2\mu)/\mu$. (Use the fact that a radial displacement of the form $u(r, t) = -\partial/\partial r\,[\psi(t - r/\alpha)/r]$ satisfies the equation of motion, where r is the distance from the center of cavity and α is the velocity of compressional waves.)

CHAPTER 4

Elastic Waves from a Point Dislocation Source

We now begin a series of six chapters that develop the basic features of seismic wave propagation. In Chapters 2 and 3, we found that seismic motions can be represented by an integration over the spatial and temporal region within which the seismic source is acting. The integration (e.g., (3.2)) is a synthesis from the displacement field of a certain Green function, introduced in Section 2.4. This chapter is concerned with properties of Green's function itself, and we shall derive a specific formula (4.23) for the displacement within a homogeneous, isotropic, unbounded medium due to a unidirectional force acting with general time-varying strength at a particular point of the medium. The formula is obtained by using the device of potentials for elastic displacement, and P-waves and S-waves are identified.

For many purposes, a body force acting at a point is an adequate source model for the seismic displacements observed from an earthquake. But, for fault slip, the point body force itself is a double couple (see (3.15)) rather than a unidirectional force. We obtain the displacement field for a double couple and discuss its radiation pattern at large distances (the "far field") and at small distances (the "near field") from the source point.

In order to adapt these exact results for practical applications, a method is needed to assess the focusing and defocusing of seismic waves by systematic inhomogeneities in the Earth. Classical ray theory can be used to obtain an approximate solution to this problem in the far field, and the chapter concludes with some general results for the amplitude and radiation pattern (P and S) due to a double couple that models slip in a specified direction on a fault surface with specified strike and dip.

4.1 Formulation: Introduction of Potentials

The first major problem before us is to solve for the displacement $\mathbf{u}(\mathbf{x}, t)$ set up by a unidirectional point body force acting with time-varying magnitude at a fixed point O in a homogeneous, unbounded, isotropic, elastic medium. Without loss of generality, we take O as the origin of Cartesian coordinates and the x_1-axis as the body-force direction. The equation to be solved for $\mathbf{u}$ is (from (2.17), (2.18), (2.3), and (2.33))

$$\rho \ddot{u}_i = f_i + (\lambda + \mu) u_{j,ji} + \mu u_{i,jj}$$

or

$$\rho \ddot{\mathbf{u}} = \mathbf{f} + (\lambda + 2\mu)\nabla(\nabla \cdot \mathbf{u}) - \mu \nabla \times (\nabla \times \mathbf{u}), \tag{4.1}$$

with body force $\mathbf{f}$ given by $f_i = X_0(t)\delta(\mathbf{x})\delta_{i1}$ and with zero initial conditions ($\dot{\mathbf{u}}(\mathbf{x}, 0) = \mathbf{0}$; $\mathbf{u}(\mathbf{x}, 0) = \mathbf{0}$ for $\mathbf{x} \neq \mathbf{0}$).

In the notation of earlier chapters, this displacement field has components

$$u_n(\mathbf{x}, t) = X_0 * G_{n1},$$

and the problem has every appearance of being complicated by details of directionality at the source ($\hat{\mathbf{x}}_1$ at O) and receiver ($\mathbf{u}$ at $\mathbf{x}$). What, then, is a similar scalar problem—but without such distracting complications—in whose terms we can begin to study the general properties of waves propagating in three dimensions away from a point source? To avoid directionality at the source, the scalar problem must be spherically symmetric, hence the problem that suggests itself is to find $g = g(\mathbf{x}, t)$ such that

$$\ddot{g} = \delta(\mathbf{x})\delta(t) + c^2 \nabla^2 g, \tag{4.2}$$

with zero initial conditions.

The solution of (4.2) is

$$g(\mathbf{x}, t) = \frac{1}{4\pi c^2} \frac{\delta(t - |\mathbf{x}|/c)}{|\mathbf{x}|}. \tag{4.3}$$

This amazingly simple result, proved in Box 4.1, is very informative as to the nature of three-dimensional waves, and from it we shall construct a hierarchy of ever-more-useful Green functions for elastodynamic problems. Solution (4.3) is the first explicit wave solution we have given in this book, and three major properties should be remembered for future use. First, the answer is the product of one factor (the delta function) whose spatial fluctuation is rapid and another factor (the reciprocal distance function) whose fluctuation is relatively slow, at least away from the source. In general, there would be a radiation-pattern factor, dependent on the direction of $\mathbf{x}$ from the source, but that factor is constant in this case because of the spherical symmetry. Second, the rapidly varying function depends, at any given $|\mathbf{x}|$, only on time relative to an "arrival time," here $|\mathbf{x}|/c$, at which motion begins. Clearly, c is the velocity of wave propagation. Third, the wave shape is the same in time (at any fixed receiver) as the time history of the inhomogeneous term in (4.2). This turns out to be true only for cases in which the spatial singularity at the source is like $\delta(\mathbf{x})$ (and not, for example, a dipole), but the elastodynamic problem (4.1) is just such a case, and we shall find there too that particle displacements are often dominated by the same pulse shape as is present in the applied body force. A related property for (4.2) and (4.3) is that g returns to zero at a given receiver after the wave (in this case, with shape $\delta(t - |\mathbf{x}|/c)$) has gone by.

BOX **4.1**

Proof that

$$g(\mathbf{x}, t) = \frac{1}{4\pi c^2|\mathbf{x}|}\,\delta\left(t - \frac{|\mathbf{x}|}{c}\right)$$

is the solution of $\ddot{g} = \delta(\mathbf{x})\delta(t) + c^2\nabla^2 g$ *with zero initial conditions*

By symmetry, the spatial dependence of the solution can be only on the distance $r = |\mathbf{x}|$ from the source, so we seek the functional form of $g = g(r, t)$. Expressing ∇^2 as a differential operator in spherical polar coordinates, it follows that

$$\nabla^2 g = \frac{1}{r^2}\frac{\partial}{\partial r}\left(r^2\frac{\partial g}{\partial r}\right) = \frac{1}{r}\frac{\partial^2}{\partial r^2}(rg).$$

Therefore, everywhere except at $r = 0$, rg satisfies the one-dimensional wave equation $(rg)'' = r\ddot{g}/c^2$ (a prime here denoting $\partial/\partial r$), and this has the well-known general solution $rg = f(t - r/c) + h(t + r/c)$. We know that $h \equiv 0$, because the required solution is outgoing, hence it remains to prove that $f(\tau) = \delta(\tau)/(4\pi c^2)$, i.e., that $4\pi c^2 f(\tau)$ has the same properties as $\delta(\tau)$ when integrated over ranges of time.

We can establish this required result by investigating the function

$$F(r, \varepsilon_1, \varepsilon_2) \equiv 4\pi c^2 \int_{r/c-\varepsilon_1}^{r/c+\varepsilon_2} g(r, t)\,dt = \frac{4\pi c^2}{r}\int_{-\varepsilon_1}^{\varepsilon_2} f(\tau)\,d\tau.$$

Operating with ∇^2 on F, we have to differentiate the limits and the integrand g with respect to r, finding

$$\nabla^2 F = 4\pi c\left[2g'(r, t) + \frac{2}{r}g(r, t) + \frac{\dot{g}}{c}(r, t)\right]_{t=r/c-\varepsilon_1}^{t=r/c+\varepsilon_2} + 4\pi c^2 \int_{r/c-\varepsilon_1}^{r/c+\varepsilon_2} \nabla^2 g(r, t)\, dt.$$

Substituting $c^2\nabla^2 g = \ddot{g} - \delta(\mathbf{x})\delta(t)$, one can carry out the above integral of $\ddot{g}$ to give another term in $\dot{g}/c$ in the square bracket above. All these terms then cancel out, since $rg = f(t - r/c)$ implies $g' = -g/r - \dot{g}/c$, which leaves

$$\nabla^2 F = -4\pi\delta(\mathbf{x}) \int_{r/c-\varepsilon_1}^{r/c+\varepsilon_2} \delta(t)\, dt.$$

When the right-hand side is integrated over any volume V, whether the origin is in V or not, it yields the same result as the volume integral of $-4\pi\delta(\mathbf{x}) \int_{-\varepsilon_1}^{\varepsilon_2} \delta(t)\, dt$. Using the property $\nabla^2(1/r) = -4\pi\delta(\mathbf{x})$, it follows that

$$F(r, \varepsilon_1, \varepsilon_2) = \frac{1}{r}\int_{-\varepsilon_1}^{\varepsilon_2} \delta(t)\, dt.$$

(F does not involve an additional harmonic function, since such a function would either add another singularity at $r = 0$ or violate the property $F \to 0$ as $r \to \infty$.) From the second equality given in the definition of F, we can now see that $4\pi c^2 \int_{-\varepsilon_1}^{\varepsilon_2} f(\tau)\, d\tau = \int_{-\varepsilon_1}^{\varepsilon_2} \delta(t)\, dt$ for all $(\varepsilon_1, \varepsilon_2)$, and hence $f(\tau)$ is the required delta function.

At this stage, the following three problems can be stated and solved (zero initial conditions are assumed throughout):

i) If $\ddot{g}_1 = \delta(\mathbf{x} - \boldsymbol{\xi})\delta(t - \tau) + c^2\nabla^2 g_1$ then

$$g_1(\mathbf{x}, t) = \frac{1}{4\pi c^2}\frac{\delta\left(t - \tau - \dfrac{|\mathbf{x} - \boldsymbol{\xi}|}{c}\right)}{|\mathbf{x} - \boldsymbol{\xi}|}.$$

ii) If $\ddot{g}_2 = \delta(\mathbf{x} - \boldsymbol{\xi})f(t) + c^2\nabla^2 g_2$, then, from $f(t) = \int_{-\infty}^{\infty} f(\tau)\delta(t - \tau)\, d\tau$, we use g_1 above to get

$$g_2(\mathbf{x}, t) = \int_{-\infty}^{\infty} f(\tau)g_1(\mathbf{x}, t)\, d\tau = \frac{1}{4\pi c^2}\frac{f\left(t - \dfrac{|\mathbf{x} - \boldsymbol{\xi}|}{c}\right)}{|\mathbf{x} - \boldsymbol{\xi}|}. \tag{4.4}$$

We shall later find solutions in a form similar to (4.4), but for wave propagation in inhomogeneous media. These, however, will be only approximations, accurate only at great distance from the point source (see Fig. 4.1).

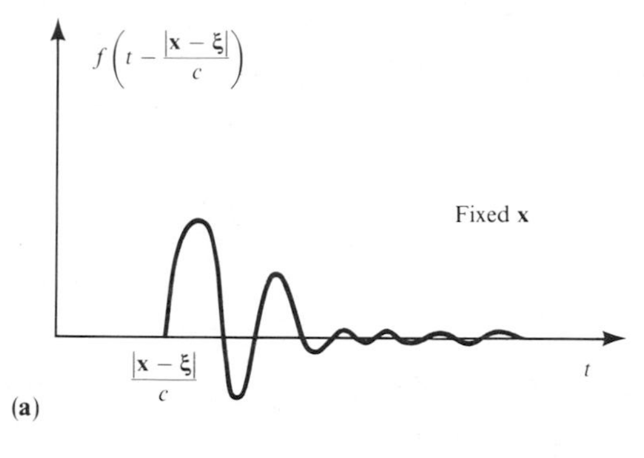

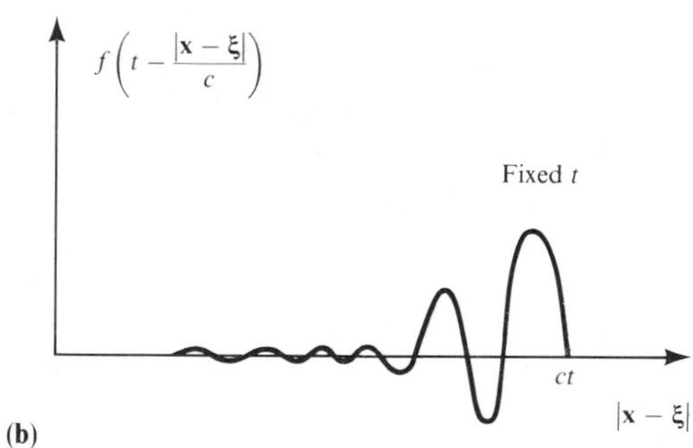

FIGURE **4.1**

In its simplest form, a wave is a propagating quantity that is a function of only a particular linear combination of spatial coordinates and time. (**a**) For some purposes, we are interested in the wave as a function of time at a fixed distance $|\mathbf{x} - \xi|$ from the source at ξ. This is the case in a seismogram. (**b**) For other purposes, we are interested in the wave as a function of position at fixed time. This is the case in a photograph.

iii) If the source is extended throughout a volume V as well as in time, say

$$\frac{\partial^2 \phi}{\partial t^2} = \frac{\Phi(\mathbf{x}, t)}{\rho} + \alpha^2 \nabla^2 \phi \tag{4.5}$$

(using ϕ now rather than g_3, to anticipate future needs), then

$$\Phi(\mathbf{x}, t) = \int_{-\infty}^{\infty} d\tau \iiint_V \Phi(\xi, \tau)\delta(\mathbf{x} - \xi)\delta(t - \tau)\, dV(\xi).$$

From superposition, the solution of (4.5) is

$$\phi(\mathbf{x}, t) = \frac{1}{4\pi\alpha^2\rho} \iiint_V \frac{\Phi\left(\xi, t - \frac{|\mathbf{x} - \xi|}{\alpha}\right)}{|\mathbf{x} - \xi|}\, dV, \tag{4.6}$$

which has the important property that the field at $(\mathbf{x}, t)$ is sensitive to source activity in the element dV (at ξ) only at the so-called *retarded time*, $t - |\mathbf{x} - \xi|/\alpha$. With this understood, it should again be emphasized that the solution (4.6) is remarkably simple, in view of the equation that it solves, (4.5). This equation is a second-order partial differential equation in four variables (three of space, one of time), and has a general inhomogeneous term.

The equation

$$\nabla^2 \phi = -\frac{\Phi(\mathbf{x})}{\alpha^2 \rho} \tag{4.7}$$

is a special example of (4.5), called a Poisson equation, and it has no time dependence. From (4.6), we see that it has the time-independent solution

$$\phi(\mathbf{x}) = \frac{1}{4\pi\alpha^2\rho} \iiint_V \frac{\Phi(\boldsymbol{\xi})}{|\mathbf{x} - \boldsymbol{\xi}|} \, dV. \tag{4.8}$$

The problem now facing us is to find a way of breaking up the elastodynamic equation (4.1) into soluble equations of the type (4.5). In what follows, we shall state a way in which this breakup can be achieved (Lamé's theorem), and in Box 4.2 we give some perspective on what is actually a rather subtle result.

4.1.1 Lamé's Theorem

If the displacement field $\mathbf{u} = \mathbf{u}(\mathbf{x}, t)$ satisfies

$$\rho \ddot{\mathbf{u}} = \mathbf{f} + (\lambda + 2\mu)\nabla(\nabla \cdot \mathbf{u}) - \mu \nabla \times (\nabla \times \mathbf{u}) \tag{4.1 again}$$

and if the body force and initial values of $\dot{\mathbf{u}}$ and $\mathbf{u}$ are expressed in terms of Helmholtz potentials via

$$\mathbf{f} = \nabla\Phi + \nabla \times \mathbf{\Psi}; \, \dot{\mathbf{u}}(\mathbf{x}, 0) = \nabla A + \nabla \times \mathbf{B}; \, \mathbf{u}(\mathbf{x}, 0) = \nabla C + \nabla \times \mathbf{D}, \tag{4.9}$$

with

$$\nabla \cdot \mathbf{\Psi}, \quad \nabla \cdot \mathbf{B}, \quad \nabla \cdot \mathbf{D} \qquad \text{all zero,} \tag{4.10}$$

then there exist potentials ϕ and $\boldsymbol{\psi}$ for $\mathbf{u}$ with all of the following four properties:

i) $$\mathbf{u} = \nabla\phi + \nabla \times \boldsymbol{\psi}, \tag{4.11}$$

ii) $$\nabla \cdot \boldsymbol{\psi} = 0, \tag{4.12}$$

iii) $$\ddot{\phi} = \frac{\Phi}{\rho} + \alpha^2 \nabla^2 \phi \quad \left(\text{with } \alpha^2 = \frac{\lambda + 2\mu}{\rho}\right), \tag{4.5 again}$$

iv) $$\ddot{\boldsymbol{\psi}} = \frac{\mathbf{\Psi}}{\rho} + \beta^2 \nabla^2 \boldsymbol{\psi} \quad \left(\text{with } \beta^2 = \frac{\mu}{\rho}\right) \tag{4.13}$$

($\nabla\phi$ and $\nabla \times \boldsymbol{\psi}$ are called the *P-wave* and *S-wave components* of $\mathbf{u}$, respectively).

The proof involves constructing ϕ and $\boldsymbol{\psi}$ by integrations,

$$\phi(x, t) = \rho^{-1} \int_0^t (t - \tau)\{\Phi(\mathbf{x}, \tau) + (\lambda + 2\mu)\nabla \cdot \mathbf{u}(\mathbf{x}, \tau)\}\, d\tau + tA + C \quad (4.14)$$

$$\boldsymbol{\psi}(\mathbf{x}, t) = \rho^{-1} \int_0^t (t - \tau)\{\boldsymbol{\Psi}(\mathbf{x}, \tau) - \mu\nabla \times \mathbf{u}(\mathbf{x}, \tau)\}\, d\tau + t\mathbf{B} + \mathbf{D}, \quad (4.15)$$

and verifying that all the properties (i)–(iv) are indeed satisfied by these definitions of ϕ and $\boldsymbol{\psi}$. Properties (i) and (ii) are easy to verify. To obtain (iii), note from (4.14) that the left-hand side of (iii) is $\{\Phi + (\lambda + 2\mu)\nabla \cdot \mathbf{u}\}/\rho$. This does equal the right-hand side of (iii), because from (i) it follows that $\nabla^2\phi = \nabla \cdot \mathbf{u}$. The final property (iv) follows in a similar fashion, making repeated use of the vector identities

$$\nabla^2\mathbf{V} \equiv \nabla(\nabla \cdot \mathbf{V}) - \nabla \times (\nabla \times \mathbf{V}), \qquad \nabla \times (\nabla\Phi) \equiv \mathbf{0}, \qquad \nabla \cdot (\nabla \times \mathbf{V}) \equiv 0.$$

BOX **4.2**

On potentials

Helmholtz potentials for the vector field $\mathbf{Z} = \mathbf{Z}(\mathbf{x})$ are fields X, $\mathbf{Y}$ such that $\mathbf{Z} = \nabla X + \nabla \times \mathbf{Y}$, with $\nabla \cdot \mathbf{Y} = 0$. To construct X and $\mathbf{Y}$ (given $\mathbf{Z}$), it is enough to solve the vector Poisson equation $\nabla^2\mathbf{W} = \mathbf{Z}$, since then the identity $\nabla^2\mathbf{W} \equiv \nabla(\nabla \cdot \mathbf{W}) - \nabla \times (\nabla \times \mathbf{W})$ tells us that we can choose potentials $X = \nabla \cdot \mathbf{W}$ and $\mathbf{Y} = -\nabla \times \mathbf{W}$. The solution for the vector Poisson equation is a simple extension of (4.8) and (4.7), giving here

$$\mathbf{W}(\mathbf{x}) = -\iiint_V \frac{\mathbf{Z}(\boldsymbol{\xi})}{4\pi|\mathbf{x} - \boldsymbol{\xi}|}\, dV(\boldsymbol{\xi}).$$

Why, then, do we not define elastic potentials ϕ and $\boldsymbol{\psi}$ to be Helmholtz potentials for $\mathbf{u}$? The reason is that a substitution of (4.11) and (4.12) into the elastic-wave equation (4.1) yields a *third*-order partial differential equation in ϕ and $\boldsymbol{\psi}$. This has to be operated on with $\nabla \cdot (\ \ )$ and $\nabla \times (\ \ )$ to give separated equations for the potentials, which then satisfy *fourth*-order wave equations,

$$\nabla^2\{\rho\ddot{\phi} - \Phi - (\lambda + 2\mu)\nabla^2\phi\} = 0, \qquad \nabla^2\{\rho\ddot{\boldsymbol{\psi}} - \boldsymbol{\Psi} - \mu\nabla^2\boldsymbol{\psi}\} = \mathbf{0}.$$

Lamé's theorem gives us a much better result, to the effect that we need seek potentials satisfying only *second*-order wave equations.

Finally, two remarkable facts: Lamé's theorem was conjectured and used for almost 100 years before a proof was ever given; and the theorem remains true even for static fields $\mathbf{u} = \mathbf{u}(\mathbf{x})$. In this case, however, ϕ and $\boldsymbol{\psi}$ are still functions of both $(\mathbf{x}, t)$, and they still satisfy wave equations, although the time dependence cancels out for the combination $\nabla\phi + \nabla \times \boldsymbol{\psi}$.

Now that we have found the solution for the scalar-wave equation and have found how to turn the elastic-wave equation into simpler equations for potentials, we can return to the main theme of this chapter.

4.2 Solution for the Elastodynamic Green Function in a Homogeneous, Isotropic, Unbounded Medium

Recall that we are seeking to solve for the displacement $\mathbf{u}(\mathbf{x}, t)$ that satisfies the elastic-wave equation (4.1) with a body force $\mathbf{f}$, which is $X_0(t)$ applied in the x_1-direction (i.e., $\hat{\mathbf{x}}_1$) at the origin.

The first step is to find body-force potentials Φ and $\mathbf{\Psi}$ such that

$$X_0(t)\delta(\mathbf{x})\hat{\mathbf{x}}_1 = \nabla\Phi + \nabla \times \mathbf{\Psi} \text{ and } \nabla \cdot \mathbf{\Psi} = 0. \tag{4.16}$$

This is a problem of the type solved in Box 4.2, since Φ and $\mathbf{\Psi}$ are Helmholtz potentials for $\mathbf{f}(\mathbf{x}, t)$ at each fixed moment of time. One first constructs

$$\mathbf{W} = -\frac{X_0(t)}{4\pi}\iiint_V (1, 0, 0)\frac{\delta(\boldsymbol{\xi})\,dV}{|\mathbf{x} - \boldsymbol{\xi}|} = -\frac{X_0(t)}{4\pi|\mathbf{x}|}\hat{\mathbf{x}}_1,$$

which then gives

$$\Phi(\mathbf{x}, t) = \nabla \cdot \mathbf{W} = -\frac{X_0(t)}{4\pi}\frac{\partial}{\partial x_1}\frac{1}{|\mathbf{x}|}$$

$$\mathbf{\Psi}(\mathbf{x}, t) = -\nabla \times \mathbf{W} = \frac{X_0(t)}{4\pi}\left(0, \frac{\partial}{\partial x_3}\frac{1}{|\mathbf{x}|}, -\frac{\partial}{\partial x_2}\frac{1}{|\mathbf{x}|}\right). \tag{4.17}$$

At first sight, it is somewhat surprising that our spatially concentrated body force (proportional to $\delta(\mathbf{x})$) has potentials (4.17) that are nonzero outside the source region. This often happens in elasticity, and it brings out the artificiality of the potential method.

The second step in finding displacements is to solve wave equations for the Lamé potentials ϕ and $\boldsymbol{\psi}$. From (4.5), (4.13), and (4.17), we get

$$\ddot{\phi} = -\frac{X_0(t)}{4\pi\rho}\frac{\partial}{\partial x_1}\frac{1}{|\mathbf{x}|} + \alpha^2\nabla^2\phi \tag{4.18}$$

and

$$\ddot{\boldsymbol{\psi}} = \frac{X_0(t)}{4\pi\rho}\left(0, \frac{\partial}{\partial x_3}\frac{1}{|\mathbf{x}|}, -\frac{\partial}{\partial x_2}\frac{1}{|\mathbf{x}|}\right) + \beta^2\nabla^2\boldsymbol{\psi}. \tag{4.19}$$

The solution of (4.18) follows by comparison with (4.5) and (4.6), so that here

$$\phi(\mathbf{x}, t) = -\frac{1}{(4\pi\alpha)^2\rho} \iiint_V \frac{X_0\left(t - \dfrac{|\mathbf{x} - \boldsymbol{\xi}|}{\alpha}\right)}{|\mathbf{x} - \boldsymbol{\xi}|} \frac{\partial}{\partial \xi_1} \frac{1}{|\boldsymbol{\xi}|} \, dV(\boldsymbol{\xi}). \tag{4.20}$$

Fortunately, this integral can be simplified by integrating over the volume V via the system of concentric spherical shells centered on $\mathbf{x}$. If $\alpha\tau$ is the radius of a typical shell S, so that $|\mathbf{x} - \boldsymbol{\xi}| = \alpha\tau$ and the shell thickness is $\alpha\, d\tau$, then

$$\phi(\mathbf{x}, t) = -\frac{1}{(4\pi\alpha)^2\rho} \int_0^\infty \frac{X_0(t - \tau)}{\tau} \left(\iint_S \frac{\partial}{\partial \xi_1} \frac{1}{|\boldsymbol{\xi}|} \, dS \right) d\tau.$$

In Box 4.3, it is shown that the integral over S is a simple explicit function of $\mathbf{x}$ and τ, and it follows that

$$\phi(\mathbf{x}, t) = -\frac{1}{4\pi\rho} \left(\frac{\partial}{\partial x_1} \frac{1}{|\mathbf{x}|} \right) \int_0^{|\mathbf{x}|/\alpha} \tau X_0(t - \tau) \, d\tau. \tag{4.21}$$

Similarly, for the vector Lamé potential, one finds

$$\boldsymbol{\psi}(\mathbf{x}, t) = \frac{1}{4\pi\rho} \left(0, \frac{\partial}{\partial x_3} \frac{1}{|\mathbf{x}|}, -\frac{\partial}{\partial x_2} \frac{1}{|\mathbf{x}|} \right) \int_0^{|\mathbf{x}|/\beta} \tau X_0(t - \tau) \, d\tau. \tag{4.22}$$

BOX **4.3**

Evaluation of a surface integral

We define

$$h(\mathbf{x}, \tau) \equiv \iint_{|\mathbf{x} - \boldsymbol{\xi}| = \alpha\tau} \frac{\partial}{\partial \xi_1} \frac{1}{|\boldsymbol{\xi}|} \, dS(\boldsymbol{\xi})$$

and show here that

$$h(\mathbf{x}, \tau) = 0 \qquad \text{for } \tau > |\mathbf{x}|/\alpha,$$

but

$$h(\mathbf{x}, \tau) = 4\pi\alpha^2\tau^2 \frac{\partial}{\partial x_1} \frac{1}{|\mathbf{x}|} \qquad \text{for } \tau < |\mathbf{x}|/\alpha.$$

i) Note the physical meaning of the result: suppose there is a uniform surface density on S. Then $|\boldsymbol{\xi}|^{-1}\, dS$ is proportional to the gravitational potential of dS at O, and $\partial|\boldsymbol{\xi}|^{-1}/\partial\xi_1 dS$

is the component of the force in the $\hat{\mathbf{x}}_1$ direction. The desired result follows from finding the total potential at O due to the shell and then differentiating to get the total force component along $\hat{\mathbf{x}}_1$. The potential inside a spherical shell is constant, and outside the spherical shell one can find the potential by lumping all the mass into a point at the center, i.e., at $\mathbf{x}$.

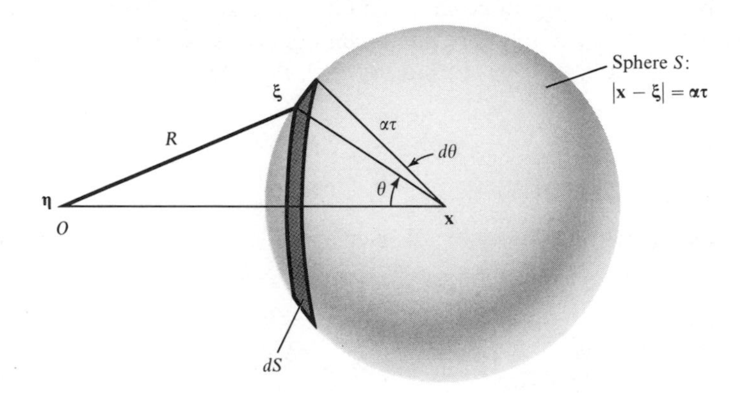

ii) Detailed proof: suppose that O is at $\boldsymbol{\eta}$, so that we can differentiate with respect to varying $\boldsymbol{\eta}$ and subsequently set $\boldsymbol{\eta} = \mathbf{0}$. Also take $r = |\mathbf{x} - \boldsymbol{\eta}|$, $R = |\boldsymbol{\xi} - \boldsymbol{\eta}|$, and θ as the angle between $\mathbf{x} - \boldsymbol{\eta}$ and $\mathbf{x} - \boldsymbol{\xi}$. Then

$$h = -\frac{\partial}{\partial \eta_1} \iint_S \frac{dS}{R} \qquad \text{(since } \boldsymbol{\eta} \text{ is fixed for all } \boldsymbol{\xi} \text{ on } S\text{)}.$$

Now choose $dS = 2\pi\alpha^2\tau^2 \sin\theta\, d\theta$:

$$\iint_S \frac{dS}{R} = 2\pi\alpha^2\tau^2 \int_0^\pi \frac{\sin\theta\, d\theta}{R}.$$

But $R^2 = r^2 + \alpha^2\tau^2 - 2r\alpha\tau\cos\theta$, so that $2R\, dR = 2r\alpha\tau \sin\theta\, d\theta$, and

$$\iint_S \frac{dS}{R} = \frac{2\pi\alpha\tau}{r} \int_{R(\theta=0)}^{R(\theta=\pi)} dR = \frac{2\pi\alpha\tau}{r} \int_{|\alpha\tau - r|}^{\alpha\tau + r} dR = \begin{cases} 4\pi\alpha\tau & \text{if } O \text{ is inside } S \qquad (\tau > r/\alpha) \\ \dfrac{4\pi\alpha^2\tau^2}{r} & \text{if } O \text{ is outside } S \qquad (\tau < r/\alpha). \end{cases}$$

Hence, if O is inside S,

$$h = -\frac{\partial}{\partial \eta_1} 4\pi\alpha\tau = 0 \qquad (\tau > r/\alpha),$$

and if O is outside S,

$$h = -\frac{\partial}{\partial \eta_1} \frac{4\pi\alpha^2\tau^2}{r} = 4\pi\alpha^2\tau^2 \frac{\partial}{\partial x_1} \frac{1}{r} \qquad (\tau < r/\alpha).$$

The third and final step in obtaining the Green function for displacement due to body force $X_0(t)$ applied in the x_1-direction at the origin is to form $\nabla\phi + \nabla \times \boldsymbol{\psi}$ from (4.21) and (4.22). Using $r = |\mathbf{x}|$, this gives

$$u_i(\mathbf{x}, t) = \frac{1}{4\pi\rho}\left(\frac{\partial^2}{\partial x_i\,\partial x_1}\frac{1}{r}\right)\int_{r/\alpha}^{r/\beta} \tau X_0(t - \tau)\,d\tau + \frac{1}{4\pi\rho\alpha^2 r}\left(\frac{\partial r}{\partial x_i}\frac{\partial r}{\partial x_1}\right)X_0\left(t - \frac{r}{\alpha}\right) + \frac{1}{4\pi\rho\beta^2 r}\left(\delta_{i1} - \frac{\partial r}{\partial x_i}\frac{\partial r}{\partial x_1}\right)X_0\left(t - \frac{r}{\beta}\right).$$

If we change the subscript 1 to j throughout this formula, the result corresponds to the displacement set up by a point force in the x_j-direction.

Using direction cosines γ_i for the vector $\mathbf{x}$, so that $\gamma_i = x_i/r = \partial r/\partial x_i$, we can write

$$\frac{\partial^2}{\partial x_i\,\partial x_j}\frac{1}{r} = \frac{3\gamma_i\gamma_j - \delta_{ij}}{r^3}.$$

Then for a point force X_0(t) in the x_j-direction at the origin, we have

$$\begin{aligned} u_i(\mathbf{x}, t) &= X_0 * G_{ij} \qquad \text{(in the notation of Chapter 3)} \\ &= \frac{1}{4\pi\rho}(3\gamma_i\gamma_j - \delta_{ij})\frac{1}{r^3}\int_{r/\alpha}^{r/\beta} \tau X_0(t - \tau)\,d\tau \\ &\quad + \frac{1}{4\pi\rho\alpha^2}\gamma_i\gamma_j\frac{1}{r}X_0\left(t - \frac{r}{\alpha}\right) \\ &\quad - \frac{1}{4\pi\rho\beta^2}(\gamma_i\gamma_j - \delta_{ij})\frac{1}{r}X_0\left(t - \frac{r}{\beta}\right). \end{aligned} \tag{4.23}$$

This is the formula we wanted to find; an equivalent version was first given by Stokes in 1849. It is one of the most important solutions in elastic wave radiation, and we next examine its main properties.

The relative magnitude of different terms in the Green function depends upon the source-receiver distance r. Thus $r^{-3}\int_{r/\alpha}^{r/\beta} \tau X_0(t - \tau)\,d\tau$ behaves like r^{-2} for sources in which X_0 is nonzero for times that are short compared to $r/\beta - r/\alpha$ (e.g., for the impulsive source of Green's function itself). But the remaining terms in (4.23) behave like r^{-1}, becoming dominant (over r^{-2}) as

$r \to \infty$. The terms including $r^{-1}X_0(t - r/\alpha)$ and $r^{-1}X_0(t - r/\beta)$ are therefore called *far-field* terms. Since r^{-2} dominates over r^{-1} as $r \to 0$, the term including $r^{-3}\int \tau X_0(t - \tau)\,d\tau$ is called a *near-field* term. Almost all seismic data used in geophysics are collected in the far-field (i.e., at a distance at which far-field terms in (4.23) are dominant). There are important exceptions, however, such as observations of the final static offset due to faulting, which is a near-field effect. The seismic data used in earthquake engineering are occasionally collected in the near-field. But when one takes up in more detail the question of where the near-field ends and where the far-field begins, it becomes apparent that far-field terms also can be big enough to cause earthquake damage to engineering structures. (See Problem 4.1.)

4.2.1 Properties of the far-field P-wave

We introduce here the *far-field P-wave*, which for (4.23) has the displacement $\mathbf{u}^P$ given by

$$u_i^P(\mathbf{x}, t) = \frac{1}{4\pi\rho\alpha^2}\gamma_i\gamma_j\frac{1}{r}X_0\left(t - \frac{r}{\alpha}\right). \tag{4.24}$$

Then, along a given direction γ from the source, this wave

i) attenuates as r^{-1};
ii) has a waveform that depends on the time-space combination $t - r/\alpha$, and therefore propagates with speed α (recall that $\alpha^2 = (\lambda + 2\mu)/\rho$). (If $t = 0$ is chosen as the time at which $X_0(t)$ first becomes nonzero, then r/α is the *arrival time* of the P-wave at r.);
iii) the waveform is proportional to the applied force at retarded time; and
iv) the direction of displacement $\mathbf{u}^P$ at $\mathbf{x}$ is parallel to the direction γ from the source. This follows from the property $\mathbf{u}^P \times \gamma = \mathbf{0}$, which is easily shown from (4.24). The far-field P-wave is therefore *longitudinal* (sometimes called *radial*) in that its direction of particle motion is the same as the direction of propagation.

4.2.2 Properties of the far-field S-wave

The *far-field S-wave* in (4.23) has displacement $\mathbf{u}^S$ given by

$$u_i^S(x, t) = \frac{1}{4\pi\rho\beta^2}(\delta_{ij} - \gamma_i\gamma_j)\frac{1}{r}X_0\left(t - \frac{r}{\beta}\right). \tag{4.25}$$

Recall that γ is the unit vector directed from the source to the receiver. Along

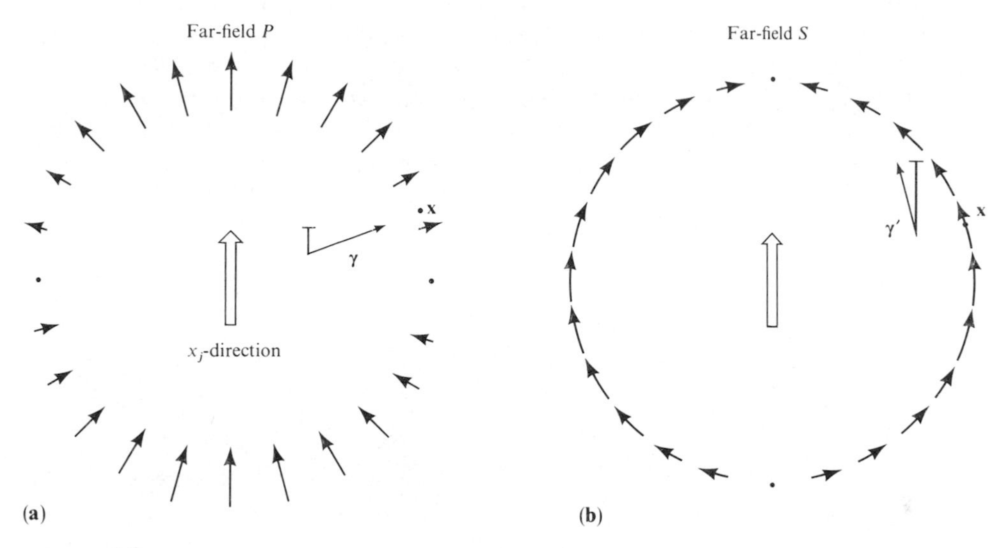

FIGURE **4.2**
Radiation patterns for P and S in the far-field, for a point force ⇧ in the x_j-direction within an infinite homogeneous isotropic medium. Directions for P and S are given by properties (iv) for each wave (see text), the particular choice of transverse direction for S being determined by requiring axial symmetry. **(a)** The magnitude of $\mathbf{u}^P$ is given by $\mathbf{u}^P \cdot \boldsymbol{\gamma} \propto \gamma_j$, where γ_j is the cosine of the angle between the force direction and the direction of $\mathbf{u}^P$. **(b)** The magnitude of $\mathbf{u}^S$ is given by $\mathbf{u}^S \cdot \boldsymbol{\gamma}' \propto \gamma'_j$, where γ'_j is the cosine of the angle between the force direction and the direction of $\mathbf{u}^S$.

a given direction $\boldsymbol{\gamma}$, this wave

i) attenuates as r^{-1};
ii) has arrival time r/β at $\mathbf{x}$ and propagates with speed β;
iii) has a displacement waveform that is proportional to the applied force at retarded time; and
iv) has a direction of displacement $\mathbf{u}^S$ at $\mathbf{x}$ that is perpendicular to the direction $\boldsymbol{\gamma}$ from the source. (From (4.25) it is easy to show that $\mathbf{u}^S \cdot \boldsymbol{\gamma} = 0$.) The far-field S-wave is therefore a *transverse* wave, in that its direction of particle motion is normal to the direction of propagation.

Radiation patterns for $\mathbf{u}^P$ and $\mathbf{u}^S$ are given in Figure 4.2.

4.2.3 Properties of the near-field term

We define the near-field displacement $\mathbf{u}^N$ in (4.23) by

$$u_i^N(\mathbf{x}, t) = \frac{1}{4\pi\rho}(3\gamma_i\gamma_j - \delta_{ij})\frac{1}{r^3}\int_{r/\alpha}^{r/\beta} \tau X_0(t - \tau)\, d\tau. \qquad (4.26)$$

In our derivation (see above) of this near-field component, we see that there are contributions both from the gradient of the P-wave potential (ϕ) and from the curl of the S-wave potential ($\boldsymbol{\psi}$). In this sense, $\mathbf{u}^N$ is composed of both P-wave and S-wave motions. It is neither irrotational (i.e., having zero curl), nor solenoidal (i.e., having zero divergence), and this indicates that it is not always

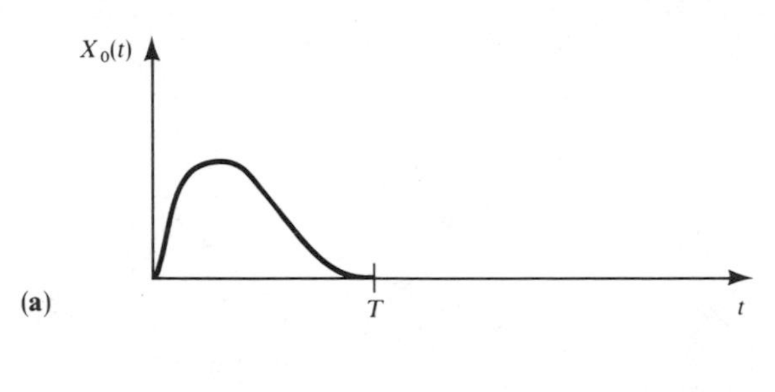

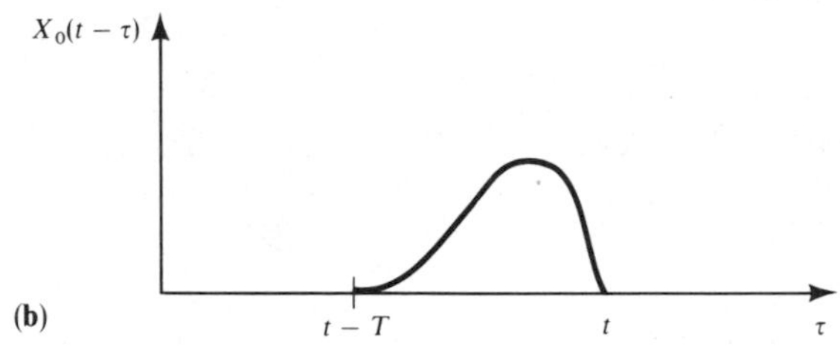

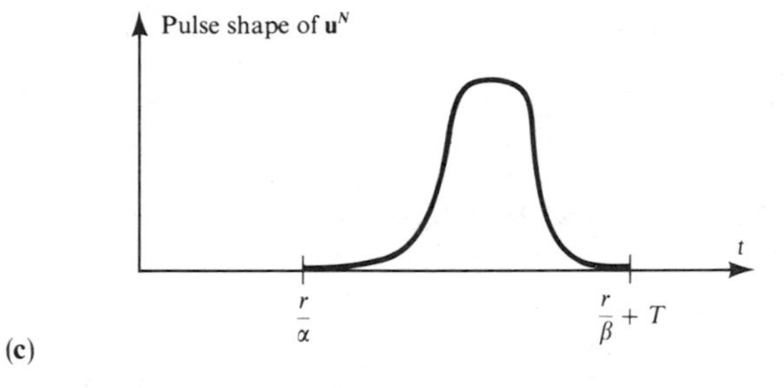

FIGURE **4.3**

Diagrams to interpret the "arrival time" and "duration" of the near-field motion, $\mathbf{u}^N$. (**a**) Body-force time function, nonzero only for t between 0 and T. (**b**) Variation of $X_0(t - \tau)$ against τ for fixed t. This function is nonzero for τ in the range r/α to r/β only if t lies between r/α and $r/\beta + T$. (**c**) Variation of $\int_{r/\alpha}^{r/\beta} \tau X_0(t - \tau)\, d\tau$ with time. At great distance, $r/\alpha \gg T$, $X_0(t - \tau)$ in the integrand for $\mathbf{u}^N$ is effectively a delta function, and $\mathbf{u}^N$ has nonzero values only between r/α and r/β, with height proportional to time. At such distances, however, $\mathbf{u}^N$ is usually negligible compared to the far-field terms.

fruitful to decompose an elastic displacement field into its P-wave and S-wave components. Furthermore, $\mathbf{u}^N$ has both longitudinal and transverse motions, since the longitudinal component is

$$\mathbf{u}^N \cdot \boldsymbol{\gamma} = \gamma_j \frac{1}{2\pi\rho r^3} \int_{r/\alpha}^{r/\beta} \tau X_0(t - \tau)\, d\tau$$

and the transverse component is

$$\mathbf{u}^N \cdot \boldsymbol{\gamma}' = -\gamma_j' \frac{1}{4\pi\rho r^3} \int_{r/\alpha}^{r/\beta} \tau X_0(t - \tau)\, d\tau$$

(see Fig. 4.2 for definitions of γ and γ').

Although it is not possible to identify, for the near-field displacement, the simple properties corresponding to (i)–(iv) found for the far-field, we can identify an arrival time and the duration of displacement $\mathbf{u}^N$ at a fixed receiver. If $t = 0$ is chosen as the time when $X_0(t)$ first becomes nonzero, let us suppose that $X_0(t)$ returns again to zero for all times $t > T$. A function of this type is shown in Figure 4.3a, and $X_0(t - \tau)$ is shown (against τ) in Figure 4.3b. It then follows from (4.26) that $\mathbf{u}^N$ is a motion that arrives at $\mathbf{x}$ at the P-wave arrival time (r/α) and which is active until the time $r/\beta + T$, so that it has duration $(r/\beta - r/\alpha) + T$. If $X_0(t)$ never does return permanently to zero (i.e., if T is not finite), then the near-field term persists indefinitely.

Several further properties of $X_0 * \mathbf{G}$ are brought out in the Problems. We have seen in Chapters 2 and 3 that in seismology the Green function is more appropriately studied in the form $G_{np,q}$, with an active couple at the source, rather than a single force. We now develop some special properties of this more directly relevant displacement field.

4.3 The Double-couple Solution in an Infinite Homogeneous Medium

This section describes one more step in the sequence of wave solutions that we initiated in Box 4.1 with a scalar wave, spreading in spherically symmetric fashion from a point source. We have seen how to synthesize vector waves and second-order tensor fields G_{ij}. Now we wish to study the third-order tensor field $G_{np,q}$ in order to understand the radiation pattern (in both near-field and far-field) for the waves set up by a displacement discontinuity (see (3.17), or, for a point source, (3.22)). It is only fair to acknowledge that this may present a formidable prospect to some readers, and so we state here, at the outset, that we shall conclude with remarkably straight-forward formulas (4.32)–(4.33) for the displacement due to a point source with moment tensor $\mathbf{M}$ having zero trace. This is the case of a point shear dislocation and for it we shall obtain the far-field radiation pattern that has actually been observed in many thousands of earthquakes. Since this double-couple source is of such practical interest, it might be thought that we should restrict our analysis to the particular combination $M_0 * (\partial G_{n1}/\partial \xi_3) + M_0 * (\partial G_{n3}/\partial \xi_1)$ (see (3.22) and (3.24)). With rather less effort, however, using the summation convention will allow us to work with the fully general nine couples in $M_{pq} * G_{np,q}$.

Our principal findings will be that far-field displacements still attenuate as r^{-1} and are proportional to particle velocity at the source; that certain remarkable similarities are found between far-field and near-field displacements; and that the final static displacement, set up throughout the medium by a displacement dislocation that eventually reaches a final fixed offset, attenuates as r^{-2}.

We start with an application of Stokes' solution, (4.23), to obtain the n-component of displacement due to a body-force distribution $\mathbf{f}(\mathbf{x}, t) = \mathbf{F}(t)\delta(\mathbf{x} - \boldsymbol{\xi})$, i.e., the force $\mathbf{F}(t)$ applied at $\boldsymbol{\xi}$. This displacement, at $(\mathbf{x}, t)$, is

$$
\begin{aligned}
F_p * G_{np} = {} & \frac{1}{4\pi\rho}(3\gamma_n\gamma_p - \delta_{np})\frac{1}{r^3}\int_{r/\alpha}^{r/\beta} \tau F_p(t - \tau)\, d\tau \\
& + \frac{1}{4\pi\rho\alpha^2}\gamma_n\gamma_p \frac{1}{r} F_p\left(t - \frac{r}{\alpha}\right) \\
& - \frac{1}{4\pi\rho\beta^2}(\gamma_n\gamma_p - \delta_{np})\frac{1}{r} F_p\left(t - \frac{r}{\beta}\right),
\end{aligned}
\tag{4.27}
$$

in which $r = |\mathbf{x} - \boldsymbol{\xi}|$ is the source-receiver distance and direction cosines $\gamma_i = (x_i - \xi_i)/r$ are referred to a source at $\boldsymbol{\xi}$. Formula (4.27) has the same apparent form as (4.23), but now a summation over p is present, since $\mathbf{F}$ in general is not along a particular coordinate direction.

In order to obtain the total effect of nine couples, of the type shown in Figure 3.7, we can evaluate (4.27) for $\mathbf{F}(t)$ applied at $\boldsymbol{\xi} + \Delta\mathbf{l}_q$ (where $\Delta\mathbf{l}_q$ is a small distance in the ξ_q-direction) and subtract the value of (4.27) for $\mathbf{F}(t)$ applied at $\boldsymbol{\xi}$. This difference gives the displacement field (at $\mathbf{x}$) due to a couple with moment $|\Delta l_q|\,|\mathbf{F}|$, and it is a difference which, to first order in Δl_q, is given directly by the calculus operation $\Delta l_q(\partial/\partial\xi_q)$. (Note here that this is a dimensionless operation: the result is still a displacement.) The final step is to equate the product $\Delta l_q F_p(t)$, in which $\Delta l_q \to 0$ and $F_p \to \infty$ such that the product remains finite, with the moment tensor component $M_{pq}(t)$. The above procedure is expressed by the equality

$$
M_{pq} * G_{np,q} = \lim_{\substack{\Delta l_q \to 0 \\ F_p \to \infty \\ \Delta l_q F_p = M_{pq}}} \Delta l_q F_p * \frac{\partial}{\partial \xi_q} G_{np}.
\tag{4.28}
$$

Note here that the summation is over both p and q. The operation given in

(4.28) is quite straightforward to apply to (4.27), using the two rules

$$\frac{\partial r}{\partial \xi_q} = -\gamma_q \quad \text{and} \quad \frac{\partial \gamma_j}{\partial \xi_q} = \frac{\gamma_j \gamma_q - \delta_{jq}}{r},$$

and the outcome is a displacement field (see (3.22)) having the nth component

$$\begin{aligned} M_{pq} * G_{np,q} = & \left(\frac{15\gamma_n\gamma_p\gamma_q - 3\gamma_n\delta_{pq} - 3\gamma_p\delta_{nq} - 3\gamma_q\delta_{np}}{4\pi\rho}\right)\frac{1}{r^4}\int_{r/\alpha}^{r/\beta} \tau M_{pq}(t-\tau)\,d\tau \\ & + \left(\frac{6\gamma_n\gamma_p\gamma_q - \gamma_n\delta_{pq} - \gamma_p\delta_{nq} - \gamma_q\delta_{np}}{4\pi\rho\alpha^2}\right)\frac{1}{r^2} M_{pq}\left(t - \frac{r}{\alpha}\right) \\ & - \left(\frac{6\gamma_n\gamma_p\gamma_q - \gamma_n\delta_{pq} - \gamma_p\delta_{nq} - 2\gamma_q\delta_{np}}{4\pi\rho\beta^2}\right)\frac{1}{r^2} M_{pq}\left(t - \frac{r}{\beta}\right) \\ & + \frac{\gamma_n\gamma_p\gamma_q}{4\pi\rho\alpha^3}\frac{1}{r}\dot{M}_{pq}\left(t - \frac{r}{\alpha}\right) \\ & - \left(\frac{\gamma_n\gamma_p - \delta_{np}}{4\pi\rho\beta^3}\right)\gamma_q \frac{1}{r}\dot{M}_{pq}\left(t - \frac{r}{\beta}\right). \end{aligned} \tag{4.29}$$

The near-field terms in this displacement field for a dislocation are proportional to $r^{-4}\int_{r/\alpha}^{r/\beta} \tau M_{pq}(t-\tau)\,d\tau$, and the far-field terms are proportional to $r^{-1}\dot{M}_{pq}(t - r/\alpha)$ (P-waves) or to $r^{-1}\dot{M}_{pq}(t - r/\beta)$ (S-waves). Recall from Chapter 3 (equation (3.22)) that $M_{pq} * G_{np,q}$ is the n-component of displacement, at $\mathbf{x}$, from a displacement discontinuity over a fault plane with linear dimensions much smaller than the wavelength of radiated waves of interest at the receiver, so that components of the moment tensor $\mathbf{M}$ are proportional to particle displacements averaged over the fault plane. It follows that $\dot{M}_{pq}(t - r/\alpha)$ and $\dot{M}_{pq}(t - r/\beta)$, giving the pulse shape of displacement in the far-field, are proportional to particle *velocities* at the source, averaged over the fault plane.

Present in (4.29) are some terms proportional to $r^{-2}M_{pq}(t - r/\alpha)$ and $r^{-2}M_{pq}(t - r/\beta)$. Since their asymptotic properties, at small and large values of r, are intermediate to the asymptotic properties of the near-field and far-field displacements, we naturally call these the *intermediate-field terms*. This is, however, a slightly misleading name, since there is no intermediate range of distances in which these terms dominate. In practice, they are found to be small in the far-field and are (often) of comparable importance to the near-field displacements at distances where the latter are appreciable.

From the generality of formula (4.29), which gives the radiation from any moment tensor $\mathbf{M}$, we shall often specialize to cases where $\mathbf{M}$ has zero trace: $M_{kk} = 0$. This can happen with three vector dipoles that create no net volume change. But we shall more often be interested in the zero-trace $\mathbf{M}$ that can arise from a discontinuity in displacement. From (3.23) and (3.20), we see that the averaged displacement discontinuity, $\bar{\mathbf{u}}$, is parallel to the fault surface: $\bar{\mathbf{u}} \cdot \boldsymbol{\nu} = 0$, where $\boldsymbol{\nu}$ is normal to the fault surface and $M_{pq} = \mu(\bar{u}_p \nu_q + \bar{u}_q \nu_p) A$ for a fault with area A. Then

$$\mu(\bar{u}_p \nu_q + \bar{u}_q \nu_p) A * G_{np,q}$$

$$= \left(\frac{30\gamma_n\gamma_p\gamma_q\nu_q - 6\nu_n\gamma_p - 6\delta_{np}\gamma_q\nu_q}{4\pi\rho r^4}\right)\mu A \int_{r/\alpha}^{r/\beta} \tau \bar{u}_p(t - \tau)\, d\tau$$

$$+ \left(\frac{12\gamma_n\gamma_p\gamma_q\nu_q - 2\nu_n\gamma_p - 2\delta_{np}\gamma_q\nu_q}{4\pi\rho\alpha^2 r^2}\right)\mu A \bar{u}_p\left(t - \frac{r}{\alpha}\right)$$

$$- \left(\frac{12\gamma_n\gamma_p\gamma_q\nu_q - 3\nu_n\gamma_p - 3\delta_{np}\gamma_q\nu_q}{4\pi\rho\beta^2 r^2}\right)\mu A \bar{u}_p\left(t - \frac{r}{\beta}\right)$$

$$+ \frac{2\gamma_n\gamma_p\gamma_q\nu_q}{4\pi\rho\alpha^3 r}\mu A \dot{\bar{u}}_p\left(t - \frac{r}{\alpha}\right)$$

$$- \left(\frac{2\gamma_n\gamma_p\gamma_q\nu_q - \nu_n\gamma_p - \delta_{np}\gamma_q\nu_q}{4\pi\rho\beta^3 r}\right)\mu A \dot{\bar{u}}_p\left(t - \frac{r}{\beta}\right). \quad (4.30)$$

Our next goal is to turn this expression for the displacement field radiated by a shear dislocation from its Cartesian form into a form that naturally brings out the radial and transverse components of motion. This can be accomplished by choosing axes, so that the fault lies in the (x_1, x_2) plane, i.e., $\boldsymbol{\nu} = (0, 0, 1)$, with $\boldsymbol{\xi} = \mathbf{0}$, and introducing spherical polar coordinates r, θ, and ϕ centered on the source. We measure θ from the x_3-direction (see Fig. 4.4); choose the

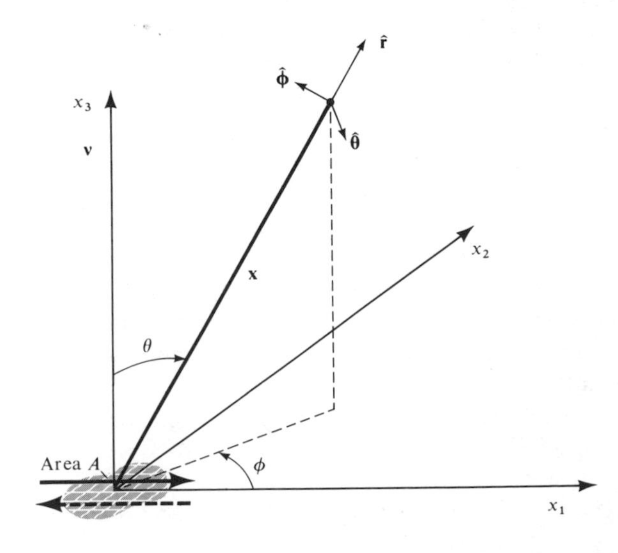

FIGURE **4.4**
Cartesian and spherical polar coordinates for analysis of radial and transverse components of displacement radiated by a shear dislocation of area A and average slip $\bar{\mathbf{u}}$. See (4.31).

x_1-axis to be the direction of slip, so that $\bar{\mathbf{u}} = (\bar{u}, 0\ 0)$; and take $\phi = 0$ as the plane containing $\mathbf{v}$ and $\bar{\mathbf{u}}$. Unit vectors $\hat{\mathbf{r}}$, $\hat{\boldsymbol{\theta}}$, $\hat{\boldsymbol{\phi}}$ are in the directions of r, θ, ϕ increasing (respectively), so that the "radial" direction is along $\hat{\mathbf{r}}$, and $\hat{\boldsymbol{\theta}}$ and $\hat{\boldsymbol{\phi}}$ are both "transverse" directions.

We seek to express the displacement vector at $\mathbf{x}$ (for which the nth Cartesian component is given in (4.30)) as a sum of vectors in the three directions $\hat{\mathbf{r}}$, $\hat{\boldsymbol{\theta}}$ and $\hat{\boldsymbol{\phi}}$. Fortunately, (4.30) is composed of vectors of only three types—namely, $\gamma_n\gamma_p\bar{u}_p\gamma_q v_q$, $v_n\gamma_p\bar{u}_p$, and $\delta_{np}\bar{u}_p\gamma_q v_q$. These three types can be recognized, respectively, as follows:

$2\gamma_n\gamma_p\bar{u}_p\gamma_q v_q$

is the nth component of $\hat{\mathbf{r}} \sin 2\theta \cos\phi\, \bar{u}$,

$2v_n\gamma_p\bar{u}_p$

is the nth component of $\hat{\mathbf{r}} \sin 2\theta \cos\phi\, \bar{u} - \hat{\boldsymbol{\theta}} 2 \sin^2\theta \cos\phi\, \bar{u}$, (4.31)

$2\bar{u}_n\gamma_q v_q$

is the nth component of $\hat{\mathbf{r}} \sin 2\theta \cos\phi\, \bar{u} + \hat{\boldsymbol{\theta}} 2 \cos^2\theta \cos\phi\, \bar{u}$
$- \hat{\boldsymbol{\phi}}\, 2 \cos\theta \sin\phi\, \bar{u}$.

(These results follow from relations $\hat{\mathbf{r}} = \gamma = (\sin\theta\cos\phi, \sin\theta\sin\phi, \cos\theta)$, $\hat{\boldsymbol{\theta}} = (\cos\theta\cos\phi, \cos\theta\sin\phi, -\sin\theta)$, and $\hat{\boldsymbol{\phi}} = (-\sin\phi, \cos\phi, 0)$.)

With the identification of vector components in (4.31), it now becomes possible to write our displacement field $u_n = M_{pq} * G_{np,q}$ in a concise vector form, using the time-dependent seismic moment $M_0(t) = \mu\bar{u}(t)A$. We find

$$\begin{aligned}
\mathbf{u}(\mathbf{x}, t) = {} & \frac{1}{4\pi\rho}\mathbf{A}^N \frac{1}{r^4}\int_{r/\alpha}^{r/\beta} \tau M_0(t-\tau)\, d\tau \\
& + \frac{1}{4\pi\rho\alpha^2}\mathbf{A}^{IP}\frac{1}{r^2}M_0\left(t - \frac{r}{\alpha}\right) + \frac{1}{4\pi\rho\beta^2}\mathbf{A}^{IS}\frac{1}{r^2}M_0\left(t-\frac{r}{\beta}\right) \\
& + \frac{1}{4\pi\rho\alpha^3}\mathbf{A}^{FP}\frac{1}{r}\dot{M}_0\left(t-\frac{r}{\alpha}\right) + \frac{1}{4\pi\rho\beta^3}\mathbf{A}^{FS}\frac{1}{r}\dot{M}_0\left(t-\frac{r}{\beta}\right),
\end{aligned} \tag{4.32}$$

in which the near-field, intermediate-field P and S, and far-field P and S have radiation patterns given, respectively, by

$$\begin{aligned}
\mathbf{A}^N &= 9\sin 2\theta\cos\phi\hat{\mathbf{r}} - 6(\cos 2\theta\cos\phi\hat{\boldsymbol{\theta}} - \cos\theta\sin\phi\hat{\boldsymbol{\phi}}) \\
\mathbf{A}^{IP} &= 4\sin 2\theta\cos\phi\hat{\mathbf{r}} - 2(\cos 2\theta\cos\phi\hat{\boldsymbol{\theta}} - \cos\theta\sin\phi\hat{\boldsymbol{\phi}}) \\
\mathbf{A}^{IS} &= -3\sin 2\theta\cos\phi\hat{\mathbf{r}} + 3(\cos 2\theta\cos\phi\hat{\boldsymbol{\theta}} - \cos\theta\sin\phi\hat{\boldsymbol{\phi}}) \\
\mathbf{A}^{FP} &= \sin 2\theta\cos\phi\hat{\mathbf{r}} \\
\mathbf{A}^{FS} &= \cos 2\theta\cos\phi\hat{\boldsymbol{\theta}} - \cos\theta\sin\phi\hat{\boldsymbol{\phi}}.
\end{aligned} \tag{4.33}$$

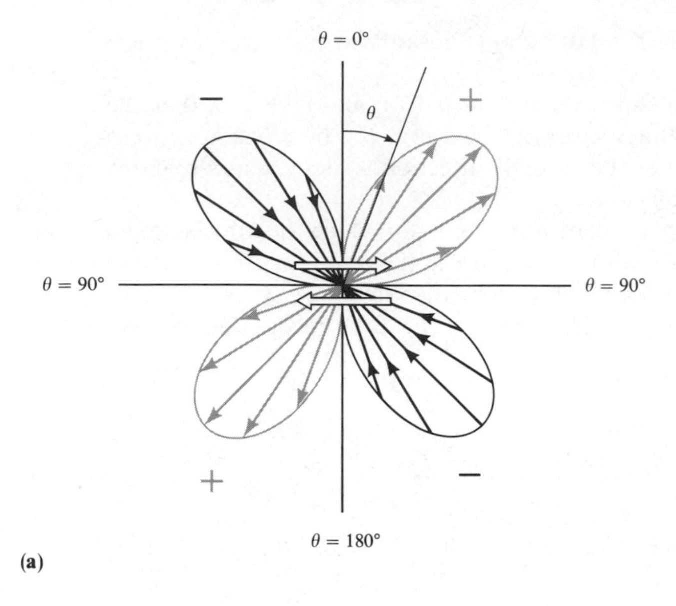

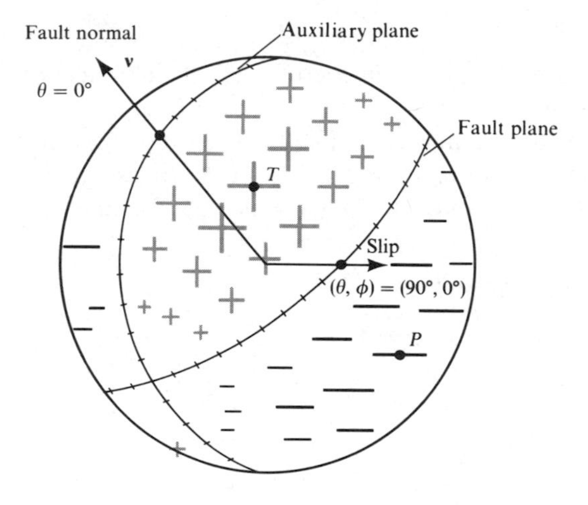

FIGURE **4.5**

Diagrams for the radiation pattern of the radial component of displacement due to a double couple, i.e., $\sin 2\theta \cos \phi \hat{\mathbf{r}}$. (**a**) The lobes are a locus of points having a distance from the origin that is proportional to $\sin 2\theta$. The diagram is for a plane of constant azimuth, and the pair of arrows at the center denotes the shear dislocation. Note the alternating quadrants of inward and outward directions. In terms of far-field P-wave displacement, plus signs denote outward displacement (if $M_0(t - r/\alpha)$ is positive), and minus signs denote inward displacement. (**b**) View of the radiation pattern over a sphere centered on the origin. Plus and minus signs of various sizes denote variation (with θ, ϕ) of outward and inward motions. The fault plane and the auxiliary plane are nodal lines (on which $\sin 2\theta \cos \phi = 0$). An equal-area projection has been used (see Fig. 4.17). Point P marks the pressure axis, and T the tension axis.

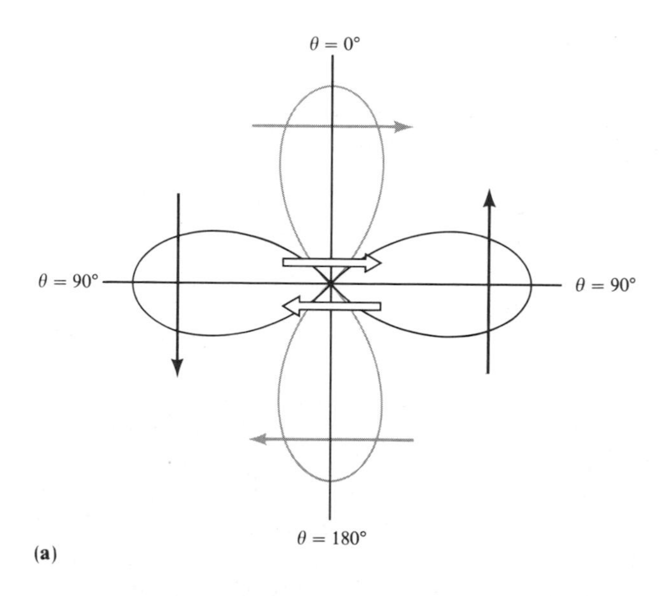

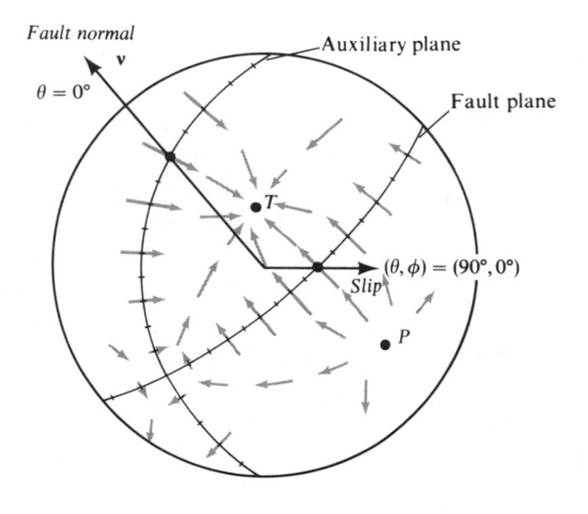

FIGURE **4.6**

Diagrams for the radiation pattern of the transverse component of displacement due to a double couple, i.e., $\cos 2\theta \cos \phi \hat{\boldsymbol{\theta}} - \cos \theta \sin \phi \hat{\boldsymbol{\phi}}$. **(a)** The four-lobed pattern in plane $\{\phi = 0, \phi = \pi\}$. The central pair of arrows shows the sense of shear dislocation, and arrows imposed on each lobe show the direction of particle displacement associated with the lobe. If applied to the far-field *S*-wave displacement, it is assumed that $M_0(t - r/\beta)$ is positive. **(b)** Off the two planes $\theta = \pi/2$ and $\{\phi = 0, \phi = \pi\}$, the $\hat{\boldsymbol{\phi}}$ component is nonzero, hence (a) is of limited use. This diagram is a view of the radiation pattern over a whole sphere centered on the origin, and arrows (with varying size and direction) in the spherical surface denote the variation (with θ, ϕ) of the transverse motions. The stereographic (equal-angle) projection has been used (see Fig. 4.16). There are no nodal lines (where there is zero motion), but nodal points do occur. Note that the nodal point for transverse motion at $(\theta, \phi) = (45°, 0)$ is a maximum in the radiation pattern for longitudinal motion (Fig. 4.5b). But the maximum transverse motion (e.g., at $\theta = 0$) occurs on a nodal line for the longitudinal motion.

These radiation patterns explicitly display a radial component, proportional to $\sin 2\theta \cos \phi \hat{\mathbf{r}}$, and a transverse component, proportional to $(\cos 2\theta \cos \phi \hat{\boldsymbol{\theta}} - \cos \theta \sin \phi \hat{\boldsymbol{\phi}})$. The important property brought out by (4.33) is that these are the only two radiation patterns needed to obtain a complete picture of all the different terms in the displacement field radiated from a shear dislocation (double couple). Figure 4.5 shows the way in which the radial component varies in magnitude for different directions (θ, ϕ), and Figure 4.6 shows how the transverse component varies in both magnitude and direction. Only the radial component is present for the far-field P-wave, and only the transverse component is present for the far-field S-wave. However, the intermediate-field displacements, both P and S, involve both radial and transverse components, as does the near-field displacement.

The surprisingly simple dependence on (θ, ϕ), which we have found in (4.32) and (4.33) and shown in Figures 4.5 and 4.6, prompts one to ask if a more direct method can be used in the derivation. Indeed this is the case, vector surface harmonics (see Chapter 8) providing the necessary analytical framework and demonstrating the simple dependence on (θ, ϕ) from the outset, but the associated algebraic manipulations for this more sophisticated method are (in our opinion) as laborious as the Cartesian analysis we have chosen to use.

In Section 4.4, we shall develop properties of the far-field displacement for a medium that, like the Earth, is inhomogeneous. In Section 4.5 we shall re-examine the radiation patterns (4.33), showing how they may be used in practice to obtain (from seismic data) the fault-plane orientation and the direction of slip. To conclude the present section, we obtain from (4.32) the final static displacement field for a shear dislocation of strength M_0. This involves taking the limit of $\dot{M}_0(t - \tau)$, $M_0(t - \tau)$, and $\int_{r/\alpha}^{r/\beta} \tau M_0(t - \tau)\, d\tau$ as $t \to \infty$, assuming that the seismic moment itself has a final constant value, $M_0(\infty)$. The result is

$$\mathbf{u}(\mathbf{x}, \infty) = \frac{M_0(\infty)}{4\pi\rho r^2}\left[\mathbf{A}^N\left(\frac{1}{2\beta^2} - \frac{1}{2\alpha^2}\right) + \frac{\mathbf{A}^{IP}}{\alpha^2} + \frac{\mathbf{A}^{IS}}{\beta^2}\right] \tag{4.34}$$

$$= \frac{M_0(\infty)}{4\pi\rho r^2}\left[\frac{1}{2}\left(\frac{3}{\beta^2} - \frac{1}{\alpha^2}\right)\sin 2\theta \cos \phi \hat{\mathbf{r}} + \frac{1}{\alpha^2}(\cos 2\theta \cos \phi \hat{\boldsymbol{\theta}} - \cos \theta \sin \phi \hat{\boldsymbol{\phi}})\right],$$

which attenuates (along any given direction (θ, ϕ)) as r^{-2}.

4.4 Ray Theory for Far-field *P*-waves and *S*-waves from a Point Source

Books and papers on the theory of elastic wave propagation are for the most part concerned with homogeneous media. Seismologists often require a good grasp of the properties of waves in such simple media and of the exact solutions that may be obtained (e.g., Chapters 5 and 6). However, the practical analysis of seismograms requires also a good grasp of approximate solutions for the waves that propagate in inhomogeneous media. Not only does the Earth have material boundaries across which the elastic properties are discontinuous, but also it contains vast regions within which there is a systematic and continuous change of bulk modulus, rigidity, and density. Thus *P*-wave and *S*-wave velocities both increase by a factor of about two from the top to the bottom of the mantle, and this is enough to distort beyond recognition the radiation patterns we have described above, unless the effects of inhomogeneity are accounted for. In this section and the next, we shall show how to remove the distortion, and demonstrate how amplitudes of body waves can be substantially changed by focusing or defocusing effects of the type exhibited by light propagating in media of varying refractive index. The approximate solution we shall obtain, called the *geometric ray solution*, provides the basis for routine interpretation of most seismic body waves, and it always provides a guide to more sophisticated methods, should they be necessary.

Ray theory can perhaps best be remembered as a collection of verifiable intuitive ideas and approximations. Thus body waves travel with a local propagation speed along "ray paths" determined by Snell's law, arriving (as a "wavefront") with an amplitude determined by the geometrical spreading of rays from the source to the receiver. These are statements that we shall prove, but intuition enters strongly at the initial stage of setting up a trial form for the solution.

Our approach will be to generalize the form of the far-field solution we have already found for *P*- and *S*-waves, and to use a system of coordinates that provides a natural way to describe the spreading of a wavefront due to a point source in an isotropic inhomogeneous medium. For spherically symmetric media, orthogonal coordinates based on the rays have the property that *P* and two different components of *S* (known as *SV* and *SH*) are separated out. These three different motions are parallel to three different coordinate directions.

A *wavefront* is a propagating discontinuity in some dependent variable of physical interest (such as the particle acceleration). In this connection, the word "discontinuity" is taken to mean "discontinuity in the variable or one of its derivatives." Thus the ramp shown in Figure 4.7a is actually continuous at $t = T$, but such a function is an eligible candidate for describing the behavior near a wavefront, since the ramp has a discontinuous derivative. Where necessary, one speaks of a function whose $(n - 1)$th derivative is discontinuous at $t = T$ (but with all lower-order derivatives continuous there) as having an *nth order discontinuity at* $t = T$. A review of the Green function solution (4.23) will show that a (temporal) discontinuity in the body force $X_0(t)$ leads to a propagating discontinuity in the far-field terms which is of the same order as that occurring in the body force: $X_0(t)$ acting at $r = 0$ leads to displacement pulse shapes $X_0(t - r/\alpha)$ and $X_0(t - r/\beta)$ at large r. (There has to be a discontinuity of some order at the source, otherwise $X_0(t)$ and all its derivatives are zero for all time

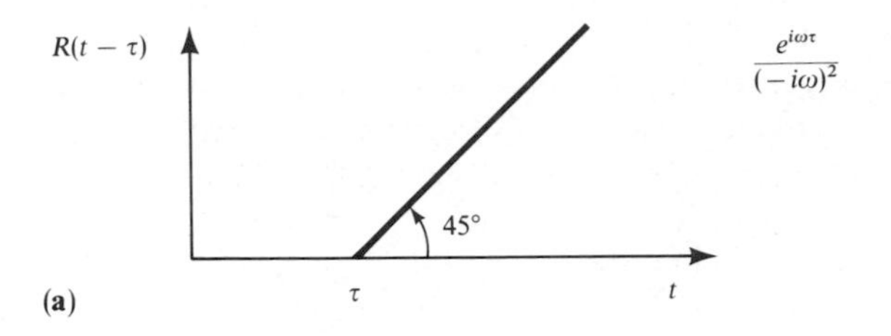

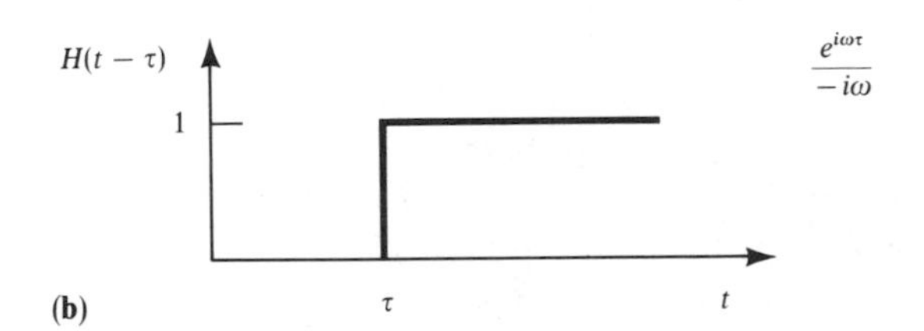

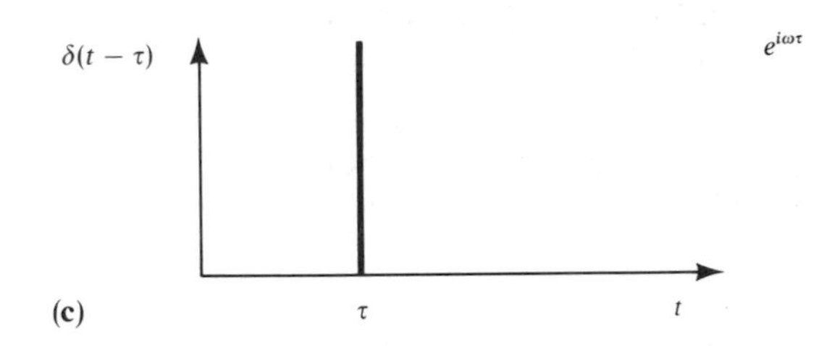

FIGURE **4.7**

A hierarchy of discontinuous functions, with formulas for their associated high-frequency spectra. The ramp values are given by $R(t) = tH(t)$.

and no waves are generated.) There is also a propagating discontinuity in the near-field terms, $\int_{r/\alpha}^{r/\beta} \tau X_0(t - \tau)\, d\tau$, but it is less severe. For example, if $X_0(t) \propto H(t)$ (the Heaviside step function, which has a first-order discontinuity), then propagating steps in displacement are radiated to the far-field, but the near-field terms carry second-order discontinuities, since $\int_{r/\alpha}^{r/\beta} \tau H(t - \tau)\, d\tau$ is continuous at $t = r/\alpha$ and $t = r/\beta$ but has discontinuous first derivatives there (see Fig. 4.8).

This observation, that the strongest discontinuities are carried by what we have called the far-field terms, is of fundamental importance. In fact, it is sometimes better to use it as the defining property of these terms, since the commonly used labels "far-field" and "near-field" can be misleading. At *any* fixed distance r, whether large or small, the behavior of the radiated wave at times sufficiently near r/α and r/β will be dominated by the strongest discontinuity (strongest wavefront) arriving at these times, and in general this is contained in a far-field term. At large enough r, there are therefore two reasons for the far-field terms to dominate: strength of discontinuity and relatively weak attenuation with distance. At fixed small r, however, the far-field terms will dominate or not dominate according as the strength of their discontinuity is more or less important than the weakness of the singularity in r^{-1}.

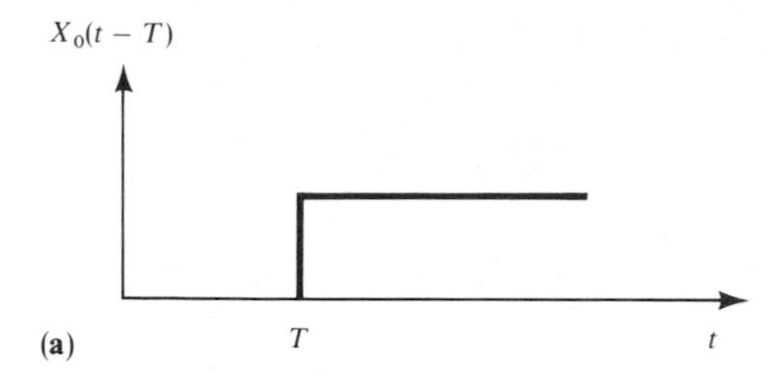

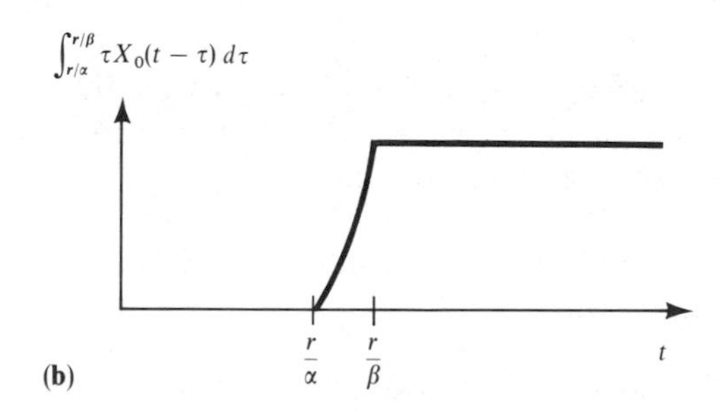

FIGURE **4.8**

Comparison of far-field and near-field displacement pulse shapes for the Green function convolved with $X_0(t) \propto H(t)$. (**a**) The far field: time $t = T$ is either $t = r/\alpha$ or r/β. (**b**) The near field: between times r/α and r/β, the displacement grows parabolically, but particle velocity is discontinuous at both these wavefronts.

Fortunately, by turning to the frequency domain, a much simpler way can be found to compare the displacements that we have labeled as near-field and far-field. Throughout this book, we shall use the notation and conventions

$$f(\omega) = \int_{-\infty}^{\infty} f(t) e^{+i\omega t}\, dt, \qquad f(t) = \frac{1}{2\pi} \int_{-\infty}^{\infty} f(\omega) e^{-i\omega t}\, d\omega$$

for a Fourier transform on functions of time and frequency, letting the context determine whether f is evaluated in the time or frequency domain. (See Box 5.2 for our reasons in choosing this convention). Then $X_0(t - r/\alpha)$ transforms to $e^{+i\omega r/\alpha} X_0(\omega)$, and $\int_{r/\alpha}^{r/\beta} \tau X_0(t - \tau)\, d\tau$ can be seen as the convolution of $X_0(t)$ with a function shaped like

The convolution therefore transforms to the product

$$X_0(\omega) \int_{r/\alpha}^{r/\beta} t e^{+i\omega t}\, dt = \left[-\left(\frac{r}{\alpha} - \frac{1}{i\omega}\right) \frac{e^{+i\omega r/\alpha}}{i\omega} + \left(\frac{r}{\beta} - \frac{1}{i\omega}\right) \frac{e^{+i\omega r/\beta}}{i\omega} \right] X_0(\omega),$$

and from (4.23) we obtain the Fourier transform of $X_0 * G_{ij}$ as

$$\frac{X_0(\omega)e^{+i\omega r/\alpha}}{4\pi\rho\alpha^2 r}\left[\gamma_i\gamma_j + (3\gamma_i\gamma_j - \delta_{ij})\left(-\frac{\alpha}{i\omega r}\right) + (3\gamma_i\gamma_j - \delta_{ij})\left(-\frac{\alpha}{i\omega r}\right)^2\right]$$
$$- \frac{X_0(\omega)e^{+i\omega r/\beta}}{4\pi\rho\beta^2 r}\left[(\gamma_i\gamma_j - \delta_{ij}) + (3\gamma_i\gamma_j - \delta_{ij})\left(-\frac{\beta}{i\omega r}\right) + (3\gamma_i\gamma_j - \delta_{ij})\left(-\frac{\beta}{i\omega r}\right)^2\right]. \tag{4.35}$$

The dimensionless terms within the square brackets of (4.35) give relative magnitudes of different far-field and near-field displacements. Of importance here are the dimensionless ratios $\omega r/\alpha$, $\omega r/\beta$, and we note that

$$\begin{cases} \omega r/\alpha = 2\pi \times \text{(number of wavelengths, for a } P\text{-wave of frequency } \omega, \\ \qquad\qquad \text{between source and receiver), and} \\ \omega r/\beta = 2\pi \times \text{(number of wavelengths, for an S-wave of frequency } \omega, \\ \qquad\qquad \text{between source and receiver).} \end{cases}$$

With this interpretation, we can at last get a clear picture of where different terms of the Green function dominate. The far-field is simply "all positions that are more than a few wavelengths away from the source," and the near-field is "all positions within a small fraction of a wavelength from the source." At near and intermediate distances, one must assess the relative magnitude of each term in (4.35), perhaps concluding that no single term is dominant (see Problem 4.1).

Our above discussion of the discontinuities that can propagate in the time domain translates formally to properties of the high-frequency spectra. If $g(t)$ has a unit jump at the origin, then the spectrum $g(\omega)$ has a term behaving asymptotically like $(-i\omega)^{-1}$ as $\omega \to \infty$ (Bracewell, 1965). Integrating this result repeatedly and applying the shift theorem,

$$g(t - T) \to g(\omega)e^{i\omega T},$$

we obtain the following: if $f(t)$ has an nth-order discontinuity at $t = T$, such that

$$\left.\frac{\partial^{n-1} f(t)}{\partial t^{n-1}}\right|_{t=T^-}^{t=T^+} = A \text{ (where } A \text{ is a constant)},$$

then the Fourier transform $f(\omega)$ behaves asymptotically like $A(-i\omega)^{-n}e^{+i\omega T}$ as $\omega \to \infty$. This does enable us to see how the different terms in (4.35) are associated correctly with different discontinuities in the time domain, but it can be a misleading result, since the behavior of $f(\omega)$ may be like that of $A(-i\omega)^{-n}e^{+i\omega T}$ only at frequencies much higher than can be observed. Putting it another way: the observed trend of a physical variable $f(\omega)$ at high frequencies may not be indicative of the properties of any underlying wavefront, since the latter might require kiloherz or megaherz frequencies to be observed; and these frequencies are usually unavailable in seismology. We can separate near-field and far-field terms, even at seismic frequencies, on the basis of the different frequency dependencies shown in (4.35), although the abstract quantities "wavefront" and "strength of discontinuity" may be unobservable.

These results are brought out in some detail because formal texts on ray theory tend to emphasize the analysis of different levels of discontinuity, and we cannot appeal to this method in seismology. Rather, we shall adopt a less rigorous approach, though with clearly stated assumptions, in which the derivative $\dot{X}_0(t - r/\alpha)$ is presumed to dominate over $(\alpha/r)X_0(t - r/\alpha)$, because this would be true at the high frequencies we commonly observe in far-field body waves (i.e., $\omega r/\alpha \gg 1$).

An elegant way to study the longitudinal and two transverse components of body-wave displacement is to set up a system of curvilinear coordinates in which these three components are either along a coordinate axis or lie within a coordinate plane. Suppose a point source at position $\boldsymbol{\xi}$ becomes active at a time chosen to be the origin, $t = 0$. In a homogeneous medium, wavefronts emanate from the source as ever-expanding spheres, with radius αt (for P-waves) and βt (for S), arriving at the general position $\mathbf{x}$ at times r/α and r/β, where $r = |\mathbf{x} - \boldsymbol{\xi}|$. We generalize this for inhomogeneous media by introducing $T(\mathbf{x}, \boldsymbol{\xi})$ as the position-dependent travel time required for the wavefront to reach $\mathbf{x}$ from $\boldsymbol{\xi}$. If more than one wavefront can exist for the particular source in question, as will be the case if both longitudinal and transverse waves are present, then more than one travel-time function will be needed. Often, however, it is enough to refer to $T(\mathbf{x})$ and let the context indicate which wavefront and which point source are of interest.

Our first *ansatz* (i.e., trial form of solution) is therefore

$$\mathbf{u}(\mathbf{x}, t) = \mathbf{U}(t - T)f(\mathbf{x}), \tag{4.36}$$

which we shall examine as an approximate solution for the wave equation $\rho\ddot{u}_i = \tau_{ij,j}$ (valid everywhere except at the source singularity $\mathbf{x} = \boldsymbol{\xi}$). The stress-displacement relations for a general anisotropic inhomogeneous medium are $\tau_{ij} = c_{ijkl}(\mathbf{x})u_{k,l}(\mathbf{x}, t)$, and direct substitution yields

$$\rho\ddot{U}_i f = [c_{ijkl}(U_k f)_{,l}]_{,j}. \tag{4.37}$$

Near a wavefront, we assume that $\mathbf{U}$ is fluctuating much more rapidly than f or c_{ijkl}, and the successive derivatives $\dot{\mathbf{U}}$ and $\ddot{\mathbf{U}}$ are fluctuating still more rapidly. Equation (4.37) contains both temporal and spatial derivatives of

U_k, but the dependence of $\mathbf{U}$ on the space-time combination $(t - T(\mathbf{x}))$ makes possible such relations as

$$U_{k,lj} = \ddot{U}_k \frac{\partial T}{\partial x_l} \frac{\partial T}{\partial x_j} - \dot{U}_k \frac{\partial^2 T}{\partial x_l \, \partial x_j},$$

so that the second derivatives in (4.37) can be gathered together as

$$\left(\rho \delta_{ik} - c_{ijkl} \frac{\partial T}{\partial x_j} \frac{\partial T}{\partial x_l} \right) \ddot{U}_k f = E_i(\mathbf{U}f), \tag{4.38}$$

where $\mathbf{E}$ includes merely (i) first-order derivatives of $\mathbf{U}$, (ii) $\mathbf{U}$ itself, and (iii) the elastic constants and amplitude function $f(\mathbf{x})$ and gradients of these. Thus $\mathbf{E}$ must be much smaller than $\ddot{\mathbf{U}}$, so we conclude that the matrix of coefficients of $\ddot{U}_k f$ must be singular:

$$\det \left| \rho \delta_{ik} - c_{ijkl} \frac{\partial T}{\partial x_j} \frac{\partial T}{\partial x_l} \right| = 0. \tag{4.39}$$

This equation determines the possible wavefronts in an elastic medium, since it gives a constraint on the function $T(\mathbf{x})$.

In an inhomogeneous isotropic medium, the special form of c_{ijkl} (see (2.33)) makes it possible to factor (4.39) in the form

$$\left(\nabla T \cdot \nabla T - \frac{\rho}{\lambda + 2\mu} \right) \left(\nabla T \cdot \nabla T - \frac{\rho}{\mu} \right)^2 = 0. \tag{4.40}$$

That is, T satisfies the *eikonal equation*

$$(\nabla T)^2 = \frac{1}{c^2}, \tag{4.41}$$

where c is either the local P-wave speed, $\sqrt{(\lambda(\mathbf{x}) + 2\mu(\mathbf{x}))/\rho(\mathbf{x})} = \alpha$, or the local S-wave speed, $\sqrt{\mu(\mathbf{x})/\rho(\mathbf{x})} = \beta$.

We next develop the consequences of T satisfying an eikonal equation with velocity c—an equation from which rays can be introduced.

4.4.1 Properties of the travel-time function $T(\mathbf{x})$ associated with velocity field $c(\mathbf{x})$

Suppose that a wavefront S is given by $t = T(\mathbf{x})$ and that S reaches the point $\mathbf{x} + d\mathbf{x}$ at a time dt later than it reaches the point $\mathbf{x}$. Then $t + dt = T(\mathbf{x} + d\mathbf{x})$, so that $dt = \nabla T \cdot d\mathbf{x}$. If $\mathbf{V}$ is the velocity of the advancing wavefront, in the direction $d\mathbf{x}$, it follows that $\mathbf{V} = d\mathbf{x}/dt$, and $\nabla T \cdot \mathbf{V} = 1$. Hence

$$V^2 \geq \frac{1}{(\nabla T)^2} = c^2 \quad \text{(using (4.41))},$$

with equality only when vectors $\mathbf{V}$ and ∇T are parallel, i.e., when $d\mathbf{x}$ is perpendicular to S. It follows that c is the velocity of S, normal (perpendicular) to itself.

For a given wavefront S, we introduce *rays* as the normals to S as the wavefront propagates. If a ray is parameterized in the form $\mathbf{x} = \mathbf{x}(\zeta)$, with ζ changing monotonically along the ray, it follows that $d\mathbf{x}/d\zeta = g(\mathbf{x})\,\nabla T$ is the equation of the ray, where g, a scalar function relating parallel vectors, is determined by the particular choice one makes for ζ.

For example, choosing

$$\frac{d\mathbf{x}}{d\zeta} = c^2\,\nabla T = (c\,\nabla T)c \tag{4.42}$$

and noting that $c\,\nabla T$ is the unit normal to S and that c is the velocity of S normal to itself, it follows that ζ has the interpretation of travel time along the ray. As S propagates normal to a particular ray R, the intersection of S and R will occur at different values of T. To examine the relation between this T and the ζ of (4.42), we note that

$$\frac{dT}{d\zeta} = \nabla T \cdot \frac{d\mathbf{x}}{d\zeta} = \nabla T \cdot (c^2\,\nabla T) = 1,$$

and hence T and ζ increase in precisely the same manner along the ray direction. The ray itself may thus be parameterized by T, and T has the physical interpretation of travel time along the ray.

In working with the equation of the ray path, it is occasionally convenient to use as a parameter the distance s, measured along the ray from some reference point. It follows that

$$\frac{d\mathbf{x}}{ds} = c\,\nabla T \tag{4.43}$$

(since $d\mathbf{x}/ds$ and ∇T arc parallel, but the left-hand side of (4.43) is a unit vector).

The rays themselves are fixed curves in space, and so to examine their geometrical properties we often need to eliminate the time quantity T in (4.43), obtaining an equation for $\mathbf{x} = \mathbf{x}(s)$ that involves only $c(\mathbf{x})$. Several stages are needed in such an elimination (which uses the eikonal equation), and step by step we have

$$\frac{d}{ds}\left(\frac{1}{c}\frac{d\mathbf{x}}{ds}\right) = \frac{d}{ds}\nabla T = \left(\frac{d\mathbf{x}}{ds}\cdot\nabla\right)\nabla T = (c\,\nabla T\cdot\nabla)\,\nabla T$$

$$= \frac{1}{2}c\nabla[(\nabla T)^2] = \frac{1}{2}c\nabla\left(\frac{1}{c^2}\right) = -\frac{1}{c^2}\nabla c.$$

The differential equation for a ray, in spatial coordinates only, is thus

$$\frac{d}{ds}\left(\frac{1}{c}\frac{d\mathbf{x}}{ds}\right) = \nabla\left(\frac{1}{c}\right). \tag{4.44}$$

In a homogeneous region, this reduces to $d^2\mathbf{x}/ds^2 = \mathbf{0}$, with the general solution $\mathbf{x} = \mathbf{a}s + \mathbf{b}$ ($\mathbf{a}$ and $\mathbf{b}$ constant), which is a straight line.

In a medium where c depends only on depth z (in a Cartesian coordinate system), the quantity

$$\hat{\mathbf{z}} \times \frac{1}{c}\frac{d\mathbf{x}}{ds} \equiv \mathbf{Q} \tag{4.45a}$$

is constant along a ray, since

$$\frac{d\mathbf{Q}}{ds} = \hat{\mathbf{z}} \times \frac{d}{ds}\left(\frac{1}{c}\frac{d\mathbf{x}}{ds}\right) = \hat{\mathbf{z}} \times \nabla\left(\frac{1}{c}\right) = \mathbf{0}$$

($\hat{\mathbf{z}}$ being parallel to $\nabla(1/c)$). It follows that rays are confined to planes parallel to the z-axis and that $\sin i(z)/c(z) \equiv p$ is constant along a ray, where i is the angle between the increasing z-direction and the ray (see Fig. 4.9a). This is *Snell's law*, and p is known as the *ray parameter*.

In a spherically symmetric medium, $c = c(r)$, it is the quantity

$$\mathbf{r} \times \frac{1}{c}\frac{d\mathbf{r}}{ds} \equiv \mathbf{Q} \tag{4.45b}$$

that is constant along a ray. (Here we are using $\mathbf{r} = (r, \theta, \phi)$ and spherical polars with origin at the center of symmetry.) This result again uses (4.44), as

$$\frac{d\mathbf{Q}}{ds} = \frac{d\mathbf{r}}{ds} \times \frac{1}{c}\frac{d\mathbf{r}}{ds} + \mathbf{r} \times \frac{d}{ds}\left(\frac{1}{c}\frac{d\mathbf{r}}{ds}\right) = \mathbf{0} + \mathbf{r} \times \nabla\left(\frac{1}{c}\right) = \mathbf{0}.$$

Rays stay in a vertical plane, and now the quantity $r \sin i(r)/c(r) \equiv p$ (still called the ray parameter) is constant along a ray path (see Fig. 4.9b). This is Snell's law for spherically symmetric media. We shall make frequent use of the quantity p in later chapters, often treating it as a variable, with different values

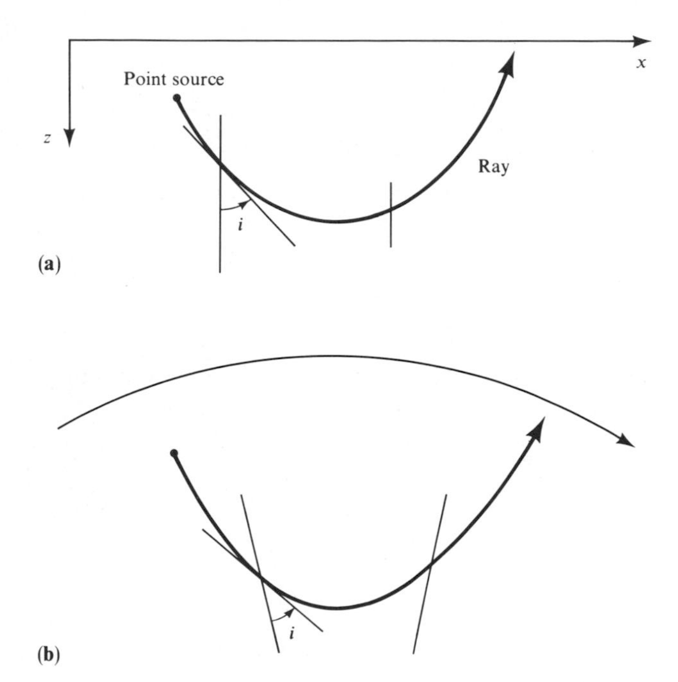

FIGURE **4.9**

(**a**) A ray path in a medium where velocity depends only on depth: $\sin i(z)/c(z)$ is constant, and the ray is confined to a vertical plane. (**b**) A ray path in a spherically symmetric medium where velocity depends only on the distance r from the center of symmetry: $r \sin i(r)/c(r)$ is constant, and the ray stays in a vertical plane. The angle i evaluated at the source is known as the *take-off angle.*

corresponding to different rays from a point source. Note that p has a different physical dimension in depth-dependent media (4.45a) than is the case in spherically symmetric media (4.45b).

There is one more general property of rays to be brought out before returning to the study of amplitudes of radiated particle displacement. This result is known as Fermat's Principle, which states that for two points A and B on a ray R, the ray itself is a path along which, in the velocity field $c(\mathbf{x})$, the travel-time from A to B is stationary. We can prove this by using further properties of the parameter ζ in the ray equation $d\mathbf{x}/d\zeta = g(\mathbf{x})\,\nabla T = g(\zeta)\,\nabla T$ (recall that we have already used ζ for travel time and for distance). The quantities $g(\zeta)$ and ζ are constrained by the eikonal equation (4.41), which implies

$$\frac{d\mathbf{x}}{d\zeta}\cdot\frac{d\mathbf{x}}{d\zeta} = \left(\frac{g}{c}\right)^2.$$

Given ζ, we then require that

$$g = c\left(\frac{d\mathbf{x}}{d\zeta}\cdot\frac{d\mathbf{x}}{d\zeta}\right)^{1/2}. \tag{4.46}$$

In general $d/ds = (c/g)d/d\zeta$, hence from (4.44) we have the ray equation for any choice of ζ as

$$\frac{c}{g}\frac{d}{d\zeta}\left[\frac{1}{g}\frac{d\mathbf{x}}{d\zeta}\right] = \nabla\left(\frac{1}{c}\right).$$

This form can be identified with the Euler equations of variational calculus by first eliminating g via (4.46) to obtain

$$\frac{d}{d\zeta}\left[\frac{1}{c(\mathbf{x})}\frac{\dfrac{d\mathbf{x}}{d\zeta}}{\left(\dfrac{d\mathbf{x}}{d\zeta}\cdot\dfrac{d\mathbf{x}}{d\zeta}\right)^{1/2}}\right] = \left(\frac{d\mathbf{x}}{d\zeta}\cdot\frac{d\mathbf{x}}{d\zeta}\right)^{1/2}\nabla\left(\frac{1}{c}\right).$$

This in turn can be written as the three scalar equations ($i = 1, 2, 3$):

$$\frac{d}{d\zeta}\left\{\frac{\partial}{\partial\left(\dfrac{\partial x_i}{\partial\zeta}\right)}\left[\frac{1}{c(\mathbf{x})}\left(\frac{d\mathbf{x}}{d\zeta}\cdot\frac{d\mathbf{x}}{d\zeta}\right)^{1/2}\right]\right\} = \frac{\partial}{\partial x_i}\left[\frac{1}{c(\mathbf{x})}\left(\frac{d\mathbf{x}}{d\zeta}\cdot\frac{d\mathbf{x}}{d\zeta}\right)^{1/2}\right].$$

These are the Euler equations we seek, since they tell us that the travel time is stationary for a ray path. That is,

$$T(B, A) = \int_A^B \frac{|d\mathbf{x}|}{c(\mathbf{x})} = \int_A^B \frac{1}{c(\mathbf{x})}\left(\frac{d\mathbf{x}}{d\zeta}\cdot\frac{d\mathbf{x}}{d\zeta}\right)^{1/2} d\zeta \tag{4.47}$$

is an integral that, for all possible paths connecting A and B, is stationary if and only if $\mathbf{x} = \mathbf{x}(\zeta)$ is a ray path.

4.4.2 Ray coordinates

The wavefront S is orthogonal to its associated rays, and we have shown how points along a ray are parameterized by values of T. We now introduce (c^2, c^3) as some coordinate system within the wavefront itself (or, equivalently, recognize that rays associated with a point source form a 2-parameter family), and then each point reached by rays in the velocity field $c(\mathbf{x})$ ($c = \alpha$ or β) is described by values in a ray coordinate system (T, c^2, c^3). Our notation for curvilinear coordinates is introduced in Box 2.6. The T-direction (which is along the ray) is orthogonal to the c^2 and c^3 directions at every point, but in general inhomogeneous media it is not possible to choose coordinates so that the c^2 and c^3 directions are always orthogonal. Fortunately, the special case of a spherically symmetric medium does permit a choice for orthogonal c^2 and c^3, allowing us then to apply directly the results for general orthogonal systems developed in Chapter 2. Note here that the scale factor $h_T = h^1 = c$, since a distance increment ds along the ray is $c\,dT$.

In homogeneous media, we found that far-field *P*-waves are longitudinal and *S*-waves transverse. The same holds true for general inhomogeneous (but still isotropic) media, as may be seen from taking the vector product and the scalar product of equation (4.38) with ∇T. In isotropic media, the results are

$$[\rho - \mu\nabla T \cdot \nabla T]\ddot{\mathbf{U}} \times \nabla T = \mathbf{E}(\mathbf{U}) \times \nabla T \tag{4.48}$$

and

$$[\rho - (\lambda + 2\mu)\nabla T \cdot \nabla T]\ddot{\mathbf{U}} \cdot \nabla T = \mathbf{E}(\mathbf{U}) \cdot \nabla T. \tag{4.49}$$

Recalling that $|\mathbf{E}|$ is of order $|\dot{\mathbf{U}}|$, that the ansatz $\mathbf{u}(\mathbf{x}, \mathbf{t}) = f(\mathbf{x})\mathbf{U}(t - T(\mathbf{x}))$ has been used, and that terms of order $|\ddot{\mathbf{U}}|$ are much greater than terms of order $|\dot{\mathbf{U}}|$, it follows that either

$$\ddot{\mathbf{U}} \times \nabla T = \mathbf{0} \quad \text{and} \quad (\nabla T)^2 = \frac{\rho}{\lambda + 2\mu} = \frac{1}{\alpha(\mathbf{x})^2} \tag{4.50}$$

or

$$\ddot{\mathbf{U}} \cdot \nabla T = 0 \quad \text{and} \quad (\nabla T)^2 = \frac{\rho}{\mu} = \frac{1}{\beta(\mathbf{x})^2}. \tag{4.51}$$

The ray direction is given by ∇T, so that equation (4.50) describes a longitudinal wave propagating along rays defined by the local *P*-wave speed, and (4.51) describes a transverse wave propagating along rays defined by the local *S*-wave speed.

Our next goal is to find the amplitude variation of *P*- and *S*-waves as they propagate along rays in an inhomogeneous medium. From our results (4.50) and (4.51), we shall assume that:

i) if (T, c^2, c^3) are defined in terms of the *P*-wave speed $\alpha(\mathbf{x})$, then the equations of motion have an approximate solution in which u^1 (the component of $\mathbf{u}$ along direction ∇T) is dominant, i.e., that the transverse components u^2 and u^3 are negligible. This is the longitudinal solution, with ansatz

$$\mathbf{u}(\mathbf{x}, t) = f^1(\mathbf{x})(U^1(t - T(\mathbf{x})), 0, 0). \tag{4.52}$$

However, we assume that

ii) if (T, c^2, c^3) are defined in terms of the S-wave speed $\beta(\mathbf{x})$, then the displacements u^2 and u^3 are dominant. This is the transverse solution, with ansatz

$$\mathbf{u}(\mathbf{x}, t) = f^2(\mathbf{x})(0, U^2(t - T(\mathbf{x})), 0) + f^3(\mathbf{x})(0, 0, U^3(t - T(\mathbf{x})). \quad (4.53)$$

We shall examine first the special case of a spherically symmetric medium, this being a good approximation to the Earth's structure. For this case, c^2 and c^3 are chosen as the orthogonal coordinates p and ϕ described in Figure 4.10.

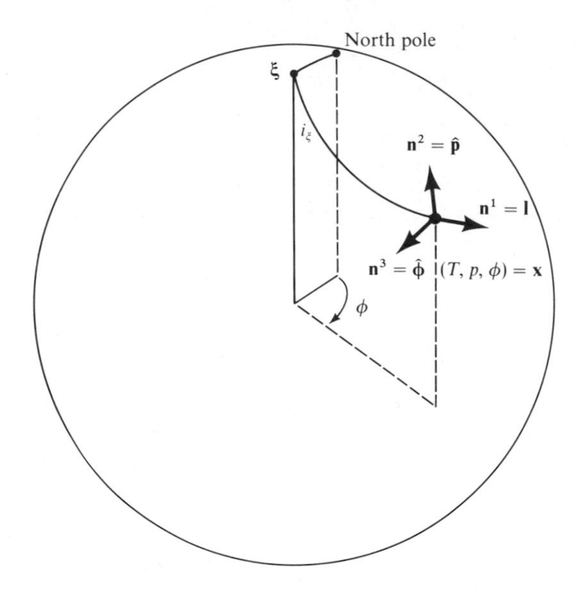

FIGURE **4.10**

General orthogonal ray coordinates for a spherically symmetric medium. We choose $(c^1, c^2, c^3) = (T, p, \phi)$, where p is the ray parameter $r \sin i(r)/c(r)$ (which is constant along a given ray), ϕ is the geographical azimuthal angle, and T is the travel time taken for a wavefront to go from $\boldsymbol{\xi}$ to $\mathbf{x}$ along the ray path. Thus $(c^2, c^3) = (p, \phi)$ specify a particular ray leaving the source, and T specifies a particular point along the ray. Associated normals at $\mathbf{x}$ are $\mathbf{n}^1$ (along the ray), $\mathbf{n}^2$ (transverse and in a vertical plane), and $\mathbf{n}^3$ (transverse and horizontal). We shall often use the symbols $\mathbf{l}$, $\hat{\mathbf{p}}$, $\hat{\boldsymbol{\phi}}$ for these respective unit normals.

Another possible choice for c^2 is i_ξ, often called the take-off angle: in this case i_ξ and ϕ would be the angles for a system of spherical polar coordinates centered on the source, with direction $i_\xi = 0$ taken downward and the direction $\phi = 0$ being due North from the source. The reason for taking $i_\xi = 0$ as the *downward* vertical is that ϕ then has the ordinary geographical definition of azimuth from the source, and a right-handed coordinate system is indeed obtained with ϕ measured *clockwise* round from North.

We shall find that the amplitude functions f^1, f^2, and f^3 of (4.52) and (4.53) can be obtained quickly, and two independent components of the S-wave are identified.

4.4.3 *The geometrical solution for P-waves in spherically symmetric media*

The exact equation satisfied by $u^1(\mathbf{x}, t)$ is now

$$\rho \frac{\partial^2 u^1}{\partial t^2} = \frac{1}{\alpha h^2 h^3}\left[\frac{\partial}{\partial T}(h^2 h^3 \tau^{11}) + \frac{\partial}{\partial c^2}(h^3 \alpha \tau^{12}) + \frac{\partial}{\partial c^3}(\alpha h^2 \tau^{31})\right]$$
$$+ \frac{\tau^{12}}{\alpha h^2}\frac{\partial \alpha}{\partial c^2} + \frac{\tau^{31}}{h^3 \alpha}\frac{\partial \alpha}{\partial c^3} - \frac{\tau^{22}}{\alpha h^2}\frac{\partial h^2}{\partial T} - \frac{\tau^{33}}{h^3 \alpha}\frac{\partial h^3}{\partial T} \tag{4.54}$$

(see (2.48) and Figure 4.10), except at the source $\mathbf{x} = \boldsymbol{\xi}$. However, from the strain-displacement formulas ((2.45), (2.46)) and our assumption

iii) that wavefunction derivatives perpendicular to the wavefront are much greater than derivatives parallel to the wavefront,

it follows that most of the strain components in this coordinate system are negligible. The stress-strain relation (2.5) then gives τ^{12}, τ^{13}, τ^{22}, τ^{33} effectively equal to zero, whereas

$$\tau^{11} = \frac{\lambda + 2\mu}{\alpha}\frac{\partial u^1}{\partial T},$$

so that (4.54) reduces to the approximate form

$$\rho \frac{\partial^2 u^1}{\partial t^2} = \frac{1}{\alpha h^2 h^3}\left[\frac{\partial}{\partial T}\left(h^2 h^3 \frac{\lambda + 2\mu}{\alpha}\frac{\partial u^1}{\partial T}\right)\right].$$

To the same order of approximation (i.e., retaining $\ddot{u}^1$ and $\dot{u}^1$), this is

$$\frac{\partial^2}{\partial t^2}[(\rho \alpha h^2 h^3)^{1/2} u^1] = \frac{\partial^2}{\partial T^2}[(\rho \alpha h^2 h^3)^{1/2} u^1], \tag{4.55}$$

which is merely a one-dimensional wave equation for a wave propagating along the direction of varying T. The general solution for $(\rho \alpha h^2 h^3)^{1/2} u^1$ is a function of $t - T$ plus a function of $t + T$. The P-wave solution for longitudinal motion along *increasing* T is therefore

$$\mathbf{u}^P(\mathbf{x}, t) = (u^1, 0, 0) = \left(\frac{1}{\rho \alpha h^2 h^3}\right)^{1/2} \mathscr{F}^P(c^2, c^3)(U^1(t - T(\mathbf{x}), 0, 0). \tag{4.56}$$

The factors in the right-hand side of this expression have several of the properties that were noted in Section 4.1 (following (4.3)) for the elementary solution of a wave spreading from a point source in a homogeneous medium. The quantity $(h^2h^3)^{-1/2}$ describes the attenuation of the wave due to geometrical spreading (see Fig. 4.11); the factor $\mathscr{F}^P(c^2, c^3)$ describes the radiation pattern of P-waves emanating in different directions (c^2, c^3) from the source, and we can expect that the function $U^1(t - T(\mathbf{x}))$, giving the displacement pulse shape in the longitudinal direction, must be related to the time function of the operative body force at the source.

The most cryptic factor in (4.56) is $(h^2h^3)^{-1/2}$, which in Figure 4.11 we have associated with geometrical spreading, and which in homogeneous media is simply $1/|\mathbf{x} - \boldsymbol{\xi}|$. The geometrical spreading factor in inhomogeneous media is of great practical importance, since it describes the focusing and defocusing of

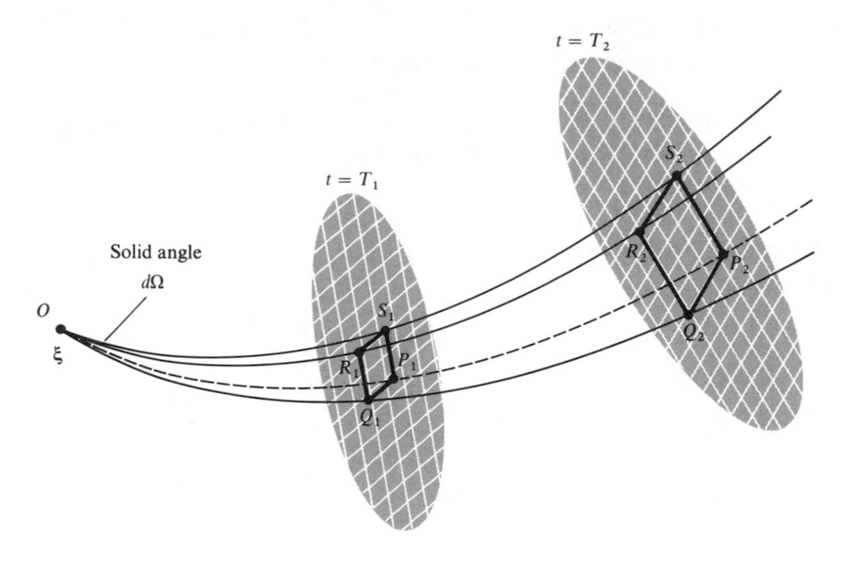

FIGURE **4.11**

Geometrical spreading of four rays. Suppose OP_1P_2 is the ray with coordinates (c^2, c^3) in the general orthogonal system (T, c^2, c^3), and that T_1 and T_2 give the position of the same wavefront at successive times $t = T_1, t = T_2$. Suppose that rays OQ_1Q_2, OR_1R_2, OS_1S have coordinates $(c^2, c^3 + dc^3)$, $(c^2 + dc^2, c^3 + dc^3)$, and $(c^2 + dc^2, c^3)$, respectively, and that the solid angle of this pencil at O is $d\Omega$. Then the area of $P_1Q_1R_1S_1$ is $h^2h^3\ dc^2\ dc^3$, with (h^2h^3) evaluated at $T = T_1$, and $P_2Q_2R_2S_2$ is $h^2h^3\ dc^2\ dc^3$, with (h^2h^3) evaluated at $T = T_2$. In this sense, h^2h^3 is proportional to the areal cross section of a pencil of rays emanating from the point source.

If coordinates c^2 and c^3 are not orthogonal (this is the case for general inhomogeneous media), then the areas $P_1Q_1R_1S_1$ and $P_2Q_2R_2S_2$ are not rectangles, but are given by $|(\partial\mathbf{x}/\partial c^2) \times (\partial\mathbf{x}/\partial c^3)|\ dc^2\ dc^3$. Such areas are still proportional to $d\Omega$ on any given wavefront, and we define a geometrical spreading function $\mathscr{R}(\mathbf{x}, \boldsymbol{\xi})$ by equating the ray-tube cross-sectional area at $\mathbf{x}$ to $\mathscr{R}^2(\mathbf{x}, \boldsymbol{\xi})\ d\Omega$.

rays. We therefore introduce a new function, $\mathscr{R}^P(\mathbf{x}, \xi)$, described in the following specific way: the pencil of P-wave rays spreading from the source ξ within solid angle $d\Omega$ has, on the wavefront coinciding with $\mathbf{x}$, the area $[\mathscr{R}^P(\mathbf{x}, \xi)]^2\, d\Omega$. If the rays are straight lines, then $\mathscr{R}^P(\mathbf{x}, \xi) = |\mathbf{x} - \xi|$, but in the more general case of curved rays shown in Figure 4.11, with $\mathbf{x}$ varying along a particular ray tube, $\mathscr{R}^P(\mathbf{x}, \xi) \propto (h^2 h^3)^{1/2}$. The constant of proportionality here can depend only on ξ and on the ray direction at ξ specified by (c^2, c^3). Since it is merely a scale factor, it can be absorbed within $\mathscr{F}^P$ and U^1 to give

$$\mathbf{u}^P(\mathbf{x}, t) = \left(\frac{1}{\rho(\mathbf{x})\alpha(\mathbf{x})}\right)^{1/2} \frac{1}{\mathscr{R}^P(\mathbf{x}, \xi)} \mathscr{F}^P(c^2, c^3)(U^1(t - T^P(\mathbf{x}, \xi)), 0, 0) \tag{4.57}$$

for the geometrical ray solution of P-wave displacement. A superscript is also used here for the travel time, and $T^P(\mathbf{x}, \xi) = \int_\xi^\mathbf{x} |ds|/\alpha$, in which the integral (see (4.47)) is taken along the ray. In Section 4.5 we shall evaluate $\mathscr{F}^P(c^2, c^3)$ and $U^1(t)$ explicitly for a point shear dislocation of arbitrary orientation.

4.4.4 *The geometrical solution for S-waves in spherically symmetric media: Introduction of the components SV and SH*

The ray coordinates of Figure 4.10, with travel time T defined for the shear-wave speed $\beta(\mathbf{x}) = \beta(r)$, are also appropriate for analyzing transverse motions, $\mathbf{u} = (0, u^2, u^3)$. Using assumptions (ii) and (iii) above and the stress-strain and strain-displacement formulas of Section 2.6, we find that all stress components are negligible, except for the shearing stresses

$$\tau^{12} = \frac{\mu}{\beta}\frac{\partial u^2}{\partial T}, \quad \tau^{31} = \frac{\mu}{\beta}\frac{\partial u^3}{\partial T}.$$

These may be substituted into the equations for $\ddot{u}^2$ and $\ddot{u}^3$ (see (2.48)), and by steps similar to the derivation of (4.55) we now find

$$\frac{\partial^2}{\partial t^2}[(\rho\beta h^2 h^3)^{1/2} u^2] = \frac{\partial^2}{\partial T^2}[(\rho\beta h^2 h^3)^{1/2} u^2] \tag{4.58}$$

$$\frac{\partial^2}{\partial t^2}[(\rho\beta h^2 h^3)^{1/2} u^3] = \frac{\partial^2}{\partial T^2}[(\rho\beta h^2 h^3)^{1/2} u^3]. \tag{4.59}$$

Again, these are one-dimensional wave equations for which we know the general solution; equations (4.58) and (4.59) also tell us that u^2 and u^3 propagate independently. Referring to Figure 4.10, we see that u^3 is a motion in the horizontal direction at $\mathbf{x}$, hence it is called the SH-component. The motion u^2

is also transverse to the ray, but is in the vertical plane, hence it is called the SV-component. Note that the SV-component is not necessarily in a vertical direction.

The geometrical S-wave solutions in spherically symmetric media can now be stated as

$$\mathbf{u}^{SV}(\mathbf{x}, t) = \left(\frac{1}{\rho(\mathbf{x})\beta(\mathbf{x})}\right)^{1/2} \frac{1}{\mathscr{R}^S(\mathbf{x}, \xi)} \mathscr{F}^{SV}(p, \phi)(0, U^2(t - T^S(\mathbf{x}, \xi)), 0) \quad (4.60)$$

and

$$\mathbf{u}^{SH}(\mathbf{x}, t) = \left(\frac{1}{\rho(\mathbf{x})\beta(\mathbf{x})}\right)^{1/2} \frac{1}{\mathscr{R}^S(\mathbf{x}, \xi)} \mathscr{F}^{SH}(p, \phi)(0, 0, U^3(t - T^S(\mathbf{x}, \xi))). \quad (4.61)$$

Although the radiation patterns $\mathscr{F}^{SV}$ and $\mathscr{F}^{SH}$ may be different, the two components of S share the same geometrical spreading factor $\mathscr{R}^S(\mathbf{x}, \xi)$ and the same travel-time function T^S. The pulse shapes U^2 and U^3 may, in general, be different, although we shall find in the next section that they are the same for far-field S-waves from a point shear dislocation. The relative amount of motion partitioned into SV and SH is known as the *polarization* of the S-wave, and it does not change along a given ray: this is a property of great importance in interpreting body-wave seismograms.

4.4.5 The geometrical ray solutions in general inhomogeneous media

The precision of modern seismology makes apparent the fact that the Earth's internal structure is not quite spherically symmetric. Thus travel-time curves (plots of the time T against distance over the Earth's surface) for body waves from a surface source show systematic differences in different regions. In studies such as that by Julian and Sengupta (1973), significant differences of around 0.2 sec (in travel-times amounting to several minutes) have been used to estimate the departure from heterogeneity in the deep mantle. Part of the departure from symmetry is due to the Earth's ellipticity, but corrections for this are now well understood (Bullen, 1937; Dziewonski and Gilbert, 1976). More interesting are the effects introduced by major laterally varying structures such as descending slabs of lithosphere, mantle plumes, continental margins, and a wide variety of crustal inhomogeneities. On body waves, there are effects that may sometimes be interpreted in terms of reflection and refraction at oblique (i.e., nonhorizontal) boundaries, but our present concern is with ray theory for media that vary smoothly, i.e., without discontinuities in the structure.

If the assumption of spherical symmetry is dropped, then the principal complication is that a given ray is no longer confined to lie in a plane. At worst, it may twist around like a bent corkscrew, and it is therefore surprising that the results (4.57), (4.60), (4.61), already obtained for spherical symmetry, require very little adaption to the general case. In fact, the geometrical ray solution (4.57) for *P*-waves is unchanged in general isotropic media. For *S*-waves, the displacement amplitude still changes like $1/[(\rho\beta)^{1/2}\mathscr{R}^S(\mathbf{x}, \xi)]$ (cf. (4.60), (4.61)), and there is still a sense in which the polarization is unchanged along the ray path. Because it is no longer possible to find coordinates (c^2, c^3) that are orthogonal in every wavefront, the results we have just stated require a different and more intricate method of proof.

Thus, for *P*-waves, we can try the ansatz

$$\mathbf{u}^P(\mathbf{x}, t) = f(\mathbf{x})U(t - T(\mathbf{x}))\alpha\,\nabla T, \tag{4.62}$$

which is suggested by (4.50). Here $\alpha\,\nabla T$ is a unit vector in the direction of longitudinal particle motion, and we substitute (4.62) into the general displacement equation for inhomogeneous isotropic media,

$$\rho\frac{\partial^2 u_i}{\partial t^2} = \frac{\partial}{\partial x_j}[\lambda\delta_{ij}u_{k,k} + \mu(u_{i,j} + u_{j,i})], \tag{4.63}$$

obtained from (2.17), (2.49), and (2.3). The terms of order $\ddot{U}$ all cancel, because $(\nabla T)^2 = 1/\alpha^2$, and if the terms of order $\dot{U}$ are also required to vanish, there results an equation for the *P*-wave amplitude, $f(\mathbf{x})$. Taking the scalar product with ∇T (cf. (4.49)), this result is obtained after some manipulation in the form

$$(\lambda + 2\mu)[2\,\nabla T\cdot\nabla(\alpha f) + 2\alpha^3 f\,T_{,k}T_{,ik}T_{,i} + \alpha f\nabla^2 T] + \alpha f\nabla(\lambda + 2\mu)\cdot\nabla T = 0. \tag{4.64}$$

Most of these terms can immediately be converted to spatial derivatives with respect to distance s along the ray, since $\alpha(\partial T/\partial x_i)(\partial/\partial x_i) = \partial/\partial s$. For example,

$$T_{,k}T_{,ik}T_{,i} = \frac{\partial T}{\partial x_k}\frac{1}{\alpha}\frac{\partial}{\partial s}\frac{\partial T}{\partial x_k} = \frac{1}{2\alpha}\frac{\partial}{\partial s}\left(\frac{\partial T}{\partial x_k}\frac{\partial T}{\partial x_k}\right) = \frac{1}{2\alpha}\frac{\partial}{\partial s}\left(\frac{1}{\alpha^2}\right).$$

The effect of geometrical spreading is present in (4.64) in the term $\nabla^2 T$. This is the divergence of ∇T, and using an elementary volume with corners at $T \pm \frac{1}{2}\delta T$, $c^2 \pm \frac{1}{2}\delta c^2$, $c^3 \pm \frac{1}{2}\delta c^3$, we see that the flux of ∇T out of the volume is nonzero only for the sides with area $\mathscr{R}^2(\mathbf{x}, \xi)\,d\Omega$. (See Fig. 4.11 and its legend.) It follows from the fundamental definition of divergence (flux per unit volume) that

$$\nabla^2 T = \frac{1}{\mathscr{R}^2(\mathbf{x}, \xi)\,d\Omega\alpha\,\delta T}\frac{\partial}{\partial T}\left(\frac{\mathscr{R}^2(\mathbf{x}, \xi)\,d\Omega}{\alpha}\right)\delta T = \frac{1}{\mathscr{R}^2}\frac{\partial}{\partial s}\left(\frac{\mathscr{R}^2}{\alpha}\right), \tag{4.65}$$

and (4.64) reduces to

$$\frac{1}{f}\frac{\partial f}{\partial s} + \frac{1}{2\rho}\frac{\partial \rho}{\partial s} + \frac{1}{2\alpha}\frac{\partial \alpha}{\partial s} + \frac{1}{\mathscr{R}}\frac{\partial \mathscr{R}}{\partial s} = 0. \tag{4.66}$$

This integrates to give $f(\mathbf{x}) \propto 1/[(\rho(\mathbf{x})\alpha(\mathbf{x}))^{1/2}\mathscr{R}(\mathbf{x}, \boldsymbol{\xi})]$, and the constant of proportionality is dependent on the particular ray, i.e., on (c^2, c^3). The ansatz (4.62) therefore leads to exactly the same geometrical solution for *P*-waves, as we found in (4.57) for spherically symmetric media.

For *S*-waves, we know from (4.51) that particle displacement is still, in general, transverse to the ray, but there is now a phenomenon that cannot be quantified from experience in the study of homogeneous or spherically symmetric media. Namely, the phenomenon that the direction of particle motion may rotate around the ray as the wavefront propagates.

Our ansatz for *S*-wave motion is

$$\mathbf{u}^S(\mathbf{x}, t) = f(\mathbf{x})U(t - T(\mathbf{x}))\boldsymbol{v}(\mathbf{x}), \tag{4.67}$$

where $f(\mathbf{x})$ is the amplitude, $U(t)$ the pulse shape, and unit vector $\boldsymbol{v}$ is the direction of particle motion, so that $\boldsymbol{v}\cdot\nabla T = 0$. Again, we substitute our ansatz into the displacement equation (4.63), but this time form the vector product with ∇T. The largest terms in $\ddot{U}$ cancel in view of the property $(\nabla T)^2 = 1/\beta^2$, and by requiring the terms in $\dot{U}$ to vanish we obtain

$$\mu[2(\nabla f\cdot\nabla T)\boldsymbol{v}\times\nabla T + 2f((\nabla T\cdot\nabla)\boldsymbol{v})\times\nabla T + f(\nabla^2 T)\boldsymbol{v}\times\nabla T] + f(\nabla\mu\cdot\nabla T)\boldsymbol{v}\times\nabla T = \mathbf{0}. \tag{4.68}$$

Using the rule $\beta(\partial T/\partial x_i)(\partial/\partial x_i) = \partial/\partial s$ to convert to derivatives with respect to distance s along the ray, and also $\nabla^2 T = (1/\mathscr{R}^2)(\partial/\partial s)(\mathscr{R}^2/\beta)$, where $\mathscr{R}$ is the geometrical spreading factor for *S*-waves (cf. (4.65)), (4.68) becomes

$$\left(\frac{1}{f}\frac{\partial f}{\partial s} + \frac{1}{2\rho}\frac{\partial \rho}{\partial s} + \frac{1}{2\beta}\frac{\partial \beta}{\partial s} + \frac{1}{\mathscr{R}}\frac{\partial \mathscr{R}}{\partial s}\right)(\boldsymbol{v}\times\nabla T) + \left(\frac{\partial \boldsymbol{v}}{\partial s}\times\nabla T\right) = \mathbf{0}. \tag{4.69}$$

However, the two vector directions here, $\boldsymbol{v}\times\nabla T$ and $\partial\boldsymbol{v}/\partial s\times\nabla T$, are orthogonal. (This follows from the mutual orthogonality of ∇T, $\boldsymbol{v}$, and $\boldsymbol{v}\times\nabla T$; and from the fact that $0 = \partial/\partial s(\boldsymbol{v}\cdot\boldsymbol{v}) = 2\boldsymbol{v}\cdot\partial\boldsymbol{v}/\partial s$, so that $\partial\boldsymbol{v}/\partial s$ lies in the plane of ∇T and $\boldsymbol{v}\times\nabla T$.) Hence the amplitude function f is constrained by

$$\frac{1}{f}\frac{\partial f}{\partial s} + \frac{1}{2\rho}\frac{\partial \rho}{\partial s} + \frac{1}{2\beta}\frac{\partial \beta}{\partial s} + \frac{1}{\mathscr{R}}\frac{\partial \mathscr{R}}{\partial s} = 0, \tag{4.70}$$

and also

$$\partial \mathbf{v}/\partial s \times \nabla T = \mathbf{0}, \tag{4.71}$$

implying that $\partial \mathbf{v}/\partial s$ must be along the longitudinal direction, ∇T.

From (4.70) we see that $f(\mathbf{x}) \propto 1/[(\rho(\mathbf{x})\beta(\mathbf{x}))^{1/2}\mathscr{R}(\mathbf{x}, \boldsymbol{\xi})]$, and the amplitude has just the property found for both the *SV*- and *SH*-components in spherically symmetric media (cf. (4.60), (4.61)).

From (4.71) we can show that there is a sense in which the *S*-wave retains the same polarization at different points along the ray, even in general inhomogeneous media. This effect may be analyzed using the intrinsic system of unit vectors associated with each point along a ray. These are the unit tangent $\mathbf{l} = \beta\nabla T$, the unit normal $\mathbf{n} = (d\mathbf{l}/ds)/|d\mathbf{l}/ds|$, and the unit binormal $\mathbf{b} = \mathbf{l} \times \mathbf{n}$. The local curvature of the ray, $\kappa(s)$, measures the tendency to change direction, and is simply $|d\mathbf{l}/ds|$. Note that this direction change, which is normal to $\mathbf{l}$, is related to $\mathbf{n}$ by

$$\frac{d\mathbf{l}}{ds} = \kappa(s)\mathbf{n}. \tag{4.72}$$

For each point on the ray, the vectors $\mathbf{l}$ and $\mathbf{n}$ determine a plane known as the *osculating plane* (see Fig. 4.12), which coincides with the plane of the ray in cases where the ray is confined to one plane (e.g., rays in a spherically symmetric Earth). More generally, the osculating plane (which is normal to $\mathbf{b}$) changes

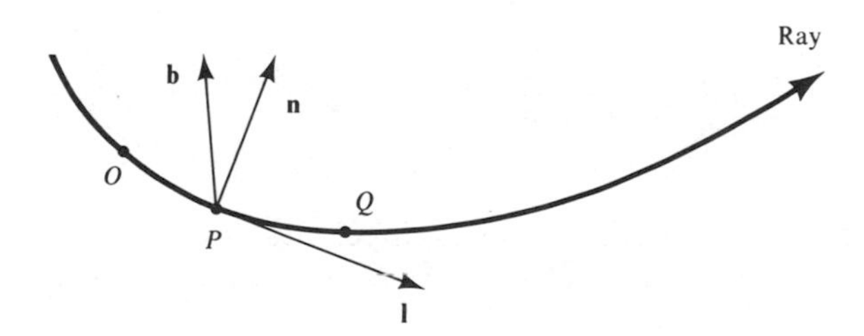

FIGURE **4.12**

The system of three mutually orthogonal unit vectors at point P, intrinsic to the geometry of a twisting ray: $\mathbf{l}$ is along the ray; $\mathbf{n}$ is normal to the ray and lies in the plane in which the ray is changing direction ($d\mathbf{l}/ds$). (This is the osculating plane, defined as the limiting plane containing points OPQ lying on the curved ray, as O and Q tend to P.) The unit binormal $\mathbf{b}$ is taken in the direction that makes ($\mathbf{l}$, $\mathbf{n}$, $\mathbf{b}$) a right-handed triple. If the ray does not twist, but merely curves in a plane, then $\mathbf{b}$ is normal to that plane.

direction along the ray, and the derivative $d\mathbf{b}/ds$ gives a measure of the tendency of the ray to "twist" out of the osculating plane.

Writing $d\mathbf{b}/ds = c_1\mathbf{l} + c_2\mathbf{n} + c_3\mathbf{b}$, it is simple to show from the fixed length of $\mathbf{b}$ that $c_3 = 0$, and from the definitions of $\mathbf{b}$ and $\mathbf{n}$ that $c_1 = 0$. Thus $d\mathbf{b}/ds$ is in direction $\mathbf{n}$, and the local torsion T of the ray, or second curvature, is defined by

$$\frac{d\mathbf{b}}{ds} = -T\mathbf{n}. \tag{4.73}$$

This equation means that the axes $(\mathbf{n}, \mathbf{b})$ rotate around the ray in a right-handed fashion at rate T per unit distance.

Since $\mathbf{n} = \mathbf{b} \times \mathbf{l}$, it follows that

$$\frac{d\mathbf{n}}{ds} = -\kappa\mathbf{l} + T\mathbf{b}. \tag{4.74}$$

Equations (4.72)–(4.74) are known as the Frénet formulas, and we now return to the problem of finding out how the direction of transverse motion changes during propagation.

Suppose that, at distance s along the ray, the transverse direction $\boldsymbol{\nu}$ of S-wave motion makes an angle γ with the local normal:

$$\boldsymbol{\nu} = \cos\gamma\mathbf{n} + \sin\gamma\mathbf{b}. \tag{4.75}$$

Then, from (4.73) and (4.74),

$$\frac{\partial\boldsymbol{\nu}}{\partial s} = -\kappa\cos\gamma\mathbf{l} - \left(\frac{d\gamma}{ds} + T\right)(\sin\gamma\mathbf{n} - \cos\gamma\mathbf{b}). \tag{4.76}$$

But we already concluded from (4.71) that $\partial\boldsymbol{\nu}/\partial s$ is directed along ∇T, i.e., along $\mathbf{l}$, hence (4.76) implies

$$\frac{d\gamma}{ds} = -T. \tag{4.77}$$

This means that the S-wave motion $\mathbf{u}^S$, as referred to the $(\mathbf{n}, \mathbf{b})$ directions, rotates around the ray at exactly the same rate T as the axes themselves rotate (see (4.73)), *but in the opposite direction.* Because of this cancellation, the S-wave may be said to retain its polarization. In practice, it appears that spatial gradients of the S-wave speed do often vary (both vertically and horizontally) in such a way as to require full three-dimensional ray tracing for the analysis of travel times and radiation patterns from a particular point source. [This involves numerical solution of the spatial equation (4.44) for ray trajectories. However, because of the two cancelling effects described above, (4.73) and (4.77), it may be true that the polarization (expressed now as an SH/SV ratio) is not substantially changed during propagation through smoothly varying structures. The subject has not been thoroughly investigated.]

These results conclude our present development of ray theory. In Chapter 5, we shall relate the above results to the conservation of energy in ray tubes; in Chapter 9 we shall develop further properties of rays in a spherically symmetric medium (finding for caustics, where $\mathscr{R}(\mathbf{x}, \xi) = 0$, and for shadows that the results of this section must be modified); and in Chapter 13 we shall show how three-dimensional ray-tracing is used to interpret the effects of laterally varying structures in the Earth.

The methods of this section have been carried much further by V. M. Babich and by J. B. Keller and their colleagues, using trial solutions such as

$$\mathbf{u}(\mathbf{x}, \omega) = \exp(i\omega T) \sum_{n=0}^{\infty} \mathbf{A}_n(\mathbf{x})(i\omega)^{-n}. \tag{4.78}$$

Recursion relations between the $\mathbf{A}_n$ can be found that make it possible to find as many terms in the infinite series as may be required (Karal and Keller, 1959). Our work above is for $\mathbf{A}_0$, which dominates at sufficiently high frequencies. Occasionally, $\mathbf{A}_1$ is useful (e.g., if material boundaries are present, and the source-receiver geometry is such as to permit head waves, then their dominant effect is contained within $\mathbf{A}_1$). More commonly, it may be said that where $\mathbf{A}_0$ alone is inadequate, then one must either take many more terms in the series (4.78) (which is impractical), or use a different form of trial solution. (In the time domain, $\mathbf{A}_n$ in (4.78) weights a term proportional to $(t - T(x))^{n-1}$, so that this solution "blows up" away from wavefronts; see Okal & Mechler, 1973).

4.5 The Radiation Pattern of Body Waves in the Far Field for a Point Shear Dislocation of Arbitrary Orientation in a Spherically Symmetric Medium

A thorough understanding of this subject-title has played a major role in the discovery and measurement of many lithospheric plate motions (Sykes, 1967; Isacks, Oliver, and Sykes, 1968). This radiation pattern, once it has been recognized and its orientation obtained for a given earthquake, gives information about the stress regime that caused the earthquake to occur.

We have already given a description of the radiation pattern in source coordinates in Section 4.3 (see Figs. 4.5b, 4.6b). For P-waves, there are two nodal planes. These are orthogonal, and one of them is the fault plane. This source-coordinate system, however, is unavailable to us until we have learned the orientation of the causative fault and the direction of slip on that fault. Fault orientation is specified by *strike* and *dip*, and then either the *rake* or the *plunge* is used to specify the direction of slip (see Fig. 4.13). A fault has two surfaces, and the one illustrated is the surface of the *foot wall*. The other surface is known as the *hanging wall*. Slip $\bar{\mathbf{u}}$ is taken as the direction of the hanging wall, relative to the foot wall. Rake λ is the angle between strike direction and slip: $-\pi < \lambda \leq \pi$. If δ is neither 0 nor $\pi/2$ and λ is within the range $(0, \pi)$, the fault is termed a *reverse fault* or a *thrust fault*. However, if λ is within the range $(-\pi, 0)$, the fault is termed a *normal fault*.

Instead of the rake (which is measured within the fault plane), some authors use the *plunge*, which is measured in the vertical plane. Measuring the plunge from the horizontal down to the direction of $\bar{\mathbf{u}}$, we find the sine of the plunge is equal to $-\sin \lambda \sin \delta$.

A *strike-slip fault* is one for which $\delta = \pi/2$ and $\lambda = 0$ or π; the choice of hanging wall and foot wall is then arbitrary, and two possible choices can be made for the strike direction. It is useful, however, to establish a convention, so that right-lateral and left-lateral strike-slip faults can immediately be distinguished by λ values alone. (A *right-lateral fault* is one for which an observer standing on one block sees the other block as having moved to the right.) Our convention is to fix on either one of the two possible strike directions and label the right-hand block (as viewed by an observer looking along the strike) as the hanging wall. This decides which of the two fault surfaces is used to define λ, and clearly $\lambda = 0$ is a left-lateral strike-slip fault and $\lambda = \pi$ is right-lateral.

A *dip-slip fault* is also one for which $\delta = \pi/2$, but $\lambda = \pi/2$ or $-\pi/2$. Again, there is ambiguity in strike direction. We take the foot wall to lie in the down-dropped block, and the strike direction as again that for which the hanging wall is on the right. Then a dip-slip fault always has $\lambda = \pi/2$.

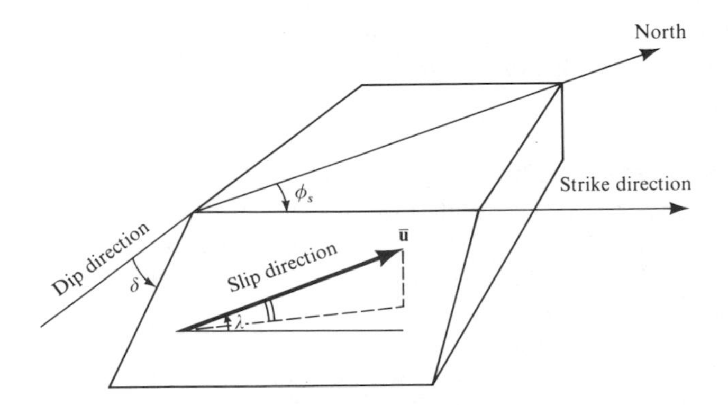

FIGURE **4.13**
Definition of the fault-orientation parameters (strike ϕ_S, dip δ), and slip-direction. ϕ_S is measured clockwise round from north, with the fault dipping down to the right of the strike direction: $0 \leq \phi_S < 2\pi$. δ is measured down from the horizontal: $0 \leq \delta \leq \pi/2$.

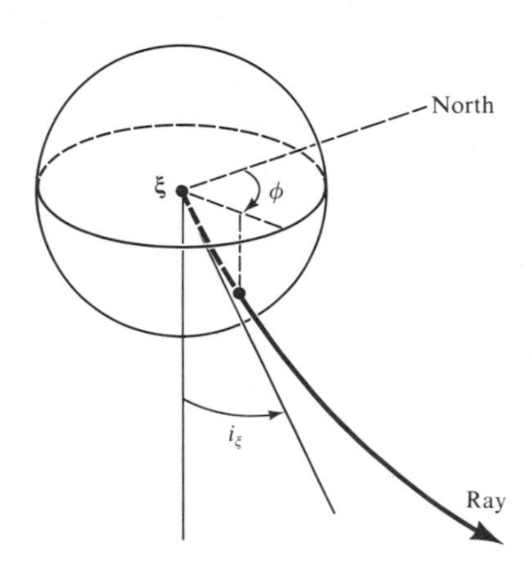

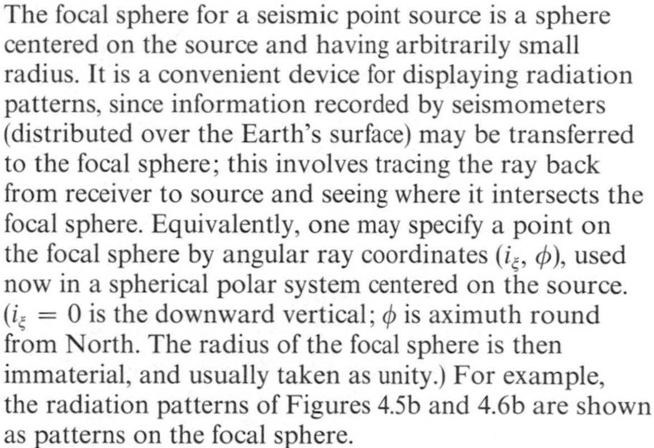

FIGURE **4.14**

The focal sphere for a seismic point source is a sphere centered on the source and having arbitrarily small radius. It is a convenient device for displaying radiation patterns, since information recorded by seismometers (distributed over the Earth's surface) may be transferred to the focal sphere; this involves tracing the ray back from receiver to source and seeing where it intersects the focal sphere. Equivalently, one may specify a point on the focal sphere by angular ray coordinates (i_ξ, ϕ), used now in a spherical polar system centered on the source. ($i_\xi = 0$ is the downward vertical; ϕ is aximuth round from North. The radius of the focal sphere is then immaterial, and usually taken as unity.) For example, the radiation patterns of Figures 4.5b and 4.6b are shown as patterns on the focal sphere.

Since the focal sphere lies within the near-field of the source, it is not obvious that far-field radiation patterns mapped onto the focal sphere would directly indicate the displacements actually occurring in the source region. However, for a shear dislocation with final offset in the same direction as averaged particle velocity, confusion does not occur. This is because the far-field radiation pattern (the terms $\mathbf{A}^{FP}$ and $\mathbf{A}^{FS}$ in (4.32) and (4.33)) is then the same as the radiation pattern for static displacements (4.34) at all distances from the source (and, in particular, on the focal sphere).

In this section, we introduce the *focal sphere* (see Fig. 4.14) as a surface on which to describe the radiation pattern, and shall show how nodal planes are obtained from global observation of far-field body waves. We then augment this descriptive approach by deriving specific formulas for the radiation pattern of *P*-, *SV*-, and *SH*-waves as a function of general strike, dip, and rake, and also as a function of ray parameter p and azimuth ϕ from source to receiver. Such formulas are required for an accurate estimation (from body waves) of scismic moment M_0.

4.5.1 A method for obtaining the fault-plane orientation of an earthquake and the direction of slip using teleseismic body-wave observations

The radiation pattern of a shear dislocation is characterized, for *P*-waves (see Fig. 4.5), by a quadrantal distribution. This effect is most commonly studied in terms of the first motion of *P*-wave displacement. For a given receiver, the longitudinal particle motion is either toward or away from the source: the first-arriving waves at a seismometer often provide the most unambiguous and most easily interpreted part of the record of ground displacement. The radiation pattern can conveniently be understood in terms of the focal sphere, defined in Figure 4.14. Thus, for a given observation of a *P*-wave at $\mathbf{x}$, we trace the ray back to the source at $\boldsymbol{\xi}$, transferring information obtained at $\mathbf{x}$ onto the corresponding point on the focal sphere (i.e., the point with coordinates (i_ξ, ϕ)).

To illustrate the distortion such a mapping can make, Figure 4.15 shows an outline of the major continents as mapped by *P*-wave rays on the lower half of a focal sphere centered on a point 42 km below the Kurile Islands (a part of the North Pacific). Because the *P*-wave speed increases with depth throughout the mantle (in the Earth model used to obtain this figure), very strong areal distortion occurs in mapping the Earth's surface onto the focal sphere.

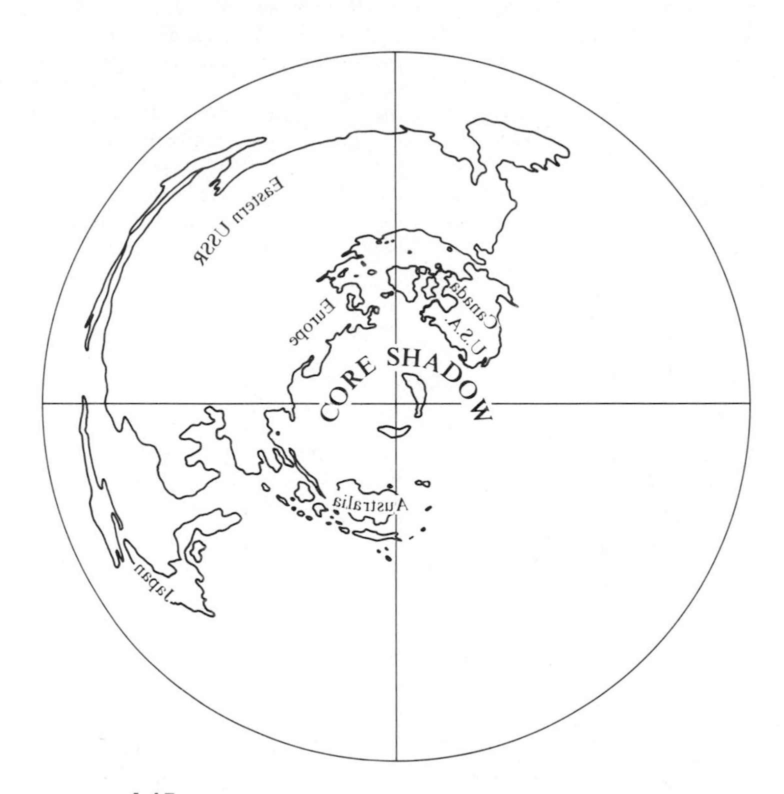

FIGURE **4.15**

The focal sphere is shown, in equal area projection, for a point beneath the Kurile Islands (43°26′N, 147°03′E, depth 42 km). For seismological purposes, the eastern U.S.S.R. could contribute more information than all other continental areas combined. The lack of coverage in the Pacific Ocean would make it very difficult to obtain a fault-plane solution. [From Davies and McKenzie, *Geophysical Journal*, v. 18. p. 60, 1969.]

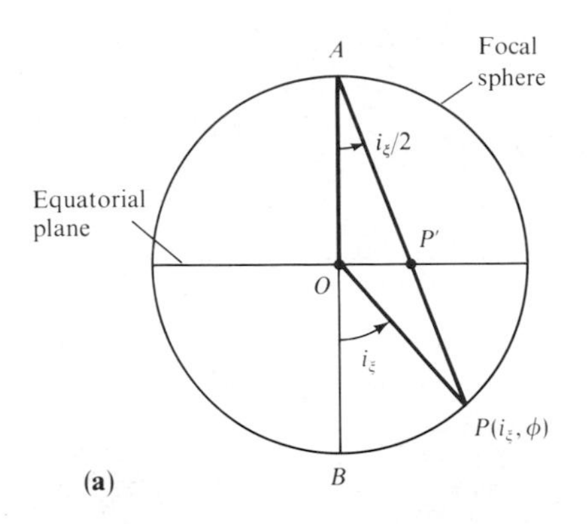

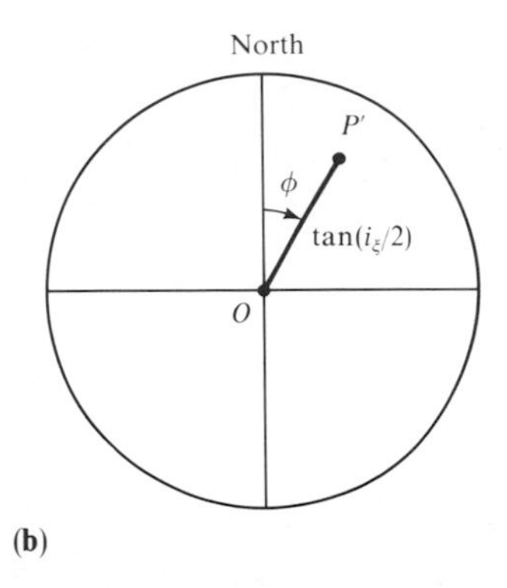

FIGURE **4.16**

The stereographic projection of the focal sphere maps point P to the point P' as shown in the equatorial plane of the focal sphere. (**a**) Shown here is a vertical plane through the center of the focal sphere and the point P. (**b**) A plan view of the horizontal plane viewed on edge in (a). This is the plane of the projection, and point P' is at distance $\tan \frac{1}{2}i_\xi$ from the point O. The figure shows projection only of the lower half of the focal sphere, but by projecting from B instead of A, the upper half of the focal sphere can also be mapped inside a circle of unit radius.

Since this projection preserves angles, the scale is locally the same in all directions at position P'. However, this scale at P' is proportional to $\sec^2 \frac{1}{2}i_\xi$, so that area elements at P are scaled at P' by a factor $\sec^4 \frac{1}{2}i_\xi$. This is highly undesirable, since it gives a large amplification of areas for points on the focal sphere that are not near A and B. The effect appears in (b) mainly as a relative compression of the area associated with down-going rays (i.e., with small i_ξ).

The stereographic projection is used in association with a Wulff net, which provides a template for drawing in (b) the possible curves that in (a) represent the intersection of the focal sphere with fault planes of arbitrary orientation.

Note that another mapping is required to show the focal sphere on a plane surface (such as the printed page). The two most commonly used are the stereographic projection (Fig. 4.16) and the zenithal equal-area projection (sometimes called the Schmidt-Lambert equal-area projection; see Fig. 4.17).

The process of obtaining a fault-plane solution from P-wave first motions involves (a) marking particular positions on the focal sphere that correspond to P-wave rays for which one has data, using a different symbol for compressional arrivals and for dilatational arrivals, and (b) partitioning the focal sphere by two perpendicular great circles, with each quadrant having either all compressional or all dilatational arrivals. An example is given in Figure 4.18.

There are two principal points to note. The first is the ambiguity in choosing which of two nodal planes is the fault plane. This difficulty was foreseen in Section 3.2, from the symmetry between the two components of the double-couple body-force equivalent (3.15). For the event shown in Figure 4.18, a surface break was observed whose strike differed by only 8° from that of one

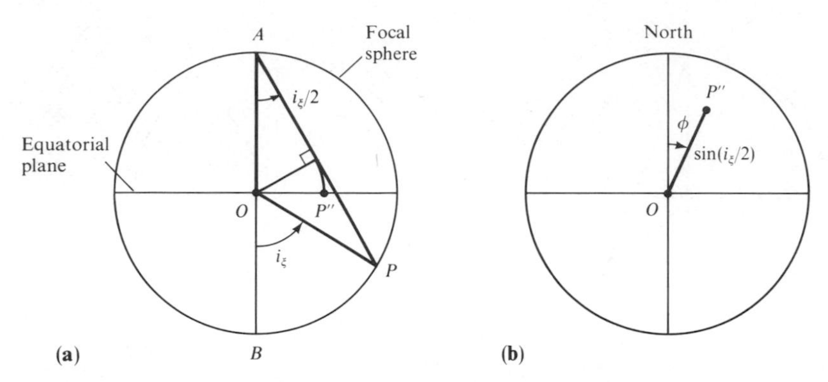

FIGURE **4.17**
The zenithal equal-area projection of the focal sphere maps point P to the point P''. The azimuthal angle ϕ is unchanged (this is a property of all zenithal projections), and distance OP'' is $\sin(\frac{1}{2}i_\xi)$. **(a)** Shown here is a vertical plane through the center of the focal sphere and the point P. **(b)** A plan view of the horizontal plane that is viewed on edge in (a). If the area element $\sin i_\xi \, di_\xi \, d\phi$ at P is mapped into an element at P'', its new area is

$$\sin(\tfrac{1}{2}i_\xi)\, d\phi[\sin(\tfrac{1}{2}i_\xi + \tfrac{1}{2}di_\xi) - \sin(\tfrac{1}{2}i_\xi)] = \tfrac{1}{4}\sin i_\xi \, di_\xi \, d\phi,$$

i.e., a reduction by the constant factor $\frac{1}{4}$. The whole focal sphere (area 4π) maps into the unit circle (area π), but it is conventional to plot the upper half (projected from B, i.e., using $\sin(\frac{1}{2}\pi - \frac{1}{2}i_\xi)$) separately from the lower half in order to avoid the large angle-distortions of the upper focal sphere that would otherwise occur. Again, a system of templates is needed to draw possible fault-plane positions in (b). Differences between the equal-angle and equal-area projections are apparent in Figures 4.5b and 4.6b, which show the fault plane and the auxiliary plane (for the same focal mechanism) in the two different ways.

of the two nodal planes, hence the choice $\phi_s = 301°$, $\delta = 67°$ can be made with confidence. The second is that a so-called *pressure axis* is located in the center of the dilatational quadrant and a *tension axis* in the center of the compressional quadrant. These are the principal axes of the moment tensor, and are also principal stress axes if the fault plane is a plane of maximum shear. The fault plane for a given earthquake, however, is established by tectonic processes extending back into geologic time. Rather than a plane of maximum shear stress, it is often the fault plane that is reactivated in a given earthquake. In these cases, McKenzie (1969) showed that the only restriction on the direction of greatest compressive stress is that it lie in a dilatational quadrant on the local sphere.

The use of S-wave radiation patterns entails considerably greater effort than for P-waves, since more than one instrument must be used at each receiving station to obtain the polarization; moreover, a correction must be made for the effect of observation at the Earth's free surface. The principal features

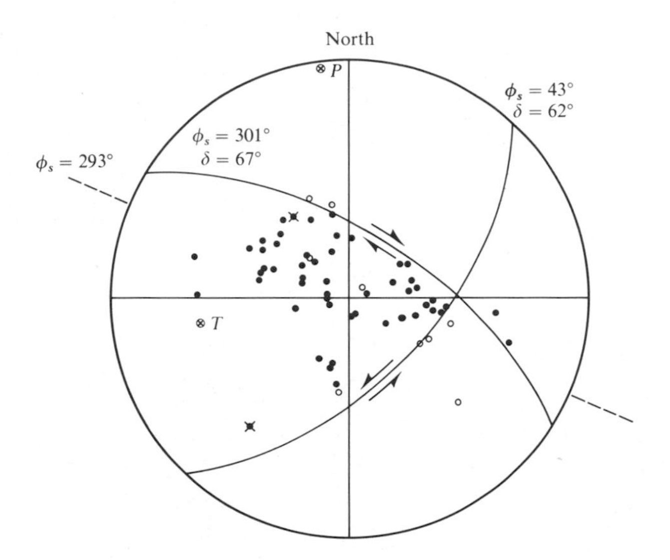

FIGURE **4.18**

P-wave first motions obtained by McKenzie for the shock of 1966 August 19 in eastern Turkey. Diagram is an equal-area projection of the lower hemisphere of the focal sphere. Solid circles represent compressions, open circles dilatations; crosses indicate stations near a nodal plane. ϕ_s and δ are the strike and dip of the nodal planes, and arrows indicate the sense of motion. The broken line shows the strike of the major right-lateral surface break that accompanied the earthquake. [From McKenzie, *Bulletin of the Seismological Society of America*, v. 59, p. 593, 1969.]

are shown in the example of Figure 4.19: particle motion diverges from the pressure axis, converges on the tension axis; and is perpendicular to the *P*-wave nodal planes.

Once the strike, dip, and rake have been determined, it is possible to set up the system of spherical polar (source) coordinates used in Figures 4.5 and 4.6 to analyze amplitude dependence within the radiation pattern in detail. However, in i_ξ and ϕ (Figs. 4.10 and 4.14) we already have a system of spherical polar angles at the source. Furthermore, these are the natural source coordinates to use, since we know their properties also as ray coordinates for *SV*- and *SH*-waves (see Section 4.4). Our next goal is therefore to obtain the radiation pattern of *P*-, *SV*- and *SH*-waves as a function of i_ξ and ϕ, and also as a function of general strike, dip, and rake.

Two different stages may be distinguished in achieving this goal. First, the problem of identifying *P*, *SV*, and *SH* in the far-field for a source within a homogeneous medium. Second, the adaptation of this result to the case of a spherically symmetric medium, like the Earth.

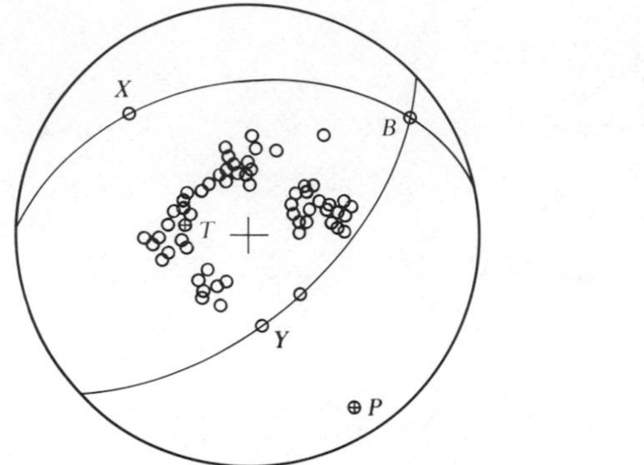

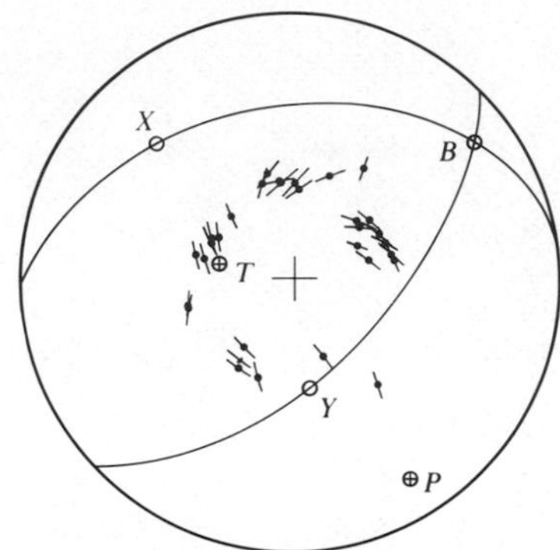

1966, May 15, 14 hr 46 min 06 sec

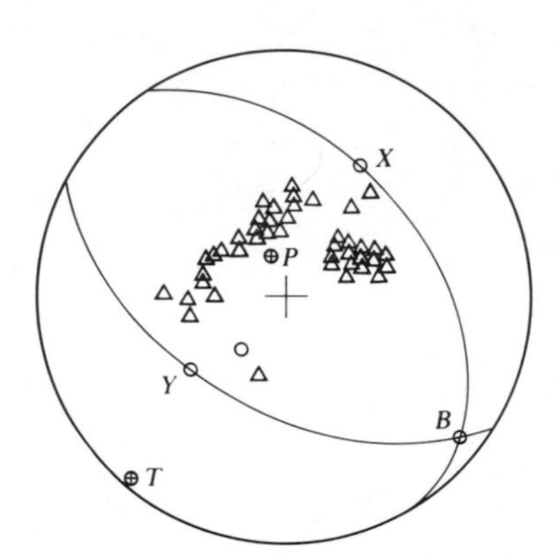

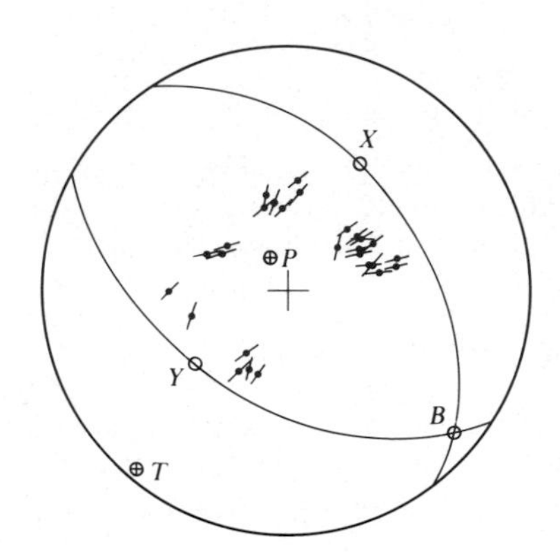

1966, June 2, 03 hr 27 min 53 sec

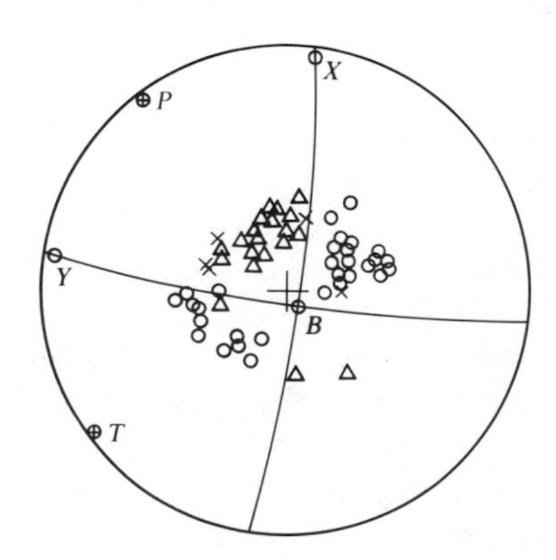

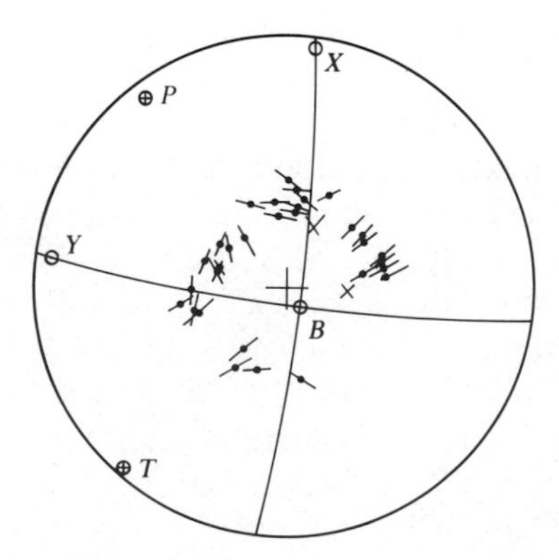

1966, July 4, 18 hr 33 min 36 sec

4.5.2 *Arbitrary orientation of the double couple in a homogeneous medium*

From the Cartesian components of displacement given in (4.30) for an effective point source with moment tensor M_{pq} having zero trace, we can identify the far-field P-wave in vector form as

$$\mathbf{u}^P(\mathbf{x}, t) = \frac{2(\boldsymbol{\gamma} \cdot \boldsymbol{\nu})(\boldsymbol{\gamma} \cdot \dot{\bar{\mathbf{u}}})\mu A \boldsymbol{\gamma}}{4\pi\rho\alpha^3 r}. \tag{4.79}$$

Here $\boldsymbol{\gamma}$ is the longitudinal direction from source at $\boldsymbol{\xi}$ to receiver at $\mathbf{x}$; $\boldsymbol{\nu}$ is the fault normal; $\dot{\bar{\mathbf{u}}}$ is the particle velocity at the source, averaged over fault area A and evaluated at retarded time $t - r/\alpha$; and r is the distance $|\mathbf{x} - \boldsymbol{\xi}|$. The corresponding result for the far-field S-wave is

$$\mathbf{u}^S(\mathbf{x}, t) = \frac{[(\boldsymbol{\gamma} \cdot \boldsymbol{\nu})\dot{\bar{\mathbf{u}}} + (\boldsymbol{\gamma} \cdot \dot{\bar{\mathbf{u}}})\boldsymbol{\nu} - 2(\boldsymbol{\gamma} \cdot \boldsymbol{\nu})(\boldsymbol{\gamma} \cdot \dot{\bar{\mathbf{u}}})\boldsymbol{\gamma}]\mu A}{4\pi\rho\beta^3 r}, \tag{4.80}$$

in which $\dot{\bar{\mathbf{u}}}$ is evaluated at $t - r/\beta$.

Since $\boldsymbol{\gamma}$ is a unit vector at the source, directed along the ray to $\mathbf{x}$, the problem of obtaining the radiation pattern of the P-wave is simply a matter of expressing $\boldsymbol{\gamma} \cdot \boldsymbol{\nu}$ and $\boldsymbol{\gamma} \cdot \dot{\bar{\mathbf{u}}}$ in terms of strike ϕ_S, dip δ, rake λ, take-off angle i_ξ, and source-receiver azimuth ϕ. The radiation patterns for SV and SH are slightly more complicated, because the separation into SV and SH is not immediately apparent in (4.80). Clearly, this formula does indicate that $\mathbf{u}^S$ is a transverse motion, because $\mathbf{u}^S \cdot \boldsymbol{\gamma} = 0$. It follows that SV and SH motions, which are (respectively) in the directions $\hat{\mathbf{p}}$ and $\hat{\boldsymbol{\phi}}$ of Figure 4.10, are given by

$$\mathbf{u}^{SV}(\mathbf{x}, t) = (\mathbf{u}^S \cdot \hat{\mathbf{p}})\hat{\mathbf{p}} = \frac{[(\boldsymbol{\gamma} \cdot \boldsymbol{\nu})(\dot{\bar{\mathbf{u}}} \cdot \hat{\mathbf{p}}) + (\boldsymbol{\gamma} \cdot \dot{\bar{\mathbf{u}}})(\boldsymbol{\nu} \cdot \hat{\mathbf{p}})]\mu A \hat{\mathbf{p}}}{4\pi\rho\beta^3 r} \tag{4.81}$$

FIGURE **4.19**

Individual focal mechanism diagrams are shown, using the equal-area projection. In the P-wave diagrams (the left figure of each pair) the triangles represent dilatational first motions, cricles represent compressions, and crosses represent P-wave arrivals that appear to be near nodal lines. (Such arrivals are identified by their emergent nature.) Symbols P, T, B are the axes of the moment tensor, and X and Y are the poles of the nodal planes. [From Stauder, 1968; copyrighted by American Geophysical Union.]

and

$$\mathbf{u}^{SH}(\mathbf{x}, t) = (\mathbf{u}^S \cdot \hat{\boldsymbol{\phi}})\hat{\boldsymbol{\phi}} = \frac{[(\boldsymbol{\gamma} \cdot \mathbf{v})(\dot{\bar{\mathbf{u}}} \cdot \hat{\boldsymbol{\phi}}) + (\boldsymbol{\gamma} \cdot \dot{\bar{\mathbf{u}}})(\mathbf{v} \cdot \hat{\boldsymbol{\phi}})]\mu A \hat{\boldsymbol{\phi}}}{4\pi\rho\beta^3 r}. \tag{4.82}$$

(See Ben-Menahem et al., 1965.)

To obtain all three radiation patterns in terms of (ϕ_s, δ, λ, i_ξ, ϕ), we introduce Cartesian coordinate directions $\hat{\mathbf{x}}, \hat{\mathbf{y}}, \hat{\mathbf{z}}$ at the epicenter. Our choice is $\hat{\mathbf{x}}$ =

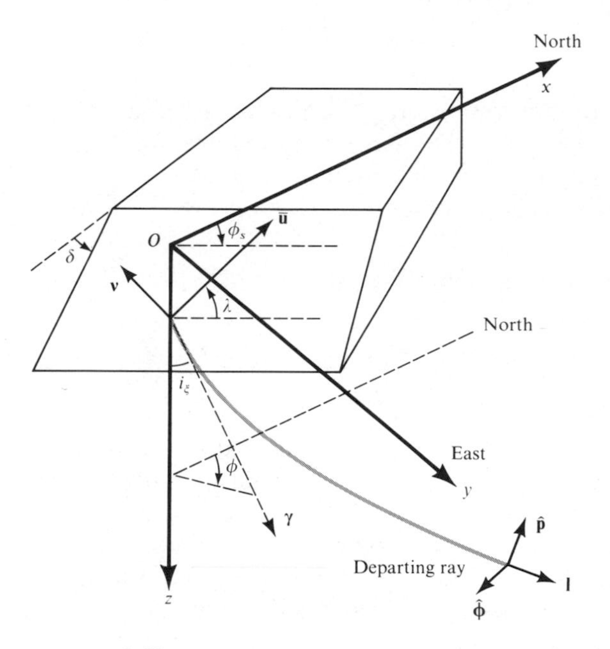

FIGURE **4.20**

Definition of Cartesian coordinates (x, y, z) used to obtain the explicit dependence of P, SV, and SH radiation patterns on (ϕ_s, δ, λ, i_ξ, ϕ). The origin is taken at the *epicenter* (the point of the Earth's surface vertically above the seismic source of interest). Advantages of this coordinate system are that z follows convention in being the depth direction, and azimuthal angles ϕ, measured round from the x-axis, follow the geographical convention. The horizontal components of $\bar{\mathbf{u}}$ can easily be resolved into $\bar{u} \cos \lambda$ (along strike) and $\bar{u} \cos \delta \sin \lambda$ (in the negative dip direction). Components given in the text (4.83) are for North (x) and East (y). This is a natural coordinate system in which to give components of a moment tensor (Box 4.4). The take-off angle i_ξ can be used interchangeably with ray parameter p as a ray coordinate. Since $\mathbf{l}$ is the longitudinal direction along the ray, $\mathbf{l} = \boldsymbol{\gamma}$ for a homogeneous medium.

North, $\hat{\mathbf{y}}$ = East, and $\hat{\mathbf{z}}$ = vertically downward, as shown in Figure 4.20. In terms of these three unit vectors,

$$
\begin{aligned}
\text{slip } \bar{\mathbf{u}} &= \bar{u}(\cos\lambda\cos\phi_s + \cos\delta\sin\lambda\sin\phi_s)\hat{\mathbf{x}} \\
&\quad + \bar{u}(\cos\lambda\sin\phi_s - \cos\delta\sin\lambda\cos\phi_s)\hat{\mathbf{y}} \\
&\quad - \bar{u}\sin\lambda\sin\delta\hat{\mathbf{z}}, \\
\text{fault normal } \boldsymbol{\nu} &= -\sin\delta\sin\phi_s\hat{\mathbf{x}} + \sin\delta\cos\phi_s\hat{\mathbf{y}} - \cos\delta\hat{\mathbf{z}}, \\
P\text{-wave direction } \mathbf{l} &= \boldsymbol{\gamma} = \sin i_\xi\cos\phi\hat{\mathbf{x}} + \sin i_\xi\sin\phi\hat{\mathbf{y}} + \cos i_\xi\hat{\mathbf{z}}, \\
SV\text{-wave direction } \hat{\mathbf{p}} &= \cos i_\xi\cos\phi\hat{\mathbf{x}} + \cos i_\xi\sin\phi\hat{\mathbf{y}} - \sin i_\xi\hat{\mathbf{z}}, \\
SH\text{-wave direction } \hat{\boldsymbol{\phi}} &= -\sin\phi\hat{\mathbf{x}} + \cos\phi\hat{\mathbf{y}}.
\end{aligned}
\tag{4.83}
$$

Note here that $\mathbf{l}$, $\hat{\mathbf{p}}$, $\hat{\boldsymbol{\phi}}$ are given as for a homogeneous medium.

Six different scalar products are needed in the radiation pattern formulas (4.79), (4.81), (4.82), and these can readily be obtained from (4.83). In dimensionless form, the radiation patterns F^P, F^{SV}, and F^{SH} are given by

$$
\begin{aligned}
\mathscr{F}^P &= 2(\boldsymbol{\gamma}\cdot\boldsymbol{\nu})(\boldsymbol{\gamma}\cdot\dot{\bar{\mathbf{u}}})/\dot{\bar{u}} \\
&= \cos\lambda\sin\delta\sin^2 i_\xi\sin 2(\phi - \phi_s) - \cos\lambda\cos\delta\sin 2i_\xi\cos(\phi - \phi_s) \\
&\quad + \sin\lambda\sin 2\delta(\cos^2 i_\xi - \sin^2 i_\xi\sin^2(\phi - \phi_s)) \\
&\quad + \sin\lambda\cos 2\delta\sin 2i_\xi\sin(\phi - \phi_s),
\end{aligned}
\tag{4.84}
$$

$$
\begin{aligned}
\mathscr{F}^{SV} &= [(\boldsymbol{\gamma}\cdot\boldsymbol{\nu})(\dot{\bar{\mathbf{u}}}\cdot\hat{\mathbf{p}}) + (\boldsymbol{\gamma}\cdot\dot{\bar{\mathbf{u}}})(\boldsymbol{\nu}\cdot\hat{\mathbf{p}})]/\dot{\bar{u}} \\
&= \sin\lambda\cos 2\delta\cos 2i_\xi\sin(\phi - \phi_s) - \cos\lambda\cos\delta\cos 2i_\xi\cos(\phi - \phi_s) \\
&\quad + \tfrac{1}{2}\cos\lambda\sin\delta\sin 2i_\xi\sin 2(\phi - \phi_s) \\
&\quad - \tfrac{1}{2}\sin\lambda\sin 2\delta\sin 2i_\xi(1 + \sin^2(\phi - \phi_s)),
\end{aligned}
\tag{4.85}
$$

$$
\begin{aligned}
\mathscr{F}^{SH} &= [(\boldsymbol{\gamma}\cdot\boldsymbol{\nu})(\dot{\bar{\mathbf{u}}}\cdot\hat{\boldsymbol{\phi}}) + (\boldsymbol{\gamma}\cdot\dot{\bar{\mathbf{u}}})(\boldsymbol{\nu}\cdot\hat{\boldsymbol{\phi}})]/\dot{\bar{u}} \\
&= \cos\lambda\cos\delta\cos i_\xi\sin(\phi - \phi_s) + \cos\lambda\sin\delta\sin i_\xi\cos 2(\phi - \phi_s) \\
&\quad + \sin\lambda\cos 2\delta\cos i_\xi\cos(\phi - \phi_s) \\
&\quad - \tfrac{1}{2}\sin\lambda\sin 2\delta\sin i_\xi\sin 2(\phi - \phi_s).
\end{aligned}
\tag{4.86}
$$

Associated far-field displacements are then

$$
\begin{aligned}
\mathbf{u}^P(\mathbf{x}, t) &= \frac{\mathscr{F}^P\mu A}{4\pi\rho\alpha^3 r}\dot{\bar{u}}\left(t - \frac{r}{\alpha}\right)\mathbf{l}, \\
\mathbf{u}^{SV}(\mathbf{x}, t) &= \frac{\mathscr{F}^{SV}\mu A}{4\pi\rho\beta^3 r}\dot{\bar{u}}\left(t - \frac{r}{\beta}\right)\hat{\mathbf{p}}, \\
\mathbf{u}^{SH}(\mathbf{x}, t) &= \frac{\mathscr{F}^{SH}\mu A}{4\pi\rho\beta^3 r}\dot{\bar{u}}\left(t - \frac{r}{\beta}\right)\hat{\boldsymbol{\phi}}.
\end{aligned}
\tag{4.87}
$$

4.5.3 Adapting the radiation pattern to the case of a spherically symmetric medium

We have taken care to obtain P, SV, and SH displacements (4.87) in a form comparable with the geometric ray solutions derived in Section 4.4 (see (4.57), (4.60), (4.61)). To complete the comparison for P-waves, it remains only to identify r/α as the ray travel time T^P, $1/r$ as the geometrical spreading factor $1/\mathscr{R}^P(\mathbf{x}, \xi)$, and $\mu/(\rho\alpha^3)$ as factor $\mu(\xi)/[\rho^{1/2}(\xi)\rho^{1/2}(\mathbf{x})\alpha^{5/2}(\xi)\alpha^{1/2}(\mathbf{x})]$. This last result follows from generalizing $\mu/(\rho\alpha^3)$ to a term proportional to $[\rho(\mathbf{x})\alpha(\mathbf{x})]^{-1/2}$ (see (4.57)), in which the constant of proportionality can depend only on properties at the source. Then

$$\mathbf{u}^P(\mathbf{x}, t) = \frac{F^P \mu(\xi) A\dot{\bar{u}}(t - T^P)\mathbf{l}}{4\pi\rho^{1/2}(\xi)\rho^{1/2}(\mathbf{x})\alpha^{5/2}(\xi)\alpha^{1/2}(\mathbf{x})\mathscr{R}^P(\mathbf{x}, \xi)}, \tag{4.88}$$

and similarly

$$\mathbf{u}^{SV}(\mathbf{x}, t) = \frac{F^{SV} \mu(\xi) A\dot{\bar{u}}(t - T^S)\hat{\mathbf{p}}}{4\pi\rho^{1/2}(\xi)\rho^{1/2}(\mathbf{x})\beta^{5/2}(\xi)\beta^{1/2}(\mathbf{x})\mathscr{R}^S(\mathbf{x}, \xi)}, \tag{4.89}$$

$$\mathbf{u}^{SH}(\mathbf{x}, t) = \frac{F^{SH} \mu(\xi) A\dot{\bar{u}}(t - T^S)\hat{\boldsymbol{\phi}}}{4\pi\rho^{1/2}(\xi)\rho^{1/2}(\mathbf{x})\beta^{5/2}(\xi)\beta^{1/2}(\mathbf{x})\mathscr{R}^S(\mathbf{x}, \xi)}. \tag{4.90}$$

The radiation patterns here are exactly the same as for a homogeneous medium, and are given in (4.84)–(4.86). The only noteworthy symmetry is a reversal in sign of F^P, F^{SV}, F^{SH} if the rake is changed by 180°. Particularly, one should note that there is no symmetry to changes of 180° in strike ϕ_s or takeoff azimuth ϕ, so that care must be taken to follow the definitions given in Figures 4.13 and 4.20, in which these angles increase clockwise round from North.

The principal use of our final formulas (4.88)–(4.90) lies in estimating the seismic moment. From a far-field observation of the displacement, one can obtain $\mu A\dot{\bar{u}}(t - T)$ after correction for the radiation pattern, geometrical spreading, and scaling factors at source and receiver. (In fact, correction is also required for the effects of transmission across material boundaries, for attenuation, and, in practice, for instrument response.) It often occurs that the duration of faulting is much shorter than the periods at which body waves are observed, yet these periods are still short enough to use ray theory. In this case, the averaged particle displacement $\bar{u}$ at the source is effectively a step function, $M_0(t) = M_0(\infty)H(t)$, hence the spectrum of the observed displacement pulse shape, $\mu(\xi)\,A\dot{\bar{u}}$, is just $M_0(\infty)$ at long periods.

BOX **4.4**

Cartesian components of the moment tensor for a shear dislocation of arbitrary orientation

In our final formulas for the body-wave radiation pattern from a shear dislocation, we have combined properties of the moment tensor with properties of the Green function. If it is surface waves we wish to study, or free oscillations in a more complicated medium, then we must develop further theories of wave propagation to evaluate the appropriate Green function. But the moment tensor is unchanged. We here obtain its Cartesian components for a shear fault described by strike ϕ_s, rake λ, dip δ, and moment M_0.

Referring to Figure 4.20, in which the x-direction is taken as North, the Cartesian components of fault slip $\bar{\mathbf{u}}$ and fault normal $\boldsymbol{\nu}$ have already been given in (4.83). Recall from Chapter 3 that the moment M_0 of a shear dislocation is $\mu A\bar{u}$, where μ is the rigidity in the source region (Box 3.1) and A is the area over which the displacement discontinuity has been averaged. From (3.21) and (3.23), it follows that $M_{pq} = \mu A(\bar{u}_p \nu_q + \bar{u}_q \nu_p)$, and hence the Cartesian components of moment tensor $\mathbf{M}$ are

$$
\begin{aligned}
M_{xx} &= -M_0(\sin\delta\cos\lambda\sin 2\phi_s + \sin 2\delta\sin\lambda\sin^2\phi_s),\\
M_{xy} &= M_0(\sin\delta\cos\lambda\cos 2\phi_s + \tfrac{1}{2}\sin 2\delta\sin\lambda\sin 2\phi_s) = M_{yx},\\
M_{xz} &= -M_0(\cos\delta\cos\lambda\cos\phi_s + \cos 2\delta\sin\lambda\sin\phi_s) = M_{zx},\\
M_{yy} &= M_0(\sin\delta\cos\lambda\sin 2\phi_s - \sin 2\delta\sin\lambda\cos^2\phi_s),\\
M_{yz} &= -M_0(\cos\delta\cos\lambda\sin\phi_s - \cos 2\delta\sin\lambda\cos\phi_s) = M_{zy},\\
M_{zz} &= M_0\sin 2\delta\sin\lambda.
\end{aligned}
\tag{1}
$$

This general result can be recognized as a weighted sum of four elementary moment tensors,

$$
\mathbf{M} = \cos\delta\cos\lambda\,\mathbf{M}^{(1)} + \sin\delta\cos\lambda\,\mathbf{M}^{(2)} - \cos 2\delta\sin\lambda\,\mathbf{M}^{(3)} + \sin 2\delta\sin\lambda\,\mathbf{M}^{(4)} \tag{2}
$$

where

$$
\begin{aligned}
\mathbf{M}^{(1)} &= M_0\begin{pmatrix} 0 & 0 & -\cos\phi_s \\ 0 & 0 & -\sin\phi_s \\ -\cos\phi_s & -\sin\phi_s & 0 \end{pmatrix}, \quad
\mathbf{M}^{(2)} = M_0\begin{pmatrix} -\sin 2\phi_s & \cos 2\phi_s & 0 \\ \cos 2\phi_s & \sin 2\phi_s & 0 \\ 0 & 0 & 0 \end{pmatrix},\\
\mathbf{M}^{(3)} &= M_0\begin{pmatrix} 0 & 0 & \sin\phi_s \\ 0 & 0 & -\cos\phi_s \\ \sin\phi_s & -\cos\phi_s & 0 \end{pmatrix}, \quad
\mathbf{M}^{(4)} = M_0\begin{pmatrix} -\sin^2\phi_s & \tfrac{1}{2}\sin 2\phi_s & 0 \\ \tfrac{1}{2}\sin 2\phi_s & -\cos^2\phi_s & 0 \\ 0 & 0 & 1 \end{pmatrix}.
\end{aligned}
\tag{3}
$$

Each of $\mathbf{M}^{(i)}$ ($i = 1, 2, 3, 4$) has eigenvalues M_0, $-M_0$, and zero, so that each is the moment tensor for a shear dislocation. In fact (from (2)), $\mathbf{M}^{(1)}$ is given by $\mathbf{M}$ in the case of $\delta = 0$, $\lambda = 0$, so $\mathbf{M}^{(1)}$ is appropriate for a horizontal fault plane, with slip direction defining the strike. Similarly, $\mathbf{M}^{(2)}$ is a pure strike-slip fault ($\delta = \pi/2$, $\lambda = 0$), $\mathbf{M}^{(3)}$ is a pure dip-slip fault ($\delta = \pi/2$, $\lambda = \pi/2$), and $\mathbf{M}^{(4)}$ is a fault dipping at 45° with slip being purely up-dip ($\delta = \pi/4$, $\lambda = \pi/2$).

Although our formulas for the Green function may involve some approximations, the decomposition of the radiation field implied by (2) is exact. That is, the waves radiated from a shear dislocation of arbitrary orientation can always be written as the sum of waves radiated from four different elementary shear dislocations, all sharing the same strike. In fact, only *three* elementary shear dislocations need be studied, if one drops the requirement that they share a common strike. This follows by replacing $\mathbf{M}^{(1)}$ in (2) by $\mathbf{M}^{(3)}$ evaluated at strike $\phi_s - \pi/2$.

In the above presentation, we have taken coordinates (x, y, z) at the source as (North, East, Down). Another convention, more natural in the context of normal mode studies for the whole Earth, is based on coordinates (r, Δ, ϕ), where Δ measures epicentral distance and ϕ is the epicentral longitude ($\phi = 0$ is South). Cartesian components of the moment tensor are then M_{rr}, $M_{r\Delta}$, etc., where these components are referred to a Cartesian system at the source with directions (Up, South, East). It follows that components of $\mathbf{M}$ in the two systems are related by

$$\begin{pmatrix} M_{rr} & M_{r\Delta} & M_{r\phi} \\ M_{\Delta r} & M_{\Delta\Delta} & M_{\Delta\phi} \\ M_{\phi r} & M_{\phi\Delta} & M_{\phi\phi} \end{pmatrix} = \begin{pmatrix} M_{zz} & M_{zx} & -M_{zy} \\ M_{xz} & M_{xx} & -M_{xy} \\ -M_{yz} & -M_{yx} & M_{yy} \end{pmatrix}.$$

Note, in this normal-mode convention, that the source-receiver azimuth becomes $\pi - \phi$.

For point sources specified by a general moment tensor (i.e., nine couples with symmetry $M_{pq} = M_{qp}$), the far-field body waves are given in terms of the six independent components of $\mathbf{M}$ by generalizing (4.79) and (4.81)–(4.82). We use the far-field terms in (4.29), and for homogeneous media it follows that

$$\mathbf{u}^P(\mathbf{x}, t) = \left(\frac{\boldsymbol{\gamma} \cdot \dot{\mathbf{M}}\left(t - \dfrac{r}{\alpha}\right) \cdot \boldsymbol{\gamma}}{4\pi\rho\alpha^3 r} \right) \mathbf{l},$$

$$\mathbf{u}^{SV}(\mathbf{x}, t) = \left(\frac{\hat{\mathbf{p}} \cdot \dot{\mathbf{M}}\left(t - \dfrac{r}{\beta}\right) \cdot \boldsymbol{\gamma}}{4\pi\rho\beta^3 r} \right) \hat{\mathbf{p}}, \tag{4.91}$$

$$\mathbf{u}^{SH}(\mathbf{x}, t) = \left(\frac{\hat{\boldsymbol{\phi}} \cdot \dot{\mathbf{M}}\left(t - \dfrac{r}{\beta}\right) \cdot \boldsymbol{\gamma}}{4\pi\rho\beta^3 r} \right) \hat{\boldsymbol{\phi}},$$

which are easily adapted to inhomogeneous media by using ray theory. First, the denominators in (4.91) are replaced by those of (4.88)–(4.90). Second, the unit vectors $\hat{\mathbf{p}}$ and $\hat{\boldsymbol{\phi}}$ that are contracted with $\dot{\mathbf{M}}$ in (4.91) must be recognized as being transverse to the ray at the *source*, whereas $\hat{\mathbf{p}}$ and $\hat{\boldsymbol{\phi}}$ at the *receiver* are the appropriate directions for identifying *SV*- and *SH*-components in the radiated particle motion.

SUGGESTION FOR FURTHER READING

Červený, V., I. A. Molotkov, and I. Pšenčík. *The Ray Method in Seismology*, Prague: Charles University Press, 1978

Courant, R., and D. Hilbert. *Methods of Mathematical Physics* (Vol. 2, Chap. 6 for ray theory). New York; Interscience, 1952.

Kline, M., and I. W. Kay. *Electromagnetic Theory and Geometrical Optics*. New York: Interscience, 1966.

Love, A. E. H. *A Treatise on the Mathematical Theory of Elasticity*. New York: Dover Publications, 1944.

Morse, P.M., and H. Feshbach. *Methods of Theoretical Physics* (Chap. 13 for Green's functions in elasticity), New York: McGraw-Hill, 1953.

Sternberg, E. On the integration of the equations of motion in the classical theory of elasticity. *Archive for Rational Mechanics and Analysis*, **6**, 34–50, 1960.

PROBLEMS

4.1 Sketch the "seismogram" for Green's function G_{ij} itself. That is, draw the relative positions and pulse shapes of the three different terms in (4.23) when X_0 is an impulse. Then show that the area under each one of the three pulses has a distance dependence proportional to $1/r$. The area under a pulse is equal to the limit (as $\omega \to 0$) of the Fourier transform of the pulse, and from the frequency dependence of (4.35) it therefore appears that the area is unbounded. Show that in fact there is a cancellation of the terms in $1/(\omega^2 r^3)$, $1/(\omega r^2)$ in (4.35), and that the distance dependence as $\omega \to 0$ is indeed like $1/r$.

For a seismometer that is sensitive only to periods much longer than the $S - P$ time, i.e., $(r/\beta - r/\alpha)$, note then that the near-field term of the Green function is effectively an impulse, as well as the far-field terms. To study near-field effects (such as the strong ground motion near a rupturing fault), it also follows that the so-called "far-field" and "near-field" terms are equally important, in the sense that the source time function X_0 will be nonzero for times that are long compared to $(r/\beta - r/\alpha)$, so the first term in (4.23) is also of order $1/r$.

To summarize: "far-field" terms dominate in the far-field, but "near-field" and "far-field" terms are of equal importance in the near-field after they have been convolved with a source time function.

4.2 If a constant force X_0 is applied in the j-direction at the origin within an infinite homogeneous isotropic medium, show from (4.23) that the static solution for the i-component of displacement at $\mathbf{x}$ is

$$\frac{X_0}{8\pi\rho r}\left[\left(\frac{1}{\beta^2}-\frac{1}{\alpha^2}\right)\gamma_i\gamma_j+\left(\frac{1}{\beta^2}+\frac{1}{\alpha^2}\right)\delta_{ij}\right].$$

(This expression is sometimes known as the Somigliana tensor.)

4.3 To evaluate the geometrical spreading function $\mathscr{R}(\mathbf{x}, \boldsymbol{\xi})$ between two points $\boldsymbol{\xi}$ and $\mathbf{x}$ in a spherically symmetric Earth, consider two rays departing from $\boldsymbol{\xi}$ with the same azimuth and with takeoff angles i_ξ and $i_\xi + \delta i_\xi$ (see Fig. 4.9b).

Consider also two rays obtained from the previous two by making an azimuth increment $\delta\phi_0$. The solid angle made at $\boldsymbol{\xi}$ by the four rays is $\sin i_\xi\ \delta i_\xi\ \delta\phi_0$. Show that the cross-sectional area of the ray tube at $\mathbf{x}$ is $-|\mathbf{x}|^2 \cos i_x \sin\Delta\ \delta\Delta\ \delta\phi_0$. Then use ray parameter

$$p = \frac{|\boldsymbol{\xi}| \sin i_\xi}{c(\boldsymbol{\xi})} = \frac{|\mathbf{x}| \sin i_x}{c(\mathbf{x})}$$

(see (4.45b)) to obtain

$$\mathscr{R}(\mathbf{x}, \boldsymbol{\xi})c(\boldsymbol{\xi}) = |\mathbf{x}|\ |\boldsymbol{\xi}|\left[\frac{\cos i_x \cos i_\xi \sin\Delta}{p}\left|\frac{\partial\Delta}{\partial p}\right|\right]^{1/2}.$$

Because of the symmetry of the right-hand side, we immediately find the reciprocity $\mathscr{R}(\mathbf{x}, \boldsymbol{\xi})\alpha(\boldsymbol{\xi}) = \mathscr{R}(\boldsymbol{\xi}, \mathbf{x})c(\mathbf{x})$, proved here for spherically symmetric media only.

4.4 Obtain the ray theory solution

$$\frac{l_i(\mathbf{x})l_j(\boldsymbol{\xi})\ \delta(t - T(\mathbf{x}, \boldsymbol{\xi}))}{4\pi\rho^{1/2}(\mathbf{x})\rho^{1/2}(\boldsymbol{\xi})\alpha^{3/2}(\boldsymbol{\xi})\alpha^{1/2}(\mathbf{x})\mathscr{R}(\mathbf{x}, \boldsymbol{\xi})}$$

for the far-field P-wave contribution to $G_{ij}(\mathbf{x}, t; \boldsymbol{\xi}, 0)$. (Use the solution given for (4.66), (4.57), and the far-field P-wave term in (4.23).) Then apply the reciprocity theorem (2.39) to prove that

$$\mathscr{R}(\mathbf{x}, \boldsymbol{\xi})\alpha(\boldsymbol{\xi}) = \mathscr{R}(\boldsymbol{\xi}, \mathbf{x})\alpha(\mathbf{x})$$

in general inhomogeneous isotropic media.

This reciprocity for geometrical spreading is a remarkable result in the differential geometry of rays. A direct proof is given by Richards (1971).

4.5 The principal axes of a symmetric second-order tensor are the Cartesian axes for which off-diagonal components of the tensor are zero. Starting from (3.24), giving the moment tensor referred to directions of slip $\bar{\mathbf{u}}$ and fault normal $\boldsymbol{\nu}$, rotate axes by 45° about a line perpendicular to $\bar{\mathbf{u}}$ and $\boldsymbol{\nu}$, and hence show that the pressure axis and the tension axis are principal axes for the moment tensor. Show that the seismic moment M_0 for fault slip is an invariant of the moment tensor $\mathbf{M}$ by finding the magnitude of the vector dipole terms in the principal axes coordinate system.

4.6 Show that the R.M.S. value for the radiation pattern of far-field P-wave amplitudes from a point source of fault slip, averaged over the focal sphere, is $\sqrt{\frac{4}{15}}$; for S-waves, show that the corresponding result is $\sqrt{\frac{2}{5}}$. (Start from the radiation patterns defined in (4.33).)

4.7 For the spherical polar coordinates used in (4.31)–(4.33) to describe the radiation pattern for a double couple, show that the far-field components of $F_j * G_{ij}$ are

$$\frac{1}{4\pi\rho\alpha^2}\frac{\hat{r}_i\hat{r}_j}{r}F_j\left(t-\frac{r}{\alpha}\right)+\frac{1}{4\pi\rho\beta^2}\frac{(\hat{\theta}_i\hat{\theta}_j+\hat{\phi}_i\hat{\phi}_j)}{r}F_j\left(t-\frac{r}{\beta}\right).$$

(This gives the i-component of displacement at (r, θ, ϕ) in terms of the Cartesian components of unit vectors $\hat{\mathbf{r}}$, $\hat{\boldsymbol{\theta}}$, $\hat{\boldsymbol{\phi}}$ for a force $\mathbf{F}(t)$ applied at the origin.)

4.8 Show that the problem of solving for the path of a ray in an inhomogeneous medium with wavespeed $c = c(\mathbf{x})$ is equivalent to solving for the motion of a particle moving in a force field with potential proportional to $1/c^2$. (*Hint*: Choose a scalar variable σ to define position along a ray, with σ having the property that $d\mathbf{x}/d\sigma = \nabla T$.)

CHAPTER 5

Plane Waves in Homogeneous Media and Their Reflection and Transmission at a Plane Boundary

So far, we have investigated the details of seismic wave generation by various point sources in a homogeneous, infinite, isotropic, elastic medium, and we have begun to look at transmission problems by developing geometrical ray theory for smoothly varying inhomogeneous media. We now continue to develop the theory of wave transmission, looking at basic problems involving a discontinuity of elastic properties in the medium within which waves are propagating. We analyze here only the simplest type of discontinuity, in which two homogeneous isotropic elastic media are in welded contact on a plane boundary. Details of the seismic source are avoided by considering only the case of a plane wave incident on the boundary.

The first analysis of such problems is due to George Green (1839); it appeared in the same classic paper that introduced a strain-energy function. Green was attempting to explain reflection and refraction of light in terms of elastic waves, and his work is similar in some detail to a modern analysis of *P*, *SV*, and *SH* plane waves. He did not, however, complete all the algebra necessary for the case in which the two half-spaces have entirely different elastic constants and densities. This generalization was obtained by Knott (in 1899), using potentials, and independently by Zoeppritz (in 1907).

The assumption of an *incident plane wave* may be quite good in practice for investigating waves at great distances from their source (see Fig. 5.1a).

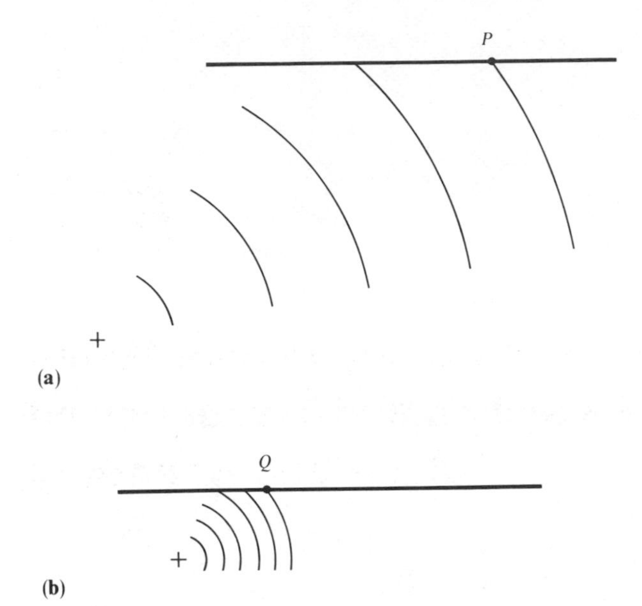

FIGURE **5.1**

(**a**) If the point P on some boundary within the Earth is sufficiently far from the localized source of waves under study, then wavefronts arriving at P may effectively be treated as incoming plane waves. (**b**) If Q is sufficiently close to the source, then curvature of the wavefronts at Q may have to be taken into account (see Chapter 6). Note that by "sufficiently far" and "sufficiently close" we refer to the number of wavelengths between the point of interest (P or Q) and the source. Thus, even for the source/receiver geometry in (b), it may be possible to use plane wave theory for the very high frequencies.

But if the reflection and refraction take place near the source (Fig. 5.1b), phenomena are observed that cannot be explained directly by Knott's theory. An outstanding example is the P_n-wave, discovered by Mohorovičic in 1909. Although this wave is of a type that can propagate in a medium composed of two welded homogeneous half-spaces (see Box 6.4), it involves a point source radiating *spherical waves*. We mention this point in order to bring out an important indirect application of Knott's theory. A method of studying spherical waves is to decompose them into a sum (or integral) of plane waves, then to apply Knott's theory to each plane wave, and finally to synthesize the required result by superimposing the results for each plane wave.

In this chapter, we shall summarize the basic properties of plane waves, which are required in many of the following chapters. After showing how to set up P, SV, SH problems with three scalar potentials, we obtain specific formulas for reflection/conversion/transmission coefficients, and relate these to unitary scattering matrices. We describe inhomogeneous waves and the associated phase shift of scattering coefficients.

The Earth is an imperfect elastic medium, in the sense that small particle motions initiated by a propagating wave lead to irreversible loss of energy from the wave due to a wide variety of dissipative mechanisms. From the point of view of the effect on the propagating wave, such attenuation is summarized conveniently by the parameter Q. We continue the chapter with a brief description of the effect on the pulse shape of propagation in an attenuating medium like the Earth, in which Q is nearly constant over the frequency range of observed seismic waves. Finally, we outline the basic theory for plane-wave propagation in anisotropic media.

5.1 Basic Properties of Plane Waves in Elastic Media

A physical quantity (such as particle acceleration or a stress component) propagates as a *plane wave* in direction **l** with speed c if

i) at a fixed time, the quantity is spatially unchanged over each plane normal to the unit vector **l**, and if
ii) the plane associated with a particular value of the quantity moves with speed c in direction **l**.

It follows that physical quantities that propagate with these two properties must have a functional dependence on space and time only via the combination $t - (\mathbf{l} \cdot \mathbf{x})/c$. We call $\mathbf{l}/c$ the *slowness vector* **s**. An advantage of using slowness (rather than velocity) to summarize the speed and direction of propagation of a wave is that slownesses may be added vectorially (but velocities, in this context, may not). Thus, using Cartesian coordinates (x, y, z), the slowness of a given wave is the vectorial sum of its components s_x, s_y, s_z along each coordinate direction: $\mathbf{s} = s_x\hat{\mathbf{x}} + s_y\hat{\mathbf{y}} + s_z\hat{\mathbf{z}}$, and the slowness in direction **n** is simply $\mathbf{s} \cdot \mathbf{n}$. In contrast, the velocity with which a plane wave advances in a particular direction is, in general, faster than its velocity in the direction of propagation (see Fig. 5.2).

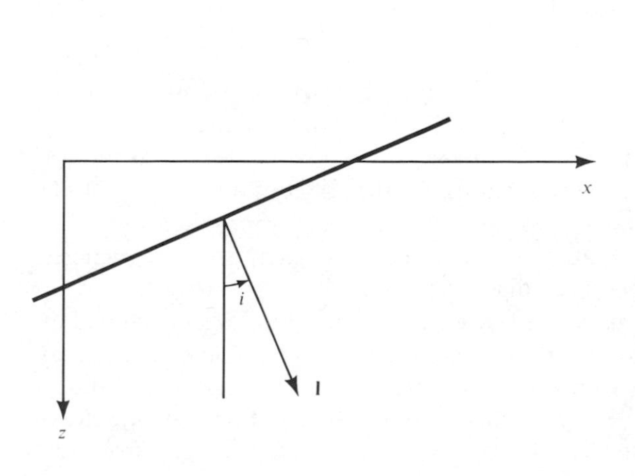

FIGURE **5.2**

For a quantity propagating with speed c in the **l** direction, the heavy oblique line marks a plane on which values of the quantity are all equal. Conventionally in seismology we take positive z as the depth direction and define the x-axis to be the direction that lies along the horizontal component of **l**. We often use the angle i, between the z-axis and direction **l**, to specify the direction of wave propagation. Then the apparent speed of propagation along the x-axis is $c/\sin i$. This apparent speed is often measured in seismology by using instruments arrayed over part of the Earth's surface. Its reciprocal is the horizontal slowness, or ray parameter.

BOX **5.1**

Notation

The advantage of using subscripts to denote vector and tensor components is considerably reduced if one is interested in the detailed properties of a particular component. In this chapter, we shall often use the notation $\mathbf{u} = (u, v, w)$ for the three Cartesian components of displacement, with $\mathbf{x} = (x, y, z)$ as the coordinates. We shall find that *P*- and *SV*-waves are coupled by boundary conditions on horizontal planes and that *P* and *SV* plane waves can each be analyzed in terms of displacement components $u = u(x, z, t)$, $w = w(x, z, t)$. *SH* plane waves involve $v = v(x, z, t)$, and they do not couple to *P* or *SV*.

Where the subscript notation is still convenient (equations (5.1), (5.2), (5.5)), we retain it with the obvious interpretations, e.g., $e_{23} = \frac{1}{2}(\partial u_2/\partial x_3 + \partial u_3/\partial x_2) = \frac{1}{2}(\partial v/\partial z + \partial w/\partial y) = e_{yz}$.

The two basic types of plane waves are easily distinguished by substituting the general form $\mathbf{u} = \mathbf{u}(t - \mathbf{s} \cdot \mathbf{x})$ for displacement into the elastic displacement equation for a homogeneous isotropic medium,

$$\rho \ddot{u}_i = (\lambda + \mu) u_{j,ji} + \mu u_{i,jj}, \tag{5.1}$$

to give

$$[\rho \delta_{ik} - (\lambda + \mu) s_i s_k - \mu s_j s_j \delta_{ik}] \ddot{u}_k = 0. \tag{5.2}$$

Forming the vector product and scalar product of (5.2) with $\mathbf{s}$, and using $s^2 = 1/c^2$, we obtain

$$\left(\rho - \frac{\mu}{c^2}\right) \ddot{\mathbf{u}} \times \mathbf{s} = \mathbf{0}, \qquad \left(\rho - \frac{\lambda + 2\mu}{c^2}\right) \ddot{\mathbf{u}} \cdot \mathbf{s} = 0. \tag{5.3}$$

Therefore, either $[\ddot{u} \times \mathbf{s} = \mathbf{0}$ and $c^2 = (\lambda + 2\mu)/\rho]$ or $[\ddot{\mathbf{u}} \cdot \mathbf{s} = 0$ and $c^2 = \mu/\rho]$. It follows that the plane wave is *either* a *P*-wave, with longitudinal motion (parallel to $\mathbf{s}$) and speed $\sqrt{(\lambda + 2\mu)/\rho}$, *or* an *S*-wave with transverse motion and speed $\sqrt{\mu/\rho}$. Our analysis here is similar to that of equations (4.48)–(4.51), the difference being now that we do not have to make approximations. The longitudinal or transverse nature of *P* or *S* motion is exact, at all frequencies, for plane waves in homogeneous media.

To describe the energy associated with elastic motion, we developed in Chapter 2 the concept of an elastic strain-energy density. The strain energy of a medium is its capacity to do work by virtue of its configuration, and in (2.32) we found the strain-energy density to be $\frac{1}{2}\tau_{ij}e_{ij}$. For a plane wave $u_i = u_i(t - \mathbf{s} \cdot \mathbf{x})$, the strain tensor is $e_{ij} = -\frac{1}{2}[\dot{u}_i s_j + \dot{u}_j s_i]$, and from the stress-strain relations for an isotropic medium (2.49) it is easy to show that

$$\tfrac{1}{2}\tau_{ij}e_{ij} = \tfrac{1}{2}[(\lambda + \mu)(\mathbf{s}\cdot\dot{\mathbf{u}})^2 + \lambda(\dot{\mathbf{u}}\cdot\dot{\mathbf{u}})(\mathbf{s}\cdot\mathbf{s})]. \tag{5.4}$$

In the case of either a P-wave (for which $\mathbf{s}$ is parallel to $\dot{\mathbf{u}}$, and $|\mathbf{s}| = \alpha^{-1}$) or an S-wave ($\mathbf{s}$ perpendicular to $\dot{\mathbf{u}}$, and $|\mathbf{s}| = \beta^{-1}$), it follows from (5.4) that

$$\tfrac{1}{2}\tau_{ij}e_{ij} = \tfrac{1}{2}\rho\dot{u}^2, \tag{5.5}$$

i.e., that the strain-energy density equals the kinetic-energy density. The quantities in (5.5) are all real, and energy densities depend on t and $\mathbf{x}$ only via the combination $t - \mathbf{s}\cdot\mathbf{x}$. Hence the speed of energy propagation is no different from the speed with which a pulse shape in particle displacement is propagated: either α for P-waves or β for S-waves.

It follows that the flux rate of energy transmission in a plane wave (i.e., the amount of energy transmitted per unit time across unit area normal to the direction of propagation) is $\rho\alpha\dot{u}^2$ for P-waves and $\rho\beta\dot{u}^2$ for S-waves. We have proved this result only for plane waves in homogeneous media, and it is a "local" property, depending on material properties and on the planar nature of the wave only at the point at which the flux rate is evaluated. We can therefore expect that flux rates are still given approximately by $\rho\dot{u}^2$ times the propagation velocity for the case of slightly curved wavefronts in a medium with some spatial fluctuation in material properties. It follows that there is a physical interpretation of the results of geometrical ray theory for displacement amplitude, which we obtained in Chapter 4. Thus, for P-waves spreading from a point source in an inhomogeneous medium, (4.57) indicates (for a particular ray) that $[\rho(\mathbf{x})\alpha(\mathbf{x})]^{1/2}\dot{u}^P$ is dependent on receiver position $\mathbf{x}$ only via the geometrical factor $1/\mathscr{R}^P(\mathbf{x}, \xi)$ and the travel time $T^P(\mathbf{x}, \xi)$. Hence the flux rate is controlled only by the ray geometry. Referring to Figure 5.3, the rate at which

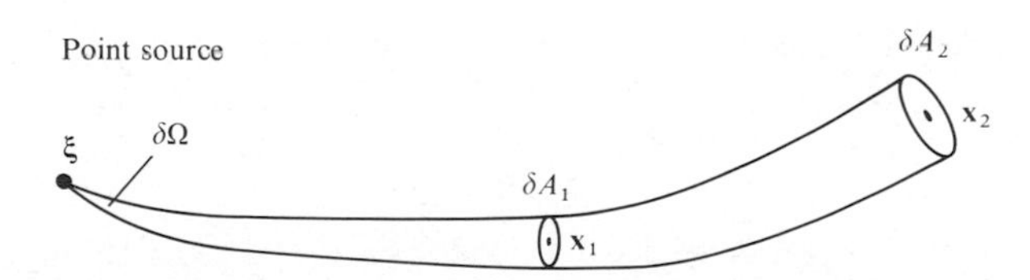

FIGURE **5.3**

Parameters for a ray tube originating on a point source at $\boldsymbol{\xi}$. The cross-sectional areas of the tube at $\mathbf{x}_1$ and $\mathbf{x}_2$ are given by $\delta A_i = \delta\Omega\mathscr{R}^2(\mathbf{x}_i, \xi)$ ($i = 1, 2$), where $\delta\Omega$ is the solid angle at the source. According to geometrical ray theory, the energy crossing δA_1 in unit time is the same as the energy crossing δA_2 in unit time, allowing for the time delay in propagation between $\mathbf{x}_1$ and $\mathbf{x}_2$.

energy crosses δA_1 is equal to $\rho(\mathbf{x}_1)\alpha(\mathbf{x}_1)\dot{u}^2\,\delta A_1$. But since $\delta A_1 \propto [\mathscr{R}^P(\mathbf{x}_1, \xi)]^2$, it follows from (4.57) that the rate at which energy crosses δA_1 at time T_1 is equal to the rate at which it crosses δA_2 at time T_2, $T_2 - T_1$ being the time taken for the wave to advance from $\mathbf{x}_1$ to $\mathbf{x}_2$. In this sense we learn from geometrical ray theory that propagating seismic energy is confined within ray tubes. This is only an approximation, becoming accurate at sufficiently high frequencies.

5.1.1 Potentials for plane waves

We saw in Chapter 4 that potentials for elastic displacement can be used to separate *P*- and *S*-components, and that this is desirable because wave equations for the separate potentials are much simpler (involving only one wave speed) than the wave equation for elastic displacement. The advantages of using potentials (ϕ and $\boldsymbol{\psi}$) appear to be offset by the fact that ϕ and $\boldsymbol{\psi}$ involve *four* unknown functions, whereas the physical quantity we are often interested in, elastic displacement, is a vector with only *three* unknown components. The extra unknown is constrained by an extra equation, usually $\nabla \cdot \boldsymbol{\psi} = 0$. In homogeneous isotropic media, however, only three scalar potential functions are necessary, corresponding separately to the *P*-, *SV*-, and *SH*-components of motion.

We shall prove this general result in Box 6.5, but our present interest is in plane waves, and for these there is a special form for the two scalar *S*-wave potentials. Thus an *S*-wave in general has displacement $\nabla \times \boldsymbol{\psi}$. If the wave is a plane wave and Cartesian coordinates are chosen as in Figure 5.2, with the x-axis taken in the direction of the horizontal slowness component, then $\boldsymbol{\psi}$ depends only on x, z, and t. It follows from the constraint $\nabla \cdot \boldsymbol{\psi} = 0$ that

$$\frac{\partial \psi_x}{\partial x} + \frac{\partial \psi_z}{\partial z} = 0. \tag{5.6}$$

If the wave is polarized purely as SV, then the y-component of displacement is zero, and

$$\frac{\partial \psi_x}{\partial z} - \frac{\partial \psi_z}{\partial \psi} = 0. \tag{5.7}$$

In the context of the theory of functions of a complex variable, (5.7) and (5.8) are Cauchy-Riemannn equations, and it follows that $\psi_z + i\psi_x$ is an analytic function of the variable $x + iz$. A function that is everywhere analytic and bounded is constant (this is Liouville's theorem). For a plane wave, $\psi_z + i\psi_x$ is certainly bounded. Furthermore, if the *SV*-wave is given by $\nabla \times \boldsymbol{\psi}$ only in a restricted depth range, one can conceptually extend this displacement to all depths, so that $\psi_z + i\psi_x$ is analytic everywhere. It follows that $\psi_z + i\psi_x$ is a constant: since only gradients of ψ_x and ψ_z are of any physical concern, the constant can be taken as zero. We therefore conclude that the most general plane *SV*-wave, propagating in a vertical plane containing the *x*-axis, can be expressed in terms of the potential $\boldsymbol{\psi} = (0, \psi, 0)$, with displacement $\nabla \times \boldsymbol{\psi} = (-\partial\psi/\partial z, 0, \partial\psi/\partial x)$. The vector wave equation reduces to the scalar form $\beta^2\nabla^2\psi = \ddot{\psi}$.

For a plane *SH*-wave, the use of a vector potential is inappropriate, since the horizontal displacement component is itself a satisfactory scalar function with which to work. For coordinates chosen as in Figure 5.2, the displacement $\mathbf{u} = (u, v, w)$ reduces for *SH* to $\mathbf{u} = (0, v, 0)$, with $v = v(x, z, t)$ for a plane wave. Already the constraint $\nabla \cdot \mathbf{u} = 0$ for a shear wave is satisfied, and it is easy to show that the elastic displacement equation reduces to $\beta^2\nabla^2 v = \ddot{v}$.

For *P*-waves, displacement is $\nabla\phi$, with ϕ satisfying $\alpha^2\nabla^2\phi = \ddot{\phi}$. The special case of a plane *P*-wave propagating as in Figure 5.2 leads to a zero component of displacement in the *y*-direction, and $\phi = \phi(x, z, t)$ (independent of the *y*-coordinate). Thus displacement is $(\partial\phi/\partial x, 0, \partial\phi/\partial z)$.

BOX **5.2**

The sign convention for Fourier transforms used in solving wave-propagation problems

Using a Cartesian coordinate system in which z increases with depth, we shall often carry out Fourier transforms of the two horizontal variables, x transforming to k_x and y to k_y. Our convention for spatial transforms here is

$$f(x) \rightarrow f(k_x) = \int_{-\infty}^{\infty} f(x)e^{-ik_x x}\,dx,$$

and similarly for functions of y. Note that we shall usually avoid special symbols (such as $\tilde{f}$ or $\bar{f}$ or F) to denote that f has been transformed. Whenever it is not clear from the context whether f denotes $f(x)$ or the transformed function, we shall write f with its specific argument, $f(x)$ or $f(k_x)$. Our only exceptions occur in Chapter 11, on data processing, where many results are made clearer with a capital letter for the transform.

We shall also be transforming the time dependence, either with a Laplace transform from t to s and the convention

$$f(t) \rightarrow f(s) = \int_0^{\infty} f(t)e^{-st}\,dt$$

or with a Fourier transform from t to ω. Although these transforms are essentially the same, if the transform variable assumes complex values, it is still useful to distinguish between them, since some methods of analysis work with real ω, so that Fourier transformation is appropriate, whereas some methods work with real s (i.e., imaginary ω), in which case the Laplace transform is more convenient.

For Fourier transformation of time dependence, our sign convention for the exponent is

$$f(t) \rightarrow f(\omega) = \int_{-\infty}^{\infty} f(t) \exp(+i\omega t)\, dt.$$

Note that this differs from the convention adopted above in this Box for spatial Fourier transformations. Of course, one would like to avoid using a mixed convention, but there are three good reasons why it is appropriate in solving wave-propagation problems relevant to seismology.

First, it permits a convenient interpretation of the inverse Fourier transforms. If $f(x, y, z, t)$ is some propagating physical variable of interest, it is often possible to obtain the triply transformed function $f(k_x, k_y, z, \omega)$. Then the required solution is

$$f(x, y, z, t) = \frac{1}{8\pi^3} \int_{-\infty}^{\infty} dk_x \int_{-\infty}^{\infty} dk_y \int_{-\infty}^{\infty} d\omega\, f(k_x, k_y, z, \omega) \exp[i(k_x x + k_y y - \omega t)].$$

For our choice of sign convention, this integrand can be interpreted, in the case of *positive* k_x, k_y, and ω, as a wave propagating in the directions of *increasing* x and y.

Second, if $f(x, y, z, t)$ satisfies a wave equation of type $c^2\nabla^2 f = \ddot{f}$, then $f(k_x, k_y, z, \omega)$ satisfies

$$\frac{\partial^2 f}{\partial z^2} = \left(k_x^2 + k_y^2 - \frac{\omega^2}{c^2}\right) f.$$

If the medium is homogeneous (i.e., c is constant), then

$$f(k_x, k_y, z, \omega) \propto e^{\pm i\omega\xi z}, \tag{1}$$

where $\omega\xi = (\omega^2/c^2 - k_x^2 - k_y^2)^{1/2}$. The choice of sign in (1) indicates whether f is a down-going wave (+) or an upcoming wave (−). However, if $\omega^2/c^2 < k_x^2 + k_y^2$, we shall find almost always that we wish to work with the *positive* imaginary value of $\omega\xi$, for then the wave $e^{+i\omega\xi z}$ attenuates correctly with depth ($z \rightarrow \infty$), and the wave $e^{-i\omega\xi z}$ attenuates correctly with height ($z \rightarrow -\infty$).

Third, anticipating our need in later chapters to use Hankel functions, we use what physicists have almost universally adopted, i.e., the convention that Hankel functions of type 1 represent outgoing waves and those of type 2 represent incoming waves. As propagating (steady-state) waves, these must then be associated with the factor $e^{-i\omega t}$. Integration (with respect to ω) over terms with this factor constitutes the inverse Fourier transform back to the time domain, and hence our sign convention is indeed correct for the standard Hankel function convention.

5.1.2 Separation of variables; steady-state plane waves

We shall show briefly that solving the wave equation

$$\alpha^2\nabla^2\phi = \ddot{\phi}, \tag{5.8}$$

by the method of separation of variables in a Cartesian coordinate system is equivalent to analyzing a type of plane-wave solution.

Thus we shall seek solutions of (5.8) in the form $X(x)Y(y)Z(z)T(t)$, each factor being a function of only one variable. It follows from (5.8) that

$$\frac{\alpha^2}{X}\frac{d^2X}{dx^2} + \frac{\alpha^2}{Y}\frac{d^2Y}{ay^2} + \frac{\alpha^2}{Z}\frac{d^2Z}{dz^2} = \frac{1}{T}\frac{d^2T}{dt^2}, \tag{5.9}$$

implying that $(1/T)/d^2T/dt^2$ is constant (differentiate (5.9) with respect to t in order to see this). For example,

$$\frac{d^2T}{dt^2} + \omega^2 T = 0, \quad \text{and thus} \quad T \propto \exp(\pm i\omega t).$$

Similarly,

$$\frac{d^2X}{dx^2} + k_x^2 X = 0 \quad \text{and} \quad X \propto \exp(\pm ik_x x)$$

$$\frac{d^2Y}{dy^2} + k_y^2 Y = 0 \quad \text{and} \quad Y \propto \exp(\pm ik_y y)$$

for some constants k_x, k_y. The z-dependence, however, is constrained in that

$$\frac{d^2Z}{dz^2} + k_z^2 Z = 0$$

(i.e., $Z \propto \exp(\pm ik_z z)$), where k_z is given by

$$k_z^2 = \frac{\omega^2}{\alpha^2} - k_x^2 - k_y^2, \tag{5.10}$$

so that the solution is characterized by only three independent numbers (ω, k_x, k_y), not four.

Separated solutions are therefore of the type

$$\exp[i(\mathbf{k}\cdot\mathbf{x} - \omega t)],$$

in which $\mathbf{k} = (k_x, k_y, k_z)$, and $|\mathbf{k}| = |\omega/\alpha|$. Clearly, this is a plane wave with a particularly simple time dependence, a steady oscillation at fixed frequency ω. The vector $\mathbf{k}$, of three separation constants, is known as the *wavenumber vector*, and it is just ω times the slowness.

General solutions of (5.8) are obtained by superposition of the separated solutions, and

$$\phi(x, y, z, t) = \int_{-\infty}^{\infty} d\omega \int_{-\infty}^{\infty} dk_x \int_{-\infty}^{\infty} dk_y \Phi(k_x, k_y, \omega, z)$$
$$\times \exp\left\{i\left[k_x x + k_y y + \left(\frac{\omega^2}{\alpha^2} - k_x^2 - k_y^2\right)^{1/2} z - \omega t\right]\right\}. \tag{5.11}$$

Here, $\Phi(k_x, k_y, \omega, z)$ is acting merely as some weighting function, giving the amount of plane wave characterized by (k_x, k_y, ω) that is present in the superposition for the required solution ϕ.

The result we have obtained in (5.11) is essentially the same as that stated by using inverse Fourier transforms in Box 5.2. The emphasis here has been on Cartesian coordinates, but for other coordinate systems it is also true that a solution given by inverse transforms can be thought of as a superposition of separated solutions.

Plane waves are directly of importance in seismology, because body waves from a distant source behave locally like plane waves. Representation (5.11) also indicates the indirect importance of steady-state plane waves, showing that they are a basis for synthesizing more general solutions. The details of this synthesis have been extensively studied, and they form the subject of Chapter 6 and parts of 7 and 9. Since these details determine the properties of plane waves that we shall need to develop, it is useful to list here the following comments on representation (5.11).

i) Because of the dependence on k_x, k_y, it will be important to study plane waves as a function of their horizontal wavenumber—or, equivalently, as a function of their horizontal slowness.
ii) Part of the process of solving a wave-propagation problem via (5.11) will be the determination of the function $\Phi(k_x, k_y, \omega, z)$ that is appropriate for the particular source under study. In the context of an integration over elements $d\omega\, dk_x\, dk_y$, Φ can be thought of as a *density function* in (ω, k_x, k_y) space. In the context of Fourier transform theory, it is related to the triple *transform* of ϕ with respect to t, x, and y. In the context of superposition, it is the *excitation function*, indicating how much of a particular plane wave is excited by the source under study.
iii) It is often useful to study waves propagating in only two spatial dimensions, x and z, in which case the y-dependence and the k_y integral are absent in (5.11).
iv) A decision must be made as to the sign of a square root appearing in the integrand in (5.11). Also, if the horizontal wavenumber is great enough, the square root of a negative number must be taken, and exponential growth or decay will occur in the z-direction. We take up this subject in Section 5.3.
v) Once the various factors appearing in the integrand of (5.11) have all been obtained, a method is required for evaluating the triple integral. We shall find that many different methods are in use, and that they often involve making approximations. In a few cases, the integrals can be inverted exactly to give a closed-form expression for ϕ. For the elastic media that are studied in problems relevant to seismology, it often occurs that a part of the evaluation of the integrals must be carried out numerically. The most fruitful methods of this type all emphasize manipulations of the integral over horizontal slowness.

With this justification for a thorough analysis of plane waves, we return to the main theme of the chapter.

5.2 Elementary Formulas for Reflection/Conversion/Transmission Coefficients

We have seen that steady-state plane wave solutions to equations of type $\alpha^2\nabla^2\phi = \ddot{\phi}$ in a homogeneous medium take the form $A\exp[i\omega(\mathbf{s}\cdot\mathbf{x} - t)]$, where A is the amplitude, and $(\partial\phi/\partial x, 0, \partial\phi/\partial z)$ is the associated P-wave displacement for propagation in a direction normal to the y-axis. In this section, we shall study the effect of the boundary between two half-spaces that are in contact along the plane $z = 0$. If the half-spaces consist of a solid, a fluid, or a vacuum, then there are five nontrivial case to consider: solid/solid, solid/fluid, solid/vacuum, fluid/fluid, and fluid/vacuum.

5.2.1 Boundary conditions

There are two types of boundary conditions: those concerning displacement (often called kinematic boundary conditions) and those concerning traction, or stress components (dynamic boundary conditions).

BOX **5.3**

The distinction between kinematics and dynamics

Kinematics is the branch of mechanics that deals purely with motion, without analyzing the underlying forces that cause or participate in the motion. Dynamics is the branch of mechanics that deals directly with force systems, and with the energy balance that governs motion. From these fundamental definitions, two useful conventions have developed for applying the words "kinematic" and "dynamic."

First, in the analysis of displacements alone, kinematic properties are those that may be derived from the eikonal equation (4.41), whereas dynamic properties are those related to displacement amplitudes. Thus the existence of particular wavefronts and ray paths is part of the kinematics of the problem in hand. As an example of a dynamic problem, we might ask if a certain approximation is adequate for the displacements observed at a given receiver at some given distance from a localized source.

Second, in those problems in which we have a direct interest both in the displacement and the associated system of stresses, then kinematic properties are properties of the displacement field and dynamic properties are related to the stresses. For example, if the relative displacement between opposite faces of a fault surface is known as a function of space and time, we say that we have a kinematic description of the fault motion. If the stresses (i.e., traction components) are known on the fault surface, we have a dynamic description. As another example, one refers to boundary conditions as being kinematic or dynamic, in the sense developed in the present section.

For two solids in welded contact, the kinematic condition is that all three components of displacement be continuous through the boundary. There would also be continuity of displacement across the boundary between a solid half-space and a viscous fluid half-space. However, if the fluid were completely inviscid, then slip could occur parallel to the boundary, but the normal component of displacement would be continuous (unless cavitation occurs or the fluid moves into interstices in the solid). For typical wavelengths and periods of seismic waves (kilometers, seconds), it appears that the two fluids of importance in seismology (the oceans and the Earth's outer core) do behave in an inviscid fashion. That is, their viscosity is so low as to confine kinematic viscous drag in the fluid to only a negligible fraction of a wavelength away from solid/fluid interfaces such as the sea floor or the core-mantle boundary. Under these circumstances, the tangential component of displacement can effectively be discontinuous, and the only hope for a boundary condition lies in the normal component. The strong compressive stresses prevailing in the Earth's interior will not permit cavitation to occur (since this would entail shock propagation with a stress discontinuity many orders of magnitude greater than the strength of rocks). Moreover, any significant diffusion of fluid into solid would require a time far greater than the period of seismic waves. It is therefore appropriate in seismology to take the kinematic boundary condition for a solid/fluid interface as continuity of normal displacement only.

The dynamic boundary condition is continuity of traction across the interface. This result can be proved along the lines of our discussion of Figure 2.4. The tractions acting across a small thin disc, with its two flat faces in different

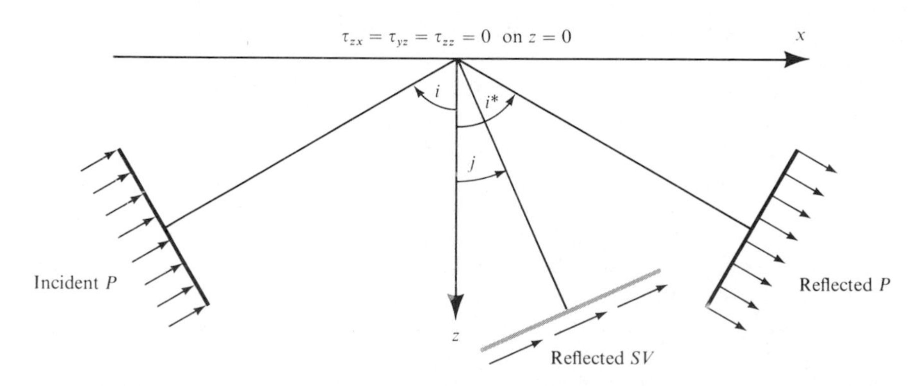

FIGURE 5.4

The wave system and coordinates for analysis of the reflected waves set up by a plane P-wave incident on the free surface of a solid elastic half-space. Portions of plane waves are shown as heavy line segments, and associated sets of arrows show the direction of particle motion. Angles i and j are defined in terms of the ray trajectories, orthogonal to the plane fronts of P and SV, respectively. Reflection angle i^* equals incident angle i.

media, are equal in magnitude but opposite in direction (see (2.7)). Reversing the direction of one of the outward normals of the disc, the traction $\mathbf{T}(\mathbf{n})$ is the same for each face of the disc, and hence is continuous through the interface (see also Problem 2.5). Since traction is a vector, this appears to give three scalar constraints. One or two of these, however, may be satisfied trivially, since a propagating plane wave does not necessarily excite all three components of traction. Note that these traction components, for our choice of the interface as a plane normal to the z-axis, are the stress tensor components τ_{zx}, τ_{zy}, τ_{zz}. At an interface with a vacuum, these three stress components are all zero, and this is effectively the case also for the surface of the Earth or the oceans, since the elastic constants for the atmosphere are several orders of magnitude less than the elastic constants of rock or the bulk modulus of sea water. (Exceptions can occur for air-coupled surface waves, as described in Appendix 1.) The case $(\tau_{zx}, \tau_{zy}, \tau_{zz}) = (0, 0, 0)$ on $z = 0$ is referred to as the "free-surface boundary condition" on $z = 0$, and this is the first reflection problem we shall examine in detail.

5.2.2 *Reflection of plane P-waves and SV-waves at a free surface*

Suppose a plane P-wave is traveling with horizontal slowness in the direction of increasing x (see Fig. 5.4). Then, for some potential ϕ, P-displacement is given by $\mathbf{u} = (\partial\phi/\partial x, 0, \partial\phi/\partial z)$ and the associated traction by

$$\mathbf{T}(\mathbf{u}, \hat{\mathbf{z}}) = (\tau_{zx}, \tau_{yz}, \tau_{zz}) = \left(2\mu \frac{\partial^2 \phi}{\partial z\, \partial x}, 0, \lambda \nabla^2 \phi + 2\mu \frac{\partial^2 \phi}{\partial z^2}\right). \tag{5.12}$$

For completeness, it is convenient to state here the corresponding results for SV- and SH-waves: for SV, there is a scalar potential ψ, and SV-displacement is given by $\mathbf{u} = (-\partial\psi/\partial z, 0, \partial\psi/\partial x)$ and traction by

$$(\tau_{zx}, \tau_{yz}, \tau_{zz}) = \left(\mu\left(\frac{\partial^2 \psi}{\partial x^2} - \frac{\partial^2 \psi}{\partial z^2}\right), 0, 2\mu \frac{\partial^2 \psi}{\partial z\, \partial x}\right); \tag{5.13}$$

for SH, only one displacement component is needed, and SH-displacement is given by $\mathbf{u} = (0, v, 0)$ and traction by

$$(\tau_{zx}, \tau_{yz}, \tau_{zz}) = \left(0, \mu \frac{\partial v}{\partial z}, 0\right). \tag{5.14}$$

In terms of the incidence angle i (see Fig. 5.4), the slowness of the incident P-wave is

$$\mathbf{s} = \left(\frac{\sin i}{\alpha}, 0, \frac{-\cos i}{\alpha}\right).$$

Since no τ_{yz} component is excited by this wave (see (5.12)), no SH-component is excited (see (5.14)), and candidates for reflected waves are P and SV only, with respective slownesses

$$\left(\frac{\sin i^*}{\alpha}, 0, \frac{\cos i^*}{\alpha}\right), \left(\frac{\sin j}{\beta}, 0, \frac{\cos j}{\beta}\right).$$

Thus the total ϕ potential is made up from an incident component ϕ^{inc}, and reflected component ϕ^{refl}, with

$$\phi = \phi^{\text{inc}} + \phi^{\text{refl}}, \tag{5.15}$$

$$\phi^{\text{inc}} = A \exp\left[i\omega\left(\frac{\sin i}{\alpha}x - \frac{\cos i}{\alpha}z - t\right)\right], \tag{5.16}$$

$$\phi^{\text{refl}} = B \exp\left[i\omega\left(\frac{\sin i^*}{\alpha}x + \frac{\cos i^*}{\alpha} - t\right)\right]. \tag{5.17}$$

The amplitudes A and B are constant in each wave, and the total SV-wave is given by

$$\psi = \psi^{\text{refl}} \tag{5.18}$$

where

$$\psi^{\text{refl}} = C \exp\left[i\omega\left(\frac{\sin j}{\beta}x + \frac{\cos j}{\beta}z - t\right)\right]. \tag{5.19}$$

There are no kinematic boundary conditions to concern us, since it is meaningless to speak of displacements above the free surface, and the displacement of the free surface of the solid is not constrained. The nontrivial dynamic boundary conditions are $\tau_{zx} = \tau_{zz} = 0$ on $z = 0$, and from equations (5.12), (5.13), (5.15)–(5.19) we find that each of τ_{zx}, τ_{zz} is a sum of three contributions involving factors of the type

$$\exp\left[i\omega\left(\frac{\sin i}{\alpha}x - t\right)\right] \quad \text{or} \quad \exp\left[i\omega\left(\frac{\sin i^*}{\alpha}x - t\right)\right] \quad \text{or} \quad \exp\left[i\omega\left(\frac{\sin j}{\beta}x - t\right)\right].$$

The boundary conditions hold on $z = 0$ for all x and t, so that these factors, which control the horizontal propagation of the wave system, must all be the same. In particular, $i^* = i$ (angles of reflected and incident P are equal), and $(\sin i)/\alpha = (\sin j)/\beta$. Thus the horizontal slowness of the incident wave is preserved on reflection, and is preserved also on conversion to SV. If there were transmission into an upper half-space, the horizontal slowness component would again be preserved, and this is a statement of Snell's law, which we have already proved (see (4.45a)) for media with smoothly varying depth-dependence of velocity. We are developing here the important concept that the whole system of waves set up by reflection and transmission of plane waves in plane-layered media, is characterized by the value of their common horizontal slowness. Often we shall call this the *ray parameter*, although this too is an inadequate name, since $(\sin i)/\alpha = (\sin j)/\beta = p$ is a parameter of a whole system of rays, not just of one ray.

BOX **5.4**

Impedance

The impedance that a given medium presents to a given motion is a measure of the amount of resistance to particle motion. Specifically, impedance in elasticity is a ratio of stress to particle velocity, so that for a given applied stress, the particle velocity is inversely proportional to impedance.

Impedance properties of different wave types can vary considerably, as we now discuss by looking at specific examples.

First, consider an *SH*-wave with displacement $v = v_0 \exp\{i\omega[px + (\beta^{-1}\cos j)z - t]\}$. Then, for horizontal planes (z = constant), the tangential stress is $\tau_{yz} = i\mu\omega(\beta^{-1}\cos j)v$ and the tangential particle velocity is $\dot{v} = -i\omega v$, so that the impedance is $\tau_{yz}/\dot{v} = -\mu(\beta^{-1}\cos j) = -\rho\beta\cos j$. For the Earth's crust, representative values of density and shear velocity are $\rho = 2.8\ \text{g/cm}^3$ and $\beta = 3.5$ km/sec, so that the impedance is of order 10^6 c.g.s. units. A stress wave with amplitude 100 bars ($=10^8$ c.g.s. units) therefore corresponds to a ground velocity of about 100 cm/sec. For *SH* waves, however, note that impedance $\tau_{yz}/\dot{v} \to 0$ as $j \to \pi/2$ (grazing incidence).

Second, consider an acoustic wave (i.e., a *P*-wave in a fluid) in which the pressure is given by $P = P_0 \exp\{i\omega[px + (\alpha^{-1}\cos i)z - t]\}$. Then, since $\rho\ddot{\mathbf{u}} = -\nabla P$, the vertical particle velocity is given by $-i\omega\rho\dot{u}_z = -\partial P/\partial z$, and the impedance is $P/\dot{u}_z = (\rho\alpha)/\cos i$. Note now that the impedance approaches infinity as $i \to \pi/2$ (grazing incidence), in contrast to the behavior of *SH*-waves.

Simplifying the equations (5.12), (5.13) giving physical variables as a function of potentials ϕ and ψ, and writing them now in terms of p, ϕ, ψ, $\partial\phi/\partial z$, and $\partial\psi/\partial z$, we find

$$\text{for } P \begin{cases} \text{displacement} & \left(i\omega p\phi,\ 0,\ \dfrac{\partial\phi}{\partial z}\right) \\ \text{traction} & \left(2\rho\beta^2 i\omega p\dfrac{\partial\phi}{\partial z},\ 0,\ -\rho(1 - 2\beta^2p^2)\omega^2\phi\right) \end{cases} \tag{5.20}$$

$$
\text{for } SV \begin{cases} \text{displacement} & \left(-\dfrac{\partial\psi}{\partial z}, 0, i\omega p\psi\right) \\ \text{traction} & \left(\rho(1 - 2\beta^2 p^2)\omega^2\psi, 0, 2\rho\beta^2 i\omega p \dfrac{\partial\psi}{\partial z}\right). \end{cases} \tag{5.21}
$$

Our immediate goal is to obtain formulas for the ratios B/A and C/A, giving the amplitude of reflected and converted waves as a fraction of the incident wave amplitude. The two equations we can use are

$$
\left.\begin{aligned} \tau_{zx} &= 2\rho\beta^2 i\omega p\left(\frac{\partial\phi^{\text{inc}}}{\partial z} + \frac{\partial\phi^{\text{refl}}}{\partial z}\right) + \rho(1 - 2\beta^2 p^2)\omega^2\psi^{\text{refl}} = 0 \\ \tau_{zz} &= -\rho(1 - 2\beta^2 p^2)\omega^2(\phi^{\text{inc}} + \phi^{\text{refl}}) + 2\rho\beta^2 i\omega p \frac{\partial\psi^{\text{refl}}}{\partial z} = 0 \end{aligned}\right\} \text{ on } z = 0.
$$

Substituting from (5.16), (5.17), (5.19), these become

$$
2\rho\beta^2 p \frac{\cos i}{\alpha}(A - B) + \rho(1 - 2\beta^2 p^2)C = 0 \tag{5.22}
$$

$$
\rho(1 - 2\beta^2 p^2)(A + B) + 2\rho\beta^2 p \frac{\cos j}{\beta} C = 0 \tag{5.23}
$$

with solutions

$$
\frac{B}{A} = \frac{4\beta^4 p^2 \dfrac{\cos i}{\alpha}\dfrac{\cos j}{\beta} - (1 - 2\beta^2 p^2)^2}{4\beta^4 p^2 \dfrac{\cos i}{\alpha}\dfrac{\cos j}{\beta} + (1 - 2\beta^2 p^2)^2} \tag{5.24}
$$

$$
\frac{C}{A} = \frac{-4\beta^2 p \dfrac{\cos i}{\alpha}(1 - 2\beta^2 p^2)}{4\beta^4 p^2 \dfrac{\cos i}{\alpha}\dfrac{\cos j}{\beta} + (1 - 2\beta^2 p^2)^2}. \tag{5.25}
$$

Using trigonometrical identities and relations between elastic constants, a large number of different forms can be derived for the above two expressions. Note, for example, that $1 - 2\beta^2 p^2$ is $\cos 2j$. We have chosen to work with p, $\alpha^{-1}\cos i$, and $\beta^{-1}\cos j$ because in Chapter 9 we shall show that the reflection coefficients (5.24) and (5.25) can then be easily generalized to study media that are vertically heterogeneous.

We have called the ratios B/A and C/A "reflection coefficients," but actually they are amplitude ratios only for potentials. In practice, we are usually interested in amplitude ratios for displacements (and, occasionally, for energy). For a propagating steady-state P-wave, the displacement amplitude is ω(potential amplitude)/α, and similarly for S displacement the amplitude is ω(potential amplitude)/β. Hence, for displacement reflection coefficients, we still need B/A for $P \to P$, but $(C\alpha)/(A\beta)$ for the conversion $P \to S$. We also need to establish a sign convention for reflection coefficients, and our choice is described in Figure 5.5.

Many, many different notations have been proposed for reflection/conversion/transmission coefficients. Fortunately, the problems one needs to solve often turn out to require only a small number (one or two) of particular coefficients. In these cases, a comprehensive notation is unnecessary, since it may be clear from its context that a symbol such as PP is a reflection coefficient, with no ambiguity as to which particular combination of reflected wave/incident wave is intended. Such is the case in the present problem of a solid half-space with a free surface: only one type of incident P-wave is present, and only one P-wave is derived from it. Nevertheless, we shall shortly have to deal with the solid/solid interface, for which P-waves (and S-waves) can be incident from above and below. Each of the four possible types of incident P-SV waves (P or SV, from above or below) can generate all four types of outgoing P-SV waves, hence 16 coefficients are involved for a complete analysis of just this one interface. In the present problem, therefore, it will be convenient to adopt a notation that can easily be extended to more complicated interfaces. We shall take $\acute{P}\grave{P}$ as the $P \to P$ reflection coefficient for Figure 5.5a and $\acute{P}\grave{S}$ as the $P \to S$ conversion coefficient. In Figure 5.5b, the $S \to P$ conversion is given by $\acute{S}\grave{P}$, and $S \to S$ reflection by $\acute{S}\grave{S}$. This use of acute and grave accents indicates directly the intended sequence of incident wave $\to$ derived wave, since all

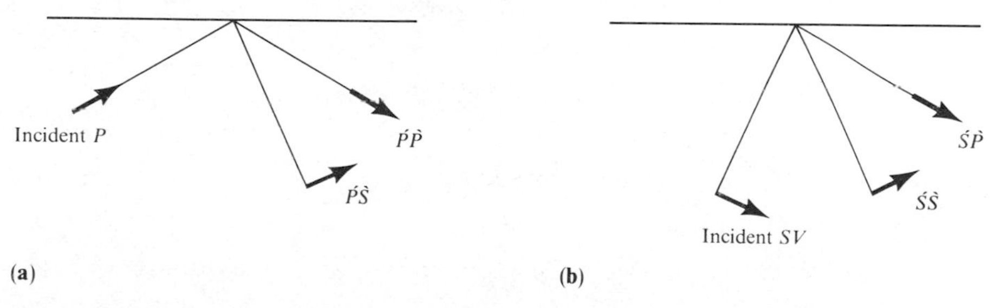

FIGURE **5.5**

Notation and sign convention for coefficients of reflection and transmission due to P or SV incidence on a free surface. A motion is taken as *positive* if its component to the right (i.e., in the horizontal direction of propagation) has the same phase as the propagation factor $\exp[i\omega(px - t)]$.

waves are moving from left to right. Thus an acute accent (e.g., $\acute{P}$) indicates an upcoming wave, and a grave accent (e.g., $\grave{S}$) a downgoing wave. Combining this notation with the sign convention of Figure 5.5, we state in Table 5.1 the exact vector form of motions corresponding to the two possible types of incident wave.

It follows from Table 5.1 and equations (5.24), (5.25), that

$$\acute{P}\grave{P} = \frac{-\left(\dfrac{1}{\beta^2} - 2p^2\right)^2 + 4p^2 \dfrac{\cos i}{\alpha} \dfrac{\cos j}{\beta}}{+\left(\dfrac{1}{\beta^2} - 2p^2\right)^2 + 4p^2 \dfrac{\cos i}{\alpha} \dfrac{\cos j}{\beta}}, \tag{5.26}$$

$$\acute{P}\grave{S} = \frac{4 \dfrac{\alpha}{\beta} p \dfrac{\cos i}{\alpha}\left(\dfrac{1}{\beta^2} - 2p^2\right)}{\left(\dfrac{1}{\beta^2} - 2p^2\right)^2 + 4p^2 \dfrac{\cos i}{\alpha} \dfrac{\cos j}{\beta}}. \tag{5.27}$$

In the case of an SV-wave incident on the free surface, we can expect a reflected P-wave ($\acute{S}\grave{P}$) and a reflected SV-wave ($\acute{S}\grave{S}$). From the vector displacements in Table 5.1, we find that the condition $\tau_{zx} = 0$ on $z = 0$ reduces to the equation

$$-2p\alpha\beta \frac{\cos i}{\alpha} \acute{S}\grave{P} + (1 - 2\beta^2 p^2)(1 - \acute{S}\grave{S}) = 0, \tag{5.28}$$

and the condition $\tau_{zz} = 0$ on $z = 0$ reduces to

$$-(1 - 2\beta^2 p^2)\acute{S}\grave{P} + \frac{2\beta^3 p}{\alpha} \frac{\cos j}{\beta}(1 + \acute{S}\grave{S}) = 0. \tag{5.29}$$

Solving (5.28) and (5.29), we obtain

$$\acute{S}\grave{P} = \frac{4 \dfrac{\beta}{\alpha} p \dfrac{\cos j}{\beta}\left(\dfrac{1}{\beta^2} - 2p^2\right)}{\left(\dfrac{1}{\beta^2} - 2p^2\right)^2 + 4p^2 \dfrac{\cos i}{\alpha} \dfrac{\cos j}{\beta}}, \tag{5.30}$$

$$\acute{S}\grave{S} = \frac{\left(\dfrac{1}{\beta^2} - 2p^2\right)^2 - 4p^2 \dfrac{\cos i}{\alpha} \dfrac{\cos j}{\beta}}{\left(\dfrac{1}{\beta^2} - 2p^2\right)^2 + 4p^2 \dfrac{\cos i}{\alpha} \dfrac{\cos j}{\beta}}. \tag{5.31}$$

TABLE **5.1**

Explicit expressions for the vector displacements involved in *P-SV* plane-wave problems of the type shown in Figure 5.5.

Incident wave		Scattered waves	
Type	*Displacement*	*Type*	*Displacement*
Upgoing P	$S(\sin i, 0, -\cos i)\exp\left[i\omega\left(\frac{\sin i}{\alpha}x - \frac{\cos i}{\alpha}z - t\right)\right]$	Downgoing P	$S(\sin i, 0, \cos i)\acute{P}\grave{P}\exp\left[i\omega\left(\frac{\sin i}{\alpha}x + \frac{\cos i}{\alpha}z - t\right)\right]$
		Downgoing SV	$S(\cos j, 0, -\sin j)\acute{P}\grave{S}\exp\left[i\omega\left(\frac{\sin j}{\beta}x + \frac{\cos j}{\beta}z - t\right)\right]$
Upgoing SV	$S(\cos j, 0, \sin j)\exp\left[i\omega\left(\frac{\sin j}{\beta}x - \frac{\cos j}{\beta}z - t\right)\right]$	Downgoing P	$S(\sin i, 0, \cos i)\acute{S}\grave{P}\exp\left[i\omega\left(\frac{\sin i}{\alpha}x + \frac{\cos i}{\alpha}z - t\right)\right]$
		Downgoing SV	$S(\cos j, 0, -\sin j)\acute{S}\grave{S}\exp\left[i\omega\left(\frac{\sin j}{\beta}x + \frac{\cos j}{\beta}z - t\right)\right]$

Note: The amplitude S of incident waves here can be thought of in two ways: either as the displacement amplitude of a steady-state wave or as the amplitude of the Fourier transform of particle velocity, in the case that the incident wave is an impulse in particle velocity associated with a step S in displacement.

Our notation with grave and acute accents was introduced for displacement amplitude ratios. It therefore can be applied also for particle-velocity amplitude ratios and for particle-acceleration amplitude ratios (since these involve scaling both scattered and incident waves by the same power of frequency). However, coefficients *are different* for ratios of potentials, or ratios of energy fluxes. Where such ratios are needed (e.g., Box 6.6, and equation (5.41)), we shall retain accented symbols for displacement amplitude ratios, and multiply by appropriate scaling corrections.

In this simple example of reflection from the free surface of a solid half-space, we have now obtained specific formulas for each component of the matrix

$$\begin{pmatrix} \acute{P}\grave{P} & \acute{S}\grave{P} \\ \acute{P}\grave{S} & \acute{S}\grave{S} \end{pmatrix}.$$

This matrix summarizes all possible reflection coefficients, and it is called a *scattering matrix*. Each of its components is plotted against slowness in Figure 5.6, and for this very simple interface, the components are found to vary quite strongly. Only the range $0 \leq p \leq 1/\alpha$ is shown. For slowness in the range 0.14 to 0.195 sec/km, note that the reflected motion is almost all opposite in type from the incident motion. That is, incident P is converted almost totally to reflected SV, and incident SV to reflected P. Far more complicated behavior can occur in other interface problems (solid/fluid, etc.), and seismologists are often forced to evaluate this behavior in great detail in order to interpret a particular piece of data. For convenience, therefore, we shall give the coefficient

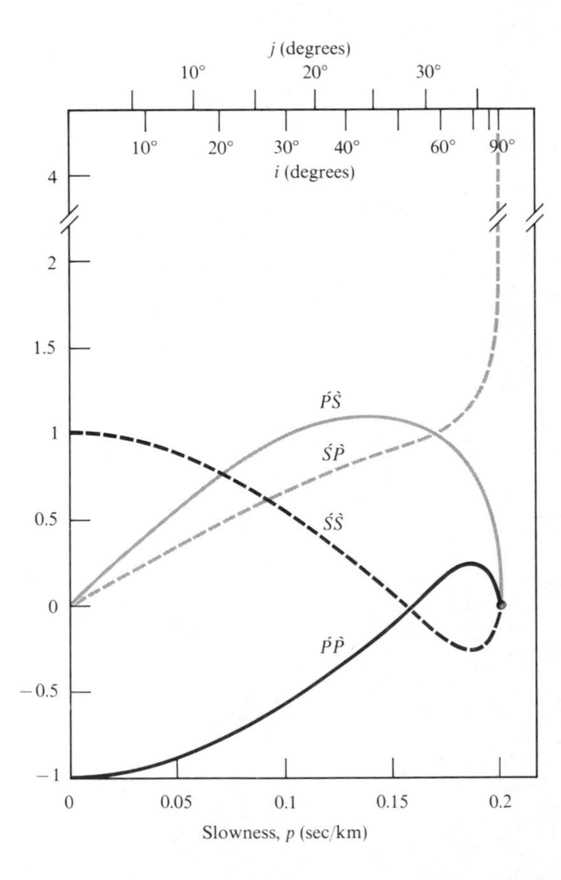

FIGURE **5.6**

The four possible P-SV reflection coefficients (displacement amplitude ratios) for a free surface are shown against horizontal slowness p. See Figure 5.5. In this case, α = 5 km/sec and β = 3 km/sec, and we restrict p to lie in the range $0 \leq p \leq 1/\alpha$ so that incidence angle i is always real. For $i = 90°$, $\acute{S}\grave{P}$ is quite large ($\sim$4.1).

TABLE **5.2**

Vector displacements for the *SH* plane wave problems shown in Figure 5.7.

Incident wave		Scattered waves	
Type	*Displacement*	*Type*	*Displacement*
Downgoing *SH*	$(0, S, 0)\exp\left[i\omega\left(px + \frac{\cos j_1}{\beta_1}z - t\right)\right]$	Upgoing *SH*	$(0, S, 0)\grave{S}\acute{S}\exp\left[i\omega\left(px - \frac{\cos j_1}{\beta_1}z - t\right)\right]$
		Downgoing *SH*	$(0, S, 0)\grave{S}\grave{S}\exp\left[i\omega\left(px + \frac{\cos j_2}{\beta_2}z - t\right)\right]$
Upgoing *SH*	$(0, S, 0)\exp\left[i\omega\left(px - \frac{\cos j_2}{\beta_2}z - t\right)\right]$	Upgoing *SH*	$(0, S, 0)\acute{S}\acute{S}\exp\left[i\omega\left(px - \frac{\cos j_1}{\beta_1}z - t\right)\right]$
		Downgoing *SH*	$(0, S, 0)\acute{S}\grave{S}\exp\left[i\omega\left(px + \frac{\cos j_2}{\beta_2}z - t\right)\right]$

formulas for two other interfaces of importance in seismology. Unfortunately, there is a long history of published misprints in these formulas (see Hales and Roberts, 1974, or Young and Braile, 1976, for a review). In order for an individual to have confidence in his or her evaluation of a particular coefficient, we shall conclude this section with a useful check to verify that the formulas have been correctly transcribed.

5.2.3 *Reflection and transmission of SH-waves*

Stress components τ_{zx} and τ_{zz} are not excited by the displacements listed in Table 5.2, so that the only nontrivial dynamic boundary condition is continuity of τ_{yz} across $x = 0$. The y-component of displacement is also continuous, and we find from (5.14) that the elements of the scattering matrix

$$\begin{pmatrix} \grave{S}\acute{S} & \acute{S}\acute{S} \\ \grave{S}\grave{S} & \acute{S}\grave{S} \end{pmatrix}$$

are

$$\grave{S}\acute{S} = \frac{\rho_1\beta_1 \cos j_1 - \rho_2\beta_2 \cos j_2}{\Delta}, \qquad \acute{S}\acute{S} = \frac{2\rho_2\beta_2 \cos j_2}{\Delta},$$
$$\grave{S}\grave{S} = \frac{2\rho_1\beta_1 \cos j_1}{\Delta}, \qquad \acute{S}\grave{S} = -\grave{S}\acute{S}, \tag{5.32}$$

where

$$\Delta = \rho_1\beta_1 \cos j_1 + \rho_2\beta_2 \cos j_2.$$

5.2.4 *Reflection and transmission of P-SV across a solid-solid interface*

The scattering matrix now is

$$\begin{pmatrix} \grave{P}\acute{P} & \grave{S}\acute{P} & \acute{P}\acute{P} & \acute{S}\acute{P} \\ \grave{P}\acute{S} & \grave{S}\acute{S} & \acute{P}\acute{S} & \acute{S}\acute{S} \\ \grave{P}\grave{P} & \grave{S}\grave{P} & \acute{P}\grave{P} & \acute{S}\grave{P} \\ \grave{P}\grave{S} & \grave{S}\grave{S} & \acute{P}\grave{S} & \acute{S}\grave{S} \end{pmatrix},$$

and it may be obtained from continuity of the x- and z-components of both displacement and traction. Each column of the scattering matrix represents the four waves scattered away from the interface by a particular type of incident

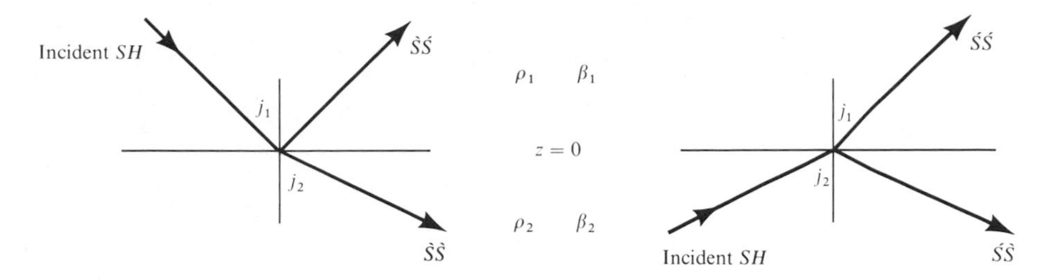

FIGURE **5.7**

Notation for the four possible reflection/transmission coefficients arising for problems of incident *SH*-waves.

wave. Hence, to evaluate all columns, it appears that we must set up four systems of four equations in four unknowns. Unless some care is taken to use all available symmetries in the problem, a complete statement of all 16 coefficients can involve extensive algebraic manipulation. This manipulation is minimized in the method we now describe, which is based on work of Nafe (1957).

We shall assume that all four possible incident waves are present together, as in Figure 5.9, with respective displacement amplitudes $\grave{P}_1$, $\grave{S}_1$, $\acute{P}_2$, $\acute{S}_2$. Subscripts are now necessary in order to distinguish the medium in which the wave is propagating. All waves are presumed to have the same horizontal slowness, and the four scattered waves have displacement amplitudes $\acute{P}_1$, $\acute{S}_1$, $\grave{P}_2$, $\grave{S}_2$. From continuity of u_x, u_z, τ_{zx}, τ_{zz}, we obtain the four equations

$$
\begin{aligned}
\sin i_1(\grave{P}_1 + \acute{P}_1) + \cos j_1(\grave{S}_1 + \acute{S}_1) &= \sin i_2(\grave{P}_2 + \acute{P}_2) + \cos j_2(\grave{S}_2 + \acute{S}_2), \\
\cos i_1(\grave{P}_1 - \acute{P}_1) - \sin j_1(\grave{S}_1 - \acute{S}_1) &= \cos i_2(\grave{P}_2 - \acute{P}_2) - \sin j_2(\grave{S}_2 - \acute{S}_2), \\
2\rho_1\beta_1^2 p \cos i_1(\grave{P}_1 - \acute{P}_1) + \rho_1\beta_1(1 - 2\beta_1^2p^2)(\grave{S}_1 - \acute{S}_1) & \\
= 2\rho_2\beta_2^2 p \cos i_2(\grave{P}_2 - \acute{P}_2) &+ \rho_2\beta_2(1 - 2\beta_2^2p^2)(\grave{S}_2 \quad \acute{S}_2), \\
\rho_1\alpha_1(1 - 2\beta_1^2p^2)(\grave{P}_1 + \acute{P}_1) - 2\rho_1\beta_1^2 p \cos j_1(\grave{S}_1 + \acute{S}_1) & \\
= \rho_2\alpha_2(1 - 2\beta_2^2p^2)(\grave{P}_2 + \acute{P}_2) &- 2\rho_2\beta_2^2 p \cos j_2(\grave{S}_2 + \acute{S}_2),
\end{aligned}
\tag{5.33}
$$

respectively. Rearranging these equations so that scattered waves are all on the left-hand side and incident waves all on the right, we find

$$
\mathbf{M}\begin{pmatrix} \acute{P}_1 \\ \acute{S}_1 \\ \grave{P}_2 \\ \grave{S}_2 \end{pmatrix} = \mathbf{N}\begin{pmatrix} \grave{P}_1 \\ \grave{S}_1 \\ \acute{P}_2 \\ \acute{S}_2 \end{pmatrix}, \tag{5.34}
$$

TABLE **5.3**

Vector displacements for the *P-SV* plane wave problems shown in Figure 5.8 (solid over solid).

Incident wave		Scattered waves	
Type	*Displacement*	*Type*	*Displacement*
Downgoing P	$S(\sin i_1, 0, \cos i_1) \exp\left[i\omega\left(px + \frac{\cos i_1}{\alpha_1} z - t\right)\right]$	Upgoing P	$S(\sin i_1, 0, -\cos i_1)\grave{P}\acute{P} \exp\left[i\omega\left(px - \frac{\cos i_1}{\alpha_1} z - t\right)\right]$
		Upgoing SV	$S(\cos j_1, 0, \sin j_1)\grave{P}\acute{S} \exp\left[i\omega\left(px - \frac{\cos j_1}{\beta_1} z - t\right)\right]$
		Downgoing P	$S(\sin i_2, 0, \cos i_2)\grave{P}\grave{P} \exp\left[i\omega\left(px + \frac{\cos i_2}{\alpha_2} z - t\right)\right]$
		Downgoing SV	$S(\cos j_2, 0, -\sin j_2)\grave{P}\grave{S} \exp\left[i\omega\left(px + \frac{\cos j_2}{\beta_2} z - t\right)\right]$
Downgoing SV	$S(\cos j_1, 0, -\sin j_1) \exp\left[i\omega\left(px + \frac{\cos j_1}{\beta_1} z - t\right)\right]$	Upgoing P	$S(\sin i_1, 0, -\cos i_1)\grave{S}\acute{P} \exp\left[i\omega\left(px - \frac{\cos i_1}{\alpha_1} z - t\right)\right]$
		Upgoing SV	$S(\cos j_1, 0, \sin j_1)\grave{S}\acute{S} \exp\left[i\omega\left(px - \frac{\cos j_1}{\beta_1} z - t\right)\right]$
		Downgoing P	$S(\sin i_2, 0, \cos i_2)\grave{S}\grave{P} \exp\left[i\omega\left(px + \frac{\cos i_2}{\alpha_2} z - t\right)\right]$
		Downgoing SV	$S(\cos j_2, 0, -\sin j_2)\grave{S}\grave{S} \exp\left[i\omega\left(px + \frac{\cos j_2}{\beta_2} z - t\right)\right]$

Upgoing P	$S(\sin i_2, 0, -\cos i_2) \exp\left[i\omega\left(px - \frac{\cos i_2}{\alpha_2} z - t\right)\right]$	Upgoing P	$S(\sin i_1, 0, -\cos i_1)\acute{P}\acute{P} \exp\left[i\omega\left(px - \frac{\cos i_1}{\alpha_1} z - t\right)\right]$
		Upgoing SV	$S(\cos j_1, 0, \sin j_1)\acute{P}\acute{S} \exp\left[i\omega\left(px - \frac{\cos j_1}{\beta_1} z - t\right)\right]$
		Downgoing P	$S(\sin i_2, 0, \cos i_2)\acute{P}\grave{P} \exp\left[i\omega\left(px + \frac{\cos i_2}{\alpha_2} z - t\right)\right]$
		Downgoing SV	$S(\cos j_2, 0, -\sin j_2)\acute{P}\grave{S} \exp\left[i\omega\left(px + \frac{\cos j_2}{\beta_2} z - t\right)\right]$
Upgoing SV	$S(\cos j_2, 0, \sin j_2) \exp\left[i\omega\left(px - \frac{\cos j_2}{\beta_2} z - t\right)\right]$	Upgoing P	$S(\sin i_1, 0, -\cos i_1)\acute{S}\acute{P} \exp\left[i\omega\left(px - \frac{\cos i_1}{\alpha_1} z - t\right)\right]$
		Upgoing SV	$S(\cos j_1, 0, \sin j_1)\acute{S}\acute{S} \exp\left[i\omega\left(px - \frac{\cos j_1}{\beta_1} z - t\right)\right]$
		Downgoing P	$S(\sin i_2, 0, \cos i_2)\acute{S}\grave{P} \exp\left[i\omega\left(px + \frac{\cos i_2}{\alpha_2} z - t\right)\right]$
		Downgoing SV	$S(\cos j_2, 0, -\sin j_2)\acute{S}\grave{S} \exp\left[i\omega\left(px + \frac{\cos j_2}{\beta_2} z - t\right)\right]$

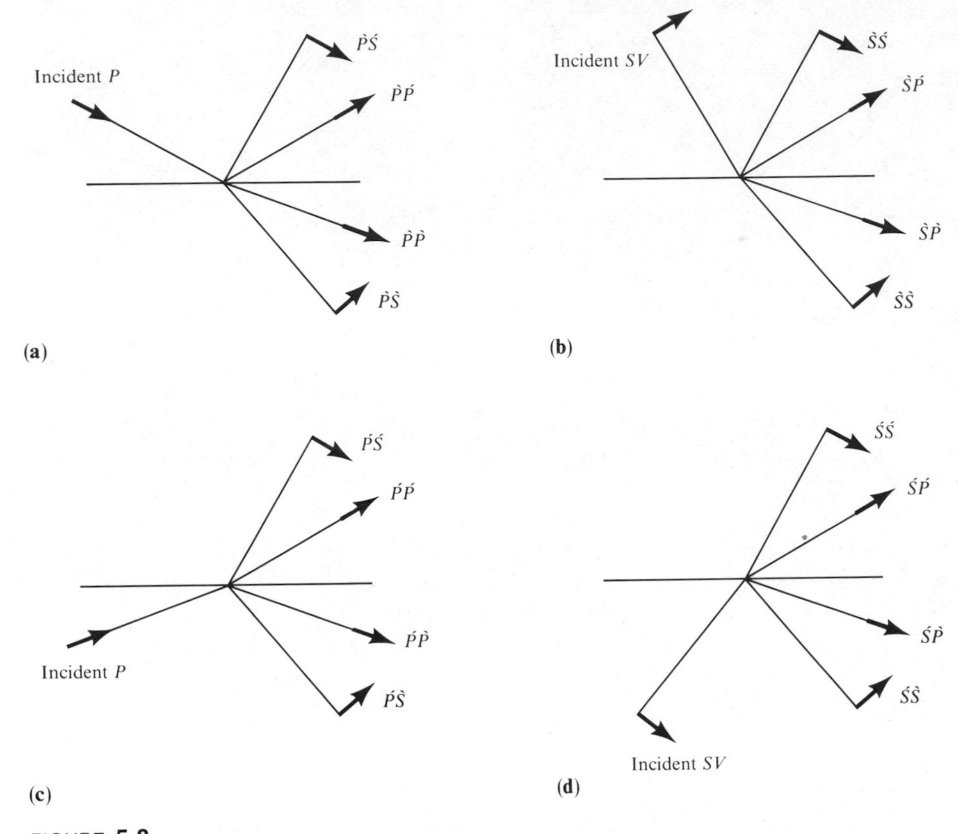

FIGURE **5.8**

Notation for the sixteen possible reflection/transmission coefficients arising for problems of *P-SV* waves at the welded interface between two different solid half-spaces.

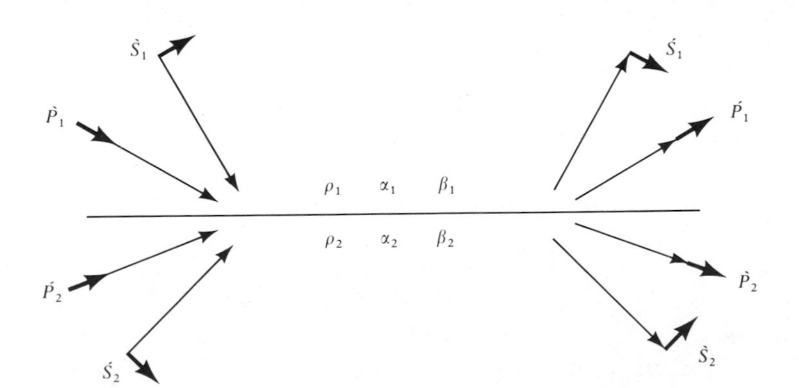

FIGURE **5.9**

The complete system of incident and scattered plane *P-SV* waves, in terms of which the scattering matrix can quickly be found. Short arrows show the direction of particle motion; long arrows show the direction of propagation.

in which the coefficient matrices are

$$\mathbf{M} = \begin{pmatrix} -\alpha_1 p & -\cos j_1 & \alpha_2 p & \cos j_2 \\ \cos i_1 & -\beta_1 p & \cos i_2 & -\beta_2 p \\ 2\rho_1\beta_1^2 p \cos i_1 & \rho_1\beta_1(1 - 2\beta_1^2 p^2) & 2\rho_2\beta_2^2 p \cos i_2 & \rho_2\beta_2(1 - 2\beta_2^2 p^2) \\ -\rho_1\alpha_1(1 - 2\beta_1^2 p^2) & 2\rho_1\beta_1^2 p \cos j_1 & \rho_2\alpha_2(1 - 2\beta_2^2 p^2) & -2\rho_2\beta_2^2 p \cos j_2 \end{pmatrix}, \tag{5.35}$$

$$\mathbf{N} = \begin{pmatrix} \alpha_1 p & \cos j_1 & -\alpha_2 p & -\cos j_2 \\ \cos i_1 & -\beta_1 p & \cos i_2 & -\beta_2 p \\ 2\rho_1\beta_1^2 p \cos i_1 & \rho_1\beta_1(1 - 2\beta_1^2 p^2) & 2\rho_2\beta_2^2 p \cos i_2 & \rho_2\beta_2(1 - 2\beta_2^2 p^2) \\ \rho_1\alpha_1(1 - 2\beta_1^2 p^2) & -2\rho_1\beta_1^2 p \cos j_1 & -\rho_2\alpha_2(1 - 2\beta_2^2 p^2) & 2\rho_2\beta_2^2 p \cos j_2 \end{pmatrix}. \tag{5.36}$$

In the case that $\grave{P}_1 = 1$ and $\grave{S}_1 = \acute{P}_2 = \acute{S}_2 = 0$, the first column of the scattering matrix becomes simply $(\acute{P}_1, \acute{S}_1, \grave{P}_2, \grave{S}_2)^T$, i.e., the first column of $\mathbf{M}^{-1}\mathbf{N}$. There are similar results for the other three columns, and it follows that the complete scattering matrix is given directly by

$$\begin{pmatrix} \grave{P}\acute{P} & \grave{S}\acute{P} & \acute{P}\acute{P} & \acute{S}\acute{P} \\ \grave{P}\acute{S} & \grave{S}\acute{S} & \acute{P}\acute{S} & \acute{S}\acute{S} \\ \grave{P}\grave{P} & \grave{S}\grave{P} & \acute{P}\grave{P} & \acute{S}\grave{P} \\ \grave{P}\grave{S} & \grave{S}\grave{S} & \acute{P}\grave{S} & \acute{S}\grave{S} \end{pmatrix} = \mathbf{M}^{-1}\mathbf{N}. \tag{5.37}$$

Fortunately, the many similarities between matrixes $\mathbf{M}$ and $\mathbf{N}$ lead to quite simple formulas for each component of the scattering matrix. In detail, these formulas make repeated use of the variables

$$a = \rho_2(1 - 2\beta_2^2 p^2) - \rho_1(1 - 2\beta_1^2 p^2), \qquad b = \rho_2(1 - 2\beta_2^2 p^2) + 2\rho_1\beta_1^2 p^2,$$

$$c = \rho_1(1 - 2\beta_1^2 p^2) + 2\rho_2\beta_2^2 p^2, \qquad d = 2(\rho_2\beta_2^2 - \rho_1\beta_1^2),$$

and repeated use also of the cosine-dependent terms

$$E = b\frac{\cos i_1}{\alpha_1} + c\frac{\cos i_2}{\alpha_2}, \qquad F = b\frac{\cos j_1}{\beta_1} + c\frac{\cos j_2}{\beta_2},$$

$$G = a - d\frac{\cos i_1}{\alpha_1}\frac{\cos j_2}{\beta_2}, \qquad H = a - d\frac{\cos i_2}{\alpha_2}\frac{\cos j_1}{\beta_1},$$

$$\mathrm{D} = EF + GHp^2 = (\det \mathbf{M})/(\alpha_1\alpha_2\beta_1\beta_2). \tag{5.38}$$

The main formulas are

$$\grave{P}\acute{P} = \left[\left(b\frac{\cos i_1}{\alpha_1} - c\frac{\cos i_2}{\alpha_2}\right)F - \left(a + d\frac{\cos i_1}{\alpha_1}\frac{\cos j_2}{\beta_2}\right)Hp^2\right]\Big/\mathsf{D},$$

$$\grave{P}\acute{S} = -2\frac{\cos i_1}{\alpha_1}\left(ab + cd\frac{\cos i_2}{\alpha_2}\frac{\cos j_2}{\beta_2}\right)p\alpha_1/(\beta_1\mathsf{D}),$$

$$\grave{P}\grave{P} = 2\rho_1\frac{\cos i_1}{\alpha_1}F\alpha_1/(\alpha_2\mathsf{D}),$$

$$\grave{P}\grave{S} = 2\rho_1\frac{\cos i_1}{\alpha_1}Hp\alpha_1/(\beta_2\mathsf{D}),$$

$$\grave{S}\acute{P} = -2\frac{\cos j_1}{\beta_1}\left(ab + cd\frac{\cos i_2}{\alpha_2}\frac{\cos j_2}{\beta_2}\right)p\beta_1/(\alpha_1\mathsf{D}),$$

$$\grave{S}\acute{S} = -\left[\left(b\frac{\cos j_1}{\beta_1} - c\frac{\cos j_2}{\beta_2}\right)E - \left(a + d\frac{\cos i_2}{\alpha_2}\frac{\cos j_1}{\beta_1}\right)Gp^2\right]\Big/\mathsf{D},$$

$$\grave{S}\grave{P} = -2\rho_1\frac{\cos j_1}{\beta_1}Gp\beta_1/(\alpha_2\mathsf{D}),$$

$$\grave{S}\grave{S} = 2\rho_1\frac{\cos j_1}{\beta_1}E\beta_1/(\beta_2\mathsf{D}),$$

$$\acute{P}\acute{P} = 2\rho_2\frac{\cos i_2}{\alpha_2}F\alpha_2/(\alpha_1\mathsf{D}),$$

$$\acute{P}\acute{S} = -2\rho_2\frac{\cos i_2}{\alpha_2}Gp\alpha_2/(\beta_1\mathsf{D}),$$

$$\acute{P}\grave{P} = -\left[\left(b\frac{\cos i_1}{\alpha_1} - c\frac{\cos i_2}{\alpha_2}\right)F + \left(a + d\frac{\cos i_2}{\alpha_2}\frac{\cos j_1}{\beta_1}\right)Gp^2\right]\Big/\mathsf{D},$$

$$\acute{P}\grave{S} = 2\frac{\cos i_2}{\alpha_2}\left(ac + bd\frac{\cos i_1}{\alpha_1}\frac{\cos j_1}{\beta_1}\right)p\alpha_2/(\beta_2\mathsf{D}),$$

$$\acute{S}\acute{P} = 2\rho_2\frac{\cos j_2}{\beta_2}Hp\beta_2/(\alpha_1\mathsf{D}),$$

$$\acute{S}\acute{S} = 2\rho_2\frac{\cos j_2}{\beta_2}E\beta_2/(\beta_1\mathsf{D}),$$

$$\acute{S}\grave{P} = 2\frac{\cos j_2}{\beta_2}\left(ac + bd\frac{\cos i_1}{\alpha_1}\frac{\cos j_1}{\beta_1}\right)p\beta_2/(\alpha_2 \mathrm{D}),$$

$$\acute{S}\grave{S} = \left[\left(b\frac{\cos j_1}{\beta_1} - c\frac{\cos j_2}{\beta_2}\right)E + \left(a + d\frac{\cos i_1}{\alpha_1}\frac{\cos j_2}{\beta_2}\right)Hp^2\right]\Big/\mathrm{D}. \quad (5.39)$$

5.2.5 *Energy flux*

For a plane wave incident on the boundary between two homogeneous half-spaces, there is no possibility of trapping energy at the interface. Hence the flux of energy leaving the boundary must equal that in the incident wave.

For a steady-state displacement P-wave having amplitude S and propagation factor $\exp[i\omega(\mathbf{s}\cdot\mathbf{x} - t)]$, the actual (real) displacement amplitude is $S\cos[\omega(\mathbf{s}\cdot\mathbf{x} - t)]$. Then $\rho\alpha S^2\omega^2 \sin^2[\omega(\mathbf{s}\cdot\mathbf{x} - t)]$ is the flux rate of energy transmission across unit area of wavefront (see Section 5.1). One must multiply this result by $\cos i$, to obtain the flux rate across unit area of a horizontal boundary upon which the wave is incident at angle i (since only $\cos i$ of wavefront area is involved). Similarly, for S-waves the flux rate is $\rho\beta\cos j S^2\omega^2\sin^2[\omega(\mathbf{s}\cdot\mathbf{x} - t)]$. Since the reflection/transmission coefficients we have found above were for displacements, it follows that the equality of incoming and outgoing energy flux will give, for example,

$$\begin{aligned}\rho_1\alpha_1\cos i_1 = {} & \rho_1\alpha_1\cos i_1(\grave{P}\acute{P})^2 + \rho_1\beta_1\cos j_1(\grave{P}\acute{S})^2 \\ & + \rho_2\alpha_2\cos i_2(\grave{P}\grave{P})^2 + \rho_2\beta_2\cos j_2(\grave{P}\grave{S})^2 \end{aligned} \quad (5.40)$$

for the scattered wave system shown in Figure 5.8a.

Equation (5.40) is a constraint on the first column of the scattering matrix for a solid/solid interface. It can be simplified by working with new dependent variables: namely, displacement × (density × propagation velocity × cosine of angle of incidence)$^{1/2}$. In terms of these scaled displacements (which are proportional to the square root of energy flux), the scattering matrix $\mathbf{S}$ is a unitary Hermitian matrix. This property can be shown from equations given by Frasier (1970), and has been extensively studied by Kennett et al. (1978).

For example, the new reflection coefficient corresponding to the previous $\acute{P}\grave{S}$ is

$$\frac{\text{displacement amplitude of downgoing } SV\text{-wave} \times (\rho_2\beta_2\cos j_2)^{1/2}}{\text{displacement amplitude of incident upgoing } P\text{-wave} \times (\rho_2\alpha_2\cos i_2)^{1/2}}$$

(see Figure 5.8c), which is $\acute{P}\grave{S}(\beta_2\cos j_2)^{1/2}/(\alpha_2\cos i_2)^{1/2}$. The complete form of this scattering matrix is

$$\mathbf{S} = \begin{pmatrix} \grave{P}\acute{P} & \grave{S}\acute{P}\left(\dfrac{\alpha_1 \cos i_1}{\beta_1 \cos j_1}\right)^{1/2} & \acute{P}\acute{P}\left(\dfrac{\rho_1\alpha_1 \cos i_1}{\rho_2\alpha_2 \cos i_2}\right)^{1/2} & \acute{S}\acute{P}\left(\dfrac{\rho_1\alpha_1 \cos i_1}{\rho_2\beta_2 \cos j_2}\right)^{1/2} \\ \grave{P}\acute{S}\left(\dfrac{\beta_1 \cos j_1}{\alpha_1 \cos i_1}\right)^{1/2} & \grave{S}\acute{S} & \acute{P}\acute{S}\left(\dfrac{\rho_1\beta_1 \cos j_1}{\rho_2\alpha_2 \cos i_2}\right)^{1/2} & \acute{S}\acute{S}\left(\dfrac{\rho_1\beta_1 \cos j_1}{\rho_2\beta_2 \cos j_2}\right)^{1/2} \\ \grave{P}\grave{P}\left(\dfrac{\rho_2\alpha_2 \cos i_2}{\rho_1\alpha_1 \cos i_1}\right)^{1/2} & \grave{S}\grave{P}\left(\dfrac{\rho_2\alpha_2 \cos i_2}{\rho_1\beta_1 \cos j_1}\right)^{1/2} & \acute{P}\grave{P} & \acute{S}\grave{P}\left(\dfrac{\alpha_2 \cos i_2}{\beta_2 \cos j_2}\right)^{1/2} \\ \grave{P}\grave{S}\left(\dfrac{\rho_2\beta_2 \cos j_2}{\rho_1\alpha_1 \cos i_1}\right)^{1/2} & \grave{S}\grave{S}\left(\dfrac{\rho_2\beta_2 \cos j_2}{\rho_1\beta_1 \cos j_1}\right)^{1/2} & \acute{P}\grave{S}\left(\dfrac{\beta_2 \cos j_2}{\alpha_2 \cos i_2}\right)^{1/2} & \acute{S}\grave{S} \end{pmatrix} \tag{5.41}$$

and

$$\mathbf{S} = \mathbf{S}^H = \mathbf{S}^{-1}. \tag{5.42}$$

(By $\mathbf{S}^H$ we mean the complex conjugate of the transpose of $\mathbf{S}$.)

Once the 16 coefficients $\grave{P}\acute{P}$, $\grave{P}\acute{S}$, etc. have been obtained from (5.37) or (5.39), they can be checked by verifying that $\mathbf{S}$ satisfies (5.42). In fact, provided that the slowness is low enough for all the angles i_1, j_1, i_2, j_2 to be real, the entries of $\mathbf{S}$ (as we have defined them) are all real, so that the transpose of $\mathbf{S}$ is also the inverse of $\mathbf{S}$. As a specific example, we find that

$$\mathbf{S} = \begin{pmatrix} 0.1065 & -0.1766 & 0.9701 & -0.1277 \\ -0.1766 & -0.0807 & 0.1326 & 0.9720 \\ 0.9701 & 0.1326 & -0.0567 & 0.1950 \\ -0.1277 & 0.9720 & 0.1950 & 0.0309 \end{pmatrix}$$

in the case $(\rho_1, \alpha_1, \beta_1, \rho_2, \alpha_2, \beta_2, p) = (3, 6, 3.5, 4, 7, 4.2, 0.1)$, and this particular matrix does indeed have the properties $\mathbf{S} = \mathbf{S}^T = \mathbf{S}^{-1}$.

A matrix $\mathbf{S}$ with complex entries would be obtained if, in forming the 16 coefficients, reference levels $z_1 < 0$ in the upper half-space and $z_2 > 0$ in the lower half-space were used. Extra phase factors must then be introduced to account for the shift in vertical reference. In this case, it is the complex conjugate transpose of $\mathbf{S}$ (i.e., $\mathbf{S}^H$), which is also the inverse of $\mathbf{S}$. Finally, we remark that if there is some more complicated transition zone in the range $z_1 < z < z_2$ (e.g., a continuous variation of elastic properties or a stack of welded homogeneous layers having different elastic constants), but with the regions above z_1 and below z_2 still homogeneous, then a scattering matrix $\mathbf{S}$ can still be defined for the whole transition zone, and $\mathbf{S}$ is still Hermitian and unitary. These properties are a consequence of energy conservation, reciprocity, and causality.

5.2.6 *A useful approximation for reflection/transmission coefficients between two similar half-spaces*

If the two half-spaces under consideration have similar properties, then one may expect that transmission coefficients will be large for waves that retain the same mode of propagation (e.g., $\grave{P}\grave{P}$, in which the mode of both the incident and the transmitted wave is downgoing P), but all other types of scattering coefficient will be small. Thus, if there is a jump in properties amounting to $\Delta\rho = \rho_2 - \rho_1$, $\Delta\alpha = \alpha_2 - \alpha_1, \Delta\beta = \beta_2 - \beta_1$, and the ratios $\Delta\rho/\rho, \Delta\alpha/\alpha, \Delta\beta/\beta$ have magnitudes much less than one (where ρ, α, β are the mean values of density and velocities for the two half-spaces), we may expect that quantities such as $\grave{S}\acute{P}$, $\acute{P}\grave{S}$ will be small, but that transmissions $\grave{P}\grave{P}$ and $\acute{S}\acute{S}$ will be of order one. We shall here derive the first-order effect of small jumps in density and velocities for the P-SV problem of two solid half-spaces, since the resulting approximate formulas (5.44) are often remarkably accurate. They give some insight into the separate contributions made by $\Delta\rho$, $\Delta\alpha$, $\Delta\beta$, and Chapman (1976a) has shown that these formulas are important in the analysis of waves in inhomogeneous media.

We shall assume that all angles i_1, i_2, j_1, j_2 are real and that none of these angles is near 90°. Then from Snell's law relating i_1 and i_2, j_1 and j_2, it follows that

$$\Delta i = i_2 - i_1 = \tan i(\Delta\alpha/\alpha), \qquad \Delta j = j_2 - j_1 = \tan j(\Delta\beta/\beta) \tag{5.43}$$

(correct to first order in the velocity jumps). After expanding the terms defined in (5.38), correct to first order in the jumps $\Delta\rho$, $\Delta\alpha$, $\Delta\beta$, we can substitute into (5.39), finding that

$$\begin{aligned}
\grave{P}\acute{P} &= \tfrac{1}{2}(1 - 4\beta^2p^2)\frac{\Delta\rho}{\rho} + \frac{1}{2\cos^2 i}\frac{\Delta\alpha}{\alpha} - 4\beta^2p^2\frac{\Delta\beta}{\beta},\\
\grave{P}\acute{S} &= \frac{-p\alpha}{2\cos j}\left[\left(1 - 2\beta^2p^2 + 2\beta^2\frac{\cos i}{\alpha}\frac{\cos j}{\beta}\right)\frac{\Delta\rho}{\rho}\right.\\
&\qquad \left. - \left(4\beta^2p^2 - 4\beta^2\frac{\cos i}{\alpha}\frac{\cos j}{\beta}\right)\frac{\Delta\beta}{\beta}\right],\\
\grave{P}\grave{P} &= 1 - \tfrac{1}{2}\frac{\Delta\rho}{\rho} + \left(\frac{1}{2\cos^2 i} - 1\right)\frac{\Delta\alpha}{\alpha},\\
\grave{P}\grave{S} &= \frac{p\alpha}{2\cos j}\left[\left(1 - 2\beta^2p^2 - 2\beta^2\frac{\cos i}{\alpha}\frac{\cos j}{\beta}\right)\frac{\Delta\rho}{\rho}\right.\\
&\qquad \left. - \left(4\beta^2p^2 + 4\beta^2\frac{\cos i}{\alpha}\frac{\cos j}{\beta}\right)\frac{\Delta\beta}{\beta}\right],
\end{aligned} \tag{5.44}$$

$$\grave{S}\acute{P} = \frac{\cos j}{\alpha} \frac{\beta}{\cos i} \grave{P}\acute{S},$$

$$\grave{S}\acute{S} = -\tfrac{1}{2}(1 - 4\beta^2 p^2)\frac{\Delta\rho}{\rho} - \left(\frac{1}{2\cos^2 j} - 4\beta^2 p^2\right)\frac{\Delta\beta}{\beta},$$

$$\grave{S}\grave{P} = \frac{-\cos j}{\alpha} \frac{\beta}{\cos i} \grave{P}\grave{S},$$

$$\grave{S}\grave{S} = 1 - \tfrac{1}{2}\frac{\Delta\rho}{\rho} + \left(\frac{1}{2\cos^2 j} - 1\right)\frac{\Delta\beta}{\beta}.$$

The remaining eight coefficients are easily deduced from the eight given above, merely by making a change in sign in the jumps $\Delta\rho$, $\Delta\alpha$, $\Delta\beta$.

There is a tendency for the coefficients of $\Delta\beta/\beta$ in (5.44) to be larger than the coefficients of $\Delta\rho/\rho$ and $\Delta\alpha/\alpha$, implying that fluctuations in shear velocity are relatively efficient in scattering elastic waves. The approximate formulas (5.44) will fail if angles i (or j) are near 90°, since then only a small jump in velocity can lead to a large change in i (or j). It can even happen that the wave undergoes total internal reflection. Uses of the first-order approximations (5.44) are given by Bortfeld (1961) and by Richards and Frasier (1976). Chapman (1976a), in a major study of the waves set up by a point source in media with depth-dependent properties, has shown how to handle the singularities in (5.44) at $i = 90°$ and $j = 90°$.

5.2.7 *Frequency independence of plane-wave reflection/transmission coefficients*

So far in this section, we have analyzed only steady-state plane waves with frequency ω (see Tables 5.1–5.3), so that reflection and transmission coefficients have been obtained in the frequency domain. But in formulas for the coefficients (e.g., (5.26), (5.32), (5.39)), there does not appear to be any frequency dependence. We phrase this cautiously, using the word "appear," because in Section 5.3 we shall find that there can be a type of frequency dependence (involving sign-dependent phase shifts) in the coefficients we have obtained. However, the coefficients *are* frequency independent when the angles of incidence (referred to above variously as i, j, j_1, j_2, i_1, i_2) are all real. In this case, p^{-1}, the horizontal phase velocity, is greater than propagation speeds α and β of P- and S-waves. The reflection/transmission coefficients are then all real, and the fact that they are independent of frequency means that waves leaving the boundary must have the same pulse shape (but with different amplitudes) as that of the incident wave. This is a very special property, and it is not true in general if the boundary is not plane or if the waves are not plane waves. One of the important phenomena we shall describe later is the way in which the pulse shape of a reflected or transmitted wave can differ from that of an incident wave when the incident wavefront is curved (Chapter 6) or if the boundary is slightly curved (Chapter 9).

5.3 Inhomogeneous Waves, Phase Shifts, and Interface Waves

For the plane waves we have described so far, it has been implied that the direction $\mathbf{l}$ of propagation is a vector (parallel to slowness $\mathbf{s}$) that has real Cartesian components. In this section, we investigate the consequences of allowing the components of $\mathbf{l}$ (or $\mathbf{s}$) in the depth direction to be imaginary. We shall find that waves of this type are possible, and that their amplitude grows or decays exponentially with depth. They are examples of *inhomogeneous waves* (formally defined on p. 183, following (5.92)), and again we shall use their horizontal slowness p as an independent variable to study their properties.

Previously, we found for P-waves that $\mathbf{s} = (\alpha^{-1} \sin i, 0, \pm\alpha^{-1} \cos i) = (p, 0, \pm\sqrt{1/\alpha^2 - p^2})$. If the depth component of $\mathbf{s}$ is imaginary, then $p > 1/\alpha$. That is, the horizontal phase velocity, p^{-1}, is less than the propagation speed of P-waves. The quantity we have previously called "$\sin i$" is still real but is now greater than one, so that angle i is no longer real. If p is even greater, so that $p > 1/\beta > 1/\alpha$, then the associated S-wave is also an inhomogeneous wave, its slowness $\mathbf{s}$ being $(p, 0, \pm\sqrt{1/\beta^2 - p^2})$. Previously, we studied steady-state plane waves with phase factor $\exp[i\omega(\mathbf{s} \cdot \mathbf{x} - t)]$, it being understood that the actual fluctuation of the wave, in space and time, is given by the real part of this phase factor, i.e., $\cos \omega(\mathbf{s} \cdot \mathbf{x} - t)$. Now, however, with imaginary s_z, the actual fluctuation of the wave is expressed by a factor like $\exp(i\omega s_z z) \cos \omega(px - t)$, in which the real quantity $\exp(i\omega s_z z)$ gives exponential growth or decay with depth, according as ωs_z is negative or positive imaginary. Previously, we found that the requirement of satisfying a wave equation reduced to the requirement that $s^2 = 1/c^2$, where $c = \alpha = \sqrt{(\lambda + 2\mu)/\rho}$ for P-waves and $c = \beta = \sqrt{\mu/\rho}$ for S. This requirement still holds for inhomogeneous waves, and since now $p = s_x > 1/c$ and $s^2 = s_x^2 + s_z^2 = 1/c^2$, we see that the wave equation provides the fundamental reason for having an imaginary vertical slowness component when the horizontal component of slowness is sufficiently large.

As an example of the need for considering inhomogeneous waves in seismology, consider again the problem shown in Figure 5.5b of a plane SV-wave incident upon a free surface. We obtained formulas in (5.30) and (5.31) for conversion coefficient $\acute{S}\grave{P}$ and reflection coefficient $\acute{S}\grave{S}$ and showed values of these in Figure 5.6 as functions of p in the range $0 \le p \le 1/\alpha$. What happens when $1/\alpha < p$? If p becomes so great that $1/\beta < p$, then even the SV-wave is inhomogeneous, and it is not clear what we mean by an "incident wave." But in the range $1/\alpha < p < 1/\beta$, the SV-wave does have a real angle of incidence $j = \sin^{-1}(\beta p)$. In this range, we must regard $\cos j$ as real and positive, but

$(\cos i)/\alpha = \sqrt{1/\alpha^2 - p^2}$ is pure imaginary. Since the entry in Table 5.1 involving $\acute{S}\grave{P}$ gives the P-wave displacement as

$$S(\sin i, 0, \cos i)\acute{S}\grave{P} \exp\left(i\omega \frac{\cos i}{\alpha} z\right) \exp\left[i\omega\left(\frac{\sin i}{\alpha} - t\right)\right], \tag{5.45}$$

and since we cannot allow the P-wave to grow exponentially with depth, it follows that

$$\frac{\cos i}{\alpha} = \pm i \sqrt{p^2 - \frac{1}{\alpha^2}}, \qquad \text{according as } \omega \gtrless 0. \tag{5.46}$$

Thus, for positive frequencies, we find that (5.45) becomes

$$S(\alpha p, 0, i\sqrt{\alpha^2 p^2 - 1})\acute{S}\grave{P} \exp\left(-\omega \sqrt{p^2 - \frac{1}{\alpha^2}} z\right) \exp[i\omega(px - t)]. \tag{5.47}$$

BOX **5.5**

Phase shifts: phase delay and phase advance

Phase shifts in a propagating wave are due to a variety of mechanisms. Thus, in the phase factor $\exp[i\omega(\mathbf{s} \cdot \mathbf{x} - t)]$, we can define a phase shift due to propagation as $\phi(\omega) = \omega\mathbf{s} \cdot \mathbf{x}$. Such a shift is termed a *phase delay*, since it is the equivalent of evaluating the basic time variation $\exp(-i\omega t)$ at a time delayed by $\mathbf{s} \cdot \mathbf{x}$, i.e., as $\exp[-i\omega(t - \mathbf{s} \cdot \mathbf{x})]$.

Another example of a phase shift occurs in equation (5.47), showing that the downward vertical component of motion in an inhomogeneous P-wave has phase $\pi/2$ greater than the horizontal component in the direction of propagation. This too is a phase delay: the maximum downward displacement is delayed a time $\pi/(2\omega)$ with respect to the maximum horizontal component (in the direction of propagation). This is a time delay by a quarter of a period, leading to prograde particle motion (see Problem 5.4).

In contrast, the phase shift occurring in $\acute{S}\grave{S}$, due to an SV-wave incident on a free surface at angle j greater than $\sin^{-1}(\beta/\alpha)$, is a *phase advance*. This can be seen from (5.31) with $\cos i$ having a positive imaginary part, so that the phase of $\acute{S}\grave{S}$ is negative (see also Fig. 5.10), corresponding to an evaluation of $\exp(-i\omega t)$ at an *earlier* time.

Our overall point here is that the sign of a phase shift can depend on the sign of frequency (see (5.46)) and on our convention of signs in the Fourier time transform, but the designations "phase delay" and "phase advance" are more fundamental, often giving a better indication of the physical significance of the phenomenon causing the phase shift. Thus, in Section 5.5, we shall discuss the phase delay due to attenuation, and in Chapter 9 we shall find that a phase advance occurs when a body wave touches a caustic.

The steps taken in (5.28)–(5.31) to derive $\acute{S}\grave{P}$ and $\acute{S}\grave{S}$ are unchanged, provided we interpret "cos i" as an imaginary quantity given by (5.46), so that now the coefficients $\acute{S}\grave{P}$ and $\acute{S}\grave{S}$ are no longer real. Consequently, when the real part is taken of the vector components in (5.47) to obtain the physical displacement, we find there is a phase shift of amount ϕ, where ϕ = phase ($\acute{S}\grave{P}$) for the horizontal component and $\phi = \pi/2$ + phase ($\acute{S}\grave{P}$) for the vertical. At fixed P, the phase shift is the same for all positive frequencies, and it is easy to see that the shift is in the opposite direction for all negative frequencies.

Although the derived P-wave (5.47) is an inhomogeneous wave, the reflected S-wave does not decay with depth. From Table 5.1, with cos j real and equal to $\sqrt{1 - \beta^2 p^2}$, the reflected S-wave displacement is

$$S(\sqrt{1 - \beta^2 p^2}, 0, -\beta p)\acute{S}\grave{S} \exp\left[i\omega\left(px + \sqrt{\frac{1}{\beta^2} - p^2}z - t\right)\right]. \tag{5.48}$$

But because of the phase shift in $\acute{S}\grave{S}$, it is no longer true that the pulse shape of the reflected S-wave is the same as that of the incident wave. From the argument in Box 5.6, the reflected pulse shape is a linear combination of the incident shape and its Hilbert transform.

BOX **5.6**

The Hilbert transform and the frequency-independent phase advance

If the Fourier components of a function $f = f(t)$ are all advanced in phase by $\pi/2$, then the resulting phase-distorted function in the time domain is

$$\frac{1}{2\pi}\int_{-\infty}^{\infty} f(\omega) \exp\left[-i\omega\left(t + \left|\frac{\pi}{2\omega}\right|\right)\right] d\omega$$

$$= \frac{1}{2\pi}\int_{-\infty}^{\infty} f(\omega) \exp\left(-i\frac{\pi}{2}\operatorname{sgn}(\omega)\right) \exp(-i\omega t)\, d\omega, \tag{1}$$

where $\operatorname{sgn}(\omega) = \pm 1$ according as $\omega \gtrless 1$. Substituting $f(\omega) = \int_{-\infty}^{\infty} f(\tau) \exp(i\omega\tau)\, d\tau$ in (1), we obtain the equivalent formula

$$\frac{1}{\pi}\int_0^{\infty} d\omega \int_{-\infty}^{\infty} f(\tau) \sin[\omega(\tau - t)]\, d\tau, \tag{2}$$

which Titchmarsh (1926) and Jeffreys and Jeffreys (1972) have called the *allied function* of $f(t)$. Integrating over ω in (2), one finds also the form

$$\frac{1}{\pi}\int_{-\infty}^{\infty} \frac{f(\tau)}{\tau - t}\, d\tau, \tag{3}$$

in which the singularity at $\tau = t$ is handled by taking the principal value of the integral, i.e., by cancelling out the contributions from τ just greater and just less than t.

The form (3) is one common definition of the Hilbert transform of $f(t)$, which we symbolize by $\mathscr{H}[f(t)]$. It can also be seen as a convolution (denoted by $*$), so that the final equivalent form for the distorted signal is

$$f(t) * (-1/\pi t). \tag{4}$$

We shall loosely refer to any one of the versions (1)–(4) as the Hilbert transform of $f(t)$. In practice, when this transform is to be computed, the original form (1) is most straightforward: one Fourier transform gives $f(\omega)$, and the $\pi/2$ phase advance reduces to an interchange of real and imaginary parts of $f(\omega)$ (with a sign change in the resulting imaginary part). An inverse Fourier transform then returns the required $\mathscr{H}[f(t)]$.

Note that

$$\mathscr{H}[f(t)] = \frac{1}{\pi}\int_0^\infty d\omega \int_{-\infty}^{\infty} f(\tau)\sin[\omega(\tau - t)]\, d\tau,$$

whereas

$$f(t) = \frac{1}{\pi}\int_0^\infty d\omega \int_{-\infty}^{\infty} f(\tau)\cos[\omega(\tau - t)]\, d\tau.$$

If the Hilbert transform pair, $f(t)$ and $\mathscr{H}[f(t)]$, is itself Hilbert-transformed, the resulting pair is $\mathscr{H}[f(t)]$ and $-f(t)$. This polarity reversal is simply a result of two $\pi/2$ phase advances.

Arons and Yennie (1950) have pointed out that if a function $f(t)$ undergoes a phase advance ε, then the resulting function may be calculated from a linear combination of $f(t)$ and its Hilbert transform. The large class of seismological examples of such phase shifts includes all plane waves that are supercritically reflected or transmitted at a discontinuity such as the Earth's free surface, ocean bottom, crust-mantle boundary, or core-mantle boundary. Choy and Richards (1975) have pointed out several examples in seismograms of SH-waves and SV-waves. Constants in the linear combination may be derived as follows:

$$\begin{aligned}
&\frac{1}{2\pi}\int_{-\infty}^{\infty} f(\omega)\exp[-i\varepsilon\,\mathrm{sgn}(\omega)]\exp(-i\omega t)\, d\omega \\
&\quad= \frac{1}{2\pi}\int_{-\infty}^{\infty} f(\omega)[\cos\varepsilon - i\,\mathrm{sgn}(\omega)\sin\varepsilon]\exp(-i\omega t)\, d\omega \\
&\quad= \cos\varepsilon f(t) + \sin\varepsilon\mathscr{H}[f(t)].
\end{aligned} \tag{5}$$

The most important Hilbert transform is that of the Dirac delta function $\delta(t)$, which is $(-1/\pi t)$. A box function that is unity for $0 < t < T$ and zero elsewhere has the Hilbert transform $(-1/\pi)(\ln|t| - \ln|t - T|)$. If $t \ll T$, the box function approaches the Heaviside function $H(t)$, and its Hilbert transform approaches $(-1/\pi)\ln|t/T|$.

Note that a Hilbert-transformed function has the same Fourier amplitude spectrum as the original function.

In Figure 5.10, we show $\acute{S}\grave{P}$ and $\acute{S}\grave{S}$ as functions of p in the range $0 \leq p \leq 1/\beta$, giving both amplitude and phase.

As another example of the need for using inhomogeneous waves, consider the reflection and transmission of SH-waves, as described in Figure 5.7 and equations (5.32). Inhomogeneous waves will be present in the upper medium if $1/\beta_1 < p$, and then this wave attenuates away from the interface, provided that we again choose

$$\frac{\cos j_1}{\beta_1} = +i\sqrt{p^2 - \frac{1}{\beta_1^2}} \quad \text{(in the case } \omega > 0\text{)}, \tag{5.49}$$

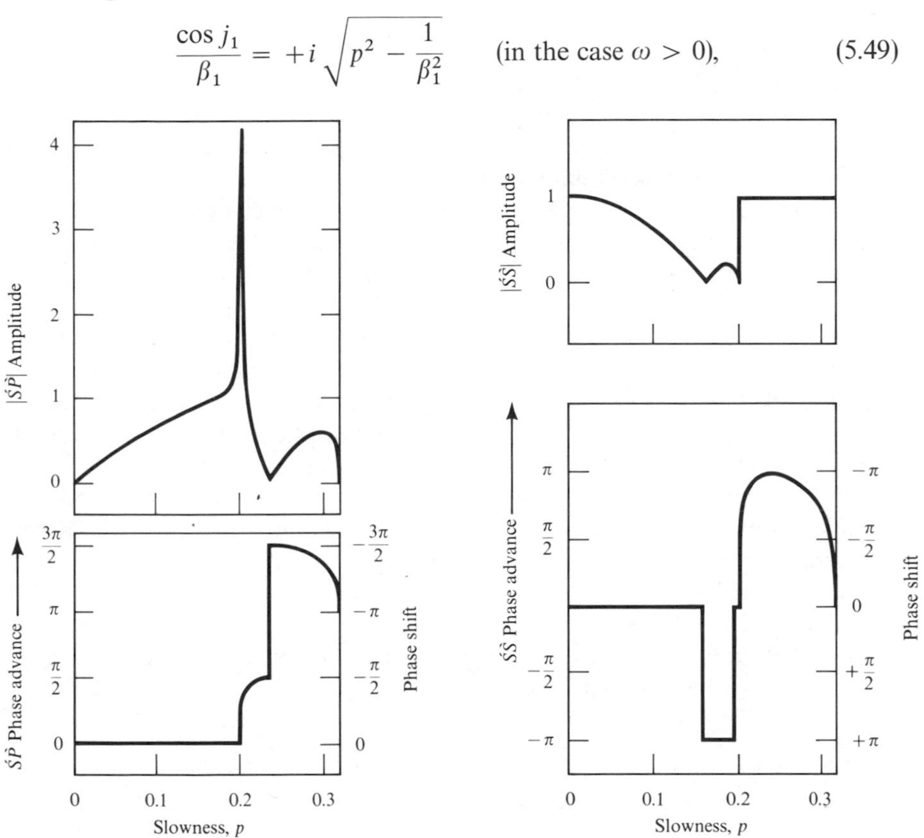

FIGURE **5.10**

The amplitude and phase of two reflection coefficients are shown as a function of horizontal slowness p. The coefficients are $\acute{S}\grave{P}$ and $\acute{S}\grave{S}$ for an SV-wave incident on the free surface, and we have taken $\alpha = 5$ km/sec, $\beta = 3$ km/sec, so that these coefficients have already been shown in Figure 5.6 for the range $0 \leq p \leq 1/\alpha$. Here we extend the range to $0 \leq p \leq 1/\beta$, so that an inhomogeneous P-wave is present for the range $1/\alpha < p \leq 1/\beta$. We have chosen to emphasize phase advance rather than phase shift, since the former is a quantity independent of the sign of frequency and independent of our Fourier sign convention. The phases actually plotted are those of $\acute{S}\grave{P}$ and $\acute{S}\grave{S}$ as determined by (5.30) and (5.31). Note that zeros in the coefficients are now associated with jumps of amount π in phase.

for then $\exp[-i\omega(\beta^{-1}\cos j)z] \to 0$ as $z \to -\infty$, when $1/\beta < p$. The phenomenon of *total internal reflection* can occur if the lower medium (velocity β_2) is the slower and an *SH*-wave is incident upward from below. For slowness in the range $1/\beta_1 < p < 1/\beta_2$, the incident wave is an ordinary traveling plane wave, and the scattered waves are still correctly given by Table 5.2 and formulas (5.32), provided that $\cos j_1$ is interpreted by (5.49). In particular, note that $|\acute{S}\grave{S}| = 1$, but there is a phase shift in the reflection. (See Hudson (1962) for an analysis of energy flux in this problem.) Since the behavior for p in the range $0 \le p < 1/\beta_1$ is so different, it is natural to refer to $p = 1/\beta_1$ as a *critical value* and $j_2 = \sin^{-1}(\beta_2/\beta_1)$ as the *critical angle of incidence*. We shall refer to a wave whose slowness components are all real as a *homogeneous* wave, so that in this *SH* example, the transmitted wave is homogeneous or inhomogeneous according as p is less than or greater than a critical value.

As a general rule, *all* the reflection/transmission coefficients associated with a particular interface will become complex if *at least one* of the waves set up at the interface by an incident wave is an inhomogeneous wave. So far, we have looked at examples in which the incident wave is homogeneous, in which case at least one of the scattered waves must also be a homogeneous wave; the overall picture is still one of energy propagating toward the boundary, then interacting with the other half-space and finally propagating away from the interface.

The next possibility we must consider is that *all* the plane waves interacting with the boundary are inhomogeneous waves. Thus, for inhomogeneous plane *P*- and *SV*-waves at the free surface of a half-space (Fig. 5.5, Table 5.1), we consider slowness in the range $1/\alpha < 1/\beta < p$. In this case, we have a very different overall picture from that considered so far, since now the energy is no longer transmitted toward the boundary and scattered from it. Rather, it can be channelled only along the boundary itself. For a half-space, we cannot permit unbounded waves, so that the only permissible wave types are exponentially decaying away from the surface:

$$(\alpha p, 0, i\sqrt{\alpha^2 p^2 - 1}) \exp\left(-\omega\sqrt{p^2 - \frac{1}{\alpha^2}}\, z\right) \exp[i\omega(px - t)] \quad (5.50)$$

for the inhomogeneous *P*-wave (the first factor is a unit vector), and

$$(i\sqrt{\beta^2 p^2 - 1}, 0, -\beta p) \exp\left(-\omega\sqrt{p^2 - \frac{1}{\beta^2}}\, z\right) \exp[i\omega(px + t)] \quad (5.51)$$

for the inhomogeneous *SV*-wave (here again, note the unit vector). These two waves are coupled by boundary conditions ($\tau_{zx} = \tau_{zz}$ on $z = 0$), and, just as we have already found for homogeneous waves, the amplitude ratio of the waves is determined by these conditions. The difference this time is that we have not identified a specific incident wave in the coupled system, so that there is one less amplitude ratio to be determined, with no reduction in the number of boundary conditions. If wave types (5.50) and (5.51) are present, in amounts given by $\grave{P}$ and $\grave{S}$, respectively, then

$$2p\alpha\beta i\sqrt{p^2 - 1/\alpha^2}\grave{P} + (1 - 2\beta^2 p^2)\grave{S} = 0 \quad (\text{from } \tau_{zx} = 0 \text{ on } z = 0) \quad (5.52)$$

and

$$(1 - 2\beta^2 p^2)\grave{P} - 2(\beta^3 p/\alpha)i\sqrt{p^2 - 1/\beta^2}\grave{S} = 0 \quad (\text{from } \tau_{zz} = 0 \text{ on } z = 0). \quad (5.53)$$

Since these are two equations for the same amplitude ratio ($\grave{S}/\grave{P}$), the determinant of coefficients must vanish and p must satisfy $\mathsf{R}(p) = 0$, where

$$\mathsf{R}(p) \equiv \left(\frac{1}{\beta^2} - 2p^2\right)^2 - 4p^2\left(p^2 - \frac{1}{\alpha^2}\right)^{1/2}\left(p^2 - \frac{1}{\beta^2}\right)^{1/2}$$
$$= \left(\frac{1}{\beta^2} - 2p^2\right)^2 + 4p^2\frac{\cos i}{\alpha}\frac{\cos j}{\beta}. \quad (5.54)$$

This function of p^2 has just one zero (for positive imaginary "cosines"), and it is real and positive. Since the corresponding positive value of p is slightly (4–14%) greater than $1/\beta$ for all elastic solids, it is indeed possible for a coupled pair of inhomogeneous waves, P and SV, to propagate along the surface of a half-space. This surface wave is named for Rayleigh, who described its properties in 1886, and $\mathsf{R}(p)$ defined in (5.54) is known as the *Rayleigh function*. Note that the Rayleigh function is just the denominator of the plane-wave coefficients given in (5.26), (5.27), (5.30), (5.31). Rayleigh waves are widely observed in seismology, and as a surface wave for a homogeneous half-space we note the following main properties: (i) their propagation speed c_R, where $\mathsf{R}(1/c_R) = 0$, is a few percent less than the shear-wave speed; (ii) c_R is independent of frequency, hence the wave does not disperse (this property arises physically because there is no intrinsic length scale in a homogeneous half-space, though we shall find in Chapter 7 that surface waves in media with depth-dependent properties are necessarily dispersed); (iii) the particles move in an elliptical trajectory (see Fig. 5.11) that is retrograde at the free surface (this result, developed in the Problems, is remarkable in that for each of the component inhomogeneous waves, P and SV, the particle motion is prograde elliptical); (iv) below a certain depth, which depends on frequency, the particle motion in a Rayleigh wave is dominated by the SV-component, and hence becomes prograde elliptical.

Once the above theory for Rayleigh waves is thoroughly grasped, the generalization to interface waves ($P - SV$) for two homogeneous half-spaces can be readily obtained. Such a wave is composed of inhomogeneous waves decaying upward in the upper medium and decaying downward in the lower medium, so

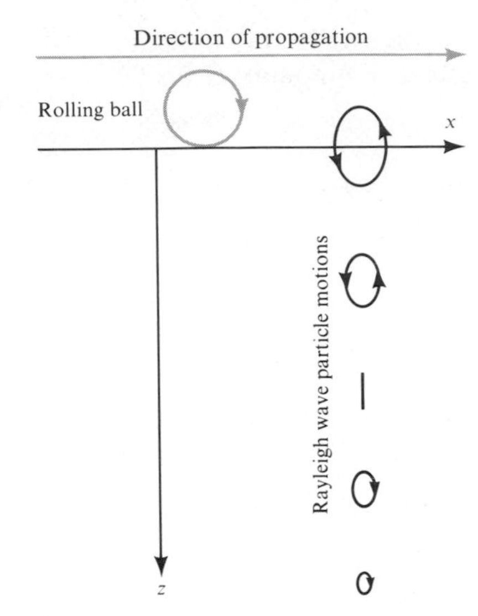

FIGURE **5.11**
The sense of particle motion in a Rayleigh wave as a function of depth. The convention for distinguishing between prograde and retrograde motion can perhaps best be remembered in terms of a ball rolling along the top of the half-space from source to receiver. The sense of rotation of the ball is prograde, which is the direction of motion for inhomogeneous *P*- and *SV*-waves, and also for the Rayleigh wave at sufficient depth. For the Rayleigh wave, however, particle motion at the surface is retrograde.

that particle motions are effectively confined to the vicinity of the interface. In terms of the "scattered" waves listed in Table 5.3, we therefore make the following interpretations (for positive frequency):

$$\begin{aligned} \frac{\cos i_1}{\alpha_1} &\to +i\sqrt{p^2 - \frac{1}{\alpha_1^2}}; \quad \frac{\cos j_1}{\beta_1} \to +i\sqrt{p^2 - \frac{1}{\beta_1^2}}, \\ \frac{\cos i_2}{\alpha_2} &\to +i\sqrt{p^2 - \frac{1}{\alpha_2^2}}; \quad \frac{\cos j_2}{\beta_2} \to +i\sqrt{p^2 - \frac{1}{\beta_2^2}} \end{aligned} \tag{5.55}$$

(compare with (5.49) and (5.46)), and the possible values of horizontal slowness p are determined by the requirement that $\mathsf{D}(p) = 0$, where D is defined in (5.38). Of course, we are interested only in real roots, for which p is greater than the largest of $(1/\alpha_1, 1/\beta_1, 1/\alpha_2, 1/\beta_2)$, so that all the "cosines" in (5.55) are pure imaginary. It is shown by Scholte (1947) that $\mathsf{D}(p) = 0$ always has four roots (i.e., two values of p^2, with four roots for p), but that these are real and appropriately large only when the ratios ρ_1/ρ_2 and β_1/β_2 lie in certain ranges. These waves are named for Stoneley, who discovered their main properties in 1924. As a special case, Stoneley waves can always exist at the interface between a fluid and a solid. Note that since frequency does not enter the definition of D, Stoneley waves, like Rayleigh waves, are not dispersive (for homogeneous half-spaces).

In this section, we have chosen to emphasize the differences between homogeneous and inhomogeneous waves, although we have unified the theory by studying these waves as a function of their horizontal slowness. In later work, we shall refer to both these wave types as "plane waves," finding that both types are required to synthesize the wave system set up by a localized source.

5.4 A Matrix Method for Analyzing Plane Waves in Homogeneous Media

For all the plane waves that were considered in Section 5.3, the equations of motion and the constitutive relation can be combined in such a way that only first-order depth derivatives of stress and displacement are needed. By this statement, we mean that plane waves can be studied in terms of an equation of the type

$$\frac{\partial \mathbf{f}}{\partial z} = \mathbf{A}\mathbf{f}, \tag{5.56}$$

in which $\mathbf{f} = \mathbf{f}(z)$ is a column-vector giving the depth dependence of particle displacement and stress. $\mathbf{A}$ is a constant matrix, with entries depending on elastic properties of the (homogeneous) medium and on horizontal slowness p and frequency ω.

For example, the SH-waves considered in Table 5.2 are all a result of analyzing the displacement $u_y = v = v(x, z, t)$ in which the dependence on (x, t) is only via a factor $\exp[i\omega(px - t)]$. The equation of motion is $\rho\ddot{v} = \tau_{yz,z} + \tau_{yx,x}$, and the constitutive relation (2.18) becomes merely $\tau_{yz} = \mu\partial v/\partial z$, $\tau_{yx} = \mu\partial v/\partial x$. It follows that such SH problems can be discussed in terms of the vector $\mathbf{f}$ given by

$$\mathbf{f}(z)\exp[i\omega(px - t)] = \begin{pmatrix} v \\ \tau_{yz} \end{pmatrix}, \tag{5.57}$$

and the depth dependence of $\mathbf{f}$ is given by solving the standard equation (5.56) with coefficient matrix

$$\mathbf{A} = \begin{pmatrix} 0 & \mu^{-1} \\ \omega^2(\mu p^2 - \rho) & 0 \end{pmatrix}. \tag{5.58}$$

For acoustic waves (see Box 5.4), with pressure P and vertical particle displacement u_z, we take

$$\mathbf{f}(z)\exp[i\omega(px - t)] = \begin{pmatrix} u_z \\ P \end{pmatrix} \quad \text{and} \quad \mathbf{A} = \begin{pmatrix} 0 & \dfrac{p^2}{\rho} - \dfrac{1}{\rho\alpha^2} \\ \rho\omega^2 & 0 \end{pmatrix}. \tag{5.59}$$

For *P-SV* waves, we take

$$\mathbf{f}\exp[i\omega(px - t)] = \begin{pmatrix} u_x \\ u_z \\ \tau_{zx} \\ \tau_{zz} \end{pmatrix} \quad \text{and}$$

$$\mathbf{A} = \begin{pmatrix} 0 & -i\omega p & \frac{1}{\mu} & 0 \\ \frac{-i\omega p\lambda}{\lambda + 2\mu} & 0 & 0 & \frac{1}{\lambda + 2\mu} \\ \frac{4\omega^2 p^2 \mu(\lambda + \mu)}{\lambda + 2\mu} - \rho\omega^2 & 0 & 0 & \frac{-i\omega p\lambda}{\lambda + 2\mu} \\ 0 & -\rho\omega^2 & -i\omega p & 0 \end{pmatrix} \tag{5.60}$$

A great deal is known about the solutions of coupled first-order differential equations like $\partial\mathbf{f}/\partial z = \mathbf{A}\mathbf{f}$ (see, e.g., Coddington and Levinson, 1955; Gantmacher, 1959), and we shall make extensive use of this theory in Chapters 7–9 to study wave propagation in inhomogeneous media. Our present discussion, however, concerns homogeneous media, for which $\mathbf{A}$ is independent of z.

The first property to notice is that, if $\mathbf{v}^\alpha$ is an eigenvector of $\mathbf{A}$ and λ^α is the associated eigenvalue, a solution to (5.56) is given by

$$\mathbf{f} = \mathbf{v}^\alpha \exp[\lambda^\alpha(z - z_{\text{ref}})] \tag{5.61}$$

(no summation over superscript α), where z_{ref} is a reference level for the phase.

Now let $\mathbf{F}$ be the matrix whose columns consist of different solutions of type (5.61). Thus, if $\mathbf{A}$ is an $n \times n$ matrix, we can find n eigenvalues and n linearly independent eigenvectors ($\alpha = 1, \ldots, n$), so that $\mathbf{F}$ is also an $n \times n$ matrix, with the αth column given by $\mathbf{v}^\alpha \exp[\lambda^\alpha(z - z_{\text{ref}})]$. It follows that the most general solution $\mathbf{f}$ to the equation $\partial\mathbf{f}/\partial z = \mathbf{A}\mathbf{f}$ is some linear combination of the columns of $\mathbf{F}$, i.e.,

$$\mathbf{f} = \mathbf{F}\mathbf{w}, \tag{5.62}$$

where $\mathbf{w}$ is a vector of constants, weighting the columns of $\mathbf{F}$ that are to appear in solution $\mathbf{f}$.

The reason why an analysis of plane waves in terms of such eigenvalues and eigenvectors can be powerful lies in the physical interpretation of (5.61) and

(5.62). Thus consider the case of *SH*-waves, for which $\partial \mathbf{f}/\partial z = \mathbf{A}\mathbf{f}$, with $\mathbf{f}$ and $\mathbf{A}$ given in (5.57) and (5.58). The two eigenvalues of $\mathbf{A}$ are $\pm i\omega\sqrt{1/\beta^2 - p^2} = \pm i\omega\beta^{-1}\cos j$, which are merely i times the vertical wavenumbers for downward and upward traveling *SH*-waves. The quantity $\beta^{-1}\cos j$ is the vertical slowness for *S* waves, and we introduce the notation

$$\eta = \frac{\cos j}{\beta} = \left(\frac{1}{\beta^2} - p^2\right)^{1/2}.$$

Corresponding to eigenvalues $\pm i\omega\eta$ are the eigenvectors

$$\begin{pmatrix} 1 \\ \pm i\omega\mu\eta \end{pmatrix}.$$

Hence, following (5.61), the basic solutions for *SH* motion are either of type

$$\begin{pmatrix} 1 \\ i\omega\mu\eta \end{pmatrix} \exp[i\omega\eta(z - z_{\text{ref}})]$$

or of type

$$\begin{pmatrix} 1 \\ -i\omega\mu\eta \end{pmatrix} \exp[-i\omega\eta(z - z_{\text{ref}})].$$

Clearly, the first of these gives particle displacement and shearing stress for a downgoing *SH*-wave, and the second is for an upgoing wave. The most general type of plane *SH*-wave is a linear combination of these, so that the interpretation of vector $\mathbf{w}$ (appearing in (5.62)) is that its first component gives the amount of downgoing wave, and its second component gives the amount of upgoing wave present in the total wave system $\mathbf{f}$.

Note that the matrix $\mathbf{F}$ can be explicitly stated as

$$\begin{aligned} \mathbf{F} &= \begin{pmatrix} 1 & 1 \\ i\omega\mu\eta & -i\omega\mu\eta \end{pmatrix} \begin{pmatrix} \exp[i\omega\eta(z - z_{\text{ref}})] & 0 \\ 0 & \exp[-i\omega\eta(z - z_{\text{ref}})] \end{pmatrix} \\ &= \mathbf{E}\mathbf{\Lambda}, \end{aligned} \tag{5.63}$$

and $\mathbf{F}$ itself has been factored into a matrix $\mathbf{E}$, made up from eigenvectors of $\mathbf{A}$, times a diagonal matrix $\mathbf{\Lambda}$ containing the vertical phase factors, which have been referred to the level z_{ref}. $\mathbf{F}$ is known as the *solution matrix*, or (within the context in which it has been most used in seismology) the *layer matrix*. The inverse of **F** for SH-waves is

$$\mathbf{F}^{-1} = \mathbf{\Lambda}^{-1}\mathbf{E}^{-1} = \begin{pmatrix} \exp[-i\omega\eta(z - z_{\text{ref}})] & 0 \\ 0 & \exp[i\omega\eta(z - z_{\text{ref}})] \end{pmatrix} \begin{pmatrix} \frac{1}{2} & \frac{-i}{2\omega\mu\eta} \\ \frac{1}{2} & \frac{i}{2\omega\mu\eta} \end{pmatrix}. \tag{5.64}$$

For P-SV waves, eigenvalues of **A** are $\pm i\omega\xi$ (where $\xi = \alpha^{-1}\cos i$ is the P-wave vertical slowness) and $\pm i\omega\eta$. The general solution of $\partial\mathbf{f}/\partial z = \mathbf{Af}$ (with **f**, **A** given in (5.60)) is again $\mathbf{f} = \mathbf{Fw}$, with **F** factored into $\mathbf{E\Lambda}$, and

$$\mathbf{E} = \begin{pmatrix} \alpha p & \beta\eta & \alpha p & \beta\eta \\ \alpha\xi & -\beta p & -\alpha\xi & \beta p \\ 2i\omega\rho\alpha\beta^2 p\xi & i\omega\rho\beta(1 - 2\beta^2p^2) & -2i\omega\rho\alpha\beta^2p\xi & -i\omega\rho\beta(1 - 2\beta^2p^2) \\ i\omega\rho\alpha(1 - 2\beta^2p^2) & -2i\omega\rho\beta^3p\eta & i\omega\rho\alpha(1 - 2\beta^2p^2) & -2i\omega\rho\beta^3p\eta \end{pmatrix}, \tag{5.65}$$

$$\mathbf{\Lambda} = \begin{pmatrix} \exp[i\omega\xi(z - z_{\text{ref}})] & 0 & 0 & 0 \\ 0 & \exp[i\omega\eta(z - z_{\text{ref}})] & 0 & 0 \\ 0 & 0 & \exp[-i\omega\xi(z - z_{\text{ref}})] & 0 \\ 0 & 0 & 0 & \exp[-i\omega\eta(z - z_{\text{ref}})] \end{pmatrix}.$$

Note that components of **f** are just those physical quantities that are continuous across a welded interface. Thus, for $P - SV$ problems, we have seen that u_x, u_z, τ_{zx}, τ_{zz} are continuous across a horizontal boundary ($z = 0$, say) between two different half-spaces. Hence, from (5.60), **f** is continuous, although **A** is not.

The matrix method we are developing plays a major role in the study of plane waves propagating in a stack of homogeneous layers. As a very simple example, we shall indicate a derivation of formulas (5.32) that gives all reflection/transmission coefficients for SH-waves incident on the boundary between two half-spaces.

In the upper medium ($z < 0$; see Fig. 5.7), we have $\mathbf{f} = \mathbf{F}_1(z)\mathbf{w}_1$, where $\mathbf{F}_1$ is given by (5.63) with values of ρ, β, j appropriate to the upper medium, and $\mathbf{w}_1$ is a vector giving the amounts of downgoing and upgoing wave in the upper medium. Similarly, in the lower medium ($z > 0$), $\mathbf{f} = \mathbf{F}_2(z)\mathbf{w}_2$. Since **f** is continuous across $z = 0$, it follows that

$$\mathbf{F}_1(0)\mathbf{w}_1 = \mathbf{F}_2(0)\mathbf{w}_2, \tag{5.66}$$

giving a general relation between the upgoing and downgoing wave systems in the two media, whatever these systems may be.

To solve for $\grave{S}\acute{S}$, etc., we apply (5.66) to the two cases shown in Figure 5.7. Thus, for *SH* incident from above, we have

$$\mathbf{w}_1 = \begin{pmatrix} 1 \\ \grave{S}\acute{S} \end{pmatrix} \quad \text{and} \quad \mathbf{w}_2 = \begin{pmatrix} \grave{S}\grave{S} \\ 0 \end{pmatrix}.$$

Substitution of these into (5.66) gives two scalar equations in the two unknowns $\grave{S}\acute{S}$ and $\grave{S}\grave{S}$. For *SH* incident from below, we have

$$\mathbf{w}_1 = \begin{pmatrix} 0 \\ \acute{S}\acute{S} \end{pmatrix} \quad \text{and} \quad \mathbf{w}_2 = \begin{pmatrix} \acute{S}\grave{S} \\ 1 \end{pmatrix},$$

which again we can substitute into (5.66), to obtain two equations for $\acute{S}\acute{S}$ and $\acute{S}\grave{S}$.

In practice, it is often worthwhile to normalize the eigenvectors $\mathbf{v}^\alpha$ so that the associated particle displacement has unit amplitude. We did this above, for our discussion both of *SH* and $P - SV$. As we shall find in Chapters 7 and 9, a major reason why the method is computationally attractive is that the matrix $\mathbf{F}(z)$ has an inverse $\mathbf{F}^{-1} = \mathbf{\Lambda}^{-1}\mathbf{E}^{-1}$ that can be stated in closed form. In fact, for $P - SV$ waves, the matrix $\mathbf{E}$ of (5.65) has inverse

$$\mathbf{E}^{-1} = \begin{pmatrix} \dfrac{\beta^2 p}{\alpha} & \dfrac{1 - 2\beta^2 p^2}{2\alpha\xi} & \dfrac{-ip}{2\omega\rho\alpha\xi} & \dfrac{-i}{2\omega\rho\alpha} \\ \dfrac{1 - 2\beta^2 p^2}{2\beta\eta} & -\beta p & \dfrac{-i}{2\omega\rho\beta} & \dfrac{ip}{2\omega\rho\beta\eta} \\ \dfrac{\beta^2 p}{\alpha} & \dfrac{-(1 - 2\beta^2 p^2)}{2\alpha\xi} & \dfrac{ip}{2\omega\rho\alpha\xi} & \dfrac{-i}{2\omega\rho\alpha} \\ \dfrac{1 - 2\beta^2 p^2}{2\beta\eta} & \beta p & \dfrac{i}{2\omega\rho\beta} & \dfrac{ip}{2\omega\rho\beta\eta} \end{pmatrix}. \tag{5.67}$$

5.5 Wave Propagation in an Attenuating Medium: Basic Theory for Plane Waves

For the adiabatic wave propagation we have so far been describing, wave motion will continue indefinitely, once it has been initiated by some specific source. Thus the wave motion may be spatially attenuated, as waves spread away from a localized source region, but the total energy of particle motion (i.e., the volume integral of kinetic energy and elastic strain energy throughout the medium) has been held constant. This follows from the discussion in Section 2.2. In contrast to such idealized behavior, however, it is common experience that as a wave is propagated through real materials, wave amplitudes

BOX **5.7**

Different definitions of Q

If a volume of material is cycled in stress at a frequency ω, then a dimensionless measure of the internal friction (or the anelasticity) is given by

$$\frac{1}{Q(\omega)} = -\frac{\Delta E}{2\pi E}, \tag{1}$$

where E is the peak strain energy stored in the volume and $-\Delta E$ is the energy lost in each cycle because of imperfections in the elasticity of the material.

This definition is rarely of direct use, since only in special experiments is it possible to drive a material element with stress waves of unchanging amplitude and period. More commonly, one observes either (i) the temporal decay of amplitude in a standing wave at fixed wavenumber or (ii) the spatial decay in a propagating wave at fixed frequency. The commonest situation in seismology involves attenuation of a signal composed of a range of frequencies, and we shall make the strong assumption (which must be checked later) that attenuation is a linear phenomenon, in the sense that a wave may be resolved into its Fourier components, each of which can be studied in terms of (i) or (ii), and that subsequent Fourier synthesis gives the correct effect of attenuation on actual seismic signals.

In the case of either (i) or (ii), for a medium with linear stress-strain relation, wave amplitude A is proportional to $E^{1/2}$. (For example, A may represent a maximum particle velocity, or a stress component in the wave. We assume also that $Q \gg 1$, so that successive peaks have almost the same strain energy.) Hence

$$\frac{1}{Q(\omega)} = -\frac{1}{\pi}\frac{\Delta A}{A}, \tag{2}$$

from which we can obtain the amplitude fluctuations due to attenuation.

Thus, in case (i), we ask: What is $A = A(t)$, given that initially $A = A_0$ and A decreases a fraction π/Q at successive times $2\pi/\omega, 4\pi/\omega, \ldots, 2n\pi/\omega \ldots$? Clearly,

$$A(t) = A_0(1 - \pi/Q)^n \qquad (\text{for } t = 2n\pi/\omega).$$

Thus

$$A(t) = A_0\left[1 - \frac{\omega t}{2Qn}\right]^n \to A_0 \exp\left[-\frac{\omega t}{2Q}\right] \qquad (\text{for large } n\text{; i.e., for large times}). \tag{3}$$

From observations of exponentially decaying values of $A(t)$, we use (3) to define the value of a *temporal Q*. It is this value that is used in describing the attenuation of the Earth's free oscillations (see Chapter 8).

We were forced to obtain the above result (3) by using discrete times, since this is the nature of experiments based on case (i). For case (ii), however, the derivation of the form $A = A(x)$ for distance x is easier, since a particular wave peak can be followed along a distance dx, and the gradual spatial decay of A can be observed. (We assume here that the direction of maximum attenuation is along the x-axis, which is also the direction of propagation.) Then $\Delta A = (dA/dx)\lambda$, where λ is the wavelength given in terms of ω and phase velocity c by $\lambda = 2\pi c/\omega$. Equation (2) becomes $dA/dx = -(\omega/2cQ)A$, with the exponentially decaying solution

$$A(x) = A_0 \exp\left[-\frac{\omega x}{2cQ}\right]. \tag{4}$$

From observations of exponentially decaying values of $A(x)$, we use (4) to define the value of a *spatial* Q. Of course, any spatial decay due to geometrical spreading effects must be studied too. In Section 5.5, we shall avoid this problem by confining our attention to plane waves propagating in homogeneous media. Such simple problems also avoid the complication of geometric dispersion due to material heterogeneity, which involves differences between Q measured via (3) and Q measured by (4). These additional effects of dispersion (which are present even in perfectly elastic media) are taken up in Chapter 7, in connection with surface waves.

Note that to obtain the effect of attenuation on a wave solution such as $e^{i(kx-\omega t)}$, an experiment to measure temporal Q can be thought of as replacing ω by a complex-valued frequency, and an experiment to measure spatial Q can be thought of as replacing k by a complex-valued wavenumber. In the first case, ω acquires an imaginary part, $-i|\omega|/2Q^{\text{temporal}}$; in the second case, k acquires an imaginary part, $i|k|/2Q^{\text{spatial}}$.

attenuate as a result of a variety of processes, which we can summarize macroscopically as "internal friction."

To give but two examples: the strains and stresses occurring within a propagating wave can lead to irreversible changes in the crystal defect structures of the medium, and work may also be done on grain boundaries within the medium if adjacent material grains are not elastically bonded. Such media are said to be *anelastic*, since the configuration of material particles is to some extent dependent on the history of applied stress. The gross effect of internal friction is summarized by the dimensionless quantity Q, defined in various ways (for slightly anelastic media) in Box 5.7.

The general subject of wave attenuation by internal friction (sometimes called intrinsic attenuation) is very large, as can be seen from identifying three different aspects:

i) As a branch of materials science, studies are made of the fundamental (microscopic) processes that cause attenuation. The effects of a variety of crystal defects, grain-boundary processes, and some thermoelastic processes have been reviewed by Mason (1958), Jackson and Anderson (1970), and Nowick and Berry (1972).

ii) The frequency dependence of Q is studied as a macroscopic phenomenon in a variety of materials. From observations of the frequency dependence of Q in seismic waves (Archambeau et al., 1969, Solomon, 1972, 1973), one may then constrain the possibilities for Earth composition. Solomon has used this method to try to quantify the extent of partial melting in the upper mantle.

iii) Many authors have developed macroscopic stress-strain relations to replace Hooke's law, and hence (with $\rho \ddot{u}_i = \tau_{ij,j}$) have obtained equations of motion for materials with some particular $Q = Q(\omega)$. Critical here is the use of a causality constraint, which leads to the phenomenon of dispersion for an attenuating medium. Key papers are those of Lomnitz (1957), Jeffreys (1958), Futterman (1962), Azimi et al. (1968), and Liu et al. (1976). The reviews of Knopoff (1964) and Stacey et al. (1975, with a correction by Savage, 1976) show that there are still some major areas of ignorance in our understanding of how attenuation occurs and how it affects the pulse shape of a propagating wave. In particular, the amount of intrinsic dispersion present in seismic waves is not yet accurately known, and there is a lack of direct evidence to support the law of linear superposition (introduced in Box 5.7).

In the remainder of this chapter, we shall cover some of the ground mentioned in (iii), since pulse shapes observed in seismology are certainly influenced by internal friction. The principal result is that, if Q is effectively constant over the wide range of frequencies observed in seismology (0.001–100 Hz), then wave propagation in the attenuating Earth can be studied by considering body-wave velocity (α and/or β, which we have hitherto regarded as real and independent of frequency) as the complex-valued and frequency-dependent quantity defined in (5.88). But if Q varies with frequency by more than a few percent in the seismic range, then (5.88) may be inadequate.

5.5.1 *The necessity for material dispersion in an attenuating medium*

Consider a plane wave $\delta(t - x/c)$ propagating with speed c along the positive x-direction in a perfectly elastic homogeneous medium. As we have shown in earlier sections of this chapter, such a wave retains its shape exactly, and all frequency components travel coherently with the same velocity c.

Now consider such a wave as the input at $x = 0$ to an attenuating medium. Knopoff (1964) has summarized the experimental measurements of attenuation by concluding that $Q(\omega) \propto \omega^{-1}$ for a variety of fluids, but that Q is approximately constant for the frequency range of observation in solids. It is this type of behavior, involving frequency-independent Q, with which we are most concerned in seismology.

Each Fourier component of the impulse,

$$\int_{-\infty}^{\infty} \delta(t - x/c)e^{i\omega t}\, dt = \exp[i\omega x/c],$$

will now be attenuated by a factor $\exp[-\alpha(\omega)x]$, and from equation (4) of Box 5.7 we find that this attenuation rate is $\alpha = \omega/(2cQ)$.

At distance x, what then will be the pulse shape $p(x, t)$ (say) of the attenuated wave? As a first attempt at answering this question, let us suppose that no dispersion is present, so that the pulse is a synthesis from its Fourier components, all having the same speed c:

$$p(x, t) = \frac{1}{2\pi} \int_{-\infty}^{\infty} \exp\left[\frac{-|\omega|x}{2cQ}\right] \exp\left[i\omega\left(\frac{x}{c} - t\right)\right] d\omega \tag{5.68}$$

(clearly, $|\omega|$ must replace ω for the attenuation factor at negative frequencies, if $x > 0$). If Q is constant, the integral here is easily shown to be

$$p(x, t) = \frac{1}{\pi}\left[\frac{\dfrac{x}{2cQ}}{\left(\dfrac{x}{2cQ}\right)^2 + \left(\dfrac{x}{c} - t\right)^2}\right], \tag{5.69}$$

and this pulse shape is plotted against time at fixed distance x in Figure 5.12.

Unfortunately, Figure 5.12 shows this pulse shape to have several unacceptable features. It shows a (small) arrival at $x = 0$ *even before time* $t = 0$, which violates our most elementary notions of causality. It shows a rise time (see figure legend) that is too large (Stacey et al., 1975, report that the measured rate of increase of τ with distance is only $(2cQ)^{-1}$, but Fig. 5.12 and equation (5.69) imply about 1.5 times this rate), and the symmetrical shape of the pulse is also at odds with experiments, which show an asymmetric shape having a decay time much greater than the rise time.

In questioning the three assumptions that went into our derivation of this pulse shape (5.69), i.e., linearity, constant Q, and no dispersion, it is natural to try to retain the first two and relax the third. (The indirect evidence for retaining linearity is simply that it has led to self-consistent results in so many careful analyses of seismic data. Nowhere is this more clear than in studies of the

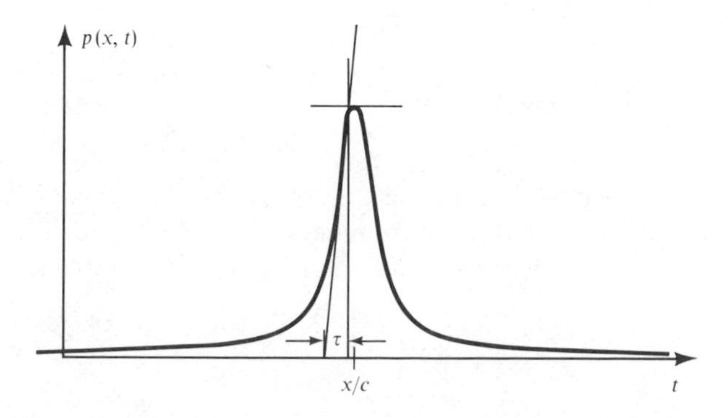

FIGURE **5.12**

The pulse shape $p(x, t)$ given by (5.69) is shown against time at a fixed distance. The rise time τ, defined by extrapolating the tangent to $p(x, t)$ at the point with maximum gradient, can be found in terms of x, c, Q as $\tau = (4/3\sqrt{3})(x/cQ)$ for this pulse shape, whereas the experimetnal rise time is found to be considerably shorter.

Earth's normal modes, which have been excited to observational levels by hundreds of earthquakes, with periods of oscillation that, to a high degree of accuracy, are found to be unchanged for different sources and different seismographs.) Although we shall shortly drop the assumption of constant Q, one of our points here is that we *must* permit some dispersion, since the symmetry of an attenuated impulse is otherwise unavoidable (Problem 5.9).

5.5.2 *Some suggested values for material dispersion in an attenuating medium*

Our first attempts at quantifying material dispersion will invoke the concept of a wavefront for anelastic media. We shall assume that a plane wave $u(x, t)$ propagates in the direction of increasing x and that the wavefront first arrives at position $x = 0$ at time $t = 0$. Thus

$$u(0, t) = 0 \quad \text{for} \quad t < 0. \tag{5.70}$$

At $x > 0$, each Fourier component of u can be factored as

$$u(x, \omega) = u(0, \omega) \exp(iKx),$$

where the complex wavenumber K is now given in terms of phase velocity $c(\omega)$ and attenuation factor $\alpha(\omega)$ by

$$K = \frac{\omega}{c(\omega)} + i\alpha(\omega). \tag{5.71}$$

With the assumption of linear superposition, the wave at (x, t) is

$$u(x, t) = \frac{1}{2\pi} \int_{-\infty}^{\infty} u(0, \omega) \exp[i(Kx - \omega t)] \, d\omega,$$

which is equivalent to the convolution of $u(0, t)$ with

$$p(x, t) = \frac{1}{2\pi} \int_{-\infty}^{\infty} \exp[i(Kx - \omega t)] \, d\omega, \tag{5.72}$$

$p(x, t)$ being the attenuated response to an impulse that was described earlier in (5.68). Now, however, we are allowing c to be frequency dependent, and in Box 5.8 we show that requiring $p(x, t)$ to be zero until an arrival time x/c_∞ implies

$$\frac{\omega}{c(\omega)} = \frac{\omega}{c_\infty} + \mathscr{H}[\alpha(\omega)]. \tag{5.73}$$

BOX **5.8**

Relations between the amplitude spectrum and phase spectrum of a causal propagating pulse shape

We assume that the propagating pulse shape $p(x, t)$ has the Fourier transform e^{iKx} with real and imaginary parts of K given by

$$K = \frac{\omega}{c(\omega)} + i\alpha(\omega). \tag{1}$$

Then if

$$p(x, t) = 0 \qquad (\text{for } t < x/c_\infty), \tag{2}$$

we shall show that

$$\frac{\omega}{c(\omega)} = \frac{\omega}{c_\infty} + \mathscr{H}[\alpha(\omega)]. \tag{3}$$

First, we note that the causality described in (2) is completely equivalent to the statement that

$$F(\omega), \text{ defined by } F(\omega) = \exp[i(K - \omega/c_\infty)x], \text{ is analytic in the upper half-plane (i.e., for Im } \omega \geq 0). \tag{4}$$

For, if (2) is true, then

$$F(\omega) = p(x, \omega)\exp(-i\omega x/c_\infty) = \int_{x/c_\infty}^{\infty} p(x, t)\exp[i\omega(t - x/c_\infty)]\,dt, \tag{5}$$

which, by Jordan's Lemma, indeed *is* a convergent integral in the upper half-plane. F and all its derivatives exist, hence (4) is obtained. On the other hand, if (4) is true, we write

$$p(x, t) = \frac{1}{2\pi}\int_{-\infty}^{\infty} F(\omega)\exp[-i\omega(t - x/c_\infty)]\,d\omega \tag{6}$$

and add a semicircular path at infinity in the upper half-plane (from which there is no contribution if $t < x/c_\infty$, since the exponential vanishes). The integrand in (6) is then analytic everywhere inside a closed contour, and so (2) is obtained.

Second, we show also that $\log F(\omega)$ is analytic in the upper half-plane. To prove this, we need show only that $F(\omega)$ has no zeros for which ω has a nonnegative imaginary part. Suppose, in fact, that $F(\omega_0) = 0$ for $\text{Im}\,\omega_0 \geq 0$. Then $F(\omega) = (\omega - \omega_0)^\lambda f(\omega)$ for some $\lambda > 0$ and some f with $f(\omega_0) \neq 0$. However, if x is increased to $x + \Delta x$, it follows from (4) that

$$(\omega - \omega_0)^{\lambda(1 + \Delta x/x)} f(\omega)^{1 + \Delta x/x}$$

is analytic in the upper half-plane—which is impossible, because in this form it is seen to have a branch point at ω_0.

Third, we note a special form of Cauchy's theorem: if $g(z)$ is analytic on and inside a closed circuit C, then for a point x *on C itself*, we have

$$g(x) = \frac{1}{\pi i} P \int_C \frac{g(z)\, dz}{z - x} \tag{7}$$

(where P indicates the process of taking Cauchy's principal value).

Finally, we apply (7) to the analytic function

$$\begin{aligned}\log F(\omega) &= \log|F(\omega)| + i \text{ phase } F(\omega) \\ &= -\alpha(\omega)x + i\omega\left(\frac{1}{c(\omega)} - \frac{1}{c_\infty}\right)x,\end{aligned} \tag{8}$$

using for C the infinite semicircle encompassing the upper half of the complex ω plane, with diameter as the real ω-axis. We find for a real value of ω that

$$\log F(\omega) = \frac{1}{\pi i} P \int_{-\infty}^{\infty} \frac{\log F(\zeta)\, d\zeta}{\zeta - \omega} \tag{9}$$

(unless there is a contribution from the arc of the semicircle; see below). Substituting from (8) into (9), and separating real and imaginary parts, we find

$$\alpha(\omega) = -\frac{1}{\pi} P \int_{-\infty}^{\infty} \zeta\left(\frac{1}{c(\zeta)} - \frac{1}{c_\infty}\right)\frac{d\zeta}{\zeta - \omega} = -\mathscr{H}\left[\omega\left(\frac{1}{c(\omega)} - \frac{1}{c_\infty}\right)\right] \tag{10}$$

and

$$\omega\left(\frac{1}{c(\omega)} - \frac{1}{c_\infty}\right) = \frac{1}{\pi} P \int_{-\infty}^{\infty} \frac{\alpha(\zeta)\, d\zeta}{\zeta - \omega} = \mathscr{H}[\alpha(\omega)], \tag{3) again}$$

which is the required result. A result equivalent to (3) in electromagnetic theory is known as the Kramers-Krönig relation.

It can happen that $\log F(\omega)$ for large $|\omega|$ does not behave in a fashion that permits (9) to be valid, because of a contribution from the semicircular arc (see, e.g., Problem 5.10). However, one can modify the above analysis by applying (7) to the function $\log F(\omega)/\omega$. The same infinite semicircle is used, leading now to two semiresidues; at $\zeta = 0$ and $\zeta = \omega$,

$$\frac{\log F(\omega)}{-\omega} + \frac{\log F(\omega)}{\omega} = \frac{1}{\pi i} P \int_{-\infty}^{\infty} \frac{\log F(\zeta)}{\zeta(\zeta - \omega)}\, d\zeta. \tag{11}$$

[There is no contribution from the large arc if, for example,

$$\log F(\omega) = O(\log \omega) \qquad \text{as } |\omega| \to \infty. \tag{12}$$

An nth order discontinuity of $p(x, t)$ at the wavefront $t = x/c_\infty$ will lead to $F(\omega) = O(\omega^{-n})$, and hence to (12) and (11).]

It follows from (11) and (8) that

$$\frac{1}{c(\omega)} = \frac{1}{c_\infty} + \mathscr{H}\left[\frac{\alpha(\omega)}{\omega}\right] \quad \text{and} \quad \frac{\alpha(\omega) - \alpha(0)}{\omega} = -\mathscr{H}\left[\frac{1}{c(\omega)} - \frac{1}{c_\infty}\right]. \tag{13}$$

This is a simple example of a more general result called a *dispersion relation with one subtraction*, and $\alpha(0)$ is called the *subtraction constant* (Nussenzveig, 1972).

There are circumstances in which it may be known that a function $f(t)$ is such that $f(t) = 0$ for $t < 0$, and hence $f(\omega)$ is analytic in the upper half-plane, but nothing else is known about $f(\omega)$. (In the special problem above, a specific formula given in (4) is available for the function in the frequency domain.) In the general case, we cannot reject the possibility that $f(\omega)$ has zeros.

Suppose, then, that $f(\omega)$ has zeros at $\omega = \omega_i$ $(i = 1, 2, \ldots, n)$ in the upper half-plane. We construct a function $f_0(\omega)$ via

$$f(\omega) = f_0(\omega) \prod_{i=1}^{n} \frac{\omega - \omega_i}{\omega - \omega_i^*}. \tag{14}$$

Since $f_0(\omega)$ has no zeros, its amplitude $f_0(\omega)$ and phase $\phi_0(\omega)$ are related by

$$\phi_0(\omega) = \mathscr{H}[\log|f(\omega)|] \tag{15}$$

(proved by steps parallel to the derivation of (3) above; note that f and f_0 have the same amplitude spectra). But, from (14), the phase $\phi(\omega)$ of $f(\omega)$ is related to $\phi_0(\omega)$ by

$$\phi(\omega) = \phi_0(\omega) - \sum_{i=1}^{n} 2 \tan^{-1}\left(\frac{\operatorname{Im} \omega_i}{\omega - \operatorname{Re} \omega_i}\right).$$

Now, given the various possible pulse shapes in the time domain that have the same amplitude spectrum $|f(\omega)|$, the pulse shape associated with phase spectrum $\phi_0(\omega)$ is known as the *minimum-delay* pulse shape, because the group delay due to the zeros,

$$\frac{d}{d\omega}\{\phi(\omega) - \phi_0(\omega)\} = +2 \sum_{i=1}^{n} \frac{\operatorname{Im} \omega_i}{\{(\operatorname{Im} \omega_i)^2 + (\omega - \operatorname{Re} \omega_i)^2\}},$$

is always positive (since $\operatorname{Im} \omega_i > 0$).

The special pulse shape considered above (e.g., in equation (6)), which is due to attenuation after propagation a distance x, is therefore a minimum-delay pulse shape after the arrival time x/c_∞.

Here, c_∞ is the limit of $c(\omega)$ as $\omega \to \infty$, and $\mathscr{H}[\alpha(\omega)]$ is the Hilbert transform of the attenuation factor.

It appears at this stage that the problem of finding values of $c(\omega)$ is essentially solved. We have merely to take the Hilbert transform of a constant-Q attenuation factor and then use (5.73). However, from equation (4) of Box 5.7, we require

$$\frac{\omega}{c_\infty} + \mathscr{H}[\alpha(\omega)] = 2Q\alpha(\omega), \tag{5.74}$$

and there is no Hilbert transform pair for which this relation is satisfied with constant Q. (If it *were* satisfied, we could Hilbert-transform (5.74). But this cannot be done, since the transform for ω/c_∞ is a divergent integral.) Instead, we must tolerate a frequency-dependent Q, satisfying (5.74), but with an attenuation chosen to make Q effectively constant over the seismic frequency range. This approach has been taken by Azimi et al. (1968), and one of the Hilbert transform pairs they have suggested using is

$$\alpha(\omega) = \frac{\alpha_0 \omega}{1 + \alpha_1 \omega}, \qquad \mathscr{H}[\alpha(\omega)] = \frac{2\alpha_0 \omega}{\pi(1 - \alpha_1^2 \omega^2)} \ln\left(\frac{1}{\alpha_1 \omega}\right), \tag{5.75}$$

in which α_0, α_1 are constants and ln is the natural logarithm.

In Figure 5.13, we show the pulse shape (5.72) that results from Azimi's attenuation law (5.75). Note the asymmetry of the pulse and its delay with respect to the nondispersed pulse shape described earlier. Stacey et al. (1975) have found that the rise time for Azimi's pulse shape does change with distance at the rate determined from measurement in a variety of solid materials, so that the dispersion corresponding to (5.75) is likely to be a good approximation for our needs in seismology. The presumption behind (5.75) is that $1 \gg \alpha_1\omega$ for all seismic frequencies, so that the departure of attenuation from a law $\alpha(\omega) \propto \omega$ is not apparent until ω is very large. We can then neglect $\alpha_1^2\omega^2$ in the expression for $\mathscr{H}[\alpha(\omega)]$, finding from (5.73) that the phase velocity is given effectively by

$$\frac{1}{c(\omega)} = \frac{1}{c_\infty} + \frac{2\alpha_0}{\pi} \ln\left(\frac{1}{\alpha_1 \omega}\right). \tag{5.76}$$

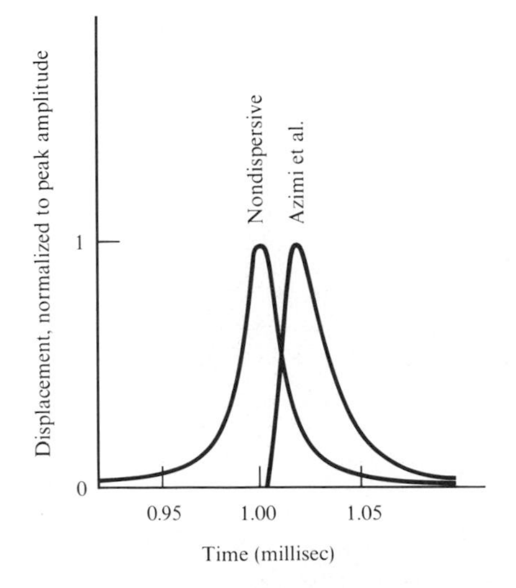

FIGURE **5.13**

Comparison of two attenuated pulse shapes $p(x, t)$, each for x fixed at 5 meters. In the nondispersive case, $c = 5$ km/sec and $Q = 50$ at all frequencies. In the dispersive case (using Azimi's attenuation law (5.75), with $c(\omega)$ and $Q(\omega)$ given by (5.73)–(5.75), $c_\infty = 5$ km/sec and $(2\alpha_0 c_\infty)^{-1} = 50$. [After Gladwin and Stacey, 1974.]

From (5.74) and (5.75), Q has approximately the value $(2c_\infty\alpha_0)^{-1}$, enabling us to obtain from (5.76) the ratio of phase velocities at two different seismic frequencies ω_1 and ω_2:

$$\frac{c(\omega_1)}{c(\omega_2)} = 1 + \frac{1}{\pi Q}\ln\left(\frac{\omega_1}{\omega_2}\right). \tag{5.77}$$

This is an important result, since it appears to be a good approximation for a variety of attenuation laws in which Q is effectively constant over the seismic frequency range (see Problem 5.10). However, specific attenuation laws have been proposed that do have an effectively constant Q, but for which the dispersion is *not* given accurately by (5.77) (see, e.g., Strick, 1970, and Problem 5.11). To see how this can happen, note from (5.73) that the phase delay at frequency ω, i.e., $x/c(\omega)$, is x/c_∞ plus an amount $x\omega^{-1}\,\mathscr{H}[\alpha(\omega)]$. Thus the phase delay at a *specific* frequency depends on an integral (the Hilbert transform) of the attenuation spectrum over *all* frequencies. If the asymptotic behavior of $\alpha(\omega)$ as $\omega \to 0$ and $\omega \to \infty$ is chosen so that the integral converges slowly, a very large phase delay can occur at seismic frequencies. In this circumstance, it is advisable for us to constrain the attenuation factor $\alpha(\omega)$ at very high and very low frequencies by returning to a discussion of the physical nature of anelasticity.

Suppose that a step-function stress $\sigma = \sigma_0 H(t)$ is applied to a solid that is initially in a state of zero stress and zero strain. For a linear medium, the resulting strain $\varepsilon(t)$ can be written as

$$M_U\varepsilon(t) = \sigma_0[1 + \phi(t)], \tag{5.78}$$

where M_U is an elastic modulus and $\phi(t)$ is known as the *creep function* for this modulus. (We shall discuss (5.78) as a relation between scalars, since in this section we are interested only in one-dimensional wave propagation.) For an isotropic elastic medium, $\phi = 0$ with $M_U = \mu$ for transverse waves, and $M_U = \lambda + 2\mu$ for longitudinal waves. For anelastic materials, there is an instantaneous strain σ_0/M_U in response to the applied stress (note that $\phi(t) = 0$ for $t \leq 0$). This is the reason for our use of a subscript U on the modulus: U stands for *unrelaxed* in the sense that M_U gives the proportionality between stress and strain as soon as the stress has been applied, before the material has started to relax (via creep) to some new configuration.

The stress-strain relation (5.78) is easily generalized to the case of general loading $\sigma = \sigma(t)$, to give

$$M_U\varepsilon(t) = \left[\sigma(t) + \int_{-\infty}^{t} \sigma(\tau)\dot{\phi}(t-\tau)\,d\tau\right],$$

as shown by Boltzmann in 1876. Since $\dot{\phi} = 0$ for $t \leq 0$, the above can conveniently be written as a convolution,

$$M_U \varepsilon(t) = \sigma(t) + \sigma(t) * \dot{\phi}(t). \tag{5.79}$$

Consider now the situation in which σ and ε are due to a plane wave propagating with fixed frequency ω in the x-direction: $\sigma = \sigma_0 \exp i(Kx - \omega t)$, with complex wavenumber K related to the phase velocity $c(\omega)$ and the attenuation factor $\alpha(\omega)$ by (5.71). The stress-strain relation then becomes

$$M_U \varepsilon(t) = \sigma(t)\left[1 + \int_0^\infty \dot{\phi}(\tau) \exp(i\omega\tau)\, d\tau\right] \tag{5.80}$$

(using the special form of $\sigma(t) \propto \exp(i\omega t)$ and $\dot{\phi}(t) = 0$ for $t \leq 0$). It follows that stress is proportional to strain via a complex modulus $M(\omega)$, where

$$M(\omega) = M_U \bigg/ \left[1 + \int_0^\infty \dot{\phi}(\tau) \exp(i\omega\tau)\, d\tau\right]. \tag{5.81}$$

The continuity of ϕ at the origin implies that the integral here tends to zero as $\omega \to \infty$, showing that $M(\omega) \to M_U$ as $\omega \to \infty$.

From the equation of motion $\rho\ddot{u} = \partial\sigma/\partial x$, we obtain $\rho\omega^2 = K^2 M(\omega)$, from which follow the relations

$$\begin{aligned} \frac{\omega}{c(\omega)} + i\alpha(\omega) &= \omega\sqrt{\frac{\rho}{M(\omega)}} = \frac{\omega}{c_\infty}\left[1 + \int_0^\infty \dot{\phi}(t) \exp(i\omega t)\, dt\right]^{1/2}, \\ \frac{1}{Q(\omega)} = \frac{2c(\omega)\alpha(\omega)}{\omega} &= -\frac{\operatorname{Im} M(\omega)}{\operatorname{Re} M(\omega)} = \frac{\operatorname{Im}\left\{\int_0^\infty \dot{\phi}(t) \exp(i\omega t)\, dt\right\}}{1 + \operatorname{Re}\left\{\int_0^\infty \dot{\phi} \exp(i\omega t)\, dt\right\}}. \end{aligned} \tag{5.82}$$

We have here identified a velocity c_∞ in terms of the unrelaxed modulus via $c_\infty = (M_U/\rho)^{1/2}$, and (5.82) is now in a form that permits us to translate various creep laws into laws of attenuation and dispersion.

Thus Lomnitz (1956, 1957) summarized his laboratory observations of creep in rocks by the logarithmic law

$$\phi(t) = \begin{cases} 0 & t \leq 0 \\ q \ln(1 + at) & t \geq 0 \end{cases}$$

in which the fundamental frequency a may be as high as the vibration frequency of a vacancy in the crystal lattice (i.e., of the order of 10^{10} Hz; see Savage and

O'Neill, 1975). The Fourier transform of $\dot{\phi}$ is then

$$aq \int_0^\infty \frac{\exp(i\omega t)}{1 + at}\, dt \sim -q\left[\gamma + \ln\left(\frac{\omega}{a}\right) - \frac{i\pi}{2}\right] e^{-i\omega/a} \tag{5.83}$$

for $\omega \ll a$, where $\gamma = 0.577\ldots$ is Euler's constant. Since $q \ll 1$, we find from (5.82) and (5.83) that

$$\frac{\omega}{c(\omega)} + i\alpha(\omega) = \frac{\omega}{c_\infty}\left\{1 - \frac{q}{2}\left[\gamma + \ln\left(\frac{\omega}{a}\right)\right] + \frac{i\pi q}{4}\right\} \tag{5.84}$$

(correct to first order). The imaginary part of this expression yields

$$q = \frac{2}{\pi Q}\frac{c_\infty}{c(\omega)} \sim \frac{2}{\pi Q}$$

for one of Lomnitz' constants. From this result and the real part of (5.84), the frequency dependence of c is found via

$$\frac{1}{c(\omega)} = \frac{1}{c_\infty}\left\{1 - \frac{1}{\pi Q}\left[\gamma + \ln\left(\frac{\omega}{a}\right)\right]\right\},$$

which is in agreement with the relation

$$\frac{c(\omega_1)}{c(\omega_2)} = 1 + \frac{1}{\pi Q}\ln\left(\frac{\omega_1}{\omega_2}\right) \qquad \text{(5.77 again)}$$

established above for Azimi's attenuation law.

Having obtained this same dispersion law via two such completely different methods (the empirical form (5.75) for attenuation, and the laboratory observation of logarithmic creep), it is satisfying also to know that Boltzmann's approach does successfully yield (5.77) again if a creep function is synthesized from well-understood mechanisms so as to give an effectively constant Q over seismic frequencies (Liu et al., 1976). Thus a physical basis for (5.77) is available directly from the observations of constant Q. The assumption made by Liu is that attenuation is due to a superposition of different relaxation phenomena, each one of which (if it acted alone) would be represented by the stress-strain relation

$$\sigma + \tau_\sigma \dot{\sigma} = M_R(\varepsilon + \tau_\varepsilon \dot{\varepsilon}).$$

(Many of the physical mechanisms proposed to explain attenuation appear to satisfy this relation, which is the stress-strain law for a *standard linear solid*, first studied in detail by Zener, 1948.) Here τ_ε is the characteristic relaxation time of strain under an applied step in stress, and τ_σ is the relaxation time for stress corresponding to a step change in strain. M_R is an elastic modulus, and note that the response to a step-function load $\sigma(t) = \sigma_0 H(t)$ is

$$\varepsilon(t) = \frac{\sigma_0}{M_R}\left[1 - \left(1 - \frac{\tau_\sigma}{\tau_\varepsilon}\right)e^{-t/\tau_\varepsilon}\right]. \tag{5.85}$$

Comparing this with (5.78) and giving special consideration to the instantaneous response at $t = 0$, we can identify

$$M_R = M_U \frac{\tau_\sigma}{\tau_\varepsilon}$$

and

$$\phi(t) = \left(\frac{\tau_\varepsilon}{\tau_\sigma} - 1\right)(1 - e^{-t/\tau_\varepsilon}). \tag{5.86}$$

M_R is known as the *relaxed modulus*, since (from (5.85)) it gives the ratio of stress to strain in the limit as $t \to \infty$. The transform

$$\int_0^\infty \dot{\phi}(t) \exp(i\omega t)\, dt$$

is needed in (5.82) to obtain Q values and the dispersion, and it is simple to show from the creep function (5.86) that

$$\frac{1}{Q} = \frac{\omega(\tau_\varepsilon - \tau_\sigma)}{1 + \omega^2 \tau_\sigma \tau_\varepsilon}$$

and

$$[c(\omega)]^2 = \frac{M_U}{\rho}\left[1 + \left(\frac{M_U}{M_R} - 1\right)\frac{1}{1 + \omega^2 \tau_\varepsilon^2}\right]^{-1}.$$

These two quantities are plotted in Figure 5.14, and note that attenuation is concentrated near the frequency $(\tau_\sigma \tau_\varepsilon)^{-1/2}$, with Q^{-1} behaving like ω for frequencies below this central peak and like ω^{-1} for frequencies above it. Moreover, the phase velocity $c(\omega)$ increases monotonically with frequency, the lower limit at $\omega = 0$ being $(M_R/\rho)^{1/2}$.

In order to reproduce the effectively constant Q values observed at seismic frequencies, Liu et al. (1976) have used a discrete superposition of twelve relaxation peaks of the type shown in Figure 5.14, all sharing the same relaxed

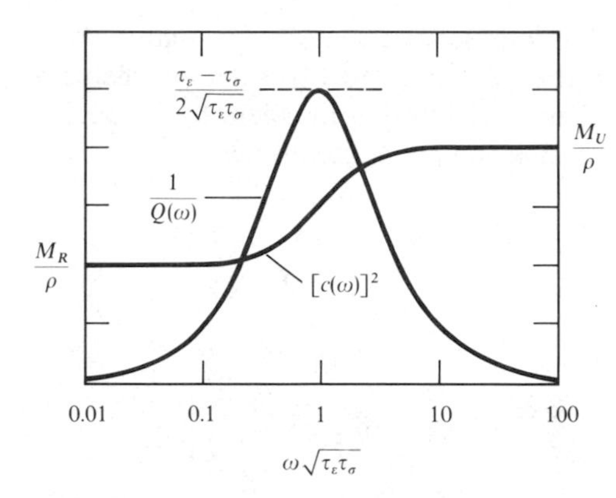

FIGURE **5.14**

The internal friction, Q^{-1}, and the square of the phase velocity, $[c(\omega)]^2$, as a function of frequency in a standard linear solid (i.e., a solid with a single relaxation mechanism). [After Liu et al., 1976.]

modulus M_R. They have chosen the 24 available anelastic parameters (τ_ε and τ_σ for each mechanism) to give Q^{-1} and phase velocity as shown in Figure 5.15. Over the range 0.0001–10 Hz, note that Q^{-1} is effectively constant within 1% and that phase velocity has a linear dependence on $\ln \omega$. Carrying the analysis one stage further, they have also considered a continuous superposition of relaxations, specified by a density function, and again find the linear dependence of $c(\omega)$ on $\ln \omega$ to be like that shown in Figure 5.15. In fact, they are able to find the gradient of $c(\omega)$ analytically, and show that the ratio $c(\omega_1)/c(\omega_2)$ has the now-familiar value $1 + (\pi Q)^{-1} \ln(\omega_1/\omega_2)$.

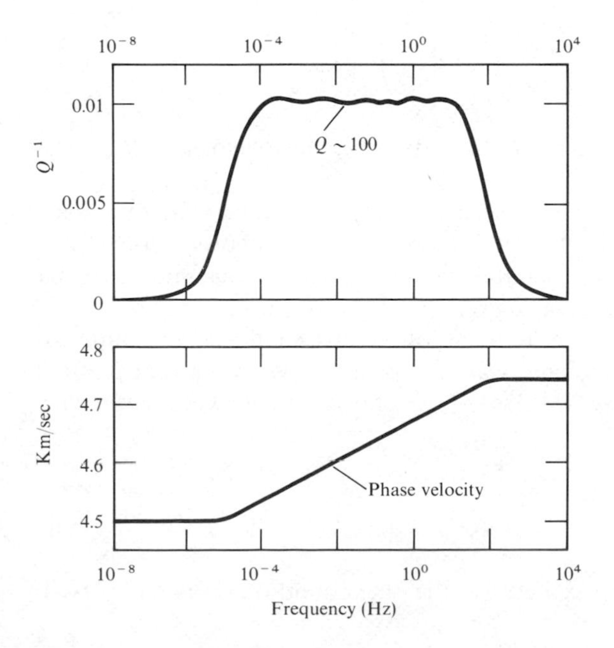

FIGURE **5.15**

Top. Internal friction values. Bottom. Phase velocities obtained (via (5.82)) from a superposition of 12 different relaxation peaks. [After Liu et al., 1976.]

This logarithmic dispersion now has ample justification for materials in which Q is observed to be constant, so that it is clear how to solve problems of wave propagation in anelastic media once the parallel problem for an elastic medium has been solved. We have seen that the propagation factor

$$\exp[i(kx - \omega t)] = \exp[i\omega(x/c - t)]$$

for elastic media is to be replaced by

$$\exp i(Kx - \omega t) = \exp\left[\frac{(-\omega x)}{2c(\omega)Q(\omega)}\right] \exp\left[i\omega\left(\frac{x}{c(\omega)} - t\right)\right].$$

Although c appears as a frequency-independent constant in the elastic solution, $c = c_e$, that solution will be an analytic function of c_e. It then follows by analytic continuation that the anelastic solution is given by replacing c_e via the rule

$$\frac{1}{c_e} \to \frac{1}{c(\omega)}\left(1 + \frac{i}{2Q(\omega)}\right). \tag{5.87}$$

If Q is effectively constant, then we can expect to be able to use the logarithmic dispersion law (5.77). We need here to fix a reference frequency against which dispersion can be recognized, and we shall choose 1 Hz (i.e., $\omega = 2\pi$). Then the elastic velocity c_e is replaced via the rule

$$c_e \to c_1\left[1 + \frac{1}{\pi Q}\ln\left(\frac{\omega}{2\pi}\right) - \frac{i}{2Q}\right] \tag{5.88}$$

(correct to first order in Q), where c_1 is the body-wave phase velocity at frequency 1 Hz.

Throughout this section, we have so far assumed that the attenuating plane wave of interest is a one-dimensional wave. Thus, for a propagating wave described by $\exp(-\alpha x) \exp i\omega[x/c(\omega) - t]$, the direction of maximum attenuation is also the direction of increasing phase delay. For plane waves in general, these two directions are different. In what follows, we shall give an example of the rule (5.87) for two-dimensional wave propagation, showing that it is still fruitful to analyze plane waves in terms of their horizontal slowness, even when the wave is attenuating.

We consider solutions to

$$\nabla^2\psi + K^2\psi = 0, \tag{5.89}$$

where $K^2 = \omega^2\rho/\mu$ is complex because the shear modulus μ is complex. In general, the steady-state plane wave solution to (5.89) is a constant times

$$\exp(-\mathbf{A} \cdot \mathbf{x}) \exp[i(\mathbf{P} \cdot \mathbf{x} - \omega t)], \tag{5.90}$$

where **A** and **P** are vectors with real Cartesian components. **A** is the direction of maximum attenuation, and **P** is the direction of wave propagation (i.e., of increasing phase delay). Substituting (5.90) into (5.89) yields

$$P^2 - A^2 = \mathrm{Re}\{K^2\}, \tag{5.91}$$

$$PA \cos \gamma = \tfrac{1}{2} \mathrm{Im}\{K^2\}, \tag{5.92}$$

where P and A are the amplitudes of **P** and **A**, and γ is the angle between **P** and **A**. The wave is said to be *homogeneous* or *inhomogeneous* according to whether $\gamma = 0$ or $\gamma \neq 0$. We met these terms before in Section 5.3 in the context of elastic media. As shown by Buchen (1971) and Borcherdt (1973), for elastic media one must either have $A = 0$ or $\gamma = \pi/2$. This follows from (5.92), with $\mathrm{Im}\{K^2\} = 0$. But, for anelastic media, necessarily $A \neq 0$ and $\gamma \neq \pi/2$. Previously in this section, we have considered only $A \neq 0$ and $\gamma = 0$, which is a homogeneous attenuating wave. But to handle the effect of an interface between anelastic half-spaces, we must consider inhomogeneous plane waves, with $0 < \gamma < \pi/2$.

Fortunately, these more general plane waves can still be studied as a function of the independent variable p, because

$$\exp[i\omega px + i(K^2 - \omega^2 p^2)^{1/2} z - i\omega t] \tag{5.93}$$

is also a general plane-wave solution to the wave equation. Solutions (5.93) and (5.90) are equivalent, with the y-axis perpendicular to **P** and **A**. The Cartesian components of **P** and **A** are

$$\begin{aligned} \mathbf{P} &= (\mathrm{Re}\{\omega p\}, 0, \mathrm{Re}\{(K^2 - \omega^2 p^2)^{1/2}\}), \\ \mathbf{A} &= (\mathrm{Im}\{\omega p\}, 0, \mathrm{Im}\{(K^2 - \omega^2 p^2)^{1/2}\}). \end{aligned} \tag{5.94}$$

These expressions allow for the possibility that p might be complex, and this we shall find is desirable in Chapters 6–9. The (physical) horizontal slowness is $\mathrm{Re}\{p\}$.

Buchen (1971) and Borcherdt (1973, 1977) have extensively studied the properties of plane waves in terms of the vectors **P** and **A**. Physical properties of individual plane waves are then easy to identify. For example, the angle of incidence to the vertical is simply θ, where

$$\mathbf{P} \cdot \hat{\mathbf{z}} = P \cos \theta. \tag{5.95}$$

The phase velocity is ω/P, and it follows from $K^2 = \omega^2\rho/\mu$ and (5.91)–(5.92) that this velocity is exactly

$$\left(\frac{\text{Re}\{\mu\}}{\rho}\right)^{1/2}\left(\frac{2(1 + Q^{-2})}{1 + (1 + \sec^2\gamma Q^{-2})^{1/2}}\right)^{1/2}. \tag{5.96}$$

Note that (5.96) is based on a Q such that $Q^{-1} = -\text{Im}\{\mu\}/\text{Re}\{\mu\}$; see (5.82). This is probably the best of several different definitions that scientists have used for Q (see O'Connell and Budiansky, 1978). In Box 5.7, we noted several properties, derived when $Q \gg 1$, that can be used as the definition of Q. But these definitions will disagree with each other if they are applied for strongly attenuating media.

Because two angles (θ and γ) are now needed to characterize a plane wave incident upon an interface, Snell's law takes an extremely complicated form in terms of physical quantities such as **P** and **A** for each wave at the interface (Borcherdt, 1977). But in terms of plane waves expressed via (5.93), with K taking different (complex) values on either side of the interface, Snell's law reduces to the familiar rule that parameter p is the same for all plane waves coupled by the interface. This allows us very quickly to generalize all the formulas we previously obtained for reflection/conversion/transmission coefficients in elastic media, making them applicable in the anelastic case. Thus, to obtain the $\grave{S}\grave{S}$ transmission coefficient for SH-waves crossing between two anelastic half-spaces (see Fig. 5.7a), we must first interpret the angles j_1 and j_2 in terms of the independent parameter p:

$$\frac{\sin j_1}{\beta_1} = p = \frac{\sin j_2}{\beta_2},$$
$$\frac{\cos j_1}{\beta_1} = \left(\frac{1}{\beta_1^2} - p^2\right)^{1/2} = \eta_1, \tag{5.97}$$
$$\frac{\cos j_2}{\beta_2} = \left(\frac{1}{\beta_2^2} - p^2\right)^{1/2} = \eta_2.$$

The incident wave here is proportional to $\exp[i\omega(px + \eta_1 z - t)]$, and the algebra that led to (5.32) gives us here

$$\grave{S}\grave{S} = \frac{2\rho_1\beta_1^2\eta_1}{\rho_1\beta_1^2\eta_1 + \rho_2\beta_2^2\eta_2} = \frac{2\mu_1\eta_1}{\mu_1\eta_1 + \mu_2\eta_2}. \tag{5.98}$$

Care must be taken in working out the angles of incidence and emergence of the transmitted wave $\grave{S}\grave{S}\exp[i\omega(px + \eta_2 z - t)]$. For this we must use (5.94) and the angle θ in (5.95), determined for each individual plane wave. Note in general that the angles of incidence and emergence will not be j_1 and j_2, since these angles have meaning only via (5.97), and we have chosen to use p as the independent variable (rather than, say, j_1).

Inspection will show that our coefficient $\grave{S}\acute{S}$ in (5.98) depends directly upon complex body-wave velocities β_1 and β_2 (or upon μ_1 and μ_2) and p. Our earlier description of attenuation was limited to one-dimensional homogeneous waves. It is satisfying that the complex elastic constants we introduced for these simple waves are all that we need to evaluate two-dimensional plane waves, even though the latter may be inhomogeneous waves. Thus reflection/transmission/conversion coefficients for an interface in attenuating media may be obtained from the analogue in elastic media (e.g., (5.32)) merely by substituting complex elastic constants, provided we are careful to interpret such concepts as "angle of incidence" and "angle of emergence."

5.6 Wave Propagation in an Elastic Anisotropic Medium: Basic Theory for Plane Waves

Steady-state plane waves in a homogeneous anisotropic medium will propagate with slowness $\mathbf{s}$, provided that the displacement $\mathbf{u} = \mathbf{U} \exp[-i\omega(t - \mathbf{s} \cdot \mathbf{x})]$ satisfies $\rho \ddot{u}_i = c_{ijkl} u_{k,lj}$. $\mathbf{U}$ is a constant, the polarization vector, giving the direction of particle motion, and for the wave equation to be satisfied we require

$$\det|\rho\delta_{ik} - c_{ijkl}s_j s_l| = 0. \tag{5.99}$$

In isotropic media, (5.99) reduces to three separate second-order equations for slowness components, corresponding to P-waves and two types of S-wave (see (4.36)–(4.39) and (5.2)). But for the general anisotropic case, it is not possible to solve analytically for $\mathbf{s}$, or for $|\mathbf{s}|$, directly in terms of ρ and components of $\mathbf{c}$.

A useful aid in visualizing the allowed slowness of wavefronts is the *slowness surface*, S. Using Cartesian axes (s_1, s_2, s_3) for points in slowness space, S is composed of points that satisfy (5.99). For isotropic media, (4.40) indicates that S consists of three concentric spheres, one with radius $[\rho/(\lambda + 2\mu)]^{1/2}$ and two, which are coincident, with radius $[\rho/\mu]^{1/2}$. But more generally, S has three sheets that are separate and nonspherical.

From the slowness surface, it is possible to construct another three-sheeted surface, $W(t)$, for the position (at a given time t) of the wavefronts associated with a point source. Clearly, W is related to the solutions (x_1, x_2, x_3) that satisfy $t = \mathbf{s} \cdot \mathbf{x}$. The problem here is that, for each $\mathbf{x}$ in the wavefront, there will be a different $\mathbf{s}$ on the slowness surface. We define W to be the envelope of planes $t = \mathbf{s} \cdot \mathbf{x}$, as $\mathbf{s}$ varies over the slowness surface. Geometrical relationships between a point $\mathbf{s}$ on S and the corresponding point $\mathbf{x}$ on W are described by Musgrave (1970). The normal to S at $\mathbf{s}$ is parallel to $\mathbf{x}$. From this a graphical construction of W can be devised (since the direction of $\mathbf{x}$ is now known, and $|\mathbf{x}|$ is determined by $t = \mathbf{s} \cdot \mathbf{x}$). The reciprocal relation also holds, and perhaps

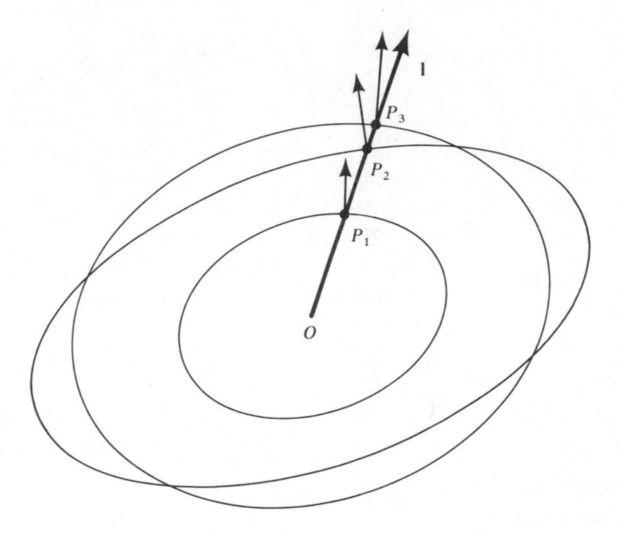

FIGURE **5.16**

The three-sheeted slowness surface S. There are three possible slownesses in in the $\mathbf{l}$ direction, with magnitudes OP_1, OP_2, and OP_3. Normals to S at P_1, P_2, P_3 give the directions of energy propagation (rays) for each of the three slownesses.

this is more important from the seismological point of view: the normal to W at $\mathbf{x}$ is parallel to $\mathbf{s}$. The simplest observable quality of a wavefront is likely to be the direction (normal to itself) in which it advances, and this is given by $\mathbf{s}$. However, as described by Vlaar (1968), the ray direction, which is the direction of energy transport, is not in general along $\mathbf{s}$, but lies along the normal to the slowness surface at $\mathbf{s}$. A general method for computing the geometrical spreading for wavefronts in anisotropic inhomogeneous media is given by Červený (1972).

For a given unit direction $\mathbf{l}$ of slowness, there are three different values of the magnitude $|\mathbf{s}|$ that make $\mathbf{s} = |\mathbf{s}|\mathbf{l}$ lie on the slowness surface (see Fig. 5.16). These correspond to three different body waves, and the magnitudes of their slowness are found by solving the eigenvalue problem

$$\det[\mathbf{M} - (\rho/|\mathbf{s}|^2)\mathbf{I}] = 0, \tag{5.100}$$

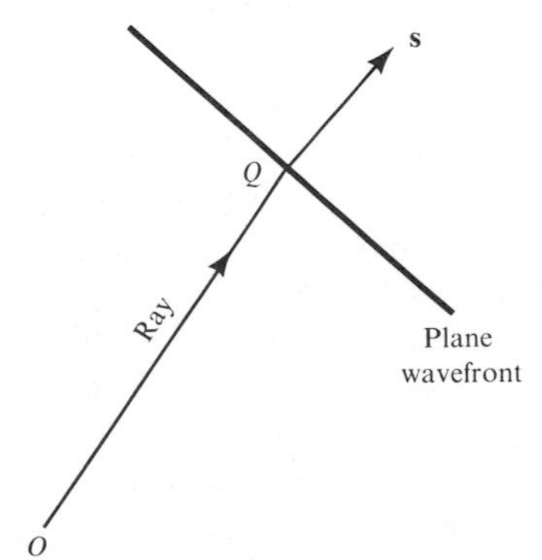

FIGURE **5.17**

For a plane wavefront, the ray direction (given by the energy flux) is not, in general, perpendicular to the wavefront, nor does it necessarily lie in the sagittal plane.

where $M_{ip} = c_{ijpq}l_jl_q$ and $I_{ip} = \delta_{ip}$. **M** is symmetric, and $M_{ip}U_p = (\rho/|\mathbf{s}|^2)U_i$, so that the three eigenvectors for the polarization **U** are mutually orthogonal.

To prepare for an analysis of horizontal interfaces between homogeneous half-spaces, we have previously taken an x_2-direction so that $\mathbf{s} = (p, 0, s_3)$. Snell's law for interfaces normal to x_3 reduces to $s_1 = p$, $s_2 = 0$ for all plane waves coupled at the interface, and s_3 is different for each wave. The same approach is helpful in anisotropic media, and Figure 5.18 gives a construction for the s_3-components for plane waves on either side of a horizontal interface between two half-spaces. The plane portrayed in Figure 5.18, containing all the coupled slowness vectors, is known as the *sagittal plane*. This also is the plane of incidence, and if it is varied then one may rotate the (x_1, x_2) axes to keep $s_2 = 0$. The penalty for this is that Cartesian components of **c** must be transformed to the new coordinate system.

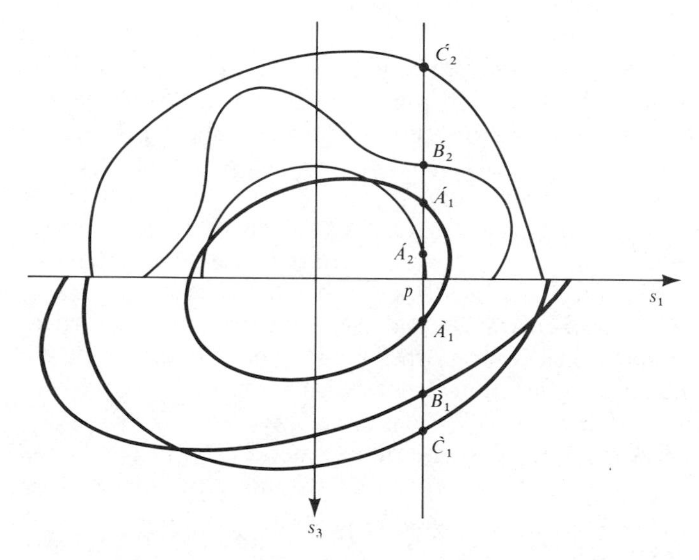

FIGURE **5.18**

Graphical method for determining the vector slowness for each plane wave scattered from a horizontal interface between two anisotropic half-spaces. Heavy curves are part of the three-sheeted slowness surface for the lower medium, and an incident wave (propagating upward) has slowness coordinates given by the point $\acute{A}_1$. The six scattered waves all share the same horizontal slowness p. Positive values of s_3 are for downgoing waves in the lower medium, and negative values (lying on medium-weight curves, which are part of the slowness surface for the upper medium) are for the transmitted waves. If p is large enough for some or all of the s_3 values to be imaginary, then corresponding plane waves are exponentially growing or decaying with depth.

A good way to quantify the reflection and transmission properties for an interface is to develop the equations of motion (on either side of the interface) in the form $d\mathbf{f}/dz = \mathbf{Af}$, where now

$$\mathbf{f}\exp[i\omega(px - t)] = (u_x, u_y, u_z, \tau_{zx}, \tau_{yz}, \tau_{zz})^T. \tag{5.101}$$

$\mathbf{A}$ is a 6×6 matrix, and one finds its six eigenvalues are the six values of $i\omega s_3$ where roots s_3 are shown graphically (for given p and a given half-space) in Figure 5.18. We form a matrix $\mathbf{E}$ whose columns are eigenvectors of $\mathbf{A}$ and a phase matrix $\mathbf{\Lambda}$ that is diagonal with six entries $\exp(i\omega s_3 z)$ for the six different s_3. With an appropriate ordering of columns of $\mathbf{E}$ and $\mathbf{\Lambda}$, the general plane wave solution in each homogeneous medium is given by $\mathbf{f} = \mathbf{E\Lambda}\,\mathbf{w}$, where w is a 6-vector of constants. The methods of Section 5.4 can now be used to obtain the 36 different reflection/transmission coefficients as a function both of p and the orientation of the sagittal plane. (The matrix $\mathbf{A}$ depends in general upon this orientation.) Some numerical examples are given by Keith and Crampin (1977).

There are several special types of anisotropy that have some importance in seismology. The simplest departure from isotropy is that for which the tensor of elasticity ($\mathbf{c}$) is isotropic in all directions normal to one particular direction, x_3. Such a medium is said to be *transversely isotropic*, and is characterized by five independent constants. The coefficient matrix for $d\mathbf{f}/dz = \mathbf{Af}$ in transversely isotropic material has been extensively analyzed by Takeuchi and Saito (1972) with the anomalous direction x_3 taken as vertical. In this case there is remarkably little difference from isotropic media. SH decouples from quasi P and quasi SV, which are often written qP and qSV, and the symmetry about a vertical axis means that wave solutions do not depend upon the orientation of the sagittal plane. This major simplification is not possible for any other type of anisotropy.

Where anisotropy is weak but otherwise quite general, Backus (1965) has shown that the velocity of body waves propagating within a particular plane has a simple dependence upon direction of propagation within that plane. For example, qP waves propagating in the (x_1, x_2) plane with slowness $(\cos\phi, \sin\phi, 0)/\alpha$ have phase velocity $\alpha(\phi)$ given approximately by

$$\rho\alpha^2 = A + B\cos 2\phi + C\sin 2\phi + D\cos 4\phi + E\sin 4\phi. \tag{5.102}$$

Backus's discovery of this form for $\alpha = \alpha(\phi)$ was prompted by Hess's (1964) observation from published data that P-waves refracted along the underside of the Moho beneath oceans appeared to have several percent velocity anisotropy. In turn, the specific form (5.102) has stimulated attempts (e.g., Crampin and Bamford, 1977) to invert observations of α in terms of five other elastic parameters (the coefficients A, B, C, D, E, which themselves are linear combinations of components of $\mathbf{c}$).

SUGGESTIONS FOR FURTHER READING

Coddington, E. A., and N. Levinson. *Theory of Ordinary Differential Equations* (Chap. 6). New York: McGraw-Hill, 1955.

Gantmacher, F. R. *The Theory of Matrices* (2 vols.). New York: Chelsea Publishing Co., 1959.

Kraut, E. A. Advances in the theory of anisotropic elastic wave propagation. *Reviews of Geophysics*, **1**, 401–448, 1963.

Liu, H. -P., D. L. Anderson, and H. Kanamori. Velocity dispersion due to anelasticity; implications for seismology and mantle composition. *Geophysical Journal of the Royal Astronomical Society*, **47**, 41–58, 1976.

Randall, M. J. Attenuative dispersion and frequency shifts of the earth's free oscillations. *Physics of the Earth and Planetary Interiors*, **12**, 1–4, 1976.

White, J. E. *Seismic Waves*. New York: McGraw-Hill, 1965.

Zener, C. *Elasticity and Anelasticity of Metals*. University of Chicago Press, 1948.

PROBLEMS

5.1 If the surface of a solid elastic half-space is traction-free, we found that there can exist a surface wave of displacement (the Rayleigh wave). If the surface is rigid, show that there can be no corresponding surface wave of tractions. (*Hint*: Find a determinant that does not vanish for appropriate values of p.)

5.2 Why can there be no interface SH-wave at the boundary between two homogeneous half-spaces?

5.3 For a solid half-space $z > 0$ (see Fig. 5.5 and Table 5.1), obtain the P–SV scattering matrices

$$\begin{pmatrix} \acute{P}\grave{P} & \acute{S}\grave{P} \\ \acute{P}\grave{S} & \acute{S}\grave{S} \end{pmatrix}$$

when the boundary condition on $z = 0$ is given (a) by $u_z = 0$ and $\tau_{zx} = 0$, and (b) by $u_x = 0$ and $\tau_{zz} = 0$. (You can reduce the algebra by working from equations (5.33).)

If a combination of P-wave and SV-wave energy is propagating toward the surface $z = 0$, show that by adding the reflections derived from boundary conditions (a) to the reflections corresponding to (b), all downward reflections are eliminated. (Smith, 1974, 1975, has shown that this is a useful way to eliminate unwanted reflections from grid boundaries in numerical work with the Finite Element and Finite Difference methods.)

5.4 a) Show briefly that the inhomogeneous P-wave (5.50) and the inhomogeneous SV-wave (5.51) each have *prograde* elliptical particle motion.
b) From (5.50)–(5.54), show for Rayleigh waves that the particle displacement vector at the free surface ($z = 0$) is proportional to the vector

$$\left(\frac{2i}{c_R}\sqrt{\frac{1}{c_R^2} - \frac{1}{\beta^2}},\ 0,\ \frac{2}{c_R^2} - \frac{1}{\beta^2}\right).$$

c) Show then that particle motion for the free surface is *retrograde* elliptical.
d) Give a brief argument to show that particle motion at sufficient depth is again prograde elliptical.

5.5 Show that it is the *horizontal* (and not the *vertical*) component of the Rayleigh wave that goes through zero as depth increases, as shown in Figure 5.11.

5.6 Since seismometers are often placed on (or very near to) the traction-free surface of the Earth, it is of interest to obtain the total displacement of the free surface of a half-space due to a P, SV, or SH plane wave incident from below. Using the notation of Table 5.1 and Figure 5.5, show that this total displacement at the surface is

$$\frac{\acute{P}\left[\dfrac{4\alpha p}{\beta^2}\dfrac{\cos i}{\alpha}\dfrac{\cos j}{\beta},\ 0,\ \dfrac{-2\alpha}{\beta^2}\dfrac{\cos i}{\alpha}\left(\dfrac{1}{\beta^2} - 2p^2\right)\right]\exp[i\omega(px - t)]}{\left(\dfrac{1}{\beta^2} - 2p^2\right)^2 + 4p^2\dfrac{\cos i}{\alpha}\dfrac{\cos j}{\beta}}$$

for a P-wave (of amplitude $\acute{P}$) incident from below; and is

$$\frac{\acute{S}\left[\dfrac{2}{\beta}\dfrac{\cos j}{\beta}\left(\dfrac{1}{\beta^2} - 2p^2\right),\ 0,\ \dfrac{4p}{\beta}\dfrac{\cos i}{\alpha}\dfrac{\cos j}{\beta}\right]\exp[i\omega(px - t)]}{\left(\dfrac{1}{\beta^2} - 2p^2\right)^2 + 4p^2\dfrac{\cos i}{\alpha}\dfrac{\cos j}{\beta}}$$

for an SV-wave (of amplitude $\acute{S}$) incident from below. (In the range $1/\alpha < p < 1/\beta$, this last formula can be used with the positive imaginary value $(\cos i)/\alpha = i\sqrt{p^2 - 1/\alpha^2}$ to give the phase shift in the surface displacement when SV is incident at a supercritical angle $j > j_c = \sin^{-1}(\beta/\alpha)$.)

For incident SH, show that particle motion in the free surface is simply double the particle motion in the incident wave.

5.7 Given the form of $\mathbf{f}$ defined in (5.60), show that indeed $\partial\mathbf{f}/\partial z = \mathbf{Af}$ for P–SV waves, provided that $\mathbf{A}$ has the form given in (5.60). Show that $\mathbf{A}$ still has this form when ρ, λ, and μ are functions of z, but that (5.61) does not then in general solve (5.56).

5.8 Consider the P-wave potential $\phi = A\exp[i\omega(px + \xi z - t)]$, which, for a homogeneous medium, satisfies $\alpha^2\nabla^2\phi = \ddot{\phi}$ provided that $\xi = (\alpha^{-2} - p^2)^{1/2}$. Show that u_x, u_z, τ_{zx}, τ_{zz} constructed from this potential forms essentially the first column of $\mathbf{F} = \mathbf{E\Lambda}$, given in (5.65). (The difference is due only to a different normalization of this downgoing P-wave.)

Repeat the discussion for SV-potential $B\exp[i\omega(px + \eta z - t)]$ (with $\eta = (\beta^{-2} - p^2)^{1/2}$), P-potential $C\exp[i\omega(px - \xi z - t)]$, and SV-potential $D\exp[i\omega(px - \eta z - t)]$. (The point here is that different columns of $\mathbf{F}$ in (5.65) are downgoing and upgoing P- and SV-waves in which the displacement and stress components are given explicitly for each wave.)

Show that these four waves, described by potentials, are equivalent to the wave system $\mathbf{f} = \mathbf{F}\mathbf{w}$, in which

$$\mathbf{w} = \begin{pmatrix} \dfrac{i\omega A}{\alpha} \\[2mm] -\dfrac{i\omega B}{\beta} \\[2mm] \dfrac{i\omega C}{\alpha} \\[2mm] \dfrac{i\omega D}{\beta} \end{pmatrix}.$$

5.9 Show that Q is in general an even function ω (i.e., $Q(-\omega) = Q(\omega)$), and hence that the attentuated impulse described by (5.68) is always symmetric about $t = x/c$ if there is no dispersion.

5.10 Obtain the approximation (5.77) in the cases:

a) $\alpha(\omega) = \dfrac{\alpha_0\omega}{1 + \alpha_2\omega^2}$,

b) $\alpha(\omega) = \alpha_0\omega\{H(\omega - \omega_l) - H(\omega - \omega_h)\}$

(ω_l and ω_h being low and high frequency cut-offs and H the Heaviside step function); use $0 < \omega_l \ll \omega \ll \omega_h$ in deriving (5.77). Hilbert transforms of the above two attenuation functions will be needed. Note the rule $\alpha(-\omega) = \alpha(\omega)$; definitions above are for $\omega > 0$. Corresponding $\mathscr{H}[\alpha(\omega)]$ are then

$$\frac{\alpha_0}{\pi}\frac{\omega}{1 + \alpha_2\omega^2}\ln\left(\frac{1}{\alpha_2\omega^2}\right) \qquad \text{for (a)}$$

and

$$\frac{\alpha_0\omega}{\pi}\ln\left\{\frac{\omega_h^2 - \omega^2}{\omega_l^2 - \omega^2}\right\} \qquad \text{for (b)}.$$

5.11 Azimi et al. (1968) and Strick (1970) examined the attenuation law $\alpha(\omega) = \alpha_0|\omega|^s$ for s just less than one. This can give an effectively linear dependence of $\alpha(\omega)$ on ω over observed frequencies. From relation (13) of Box 5.8 and the Hilbert transform $\mathscr{H}[\operatorname{sgn}\omega|\omega|^{s-1}] = |\omega|^{s-1}\tan s\pi/2$, show here that

$$\frac{1}{c(\omega)} = \frac{1}{c_\infty} + \frac{\alpha(\omega)}{|\omega|}\tan\frac{s\pi}{2}.$$

For the above attenuation law to be effectively linear, $\alpha(\omega)/|\omega|$ must effectively be constant. Using the last equation above, show that relative dispersion between different frequency components over the seismic frequency range might be small, and hence hard to detect, though absolute dispersion (with respect to the phase velocity c_∞) may be large for s only just less than one. (This curious result is due to the slow convergence of the Hilbert transform integral. The amplitude spectrum at megahertz frequencies and higher is controlling the phase delay at seismic frequencies. This is implausible, and is avoided because more reasonable attenuation laws have different high-frequency behavior.)

5.12 If $Q^{-1} \ll 1$ and (5.96) is expanded by the binomial theorem, we find that the phase velocity in general for an attenuating plane S-wave is $(\mathrm{Re}\{\mu\}/\rho)^{1/2}$ (1 + term of order $1/Q^2$). Does this violate an important conclusion of Section 5.5; namely, that attenuation in a causal medium has an effect on velocity that introduces a correction of order $1/Q$?

5.13 An attenuating medium that has all its losses confined to shear and none to pure compression can be characterized as having a complex rigidity, but a bulk modulus that is purely real. Show for such a medium that the ratio of Q for P-waves to Q for S-waves is $3\alpha^2/(4\beta^2)$, where α and β are the two wave speeds (assume $Q \gg 1$).

5.14 For times $t \ll T$, show that one can regard the Heaviside function $H(t - T)$ as having the Hilbert transform $(-1/\pi)\ln|t/T|$.

CHAPTER 6

Reflection and Refraction of Spherical Waves; Lamb's Problem

As a general rule, it is fair to say that the elementary theory needed to explain seismic data falls into one or the other of two categories: one studies either the asymptotic approximations for the waves that propagate in realistic models of the Earth or the exact form of waves that propagate in highly idealized media. Thus the first category includes geometric ray theory (Chapter 4), and the second includes all of our work in Chapter 5 on plane waves and homogeneous half-spaces. As we develop more sophisticated methods, these two categories gradually merge, in the sense that we shall aim to calculate quite accurately the waves that propagate in fairly realistic Earth models.

The present chapter is mainly a contribution to the second category of theoretical development referred to above, for we shall be concerned with the calculation of waves emanating from a point source in a medium consisting of only two homogeneous half-spaces. In 1904, Horace Lamb gave the exact solution for a problem of this type, in which the source acted as an impulse applied normal to a point on the free surface of a solid half-space. However, since his paper contained so much more, the term "Lamb's problem" has now come to refer to more general source/medium geometries with a single interface, in which the principal interest is in the exact calculation (in the time domain) of waves emanating from an impulsive line source or point source.

We introduce this subject by expressing the spherical wave emanating from a point source as a superposition of plane waves (the Weyl integral) and then as a superposition of cylindrical waves (the Sommerfeld integral). When the spherical wave interacts with a plane boundary between two different half-spaces, the resulting wave systems can naturally be divided into three major types: (i) waves that are directly reflected from or transmitted through the boundary; (ii) waves that travel from source to receiver via a path involving refraction along the boundary at a body wave speed (head waves); (iii) and waves of Rayleigh or Stoneley type, with amplitude decaying exponentially with distance from the interface. These results are derived by manipulation of integration paths in the complex ray-parameter plane. After giving a general outline of the three major wave types, using the frequency domain, we develop exact methods of solution based on the work of Cagniard and de Hoop. These methods employ the Laplace transform of time, although final solutions are stated in the time domain. The basic division into three different wave types is again apparent, together with minor contributions from a leaking mode (another type of interface wave).

At the outset, we must emphasize that the best way to solve Lamb's problem is via Laplace transformation and the inversion methods of Cagniard. The discussion of integration paths in the complex ray-parameter plane is then relatively simple; and the actual inversion of the Laplace transform, to obtain pulse shapes in the time domain, is made trivial. A self-contained description of these methods is given here in Sections 6.4 and 6.5. We introduce this material, however, with a Fourier transform. In part, this is an acknowledgment to the vast literature on the subject, including books by Ewing et al. (1957), Brekhovskikh (1960), and Červený and Ravindra (1971), and many hundreds of papers. But the major reason for developing Fourier-transform methods in connection with Lamb's problem is to prepare the ground for Chapter 9, which gives practical methods for calculating seismograms in realistic structures. We shall find there that the reflectivity method for layered media and powerful solution methods for problems of grazing incidence are based on numerical work with the Fourier transform rather than analytic inversion of the Laplace transform.

6.1 Spherical Waves as a Superposition of Plane Waves and Cylindrical Waves

Consider an inhomogeneous wave equation with point source at the origin and time dependence $\exp(-i\omega t)$:

$$\frac{\partial^2 \phi}{\partial t^2} - c^2 \nabla^2 \phi = 4\pi c^2 \delta(\mathbf{x}) \exp(-i\omega t). \tag{6.1}$$

The solution of this equation (in an infinite homogeneous space) is obtained from (4.4) as

$$\phi(\mathbf{x}, t) = \frac{1}{R}\exp\left[i\omega\left(\frac{R}{c} - t\right)\right], \tag{6.2}$$

where $R = (x^2 + y^2 + z^2)^{1/2}$.

Equation (6.1) can also be solved by recognizing the time dependence of $\phi(\mathbf{x}, t)$ as the steady oscillation $\exp(-i\omega t)$ and then using Fourier-transform methods to derive the spatial dependence. We find that

$$\phi(\mathbf{k}, t) = [4\pi c^2/(k^2c^2 - \omega^2)]\exp(-i\omega t),$$

where $k^2 = k_x^2 + k_y^2 + k_z^2$. Then from (6.2) and the triple inverse transform of $\phi(\mathbf{k}, t)$,

$$\frac{1}{R}\exp\left[i\omega\left(\frac{R}{c} - t\right)\right] = \frac{\exp(-i\omega t)}{2\pi^2}\iiint_{-\infty}^{\infty}\frac{\exp(i\mathbf{k}\cdot\mathbf{x})}{k^2 - \dfrac{\omega^2}{c^2}}\,dk_x\,dk_y\,dk_z. \tag{6.3}$$

The left-hand side of (6.3) is a spherical wave propagating from the origin with speed c. Its amplitude is a function only of radial distance and has no directional variation. The right-hand side of (6.3) has a weight function $[2\pi^2(k^2 - \omega^2/c)^2]^{-1}$ and a superposition of plane waves $\exp[i(\mathbf{k}\cdot\mathbf{x} - \omega t]$ over the entire range of k_x, k_y, k_z. It therefore appears that we have already accomplished our aim of representing the spherical wave by superposing plane waves. However, a closer look at the plane waves in (6.3) shows that they have arbitrary velocity ($=\omega/k$) from 0 to ∞, so that they do not yet have the form of plane waves in a medium with given velocity c.

In order to obtain the Weyl integral, we must carry out an integration with respect to one of the wavenumber components; we shall do this for k_z, so that the integrals that remain will be over horizontal wavenumbers.

The integration over k_z in (6.3) is simple if we extend k_z to complex values and apply residue theory. For given k_x and k_y, the integrand has poles at $k_z = \pm[(\omega^2/c^2) - k_x^2 - k_y^2]^{1/2}$, and the only difficulty is that (for a certain range of values of k_x and k_y) these poles lie on the real k_z-axis, interfering with the integration path (see Fig. 6.1a). For convenience, we introduce a small attenuation by making $1/c$ complex. As we have seen in Chapter 5 (equations (5.87), (5.88)), the proper way of doing this is to regard $1/c$ as the sum of a certain reference value, plus a small complex correction that has a *positive* imaginary part. Thus $\text{Im}\{1/c\} = \varepsilon$, and $\varepsilon > 0$ (for $\omega > 0$). The immediate result is to change the pole locations into the first and third quadrants (see Fig. 6.1b), removing the interference with the integration path.

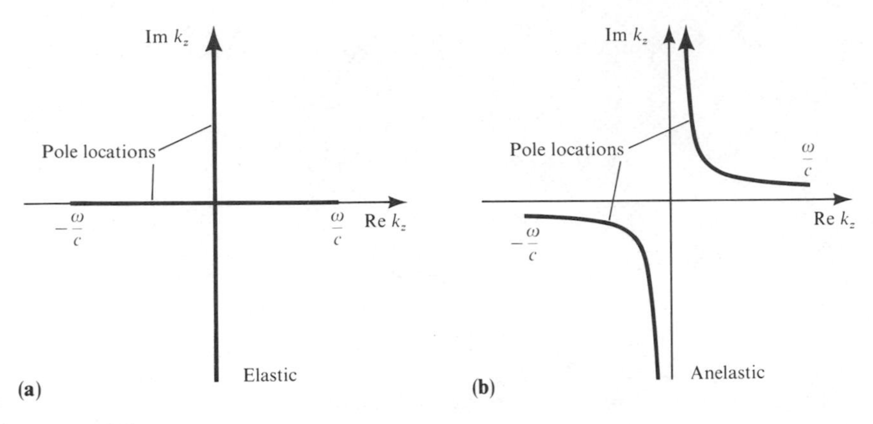

FIGURE **6.1**
Pole locations in the complex k_z-plane for different real values of k_x, k_y, and fixed $\omega > 0$. (**a**) Elastic medium. (**b**) Attentuating medium.

We define the location of the pole in the first quadrant as

$$k_z = i\gamma = +(\omega^2/c^2 - k_x^2 - k_y^2)^{1/2}.$$

Then the position of the other pole (third quadrant) is

$$k_z = -i\gamma = -(\omega^2/c^2 - k_x^2 - k_y^2)^{1/2}.$$

In both cases, we find that $\text{Re}\{\gamma\}$ is positive and $\text{Im}\{\gamma\}$ is negative.

The residue evaluation is now straightforward. For $z > 0$ a factor $\exp(ik_z z)$ suppresses the integrand in (6.3) if it is taken around a sufficiently large semicircle in the upper half-plane (see Fig. 6.2a). Adding this semicircle to the integration path along the real axis, we have a closed path going in the positive direction around a pole at $k_z = i\gamma$ in the first quadrant, so that

$$\phi = 2\pi i \times \text{residue} = \frac{\exp(-i\omega t)}{2\pi} \iint_{-\infty}^{\infty} \frac{\exp(ik_x x + ik_y y - \gamma z)}{\gamma}\, dk_x\, dk_y.$$

For $z < 0$, we add a sufficiently large semicircle in the lower half-plane (Fig. 6.2b) to obtain a closed path in the negative (i.e., clockwise) direction, which picks up a pole at $k_z = -i\gamma$ in the third quadrant:

$$\phi = -2\pi i \times \text{residue} = \frac{\exp(-i\omega t)}{2\pi} \iint_{-\infty}^{\infty} \frac{\exp(ik_x x + ik_y y + \gamma z)}{\gamma}\, dk_x\, dk_y.$$

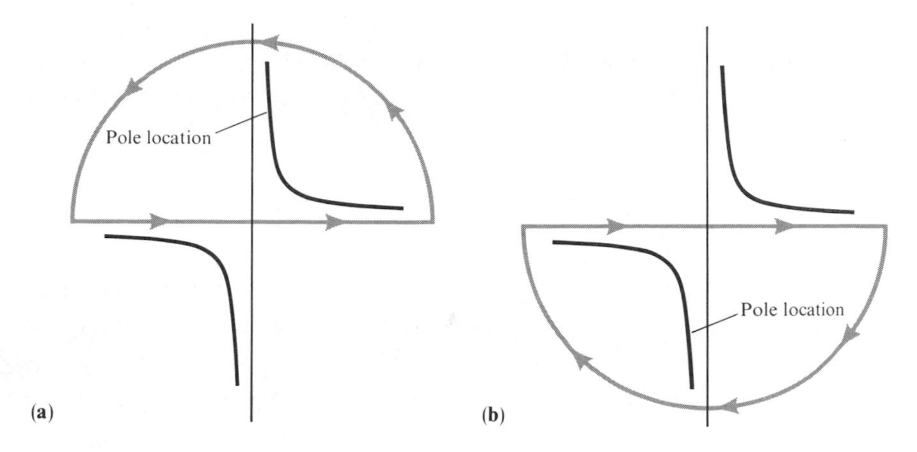

FIGURE **6.2**

Paths of integration in the complex k_z-plane for obtaining the Weyl integral. (**a**) The path when $z > 0$. (**b**) $z < 0$.

Summarizing these results, we obtain the Weyl integral

$$\frac{1}{R}\exp\left(i\omega\frac{R}{c}\right) = \frac{1}{2\pi}\iint_{-\infty}^{\infty}\frac{\exp(ik_x x + ik_y y - \gamma|z|)}{\gamma}\,dk_x\,dk_y, \qquad (6.4)$$

where $\gamma = (k_x^2 + k_y^2 - \omega^2/c^2)^{1/2}$ and the sign of γ is chosen so that $\operatorname{Re}\gamma > 0$. In the limiting case of zero attenuation, this becomes $\operatorname{Re}\gamma \geq 0$.

In the above expression, the plane waves in the integrand do satisfy the wave equation with velocity c, so that the spherical wave is indeed represented by a superposition of such plane waves. Note that for some parts of the (k_x, k_y) integration, the plane waves are inhomogeneous. This occurs for $\omega^2/c^2 < k_x^2 + k_y^2$, so that γ becomes positive real and the corresponding inhomogeneous waves propagate parallel to the xy-plane, changing amplitude most rapidly in the z-direction.

As suggested above, a different wavenumber component could have been used for the integration instead of k_z. If we had used k_x, we should find inhomogeneous waves propagating parallel to the yz-plane, equally representing the spherical wave. We chose to work with the depth component of wavenumber, because the result (6.4) is so appropriate for analyzing boundary conditions on horizontal surfaces.

To obtain the Sommerfeld integral, we change integration variables from (k_x, k_y) to (k_r, ϕ') via

$$k_x = k_r\cos\phi', \qquad k_y = k_r\sin\phi'. \qquad (6.5)$$

Thus $i\gamma = (\omega^2/c^2 - k_x^2 - k_y^2)^{1/2} = (\omega^2/c^2 - k_r^2)^{1/2}$, and the area element dk_x dk_y is replaced by the element $dk_r(k_r\, d\phi')$, with the whole horizontal wavenumber plane $(-\infty < k_x < \infty;\ -\infty < k_y < \infty)$ being covered by the ranges $0 \le k_r < \infty$, $0 \le \phi' < 2\pi$. The integral (6.4) becomes

$$\frac{1}{R}\exp\left(i\omega\frac{R}{c}\right) = \frac{1}{2\pi}\int_0^\infty dk_r \int_0^{2\pi} d\phi' \frac{k_r}{\gamma}\exp[ik_r r\cos(\phi - \phi') - \gamma|z|], \quad (6.6)$$

where we have introduced (r, ϕ) as the polar coordinates related to (x, y) via $x = r\cos\phi$, $y = r\sin\phi$. We often call r the *horizontal range*, or simply the

BOX **6.1**

Fundamental significance of Weyl and Sommerfeld integrals

The totality of solutions to a wave equation with particular homogeneous boundary conditions is a vector space, and for various coordinate systems one may establish a basis of vectors to span this space. The Weyl integral uses plane waves as a basis, summing them to give the solution for a point source. The Sommerfeld integral is the analogous result for cylindrical waves. Often we refer to the basis vectors as eigenvectors or eigenfunctions, appropriate for a particular coordinate system. Plane waves and cylindrical waves are seen to be eigenfunctions, in the context of studying solutions to the wave equation by the method of separation of variables (e.g., see Chapter 5, equations (5.9) and (5.10)).

These ideas will recur repeatedly in later chapters, where the eigenfunctions may be surface wave modes of vector displacement, where the source may have a complicated radiation pattern, where the coordinate system may involve spherical polars with origin at the center of the Earth, and where a discrete sum over eigenfunctions may be appropriate, rather than the integrations over horizontal wavenumber that we have used here.

The problem of finding the particular integral (or sum) of eigenfunctions appropriate for a particular source is then the problem of finding the coefficient of each (suitably normalized) eigenfunction in the expansion. This coefficient is a useful measure of the degree to which a particular eigenfunction has been excited.

A property common to all these source expansions, true for the Weyl integral as well as the more complicated expressions for surface waves and normal modes described in Chapters 7 and 8, is that the integrand (or each term in the sum) can be factored into a vertical eigenfunction evaluated at the source, a vertical eigenfunction evaluated at the receiver, and a transverse wavefunction that is dependent on the horizontal separation of source and receiver. (In our case, e.g., (6.4) and (6.9), vertical eigenfunctions at source and receiver are waves traveling in opposite directions, so that only the difference phase appears, proportional to the vertical separation.) The independent variable describing different terms in the integral is essentially the horizontal slowness, and a function of this variable provides a weighting factor for the integrand.

range, and ϕ is the azimuthal coordinate of cylindrical or spherical polars. Since z is measured downward, note that ϕ increases clockwise when the horizontal plane is viewed from above.

The integral over ϕ' in (6.6) can be given explicitly, using the relation $2\pi J_0(k_r r) = \int_0^{2\pi} \exp(ik_r r \cos \Phi)\, d\Phi$ to obtain the Sommerfeld integral

$$\frac{1}{R} \exp\left(i\omega \frac{R}{c}\right) = \int_0^\infty \frac{k_r J_0(k_r r) \exp(-\gamma|z|)}{\gamma} dk_r \tag{6.7}$$

where $i\gamma = (\omega^2/c^2 - k_r^2)^{1/2}$ and $\text{Re}\{\gamma\} > 0$. The integrand here is a new kind of fundamental wave, the cylindrical wave (with symmetry about a vertical axis), in which the dependence on r and z appears via separate factors.

In Chapter 5, we found it convenient to analyze plane waves by choosing a y-axis normal to the direction of propagation, and then identifying k_x with ray parameter p via $k_x = \omega \sin i/c = \omega p$ (e.g., equation (5.16)). Now, however, for a point source, the system of plane waves *departing from the source* involves all possible horizontal directions of propagation, as shown directly in (6.6) by the integration over ϕ'. The relation between horizontal wavenumber and ray parameter therefore becomes

$$k_x = \omega \sin i \cos \phi'/c = \omega p \cos \phi', \qquad k_y = \omega \sin i \sin \phi'/c = \omega p \sin \phi'.$$

By comparison with (6.5) we see that

$$k_r = \omega p. \tag{6.8}$$

Hence, from (6.7), we can express the spherical wave as an integral over all ray parameters in the form

$$\frac{1}{R} \exp\left(i\omega \frac{R}{c}\right) = i\omega \int_0^\infty \frac{p}{\xi} J_0(\omega p r) \exp(i\omega\xi|z|)\, dp. \tag{6.9}$$

Here we have taken $i\gamma = \omega(c^{-2} - p^2)^{1/2} = \omega\xi$. Recall too that $\text{Re}\{\gamma\} > 0$. Hence

$$\xi = (c^{-2} - p^2)^{1/2}, \text{ and we must choose the branch } \text{Im}\{\xi\} > 0, \tag{6.10}$$

which becomes $\text{Im}\{\xi\} \geq 0$ for perfectly elastic media. With this definition of the square root, the vertical wave function in (6.9) can readily be identified with the formulas of Chapter 5: it is $\exp[i\omega(c^{-1} \cos i)|z|]$, with $(c^{-1} \cos i) = (c^{-2} - p^2)^{1/2} = \xi$ being positive imaginary if $1/c < p$. Note that ξ is the vertical slowness and $i\gamma$ the corresponding vertical wavenumber, whereas p is the horizontal slowness. For S-waves, we shall use η as the vertical slowness and $i\nu$ for the vertical wavenumber. Thus $\xi^2 + p^2 = 1/\alpha^2$, and $\eta^2 + p^2 = 1/\beta^2$. Because this chapter is predominantly about body waves and nondispersive surface waves, it is convenient in practice to work with slownesses p, ξ, η. Therefore, we shall consistently work here with (6.9) rather than (6.7). For the surface waves described in Chapter 7, when strong dispersion can be present and the vertical wavenumber is imaginary, we shall revert to using variables such as k_r, γ, and ν.

6.2 Reflection of Spherical Waves at a Plane Boundary: Acoustic Waves

In this section and the next, we shall examine the basic phenomena that arise when a curved wavefront is incident upon the boundary between two different media. For this initial work we shall use waves at fixed frequency, since so many of the observations in seismology can be physically interpreted as a synthesis (the inverse Fourier transform) of such waves. In later work, to obtain exact wave solutions in the time domain, we shall turn to a method that is more conveniently developed via the Laplace transform.

The first problem we shall solve (see Fig. 6.3) involves acoustic waves in a body consisting of two different homogeneous semi-infinite liquids in contact along the plane $z = 0$. The source is in medium 1 (density ρ_1, velocity α_1) at depth $z_0 < 0$, and we shall also take the observation point (x, y, z) in medium 1. The source-receiver distance is given by

$$R = [x^2 + y^2 + (z - z_0)^2]^{1/2},$$

and we shall assume that the source generates waves of pressure P in the form

$$P = P^{\text{inc}} \equiv A\frac{1}{R}\exp\left[i\omega\left(\frac{R}{\alpha_1} - t\right)\right]$$

for some constant A.

Using the results of Section 6.1, we shall express P^{inc} as a superposition of cylindrical waves. Then we apply the theory of Chapter 5 to find reflected and transmitted plane waves, which must again be superposed. Thus

$$\begin{aligned} P^{\text{inc}} &= \frac{A}{R}\exp\left[i\omega\left(\frac{R}{\alpha_1} - t\right)\right] \\ &= i\omega\exp(-i\omega t)\int_0^\infty \frac{Ap}{\xi_1} J_0(\omega pr)\exp(i\omega\xi_1|z - z_0|)\,dp \qquad (6.11) \end{aligned}$$

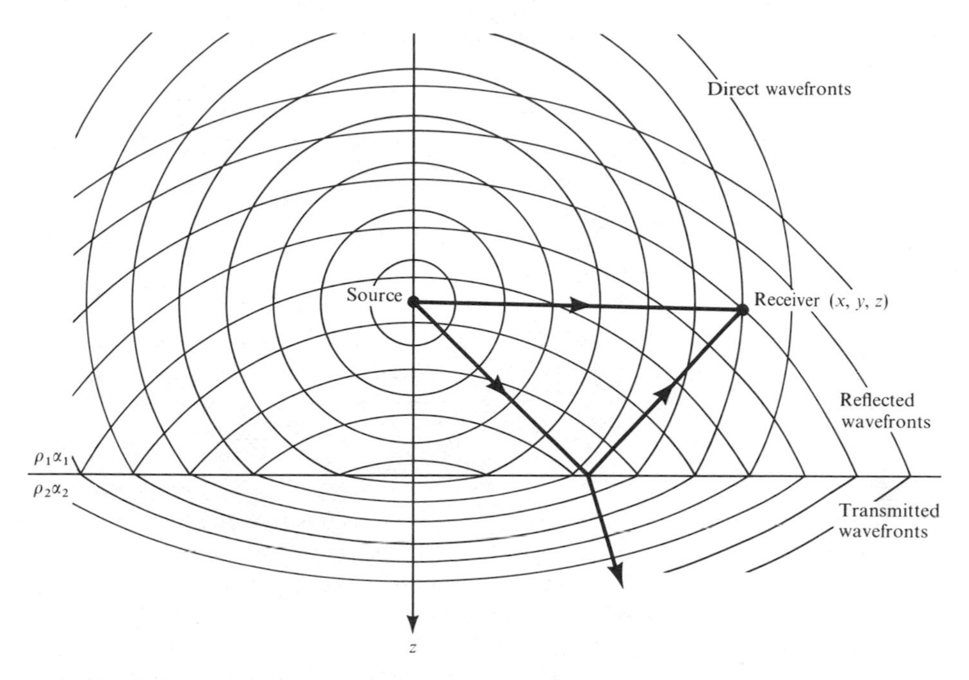

FIGURE **6.3**

Configuration of source and receiver in a fluid half-space $z < 0$ with density ρ_1 and velocity α_1. A different fluid half-space (ρ_2, α_2) occupies $z > 0$, and pressure and vertical displacement are continuous across the interface. It is because of the complexity of wavefront systems that we shall usually use rays alone.

where $\xi_1 = \alpha^{-1} \cos i_1 = (\alpha_1^{-2} - p^2)^{1/2}$ is positive imaginary if $1/\alpha_1 < p$. Seeing the integrand here as an azimuthal integral over plane waves (compare with equation (6.6)), it is natural to attempt a solution for the reflection and transmission fields as

$$P^{\text{refl}} = i\omega \exp(-i\omega t) \int_0^\infty \frac{Bp}{\xi_1} J_0(\omega pr) \exp[-i\omega\xi_1(z + z_0)]\, dp \quad \text{in } z < 0 \quad (6.12)$$

and

$$P^{\text{trans}} = i\omega \exp(-i\omega t) \int_0^\infty \frac{Cp}{\xi_1} J_0(\omega pr) \exp(-i\omega\xi_1 z_0 + i\omega\xi_2 z)\, dp \quad \text{in } z > 0. \quad (6.13)$$

Continuity of pressure and vertical displacement across $z = 0$ is assured if $A + B = C$ and $(\xi_1/\rho_1)(A - B) = (\xi_2/\rho_2)C$, which are the equations determining plane-wave reflection/transmission coefficients

$$\frac{B}{A} = -\frac{\rho_1\xi_2 - \rho_2\xi_1}{\rho_1\xi_2 + \rho_2\xi_1} = -\frac{\rho_1\dfrac{\cos i_2}{\alpha_2} - \rho_2\dfrac{\cos i_1}{\alpha_1}}{\rho_1\dfrac{\cos i_2}{\alpha_2} + \rho_2\dfrac{\cos i_1}{\alpha_1}} \tag{6.14}$$

and

$$\frac{C}{A} = \frac{2\rho_2\xi_1}{\rho_1\xi_2 + \rho_2\xi_1} = \frac{2\rho_2\dfrac{\cos i_1}{\alpha_1}}{\rho_2\dfrac{\cos i_2}{\alpha_2} + \rho_2\dfrac{\cos i_1}{\alpha_1}}.$$

Note also that $\alpha_2^{-1}\cos i_2 = \xi_2 = (\alpha_2^{-2} - p^2)^{1/2}$ is positive imaginary if $1/\alpha_2 < p$, to ensure exponential decay of (6.13) with depth.

In the remainder of this section, we shall concentrate on an approximate evaluation of the reflected wave (6.12), using the standard method of saddle-point integration and beginning with the case that $\alpha_1 > \alpha_2$.

The first step is to rewrite (6.12) using the Hankel function $H_0^{(1)}$ instead of J_0. Since $J_0(x) = \frac{1}{2}[H_0^{(1)}(x) + H_0^{(2)}(x)]$ and $H_0^{(2)}(x) = -H_0^{(1)}(-x)$, we find

$$p^{\text{refl}} = \frac{i\omega}{2}\exp(-i\omega t)\int_{-\infty}^{\infty}\frac{Bp}{\xi_1}H_0^{(1)}(\omega pr)\exp[-i\omega\xi_1(z + z_0)]\,dp \tag{6.15}$$

(using the fact that B is even in p).

Secondly, let us assume that the range r is many wavelengths, so that pr is large and we can approximate $H_0^{(1)}(\omega pr)$ by its asymptotic expansion

$$H_0^{(1)}(\omega pr) = \left(\frac{2}{\pi\omega pr}\right)^{1/2}\exp[i(\omega pr - \pi/4)]\left[1 - \frac{i}{8\omega pr} + O\left(\frac{1}{\omega^2p^2r^2}\right)\right]. \tag{6.16}$$

(The above expansion is inaccurate for p-values near zero, but we shall later distort the integration path in a way that avoids $p = 0$. It would be more correct to substitute from (6.16) after the distortion.) The approximate result is now

$$p^{\text{refl}} = \left(\frac{\omega}{2\pi r}\right)^{1/2}\exp[-i(\omega t - \pi/4)]\int_{-\infty}^{\infty}\frac{Bp^{1/2}}{\xi_1}\exp[i\omega(pr - \xi_1 z - \xi_1 z_0)]\,dp, \tag{6.17}$$

in which terms of order $1/\omega$ have been neglected in the integrand, and $B = B(p)$, involving both ξ_1 and ξ_2, is given in (6.14).

BOX **6.2**

Determining the branch cuts of $(\alpha^{-2} - p^2)^{1/2} = \xi$ *in the complex p-plane, so that* $\text{Im}\ \xi \geq 0$ *for a whole plane.*

We wish to make ξ a single-valued analytic function of p, and the ambiguity of sign for a square root requires us to consider two p-planes (Riemann sheets) to describe ξ completely. We speak of a *top sheet* and a *bottom sheet*, according as $\text{Im}\ \xi > 0$ and $\text{Im}\ \xi < 0$, connected along lines (branch cuts) determined by $\text{Im}\ \xi = 0$. Clearly the cuts include points where $\xi = 0$ (branch points), and these are $p = \pm 1/\alpha$. We shall need to understand how to take closed paths around a branch point in such a way that ξ varies smoothly with p for the whole path.

Since branch cuts are determined by $\text{Im}\ \xi = 0$, it follows that $\alpha^{-2} - p^2$ is real and nonnegative on the cut itself, and hence that

$$1/\alpha^2 - (\text{Re}\ p)^2 + (\text{Im}\ p)^2 - 2i(\text{Re}\ p)(\text{Im}\ p) \geq 0. \tag{1}$$

But since $\text{Im}(1/\alpha^2)$ is (small and) positive, the requirement that the left-hand side of (1) is real implies

$$(\text{Re}\ p)(\text{Im}\ p) = \varepsilon \qquad \text{for some } \varepsilon > 0, \quad \text{with } \varepsilon \to 0$$

in the limiting case of perfect elasticity. Hence the branch cuts lie on a hyperbola in the first and third quadrants. A further restriction, from (1), is that $(\text{Im}\ p)^2 \geq (\text{Re}\ p)^2 - \text{Re}(1/\alpha^2)$, limiting the branch cuts to positions shown schematically in Figure 6.4.

In order to see how the devices of branch cuts and Riemann sheets make a single-valued analytic function out of what ordinarily is considered a double-valued function $\xi = \xi(p)$, consider the three circuits shown below:

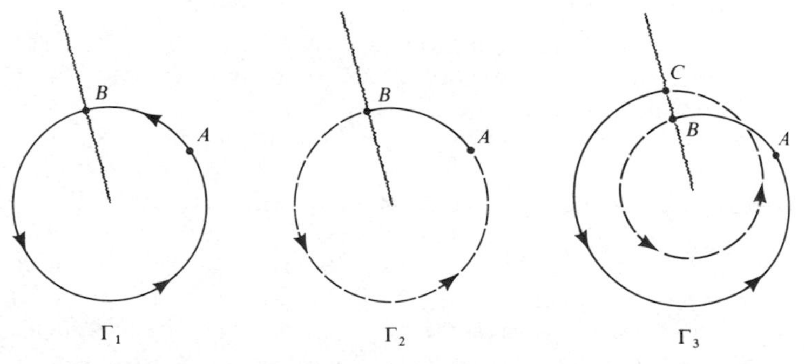

Starting from point A, the closed circuit Γ_1 is drawn entirely on the same sheet: ξ is single valued, but is not even continuous on crossing the branch cut at B. For the circuit Γ_2, ξ is analytic at B, requiring Γ_2 to go on to a different sheet (broken line); but then Γ_2 does not return ξ to the same value at A as it had initially. Finally, Γ_3 shows a path going *twice* around the branch point, with ξ varying analytically all the way and returning to its original value at A.

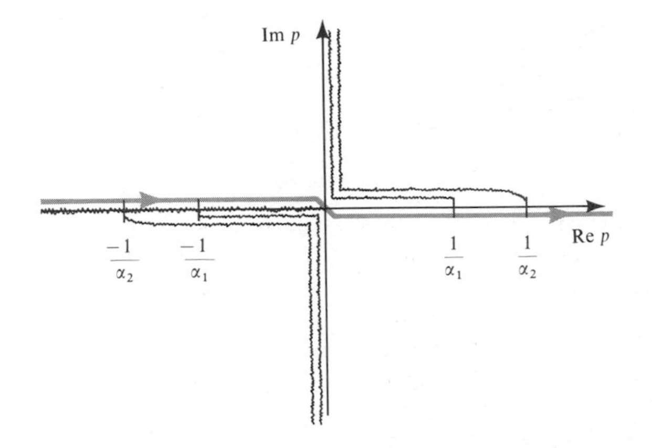

FIGURE **6.4**

Branch cuts for ξ_1, ξ_2, and $p^{1/2}$ in the complex p-plane. The cuts are given by Im $\xi_1 = 0$, Im $\xi_2 = 0$; and Re $p^{1/2} = 0$ (this being the cut assumed in (6.16). In fact, it is directly a branch cut for $H_0^{(1)}(\omega pr)$). The integration path for P^{refl} (see (6.17)) lies on the negative real axis just above three cuts, and lies on the positive real axis just below two cuts.

The location of branch cuts for ξ_1 and ξ_2 is described in Box 6.2, and the cuts for P^{refl} are shown in Figure 6.4. Our goal now is to distort the path of integration in such a way that only a limited range of p-values makes any significant contribution, and then to evaluate the integral itself. To this end, we use standard methods of saddle-point analysis to obtain an asymptotic approximation to P^{refl}, valid for sufficiently large frequencies.

BOX **6.3**

The evaluation of $I(x) = \int_C F(\zeta)\exp[xf(\zeta)]\,d\zeta$ *by the method of steepest descents*

We presume the reader has some acquaintance with this method (see, e.g., Jeffreys and Jeffreys, 1972). Our purpose here is to state notation and show some applications. To begin, suppose $f(\zeta)$ has a saddle point at $\zeta = \zeta_0$. Where f is analytic, contours of Re f = constant and contours of Im f = constant are orthogonal to each other. Near ζ_0, these contours are hyperbolas, as shown in the figure on the next page. (The angle χ is about $-32°$.)

The designations "valley" and "ridge" and the directions of rising and falling describe topography of the surface with height given by Re f, evaluated near ζ_0. The integrand for $I(x)$ is primarily controlled (in magnitude) by $\exp(x \text{ Re } f)$; therefore, since x is presumed large and positive, the terms "valley" and "ridge" apply *a fortiori* to this integrand. By taking the integration path up one valley, across the ridge at its lowest point, and down the opposite valley, one often achieves the desirable goal of minimizing the range of integration that is significant for evaluation of $I(x)$. Details of how contour C is deformed—in order to cross the saddle from valley to valley—will depend on singularities and branch cuts of the integrand and on the end points of C. We shall assume that C has been deformed to the path of ascent and descent as shown above, the path making an angle $\chi\,(-\pi < \chi \leq \pi)$ with the positive real ζ-axis. Making a Taylor series expansion of $f(\zeta)$ about ζ_0, it is easy to show that $f''(\zeta_0)e^{2i\chi}$ is real and negative. By ignoring $O[(\zeta - \zeta_0)^3]$ in the expansion for

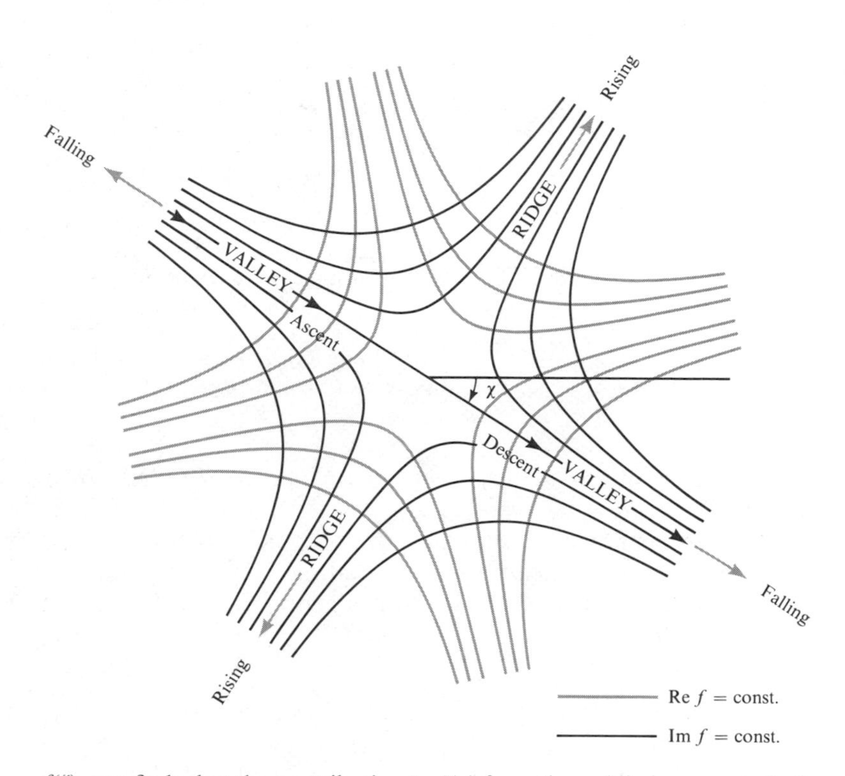

$f(\zeta)$, one finds that the contribution to $I(x)$ from the saddle is asymptotically

$$\left|\frac{2\pi}{xf''(\zeta_0)}\right|^{1/2} F(\zeta_0)e^{i\chi}\exp[xf(\zeta_0)] \qquad \text{as } x \to \infty. \tag{1}$$

This formula, however, is only of passing interest in seismology, since we shall find that it gives nothing more than the answer we would obtain by using geometrical ray theory. More important are the numerical methods of integrating over the saddle *without* making crude approximations to $f(\zeta)$. Often we shall follow the "valleys" of Re f out to a great distance from the saddle itself (see, e.g., Fig. 6.6), in which case the trend of the valleys is not orthogonal to the trend of the ridges at the saddle itself.

Following the notation of Box 6.3, with $x = \omega$, $\zeta = p$, we see that

$$\begin{aligned} f(p) &= i(pr + \xi_1|z + z_0|) \qquad \text{(recall } z < 0,\ z_0 < 0\text{)}, \\ f'(p) &= i(r - p|z + z_0|/\xi_1), \\ f''(p) &= -i|z + z_0|/(\alpha_1^2\xi_1^3), \end{aligned} \tag{6.18}$$

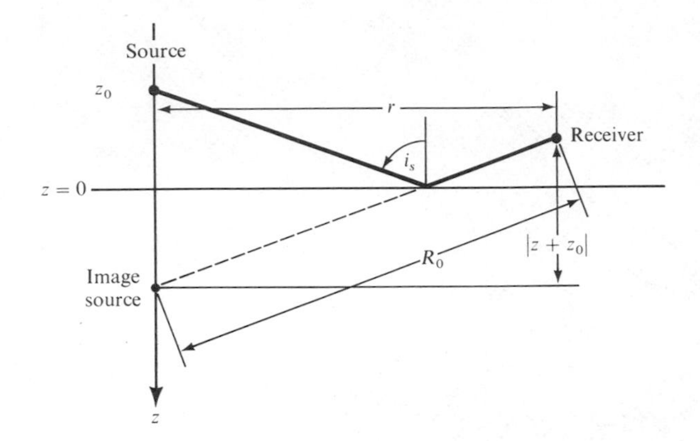

FIGURE **6.5**
Geometry for the ray reflected at the interface between source and receiver. It has the special ray parameter $p_s = \alpha_1^{-1} \sin i_s$. R_0 is the distance between image source (at $z = -z_0$) and receiver. Note the trigonometrical relationships $r = R_0 p_s \alpha_1$, $pr + \alpha_1^{-1} \cos i_s |z + z_0| = R_0/\alpha_1$, and $r \cos i_s = |z + z_0| \sin i_s$.

indicating a saddle point $p = p_s$ such that $r\xi_1 = p|z + z_0|$. In terms of an angle, $p_s = \alpha_1^{-1} \sin i_s$. Then $\xi_1 = \alpha_1^{-1} \cos i_s$, so that $r \cos i_s = |z + z_0| \sin i_s$. From Figure 6.5 it is now apparent that the saddle occurs precisely for that value of p which gives the reflected ray between source and receiver. Following through with the approximation (1) of Box 6.3, a little algebra and trigonometry gives the asymptotic result

$$P^{\text{refl}} \sim \frac{B(p_s)}{R_0} \exp\left[i\omega\left(\frac{R_0}{\alpha_1} - t\right)\right] \qquad \text{as } \omega \to \infty, \tag{6.19}$$

provided only that the real path of integration in (6.17) can be deformed onto the steepest descents path over $p = p_s$ without picking up any further contributions from singularities or branch cuts of the integrand. The saddle itself lies on two branch cuts, so that at first sight it does not seem possible to use the steepest-descents path. Indeed, it is not possible if the path is confined only to the Riemann sheet with $\text{Im}\, \xi_1 \geq 0$, $\text{Im}\, \xi_2 \geq 0$. But, by using (in the first quadrant) the sheet with $\text{Im}\, \xi_1 < 0$, $\text{Im}\, \xi_2 < 0$, the path Γ shown in Figure 6.6 is obtained with the following properties:

i) the quantity $i(pr - \xi_1 z - \xi_1 z_0)$ is an analytic function, taking value iR_0/α_1 at the saddle point;
ii) elsewhere on Γ, $i(pr - \xi_1 z - \xi_1 z_0) - iR_0/\alpha_1$ is a negative real number;
iii) the asymptotes of Γ are lines, as shown in Figure 6.6, making angles i_s with the negative and positive real p-axis.

These three properties arise from the solution of $i(pr - \xi_1 z - \xi_1 z_0) = -X^2 + iR_0/\alpha_1$ for p as a function of the real number X. We shall give further details of such solutions later (in the legend of Fig. 6.12c, and also for related paths in the context of Cagniard theory). The remaining point to make here is that when $\alpha_1 > \alpha_2$, there is no difficulty in deforming from the real-axis path of (6.17) to

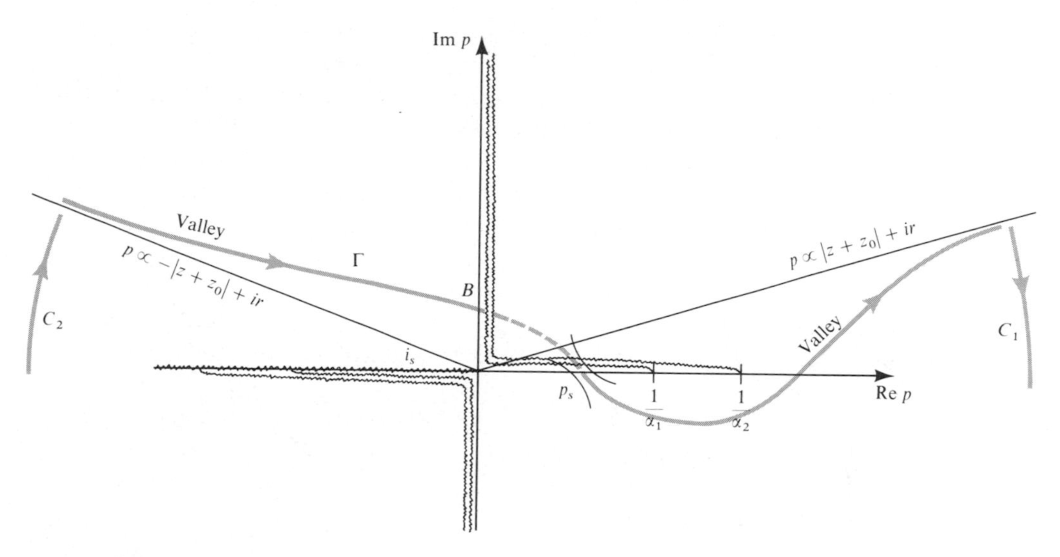

FIGURE **6.6**

The steepest-descents path Γ in the complex ray-parameter plane for obtaining P^{refl} (see (6.17)), when $\alpha_1 > \alpha_2$. We have indicated the exact path for its whole length, using the rule that $\omega f(p) - \omega f(p_s)$ is negative real, so that $\exp[\omega f(p)]$ decays exponentially away from the saddle point. Near the saddle point itself, ridges and valleys are as shown in Box 6.3, with $\chi = -\pi/4$. The dotted path is on a lower Riemann sheet, for which Im ξ_1 and Im ξ_2 are negative.

the path Γ of Figure 6.6. Connecting arcs C_1 and C_2 are required in the first and second quadrants, but on these the integrand vanishes exponentially. At point B, the path Γ must leave the top Riemann sheet if it is to avoid discontinuities in the integrand. On reaching the top of the saddle, Γ reappears on the top sheet, where it remains for the rest of the path. Having satisfied ourselves that branch cuts and singularities make no contribution, we return to a discussion of the asymptotic solution (6.19).

We note first that the geometrical spreading factor for the reflected wave would be just $1/R_0$, and its travel time would be R_0/α_1. The expectation from geometrical ray theory would then be that $P^{\text{refl}} \sim A(1/R_0) \exp[i\omega(R_0/\alpha_1 - t)]$ multiplied by a factor representing the ratio of reflected/incident pressures for the interface. We have obtained such ratios in Chapter 5 for plane waves, but here the incident wave has a spherical wavefront. Nevertheless, the plane-wave reflection coefficient B/A given in (6.14) will be adequate if the radius of curvature is sufficiently large, provided that $B = B(p)$ is evaluated at the ray-parameter value p_s (corresponding to the actual angle of reflection, i_s). It is satisfying that we have now given a physical interpretation of the solution (6.19) in terms of geometrical ray theory and plane-wave reflection coefficients. We shall find that such explanations are still available as we advance our theoretical methods to handle ever more realistic Earth models. However, if this were all we could achieve with integrations in the complex ray-parameter plane, the effort would hardly be worthwhile. Going over the derivation of (6.19), we see that several different approximations were made, beginning with the neglect of higher-order terms in (6.16) and continuing in Box 6.3 with the approximation of $f(p)$ by $f(p_s) + (p - p_s)^2 f''(p_s)$. This last approximation is particularly poor, since it is valid only in the vicinity of the saddle, yet is applied for the whole p-plane. As we shall find in Chapter 9, modern methods of quantitative seismology often make the approximation (6.16), keeping only the first term and identifying the steepest-descents path Γ for the whole complex p-plane, but then conducting the integral over a convenient path near Γ by numerical methods, evaluating $f(p)$ exactly.

By keeping additional terms in the approximations used to derive (6.19), it can formally be shown for $\alpha_1 > \alpha_2$ that

$$P^{\text{refl}} = \frac{B(p_s)}{R_0}\left[1 + \sum_{n=1}^{\infty} a_n \left(\frac{i}{\omega}\right)^n\right] \exp\left[i\omega\left(\frac{R_0}{\alpha_1} - t\right)\right] \tag{6.20}$$

as $\omega \to \infty$, in which the constants $a_1, a_2, \ldots$ depend on source-receiver geometry. In the time domain, this result becomes

$$P^{\text{refl}} = \frac{B(p_s)}{R_0}\left\{\delta\left(t - \frac{R_0}{\alpha_1}\right) + H\left(t - \frac{R_0}{\alpha_1}\right)\left[a_1 + a_2\left(t - \frac{R_0}{\alpha_1}\right) + a_3\left(t - \frac{R_0}{\alpha_1}\right)^2 + \cdots\right]\right\} \tag{6.21}$$

as $t \to R_0/\alpha_1$, so that the reflected wave has a longer tail than the incident wave in the time domain ($P^{\text{inc}} = (A/R)\,\delta(t - R/\alpha_1)$). The equivalent expressions (6.20) and (6.21) are *wavefront expansions.* As we indicated in Chapter 4 (following (4.78)), their use in seismology is limited, although the first term can be of interest in giving properties at the wavefront itself.

To continue our analysis of acoustic waves in two different half-spaces, we next discuss the case $\alpha_1 < \alpha_2$, which permits head waves to exist. Equation (6.17) is still valid, and the branch cuts are as we found them in Figure 6.4 but with $1/\alpha_1$ now to the right of $1/\alpha_2$. As before, there is a saddle point $p_s = \alpha_1^{-1} \sin i_s$, which must lie between 0 and $1/\alpha_1$, but now there is the possibility that p_s lies to the right of $1/\alpha_2$. In Figure 6.7, a critical angle $i_c = \sin^{-1}(\alpha_1/\alpha_2)$ is defined, and the spatial regions corresponding to $p_s = \alpha_1^{-1} \sin i_s < 1/\alpha_2$ and $p_s = \alpha_1^{-1} \sin i_s > 1/\alpha_2$ are identified.

First, we shall consider the case of a receiver in the light gray region of Figure 6.7, i.e., $p_s < 1/\alpha_2$. Then the reflected wave P^{refl} can be analyzed in detail by exactly the same type of steepest-descents path Γ as we found earlier. The path is shown in Figure 6.8, and again it must be taken on to the sheet $\text{Im}\,\xi_1 < 0$,

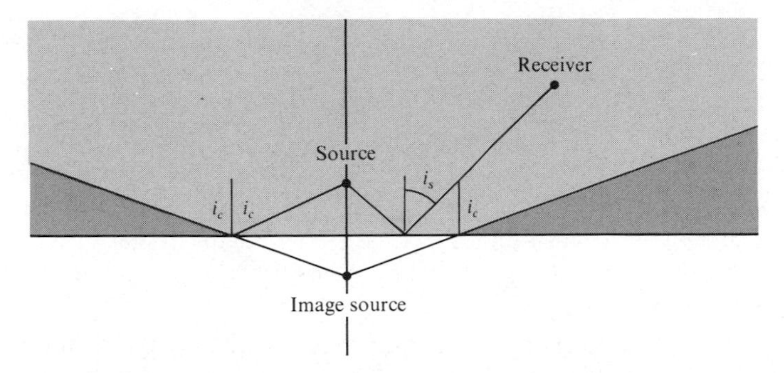

FIGURE **6.7**

For receivers in the light gray region, the reflected ray has angle of incidence $i_s < i_c$, as shown. We define $i_c = \sin^{-1}(\alpha_1/\alpha_2)$, so that, by Snell's law, rays incident at this angle emerge in the lower half-space, with $i_2 = \pi/2$, i.e., in the horizontal direction. Receivers in the dark gray region would have $i_s > i_c$, so that $p_s = \alpha_1^{-1} \sin i_s > 1/\alpha_2$. Then plane waves in the lower medium with this ray parameter would be inhomogeneous waves. It is for receivers in the dark gray region that head waves may be observed.

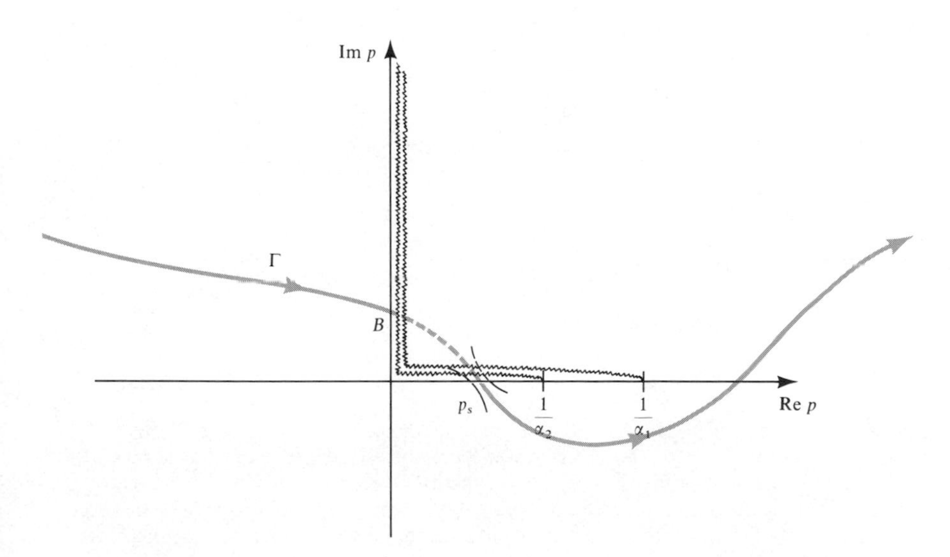

FIGURE **6.8**

The steepest-descents path Γ in the complex ray parameter plane for obtaining P^{refl} (see (6.17)), when $\alpha_1 < \alpha_2$ but $p_s < 1/\alpha_2$. The path is similar to that of Figure 6.6.

$\text{Im}\ \xi_2 < 0$ in the first quadrant. Our discussion above of the geometrical ray solution (6.19) is unchanged.

Second, we consider the case of a receiver in the dark gray region of Figure 6.7, i.e., $p_s > 1/\alpha_2$. Although p_s is still a saddle point of $i(pr - \xi_1 z - \xi_1 z_0)$, the problem now is that the valley of approach in the first quadrant lies on the sheet $\text{Im}\ \xi_1 < 0$, $\text{Im}\ \xi_2 > 0$. From this sheet, how is it possible to circumvent branch cuts and cross from one sheet to another with a path that begins (as required) in the second quadrant on the sheet $\text{Im}\ \xi_1 > 0$, $\text{Im}\ \xi_2 > 0$? A solution is shown in detail in Figure 6.9.

In considering this complicated path of integration, it must be borne in mind that the integrand is exponentially small, and hence negligible, for all points except those near the real axis. Integration over the saddle again gives a term that is approximately

$$\frac{B(p_s)}{R_0} \exp\left[i\omega\left(\frac{R_0}{\alpha_1} - t\right)\right], \tag{6.22}$$

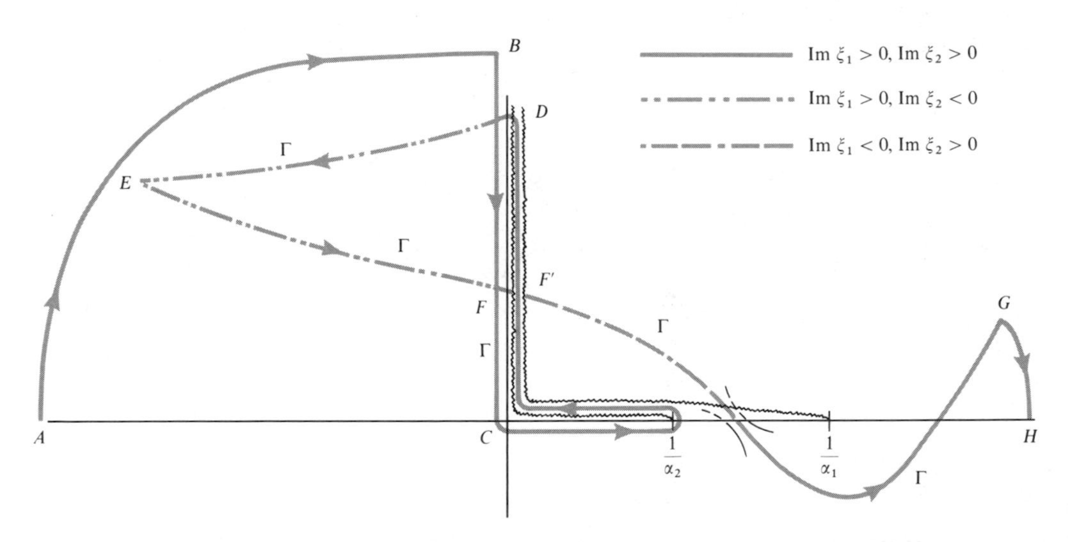

FIGURE **6.9**

The integration path Γ in the complex ray-parameter plane for obtaining P^{refl} (see (6.17)) when $\alpha_1 < \alpha_2$ but $p_s > 1/\alpha_2$. Note that three Riemann sheets are needed as shown at the upper right. Starting at point *A*, there is no contribution from arc *AB*. For *BC*, *C* to $1/\alpha_2$, and around the cut $\text{Im}\ \xi_2 = 0$, the path stays on the top sheet to *D*, which is some point sufficiently far up to give a negligible integrand. Crossing the cut at *D*, the path must descend to $\text{Im}\ \xi_1 > 0$, $\text{Im}\ \xi_2 < 0$ in order to keep the integrand analytic. From *E* to the saddle point and on to *G* is exactly the path of steepest descents; at *F*, the path crosses to $\text{Im}\ \xi_1 < 0$, $\text{Im}\ \xi_2 > 0$, and crosses back to the top sheet at the saddle. A large arc *GH* contributes nothing.

although now the reflection coefficient has a phase shift due to inhomogeneous waves in the lower medium associated with $1/\alpha_2 < p_s$. For reasons that are apparent in Figure 6.7, the wave approximated by (6.22) is known in seismology as a *wide-angle reflection* (note $i_s > i_c$). Because of the phase shift in $B(p_s)$, the reflected pulse shape is distorted in a fashion described in Section 5.3 and Box 5.6.

The other contribution from path Γ of Figure 6.9 comes from integration around the branch cut of ξ_2. On the positive real axis just below this cut, ξ_2 is positive real, equal to $\alpha_2^{-1} \cos i_2$; just above the cut, $\xi_2 = -\alpha_2^{-1} \cos i_2$. It follows from (6.17) and (6.14) that this contribution is

$$A\left(\frac{\omega}{2\pi r}\right)^{1/2} \exp[-i(\omega t - \pi/4)]\left(\int_{i\infty}^{0} + \int_{0}^{1/\alpha_2}\right)$$
$$\times \left[\frac{-\rho_1\alpha_1 \cos i_2 + \rho_2\alpha_2 \cos i_1}{+\rho_1\alpha_1 \cos i_2 + \rho_2\alpha_2 \cos i_1} - \frac{+\rho_1\alpha_1 \cos i_2 + \rho_2\alpha_2 \cos i_1}{-\rho_1\alpha_1 \cos i_2 + \rho_2\alpha_2 \cos i_1}\right]$$
$$\times \frac{\alpha_1 p^{1/2}}{\cos i_1} \exp[\omega f(p)]\, dp, \tag{6.23}$$

where the exponent function $f(p)$ is given in (6.18). Between $p = 0$ and $p = 1/\alpha_2$, $f(p)$ is imaginary, so that the integrand in (6.23) has rapid oscillations. The main contributions come from p-values for which $\text{Im}\{f'(p)\}$ is smallest, and from (6.18) one finds this smallest value occurs right at $p = 1/\alpha_2$. In fact, only the right-hand end of the branch cut contributes significantly, hence to obtain an approximation for the branch-cut integral (6.23), we use an expansion

$$f(p) = f(1/\alpha_2) + (p - 1/\alpha_2)f'(1/\alpha_2),$$

evaluate $\cos i_1$ and $p^{1/2}$ at $p = 1/\alpha_2$, and take $\cos i_2 = (2\alpha_2)^{1/2}(1/\alpha_2 - p)^{1/2}$. From (6.18) and the trigonometry of Figure 6.10, we find $f(1/\alpha_2) = it_h$, where

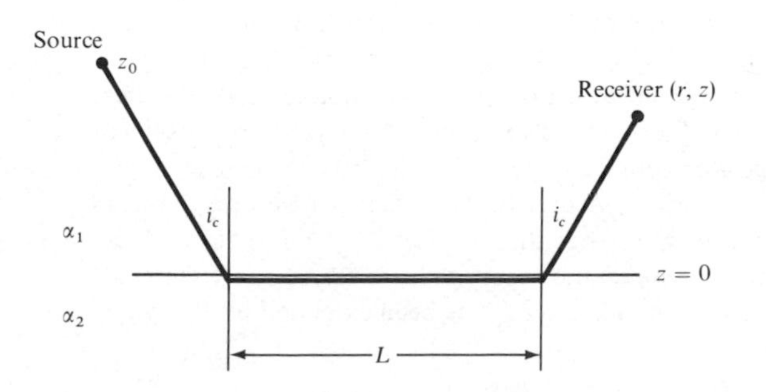

FIGURE **6.10**

Head waves have a travel time t_h corresponding to a leg of length $|z_0| \sec i_c$ in the upper medium, a leg of length L in the lower medium (traveling horizontally), and a leg of length $|z| \sec i_c$ back in the upper medium. Thus

$$t_h = (\alpha_1^{-1} \sec i_c)|z + z_0| + L/\alpha_2,$$

which can be shown to equal $r/\alpha_2 + (\alpha_1^{-1} \cos i_c)|z + z_0|$. Distance L is clearly $r - |z + z_0| \tan i_c$.

$t_h = |z + z_0|/(\alpha_1 \cos i_c) + L/\alpha_2$ is the arrival time of head waves and $f'(1/\alpha_2) = iL$. The approximation for (6.23) becomes

$$-A\left(\frac{\omega}{\pi r}\right)^{1/2} \frac{\exp[i\omega(t_h - t)]e^{i\pi/4}2\rho_1\alpha_1^2}{(1 - \alpha_1^2/\alpha_2^2)\rho_2\alpha_2} \int_0^{1/\alpha_2} \left(\frac{1}{\alpha_2} - p\right)^{1/2} \exp\left[i\omega L\left(p - \frac{1}{\alpha_2}\right)\right] dp. \tag{6.24}$$

Putting $(p - 1/\alpha_2) = iy^2$ and using $\int_0^\infty y^2 e^{-\alpha y^2}\, dy = (1/4)\pi^{1/2}/\alpha^{3/2}$, we finally get the asymptotic approximation

$$\frac{i}{\omega} A \frac{\rho_1\alpha_1^2}{\rho_2\alpha_2(1 - \alpha_1^2/\alpha_2^2)} \frac{1}{r^{1/2}L^{3/2}} \exp[i\omega(t_h - t)] \qquad \text{as } \omega \to \infty. \tag{6.25}$$

Identification of the whole branch-cut integral (6.23) as head waves is made on the basis of the approximate result we have just obtained. Again, we emphasize that numerical methods (without approximating the integrand) give more accurate results.

BOX **6.4**

Outstanding features of head waves

(i) Amplitude attenuates with distance as $r^{-1/2}L^{-3/2}$, which will be approximately r^{-2} for $r \gg |z + z_0|$, so that the attenuation is much stronger than for the spherical wave P^{inc} or the wide-angle reflection.

(ii) The factor i/ω corresponds to an integration in the time domain, making the waveform smoother, with a more emergent onset and longer tail as compared to the waveform of P^{inc}.

(iii) Near the critical distance at which head waves formally can exist, L is small (see Fig. 6.10), and our approximate formula (6.25) blows up as $L \to 0$, i.e., as $i_1 \to i_c$ from above. A corresponding failure can be shown for the reflection formula (6.20), in that $a_1 \to \infty$ as $i_1 \to i_c$ from below, so that our asymptotic formulas will fail completely at the critical distance. Different asymptotic formulas can be developed that apply specifically in the critical range $i_1 \sim i_c$ (Brekhovskikh, 1960, para. 22), but for practical purposes it is imperative to develop numerical methods applicable for all distance ranges.

Head waves are also known in the seismological literature as "conical waves" or "lateral waves," and their propagation path was first recognized by Mohorovičić during his studies of the arrival time of certain waves from a European earthquake of 1909. These waves, which we now call P_n, are refracted along the top of the mantle. They have a linear relationship between arrival time and horizontal range, unlike the travel-time curve for a reflected or direct arrival, as shown in Figures A and B. At sufficient distance, the head wave is the first-arriving wave (see receiver A_4), but this is offset by the amplitude decay noted above in point (i). Jeffreys (1926) originated the theory of head waves as recognizably distinct arrivals. The method we have used in Section 6.2 has been extended by Berry and

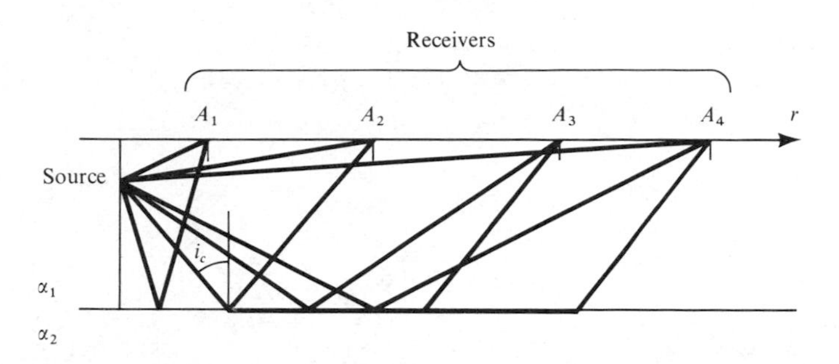

FIGURE **A**

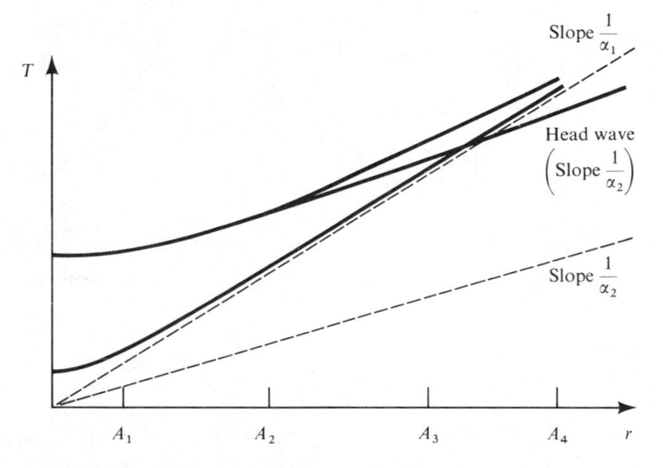

FIGURE **B**

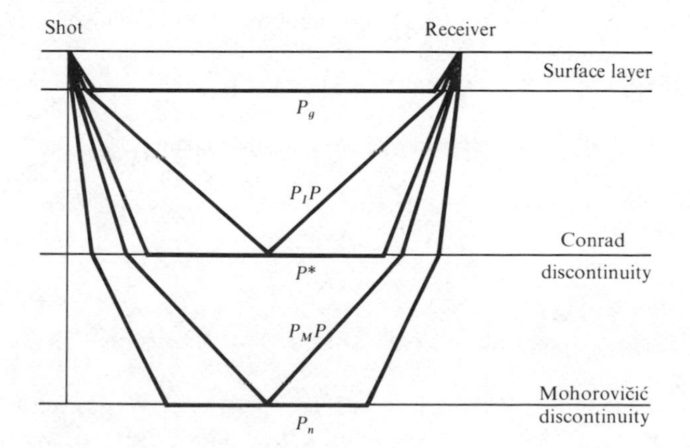

FIGURE **C**

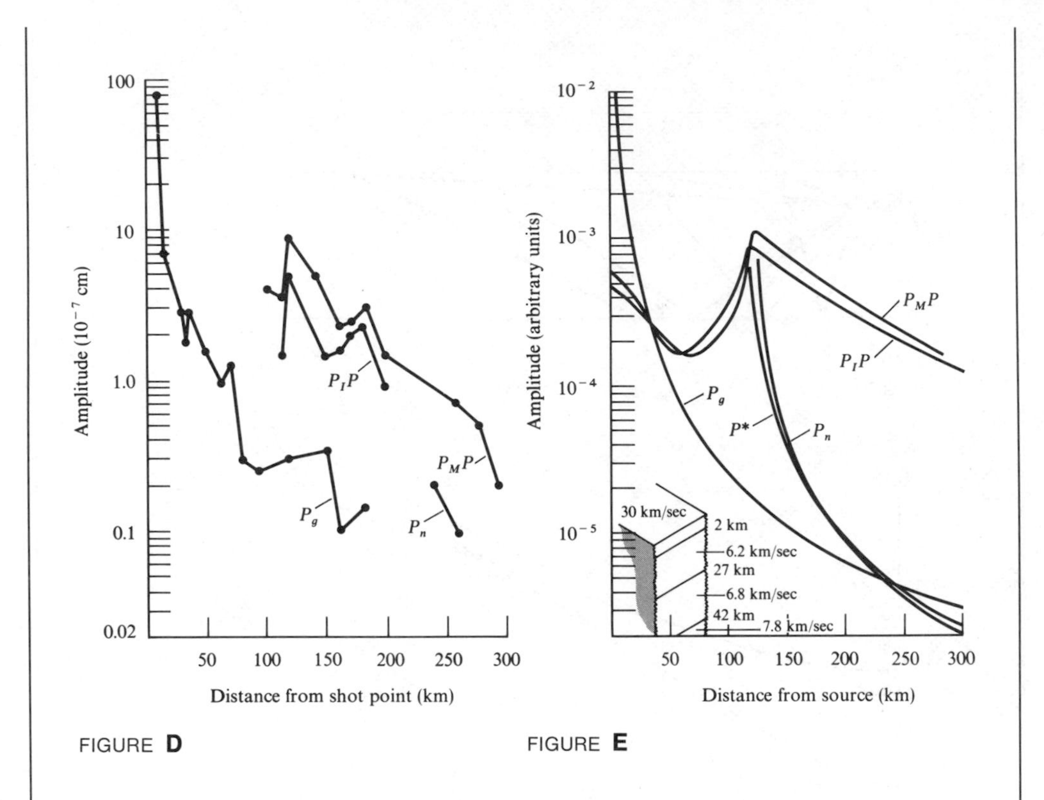

FIGURE **D**

FIGURE **E**

West (1966) for an elastic layered medium, and applied by them to interpret the waves P_g, P_n, P^*, P_MP, P_IP that had earlier been recorded on the Colorado Plateau by the U.S. Geological Survey. Figure C gives propagation paths for these waves: P_a, P^* and P_n are head waves; P_IP and P_MP are reflections. Observations are shown in D, and there is general agreement with amplitudes calculated for a particular crustal model, shown in (e). [Figures D and E here are taken from Berry and West (1966), copyrighted by American Geophysical Union.] The low amplitude of head waves makes them difficult to detect, except when they arrive ahead of *all* reflected waves. The major discrepancy is in reflected wave amplitudes, which are observed to be much smaller than those calculated. The effect is attributed by Berry and West to roughness of the Conrad and Moho discontinuities.

6.3 Spherical Waves in an Elastic Half-space: The Rayleigh Pole

In Section 5.2, we saw that a free surface can lead to very strong interactions between P and S plane waves. For example, at some angles of incidence for an incident P-wave, the reflected system consists entirely of S-waves (see Fig. 5.6). With spherical waves now recognized as an integral over plane waves, it may

BOX **6.5**

Independence of P-SV and SH motions for piecewise homogeneous media in which the material discontinuities are horizontal

A piecewise homogeneous medium is one composed of different regions, each of which is homogeneous. We have already seen in Chapter 5 that a horizontal plane interface between two homogeneous media leads to coupling between P and SV plane waves, and that SH propagates independently. Here, we generalize this result.

Beginning with Lamé's theorem (Section 4.1), we note that wave equations

$$\ddot{\phi} = \Phi/\rho + \alpha^2\nabla^2\phi, \qquad \ddot{\boldsymbol{\psi}} = \boldsymbol{\Psi}/\rho + \beta^2\nabla^2\boldsymbol{\psi} \tag{1}$$

are satisfied by potentials related to displacement $\mathbf{u}$ and body force $\mathbf{f}$ via

$$\mathbf{u} = \nabla\phi + \nabla \times \boldsymbol{\psi}, \qquad \mathbf{f} = \nabla\Phi + \nabla \times \boldsymbol{\Psi}, \quad \text{with} \quad \nabla\cdot\boldsymbol{\psi} = 0, \qquad \nabla\cdot\boldsymbol{\Psi} = 0.$$

The wave equations (1) are equivalent to three scalar equations

$$\begin{aligned} \ddot{\phi} &= \Phi/\rho + \alpha^2\nabla^2\phi, \\ \frac{d^2}{dt^2}(\nabla \times \boldsymbol{\psi})_z &= \frac{1}{\rho}(\nabla \times \boldsymbol{\Psi})_z + \nabla^2[(\nabla \times \boldsymbol{\psi})_z], \\ \ddot{\psi}_z &= \Psi_z/\rho + \nabla^2\psi_z. \end{aligned} \tag{2}$$

In the absence of body forces, it is clear from (2) that any motion can be decomposed into three kinds of motion; namely, those in which two out of the three functions ϕ, $(\nabla \times \boldsymbol{\psi})_z$, ψ_z vanish everywhere.

i) Motions of the first kind, with $(\nabla \times \boldsymbol{\psi})_z$ and ψ_z being zero, are clearly P-waves. They are characterized by nonzero $\nabla\cdot\mathbf{u}$, but $\nabla \times \mathbf{u} = \mathbf{0}$.
ii) Motions of the second kind involve $\phi = 0$, $\psi_z = 0$. Since also $\nabla\cdot\boldsymbol{\psi} = 0$, we know $\partial\psi_x/\partial x + \partial\psi_y/\partial y = 0$. This is the condition for existence of some function M such that $\psi_x = \partial M/\partial y$, $\psi_y = -\partial M/\partial x$, i.e., $\boldsymbol{\psi} = \nabla \times (0, 0, M)$. In fact, it is conventional to write M as the scalar ψ. Then

$$\mathbf{u} = \nabla \times \nabla \times (0, 0, \psi) = \left(\frac{\partial^2\psi}{\partial z\,\partial x}, \frac{\partial^2\psi}{\partial z\,\partial y}, -\frac{\partial^2\psi}{\partial x^2} - \frac{\partial^2\psi}{\partial y^2}\right).$$

For these motions, $\nabla\cdot\mathbf{u}$ and $(\nabla \times \mathbf{u})_z$ are zero, hence SV waves fall into this category.
iii) Motions of the third kind involve $\phi = 0$, $(\nabla \times \boldsymbol{\psi})_z = 0$. Then $\nabla\cdot\mathbf{u} = 0$ and $u_z = 0$, so that displacement here is like $\boldsymbol{\psi}$ in (ii): there exists a function χ such that $\mathbf{u} = \nabla \times (0, 0, \chi)$. This motion is clearly like SH, being characterized by $(\nabla \times \mathbf{u})_z \neq 0$, $u_z = 0$, $\nabla\cdot\mathbf{u} = 0$.

Similarly, the three types of body force in (2) can each be described by a scalar potential, and $\mathbf{f} = \nabla\Phi + \nabla \times \nabla \times (0, 0, \Psi) + \nabla \times (0, 0, X)$, with P-, SV-, SH-wave equations being, respectively,

$$\ddot{\phi} = \Phi/\rho + \alpha^2\nabla^2\phi, \qquad \ddot{\psi} = \Psi/\rho + \beta^2\nabla^2\psi, \qquad \ddot{\chi} = X/\rho + \beta^2\nabla^2\chi. \tag{3}$$

For a horizontal discontinuity in material properties, the three scalar conditions of continuity on components of **u** can be differentiated in horizontal directions. The three conditions are then equivalent to continuity of

$$\frac{\partial u_y}{\partial x} - \frac{\partial u_x}{\partial y}, \qquad \frac{\partial u_x}{\partial x} + \frac{\partial u_y}{\partial y}, \qquad u_z,$$

i.e., continuity of

$$(\nabla \times \mathbf{u})_z, \qquad \nabla \cdot \mathbf{u} - \partial u_z/\partial z, \qquad u_z. \tag{4}$$

Similarly, continuity of traction implies continuity of

$$\mu \frac{\partial}{\partial z}[(\nabla \times \mathbf{u})_z], \qquad \mu\left[\frac{\partial}{\partial z}\nabla \cdot \mathbf{u} - 2\frac{\partial^2 u_z}{\partial z^2} + \nabla^2 u_z\right], \qquad \lambda\nabla \cdot \mathbf{u} + 2\mu\frac{\partial u_z}{\partial z}. \tag{5}$$

For an *SH*-wave, with $(\nabla \times \mathbf{u})_z \neq 0$ but $u_z = 0$ and $\nabla \cdot \mathbf{u} = 0$, it follows that four of the six continuity conditions in (4) and (5) are satisfied trivially, and that u_z and $\nabla \cdot \mathbf{u}$ remain zero after interaction with the boundary, so that *SH*-waves remain *SH*.

Similarly, for an incident wave with $(\nabla \times \mathbf{u})_z = 0$ but with nonzero u_z and $\nabla \cdot \mathbf{u}$, there is no coupling to *SH* at a boundary. However, the coupling between u_z and $\nabla \cdot \mathbf{u}$ implies that *P* and *SV* are coupled.

In media with depth-dependent density and Lamé parameters, there is again decoupling between *SH* and *P* + *SV* motions: the coupling between *P* and *SV* is directly apparent in the wave equations, which do not separate into the form (2). Instead, one finds two equations, each involving ϕ and $(\nabla \times \boldsymbol{\psi})_z$.

thus be expected that a detailed discussion of elastic waves from a point source in a half-space will require a careful study of coupling between *P* and *S* at the boundary.

This problem was first solved by Lamb (1904), and subsequently re-examined by many others. We shall here follow Lapwood (1949) in decomposing the elastic wave field into several component waves, each with a simple physical interpretation. (The present section, however, concerns a point source, and Lapwood's work covered line sources.) In particular, we shall study the generation of Rayleigh waves, which are the dominant part of seismograms written at a distant observation point on the free surface. Later we shall solve the problem by Cagniard-de Hoop methods, giving the exact solution for the total field.

We assume that a homogeneous, isotropic, elastic body, with body-wave velocities α and β and density ρ, occupies the half-space $z > 0$. A point source is located at $z = h$, $(x^2 + y^2)^{1/2} = r = 0$. We shall work through a *P-SV* problem (see Box 6.5) that has been simplified in that axial symmetry is imposed about the vertical line through the source. In this sense, the problem is two-dimensional, with solutions dependent only on (r, z). We shall use both Cartesian and cylindrical coordinates, (x, y, z) and (r, ϕ, z), sharing the same depth direction. Thus displacement is represented as

$$\mathbf{u} = \nabla\phi + \nabla \times \nabla \times (0, 0, \psi),$$

with P-potential ϕ satisfying

$$\ddot{\phi} = \frac{\Phi}{\rho} + \alpha^2\nabla^2\phi, \tag{6.26}$$

and SV-potential ψ satisfying

$$\ddot{\psi} = \frac{\Psi}{\rho} + \beta^2\nabla^2\psi,$$

where Φ and Ψ are potentials for the body force $\mathbf{f}$ via

$$\mathbf{f} = \rho\ddot{\mathbf{u}} - (\lambda + 2\mu)\nabla(\nabla \cdot \mathbf{u}) + \mu\nabla \times (\nabla \times \mathbf{u}) = \nabla\Phi + \nabla \times \nabla \times (0, 0, \Psi). \tag{6.27}$$

We shall consider a point source of P-waves, as shown in Figure 6.11. (The figure also describes a simple point source of SV. Point sources that are more relevant for seismology are described in terms of cylindrical coordinates in the next chapter. A line source of SV is described in Section 6.4.) Thus we take here

$$\Phi = A4\pi\rho\alpha^2\, \delta(x)\, \delta(y)\, \delta(z - h) \exp(-i\omega t); \qquad \Psi = 0. \tag{6.28}$$

Via the wave equation (6.26) for potential, this source generates a spherical wave

$$\phi^{\text{inc}}(\mathbf{x}, t) = A\frac{1}{R}\exp\left[i\omega\left(\frac{R}{\alpha} - t\right)\right] \quad \text{with} \quad R = [x^2 + y^2 + (z - h)]^{1/2}, \tag{6.29}$$

which then is incident on the free surface at $z = 0$. The total potentials are

$$\phi = Ai\omega \exp(-i\omega t)\int_0^\infty \frac{p}{\xi} J_0(\omega pr) \exp[i\omega\xi|z - h|]\, dp \qquad \text{(incident wave)}$$

$$+ Ai\omega \exp(-i\omega t)\int_0^\infty \acute{P}\grave{P}\frac{p}{\xi} J_0(\omega pr) \exp[i\omega\xi(z + h)]\, dp$$

$$\text{(generalized } \acute{P}\grave{P} \text{ reflection)}, \tag{6.30}$$

$$\psi = Ai\omega \exp(-i\omega t)\int_0^\infty \left(\frac{1}{i\omega p}\frac{\beta}{\alpha}\acute{P}\grave{S}\right)\frac{p}{\xi} J_0(\omega pr) \exp[i\omega(\xi h + \eta z)]\, dp$$

$$\text{(generalized } \acute{P}\grave{S} \text{ reflection)} \tag{6.31}$$

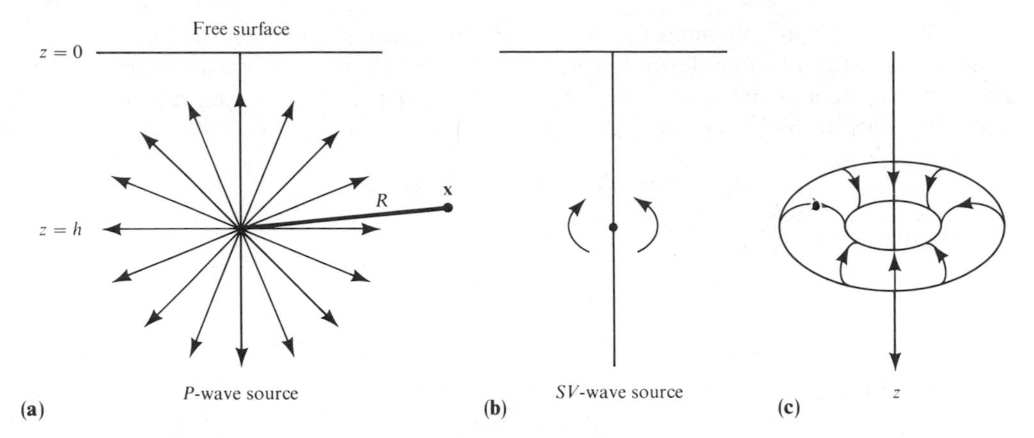

FIGURE **6.11**

Two different point sources are shown at depth h in an elastic half-space. (**a**) The P-wave source, leading to the spherical wave $\phi^{\text{inc}} = (A/R) \exp i\omega(R/\alpha - t)$, where R is the straight-line source-receiver distance. This source is a simple model of an explosion. (**b**) The SV-wave source, leading to $\psi^{\text{inc}} = (A/R) \exp i\omega(R/\beta - t)$. (**c**) To visualize this SV source, consider a small torus (doughnut) with its axis vertical and an axisymmetric motion in which particles of the torus "roll," as shown by the arrows, so that particle motions are confined to vertical planes. It is clear that SV-waves are radiated by such a torus, and it can be shown that a source of this type, in the limit as the torus tends to a point, does have a spherically symmetric ψ^{inc} as given above.

(see Box 6.6), in which $\acute{P}\grave{P}$ and $\acute{P}\grave{S}$ are the p-dependent coefficients described for the free surface in Chapter 5 (equations (5.26) and (5.27)).

We focus attention on the generalized $\acute{P}\grave{P}$ reflection in (6.30), showing how it separates out into three different types of P-waves as range r is increased.

First, we split J_0 into its traveling-wave components to obtain the generalized $\acute{P}\grave{P}$ reflection:

$$\frac{Ai\omega}{2} \exp(-i\omega t) \int_{-\infty}^{\infty} \acute{P}\grave{P} \frac{p}{\xi} H_0^{(1)}(\omega pr) \exp[i\omega\xi(z + h)]\, dp$$

$$\sim A\left(\frac{\omega}{2\pi r}\right)^{1/2} e^{-i(\omega t - \pi/4)} \int_{-\infty}^{\infty} \acute{P}\grave{P} \frac{p^{1/2}}{\xi} \exp[i\omega(pr + \xi z + \xi h)]\, dp \quad (6.32)$$

as $\omega \to \infty$ (compare with the derivation of (6.15) and (6.17)). Recall that $\text{Im}\,\xi \geq 0$, $\text{Im}\,\eta \geq 0$.

Second, we inspect the branch cuts and steepest-descents path of the exponent in (6.32). On the top Riemann sheet ($\text{Im}\,\xi \geq 0$, $\text{Im}\,\eta \geq 0$), as shown in Figure 6.12, there is a saddle point at $p_s = \alpha^{-1} \sin i_s$ and a pole where $p = 1/c_R$ (c_R being the Rayleigh wave speed and a zero of the denominator of $\acute{P}\grave{P}$). If we

BOX **6.6**

On cyclindrical coordinates

Formulas (6.30) and (6.31) can be obtained using the theory of plane waves in Cartesian coordinates, the Weyl integral (6.4), and subsequent conversion to cylindrical coordinates, parallel to our derivation of (6.9). Much quicker, however, is the following method due to Lamb (1904), using cylindrical geometry throughout.

Axisymmetric solutions of the homogeneous wave equations for potentials are

$$J_0(\omega pr)\exp(\pm i\omega\xi z - i\omega t) \quad \text{for } \phi \quad \text{and} \quad J_0(\omega pr)\exp(\pm i\omega\eta z - i\omega t) \quad \text{for } \chi,$$

where $\xi = (\alpha^{-2} - p^2)^{1/2}$ and $\eta = (\beta^{-2} - p^2)^{1/2}$. Suppose an incident upgoing P-wave and its reflections are expressed by the total fields

$$\phi = J_0(\omega pr)[Ae^{-i\omega\xi z} + Be^{i\omega\xi z}]\exp(-i\omega t),$$
$$\psi = J_0(\omega pr)Ce^{i\omega\eta z}\exp(-i\omega t).$$

Then, from $\mathbf{u} = \nabla\phi + \nabla \times \nabla \times (0, 0, \psi)$ and $dJ_0(x)/dx = -J_1(x)$, we find

$$u_r = \frac{\partial\phi}{\partial r} + \frac{\partial^2\psi}{\partial r\,\partial z} = -\omega pJ_1(\omega pr)[Ae^{-i\omega\xi z} + Be^{i\omega\xi z} + i\omega\eta Ce^{i\omega\eta z}]\exp(-i\omega t),$$

$$u_z = \frac{\partial\phi}{\partial z} - \frac{1}{r}\frac{\partial}{\partial r}\left(r\frac{\partial\psi}{\partial r}\right) = J_0(\omega pr)[-i\omega\xi Ae^{-i\omega\xi z} + i\omega\xi Be^{i\omega\xi z} + \omega^2p^2Ce^{i\omega\eta z}]\exp(-i\omega t).$$

From Section 2.6, with axisymmetry,

$$\tau_{zr} = 2\mu e_{zr} = \mu\left(\frac{\partial u_r}{\partial z} + \frac{\partial u_z}{\partial r}\right)$$
$$= i\omega^2J_1(\omega pr)[2\rho\beta^2p\xi(A - B) + \rho(1 - 2\beta^2p^2)(-i\omega pC)]\exp(-i\omega t) \qquad \text{on } z = 0, \quad (1)$$

$$\tau_{zz} = \lambda\,\text{div}\,\mathbf{u} + 2\mu\frac{\partial u_z}{\partial z} = -\frac{\lambda\omega^2}{\alpha^2} + 2\mu\frac{\partial u_z}{\partial z}$$
$$= -\omega^2J_0(\omega pr)[\rho(1 - 2\beta^2p^2)(A + B) + 2\rho\beta^2p\eta(-i\omega pC)]\exp(-i\omega t) \quad \text{on } z = 0. \quad (2)$$

Equating (1) and (2) above to zero (the free-surface boundary conditions), we find equations that differ from those obtained in our study of plane waves only in the substitution of $-i\omega pC$ for the C appearing in (5.22) and (5.23). It follows that

$$B = A\acute{P}\grave{P} \quad \text{and} \quad C = \frac{A}{i\omega p}\frac{\beta}{\alpha}\acute{P}\grave{S}. \tag{3}$$

where $\acute{P}\grave{P}$ and $\acute{P}\grave{S}$ are the reflection/conversion coefficients for displacement given in (5.26) and (5.27). Formulas (6.30) and (6.31) follow immediately from the Sommerfeld integral for the incident wave and from superposition of reflections.

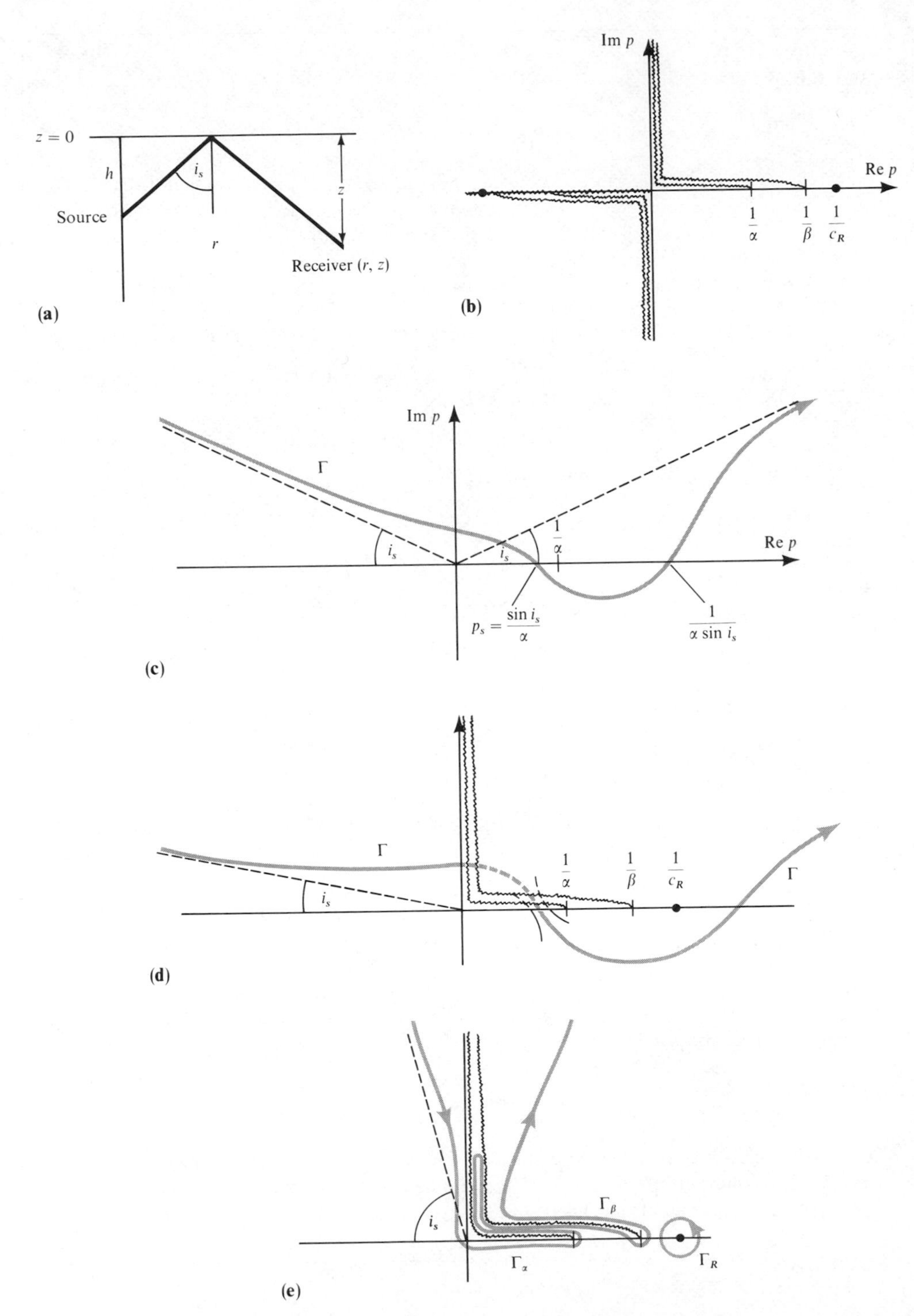

z = 0
h
i_s
z
Source
r
Receiver (r, z)
(a)
Im p
Re p
$\frac{1}{\alpha}$
$\frac{1}{\beta}$
$\frac{1}{c_R}$
(b)
Im p
Γ
i_s
i_s
$\frac{1}{\alpha}$
Re p
$p_s = \frac{\sin i_s}{\alpha}$
$\frac{1}{\alpha \sin i_s}$
(c)
Γ
i_s
$\frac{1}{\alpha}$
$\frac{1}{\beta}$
$\frac{1}{c_R}$
Γ
(d)
i_s
Γ_β
Γ_α
Γ_R
(e)

compare the steepest-descents path Γ of Figure 6.12c with the singularities of Figure 6.12b, we see that there is little interference from singularities, provided the reflection angle is small. This is the case in Figure 6.12d, and the saddle point contribution alone is important. However, if i_s becomes large (i.e., if $r \gg z + h$), a strong interference takes place, as shown in Figure 6.12e, leading to three identifiable contributions: (i) from Γ_α, the main contribution is still from the saddle point, and can be interpreted as the surface P reflection shown in Figure 6.12a; (ii) from Γ_β, the main contribution comes from the vicinity of the branch point $p = 1/\beta$. This propagation path is identifiable as an inhomogeneous P-wave, decaying upward from the source to the free surface, followed by horizontal propagation along the free surface as an SV-wave and ending as an inhomogeneous P-wave decaying downward from free surface to receiver. It is called the *surface S-wave*. Finally, (iii) from Γ_R, the P-wave component of the Rayleigh wave is obtained, with amplitude given simply by the residue of the integrand (6.32). Since

$$\acute{P}\grave{P} \sim -\left(\frac{1}{p - 1/c_R}\right)\frac{8A(c_R^{-2} - \alpha^{-2})^{1/2}(c_R^{-2} - \beta^{-2})^{1/2}}{c_R^2 \mathsf{R}'(1/c_R)}$$

as $p \to 1/c_R$, where $\mathsf{R}(p) = 4p^2\xi\eta + (1/\beta^2 - 2p^2)^2$ and $\mathsf{R}' = d\mathsf{R}/dp$, we find from (6.32) that the Rayleigh wave has P-wave component

$$\phi_R \sim -8A\left(\frac{2\pi\omega}{rc_R}\right)^{1/2} \frac{e^{i\pi/4}\left(\dfrac{1}{c_R^2} - \dfrac{1}{\beta^2}\right)^{1/2}}{c_R^2 \mathsf{R}'(1/c_R)} \exp\left[i\omega\left(\frac{r}{c_R} - t\right)\right] \exp\left[-\omega\left(\frac{1}{c_R^2} - \frac{1}{\alpha^2}\right)^{1/2}(z + h)\right] \quad (6.33)$$

FIGURE **6.12**

Diagrams for interpretation of the generalized reflection for a point source in an elastic half-space. (**a**) The source-receiver geometry and a reflected P-wave defining the reflection angle i_s. (**b**) Branch cuts for ξ and η in the complex p-plane, together with a pole at $p = 1/c_R$. (*Note*: For attenuating media, the branch cuts and pole on the positive real axis move up into the first quadrant.) (**c**) The steepest-descents path Γ for the exponent in (6.32), this being the path such that $pr + \xi z + \xi h = R_0/\alpha$ + positive imaginary quantity $= R_0/\alpha + iX^2$. One may solve for p to find

$$R_0^2 p = (iX^2 + R_0/\alpha) \pm (X^4 - 2iX^2R_0/\alpha)^{1/2}(z + h)$$

on Γ, crossing the real axis at $p = p_s = \alpha^{-1}\sin i_s$ and also at $p = 1/(\alpha \sin i_s)$. (**d**) For small i_s, the steepest-descents path can be taken (going on to the sheet Im $\xi < 0$, Im $\eta < 0$ in the first quadrant, as in Fig. 6.8). (**e**) For large i_s, the integration path (wholly on the top Riemann sheet) can be thought of as a sum of two branch-cut integrals, Γ_α and Γ_β, plus a circuit Γ_R picking up the Rayleigh-pole residue.

as $\omega \to \infty$. An estimate of the distance range at which this residue is not picked up is given by seeing when the steepest-descents path Γ (Fig. 6.12c) still lies below the Rayleigh pole. Roughly, this occurs where

$$\frac{1}{\alpha \sin i_s} > \frac{1}{c_R},$$

or, equivalently, for

$$\tan i_s = \frac{r}{z + h} < \frac{c_R}{(\alpha^2 - c_R^2)^{1/2}}. \tag{6.34}$$

The inequality (6.34) was first obtained by Nakano, and interpreted as the distance range for which Rayleigh waves have not yet built up. If

$$\frac{r}{z + h} \gg \frac{c_R}{(\alpha^2 - c_R^2)^{1/2}} \sim 0.6,$$

then the pole contribution is well isolated from branch cuts and the steepest-descents path, and (6.33) will be significant.

We can analyze the SV-wave (6.31) along similar lines, generalizing $\acute{P}\grave{S}$ and finding that an ordinary downgoing S-wave is reflected from the incident spherical P-wave. Also present, at sufficient range, is a wave with horizontal slowness $1/\beta$. Known as the *secondary S-wave*, this leaves the source as an inhomogeneous P-wave, decaying upward and then propagating horizontally as S to the receiver. A Rayleigh pole is again present, and the residue gives

$$\begin{aligned} \psi_R \sim 4A\left(\frac{2\pi\omega}{rc_R}\right)^{1/2} e^{i\pi/4}\left(\frac{c_R}{\omega}\right)\frac{\left(\frac{1}{\beta^2} - \frac{2}{c_R^2}\right)}{c_R \mathrm{R}'(1/c_R)} \\ \times \exp\left[i\omega\left(\frac{r}{c_R} - t\right)\right]\exp\left[-\omega\left(\frac{1}{c_R^2} - \frac{1}{\alpha^2}\right)^{1/2} h - \omega\left(\frac{1}{c_R^2} - \frac{1}{\beta^2}\right)^{1/2} z\right] \end{aligned} \tag{6.35}$$

as $\omega \to \infty$

Corresponding displacements can be calculated from

$$u = u_r = \frac{\partial \phi}{\partial r} + \frac{\partial^2 \psi}{\partial r\, \partial z}, \qquad w = u_z = \frac{\partial \phi}{\partial z} - \frac{1}{r}\frac{\partial}{\partial r}\left(\frac{r\, \partial \psi}{\partial r}\right).$$

(We adhere to the common convention of u for horizontal displacement, though now meaning the radial component rather than the horizontal Cartesian component.) The Rayleigh wave displacement at the surface ($z = 0$) is then

$$u_R \sim -2iQe^{i\pi/4}\left[\frac{2}{c_R}\left(\frac{1}{c_R^2} - \frac{1}{\beta^2}\right)^{1/2}\right]\exp\left[i\omega\left(\frac{r}{c_R} - t\right)\right]\exp\left[-\omega\left(\frac{1}{c_R^2} - \frac{1}{\alpha^2}\right)^{1/2} h\right],$$

$$w_R \sim -2Qe^{i\pi/4}\left[\frac{2}{c_R^2} - \frac{1}{\beta^2}\right]\exp\left[i\omega\left(\frac{r}{c_R} - t\right)\right]\exp\left[-\omega\left(\frac{1}{c_R^2} - \frac{1}{\alpha^2}\right)^{1/2} h\right], \quad (6.36)$$

where

$$Q = A\left(\frac{2\pi\omega}{rc_R}\right)^{1/2} \frac{\omega}{\beta^2 \mathrm{R}'(1/c_R)}.$$

Although the pole at $p = 1/c_R$ is a predominant feature of half-space problems, we shall find in the next section that other zeros of the Rayleigh function, occurring on different Riemann sheets, can also be significant.

BOX **6.7**

Outstanding features of Rayleigh waves from a buried point source

(i) Attenuation behaves like $r^{-1/2}$ with distance, as compared with body waves ($\sim r^{-1}$) and head waves (r^{-2}), so that Rayleigh waves must dominate the ground motion at sufficient range. Note that the ratio between horizontal and vertical motions in the *cylindrical* Rayleigh wave (6.36) is just that found for a *plane* Rayleigh wave in Problem 5.4b,

(ii) Phase delay is given by $\omega r/c_R$, and is independent of depth h, so that the travel time curve is a straight line.

(iii) Amplitude is an exponentially decaying function of h and ω:

$$\exp[-\omega(c_R^{-2} - \alpha^{-2})^{1/2}h] \sim \exp[-(0.9)^{1/2}\omega h/\beta].$$

For an S-wave source, this becomes

$$\exp[-\omega(c_R^{-2} - \beta^{-2})^{1/2}h] \sim \exp[-(0.2)^{1/2}\omega h/\beta].$$

(iv) Particle motion is retrograde elliptic (w has a phase shift of $-\pi/2$ with respect to u, and hence a phase advance of $+\pi/2$; see Box 5.5), and ellipticity is the same as for free Rayleigh waves.

(v) From (ii) and (iii) above, the slope of the phase spectrum is a function of range but not of depth, and the shape of the amplitude spectrum is a function of depth but not of range. Therefore, the amplitude and phase are independent of each other. This is a common feature of what we generally call *normal modes*, to be investigated in more detail for a layered medium in Chapter 7. This independence, of course, violates causality (see Box 5.8), hence it is meaningless to speak of the "first motion" of Rayleigh waves or individual normal modes.

In our analysis of spherical waves incident on plane boundaries, we have now seen two phenomena, head waves and Rayleigh waves, that are not manifested when the incident wavefront is plane. Our asymptotic formulas that evaluate these phenomena have enabled us quickly to assess many of their properties, but now we turn to an alternative approach that, for Lamb's problem, can be made exact.

6.4 Cagniard-De Hoop Methods for Line Sources

The time-dependence of sources we have considered so far in this chapter has been $\exp(-i\omega t)$, and we have been seeking a way to express the response (i.e., some physical variable, evaluated at a receiver) in the form $\tilde{r}(\omega)\exp(-i\omega t)$. Written in input-output form,

$$\exp(-i\omega t) \to \text{Elastic medium} \to \tilde{r}(\omega)\exp(-i\omega t).$$

For many purposes, the goal is to find the response $r(t)$ to an input $\delta(t)$. Then

$$\delta(t) = \frac{1}{2\pi}\int_{-\infty}^{\infty} \exp(-i\omega t)\,d\omega \to \text{Elastic medium} \to \frac{1}{2\pi}\int_{-\infty}^{\infty} \tilde{r}(\omega)\exp(-i\omega t)\,d\omega = r(t),$$

and we see that $r(t)$ is the inverse Fourier transform of $\tilde{r}(\omega)$. The physical causality requires $r(t) = 0$ for $t < 0$, and note that so far we have been looking at details of the Fourier transform of $r(t)$ primarily for real positive ω. Knowing $r(t)$, we can express the response to an arbitrary input $x(t)$ by the convolution

$$x(t) = \int_{-\infty}^{\infty} x(t-\tau)\,\delta(\tau)\,d\tau \to \text{Elastic medium} \to \int_0^{\infty} x(t-\tau) r(\tau)\,d\tau. \quad (6.37)$$

(Integration can be restricted to positive τ, because r is causal.)

Using Cagniard's method, we shall now obtain the response $\bar{r}(s)e^{st}$ to input $e^{st}(-\infty < t < \infty)$, where s is *real* and *positive*:

$$e^{st} \to \text{Elastic medium} \to \bar{r}(s)e^{st}.$$

Then from (6.37), $\bar{r}(s)e^{st} = \int_0^\infty r(\tau)e^{s(t-\tau)}\,d\tau$, hence $\bar{r}(s)$ is just the Laplace transform of $r(t)$:

$$\bar{r}(s) = \int_0^\infty r(t)e^{-st}\,dt \qquad (s \text{ real and positive}). \quad (6.38)$$

Supposing that a solution for $\bar{r}(s)$ can be found, the use of Cagniard's method entails manipulation of the expression for $\bar{r}(s)$ until it is in the form of (6.38), and then the integrand $r(t)$ is identified as the impulse response. (The contribution of de Hoop was to simplify the methods of manipulation. It should be noted that Cagniard attributes his method to Carson. It was developed also by Smirnov and Sobolev; Pekeris; Sauter; Garvin; and Gilbert.) Once the impulse response is known, the problem is regarded as solved, because the response to an arbitrary input can be obtained by convolution.

It is not obvious, however, that the impulse response $r(t)$ is uniquely determined by the above procedure from $\bar{r}(s)$ known only for real positive s. After all, the inverse transform of (6.38) is

$$r(t) = \frac{1}{2\pi i} \int_{c-i\infty}^{c+i\infty} \bar{r}(s) e^{st} \, ds$$

(in which the line of integration lies to the right of all singularities of $\bar{r}(s)$), requiring the analysis of $\bar{r}(s)$ for complex s. Fortunately for us, Cagniard (1962, Chapter 3) has proved that identifying $r(t)$ from the integrand of $\bar{r}(s)$ written in the form (6.38) does uniquely determine the required impulse response under fairly minor constraints on the differentiability and boundedness of $r(t)$. This is known as Lerch's lemma (or theorem). The proof is intricate and long, and is omitted here.

It is clear that $\bar{r}(s)$ can be obtained from formal substitution of $\omega = is$ in the expressions for $\tilde{r}(\omega)$ in our Sections 6.1–6.3. It is convenient, however, to derive $\bar{r}(s)$ directly from Laplace transformation of the output when the source is $\delta(t)$. For example, in two-dimensional problems involving propagation normal to the Cartesian y-direction, we seek to obtain the functional form of $r = r(x, z, t)$. With an input $\delta(t)$, we shall find it convenient to take a spatial Fourier transform over x and a Laplace transform over t, obtaining an algebraic expression for the double transform $\bar{\tilde{r}} = \bar{\tilde{r}}(k_x, z, s)$. The inverse Fourier transform is then

$$\bar{r}(x, z, s) = \frac{1}{2\pi} \int_{-\infty}^{\infty} \bar{\tilde{r}}(k_x, z, s) \exp(ik_x x) \, dk_x.$$

Since k_x here is the horizontal component of wavenumber, we extract a factor is and identify $k_x/(is) = p$ as the ray parameter. Then

$$\bar{r}(x, z, s) = -\frac{is}{2\pi} \int_{-i\infty}^{i\infty} \bar{\tilde{r}}(isp, z, s) e^{-spx} \, dp. \tag{6.39}$$

Regarding p here as a dummy variable of integration, it remains to manipulate (6.39) into the form of the integral in (6.38). The resulting integrand then permits $r(x, z, t)$ to be recognized immediately.

Note that s appears in the integral (6.38) in only one position, as a factor of the exponent. The manipulation of (6.39) often leads to a form

$$\bar{r} = s^n \int_0^{\infty} g(t) e^{-st} \, dt, \tag{6.40}$$

in which case the impulse response can be calculated from

$$r(t) = \frac{d^n}{dt^n} g(t).$$

Alternatively, taking as input a delta function that has been integrated n times, the response is directly recognized as $g(t)$.

In the remainder of this section we shall investigate the above method for line sources of SH-waves, and for a line source of P-SV waves in an elastic half-space. We shall find the medium response can be given exactly by algebraic expressions. Extension of the method to point sources is given in the next section, and we conclude the chapter with some overall comparisons between the Cagniard-de Hoop approach, and the Fourier transform methods discussed in Sections 6.1–6.3.

We begin with a simple SH problem. Using Cartesians (x, y, z), we suppose that a line source in an infinite elastic medium acts along the y-axis and consists of a body force acting impulsively in the y-direction. Thus the body force is

$$\mathbf{f} = (0, A\,\delta(x)\,\delta(z)\,\delta(t), 0),$$

where A is a constant having the dimensions of impulse per unit length. Only the y-component of displacement is excited by this source, so that displacement $\mathbf{u}$ is $(0, v(x, z, t), 0)$. The equation for v is

$$\rho\ddot{v} = A\,\delta(x)\,\delta(z)\,\delta(t) + \mu\nabla^2 v, \tag{6.41}$$

and both v and $\dot{v}$ are zero for $t < 0$.

To solve for v, the simplest method is to take the point-source solution of Box 4.1 and integrate for sources all along the y-axis:

$$v(x, z, t) = \frac{1}{4\pi\beta^2}\int_{-\infty}^{\infty} \frac{A}{\rho}\frac{1}{R_3}\,\delta\left(t - \frac{R_3}{\beta}\right) dy,$$

where $R_3 = (x^2 + y^2 + z^2)^{1/2}$. Since $dy/R_3 = dR_3/y$ and R_3 is even in y,

$$v(x, z, t) = \frac{A}{2\pi\rho\beta^2}\int_R^{\infty} \frac{1}{y}\,\delta\left(t - \frac{R_3}{\beta}\right) dR_3 = \begin{cases} 0 & \text{for } t < \dfrac{R}{\beta}, \\[2ex] \dfrac{A}{2\pi\rho\beta^2}\dfrac{1}{\left(t^2 - \dfrac{R^2}{\beta^2}\right)^{1/2}} & \text{for } t > \dfrac{R}{\beta}, \end{cases} \tag{6.42}$$

where $R = (x^2 + z^2)^{1/2}$ is the distance function for two-dimensional problems.

Let us now derive (6.42) again, using the Cagniard-de Hoop approach.

Taking the double transform

$$\int_{-\infty}^{\infty} \exp(-ik_x x)\,dx \int_0^{\infty} e^{-st}\,dt,$$

we find

$$\frac{\partial^2}{\partial z^2} v(k_x, z, s) = \frac{-A}{\rho\beta^2}\delta(z) + n^2 v(k_x, z, s), \qquad \text{where } n^2 = k_x^2 + \frac{s^2}{\beta^2}. \quad (6.43)$$

Thus everywhere except at $z = 0$, we know $\partial^2 v/\partial z^2 = n^2 v$, with solutions

$$v(k_x, z, s) = ae^{nz} + be^{-nz} \qquad \text{(fixing the choice of root by } n > 0)$$

for some constants a and b. Considering separately the two regions $z \gtrless 0$ and requiring v to be bounded as $z \to \pm\infty$, we see that $a = 0$ for $z > 0$ and $b = 0$ for $z < 0$. But (6.43) implies that v is continuous across $z = 0$, and $\partial v/\partial z$ has a step jump down of amount $A/\rho\beta^2$. It follows that $v \propto e^{-n|z|}$, and knowing the magnitude of the step in $\partial v/\partial z$ we finally find

$$v(k_x, z, s) = \frac{A}{2\rho\beta^2 n} e^{-n|z|}. \quad (6.44)$$

(Alternatively, we can obtain (6.44) by transforming the z-dependence in (6.43) and using poles in the k_z-plane, just as we did in deriving the Weyl integral, (6.4).)

The Laplace transform of the solution we seek is now

$$v(x, z, s) = \frac{A}{4\pi\rho\beta^2} \int_{-\infty}^{\infty} \frac{\exp(ik_x x - n|z|)}{n}\,dk_x \quad (6.45)$$

with $\operatorname{Re} n = \operatorname{Re}(k_x^2 + s^2/\beta^2)^{1/2} > 0$, generalized from the previous $n > 0$ to allow discussion of complex values. Observe in (6.45) that k_x is a dummy variable of integration, and write it instead as $k_x = isp$. Then

$$v(x, z, s) = \frac{A}{4\pi\rho\beta^2} \int_{-i\infty}^{i\infty} \frac{-ie^{-s(px + \eta|z|)}}{\eta}\,dp \quad (6.46)$$

where $\eta = (\beta^{-2} - p^2)^{1/2}$, $\operatorname{Re}\eta > 0$.

Already we have isolated s to just one position in the integrand (6.46); we now continue the manipulations to arrange (6.46) into the form of a forward Laplace transform. To this end, we write the real and imaginary parts of $(1/\eta)e^{-s(px + \eta|z|)}$ as $E(p)$ and $O(p)$, respectively, noting that E is even and O is odd for imaginary values of p. Then

$$\int_{-i\infty}^{i\infty} -i(E + iO)\,dp = -2i \int_0^{i\infty} E\,dp = 2\operatorname{Im}\left\{\int_0^{i\infty} (E + iO)\,dp\right\},$$

hence

$$v(x, z, s) = \frac{A}{2\pi\rho\beta^2} \operatorname{Im}\left\{\int_0^{i\infty} \frac{e^{-s(px+\eta|z|)}}{\eta}\, dp\right\}. \tag{6.47}$$

Titchmarsh (1939) gives a result similar to our derivation of (6.47) from (6.46), and calls it "the principle of reflection." The branch cuts for η are determined now by $\operatorname{Re}\eta \geq 0$, and are shown in Figure 6.13. (To analyze Fourier transforms, note that we chose $\operatorname{Im}\eta \geq 0$, for which branch cuts were much more complicated, involving parts of the real and imaginary p-axes.)

Our next step is to investigate the path C in the complex p-plane for which the quantity $px + \eta|z|$ is real. In fact, we shall label this quantity as t, so that the exponential in (6.47) becomes e^{-st}, and shall later use

$$t = px + \eta|z| \tag{6.48}$$

as an independent variable of integration. We call the path C, given by $p = p(t)$, the *Cagniard path*, and set out now to find how it is parameterized by real values of t increasing from zero to infinity.

Solving (6.48) for p as a function of t merely requires solving a quadratic equation, and one of the roots is

$$p = \begin{cases} \dfrac{xt - |z|\left(\dfrac{R^2}{\beta^2} - t^2\right)^{1/2}}{R^2} & \text{for } t < \dfrac{R}{\beta} \\[2ex] \dfrac{xt + i|z|\left(t^2 - \dfrac{R^2}{\beta^2}\right)^{1/2}}{R^2} & \text{for } t > \dfrac{R}{\beta}. \end{cases} \tag{6.49}$$

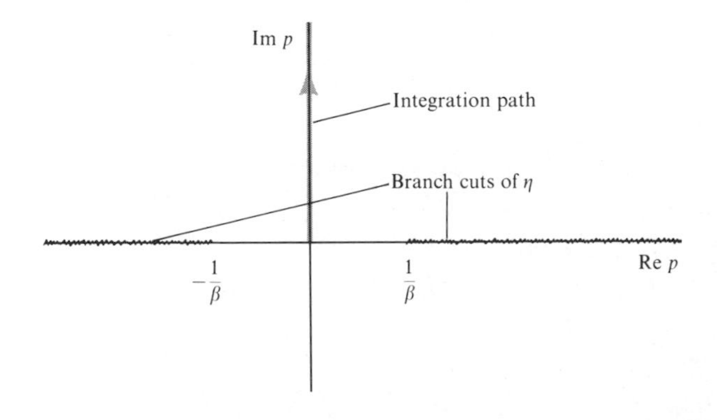

FIGURE **6.13**

Branch cuts and integration path in the complex p-plane for the Laplace transform (6.47).

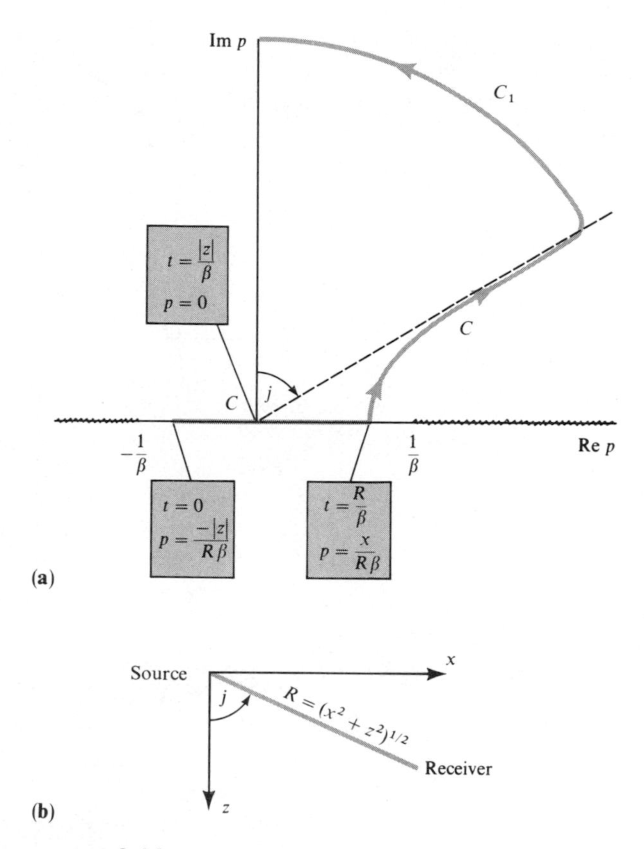

FIGURE **6.14**

(**a**) The Cagniard path C given by (6.49) is shown in the complex p-plane, beginning at $t = 0$ on the negative real p-axis, turning on to a branch of a hyperbola at $t = R/\beta$, and continuing into the first quadrant along a path with asymptote making the angle $j = \tan^{-1} x/|z|$ with the positive imaginary p-axis. Note that on the real p-axis, C cannot lie on the branch cuts. Also shown is part of a large arc in the first quadrant, connecting the imaginary p-axis and C. (**b**) Angle j is identified as the angle of incidence of the ray from source to receiver.

In Figure 6.14, we show the path $p = p(t)$ given by (6.49) in the complex p-plane for different real values of t. We have taken $x > 0$. If $x < 0$, the mirror image can be used for C.

The next stage in manipulating (6.47) is to see if the integral up the positive imaginary p-axis can instead be taken over the Cagniard path, on which t

increases from zero to infinity. In fact, there is no contribution from this integrand along that part of C between $t = 0$ and $t = R/\beta$, since p and the integrand in (6.47) are purely real. Neither is there a contribution from the large arc C_1 in the first quadrant (see Box 6.8). Since there are no singularities between C and the positive imaginary p-axis, we conclude that

$$v(x, z, s) = \frac{A}{2\pi\rho\beta^2} \operatorname{Im}\left\{\int_C \frac{e^{-s(px+\eta|x|)}}{\eta}\, dp\right\}. \tag{6.50}$$

BOX **6.8**

Jordan's Lemma

We shall show that

$$\int_{C_1} \frac{e^{-s(px+\eta|z|)}}{\eta}\, dp \to 0$$

as the radius of large arc C_1 (shown in Fig. 6.14) tends to ∞.

On C_1, $p = Le^{i\theta}$ for some large L, and $0 \le \pi/2 - j \le \theta \le \pi/2$. Then $\eta \sim -ip$ (since $\operatorname{Re}\eta > 0$), and $\eta \sim L\sin\theta - iL\cos\theta$, hence

$$\left|\int_{C_1}\right| \le \int_{\pi/2-j}^{\pi/2} 2\exp[-s(L\cos\theta x + L\sin\theta|z|)]\, d\theta \tag{1}$$

(provided L is large enough; the "2" is inserted to ensure correctness of $\le$). If neither x nor z is zero, the exponent here is always vanishingly small. If x is zero, then $j = 0$ and the Cagniard path is just the imaginary axis, so there is nothing to prove. But if z is zero, then $j = \pi/2$ and we must check that the integral in (1) still tends to zero for large L, even though the integrand (1) does not tend to zero as $\theta \to \pi/2$. We use the inequality $1 - 2\theta/\pi < \cos\theta$ for $0 < \theta < \pi/2$. Then from (1), with $|z| = 0$,

$$\left|\int_{C_1}\right| \le \int_0^{\pi/2} 2e^{-sL(1-2\theta/\pi)x}\, d\theta = \frac{\pi}{sLx}(1 - e^{-sLx}) \to 0 \qquad \text{as } L \to \infty.$$

More generally, it is true that $\int_\Gamma e^{-\lambda p} f(p)\, dp \to 0$ as $L \to \infty$, where λ is real and positive, Γ is the semicircle $p = Le^{-i\theta}$ $(-\pi/2 < \theta < \pi/2)$, and $|f(p)| \to 0$ as $|p| \to \infty$ on Γ. This general result is Jordan's Lemma, often stated for negative imaginary λ, and with Γ as the upper semicircle with $0 \le \theta \le \pi$.

It remains only to convert to using t as the variable of integration. From (6.48) and (6.49), one can show that

$$\frac{dp}{dt} = \frac{i\eta}{(t^2 - R^2/\beta^2)^{1/2}} \qquad \text{on } C, \text{ for } t > R/\beta,$$

which finally gives

$$v(x, z, s) = \frac{A}{2\pi\rho\beta^2} \int_{R/\beta}^{\infty} \frac{e^{-st}}{(t^2 - R^2/\beta^2)^{1/2}} \, dt, \tag{6.51}$$

from which we identify

$$v(x, z, t) = \frac{A}{2\pi\rho\beta^2} \frac{H(t - R/\beta)}{(t^2 - R^2/\beta^2)^{1/2}}. \qquad \text{(6.42 again)}$$

This concludes our rederivation of the delta-function response. Note that the pulse shape (6.42) has the arrival time $t = R/\beta$, which we should expect. There is a singularity in v at $t = R/\beta$, but it is integrable, so that there is no difficulty in obtaining by convolution the response to a body force $\mathbf{f}$ with general time dependence.

Next we examine a simple SH problem in which head waves can arise. We suppose that a line source given by $\mathbf{f} = (0, A\,\delta(x)\,\delta(z - z_0)\,\delta(t), 0)$ acts in a homogeneous half-space $z < 0$, so that $z_0 < 0$, with another half-space in $z > 0$ and welded contact along $z = 0$. Density and shear speed are ρ_1 and β_1 for the upper medium; ρ_2 and β_2 for the lower (see Fig. 6.15). Taking the double transform $(x, t) \to (k_x, s)$, it is clear from (6.44) that the incident wave in the upper medium is

$$v^{\text{inc}}(k_x, z, s) = \frac{A}{2\rho_1\beta_1^2 s\eta_1} \exp(-s\eta_1|z - z_0|), \tag{6.52}$$

where $\eta_1 = (1/\beta_1^2 - p^2)^{1/2}$. Continuity of v and $\tau_{yz} = \mu\,\partial v/\partial z$ across $z = 0$ determines reflection and transmission coefficients $\grave{S}\acute{S}$ and $\grave{S}\grave{S}$, and from (5.32) these are

$$\grave{S}\acute{S} = \frac{\mu_1\eta_1 - \mu_2\eta_2}{\mu_1\eta_1 + \mu_2\eta_2}, \qquad \grave{S}\grave{S} = \frac{2\mu_1\eta_1}{\mu_1\eta_1 + \mu_2\eta_2}.$$

The total fields are then:

$$\text{in} \quad z < 0, \quad v(k_x, z, s) = \frac{A}{2\mu_1 s\eta_1} \{\exp(-s\eta_1|z - z_0|) + \grave{S}\acute{S} \exp[+s\eta_1(z + z_0)]\}, \tag{6.53}$$

$$\text{in} \quad z > 0, \quad v(k_x, z, s) = \frac{A}{2\mu_1 s\eta_1} \grave{S}\grave{S} \exp[-s(\eta_2 z + \eta_1 z_0)]. \tag{6.54}$$

These are the algebraic expressions needed to start the Cagniard-de Hoop inversion. We have already examined the first term in (6.53), and shall now

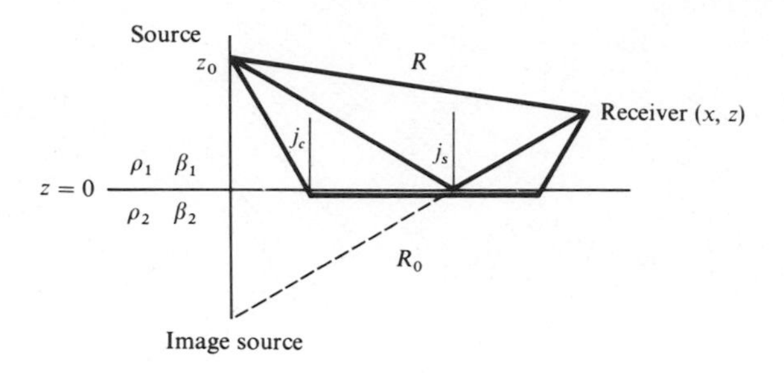

(a)

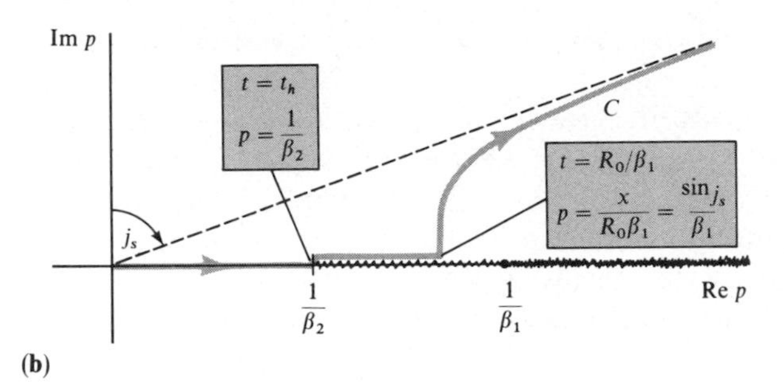

(b)

FIGURE **6.15**

(**a**) Parameters for a line source of *SH* at $x = 0$, $z = z_0 < 0$, in a medium consisting of two half-spaces in welded contact along $z = 0$. $R_0 = (x^2 + (z + z_0)^2)^{1/2}$ is the distance of the receiver from an image source in the lower medium. We presume $\beta_2 > \beta_1$, and the critical angle $j_c = \sin^{-1}(\beta_1/\beta_2)$. The wide-angle reflection has an angle of incidence j_s, and $j_s > j_c$. (**b**) Features of the complex ray-parameter plane for evaluation of the generalized reflection (6.55) at a receiver beyond the critical distance, so that $j_s > j_c$ (see (a)) and head waves can occur.

examine the second term, v^{refl}, giving the generalized reflection response in the upper medium. Following the steps that led to (6.47), we know that

$$v^{\text{refl}}(x, z, s) = \frac{A}{2\pi\rho_1\beta_1^2} \operatorname{Im}\left\{\int_0^{i\infty} \frac{1}{\eta_1}\left(\frac{\mu_1\eta_1 - \mu_2\eta_2}{\mu_1\eta_1 + \mu_2\eta_2}\right) \exp[-s(px + \eta_1|z + z_0|)]\, dp\right\} \tag{6.55}$$

involving branches defined by Re $\eta_1 \geq 0$, Re $\eta_2 \geq 0$.

The Cagniard path C for the reflection is a solution $p = p(t)$ of

$$t = px + \eta_1|z + z_0|, \tag{6.56}$$

i.e.,

$$p = \begin{cases} \dfrac{xt - |z + z_0|\left(\dfrac{R_0^2}{\beta_1^2} - t^2\right)^{1/2}}{R_0^2} & t \le \dfrac{R_0}{\beta_1}, \quad (6.56a) \\[2ex] \dfrac{xt + i|z + z_0|\left(t^2 - \dfrac{R_0^2}{\beta_1^2}\right)^{1/2}}{R_0^2} & t \ge \dfrac{R_0}{\beta_1}, \quad (6.56b) \end{cases}$$

where $R_0 = (x^2 + (z + z_0)^2)^{1/2}$ is the distance between receiver and image source (see Fig. 6.15b). It is interesting to compare this Cagniard path with the steepest-descents path of integration for (6.55). To find this latter path, we adopt the terminology of Box 6.3 with $x = s$, $\zeta = p$, and $f = -(px + \eta_1|z + z_0|)$. A saddle point $p = p_s$ must be such that $f'(p_s) = 0$, i.e., $x \cos j_s = |z + z_0| \sin i_s$, where $p_s = \beta_1^{-1} \sin j_s$, so that p_s is just the ray parameter for the reflected ray between source and receiver, having j_s as the angle of incidence in the upper medium (Fig. 6.15a). Note that some close parallels with Section 6.2 are beginning to emerge (see (6.18)). In that section, we analyzed a P-wave problem, used a Fourier transform, and considered a point source. Yet here we find essentially the same saddle-point position. A difference now is that the steepest-descents path is perpendicular to the real p-axis; i.e., angle $\chi = \pi/2$ (see Box 6.3), whereas previously we found $\chi = -\pi/4$. In fact, where the Cagniard path lies on the real axis (6.56a), it lies on a "ridge" of the integrand, the ridge descending to a saddle point at $p_s = x/(R_0\beta_1) = \beta^{-1} \sin j_s$ as t increases to R_0/β_1. There the Cagniard path turns through $\pi/2$ and follows a "valley" of the integrand, which is the ordinary steepest-descents path for t increasing from R_0/β_1 (6.56b).

If the receiver is in a position such that $x/R_0 < \beta_1/\beta_2$, then the point of departure of the Cagniard path from the real p-axis lies to the left of branch cuts emanating from $p = 1/\beta_1$ and $p = 1/\beta_2$. (The inequality implies that x is less than the critical distance at which head waves begin to be observable.) No interference with the branch cuts can occur, and since

$$dp/dt = i\eta_1(t^2 - R_0^2/\beta_1^2)^{-1/2} \qquad \text{on } C \text{ (for } t > R_0/\beta_1),$$

it follows that

$$v^{\text{refl}}(x, z, t) = \frac{A}{2\pi\rho_1\beta_1^2} \operatorname{Re}\left\{\frac{\mu_1\eta_1 - \mu_2\eta_2}{\mu_1\eta_1 + \mu_2\eta_2}\right\} \frac{H(t - R_0/\beta_1)}{(t^2 - R_0^2/\beta_1^2)^{1/2}}. \tag{6.57}$$

(There is no contribution for $t < R_0/\beta_1$, since then $p(t)$ is real (see (6.56a)), and the integrand (6.55) has zero imaginary part.) This algebraic formula (6.57) is exact, and is evaluated for $t > R_0/\beta_1$ by first using (6.56b) to obtain a corresponding point on the Cagniard path, then finding $\eta_i = (\beta_i^{-2} - p^2)^{1/2}$ $(i = 1, 2)$; and finally, substituting into (6.57).

If the receiver is beyond the critical distance, so that $1 > x/R_0 > \beta_1/\beta_2$, then the Cagniard path departs from the real p-axis at a point $p = p_s$ between $1/\beta_2$ and $1/\beta_1$, as shown in Figure 6.15b. The deformation from the positive imaginary p-axis (6.55) to the Cagniard path proceeds just as before, but now there can be a contribution from that part of the path which lies on the real p-axis. This is the head-wave contribution, which arises in the evaluation of (6.55) along the Cagniard path for real p-values between $1/\beta_2$ and $\beta^{-1} \sin j_s$; then η_2 is pure imaginary, so that the integrand has a nonzero imaginary part. Corresponding values of time are found from (6.56):

$$\text{at } p = 1/\beta_2, \qquad t = t_h = x/\beta_2 + |z + z_0|(\beta_1^{-2} - \beta_2^{-2})^{1/2},$$

t_h being the arrival time at (x, z) of head waves from $(0, z_0)$;

$$\text{at } p = \beta_1^{-1} \sin j_s, \qquad t = R_0/\beta_1,$$

the arrival time of the wide-angle reflection. Between times t_h and R_0/β_1, η_2 is a negative pure imaginary quantity. It follows that an exact formula for the generalized reflection is

$$v^{\text{refl}}(x, z, t) = \frac{A}{2\pi\rho_1\beta_1^2} \operatorname{Im}\left\{\frac{\mu_1\eta_1 - \mu_2\eta_2}{\mu_1\eta_1 + \mu_2\eta_2}\right\} \frac{H(t - t_h) - H(t - R_0/\beta_1)}{(R_0^2/\beta_1^2 - t^2)^{1/2}}$$
$$+ \frac{A}{2\pi\rho_1\beta_1^2} \operatorname{Re}\left\{\frac{\mu_1\eta_1 - \mu_2\eta_2}{\mu_1\eta_1 + \mu_2\eta_2}\right\} \frac{H(t - R_0/\beta_1)}{(t^2 - R_0^2/\beta_1^2)^{1/2}}. \tag{6.58}$$

The last term here evaluates the shape of the wide-angle reflection at time $t > R_0/\beta_1$. However, there is a phase shift, because associated waves in the lower medium are inhomogeneous ($p > 1/\beta_2$). As we showed in Box 5.6 and in Section 6.2, the pulse shape of the wide-angle reflection is a linear sum of the incident pulse shape and its Hilbert transform. The latter involves motions occurring for $t < R_0/\beta_1$, i.e., from the first term in the right-hand side of (6.58), often called the head-wave term. In this sense, one can speak of the wide-angle reflection as "emerging from the tail of the head wave." Note, however, that the attempt to separate the head-wave and reflection contributions will fail whenever there is a breakdown of the approximate (asymptotic) theory for each contribution (e.g., (6.25), if the receiver is near the critical distance, so that L is very small). Although this is a breakdown of terminology, (6.58) continues to give the exact total effect of "head wave" plus "reflection."

The final problem we shall consider in this section on the exact impulse response for two-dimensional problems is that of a line source of P-SV waves in a half-space taken as the region $z > 0$, with a free surface at $z = 0$. We closely follow Chapman (1972), obtaining exact results that have many points of similarity with Section 6.3 above. One new idea is introduced: the concept of a "leaking mode," associated with zeros of the Rayleigh function lying on the Riemann sheet $\{\text{Re}\,\xi < 0; \text{Re}\,\eta > 0\}$.

We shall consider *P-SV* motions with displacement only in the *x*- and *z*-directions. Then, from an argument given in Section 5.1, it is sufficient to work with scalar potentials ϕ and ψ related to displacement via

$$\mathbf{u} = \nabla\phi + \nabla \times (0, \psi, 0) = (\partial\phi/\partial x - \partial\psi/\partial z, 0, \partial\phi/\partial z + \partial\psi/\partial x). \quad (6.59)$$

The source of such motions is a body force $\mathbf{f}$, which (without loss of generality) is given by $\mathbf{f} = \nabla\Phi + \nabla \times (0, \Psi, 0)$. From Lamé's theorem, potentials satisfy

$$\ddot{\phi} = \alpha^2\nabla^2\phi + \Phi/\rho \quad \text{and} \quad \ddot{\psi} = \beta^2\nabla^2\psi + \Psi/\rho, \quad (6.60)$$

and to specialize to a line source at depth h, we take

$$\begin{aligned} \Phi(x, z, t) &= L_0(t)\,\delta(x)\,\delta(z - h) \qquad (P\text{-wave source}),\\ \Psi(x, z, t) &= M_0(t)\,\delta(x)\,\delta(z - h) \qquad (SV\text{-wave source}). \end{aligned} \quad (6.61)$$

Each of these two sources generates a wave incident on the free surface, and after transformation $(x, t) \to (k_x, s)$ we see by comparison with (6.52) that the incident waves have potentials

$$\phi^{\text{inc}}(k_x, z, s) = \frac{L_0(s)}{2\rho\alpha^2 s\xi}\, e^{-s\xi|z-h|}, \qquad \psi^{\text{inc}}(k_x, z, s) = \frac{M_0(s)e^{-s\eta|z-h|}}{2\rho\beta^2 s\eta}. \quad (6.62)$$

As usual, we have here used $\xi = (\alpha^{-2} - p^2)^{1/2}$, $\eta = (\beta^{-2} - p^2)^{1/2}$ and chosen the roots $\text{Re}\,\xi > 0$, $\text{Re}\,\eta > 0$ to satisfy radiation conditions. When waves (6.62) reach the free surface, they will generate P and S reflections with amplitude determined by satisfaction of boundary conditions $\tau_{zx} = \tau_{zz} = 0$ on $z = 0$. By transforming to the (k_x, z, s) domain, we have essentially reduced the calculation of reflections to a problem in plane-wave theory, so that the total potentials are just

$$\begin{aligned} \phi(k_x, z, s) &= \frac{L_0(s)}{2\rho\alpha^2 s\xi}\{\exp(-s\xi|z - h|) + \acute{P}\grave{P}\exp[-s\xi(z + h)]\}\\ &\quad + \frac{M_0(s)}{2\rho\beta^2 s\eta}\frac{\alpha\acute{S}\grave{P}}{\beta}\exp[-s(\xi z + \eta h)],\\ \psi(k_x, z, s) &= \frac{-L_0(s)}{2\rho\alpha^2 s\xi}\cdot\frac{\beta}{\alpha}\acute{P}\grave{S}\exp[-s(\eta z + \xi h)]\\ &\quad + \frac{M_0(s)}{2\rho\beta^2 s\eta}\{\exp(-s\eta|z - h|) - \acute{S}\grave{S}\exp[-s\eta(z + h)]\}, \end{aligned} \quad (6.63)$$

where reflection/conversion coefficients (see (5.26)–(5.27); (5.30)–(5.31)) are

$$\acute{P}\grave{P} = \frac{4p^2\xi\eta - (\beta^{-2} - 2p^2)^2}{\mathsf{R}(p)}, \qquad \acute{S}\grave{P} = \frac{4\beta p\eta(\beta^{-2} - 2p^2)/\alpha}{\mathsf{R}(p)},$$
$$\acute{P}\grave{S} = \frac{4\alpha p\xi(\beta^{-2} - 2p^2)/\beta}{\mathsf{R}(p)}, \qquad \acute{S}\grave{S} = \frac{-4p^2\xi\eta + (\beta^{-2} - 2p^2)^2}{\mathsf{R}(p)}, \tag{6.64}$$

BOX **6.9**

On writing down the multi-transformed solution, (6.63)

The only difficulty that prevents one from immediately writing down the formulas (6.63) concerns the detail of our sign convention and normalization of reflection/conversion coefficients. Beginning in Chapter 5, we adopted the convention of writing these coefficients for displacement-amplitude ratios, with a positive sign for motion in the increasing x-direction (see Fig. 5.5). Since (6.63) is for potentials, minus signs are introduced for $\acute{P}\grave{S}$ and $\acute{S}\grave{S}$; and factors α/β and β/α for $\acute{S}\grave{P}$ and $\acute{P}\grave{S}$, respectively.

In practice, one does not carry out the inversion of potentials to the time domain and then use (6.59) for displacement components. Rather, one obtains the transformed displacement components equivalent to (6.63), inverting these directly to the time domain. Using $\mathbf{u} = (u, 0, w)$ and recognizing that operation $\partial/\partial x$ becomes multiplication by $(-sp)$ in the transformed domain, it follows from (6.59) that, for $0 \le z \le h$,

$$\begin{pmatrix} u(k_x, z, s) \\ w(k_x, z, s) \end{pmatrix} = \frac{L_0(s)}{2\rho\beta^2\xi}\left[\begin{pmatrix} -p \\ \xi \end{pmatrix}\exp[s\xi(z - h)] + \acute{P}\grave{P}\begin{pmatrix} -p \\ -\xi \end{pmatrix}\exp[-s\xi(z + h)]\right]$$
$$+ \frac{M_0(s)}{2\rho\beta^2\eta}\frac{\alpha}{\beta}\acute{S}\grave{P}\begin{pmatrix} -p \\ -\xi \end{pmatrix}\exp[-s(\xi z + \eta h)]$$
$$+ \frac{L_0(s)}{2\rho\beta^2\xi}\frac{\beta}{\alpha}\acute{P}\grave{S}\begin{pmatrix} -\eta \\ p \end{pmatrix}\exp[-s(\eta z + \xi h)]$$
$$+ \frac{M_0(s)}{2\rho\beta^2\eta}\left[\begin{pmatrix} -\eta \\ -p \end{pmatrix}\exp[s\eta(z - h)] + \acute{S}\grave{S}\begin{pmatrix} -\eta \\ p \end{pmatrix}\exp[-s\eta(z + h)]\right]. \tag{1}$$

Since the vectors

$$\begin{pmatrix} -p \\ \xi \end{pmatrix} \quad \text{and} \quad \begin{pmatrix} -\eta \\ -p \end{pmatrix}$$

have amplitude $1/\alpha$ and $1/\beta$, respectively, a factor α/β precedes $\acute{S}\grave{P}$ (and β/α precedes $\acute{P}\grave{S}$) in order for conversion coefficients to retain their standard definition in terms of displacement-amplitude ratios.

and

$$\mathrm{R}(p) = 4p^2\xi\eta + (\beta^{-2} - 2p^2)^2 \tag{6.65}$$

is the Rayleigh function.

In the remainder of this section, to establish some general methods and obtain specific results, we shall restrict ourselves to the inversion of just one wave to the time domain. (A total of six waves is indicated by (6.63), but two of these are incident waves, which are trivial to evaluate.) We choose to evaluate the generalized $\acute{S}\grave{S}$ reflection in the case that $M_0(t) = M_0 H(t)$ (i.e., a step of constant height M_0, for which the Laplace transform is M_0/s). From the last term in (1) of Box 6.9, the horizontal component of displacement in this wave is u_{SS}, where

$$u_{SS}(k_x, z, s) = \frac{-M_0}{2\rho\beta^2 s}\acute{S}\grave{S}\exp[-s\eta(z + h)]. \tag{6.66}$$

Since $\acute{S}\grave{S}$ is even in p, the inverse k_x-transform can be written as

$$u_{SS}(x, z, s) = \frac{-M_0}{2\pi\rho\beta^2}\operatorname{Im}\left\{\int_0^{i\infty}\acute{S}\grave{S}(p)\exp[-spx - s\eta(z + h)]\,dp\right\} \tag{6.67}$$

(recall that $k_x = isp$). Defining a Cagniard path by $t = px + \eta(z + h)$, i.e.,

$$p(t) = \begin{cases} \dfrac{xt - (z + h)\left(\dfrac{R_0^2}{\beta^2} - t^2\right)^{1/2}}{R_0^2} & t \le \dfrac{R_0}{\beta}, \\[2ex] \dfrac{xt + i(z + h)\left(t^2 - \dfrac{R_0^2}{\beta^2}\right)^{1/2}}{R_0^2} & \dfrac{R_0}{\beta} \le t, \end{cases} \tag{6.68}$$

with $R_0 = [x^2 + (z + h)^2]^{1/2}$ as the distance between receiver and image source, we obtain

$$u_{SS}(x, z, s) = \frac{-M_0}{2\pi\rho\beta^2}\operatorname{Im}\left\{\int_0^{\infty}\acute{S}\grave{S}(p)e^{-st}\frac{dp}{dt}\,dt\right\},$$

and hence the exact solution in the time domain is

$$\begin{aligned} u_{SS}(x, z, t) &= \frac{-M_0}{2\pi\rho\beta^2}\operatorname{Im}\left\{\acute{S}\grave{S}(p)\frac{dp}{dt}\right\} \\ &= \frac{M_0}{2\pi\rho\beta^2}\operatorname{Im}\left\{\frac{4p^2\xi\eta - (\beta^{-2} - 2p^2)^2}{4p^2\xi\eta + (\beta^{-2} - 2p^2)^2}\frac{dp}{dt}\right\}\Bigg|_{p = p(t)} \end{aligned} \tag{6.69}$$

In this expression,

$$\frac{dp}{dt} = \begin{cases} \dfrac{\eta}{\left(\dfrac{R_0^2}{\beta^2} - t^2\right)^{1/2}} & 0 < t < \dfrac{R_0}{\beta}, \\[2ex] \dfrac{i\eta}{\left(t^2 - \dfrac{R_0^2}{\beta^2}\right)^{1/2}} & \dfrac{R_0}{\beta} < t, \end{cases}$$

so that an integrable singularity is present in (6.69) at the ray arrival time of reflected waves, $t = R_0/\beta$.

Despite the simplicity of the exact solution (6.69), it is worthwhile to go further, since insight is acquired by identifying various features in the associated complex p-plane (Fig. 6.16b) and the approximate properties of waves that are controlled by these features. Thus, the first arrival at the receiver (if it is sufficiently distant from the source) is a head wave, having ray parameter $1/\alpha$ and arrival time $t_h = x/\alpha + (z + h)(\beta^{-2} - \alpha^{-2})$.$^{1/2}$ Its pulse shape is determined

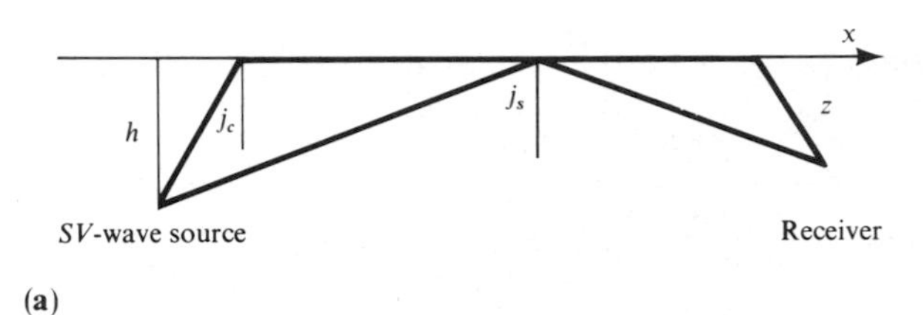

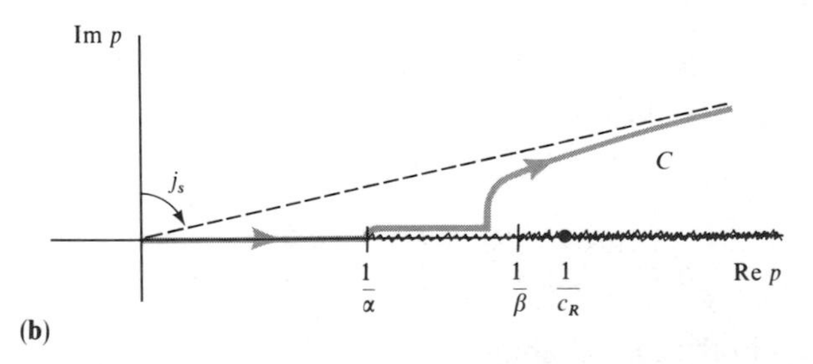

FIGURE **6.16**

The generalized $\acute{S}\grave{S}$ reflection for a line source buried in a homogeneous half-space. (**a**) The source-receiver geometry in the case $j_s > j_c$, showing both reflection and head-wave paths. (**b**) The complex p-plane, with Cagniard path on the Riemann sheet $\{\text{Re}\ \xi > 0; \text{Re}\ \eta > 0\}$. The Rayleigh pole at $p = 1/c_R$ now lies on two branch cuts.

by evaluation of (6.69) in the vicinity of $t = t_h$, $p = 1/\alpha$, values for which the Cagniard path lies on a branch cut beginning at $p = 1/\alpha$. Another major arrival is the reflection associated with a saddle point of (6.67). This occurs on the Cagniard path at $t = R_0/\beta$ and has properties similar to those we investigated in (6.57) and (6.58). The two new phenomena that occur are due to poles of the integrand in (6.67), i.e., zeros of the Rayleigh function

$$\mathsf{R}(p) = 4p^2\xi\eta + (\beta^{-2} - 2p^2)^2.$$

From Section 5.3, we know that $\mathsf{R}(p)$ has zeros at $p = \pm 1/c_R$ near $p = \pm 1/\beta$ on the sheet $\{\text{Re}\ \xi \geq 0;\ \text{Re}\ \eta \geq 0\}$, where ξ and η are in fact pure imaginary. $\mathsf{R}(p) = 0$ also has zeros near $p = \pm 1/\alpha$ on the sheet $\{\text{Re}\ \xi < 0;\ \text{Re}\ \eta \geq 0\}$. Such a sheet is often called "forbidden" or "unphysical," in the sense that an integrand evaluated at a point on this sheet would tend to infinity as $z \to \infty$, in violation of the radiation condition. However, we shall find that poles of $\acute{S}\grave{S}$ at zeros of $\mathsf{R}(p)$ that are on a forbidden sheet can yet make an identifiable contribution on a seismogram.

We shall assess first the effect of the Rayleigh pole at $p = 1/c_R$. As the source-receiver distance increases, the angle j_s increases (Fig. 6.16a) and the Cagniard-path asymptote becomes closer to the real p-axis (Fig. 6.16b). The Cagniard path itself therefore passes closer to the Rayleigh pole. If one were to contour the magnitude of $\acute{S}\grave{S}(p)$ in the vicinity of $p = 1/c_R$, as shown in Figure 6.17, it is apparent that the large values of $\acute{S}\grave{S}(p)$ in the vicinity of the pole would lead to an identifiable Rayleigh wave, provided the Cagniard path lies sufficiently close to $p = 1/c_R$. To get an estimate of when the Rayleigh pulse is well developed, we can see if, at its "arrival time" x/c_R, the Cagniard path lies closer to the Rayleigh pole than does the next-nearest feature of $\acute{S}\grave{S}$ and dp/dt that might influence u_{SS} in (6.69). Since this next-nearest feature is a branch point at $p = 1/\beta$, it is clear that the Rayleigh wave is likely to be well developed if

$$\left| p\left(\frac{x}{c_R}\right) - \frac{1}{c_R} \right| < \frac{1}{c_R} - \frac{1}{\beta}. \tag{6.70}$$

In this inequality, we can evaluate p at time x/c_R from (6.68) and find that a completely equivalent inequality is the geometrical relation

$$\tan j_s > \left(\frac{2c_R}{\beta - c_R}\right)^{1/2}. \tag{6.71}$$

In practice, this result appears to be rather better than Nakano's inequality (6.34), which we developed for a P-wave source (the inequality there being reversed, since it designated the range for which Rayleigh waves had *not*

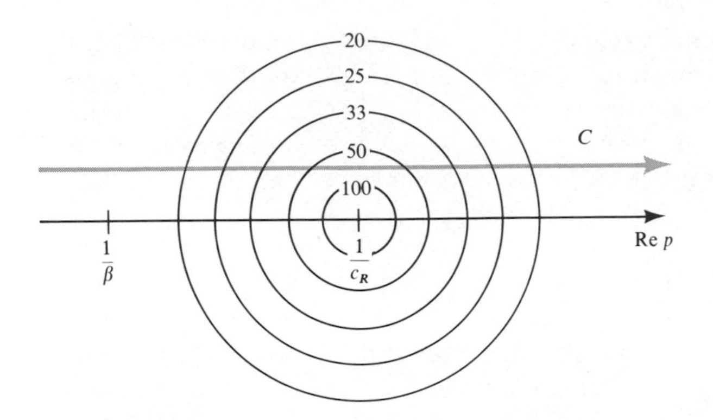

FIGURE **6.17**
The magnitude of $\acute{S}\grave{S}(p)$ is contoured in the complex p-plane for values near the Rayleigh pole. Then $\acute{S}\grave{S}(p) \propto (p - 1/c_R)^{-1}$, so that contours are essentially a system of concentric circles centered on the pole. The numbers shown are in the ratio $1:\frac{1}{2}:\frac{1}{3}:\frac{1}{4}:\frac{1}{5}:\ldots$, which is appropriate for equal increments of radius. For a large value of j_s (see Fig. 6.16ab), the Cagniard path C is shown cutting across the contours close to the pole: associated values of u_{SS} given by (6.69) will clearly be large at these times. Such values constitute the Rayleigh-wave pulse shape (i.e., its shear wave component. There is another contribution from generalized $\acute{S}\grave{P}$, giving the P-wave component.).

developed). Since $c_R \sim 0.92\beta$ for many rocks, we conclude from (6.71) that the range must be at least five times greater than the source + receiver depths in order for the Rayleigh pulse to be developed.

Gilbert and Laster (1962) labeled the pole at $p = 1/c_R$ the $\bar{S}$-pole, and the associated Rayleigh wave as the $\bar{S}$-pulse. This label was chosen because, in general terms, the Rayleigh wave is a phenomenon of diffraction associated with a nearby S-wave wavefront, which interacts with the free surface. Similarly, there is an effect associated with poles of $\mathsf{R}(p)$ which usually lie near $p = 1/\alpha$, moving away from this branch point only for values of Poisson's ratio greater than about 0.4. These poles, called $\bar{P}$-poles by Gilbert and Laster, do not lie on the physical Riemann sheet. Introducing the notation of Phinney (1961), we designate the physical sheet $\{\text{Re}\,\xi > 0;\ \text{Re}\,\eta > 0\}$ as the $(+\,+)$ sheet, and similarly for the three unphysical sheets: $(+\,-)$; $(-\,+)$, and $(-\,-)$. Whereas the Rayleigh pole ($\bar{S}$) lies on the $(+\,+)$ sheet, $\bar{P}$-poles lie on the $(-\,+)$ sheet.

It might reasonably be thought that poles of $\acute{S}\grave{S}(p)$ lying on unphysical sheets cannot influence the generalized reflection u_{SS} in (6.69), since the Cagniard path lies entirely on the $(+\,+)$ sheet. Shown schematically in Figure 6.18, however, is a system of contours of $\acute{S}\grave{S}$ in the vicinity of a $\bar{P}$-pole. Since we are

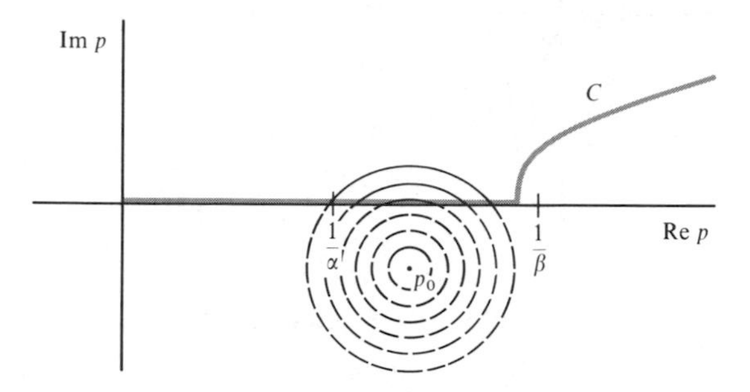

FIGURE **6.18**

Contours showing the magnitude of $\acute{S}\grave{S}(p)$ in the vicinity of a $\bar{P}$-pole $p = p_0$ lying on an "unphysical" Riemann sheet. This is the $(-\ +)$ sheet, and $\mathsf{R}(p_0) = 0$, where still $\mathsf{R}(p) = 4p^2\xi\eta + (\beta^{-2} - 2p^2)^2$. Although the contours close to p_0 are concentric circles lying entirely in the fourth quadrant of the $(-\,+)$ sheet, contours with larger radius must run into the branch cut running to the right from $p = 1/\alpha$. Such contours will then extend on to the first quadrant of the *physical* Riemann sheet, since this is the way $\acute{S}\grave{S}(p)$ is kept analytic across a branch cut (see Box 6.2). On the $(-\,+)$ sheet, contours are shown as broken lines. The extension of contours onto the $(+\,+)$ sheet is shown by solid lines. Note that the Cagniard path lies on the real p-axis for part of the segment between $1/\alpha$ and $1/\beta$.

interested in values of $\acute{S}\grave{S}$ in the first quadrant of the $(+\,+)$ sheet (this is the region in which our Cagniard path lies), it is just this sheet that is contoured for the first quadrant in Figure 6.18. Analytic continuation of $\acute{S}\grave{S}$ into the fourth quadrant (i.e., across the real p-axis for $1/\alpha < p$) then requires going on to a different Riemann sheet, the fourth quadrant of the $(-\,+)$ sheet, and in this region there *is* a pole of $\acute{S}\grave{S}$. The pole shown is in fact in the appropriate location for a Poisson's ratio of around 0.4, and the important point is that contours associated with this pole do appear on the physical sheet. For a receiver at sufficient range, so that the Cagniard path lies on or close to the real p-axis, these solid-line contours of Figure 6.18 will be crossed in the evaluation of (6.69), and a rather broad pulse shape will be the result. This is the $\bar{P}$-pulse, described in detail by Gilbert, Laster, Backus, and Schell (1962). It occurs for ray parameters lying between $1/\alpha$ and $1/\beta$, and its place in the seismogram is between the head-wave arrival and the wide-angle reflection. An example is shown in Figure 6.19a.

For values of Poisson's ratio decreasing from 0.4 to around 0.263, $\bar{P}$-poles migrate to the real p-axis just to the left of $p = 1/\alpha$ (but still on the $(-\,+)$ sheet). As Poisson's ratio decreases still further, from 0.263 to zero, $\bar{P}$-poles remain on the real p-axis, one pole actually tending to the branch point itself, at $p = 1/\alpha$ (see Problem 6.3). The presence of a pole near that point of the Cagniard path at which head waves originate (also at $p = 1/\alpha$, but on the $(+\,+)$ sheet) has the effect of distorting the head-wave arrival from its usual shape. In Figure 6.19b, we show an example of u_{SS} for an ordinary value of Poisson's ratio (0.25), and

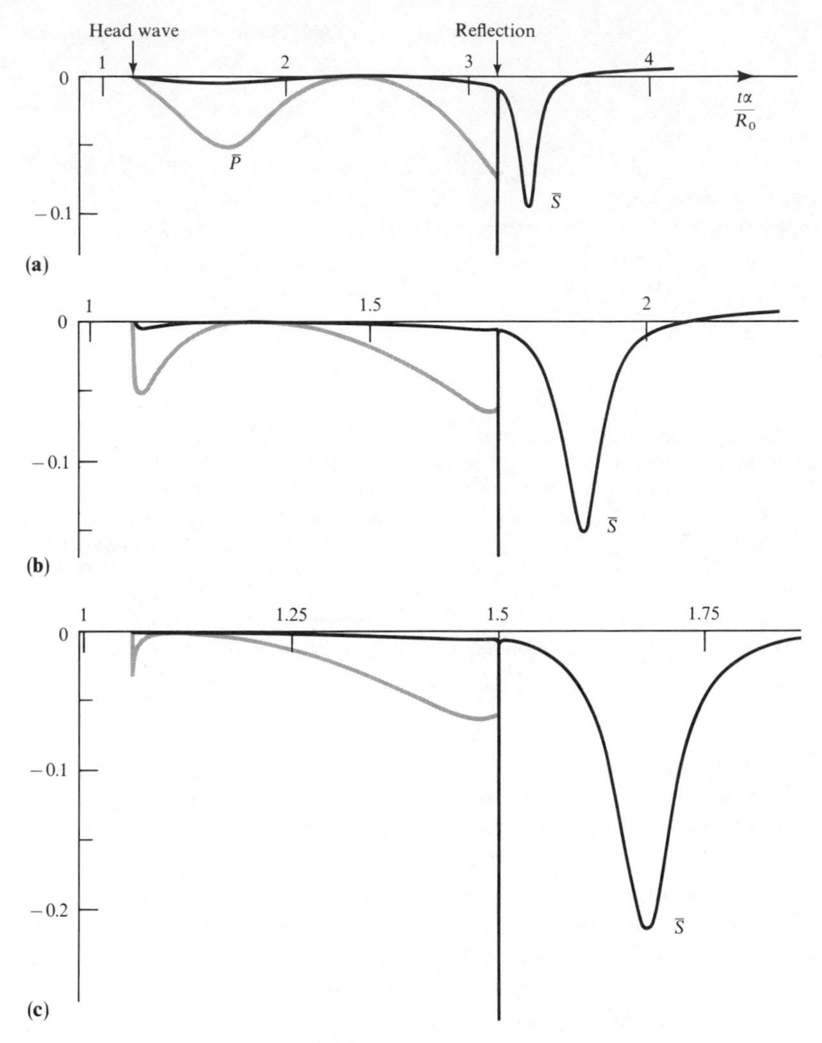

FIGURE **6.19**

The horizontal component of the generalized $\acute{S}\hat{S}$ reflection given by (6.69) is shown as a function of time for three different values of Poisson's ratio. We use dimensionless time, $t\alpha/R_0$, where R_0 is the distance between receiver and image source: $(h, x, z) = (1, 20, 0.1)$ units. Since the head wave is traveling at the P-wave speed for most of its path, it arrives just after $t\alpha/R_0 = 1$. Time scales have been chosen so that head-wave and reflection times are aligned. To bring out the detailed pulse shapes between head-wave and reflection arrival times, a tenfold increase in amplitude scale is also used for this part of the seismogram (shown in gray). (**a**) Poisson's ratio $= \frac{4}{9}$. There is a $\bar{P}$ pole away from the real p-axis, of the type shown in Figure 6.18, leading to a $\bar{P}$-wave between the head-wave arrival time and the wide-angle reflection. (**b**) Poisson's ratio $= \frac{1}{4}$. The head-wave and the wide-angle reflection are clearly separated pulses. (**c**) Poisson's ratio $= \frac{1}{10}$. A $\bar{P}$-pole lies on the real p-axis, close to the branch point $p = 1/\alpha$, although not on the $(+\,+)$ sheet. Its principal effect is to influence the shape of the head wave, giving here an impulsive arrival. Although there is a singularity at the reflection arrival time, there is insignificant area under this part of the pulse.

in Figure 6.19c we show the theoretical seismogram for a very low Poisson's ratio (0.1).

The $\bar{P}$-pulse is a simple example of a *leaking mode*. The adjective "leaking" is appropriate, because, for that part of the Cagniard path which is influenced by the $\bar{P}$-pole, η is either purely real or, for times just after the reflection arrival, has only a small imaginary part. This implies that energy is radiated downward, away from the free surface, as SV motion. Such a radiation leakage of energy does not occur for $\bar{S}$, for which both ξ and η are imaginary. We shall look further at leaking modes in Chapter 7.

6.5 Cagniard-De Hoop Methods for Point Sources

The attractive feature of Cagniard-de Hoop analysis for two-dimensional problems (line sources) was found above to be that exact algebraic solutions are possible. For three-dimensional problems, we shall find here that exact solutions in the time domain cannot usually be stated as algebraic expressions, but rather as single finite integrals. Fairly recently in seismology, these integrals have been manipulated so that they are carried out over a finite segment of the Cagniard path itself. But for many three-dimensional problems of practical interest, an adequate approximate form of solution can be found (see (6.94), (6.95)), in which the integral is readily computed because it is a convolution.

We begin again with a simple SH problem, using both Cartesian coordinates (x, y, z) and related cylindrical coordinates (r, ϕ, z). We suppose that a point source in an infinite elastic medium acts as a point torque at the origin, so that the body force is

$$\mathbf{f} = \nabla \times (0, 0, X) \tag{6.72}$$

in which the body-force potential X is axisymmetric, so that $X = X(r, z, t)$, independent of ϕ. Only the azimuthal component of displacement $\mathbf{u}$ is excited by this source, and it is independent of ϕ. Thus, in cylindrical coordinates,

$$\mathbf{u} = \nabla \times (0, 0, \chi) = (0, -\partial\chi/\partial r, 0), \tag{6.73}$$

and the equation of motion reduces to

$$\rho\ddot{\chi} = X + \mu\nabla^2\chi. \tag{6.74}$$

For a point source at the origin, with a step-like time dependence,

$$X = N_0 H(t)\,\delta(x)\,\delta(y)\,\delta(z) = N_0 H(t)\frac{\delta(r)}{2\pi r}\,\delta(z) \tag{6.75}$$

for some constant N_0, and from (4.4) it is clear that the solution for SH-potential χ is exactly

$$\chi(x, y, z, t) = \frac{N_0}{4\pi\rho\beta^2} \frac{H(t - R/\beta)}{R}, \tag{6.76}$$

using $R = |\mathbf{x}|$.

Let us now derive this result again, using instead the Cagniard-de Hoop approach for three-dimensional problems.

We start by taking the triple transform

$$\int_{-\infty}^{\infty} \exp(-ik_x x)\, dx \int_{-\infty}^{\infty} \exp(-ik_y y)\, dy \int_0^{\infty} \exp(-st)\, dt$$

operating on (6.74) and (6.75):

$$\frac{\partial^2}{\partial z^2} \chi(k_x, k_y, z, s) = -\frac{N_0}{\rho\beta^2 s} \delta(z) + n^2 \chi(k_x, k_y, z, s),$$

where $n^2 = k_x^2 + k_y^2 + s^2/\beta^2$. Comparing with (6.43) and (6.44), it follows that

$$\chi(k_x, k_y, z, s) = \frac{N_0}{2\rho\beta^2 sn} e^{-n|z|}, \tag{6.77}$$

choosing $\operatorname{Re} n > 0$.

The Laplace transform of the solution we wish to find is given by two inverse Fourier transforms of (6.77):

$$\chi(\mathbf{x}, s) = \frac{N_0}{8\pi^2\rho\beta^2 s} \int_{-\infty}^{\infty} dk_x \int_{-\infty}^{\infty} dk_y \frac{\exp(ik_x x + ik_y y - n|z|)}{n}, \tag{6.78}$$

and our goal is to rearrange this into the form of a forward Laplace transform, from which the integrand can be recognized as χ in the time domain.

To this end, we investigate the (k_x, k_y)-plane by using new variables (w, q) defined by

$$k_x = s(w \cos\phi - q \sin\phi), \qquad k_y = s(w \sin\phi + q \cos\phi). \tag{6.79}$$

This is the de Hoop transformation (de Hoop, 1960): it involves the azimuthal coordinate ϕ, and actually consists of a rotation and stretch of the whole horizontal-wavenumber plane. Since $x = r\cos\phi$, $y = r\sin\phi$, and the area element $dk_x\, dk_y$ is now replaced by $s^2 dw\, dq$, we find from (6.78) that

$$\chi(\mathbf{x}, s) = \frac{N_0}{8\pi^2\rho\beta^2}\int_{-\infty}^{\infty} dw \int_{-\infty}^{\infty} dq\, \frac{\exp(iswr - s\eta|z|)}{\eta}, \tag{6.80}$$

where $\eta = (\beta^{-2} + q^2 + w^2)^{1/2}$ and $\operatorname{Re}\eta > 0$.

One standard method of inverting (6.80) to the time domain is to convert the q-integral to a forward Laplace transform of the w-integral. This approach is often taken in the applied mechanics literature, and is used for a solution quoted in our Chapter 15 (see equations (15.30) and (15.31), in which an extra factor $1/\alpha$ has been removed from q and w to make these variables dimensionless). The resulting solution for $\chi(\mathbf{x}, t)$ is an integral over a finite range of w-values.

Another method of inverting (6.80) is to exploit the similarities with line-source problems. To bring the integrand (6.80) into a form familiar in the two-dimensional case (e.g., see (6.46)), we change the w-variable via $p = -iw$, giving

$$\chi(\mathbf{x}, s) = \frac{N_0}{2\pi^2\rho\beta^2}\int_0^{\infty} dq\, \operatorname{Im}\left[\int_0^{i\infty} \frac{\exp[-s(pr + \eta|z|)]}{\eta}\, dp\right], \tag{6.81}$$

where $\eta = (\beta^{-2} + q^2 - p^2)^{1/2}$. (To obtain (6.81) from (6.80), we also used half-ranges of integration and properties of evenness in q; and evenness and oddness in w for the real and imaginary parts of the integrand.)

Strong similarities are now apparent between the integrands of (6.47) and (6.81). Note here that the horizontal variable is r, rather than x, and $1/\beta^2 + q^2$ replaces $1/\beta^2$ in the definition of η. Previously, we found that

$$\operatorname{Im}\left[\int_0^{i\infty} \frac{\exp[-s(px + \eta|z|)]}{\eta}\, dp\right] \qquad (\text{with } \eta = (\beta^{-2} - p^2)^{1/2})$$

is the Laplace transform of

$$\frac{H\left(t - \dfrac{(x^2 + z^2)^{1/2}}{\beta}\right)}{\left(t^2 - \dfrac{x^2 + z^2}{\beta^2}\right)^{1/2}},$$

and this enables us now to write (6.81) as

$$\chi(\mathbf{x}, s) = \frac{N_0}{2\pi^2\rho\beta^2}\int_0^{\infty} dq \int_0^{\infty} \frac{H[t - R(\beta^{-2} + q^2)^{1/2}]}{[t^2 - R^2(\beta^{-2} + q^2)]^{1/2}}\, e^{-st}\, dt. \tag{6.82}$$

Here we are using $R = (x^2 + y^2 + z^2)^{1/2} = (r^2 + z^2)^{1/2}$ as the three-dimensional distance function. It remains only to interchange the order of integration in (6.82), giving

$$\chi(\mathbf{x}, s) = \frac{N_0}{2\pi^2\rho\beta^2}\int_0^\infty dt e^{-st}\left\{H\left(t - \frac{R}{\beta}\right)\int_0^{(t^2/R^2 - 1/\beta^2)^{1/2}} \frac{dq}{[t^2 - R^2(\beta^{-2} + q^2)]^{1/2}}\right\} \tag{6.83}$$

(as explained in Fig. 6.20), and then we can recognize the required solution as

$$\chi(\mathbf{x}, t) = \frac{N_0}{2\pi^2\rho\beta^2} H\left(t - \frac{R}{\beta}\right)\int_0^{(t^2/R^2 - 1/\beta^2)^{1/2}} \frac{dq}{[t^2 - R^2(\beta^{-2} + q^2)]^{1/2}}. \tag{6.84}$$

This is essentially the method of de Hoop (1960), and we list the following comments on (6.84):

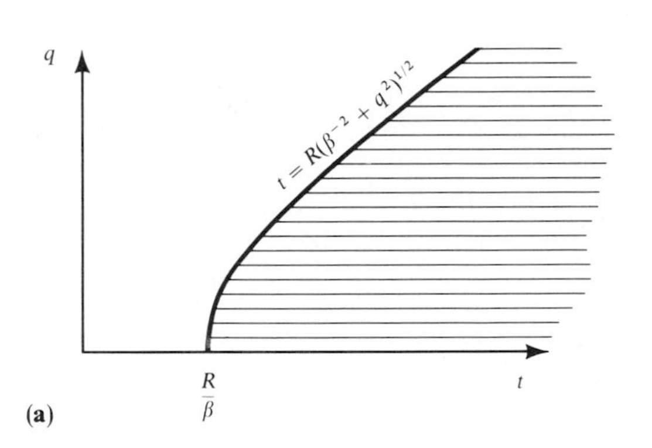

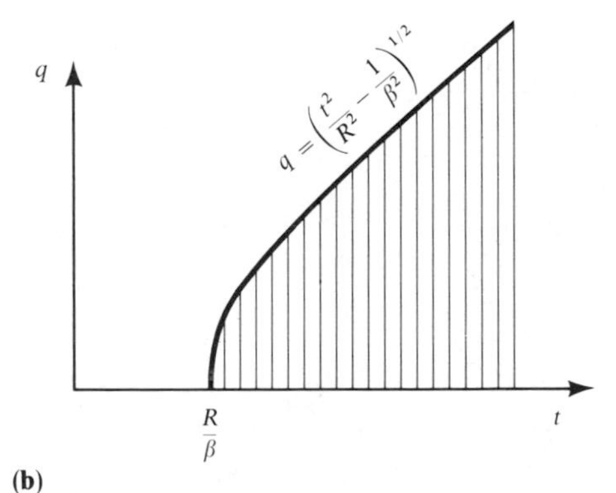

FIGURE **6.20**

Diagrams to explain equivalent integrations in the $t - q$ plane. (**a**) The t-integration at fixed q is conducted first, as in (6.82), with lower limit as a function of q. (**b**) The q-integration at fixed t is conducted first, as in (6.83), with upper limit as a function of t.

i) The solution has the typical form for point-source problems, in that the exact solution is a single finite integral.
ii) In our particular case, the integral in (6.84) is easy to evaluate analytically. The substitution $q = (t^2/R^2 - 1/\beta^2)^{1/2} \sin\theta$ immediately gives $\pi/(2R)$ for the integral, and solution (6.76) is recovered.
iii) Going back to (6.83), we find $\chi(\mathbf{x}, s) = (N_0 e^{-sR/\beta})/(4\pi\rho\beta^2 sR)$. For a general time dependence $N_0(t)$, rather than the step function we have so far assumed,

$$\chi(\mathbf{x}, s) = \frac{N_0(s)}{4\pi\rho\beta^2 R} e^{-sR/\beta} \quad \text{and} \quad \chi(\mathbf{x}, t) = \frac{N_0(t - R/\beta)}{4\pi\rho\beta^2 R}.$$

The physical displacement, from (6.73), is

$$u_\phi = -\left[\frac{\partial}{\partial r}\frac{N_0(t - R/\beta)}{4\pi\rho\beta^2 R}\right].$$

iv) When an interface is present, with its attendant head waves and interface waves, the above method is changed only in that the t-integral of (6.82) is more complicated (again, it can be written down from prior study of the corresponding two-dimensional problem).

Although the above method of solution is a brilliantly conceived and successful approach, it does not fully exploit the properties of ray parameter and of integration in the complex ray-parameter plane, which educated our intuition in Section 6.4. Therefore, modern methods of solving Lamb's problem for a point source have often taken the approach initiated by Strick (1959), and later extended by Helmberger (1968) and Gilbert and Helmberger (1972), in which the complex ray-parameter plane again plays a central role. These methods require familiarity with the Hankel transform and related results for modified Bessel functions, which are reviewed in Box 6.10. By way of illustration, we again turn to the problem (6.74)–(6.75), operating this time with the double transform $\int_0^\infty rI_0(\lambda r)\,dr \int_0^\infty e^{-st}\,dt$ to find

$$\frac{\partial^2}{\partial z^2}\chi(\lambda, z, s) = \frac{-N_0}{2\pi\rho\beta^2 s}\delta(z) + \left(\frac{s^2}{\beta^2} - \lambda^2\right)\chi(\lambda, z, s).$$

The doubly transformed solution therefore is (compare (6.44) and (6.43))

$$\chi(\lambda, z, s) = \frac{N_0}{4\pi\rho\beta^2 s^2\eta} e^{-s\eta|z|}, \tag{6.85}$$

where $\eta = (1/\beta^2 - \lambda^2/s^2)^{1/2}$ and $\operatorname{Re}\eta > 0$. Using $\lambda = sp$ (a factorization we have now done several times in one form or another) and the inverse transform

(Box 6.10, equation (4)), we see that

$$\chi(r, z, s) = \frac{N_0}{4\pi^2\rho\beta^2 i}\int_{-i\infty}^{i\infty}\frac{p}{\eta}K_0(spr)e^{-s\eta|z|}\,dp, \tag{6.86}$$

in which $\eta = (\beta^{-2} - p^2)^{1/2}$ and s, r, η are all real and positive.

BOX **6.10**

Horizontal transforms for functions symmetric about a vertical axis

Consider a function $f = f(x, y)$ in which the dependence is really only on $(x^2 + y^2)^{1/2} = r$. Then

$$f(k_x, k_y) = \iint\limits_{-\infty}^{\infty} f(x, y)\exp[-i(k_x x + k_y y)]\,dx\,dy.$$

We now introduce k_r and ϕ', just as we did in (6.5) to develop the Sommerfeld integral. Following the method that led to (6.7), we find now that

$$f(k_x, k_y) = 2\pi\int_0^\infty rf(r)J_0(k_r r)\,dr,$$

which is a function only of k_r, and not ϕ'. It is conventional to use $f(r) = f(x, y)$, but $f(k_r) = f(k_x, k_y)/2\pi$. We then find from the above and a similar treatment of the inverse transform that

$$\left.\begin{aligned} f(k_r) &= \int_0^\infty rf(r)J_0(k_r r)\,dr \\ f(r) &= \int_0^\infty k_r f(k_r)J_0(k_r r)\,dk_r \end{aligned}\right\}\quad \text{which is the Hankel transform pair of order zero.} \tag{1}$$

It follows by induction that there is a Hankel transform pair of order n:

$$\left.\begin{aligned} f^{(n)}(k_r) &= \int_0^\infty rf(r)J_n(k_r r)\,dr \\ f(r) &= \int_0^\infty k_r f^{(n)}(k_r)J_n(k_r r)\,dk_r. \end{aligned}\right\} \tag{2}$$

(We use the superscript (n) when it is necessary to emphasize the order of the Bessel function used in the transform.)

The order of Hankel transform to choose in any particular wave problem is the one for which, in the wave equation under discussion, operations with the variable r are reduced to scalar multiplication in the transform domain. For example, the derivatives in (6.74) are written

$$\rho\ddot{\chi} = X + \mu\left[\frac{1}{r}\frac{\partial}{\partial r}\left(\frac{r\,\partial\chi}{\partial r}\right) + \frac{\partial^2\chi}{\partial z^2}\right].$$

Here it is appropriate to use the zero-order Hankel transform, since

$$\int_0^\infty r\left[\frac{1}{r}\frac{\partial}{\partial r}\left(\frac{r\,\partial\chi}{\partial r}\right)\right]J_0(k_r r)\,dr = -k_r^2\chi(k_r, z, t).$$

(This equality follows from integrating by parts twice, then using the differential equation satisfied by J_0.) However, the wave equation satisfied by $u_\phi = -\partial\chi/\partial r$ is

$$\rho\ddot{u}_\phi = f_\phi + \mu(\nabla^2\mathbf{u})_\phi = f_\phi + \mu\left(\nabla^2 u_\phi - \frac{u_\phi}{r^2}\right), \tag{3}$$

and the extra last term here indicates that the zero-order Hankel transform would be unsuccessful. Instead, the first-order transform is the one that reduces all operations with r in (3) to scalar multiplication in the k_r-domain. In *P-SV* problems with a point source on $r = 0$, use of the zero-order transform is appropriate for u_z and τ_{zz}; the first-order transform, for u_r and τ_{zr}. These results are apparent from Box 6.6, which treats waves in the wavenumber-frequency domain.

In solving two-dimensional problems via the Cagniard-de Hoop method, we made early use of a relation $k_x = isp$ between horizontal wavenumber and ray parameter (see (6.39)). For point-source problems, this relation clearly becomes $k_r = isp$ (compare with (6.8) for the Fourier time transform), hence it will be important to examine (2) for pure imaginary values of k_r. We shall use $k_r = i\lambda$, and then (2) can be rewritten as

$$\left.\begin{aligned} f^{(n)}(\lambda) &= \int_0^\infty rf(r)I_n(\lambda r)\,dr \\ f(r) &= \frac{1}{\pi i}\int_{-i\infty}^{i\infty} \lambda f^{(n)}(\lambda)K_n(\lambda r)\,d\lambda. \end{aligned}\right\} \tag{4}$$

Here, I_n and K_n are modified Bessel functions. The transform pair (4) bears the same relationship to (2) as the Laplace transform does to the Fourier transform.

Noting that $K_0(\zeta^*) = [K_0(\zeta)]^*$, we can see that (6.86) gives

$$\chi(\mathbf{x}, s) = \frac{N_0}{2\pi^2\rho\beta^2}\,\mathrm{Im}\left\{\int_0^{i\infty}\frac{p}{\eta}\,K_0(spr)e^{-s\eta|z|}\,dp\right\}. \tag{6.87}$$

In view of the asymptotic result

$$K_0(spr) = \left(\frac{\pi}{2spr}\right)^{1/2} e^{-spr}\left(1 + O\left(\frac{1}{spr}\right)\right) \tag{6.88}$$

for large argument, (6.87) bears a very strong similarity to the line-source formula (6.47). Both integrands involve the complex ray-parameter plane, but now r replaces x as the horizontal variable. We define

$$\tau = pr + \eta|z|$$

and then solve for a Cagniard path $p = p(\tau)$ in the first quadrant (see Fig. 6.21). Deforming the path of integration in (6.87) to the Cagniard path gives

$$\chi(\mathbf{x}, s) = \frac{N_0}{2\pi^2\rho\beta^2}\,\mathrm{Im}\int_{p(\tau)} \frac{p}{\eta} K_0(spr)e^{-s\eta|z|}\,dp, \tag{6.89}$$

and at last we are in a position to invert to the time domain.

From tables, we know that $K_0(s\sigma_1)$ is the Laplace transform of

$$H(t - \sigma_1)(t^2 - \sigma_1^2)^{-1/2},$$

and hence $K_0(s\sigma_1)e^{-s\sigma_2}$ is the Laplace transform of

$$H(t - \sigma_1 - \sigma_2)[(t - \sigma_2)^2 - \sigma_1^2]^{-1/2}.$$

Therefore, (6.89) implies

$$\chi(\mathbf{x}, t) = \frac{N_0}{2\pi^2\rho\beta^2}\,\mathrm{Im}\int_{p(\tau)} \frac{p}{\eta} \frac{H(t - pr - \eta|z|)}{[(t - \eta|z|)^2 - p^2r^2]^{1/2}}\,dp. \tag{6.90}$$

The contribution to this integral comes only from values $R/\beta < \tau < t$ (at the lower limit, the integrand begins to have an imaginary part: the upper limit is

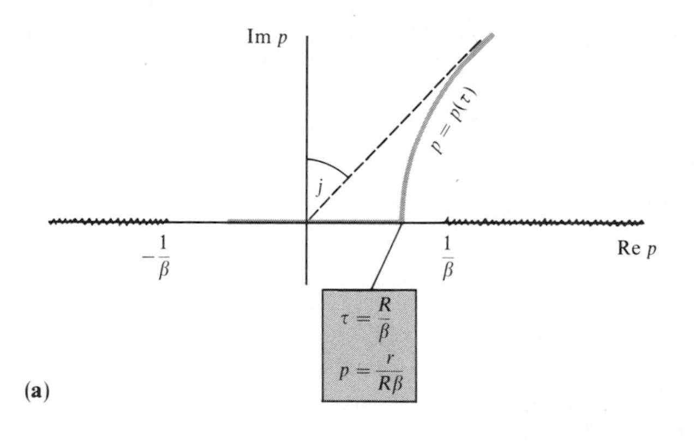

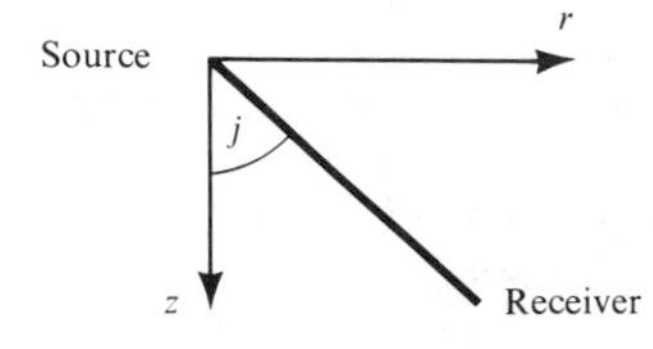

FIGURE **6.21**

The complex ray-parameter plane and features for the evaluation of (6.89)–(6.91). Note the similarity to Figure 6.14.

the cutoff from the step function), so that

$$\chi(\mathbf{x}, t) = \frac{N_0}{2\pi^2 \rho \beta^2} \operatorname{Im} \int_{R/\beta}^{t} \frac{p}{\eta} \frac{1}{(t - \tau)^{1/2}(t - \tau + 2pr)^{1/2}} \frac{dp}{d\tau}\, d\tau, \tag{6.91}$$

in which

$$p = p(\tau) = \frac{r\tau + i|z|(\tau^2 - R^2/\beta^2)^{1/2}}{R^2},$$

$$\frac{dp}{d\tau} = \frac{i\eta}{(\tau^2 - R^2/\beta^2)^{1/2}}, \quad \text{and} \quad \eta = \left(\frac{1}{\beta^2} - p^2\right)^{1/2} \quad \text{with } \operatorname{Re}\eta > 0.$$

Although it is not obvious that the integral in (6.91) has imaginary part $\pi/(2R)$, the integral is simple to evaluate numerically after taking account of integrable singularities at both upper and lower limits.

Once the derivation of (6.91) has been mastered, almost all the features of practical Cagniard-de Hoop applications are relatively simple. Thus, suppose two half-spaces are in welded contact along $z = 0$, as shown in Figure 6.15, with a *point* source of SH-waves acting at $z_0 < 0$ in the upper medium. The SH-potential in the upper medium is just the incident wave

$$\chi^{\text{inc}}(\mathbf{x}, t) = \frac{N_0}{4\pi\rho_1\beta_1^2} \frac{H(t - R/\beta_1)}{R}$$

using $R = (r^2 + (z - z_0)^2)^{1/2}$ (see (6.76)), together with a generalized reflection potential. By analogy with the line-source problem and (6.91), the reflection is exactly

$$\chi^{\text{refl}}(\mathbf{x}, t) = \frac{N_0}{2\pi^2\rho_1\beta_1^2} \operatorname{Im} \int_0^t \frac{p}{\eta_1} \left[\frac{\mu_1\eta_1 - \mu_2\eta_2}{\mu_1\eta_1 + \mu_2\eta_2}\right] \frac{1}{(t - \tau)^{1/2}(t - \tau + 2pr)^{1/2}} \frac{dp}{d\tau}\, d\tau \tag{6.92}$$

in which $p(\tau)$ is given by (6.56), but now of course with r in place of x, and τ in place of t. The integral in (6.92) is extended down to zero to pick up possible head waves in the case that $\beta_2 > \beta_1$.

We have made no approximations in deriving (6.92), but for many practical purposes an important approximation can be made that vastly reduces the computational effort, with little sacrifice in accuracy. Thus, suppose $|pr| \gg t - \tau$ throughout the range of integration. Then we can make the approximation

$$(t - \tau + 2pr)^{1/2} \sim (2pr)^{1/2} \tag{6.93}$$

and recognize (6.91) and (6.92) as convolutions of a function of time with $1/t^{1/2}$. That is, if for (6.92) we wish to calculate χ^{refl} for the whole interval $0 \le t \le T$, we first calculate the function

$$\psi(t) = \frac{1}{2\pi^2 \rho_1 \beta_1^2} \operatorname{Im}\left\{\frac{p(t)}{\eta_1}\left[\frac{\mu_1\eta_1 - \mu_2\eta_2}{\mu_1\eta_1 + \mu_2\eta_2}\right]\frac{1}{(2pr)^{1/2}}\frac{dp}{dt}\right\} \tag{6.94}$$

for $0 \le t \le T$. Then

$$\chi(\mathbf{x}, t) = N_0 \int_0^t \frac{\psi(\tau)}{(t - \tau)^{1/2}}\, d\tau = N_0 \psi(t) * \frac{1}{t^{1/2}}, \tag{6.95}$$

and the theoretical seismogram for $0 \le t \le T$ can be generated in just one convolution operation—a very efficient and fast process on many computers. By comparison, the exact result (6.92) requires a different numerical integration for each single point in the desired time series.

Note that the approximation (6.93) is good for large ranges, but fails even then for sufficiently long times. It is equivalent to using $[\pi/(2spr)]^{1/2}e^{-spr}$ for $K_0(spr)$ in (6.87), and hence is the same approximation we made at (6.17) in Section 6.2 for a point-source problem discussed in the frequency domain. Further terms can be kept in the expansion for K_0, and are equivalent to approximating $(t - \tau + 2pr)^{-1/2}$ by the binomial expansion

$$\frac{1}{(2pr)^{1/2}}\left[1 - \frac{t - \tau}{4pr} + \frac{3}{32}\left(\frac{t - \tau}{pr}\right)^2 + \cdots\right].$$

Provided $|2pr| > t - \tau$, these further terms may be expected to give improved accuracy, though the effort is rarely worthwhile. Since they depend on t only via factors of type $(t - \tau)^n$, they too yield simple convolutions (Helmberger and Harkrider, 1978).

The solutions (6.91) and (6.92) are for the case of a step-function time-dependence of the force. Other time-dependences can clearly be obtained by convolution, but there is one time-dependence that in effect "cancels out" the convolution (6.95). Suppose, for example, that

$$X(r, z, t) = N_0(t)\frac{\delta(r)}{2\pi r}\delta(z) \quad \text{with} \quad N_0(t) = \frac{H(t)}{\pi t^{1/2}}.$$

Then it is an elementary exercise in convolutions to show that

$$\chi(\mathbf{x}, t) = \frac{d\psi}{dt} * \frac{1}{t^{1/2}} * \frac{1}{\pi t^{1/2}} = \psi(t),$$

and in this sense we can actually get a useful *algebraic* expression for the seismogram, given by (6.94).

The approximation (6.93), and the resulting convolution (6.95), is now very much a part of modern seismology, as we shall find in Chapter 9 when looking at the effects of multiple layering.

6.6 Summary of Main Results and Comparison Between Different Methods

We have described two methods for solving problems of a spherical wave interacting with a plane boundary. The first method (Sections 6.1–6.3) uses the Fourier transform of time-dependence, and leads to solutions for displacement, pressure, etc. as a function of frequency. The second method (Sections 6.4–6.5) uses a Laplace transform, but (by manipulations due to Cagniard, de Hoop, and others) leads to solutions directly in the time domain. In this section we list some similarities and differences between the two methods and discuss their merits and disadvantages, so that the user can determine which method is better for a particular application.

First, we list the similarities. Both the Fourier method and the Cagniard method entail integrations in the complex ray-parameter plane: ray paths in the physical problem correspond to saddle points in the integrand under consideration; head waves correspond to branch cuts; interface waves (e.g., Rayleigh, Stoneley) correspond to poles; and leaking modes (e.g., $\bar{P}$) correspond to poles on Riemann sheets other than that on which the radiation condition is satisfied.

Second, there are several superficial differences. Thus (i) to obtain results in the time domain via the Fourier method, a numerical inverse transform is required. But, in practice, the Cagniard solution must be convolved with a source function and with the instrument response, and these operations are essentially equivalent to numerical Fourier transformations. (ii) We characterized the Cagniard methods as being exact, whereas early in the development of the Fourier method, we made an approximation to certain Hankel functions (see (6.15)–(6.17)). In practice, an equivalent approximation (6.93) is often made in the Cagniard approach. (iii) Branch cuts in the Cagniard method were chosen to make $\{\mathrm{Re}\,\xi \geq 0;\ \mathrm{Re}\,\eta \geq 0\}$, and it was found possible to keep the path of integration on this same physical Riemann sheet, without crossing branch cuts. However, we developed the Fourier theory with branch cuts fixed by $\{\mathrm{Im}\,\xi \geq 0;\ \mathrm{Im}\,\eta \geq 0\}$, in which case we found it necessary to develop complicated paths of integration (e.g., Fig. 6.9) that had segments on non-physical sheets. Many authors take this approach (e.g., Lapwood, 1949; Berry and West, 1966), and a correct discussion of the effect of leaking modes can be highly involved. Fortunately, the choice of branch cuts is quite flexible in the

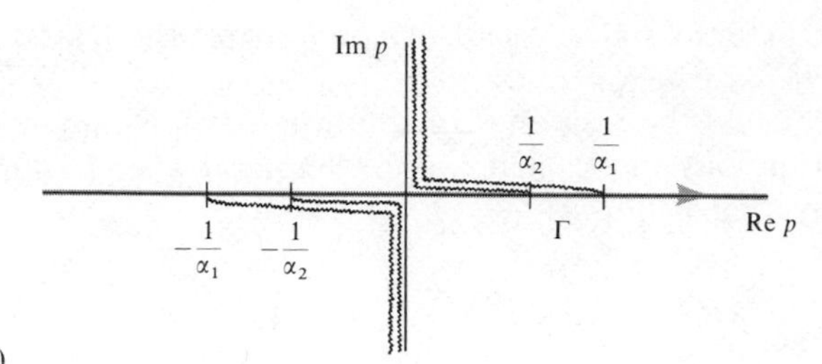
Im p
$\frac{1}{\alpha_2}$
$\frac{1}{\alpha_1}$
$-\frac{1}{\alpha_1}$
$-\frac{1}{\alpha_2}$
Γ
Re p

(a)

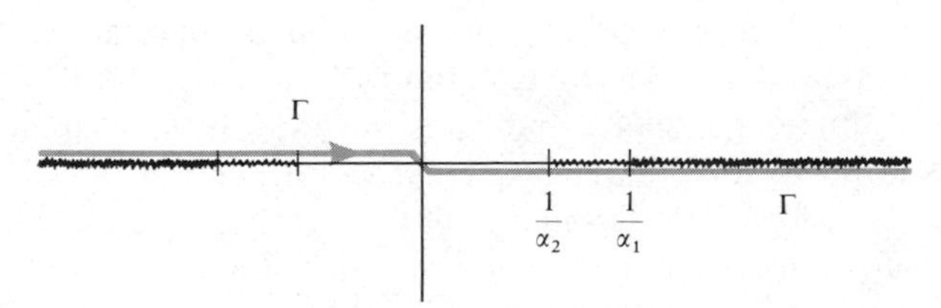
Γ
$\frac{1}{\alpha_2}$
$\frac{1}{\alpha_1}$
Γ

(b)

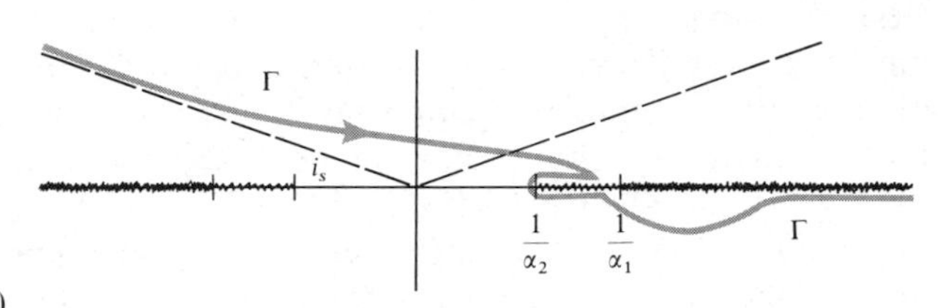
Γ
i_s
$\frac{1}{\alpha_2}$
$\frac{1}{\alpha_1}$
Γ

(c)

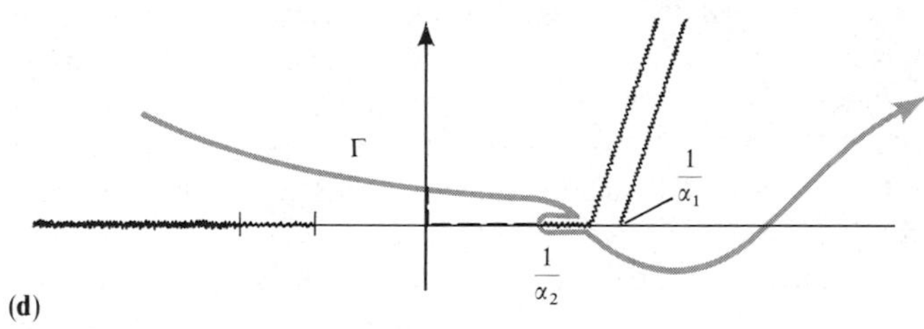
Γ
$\frac{1}{\alpha_1}$
$\frac{1}{\alpha_2}$

(d)

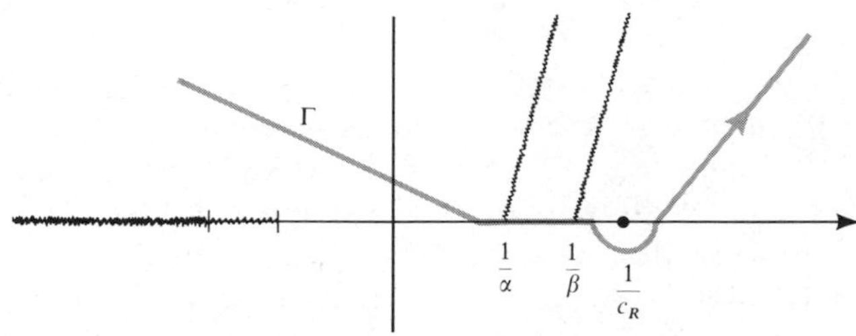
Γ
$\frac{1}{\alpha}$
$\frac{1}{\beta}$
$\frac{1}{c_R}$

(e)

Fourier method, and a path of integration can in fact be chosen that does lie close to the steepest-descents path, yet also stays on the same Riemann sheet. The main constraint, the radiation condition, requires that the integrand (e.g., (6.32)) tend to zero as $|z| \to \infty$ only for values of p on the original path of integration. Thus, in (6.32), we really require Im $\xi \geq 0$ only for p-values on the real p-axis, and do *not* need to use branch cuts for which Im $\xi \geq 0$ on the whole plane. In Figure 6.22 are shown some different choices for branch cuts, which enable one to take integration paths that are more successful for numerical applications of the Fourier method.

Finally, we point out some of the differences between our two methods in order to show the appropriateness of each for certain circumstances. Thus (i) the Cagniard path must be found exactly for each generalized ray, but the path of integration in the Fourier method need only approximately follow a steepest-descents path: as we shall see in Chapter 9, two or three straight-line segments are often adequate. This difference between the two methods is particularly important when theoretical seismograms are required for a large suite of source depths or source-receiver distances. Then, in the Fourier method, the same integration path can be used several times to generate different seismograms by changing only one factor (e.g., $e^{i\omega pr}$, to look at different ranges). The various Cagniard paths we have described have all been obtained by

---

FIGURE **6.22**

Various branch cuts and integration paths in the complex p-plane, showing the flexibility of choice. (a)–(d) are relevant to the problem solved in Section 6.2 of a point source of pressure in a medium consisting of two fluid half-spaces. (**a**) Branch cuts are chosen so that Im $\xi_1 > 0$, Im $\xi_2 > 0$ for the whole plane. Γ lies on the real axis, just above cuts in the third quadrant, and below cuts in the first. Compare with Figure 6.4. (**b**) For the same problem, we have changed branch cuts to Re $\xi_1 > 0$, Re $\xi_2 > 0$. The solution is unchanged, because Γ is unchanged and the value of an integrand (e.g., (6.17)) at any point on Γ is unchanged from (a). (**c**) Γ is distorted from the position shown in (b) to lie on a steepest-descents path in a case where head waves are possible. The path around the cut is now much simpler than that shown in Figure 6.9, although there is still a problem in that the steepest-descents path runs into the cuts at $p = 1/(\alpha_1 \sin i_s)$, hence Γ is subsequently drawn below the cut. (**d**) For cuts like those shown here, there is no difficulty in keeping Γ everywhere on the steepest-descents path (except around the branch point at $p = 1/\alpha_2$). To see that these cuts are possible, note that they can be moved from the position shown in (a) *before* Γ is distorted from the real axis. In subsequent distortion of Γ to the path shown here, Im ξ_1 and Im ξ_2 do become negative in the first quadrant to the left of the cuts shown. This is allowed because no singularities are present between this part of Γ and a path (shown as a broken line) on which Im ξ_1 and Im ξ_2 are positive. (**e**) This shows the p-plane for a solid half-space problem, e.g., for evaluating the generalized $\acute{P}\grave{P}$ reflection (see Fig. 6.12 for comparison). Branch cuts are drawn upward into the first quadrant, and a path Γ favorable for computations is made up from straight-line segments and a semicircle around $p = 1/c_R$. Branch cuts of this type have properties similar to the lines of poles found in Chapter 9 in generating theoretical seismograms when the Earth's spherical geometry is taken into account.

solving a quadratic equation, but in practice the equation to be solved is often of higher order. For example, the $\acute{S}\grave{P}$ reflection (6.63) entails solving $t = px + \eta z + \xi h$, which is a quartic in p. Fortunately, as Helmberger (1968) has shown, there is no difficulty in determining a numerical solution, even for multilayered media. (ii) The success of the Cagniard method depends on a certain property of the Laplace transform of the required solution—namely, that it can be written as a function of s times a factor in the form $\int_0^\infty g(t)e^{-st}\,dt$ (see (6.40)). Unfortunately, this property is quite easily destroyed. For example, if attenuation is introduced, then, as we saw in Chapter 5, the body-wave velocity becomes frequency dependent. The Laplace transform of the response takes a form $\int_0^\infty g(s, t)e^{-st}\,dt$, and it is not possible to factor g in a way that permits the time-domain solution to be identified directly. The same difficulty arises for media in which elastic properties vary continuously with depth, but with occasional discontinuities, as we shall see in Chapter 9.

In summary, it is clear that the Cagniard-de Hoop solution has great advantages for solving Lamb's problem with a point source or a line source. The impulse response is obtained directly and with a minimum of computational effort. The Fourier method also has merit, giving some complementary insight in that the behavior of different frequency components is studied directly and being more flexible than the Cagniard method in applications beyond Lamb's problem.

SUGGESTIONS FOR FURTHER READING

Brekhovskikh, L. M. *Waves in Layered Media*. New York: Academic Press, 1960.

Červený, V., and R. Ravindra. *Theory of Seismic Head Waves*. University of Toronto Press, 1971.

Chapman, C. H. Lamb's problem and comments on the paper 'On leaking modes' by Usha Gupta. *Pure and Applied Geophysics*, **94**, 233–247, 1972.

de Hoop, A. T. Modification of Cagniard's method for solving seismic pulse problems. *Applied Science Research*, **B8**, 349–356, 1960.

Gilbert, F., and L. Knopoff. The directivity problem for a buried line source. *Geophysics*, **26**, 626–634, 1961.

Johnson, L. R. Green's function for Lamb's problem. *Geophysical Journal of the Royal Astronomical Society*, **37**, 99–131, 1974.

Lamb, H. On the propagation of tremors over the surface of an elastic solid. *Philosophical Transactions of the Royal Society of London*, **A203**, 1–42, 1904.

Lapwood, E. R. The disturbance due to a line source in a semi-infinite elastic medium. *Philosophical Transactions of the Royal Society of London*, **A242**, 63–100, 1949.

Pekeris, C. L. The seismic buried pulse. *Proceedings of the National Academy of Sciences*, **41**, 629–639, 1955.

PROBLEMS

6.1 When head waves can occur in the solution (6.92), arriving at time t_h, show that (6.93) is equivalent to requiring

$$\frac{r}{\beta_2} \gg t - t_h.$$

When $\beta_2 < \beta_1$, show that (6.93) in the solution (6.92) is equivalent to requiring

$$\frac{r^2}{R_0\beta_1} \gg t - \frac{R_0}{\beta_1}.$$

The above constraints on the applicability of our convolution solution (6.95) might at first sight make it seem that (6.95) is nothing more than a wavefront approximation (e.g., (6.21)). In fact, (6.95) is much more powerful than a wavefront approximation. Show briefly that this is so by considering expressions for the seismograms at distances near critical, where the head wave and wide-angle reflection are superimposed.

6.2 Use a body-force equivalent like (3.8) of Chapter 3 to show that the discontinuity in $\partial v/\partial z$, used to obtain (6.44), is physically due to a discontinuity in stress.

6.3 Show that the zero crossings of $\acute{P}\grave{P}$ and $\acute{S}\grave{S}$ in Figure 5.6 occur precisely at values of p for which the Rayleigh function has zeros on nonphysical Riemann sheets. (It is the right-most zero crossing of Fig. 5.6, near $p = 1/\alpha$, that can affect the head-wave arrival contained in expression (6.69), evaluated in Fig. 6.18.)

6.4 Give an algebraic solution for the classical Lamb problem with a line source. That is, give the surface displacements (u, w) when a line source along the y-axis acts on the surface of an elastic half-space, so that there is an impulse I (per unit length) applied downward at the origin ($x = 0$, $t = 0$) on the free surface $z = 0$. Obtain a simple approximation for the shape of the Rayleigh wave.

6.5 In setting up P-SV problems with the method of potentials, note that in general the SV-displacement is represented by $\nabla \times \nabla \times (0, 0, \psi)$. See Box 6.5. Reconcile this with a conclusion reached in Section 5.1, in which the SV-displacement was represented by $\nabla \times (0, \psi, 0)$ for plane waves. (By a synthesis from such waves, $\nabla \times (0, \psi, 0)$ is appropriate for SV in all problems with fields independent of the y-coordinate,)

6.6 In Figure 6.9, is the result of integrating along the total path Γ (as shown) any different from using the path $A \to B \to C \to$ (around $1/\alpha_2$) $\to F' \to$ (saddle) $\to G \to H$? (That is, omitting $F' \to D \to E \to F'$.)

6.7 Show that the motion u_{SS} obtained in (6.69), and illustrated in Figure 6.19, diverges to infinity as $t \to \infty$. Show (without developing full algebraic details) that this behavior is cancelled out by a similar divergent expression arising from the P-wave component of particle motion (generalized $\acute{S}\grave{P}$).

CHAPTER 7

Surface Waves in a Vertically Heterogeneous Medium

Surface waves are propagated in directions parallel to the surface of the Earth. Their amplitude distribution over depth is stationary with horizontal position (apart from an overall multiplicative factor describing the horizontal spreading). Therefore, the geometrical spreading effect is much less on surface waves than on body waves, for which the energy spreads both horizontally and vertically down into the Earth's interior. In fact, surface waves are usually the main part of records on long-period seismograms of the Worldwide Standard Seismograph Network (WWSSN), and much of the reliable information on the long-period part of seismic spectra is obtained from surface waves. Long-period surface waves, with periods of 10 to 200 sec, have been a valuable source of information on both the Earth's structure and the seismic source mechanism. Their phase velocities, group velocities, and attenuation characteristics have been useful in delineating the structure of the crust and upper mantle in various regions of the Earth. Their source spectra and radiation patterns have contributed to determination of seismic moment, focal mechanism, and focal depths of remote events. They also supply crucial data for discriminating earthquakes from underground explosions.

The main purpose of this chapter is to describe the basic properties of surface waves in a vertically heterogeneous medium bounded by a free surface and to derive formulas for their amplitude and phase spectrum due to an arbitrary point source. In the final section, to complete the modal approach, we shall represent the total wave field as a sum of normal modes and leaky modes.

7.1 Basic Properties of Surface Waves

Let the z-axis be vertical, with the positive direction downward. Our vertically heterogeneous medium occupies the lower half-space, $z > 0$. As we saw in Sections 5.1 and 5.2, it is worthwhile to consider waves in which the dependence on spatial coordinates and time is given by the method of separation of variables. Thus, beginning with the simplest of problems, we shall consider Cartesian coordinates and a surface wave propagating in the horizontal direction of increasing x, with frequency ω:

$$\mathbf{u}(x, y, z, t) = \mathbf{Z}(z) \exp[i(kx - \omega t)]. \tag{7.1}$$

The all-important consequences of this equation are that only z-derivatives remain in the wave equation and boundary conditions, and such one-dimensional problems are relatively simple to solve.

BOX **7.1**

Initial assumptions

As we begin to discuss surface waves, it should be noted that a double meaning is attached to the very first equation here, (7.1). In the main text, this basic wave is introduced as a steady-state wave, having fixed horizontal wavenumber. (For brevity, we shall use k where in previous chapters we used k_x or k_r.) This permits a clear physical picture of the wave, and obviously

$$\frac{\partial}{\partial t}\mathbf{u}(x, z, t) = -i\omega\mathbf{u}(x, z, t),$$

$$\frac{\partial}{\partial x}\mathbf{u}(x, z, t) = ik\mathbf{u}(x, z, t).$$

The surface waves in seismology, however, are actually a synthesis of such basic waves. In this sense, it is appropriate for us to think of the displacement field as a function of (k, z, ω), being the double Fourier transform (over x and t) of the nonseparable surface wave $\mathbf{u}(\mathbf{x}, t)$. In this context, the double transforms of $\partial\mathbf{u}(\mathbf{x}, t)/\partial t$ and $\partial\mathbf{u}(\mathbf{x}, t)/\partial x$ are, respectively, $-i\omega\mathbf{u}(k, z, \omega)$ and $+ik\mathbf{u}(k, z, \omega)$. By two different but related approaches, we have thus reduced partial derivatives (with respect to x and t) to scalar multiplications.

In fact, the problem is to evaluate (7.1) under three general conditions. These waves must satisfy the free surface condition at $z = 0$, where the traction vanishes. At the other boundary $z = \infty$, their amplitude must vanish because no energy is supplied at infinity. In addition, they must, of course, satisfy the equation of motion. Under such restrictive conditions, a nontrivial solution of the form (7.1) does not exist for an arbitrary frequency ω and for an arbitrary wavenumber k. It can exist, however, if for a given ω, k takes a special value, say $k_n(\omega)$. This is an eigenvalue problem, and we use a subscript n because it can often happen that more than one value of k provides a surface wave (7.1) with given ω. Here $k_n(\omega)$ is an eigenvalue, and the corresponding solution $\mathbf{u}_n(z)$ is the eigenfunction. Thus, for a given frequency ω, surface waves (if they exist) have uniquely determined wave numbers $k_0(\omega)$, $k_1(\omega)$, $k_2(\omega)$ . . . . In other words, the phase velocites $c_n = \omega/k_n$ are fixed for a given frequency. The z-dependence is also fixed independent of the horizontal location.

To illustrate these concepts, we shall first study the simplest case of plane Love waves in a single homogeneous layer overlying a homogeneous half-space (Fig. 7.1). Putting the rigidities of the layer and half-space as μ_1, μ_2 and their densities as ρ_1, ρ_2 respectively, the y-component of displacement (v) satisfies the wave equation

$$\frac{\partial^2 v}{\partial t^2} = \frac{\mu_1}{\rho_1}\left(\frac{\partial^2 v}{\partial x^2} + \frac{\partial^2 v}{\partial z^2}\right) \qquad \text{in the layer } 0 < z < H$$

and

$$\frac{\partial^2 v}{\partial t^2} = \frac{\mu_2}{\rho_2}\left(\frac{\partial^2 v}{\partial x^2} + \frac{\partial^2 v}{\partial z^2}\right) \qquad \text{in the half-space } z > H.$$

Using trial solutions of the form (7.1), we find

$$v = [\grave{S}_1 e^{-\nu_1 z} + \acute{S}_1 e^{\nu_1 z}] \exp[i(kx - \omega t)] \qquad 0 \le z \le H$$

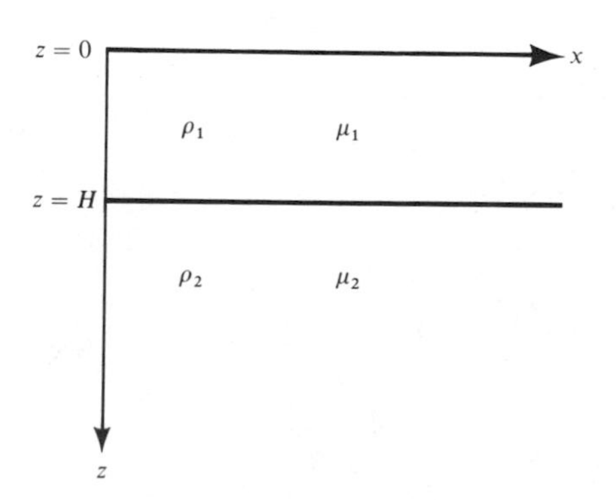

FIGURE **7.1**

A homogeneous layer over a homogeneous half-space, the simplest medium in which Love waves can be generated.

and

$$v = [\grave{S}_2 e^{-\nu_2 z} + \acute{S}_2 e^{\nu_2 z}] \exp[i(kx - \omega t)] \qquad H \leq z, \tag{7.2}$$

where

$$\nu_i = (k^2 - \omega^2/\beta_i^2)^{1/2}, \quad \mathrm{Re}\, \nu_i \geq 0, \quad \text{and} \quad \beta_i = (\mu_i/\rho_i)^{1/2}$$

for $i = 1$ and 2. Then $\mathrm{Im}\, \nu_i \leq 0$ (cf. properties of γ in (6.4)). $\grave{S}_1$, $\acute{S}_1$, $\grave{S}_2$ and $\acute{S}_2$ are constants, as yet unknown.

We shall assume that the velocity β_1 of the layer is lower than the velocity β_2 of the half-space. From the boundary conditions as $z \to \infty$, we know (if the wave is homogeneous, i.e., if ν_2 is negative imaginary) that there can be no upcoming wave, that so $\acute{S}_2 = 0$, and it is also just this wave that must vanish if ν_2 is positive real, to avoid exponential growth with depth. (The sign of root for ν_1 and ν_2 gives $i\nu_1$ and $i\nu_2$ the property of vertical wavenumber, having real and imaginary parts that are either positive or zero.) Since the free surface condition is satisfied if $\partial v/\partial z = 0$ at $z = 0$, then $\grave{S}_1 = \acute{S}_1$. Thus we are left with two unknowns, $\grave{S}_1$ and $\grave{S}_2$, which must satisfy the following equations required for continuity of displacement and traction across $z = H$:

$$\begin{aligned} 2\grave{S}_1 \cos(i\nu_1 H) &= \grave{S}_2 e^{-\nu_2 H} \\ 2i\mu_1 \nu_1 \grave{S}_1 \sin(i\nu_1 H) &= \mu_2 \nu_2 \grave{S}_2 e^{-\nu_2 H} \end{aligned} \tag{7.3}$$

or

$$\frac{\grave{S}_2}{\grave{S}_1} = \frac{2 \cos(i\nu_1 H)}{e^{-\nu_2 H}} = \frac{2i\mu_1 \nu_1 \sin(i\nu_1 H)}{\mu_2 \nu_2 e^{-\nu_2 H}}. \tag{7.4}$$

This gives eigenvalues k_n as the solutions of $F(k) = 0$, where

$$F(k) \equiv \tan(i\nu_1 H) - \frac{\mu_2 \nu_2}{i\mu_1 \nu_1}, \tag{7.5}$$

or the phase velocity $c_n = \omega/k_n$ as the solution of

$$\tan \omega H \left(\frac{1}{\beta_1^2} - \frac{1}{c^2} \right)^{1/2} = \frac{\mu_2}{\mu_1} \frac{\left(\dfrac{1}{c^2} - \dfrac{1}{\beta_2^2} \right)^{1/2}}{\left(\dfrac{1}{\beta_1^2} - \dfrac{1}{c^2} \right)^{1/2}}. \tag{7.6}$$

The corresponding eigenfunction v_n is obtained by putting (7.4), $\grave{S}_1 = \acute{S}_1$, and $\acute{S}_2 = 0$ into (7.2):

$$v_n(x, z, t) = 2\grave{S}_1 \cos\left[\omega\left(\frac{1}{\beta_1^2} - \frac{1}{c_n^2}\right)^{1/2} z\right] \times \exp[i(k_n x - \omega t)] \qquad \text{for } 0 \le z \le H$$

$$= 2\grave{S}_1 \cos\left[\omega\left(\frac{1}{\beta_1^2} - \frac{1}{c_n^2}\right)^{1/2} H\right] \exp\left[-\omega\left(\frac{1}{c_n^2} - \frac{1}{\beta_2^2}\right)^{1/2} (z - H)\right] \times \exp[i(k_n x - \omega t)] \qquad \text{for } H \le z, \tag{7.7}$$

which shows a sinusoidal oscillation in the layer and an exponential decay in the half-space.

Equation (7.6) may be solved graphically to give a rough idea about the phase velocity. Figure 7.2 shows the right side of (7.6) as a broken line and the left side as a solid line, both as functions of $(H/\beta_1)(1 - \beta_1^2/c^2)^{1/2}$. The roots are given by the intersection of two lines. From the figure, it is clear that real roots for c are limited to lie between β_1 and β_2 (we assumed $\beta_1 < \beta_2$). We find that there are only a finite number of real roots for a given frequency ω. When $\omega = 0$, there is only one root, which is a point on the fundamental mode, corresponding to the extreme left branch of the tangent curves. If we increase ω, the tangent curve marked as $n = 1$ enters the range from the right. It enters when π/ω is equal to $(H/\beta_1)(1 - \beta_1^2/\beta_2^2)^{1/2}$. If we further increase ω, more tangent curves will enter the

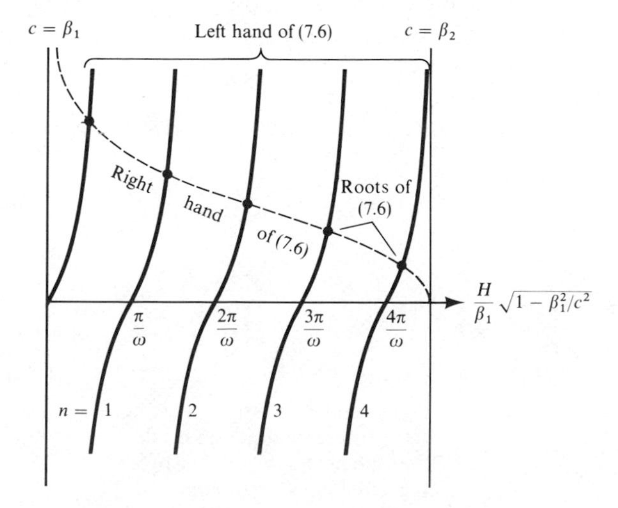

FIGURE **7.2**
Graphic solution of equation (7.6) for the dispersion of Love waves in a single layer over a half-space.

range from the right. The nth curve enters there when ω is equal to

$$\omega_{cn} = \frac{n\pi\beta_1}{H} \Bigg/ \left(1 - \frac{\beta_1^2}{\beta_2^2}\right)^{1/2}, \tag{7.8}$$

which is called the *cut-off frequency of the* nth *higher mode*, because, as shown in Figure 7.3, the nth mode exists only for $\omega > \omega_{cn}$. For example, in the case of a typical crust-mantle structure in a continent, taking $H = 35$ km, $\beta_1 = 3.5$ km/sec, and $\beta_2 = 4.5$ km/sec, the cutoff frequency of the first higher mode is 0.08 Hz or has a period of 13 sec.

Thus the nth-higher mode appears at $\omega = \omega_{cn}$ and exists for frequencies higher than ω_{cn}. At the cutoff frequency, all the modes have a phase velocity of $c = \beta_2$. As $\omega \to \infty$, the phase velocity approaches β_1 for all modes. Therefore, we can draw schematic phase-velocity curves as shown in Figure 7.3. It is now clear that for a given frequency ω, there are only a finite number of modes. The eigenfunction corresponding to each mode can be calculated by (7.7), once the phase velocity is determined.

The relative excitation of different modes depends on the depth and nature of the seismic source. For example, long-period waves from a shallow source are in general predominantly of the fundamental mode, which we designated as $n = 0$. One way of separating different modes is to record them at a great distance, where they arrive separately at different times due to propagation with different group velocities. The group velocity for a given frequency ω is the velocity at which an envelope of a wave packet with frequency around ω is transported. The peaks, troughs, and zeros of the wave packet are propagated at the phase velocity, which is in general different from the group velocity. Since the group velocity is a concept attached to a wave packet having a continuous spectrum instead of a discrete line spectrum, it is best explained by constructing a transient waveform by integrating the single-mode solution (with unit amplitude and zero initial phase) over a finite frequency band around a given frequency ω_0. Using a cosine transform, this gives

$$f_0(x, t) = \frac{1}{\pi} \int_{\omega_0 - 1/2\,\Delta\omega}^{\omega_0 + 1/2\,\Delta\omega} \cos[\omega t - k_n(\omega)x]\, d\omega. \tag{7.9}$$

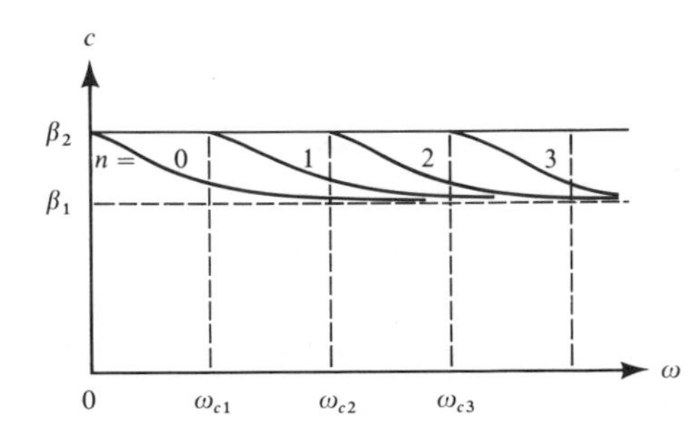

FIGURE **7.3**
Schematic diagram of phase velocities of the fundamental mode ($n = 0$) and the next three higher modes plotted against frequency.

For small $\Delta\omega$, we can expand $k_n(\omega)$ in a Taylor series,

$$k_n(\omega) = k_n(\omega_0) + \left(\frac{dk_n}{d\omega}\right)_0 (\omega - \omega_0) + \cdots. \tag{7.10}$$

Neglecting terms higher than $(\omega - \omega_0)$, we put (7.10) into (7.9) to obtain

$$f_0(x, t) \sim \frac{\Delta\omega}{\pi} \frac{\sin Y}{Y} \cos[\omega_0 t - k_n(\omega_0)x], \tag{7.11}$$

where

$$Y = \frac{\Delta\omega}{2}\left[t - \left(\frac{dk_n}{d\omega}\right)_0 x\right].$$

The above waveform is a sinusoidal oscillation of frequency ω_0 modulated by an envelope $\sin Y/Y$, which is peaked at $Y = 0$ or $x = (d\omega/dk_n)_0 t$. We see, therefore, that the envelope is transported with the group velocity $(d\omega/dk_n)_0$ and that the phase (peak, trough, or zeros of the cosine function) is moving with the phase velocity $\omega_0/k_n(\omega_0)$.

A wave packet with a given spectral density $|F(\omega)|$ and initial phase $\phi(\omega)$ composed of a single mode may be expressed as

$$f(x, t) = \frac{1}{2\pi} \int_{-\infty}^{\infty} |F(\omega)| \exp[-i\omega t + ik_n x + i\phi(\omega)]\, d\omega. \tag{7.12}$$

This integral may be evaluated by dividing the ω-axis into consecutive sections of width $\Delta\omega_i$ and summing up the approximate solutions, such as (7.11). See Problem 7.8 to find how the effect of $d|F(\omega)|/d\omega$ is included for all the sections.

When a surface wave is nondispersive, such as Rayleigh waves in a half-space, the waveform does not change with propagation, because if c_n is nondispersive

$$\begin{aligned} f(x, t) &= \frac{1}{2\pi} \int F(\omega) \exp[-i\omega(t - x/c_n)]\, d\omega \\ &= f(t - x/c_n). \end{aligned} \tag{7.13}$$

When a surface wave is strongly dispersive, one can use the stationary-phase method to approximate the waveform. Similar to the steepest-descents method, integration by the stationary-phase method also exploits behavior of an integrand near its saddle point. In the steepest descent, the integration path is taken along a path of constant phase and most rapidly changing absolute value. In the stationary-phase method, the path is along the real axis of ω, along which the phase $(-\omega t + k_n x)$ varies most rapidly. Then, for large x or t the integrand is very rapidly oscillating, with a self-canceling effect on the integral. Only at and near the saddle point, where the phase varies slowly, will there be an appreciable contribution to the integral. The saddle point, or the point of stationary phase, is given by

$$\frac{d}{d\omega}(-\omega t + k_n x) = 0$$

or

$$\frac{x}{t} = \frac{d\omega}{dk_n}. \tag{7.14}$$

This equation determines the frequency $\omega_s = \omega_s(x, t)$ expected to dominate at time t for a given distance x, and ω_s is clearly the frequency for which the group velocity is equal to x/t.

Expanding the phase as a Taylor series in the vicinity of the point of stationary phase and neglecting terms of order higher than $(\omega - \omega_s)^2$, we have

$$-\omega t + k_n x \sim -\omega_s t + k_n(\omega_s)x + \frac{x}{2}\frac{d^2k_n}{d\omega^2}(\omega - \omega_s)^2. \tag{7.15}$$

For large x, a small departure of ω from ω_s will generate a rapid oscillation of the integrand, leading to a self-canceling effect. Therefore, the integration limit can be extended beyond the limit for Taylor expansion. Thus we have

$$f(x, t) = \frac{1}{2\pi}\exp[-i\omega_s t + ik_n(\omega_s)x]|F(\omega_s)| \int_{-\infty}^{\infty} \exp\left[\frac{x}{2}i\frac{d^2k_n}{d\omega^2}(\omega - \omega_s)^2\right] d\omega. \tag{7.16}$$

Since integral tables give

$$\int_0^{\infty} \sin\left(\frac{1}{2}a\omega^2\right) d\omega = \int_0^{\infty} \cos\left(\frac{1}{2}a\omega^2\right) d\omega = \frac{1}{2}\left(\frac{\pi}{a}\right)^{1/2}, \tag{7.17}$$

we have

$$f(x, t) = \frac{|F(\omega_s)|}{2\pi}\left\{\frac{2\pi}{x\left|\frac{d^2k_n}{d\omega^2}\right|}\right\}^{1/2} \exp\left[-i\omega_s t + ik_n(\omega_s)x \pm i\frac{\pi}{4}\right], \tag{7.18}$$

where $\pm$ corresponds to $d^2k_n/d\omega^2 \gtrless 0$, respectively. For a given x and t, ω_s is obtained by solving (7.14). Thus the above formula gives the waveform of dispersive surface waves when the phase velocity $c_n = \omega/k_n$, group velocity $U_n = d\omega/dk_n$, and $dU_n^{-1}/d\omega$ are known. The sign of a phase shift by $\pi/4$ (time shift by $\frac{1}{8}$ of a period) implies a delay when the group velocity increases with

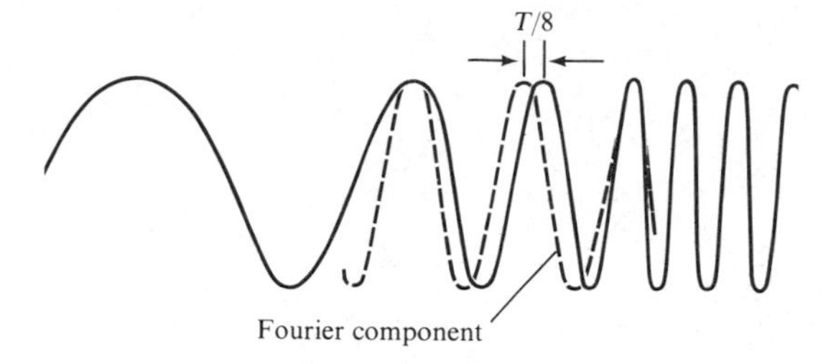

FIGURE **7.4**
The misalignment (by $T/8$) between a peak with period T in a dispersed wave train and the peak of the corresponding Fourier component.

period. A visual Fourier analysis of waves with such a dispersive property (shown in Fig. 7.4) gives an intuitive explanation of the apparent phase delay for $f(x, t)$ relative to the Fourier component (Brune et al., 1960).

When the group velocity is stationary with respect to frequency, the denominator in (7.18) vanishes and the formula is no longer valid. Inclusion of higher-order terms in the Taylor expansion of phase gives a result in terms of the Airy function (see Savage, 1966). For this reason, the arrivals associated with the group velocity maxima and minima are called Airy phases. The evaluation of an Airy phase is also easily done by a straightforward numerical integration of (7.12), because the group velocity is slowly varying and the total duration of the phase is short.

Various methods for measuring the phase and group velocities of surface waves are described in Chapter 11.

7.2 Eigenvalue Problem for the Displacement-stress Vector

Let us now consider surface waves in a vertically heterogeneous, isotropic, elastic medium occupying a half-space $z > 0$ in which elastic constants $\lambda(z)$, $\mu(z)$ and density $\rho(z)$ are arbitrary functions of z. In this section, we shall work on plane surface waves propagating in the x-direction. It will be shown later that they share the same z-dependence as the cylindrical surface waves spreading from a point.

In some aspects, this section duplicates parts of an earlier discussion of body waves, Section 5.4. However, our notation here is more suited to surface waves in that $\pi/2$ phase shifts between various displacement and stress components are recognized explicitly. Moreover, because of dispersion, the frequency dependence of horizontal wavenumber k is here more complicated, so that k is left as an explicit symbol and not factored into $\omega \times$ horizontal slowness.

For Love waves, we shall seek a solution for the equation of motion (2.13) of the form

$$
\begin{aligned}
u &= 0, \\
v &= l_1(k, z, \omega) \exp[i(kx - \omega t)], \qquad (7.19) \\
w &= 0.
\end{aligned}
$$

The stress components associated with the above displacement are

$$\begin{aligned} \tau_{xx} &= \tau_{yy} = \tau_{zz} = \tau_{zx} = 0, \\ \tau_{yz} &= \mu \frac{dl_1}{dz} \exp[i(kx - \omega t)], \\ \tau_{xy} &= ik\mu l_1 \exp[i(kx - \omega t)]. \end{aligned} \tag{7.20}$$

If we substitute (7.19) and (7.20) into (2.13), we obtain the equation of motion for $l_1(k, z, \omega)$:

$$-\omega^2 \rho(z) l_1 = \frac{d}{dz}\left[\mu(z) \frac{dl_1}{dz}\right] - k^2 \mu(z) l_1. \tag{7.21}$$

As described in the preceding section, such Love waves are a solution of the equation of motion that satisfies the source-free condition everywhere (no body force). The traction must vanish at the free surface $z = 0$, and no source exists at infinity. In addition to these conditions, the displacement and traction must be continuous at any interfaces where the elastic constants have a jump discontinuity, because otherwise the discontinuities in displacement and traction would act as a seismic source (Section 3.1). Since medium discontinuities occur only across horizontal planes, the stress component τ_{yz} is required to be continuous for Love waves. Here we shall introduce a new function to describe the z-dependence of τ_{yz} by

$$\tau_{yz} = l_2(k, z, \omega) \exp[i(kx - \omega t)]. \tag{7.22}$$

Then (7.20) and (7.21) can be rewritten as a set of first-order ordinary differential equations,

$$\begin{aligned} \frac{dl_1}{dz} &= \frac{l_2}{\mu(z)}, \\ \frac{dl_2}{dz} &= (k^2 \mu(z) - \omega^2 \rho(z)) l_1, \end{aligned} \tag{7.23}$$

or in a matrix form,

$$\frac{d}{dz}\begin{pmatrix} l_1 \\ l_2 \end{pmatrix} = \begin{pmatrix} 0 & \mu(z)^{-1} \\ k^2 \mu(z) - \omega^2 \rho(z) & 0 \end{pmatrix} \begin{pmatrix} l_1 \\ l_2 \end{pmatrix}. \tag{7.24}$$

We shall refer to (l_1, l_2) as the *motion-stress vector for Love waves* and note (Box 7.1) that it is also the double transform of the (x, t) dependence of (u_y, τ_{yx}) for general *SH* particle motion along the y-direction.

The *motion-stress vector for Rayleigh waves* can be obtained in a similar manner. We start with the following form of displacements:

$$\begin{aligned} u &= r_1(k, z, \omega) \exp[i(kx - \omega t)], \\ v &= 0, \\ w &= i r_2(k, z, \omega) \exp[i(kx - \omega t)]. \end{aligned} \tag{7.25}$$

(For real positive r_1 and r_2, the above combination of u, w represents a prograde motion.) The corresponding stress components are given by

$$\begin{aligned} \tau_{yz} &= \tau_{xy} = 0, \\ \tau_{xx} &= i\left[\lambda \frac{dr_2}{dz} + k(\lambda + 2\mu)r_1\right] \exp[i(kx - \omega t)], \\ \tau_{yy} &= i\left(\lambda \frac{dr_2}{dz} + k\lambda r_1\right) \exp[i(kx - \omega t)], \\ \tau_{zz} &= i\left[(\lambda + 2\mu) \frac{dr_2}{dz} + k\lambda r_1\right] \exp[i(kx - \omega t)], \\ \tau_{zx} &= \mu\left(\frac{dr_1}{dz} - kr_2\right) \exp[i(kx - \omega t)]. \end{aligned} \tag{7.26}$$

The stress components τ_{zx} and τ_{zz} are continuous in z. We write them as

$$\begin{aligned} \tau_{zx} &= r_3(k, z, \omega) \exp[i(kx - \omega t)], \\ \tau_{zz} &= i r_4(k, z, \omega) \exp[i(kx - \omega t)]. \end{aligned} \tag{7.27}$$

Then the differential equations for the motion-stress vector (r_1, r_2, r_3, r_4) are obtained from (7.26) and (2.13) as

$$\frac{d}{dz}\begin{pmatrix} r_1 \\ r_2 \\ r_3 \\ r_4 \end{pmatrix} = \begin{pmatrix} 0 & k & \mu^{-1}(z) & 0 \\ -k\lambda(z)[\lambda(z) + 2\mu(z)]^{-1} & 0 & 0 & [\lambda(z) + 2\mu(z)]^{-1} \\ k^2\zeta(z) - \omega^2\rho(z) & 0 & 0 & k\lambda(z)[\lambda(z) + 2\mu(z)]^{-1} \\ 0 & -\omega^2\rho(z) & -k & 0 \end{pmatrix} \begin{pmatrix} r_1 \\ r_2 \\ r_3 \\ r_4 \end{pmatrix}, \tag{7.28}$$

where $\zeta(z) = 4\mu(z)[\lambda(z) + \mu(z)]/[\lambda(z) + 2\mu(z)]$. The matrices in (7.24) and (7.28) look very simple, and in essence are the same as those obtained in Section 5.4 for a homogeneous medium. They do not contain spatial gradients of medium parameters explicitly, although $\lambda(z)$, $\mu(z)$, and $\rho(z)$ are functions of z.

The boundary conditions for surface waves—i.e., vanishing traction at the free surface $z = 0$ and no motion at infinity—require that

$$\begin{aligned} &r_1, r_2, \text{ and } l_1 \to 0 \qquad \text{as } z \to \infty, \\ &r_3 = r_4 = l_2 = 0 \qquad \text{at the free surface } (z = z_0). \end{aligned} \tag{7.29}$$

For a given ω, nonvanishing solutions of (7.24) or (7.28) under the condition (7.29) exist only for certain $k = k_n(\omega)$. In exactly the same manner as for Love waves in a single-layered half-space studied in Section 7.1, $\omega/k_n(\omega)$ gives the phase velocity and the corresponding solution of (7.24) or (7.28) gives the z-dependence of the mode.

There are many ways to solve this eigenvalue-eigenvector problem. We shall start with the numerical integration method and the propagator matrix method. Explicit results are obtained by the Thomson-Haskell method, which is a special case of the propagator matrix method applicable to a stack of homogeneous layers overlying a half-space.

7.2.1 Numerical integration

The eigenvalue problem for the motion-stress vector may be solved by a direct integration of (7.24) and (7.28) by numerical methods such as that of Runge-Kutta. Takeuchi and Saito (1972) advocate this approach for its smoother way of modeling the Earth as compared to the matrix method in which the Earth is approximated by a stack of homogeneous layers.

To illustrate how to solve the eigenvalue problem by a numerical integration, let us first suppose that the displacements are fixed at a great depth z_n (i.e., a rigid boundary). In the case of Love waves, we start integrating equation (7.24) upward from $z = z_n$, with the initial values $l_1(z_n) = 0$ and $l_2(z_n) = 1$. A trial value of k has to be chosen for a given ω.

The result of integration should give $l_2 = 0$ at $z = z_0$ if k is the eigenvalue. The process is iterated with a corrected k until $l_2(z_0)$ vanishes. In the case of Rayleigh waves, we integrate equation (7.28) for two different sets of initial values; $r_1 = r_2 = r_4 = 0$ and $r_3 = 1$ in set 1, and $r_1 = r_2 = r_3 = 0$ and $r_4 = 1$ in set 2. Writing the solutions for set 1 and set 2 as $\mathbf{r}^{(1)}$ and $\mathbf{r}^{(2)}$, respectively, a general solution of (7.28) can be expressed as

$$\mathbf{r} = A\mathbf{r}^{(1)} + B\mathbf{r}^{(2)}. \tag{7.30}$$

If $\mathbf{r}$ is an eigenvector, its stress components $r_3(z_0)$ and $r_4(z_0)$ must vanish, i.e.,

$$\begin{aligned} Ar_3^{(1)}(z_0) + Br_3^{(2)}(z_0) &= 0, \\ Ar_4^{(1)}(z_0) + Br_4^{(2)}(z_0) &= 0. \end{aligned} \tag{7.31}$$

BOX **7.2**

Runge-Kutta method

Let us consider the numerical method for solving the first-order ordinary differential equation

$$\frac{dy}{dx} = f(x, y). \tag{1}$$

Using the forward-difference formula for dy/dx (Box 13.2), the finite difference equation corresponding to (1) is written as

$$\begin{aligned} y_{i+1} &= y_i + (x_{i+1} - x_i)f_i, \\ f_i &= f(x_i, y_i). \end{aligned} \tag{2}$$

Repeated application of (2) gives y_i if the initial value y_0 is known. This is Euler's method, and (2) has a truncation error of order h^2, where $h = x_{i+1} - x_i$. On the other hand, the Runge-Kutta method of nth order is designed to give a truncation error of order h^{n+1}. We start with defining a location $(\bar{x}, \bar{y})$ in the xy-plane by

$$\begin{aligned} \bar{x} &= x_i + \alpha h \\ \bar{y} &= y_i + \beta h. \end{aligned} \tag{3}$$

We wish to determine α and β such that

$$\overline{\Delta y_i} = f(\bar{x}, \bar{y})h \tag{4}$$

is identical with the increment

$$\Delta y_i = y_i'h + y_i''\frac{h^2}{2} + \cdots + \frac{d^n y_i}{dx^n}\frac{h^n}{n!},$$

where primes denote derivatives with respect to x, and such derivatives are evaluated at x_i.

If α and β can be found with this property, then the solution can be carried forward one step, by (4), with the same accuracy as if derivatives of y up to nth order were available at the start of the step (in which case the nth-order Taylor series would be available).

For example, if $n = 2$, α and β are determined such that

$$\Delta y_i = y_i'h + y_i''\tfrac{1}{2}h^2 = \overline{\Delta y_i}. \tag{5}$$

Expanding (4) around (x_i, y_i), we have

$$\overline{\Delta y_i} = f(x_i + \alpha h, y_i + \beta h) \cdot h \sim \left[f_i + \alpha h\left(\frac{\partial f}{\partial x}\right)_i + \beta h\left(\frac{\partial f}{\partial y}\right)_i\right]h. \tag{6}$$

On the other hand, since

$$y'' = \frac{df}{dx} = \frac{\partial f}{\partial x} + \frac{\partial f}{\partial y}\frac{dy}{dx}$$

we obtain

$$y_i'h + y_i''\tfrac{1}{2}h^2 = f_ih + \tfrac{1}{2}h^2\left[\left(\frac{\partial f}{\partial x}\right)_i + \left(\frac{\partial f}{\partial y}\right)_i f_i\right]. \tag{7}$$

Putting (6) and (7) into (5), we find (5) is satisfied if we choose $\alpha = \frac{1}{2}$ and $\beta = f_i/2$, or

$$\bar{x} = x_i + \tfrac{1}{2}h, \qquad \bar{y} = y_i + \tfrac{1}{2}hf_i. \tag{8}$$

Thus, in the second-order Runge-Kutta method, the solution is given by

$$\begin{aligned} y_{i+1} &= y_i + \overline{\Delta y_i}, \\ \overline{\Delta y_i} &= hf(\bar{x}, \bar{y}) = hf(x_i + \tfrac{1}{2}h, y_i + \tfrac{1}{2}hf_i). \end{aligned} \tag{9}$$

In the case of simultaneous equations

$$\frac{dy}{dx} = f(x, y, z), \qquad \frac{dz}{dx} = g(x, y, z), \tag{10}$$

the solution by the second-order Runge-Kutta method is given by

$$\begin{aligned} y_{i+1} &= y_i + \overline{\Delta y_i}, \\ z_{i+1} &= z_i + \overline{\Delta z_i}, \\ \overline{\Delta y_i} &= hf(x_i + \tfrac{1}{2}h, y_i + \tfrac{1}{2}hf_i, z_i + \tfrac{1}{2}hg_i), \\ \overline{\Delta z_i} &= hg(x_i + \tfrac{1}{2}h, y_i + \tfrac{1}{2}hf_i, z_i + \tfrac{1}{2}hg_i). \end{aligned} \tag{11}$$

Thus nontrivial solution (7.30) is possible only when

$$\begin{vmatrix} r_3^{(1)}(z_0) & r_3^{(2)}(z_0) \\ r_4^{(1)}(z_0) & r_4^{(2)}(z_0) \end{vmatrix} = 0. \tag{7.32}$$

The eigenvalue is searched for by trial and correction until (7.32) is met. Once the eigenvalue is found, A/B can be determined from (7.31), and then the eigenvector from (7.30).

A better approximation of the lower boundary condition than rigidity is to replace the Earth below z_n by a homogeneous half-space. The solutions in the homogeneous half-space that involve exponential decay with depth are well known. For Love waves, we use the following initial values:

$$\begin{aligned} l_1(z_n) &= \exp(-\nu_{n+1}z_n), \\ l_2(z_n) &= \mu_{n+1}\frac{dl_1}{dz} = -\nu_{n+1}\mu_{n+1}\exp(-\nu_{n+1}z_n), \end{aligned} \tag{7.33}$$

where $\nu_{n+1} = (k^2 - \omega^2/\beta_{n+1}^2)^{1/2}$, chosen so that Re $\nu_{n+1} \geq 0$ (and hence Im $\nu_{n+1} \leq 0$). The subscript $n + 1$ applies to the properties of the half-space. For Rayleigh waves, the required two sets of initial values may be calculated using potentials given by

$$\phi = \exp(-\gamma_{n+1} z), \qquad \psi = 0 \tag{7.34}$$

for the first set and

$$\phi = 0, \qquad \psi = \exp(-\nu_{n+1} z) \tag{7.35}$$

for the second set, where $\gamma_{n+1} = (k^2 - \omega^2/\alpha_{n+1}^2)^{1/2}$ and $\nu_{n+1} = (k^2 - \omega^2/\beta_{n+1}^2)^{1/2}$. The corresponding motion-stress vectors $\mathbf{r}^{(1)}(z)$ and $\mathbf{r}^{(2)}(z)$ in the half-space can be calculated by putting (7.34) and (7.35) into (6.59), then using (7.25) and (7.26). The numerical integration is carried out for the two sets of initial values $\mathbf{r}^{(1)}(z_n)$ and $\mathbf{r}^{(2)}(z_n)$, and the result is iterated over k-values until (7.32) is satisfied to find the eigenvalue. For small ω/k, the two sets of initial values become numerically indistinguishable. In such cases, the sum and difference of the two solutions may be used as the initial values.

7.2.2 *Propagator-matrix method*

The matrix method due to Thomson (1950) and corrected by Haskell (1953) has been extensively used in surface-wave analysis since the advent of large-scale computers. In their method, the vertically heterogeneous medium is replaced by a stack of homogeneous layers overlying a homogeneous half-space, as shown in Figure 7.5. The Thomson-Haskell method is a special case of the propagator matrix method introduced to seismology by Gilbert and Backus (1966).

We first generalize the differential equations (7.24) and (7.28) for the motion-stress vector to a matrix form,

$$\frac{d\mathbf{f}(z)}{dz} = \mathbf{A}(z)\mathbf{f}(z), \tag{7.36}$$

where $\mathbf{f}(z)$ is an $n \times 1$ column vector and $\mathbf{A}(z)$ is an $n \times n$ matrix; $n = 2$ for Love waves, and $n = 4$ for Rayleigh waves.

The *propagator matrix* (sometimes called the *matrizant*) is defined as

$$\mathbf{P}(z, z_0) = \mathbf{I} + \int_{z_0}^{z} \mathbf{A}(\zeta_1)\, d\zeta_1 + \int_{z_0}^{z} \mathbf{A}(\zeta_1) \int_{z_0}^{\zeta_1} \mathbf{A}(\zeta_2)\, d\zeta_2\, d\zeta_1 + \cdots, \tag{7.37}$$

where $\mathbf{I}$ is the unit matrix of order n. Obviously, $\mathbf{P}(z, z_0)$ satisfies the differential equation (7.36):

$$\frac{d}{dz}\mathbf{P}(z, z_0) = \mathbf{A}(z)\mathbf{P}(z, z_0). \tag{7.38}$$

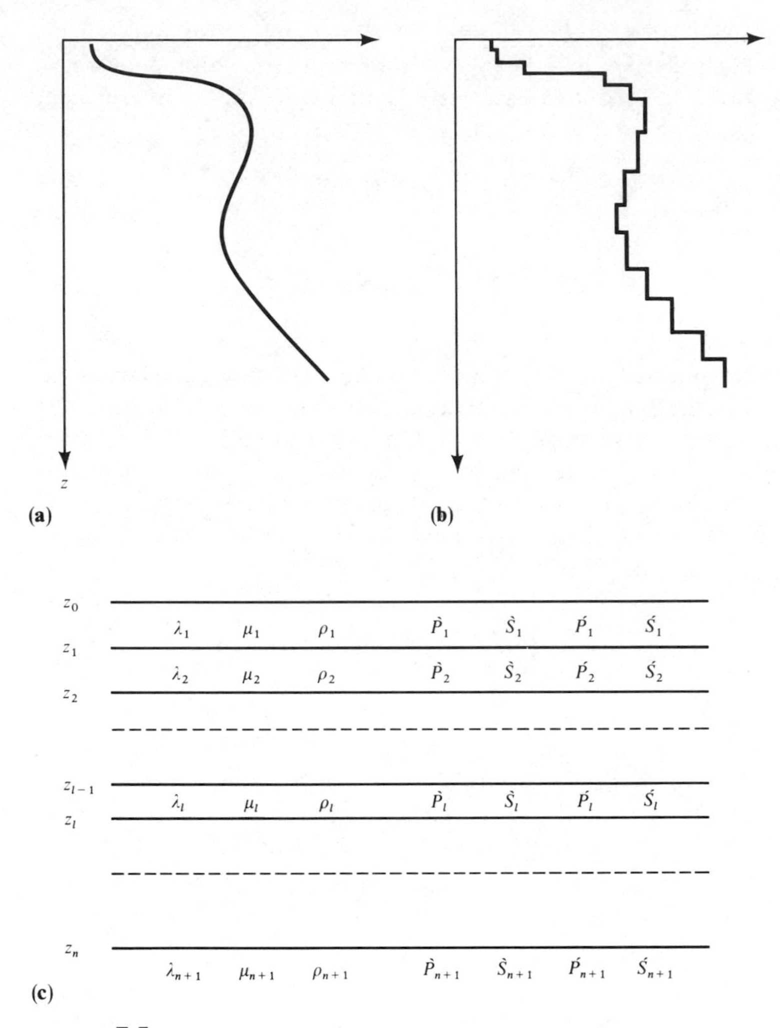

FIGURE **7.5**

(**a**) *A* continuous variation of properties (one of ρ, λ, μ, α, or β) with depth. (**b**) A piecewise constant approximation to (a). This may be regarded as a stack of homogeneous welded plates. (**c**) Numbering system for elastic constants, layers, and upgoing and downgoing waves.

Further, from (7.37), $\mathbf{P}(z_0, z_0) = \mathbf{I}$. Hence we obtain the most important property of the propagator matrix, namely

$$\mathbf{f}(z) = \mathbf{P}(z, z_0)\mathbf{f}(z_0), \tag{7.39}$$

since the right-hand side here satisfies the differential equation (7.36) and does give $\mathbf{f}(z_0)$ at $z = z_0$. Thus $\mathbf{P}(z, z_0)$ generates the motion-stress vector at z by operating on the vector at z_0.

An interesting property of $\mathbf{P}(z, z_0)$ can be found by observing that

$$\begin{aligned} \mathbf{f}(z_2) &= \mathbf{P}(z_2, z_1)\mathbf{f}(z_1) \\ &= \mathbf{P}(z_2, z_1)\mathbf{P}(z_1, z_0)\mathbf{f}(z_0). \end{aligned}$$

Choosing $z_2 = z_0$ and applying the above chain rule for any $\mathbf{f}(z_0)$, it follows that

$$\mathbf{I} = \mathbf{P}(z_0, z_1)\mathbf{P}(z_1, z_0), \tag{7.40}$$

so that the inverse of $\mathbf{P}(z_1, z_0)$ is $\mathbf{P}(z_0, z_1)$.

When $\mathbf{A}(z)$ is constant independent of z, as assumed within a given layer by Thomson and Haskell, the propagator matrix takes a simple form. From (7.37)

$$\begin{aligned} \mathbf{P}(z, z_0) &= \mathbf{I} + (z - z_0)\mathbf{A} + \tfrac{1}{2}(z - z_0)^2\mathbf{A}\mathbf{A} + \cdots \\ &= \exp[(z - z_0)\mathbf{A}]. \end{aligned} \tag{7.41}$$

For a square matrix $\mathbf{A}$ with distinct eigenvalues λ_k $(k = 1, 2, \ldots, n)$, a function of matrix $\mathbf{A}$ can be expanded by Sylvester's formula (e.g., Hildebrand, 1952):

$$F(\mathbf{A}) = \sum_{k=1}^{n} F(\lambda_k) \frac{\prod_{r \neq k} (\mathbf{A} - \lambda_r \mathbf{I})}{\prod_{r \neq k} (\lambda_k - \lambda_r)}. \tag{7.42}$$

It is this relation that assigns meaning to the last part of (7.41). For Love waves, we have from (7.24)

$$\mathbf{A} = \begin{pmatrix} 0 & \mu^{-1} \\ k^2\mu - \omega^2\rho & 0 \end{pmatrix}.$$

To find eigenvalues, we set

$$|\mathbf{A} - \lambda\mathbf{I}| = \begin{vmatrix} -\lambda & \mu^{-1} \\ k^2\mu - \omega^2\rho & -\lambda \end{vmatrix} = 0$$

and obtain $\lambda = \pm(k^2 - \omega^2/\beta^2)^{1/2} = \pm\nu$. Putting these results in (7.42), we find

$$\mathbf{P}(z, z_0) = \exp[(z - z_0)\mathbf{A}] - \begin{pmatrix} \cosh \nu(z - z_0) & (\nu\mu)^{-1} \sinh(z - z_0) \\ \nu\mu \sinh \nu(z - z_0) & \cosh \nu(z - z_0) \end{pmatrix}. \tag{7.43}$$

This matrix generates the motion-stress vector $\mathbf{l}(z)$ by operating on $\mathbf{l}(z_0)$ when both z and z_0 are in the same layer. For a layered medium as shown in Figure 7.5, the propagator matrix $\mathbf{P}(z, z_0)$ for $z_k > z > z_{k-1}$ is found from

$$\mathbf{f}(z) = \mathbf{P}(z, z_{k-1})\mathbf{P}(z_{k-1}, z_{k-2}) \cdots \mathbf{P}(z_1, z_0)\mathbf{f}(z_0) = \mathbf{P}(z, z_0)\mathbf{f}(z_0),$$

and hence

$$\mathbf{P}(z, z_0) = \exp[(z - z_{k-1})\mathbf{A}_k] \prod_{l=1}^{k-1} \exp[(z_l - z_{l-1})\mathbf{A}_l]. \tag{7.44}$$

Similarly, the layer matrix for Rayleigh waves can be obtained by putting $\mathbf{A}$ given in (7.28) into (7.41). In this case, the eigenvalues of $\mathbf{A}$ are $\pm\gamma = \pm(k^2 - \omega^2/\alpha^2)^{1/2}$ and $\pm\nu = \pm(k^2 - \omega^2/\beta^2)^{1/2}$. The resulting elements of $\mathbf{P}(z, z_0)$ are given below (with z and z_0 in the same layer):

$$\begin{aligned}
P_{11} &= P_{33} = 1 + \frac{2\mu}{\omega^2\rho}\left[2k^2 \sinh^2 \frac{\gamma(z - z_0)}{2} - (k^2 + \nu^2)\sinh^2 \frac{\nu(z - z_0)}{2}\right],\\
P_{12} &= -P_{43} = \frac{k\mu}{\omega^2\rho}\left[(k^2 + \nu^2)\frac{\sinh \gamma(z - z_0)}{\gamma} - 2\nu \sinh \nu(z - z_0)\right],\\
P_{13} &= \frac{1}{\omega^2\rho}\left[k^2 \frac{\sinh \gamma(z - z_0)}{\gamma} - \nu \sinh \nu(z - z_0)\right],\\
P_{14} &= -P_{23} = \frac{2k}{\omega^2\rho}\left[\sinh^2 \frac{\gamma(z - z_0)}{2} - \sinh^2 \frac{\nu(z - z_0)}{2}\right],\\
P_{21} &= -P_{34} = \frac{k\mu}{\omega^2\rho}\left[(k^2 + \nu^2)\frac{\sinh \gamma(z - z_0)}{\gamma} - 2\gamma \sinh \gamma(z - z_0)\right],\\
P_{22} &= P_{44} = 1 + \frac{2\mu}{\omega^2\rho}\left[2k^2 \sinh^2 \frac{\nu(z - z_0)}{2} - (k^2 + \nu^2)\sinh^2 \frac{\gamma(z - z_0)}{2}\right],\\
P_{24} &= \frac{1}{\omega^2\rho}\left[k^2 \frac{\sinh \nu(z - z_0)}{\nu} - \gamma \sinh \gamma(z - z_0)\right],\\
P_{31} &= \frac{\mu^2}{\omega^2\rho}\left[4k^2\gamma \sinh \gamma(z - z_0) - (k^2 + \nu^2)^2 \frac{\sinh \nu(z - z_0)}{\nu}\right],\\
P_{32} &= -P_{41} = 2\mu^2(k^2 + \nu^2)P_{14},\\
P_{42} &= \frac{\mu^2}{\omega^2\rho}\left[4k^2\nu \sinh \nu(z - z_0) - (k^2 + \nu^2)^2 \frac{\sinh \gamma(z - z_0)}{\gamma}\right].
\end{aligned} \tag{7.45}$$

Equations (7.43), (7.44), and (7.45) give the propagator matrix in the Thomson-Haskell method.

The propagator matrix can also be used to solve the system with a source term,

$$\frac{d\mathbf{f}(z)}{dz} = \mathbf{A}(z)\mathbf{f}(z) + \mathbf{g}(z), \tag{7.46}$$

where $\mathbf{g}(z)$ is a known $n \times 1$ matrix function of z. The solution is

$$\mathbf{f}(z) = \mathbf{P}(z, z_0)\left[\int_{z_0}^{z} \mathbf{P}^{-1}(\zeta, z_0)\mathbf{g}(\zeta)\,d\zeta + \mathbf{f}(z_0)\right]. \tag{7.47}$$

To verify this result, we use (7.38) to get

$$\begin{aligned}\frac{d\mathbf{f}}{dz} &= \frac{d\mathbf{P}}{dz}\left[\int_{z_0}^{z} \mathbf{P}^{-1}(\zeta, z_0)\mathbf{g}(\zeta)\,d\zeta + \mathbf{f}(z_0)\right] + \mathbf{P}(z, z_0)\mathbf{P}^{-1}(z, z_0)\mathbf{g}(z)\\ &= \mathbf{A}(z)\mathbf{P}(z, z_0)\mathbf{P}^{-1}(z, z_0)\mathbf{f}(z) + \mathbf{g}(z)\\ &= \mathbf{A}(z)\mathbf{f}(z) + \mathbf{g}(z).\end{aligned}$$

Thus (7.47) satisfies (7.46). Since, from (7.40),

$$\mathbf{P}(z, z_0)\mathbf{P}^{-1}(\zeta, z_0) = \mathbf{P}(z, z_0)\mathbf{P}(z_0, \zeta) = \mathbf{P}(z, \zeta),$$

the solution (7.47) simplifies to

$$\mathbf{f}(z) = \int_{z_0}^{z} \mathbf{P}(z, \zeta)\mathbf{g}(\zeta)\,d\zeta + \mathbf{P}(z, z_0)\mathbf{f}(z_0). \tag{7.48}$$

Let us now discuss some of the practical problems encountered in the application of Thomson-Haskell methods. Since the radiation condition is imposed by suppressing certain waves at infinity, rather than by a constraint on the motion-stress vector directly, we need to relate the motion-stress vector to the presence (or absence) of upgoing and downgoing wave types in the bottom half-space. Of course, for large enough k, these wave types become respectively growing or decaying exponentially with depth. From (5.62) and (5.63), with vertical wavenumber $i\nu$ replacing $\omega\eta$, the relation for SH-waves in a homogeneous body is

$$\begin{pmatrix} l_1 \\ l_2 \end{pmatrix} = \begin{pmatrix} e^{-\nu z} & e^{\nu z} \\ -\nu\mu e^{-\nu z} & \nu\mu e^{\nu z} \end{pmatrix}\begin{pmatrix} \grave{S} \\ \acute{S} \end{pmatrix} \tag{7.49}$$

or $\mathbf{l} = \mathbf{F}\mathbf{w}$, where $\nu = (k^2 - \omega^2/\beta^2)^{1/2}$ and $\beta = (\mu/\rho)^{1/2}$. $\grave{S}$ and $\acute{S}$ are constants giving the displacement amplitude of downgoing and upgoing waves, respectively. The inverse relation to (7.49) is

$$\mathbf{w} = \begin{pmatrix} \grave{S} \\ \acute{S} \end{pmatrix} = 1/(2\nu\mu) \begin{pmatrix} \nu\mu e^{\nu z} & -e^{-\nu z} \\ \nu\mu e^{\nu z} & e^{-\nu z} \end{pmatrix} \begin{pmatrix} l_1 \\ l_2 \end{pmatrix} \tag{7.50}$$

or $\mathbf{w} = \mathbf{F}^{-1}\mathbf{l}$.

Applying this relation to the motion-stress vector at $z = z_n$, the amplitudes of upgoing and downgoing waves in the half-space are expressed in terms of the motion-stress vector at $z = z_0$:

$$\mathbf{w}_{n+1} = \mathbf{F}_{n+1}^{-1}\mathbf{P}(z_n, z_0)\mathbf{l}(z_0) = \mathbf{B}\mathbf{l}(z_0). \tag{7.51}$$

Since $\acute{S}_{n+1} = 0$ and $l_2(z_0) = 0$ from the radiation/boundary conditions, we have

$$\begin{pmatrix} \grave{S}_{n+1} \\ 0 \end{pmatrix} = \begin{pmatrix} B_{11} & B_{12} \\ B_{21} & B_{22} \end{pmatrix} \begin{pmatrix} l_1(z_0) \\ 0 \end{pmatrix}. \tag{7.52}$$

BOX **7.3**

On avoiding potentials

In contrast to the methods of Haskell (1953, 1960, 1962) for wave propagation in layered media, we have minimized the use of potentials for P, SV and SH-waves in this chapter. Although it is true that potentials can clarify certain stages in the theory of elastic waves in homogeneous isotropic media, the fact is that matrix methods for discussing solutions to $\partial\mathbf{f}/\partial z = \mathbf{A}(z)\mathbf{f}$ are far more powerful. Potentials are of no direct interest, and are awkward to use in imposing boundary conditions, whereas $\mathbf{f}$ is made up from the components of particle motion and stress, which are immediately the quantities for which we wish to solve.

The reader may note that the first reflection coefficient discussed in Chapter 5 was derived by using potentials, but that we quickly turned to work with displacements directly. Potentials recur in Chapter 6, because their scalar wave equations appear simpler than vector equations containing P- and S-motions, since we had not then put wave equations into the form $\partial\mathbf{f}/\partial z = \mathbf{A}\mathbf{f}$. However, in Box 6.9 we did discuss the first two entries of the motion-stress vector $\mathbf{f}$ for P-SV waves, and did develop a result that equates $\mathbf{f}$ to a sum of the basic waves possible, i.e., $\mathbf{f} = \mathbf{F}\mathbf{w}$ (in the notation of this chapter).

Radiation conditions are easy to impose by setting appropriate components of $\mathbf{w}$ equal to zero.

All this is not to deny that understanding elastic wave potentials, and solving for them, is an important part of the learning process for seismologists. We shall occasionally use potentials in the pages that follow, but for the most part we shall emphasize an understanding of $\mathbf{f}(z) = \mathbf{P}(z, z_0)\mathbf{f}(z_0)$ and $\mathbf{f}(z) = \mathbf{F}(z)\mathbf{w}$. A connection between potentials and the layer matrix $\mathbf{F}$ is developed in Problem 5.8.

For a nontrivial solution **l**, we see from the above equation that B_{21} must vanish. Thus the eigenvalue is determined by

$$B_{21} = 0. \tag{7.53}$$

To find the root of this equation for a given ω, we start with a trial value of k. For known ω and k, one can calculate the value of B_{21} by the matrix multiplication (7.51), using given layer parameters. Then k is changed slightly, and the resulting change in B_{21} is monitored. Extrapolation and interpolation techniques may be used quickly to locate the value of k at which the zero-crossing of B_{21} occurs. Once the eigenvalue is found, the eigenfunction can be calculated via (7.44).

The above method is easily extended to Rayleigh waves by finding a relation between the motion-stress vector (r_1, r_2, r_3, r_4) and the numbers $(\grave{P}, \grave{S}, \acute{P}, \acute{S})$ describing how much of each of the four possible wave types is present in each layer. Just as we developed $\mathbf{f} = \mathbf{Fw}$ to solve $\partial\mathbf{f}/\partial z = \mathbf{Af}$ in Section 5.4, with $\mathbf{F}$ factoring into a matrix made up from eigenvectors of $\mathbf{A}$ times a diagonal matrix containing the vertical phase factors (see (5.65)), so here do we have

$$\begin{pmatrix} r_1 \\ r_2 \\ r_3 \\ r_4 \end{pmatrix} = \mathbf{Fw} = \mathbf{F}\begin{pmatrix} \grave{P} \\ \grave{S} \\ \acute{P} \\ \acute{S} \end{pmatrix}. \tag{7.54}$$

The matrix $\mathbf{F}$ is factored as

$$\mathbf{F} = \omega^{-1}\begin{pmatrix} \alpha k & \beta\nu & \alpha k & \beta\nu \\ \alpha\gamma & \beta k & -\alpha\gamma & -\beta k \\ -2\alpha\mu k\gamma & -\beta\mu(k^2+\nu^2) & 2\alpha\mu k\gamma & \beta\mu(k^2+\nu^2) \\ -\alpha\mu(k^2+\nu^2) & -2\beta\mu k\nu & -\alpha\mu(k^2+\nu^2) & -2\beta\mu k\nu \end{pmatrix}$$

$$\times\begin{pmatrix} e^{-\gamma z} & 0 & 0 & 0 \\ 0 & e^{-\nu z} & 0 & 0 \\ 0 & 0 & e^{\gamma z} & 0 \\ 0 & 0 & 0 & e^{\nu z} \end{pmatrix}. \tag{7.55}$$

We are free to choose any suitable normalization for the eigenvectors of the coefficient matrix appearing in (7.28), and for the specific forms of (7.54) and (7.55) we have been guided by two requirements. The first is that the four component waves contained in $\mathbf{Fw}$ do have displacement amplitudes given by $(\grave{P}, \grave{S}, \acute{P}, \acute{S})$ for homogeneous waves (i.e., when γ and ν are negative imaginary). The second is that for surface waves (when γ and ν are positive real), the entries in $\mathbf{F}$ are all real numbers.

BOX **7.4**

Mixture of solid and liquid layers

A liquid layer in seismology may be (i) at the top, as in an oceanic structure; (ii) sandwiched between solid layers, as in an ice-covered ocean or a volcanic area with a magma lens; or (iii) at the bottom, as in a mantle underlain by a liquid core. We shall consider here an ideal liquid with zero viscosity, for which the shear stress must vanish at the liquid-solid interface, but a discontinuity in horizontal displacements is allowed (see Section 5.2).

In the case of Love waves, a liquid-solid interface becomes a free surface, and the liquid layer behaves like a vacuum.

In the case of Rayleigh waves, the shear stress vanishes, but the normal stress and normal displacement do not vanish, and must be continuous across the liquid-solid interface. Thus the layer matrix **F** for a liquid layer becomes singular, and its inverse does not exist. We need to find a method that works as if the inverse *did* exist.

Consider the case of a liquid layer sandwiched at depths $z_f > z > z_{f-1}$ in a solid, layered half-space, as shown in the figure. Within the layer, only compressional waves exist. To determine the motion-stress vector through this mixed-layered medium, we shall start from the free surface $z = z_0$. The motion-stress vector $\mathbf{r}(z)$ at $z = z_{f-1}$ can be expressed in terms of $\mathbf{r}(z_0)$ using the layer matrices **F** given in the text (see Problem 7.3). Since the traction vanishes, $r_3 = r_4 = 0$ at z_0, and $\mathbf{r}(z_{f-1})$ is determined completely as a linear combination of two unknowns $r_1(z_0)$ and $r_2(z_0)$. On the other hand, $z = z_{f-1}$ is a a liquid-solid interface, and the shear traction $r_3(z_{f-1})$ vanishes. This imposes a linear relation between $r_1(z_0)$ and $r_2(z_0)$. Using this relation, one can eliminate $r_1(z_0)$ from the expression for $\mathbf{r}(z_{f-1})$, which is now completely determined by one unknown $r_2(z_0)$ alone. Then $r_2(z_f)$ and $r_4(z_f)$ can be determined from $r_2(z_{f-1})$ and $r_4(z_{f-1})$ through the upgoing and downgoing compressional waves $\acute{P}_f$ and $\grave{P}_f$. At $z = z_f$, $r_3(z_f)$ is known to be zero, so that the only unknown component of $\mathbf{r}(z_f)$ is $r_1(z_f)$. The motion-stress vector below z_f can therefore be expressed as a linear combination of two unknowns $r_1(z_f)$ and $r_2(z_0)$. It follows that the radiation condition on waves in the half-space leads to an equation similar to (7.58), where $r_1(z_0)$ is replaced by $r_1(z_f)$.

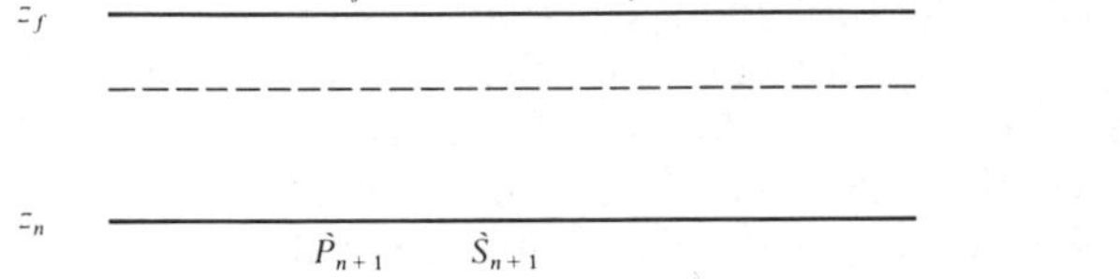

z_0

$\grave{P}_1$ $\grave{S}_1$ $\acute{P}_1$ $\acute{S}_1$

z_1

z_{f-1}

$\grave{P}_f$ $\acute{P}_f$ Fluid layer

z_f

z_n

$\grave{P}_{n+1}$ $\grave{S}_{n+1}$

In the case of a liquid layer overlying a solid medium, the unknown component of $\mathbf{r}(z_0)$ will be only $r_1(z_0)$ to start with, and therefore the steps to be taken are identical to the preceding case in which overlying solid layers are eliminated. In the case of a liquid half-space, the condition of vanishing $r_3(z_n)$ will replace that of vanishing $\acute{S}_{n+1}$.

A specific inverse can be written down for each of the matrix factors of $\mathbf{F}$ in (7.55), and

$$\mathbf{F}^{-1} = \begin{pmatrix} e^{\gamma z} & 0 & 0 & 0 \\ 0 & e^{\nu z} & 0 & 0 \\ 0 & 0 & e^{-\gamma z} & 0 \\ 0 & 0 & 0 & e^{-\nu z} \end{pmatrix} \times \frac{\beta}{2\alpha\mu\gamma\nu\omega}$$

$$\times \begin{pmatrix} 2\beta\mu k\gamma\nu & -\beta\mu\nu(k^2+\nu^2) & -\beta k\nu & \beta\gamma\nu \\ -\alpha\mu\gamma(k^2+\nu^2) & 2\alpha\mu k\gamma\nu & \alpha\gamma\nu & -\alpha k\gamma \\ 2\beta\mu k\gamma\nu & \beta\mu\nu(k^2+\nu^2) & \beta\gamma\nu & \beta\gamma\nu \\ -\alpha\mu\gamma(k^2+\nu^2) & -2\alpha\mu k\gamma\nu & -\alpha\gamma\nu & -\alpha k\gamma \end{pmatrix} \quad (7.56)$$

Each of $\mathbf{w}$, $\mathbf{F}$, $\mathbf{F}^{-1}$ is layer-dependent, and we can relate the amplitude of different wave-types in the lowermost half-space to the motion-stress vector at the free surface:

$$\mathbf{w}_{n+1} = \mathbf{F}_{n+1}^{-1}\mathbf{P}(z_n, z_0)\mathbf{r}(z_0) = \mathbf{B}\mathbf{r}(z_0). \quad (7.57)$$

Corresponding to our treatment of Love waves, (7.51)–(7.52), we find here that

$$\begin{pmatrix} \grave{P}_{n+1} \\ \grave{S}_{n+1} \\ 0 \\ 0 \end{pmatrix} = \begin{pmatrix} B_{11} & B_{12} & B_{13} & B_{14} \\ B_{21} & B_{22} & B_{23} & B_{24} \\ B_{31} & B_{32} & B_{33} & B_{34} \\ B_{41} & B_{42} & B_{43} & B_{44} \end{pmatrix} \begin{pmatrix} r_1(z_0) \\ r_2(z_0) \\ 0 \\ 0 \end{pmatrix}. \quad (7.58)$$

Thus the eigenvalue is given by

$$\begin{vmatrix} B_{31} & B_{32} \\ B_{41} & B_{42} \end{vmatrix} = 0. \quad (7.59)$$

To find the root of this determinant for a given ω, we start again with a trial value of k and iterate on values of k until (7.59) is met. In calculating the sub-determinant on the left side of (7.59) for a trial value of k, a numerical problem arises if the wavelength λ of surface waves becomes shorter than a certain limit. According to Schwab and Knopoff (1970), when using 16-decimal digits in the computations, this limit is approximated by

$$H/\lambda = 3 \times 10^{(14.3-\sigma)/12.4}, \tag{7.60}$$

where H is the depth to the homogeneous half-space and σ is the required accuracy in phase velocity given in significant figures. This limit does not exist for Love waves. The reason why it exists for Rayleigh waves is due to the following matrix $\mathbf{\Lambda}_l(z)$, present in the layer matrix (7.55):

$$\mathbf{\Lambda}_l(z) = \begin{pmatrix} \exp(-\gamma_l z) & 0 & 0 & 0 \\ 0 & \exp(-\nu_l z) & 0 & 0 \\ 0 & 0 & \exp(\gamma_l z) & 0 \\ 0 & 0 & 0 & \exp(\nu_l z) \end{pmatrix}. \tag{7.61}$$

When H/λ is large, some of the diagonal elements of $\mathbf{\Lambda}_l$ will be very large. Lumping the matrix products in $\mathbf{B}$ before and after $\mathbf{\Lambda}_l$ as $\mathbf{L}$ and $\mathbf{R}$, we write

$$\mathbf{B} = \mathbf{L}\mathbf{\Lambda}_l\mathbf{R}. \tag{7.62}$$

The subdeterminant of $\mathbf{B}$ corresponding to (7.59) will contain terms like

$$\exp(2\gamma_l z)(l_{33}r_{31}l_{43}r_{32} - l_{43}r_{31}l_{33}r_{32}),$$

which is identically zero, but will be obtained as a difference of two large numbers in the process of computation if $\exp(\gamma_l z)$ is large, causing a loss of significant figures. This problem was pointed out by Dunkin (1965) and remedied by him and Knopoff (1964) by the use of Laplace's development by minors. It can be remedied also by the layer-reduction method (Schwab and Knopoff, 1970), in which the lower part of the medium is replaced by a homogeneous half-space at increasingly shallower depths for waves of shorter period. The layer-reduction method, however, should be avoided for the cases with deep wave guides and be used only at depths for which all the layers below contain only inhomogeneous P- and SV-waves. Another approach is to use propagator-matrix methods for the second-order minors from solutions $\mathbf{r}^{(1)}$ and $\mathbf{r}^{(2)}$ (see (7.30)). These six minors are:

$$R_1 = r_1^{(1)}r_2^{(2)} - r_2^{(1)}r_1^{(2)}, \qquad R_2 = r_1^{(1)}r_3^{(2)} - r_3^{(1)}r_1^{(2)},$$

$$R_3 = r_1^{(1)}r_4^{(2)} - r_4^{(1)}r_1^{(2)}, \qquad R_4 = r_2^{(1)}r_3^{(2)} - r_3^{(1)}r_2^{(2)},$$

$$R_5 = r_2^{(1)}r_4^{(2)} - r_4^{(1)}r_2^{(2)}, \qquad R_6 = r_3^{(1)}r_4^{(2)} - r_4^{(1)}r_3^{(2)}.$$

Takeuchi and Saito (1972) showed that $\mathbf{f} = (R_1, R_2, R_3, R_4, R_5, R_6)^T$ satisfies an equation of type $d\mathbf{f}/dz = \mathbf{Cf}$, which we have already shown how to solve. This time, however, the dispersion relation (7.32) reduces to requiring one scalar component of $\mathbf{f}$ (namely, R_6) to vanish at the free surface $z = z_0$.

BOX **7.5**

Surface waves in the Gutenberg Earth model

As an example, we show in Figures A to E the phase velocity c, group velocity U, and eigenfunctions for Love and Rayleigh waves in Gutenberg's classic Earth model for a continent. The model consists of a stack of 24 homogeneous layers, for each of which the depth to bottom, density, and compressional and shear wave velocities are listed in the table below.

Gutenberg's layered model of continental structure.

Layer number	*Depth to bottom* (km)	*density* (g/cm^3)	v_p (km/sec)	v_s (km/sec)
1	19	2.74	6.14	3.55
2	38	3.00	6.58	3.80
3	50	3.32	8.20	4.65
4	60	3.34	8.17	4.62
5	70	3.35	8.14	4.57
6	80	3.36	8.10	4.51
7	90	3.37	8.07	4.46
8	100	3.38	8.02	4.41
9	125	3.39	7.93	4.37
10	150	3.41	7.85	4.35
11	175	3.43	7.89	4.36
12	200	3.46	7.98	4.38
13	225	3.48	8.10	4.42
14	250	3.50	8.21	4.46
15	300	3.53	8.38	4.54
16	350	3.58	8.62	4.68
17	400	3.62	8.87	4.85
18	450	3.69	9.15	5.04
19	500	3.82	9.45	5.21
20	600	4.01	9.88	5.45
21	700	4.21	10.30	5.76
22	800	4.40	10.71	6.03
23	900	4.56	11.10	6.23
24	1000	4.63	11.35	6.32

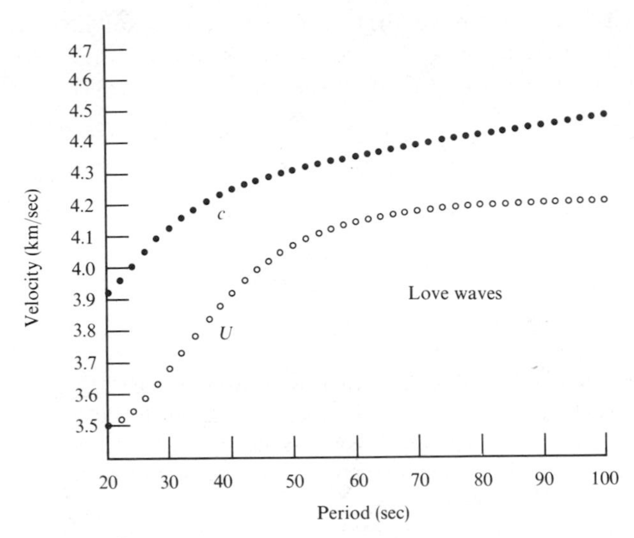

FIGURE **A**

Phase and group velocity of the fundamental-mode Love waves for the Gutenberg Earth model.

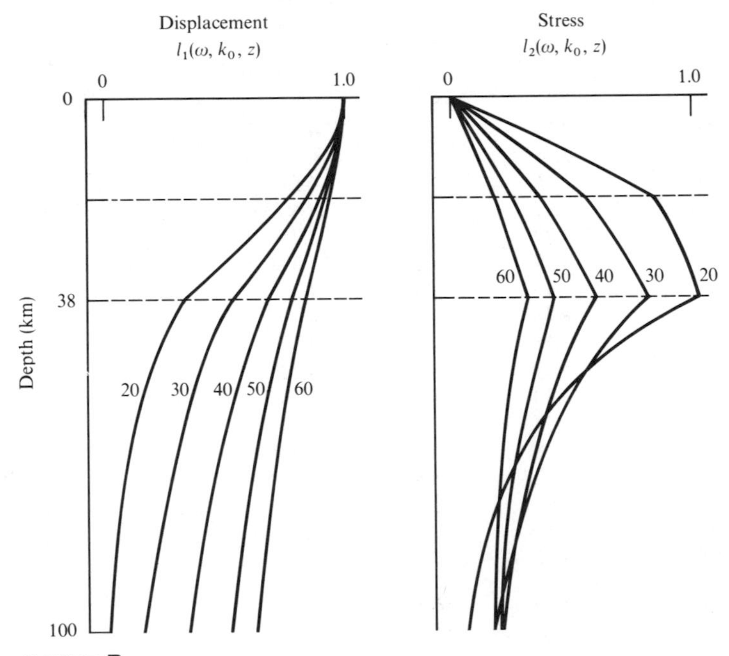

FIGURE **B**

The eigenfunctions for the fundamental-mode Love waves for various periods. The amplitude is normalized to the displacement at $z = 0$.

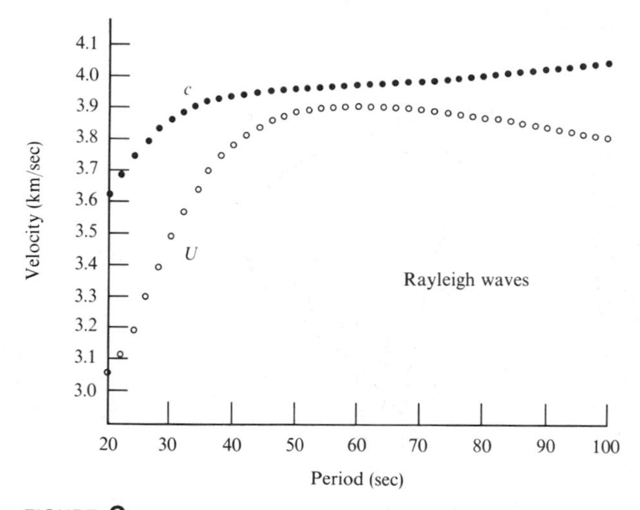

FIGURE **C**

Phase and group velocity of the fundamental-mode Rayleigh waves for the Gutenberg Earth model.

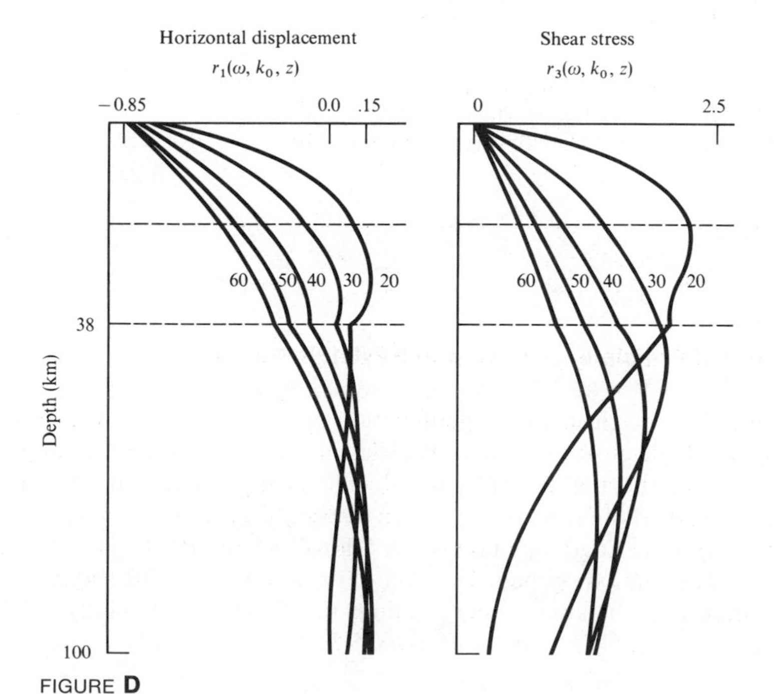

FIGURE **D**

The horizontal eigenfunctions for the fundamental-mode Rayleigh waves for various periods. The amplitude is normalized to the vertical displacement r_2 at $z = 0$.

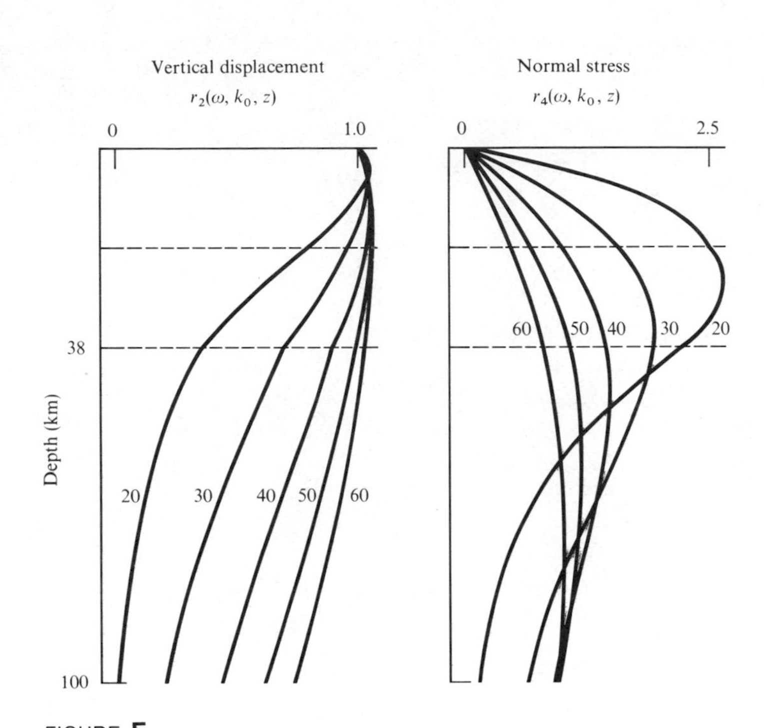

FIGURE **E**

The vertical eigenfunctions for the fundamental-mode Rayleigh waves for various periods. Again, the amplitude is normalized to the vertical displacement r_2 at $z = 0$.

7.3 Variational Principle for Love and Rayleigh Waves

We show here that the eigenvalue-eigenfunction problem discussed above can be solved by the Rayleigh-Ritz method. We shall also use variational techniques for finding several important formulas to calculate group velocity, attenuation, and partial derivatives of phase velocity with respect to medium parameters.

Let us start with the Lagrangian densities L for Love and Rayleigh waves in isotropic vertically inhomogeneous media. For a linear elastic body, the Lagrangian density is the kinetic energy minus the elastic strain energy (2.32). For an isotropic body,

$$L = \tfrac{1}{2}\rho \dot{u}_i \dot{u}_i - \left[\tfrac{1}{2}\lambda(e_{kk})^2 + \mu e_{ij} e_{ij}\right]. \tag{7.63}$$

In the case of plane Love waves given by (7.19), since $e_{ij} = (u_{i,j} + u_{j,i})/2$, we have

$$\langle L \rangle = \tfrac{1}{4}\rho\omega^2 l_1^2 - \tfrac{1}{4}\mu\left[k^2 l_1^2 + \left(\frac{dl_1}{dz}\right)^2\right]. \tag{7.64}$$

The brackets $\langle \ \rangle$ here denote an averaging process, so that terms like $\cos^2(kx - \omega t)$ and $\sin^2(kx - \omega t)$, present in (7.63), have been replaced by $\frac{1}{2}$ in the derivation of (7.64).

In the case of plane Rayleigh waves given by (7.25),

$$\begin{aligned}\langle L \rangle = \tfrac{1}{4}\rho\omega^2(r_1^2 + r_2^2) - \tfrac{1}{4}\bigg[\lambda\left(kr_1 + \frac{dr_2}{dz}\right)^2 + \mu\left(\frac{dr_1}{dz} - kr_2\right)^2 \\ + 2\mu k^2 r_1^2 + 2\mu\left(\frac{dr_2}{dz}\right)^2\bigg].\end{aligned} \tag{7.65}$$

7.3.1 *Love waves*

For these surface waves we define the energy integrals

$$\begin{aligned} I_1 &= \tfrac{1}{2}\int_0^\infty \rho l_1^2\,dz, \quad I_2 = \tfrac{1}{2}\int_0^\infty \mu l_1^2\,dz, \\ I_3 &= \tfrac{1}{2}\int_0^\infty \mu\left(\frac{dl_1}{dz}\right)^2 dz. \end{aligned} \tag{7.66}$$

When there are no body forces and no surface tractions, then, from Hamilton's principle, the integral of $\langle L \rangle$ must be stationary for perturbation of l_1 about the actual motion. To verify this, we examine the perturbation in the integral of $\langle L \rangle$, which is just one-half of

$$\begin{aligned}\omega^2\,\delta I_1 - k^2\,\delta I_2 - \delta I_3 &= \omega^2\int_0^\infty \rho l_1\,\delta l_1\,dz - \int_0^\infty \mu\frac{dl_1}{dz}\frac{d\,\delta l_1}{dz}\,dz \\ &\quad - k^2\int_0^\infty \mu l_1\,\delta l_1\,dz \\ &= \int_0^\infty\left\{\omega^2\rho l_1 - k^2\mu l_1 + \frac{d}{dz}\left(\mu\frac{dl_1}{dz}\right)\right\}\delta l_1\,dz \\ &\quad - \mu\frac{dl_1}{dz}\,\delta l_1\bigg|_0^\infty.\end{aligned}$$

If (i) the equation of motion (7.21), (ii) the free surface condition $dl_1/dz = 0$ at $z = 0$, (iii) vanishing δl_1 at $z = \infty$ and (iv) continuity of l_1 and $\mu\, dl_1/dz$ (required for the integration by parts) are all satisfied, we do indeed find that

$$\omega^2\,\delta I_1 - k^2\,\delta I_2 - \delta I_3 = 0. \tag{7.67}$$

The l_1 that satisfies these four conditions is nothing but an eigenfunction. Thus we have verified that $\omega^2 I_1 - k^2 I_2 - I_3$ is stationary for perturbation of l_1 around an eigenfunction. Furthermore, we can show that the Lagrangian integral $\omega^2 I_1 - k^2 I_2 - I_3$ vanishes at its stationary point by multiplying the equation of motion (7.21) by l_1 and integrating it with respect to z from 0 to ∞:

$$\begin{aligned} 0 &= \int_0^\infty \left\{\omega^2 \rho l_1^2 - k^2 \mu l_1^2 + l_1 \frac{d}{dz}\left(\mu \frac{dl_1}{dz}\right)\right\} dz \\ &= 2\omega^2 I_1 - 2k^2 I_2 - 2I_3 + l_1 \mu \frac{dl_1}{dz}\bigg|_0^\infty, \end{aligned} \tag{7.68}$$

and when l_1 is an eigenfunction, $l_1 = 0$ at $z = \infty$ and $dl_1/dz = 0$ at $z = 0$, so that the last term in (7.68) vanishes. Thus, for an eigenfunction,

$$\omega^2 I_1 = k^2 I_2 + I_3. \tag{7.69}$$

In other words, the total elastic energy contained in a normal mode is equal to the total kinetic energy.

There are three particularly important applications of the two relations $\omega^2\,\delta I_1 = k^2\,\delta I_2 + \delta I_3$ and $\omega^2 I_1 = k^2 I_2 + I_3$ for us to consider. Different quantities are perturbed in each application, and the subtlety and power of variational methods is well illustrated by the fact that three such different results all stem from the same root.

First, we shall suppose that in a numerical attempt to evaluate a true eigenfunction $l_1(z)$, we have made errors and have in fact obtained $l_1 + \delta l_1$. From this, we would numerically obtain $I_1 + \delta I_1$, $I_2 + \delta I_2$, and $I_3 + \delta I_3$ by using $l_1 + \delta l_1$ in (7.66). However, addition of the two relations (7.67) and (7.69) tells us that ω^2 is related to k^2 via

$$k^2 = \frac{\omega^2(I_2 + \delta I_2) - (I_3 + \delta I_3)}{I_1 + \delta I_1}.$$

This can be used to determine k^2 accurately, and the point is that first-order errors in l_1 do not lead to first-order errors in the eigenvalue.

Second, we shall obtain a useful formula for group velocity U without having to perform a numerical derivative. We suppose that $l_1(k, \omega)$ and $l_1 + \delta l_1 = l_1(k + \delta k, \omega + \delta\omega)$ are both eigenfunctions. Using this $l_1 + \delta l_1$ to evaluate energy integrals, it follows from (7.69) that

$$(\omega + \delta\omega)^2(I_1 + \delta I_1) = (k + \delta k)^2(I_2 + \delta I_2) + (I_3 + \delta I_3).$$

Subtracting (7.69), we find, to first order, that

$$\omega^2\,\delta I_1 + 2\omega\,\delta\omega I_1 = k^2\,\delta I_2 + 2k\,\delta k I_2 + \delta I_3.$$

However, using this $l_1 + \delta l_1$ does constitute a perturbation to l_1 for which (7.67) applies. Therefore, $2\omega\,\delta\omega I_1 = 2k\,\delta k I_2$, and

$$U = \frac{\delta\omega}{\delta k} = \frac{k}{\omega}\frac{I_2}{I_1} = \frac{I_2}{cI_1}. \tag{7.70}$$

This gives the group velocity in terms of integrals, and is numerically more stable than differentiation.

Third, we shall evaluate the changes in phase velocity and group velocity that arise (at fixed frequency) when small perturbations in rigidity μ and density ρ are made in the structure. Suppose that $l_1(k, \omega)$ is the eigenfunction in structure $(\rho(z), \mu(z))$. We write this as $l_1 = l_1(\rho, \mu, k, \omega)$ and consider a perturbation $l_1 + \delta l_1 = l_1(\rho + \delta\rho, \mu + \delta\mu, k + \delta k, \omega)$ that is also an eigenfunction, but in a slightly different structure. Applying (7.69) to the new eigenfunction gives

$$\omega^2 \int_0^\infty (\rho + \delta\rho)(l_1 + \delta l_1)^2\,dz = (k + \delta k)^2 \int_0^\infty (\mu + \delta\mu)(l_1 + \delta l_1)^2\,dz + \int_0^\infty (\mu + \delta\mu)\left[\frac{d}{dz}(l_1 + \delta l_1)\right]^2 dz,$$

and subtracting (7.69) for the original eigenfunction gives (to first order)

$$\omega^2 \int_0^\infty (l_1^2\,\delta\rho + 2\rho l_1\,\delta l_1)\,dz = k^2 \int_0^\infty (l_1^2\,\delta\mu + 2\mu l_1\,\delta l_1)\,dz + 2k\,\delta k \int_0^\infty \mu l_1^2\,dz + \int_0^\infty \left(\frac{dl_1}{dz}\right)^2 \delta\mu\,dz + \int_0^\infty 2\mu\frac{dl_1}{dz}\frac{d\,\delta l_1}{dz}\,dz.$$

But from (7.67) we can cancel out all integrals here that contain the terms in δl_1 and $d(\delta l_1)/dz$. We are left with a relation between δk and integrals over $\delta\rho$

and $\delta\mu$. It remains to relate δc and δk, and the final result is

$$\left(\frac{\delta c}{c}\right)_\omega = -\frac{\delta k}{k}$$
$$= \frac{\int_0^\infty \left[k^2 l_1^2 + \left(\frac{dl_1}{dz}\right)^2\right] \delta\mu \, dz - \int_0^\infty \omega^2 l_1^2 \, \delta\rho \, dz}{2k^2 \int_0^\infty \mu l_1^2 \, dz}. \tag{7.71}$$

This formula shows a linear relation between the fractional change $\delta c/c$ in phase velocity and the perturbation in model parameters, and it plays a central role in the inversion of phase-velocity data, as discussed in Chapter 12. The weight, or kernel, function for $(\delta\mu/\mu)$ is proportional to the strain-energy density and that for $(\delta\rho/\rho)$ to the kinetic-energy density, showing that the perturbation in μ and ρ at depths where the energy densities are larger affect the phase velocity more strongly.

The change of group velocity at fixed frequency, due to small perturbations in rigidity and density, is obtained from (7.70) as

$$\frac{\delta U}{U} = \frac{\delta I_2}{I_2} - \frac{\delta c}{c} - \frac{\delta I_1}{I_1}, \tag{7.72}$$

but in this case δl_1 must be computed before δI_1 and δI_2 are found, since integrals involving δl_1 and $d(\delta l_1)/dz$ do not cancel out in a manner similar to (7.71).

7.3.2 Rayleigh waves

Using the Lagrangian density given in (7.65), Hamilton's principle for Rayleigh waves can be written as

$$\omega^2 \, \delta I_1 - k^2 \, \delta I_2 - k \, \delta I_3 - \delta I_4 = 0, \tag{7.73}$$

where

$$\begin{aligned}
I_1 &= \tfrac{1}{2} \int_0^\infty \rho(r_1^2 + r_2^2) \, dz, \\
I_2 &= \tfrac{1}{2} \int_0^\infty [(\lambda + 2\mu) r_1^2 + \mu r_2^2] \, dz, \\
I_3 &= \int_0^\infty \left(\lambda r_1 \frac{dr_2}{dz} - \mu r_2 \frac{dr_1}{dz}\right) dz, \\
I_4 &= \tfrac{1}{2} \int_0^\infty \left[(\lambda + 2\mu)\left(\frac{dr_2}{dz}\right)^2 + \mu\left(\frac{dr_1}{dz}\right)^2\right] dz.
\end{aligned} \tag{7.74}$$

It can be shown that (7.73) is equivalent to the equation of motion (7.28), boundary conditions (7.29), and continuity of r_1, r_2, r_3, and r_4. Since r_1, r_2, which satisfy these conditions, are eigenfunctions for Rayleigh waves, (7.73) shows that the integral $\omega^2 I_1 - k^2 I_2 - kI_3 - I_4$ is stationary for perturbation of eigenfunctions. As is true for Love waves, we find that this integral vanishes at the stationary point:

$$\omega^2 I_1 - k^2 I_2 - kI_3 - I_4 = 0. \tag{7.75}$$

Thus this equation can be used as a check on the eigenvalue and eigenfunction calculations by the methods discussed earlier.

By perturbing k and ω in (7.75), we find the formula for group velocity to be

$$I_1\,\delta(\omega^2) = I_2\,\delta(k^2) + I_3\,\delta k$$

or

$$U = \frac{\delta\omega}{\delta k} = \frac{I_2 + \dfrac{I_3}{2k}}{cI_1} \tag{7.76}$$

where c is the phase velocity.

The change of phase velocity due to small perturbations in λ, μ, and ρ can also be obtained from (7.75). For a given ω, again using (7.73), we have

$$\begin{aligned}\delta k(2kI_2 + I_3) = {} & \tfrac{1}{2}\int_0^\infty \omega^2(r_1^2 + r_2^2)\,\delta\rho\,dz \\ & - \tfrac{1}{2}\int_0^\infty \left(kr_1 + \frac{dr_2}{dz}\right)^2 \delta\lambda\,dz \\ & - \tfrac{1}{2}\int_0^\infty \left[2k^2r_1^2 + 2\left(\frac{dr_2}{dz}\right)^2 + \left(kr_2 - \frac{dr_1}{dz}\right)^2\right]\delta\mu\,dz. \end{aligned} \tag{7.77}$$

The corresponding fractional change in phase velocity can be written as

$$\begin{aligned}\left(\frac{\delta c}{c}\right)_\omega & = -\frac{\delta k}{k} \\ & = \frac{1}{4UcI_1}\left\{\int_0^\infty \left(kr_1 + \frac{dr_2}{dz}\right)^2 \delta\lambda\,dz + \int_0^\infty \left[2k^2r_1^2 + 2\left(\frac{dr_2}{dz}\right)^2\right.\right. \\ & \quad \left.\left. + \left(kr_2 - \frac{dr_1}{dz}\right)^2\right]\delta\mu\,dz - \int_0^\infty \omega^2(r_1^2 + r_2^2)\,\delta\rho\,dz\right\}, \end{aligned} \tag{7.78}$$

BOX **7.6**

"Partial derivatives" of phase velocity

The small change in phase velocity of Love waves, due to small perturbations $\delta\rho$ and $\delta\mu$ in the structure, can naturally be written in the form

$$\left(\frac{\delta c}{c}\right)_\omega = \int_0^\infty \frac{\rho}{c}\left[\frac{\partial c}{\partial \rho}\right]_{\omega,\mu} \frac{\delta\rho}{\rho}\, dz + \int_0^\infty \frac{\mu}{c}\left[\frac{\partial c}{\partial \mu}\right]_{\omega,\rho} \frac{\delta\mu}{\mu}\, dz. \tag{1}$$

The symbols appearing here in square brackets are known as partial derivatives of phase velocity with respect to density (or rigidity) at a particular depth. Comparing with (7.71), we have

$$\frac{\rho}{c}\left[\frac{\partial c}{\partial \rho}\right]_{\omega,\mu} = -\frac{\rho\omega^2 l_1^2}{2k^2 \int_0^\infty \mu l_1^2\, dz} \tag{2}$$

and

$$\frac{\mu}{c}\left[\frac{\partial c}{\partial \mu}\right]_{\omega,\rho} = \frac{\mu\left\{k^2 l_1^2 + \left(\frac{dl_1}{dz}\right)^2\right\}}{2k^2 \int_0^\infty \mu l_1^2\, dz}. \tag{3}$$

The notation here is deceptive, in that the left-hand sides of (2) and (3) appear to be dimensionless. As is made clear from the right-hand sides, the unit is in fact reciprocal length (this is apparent directly from (1)).

The first computations of partial derivatives of phase velocity were conducted by Dorman and Ewing (1962) and Brune and Dorman (1963), who simply found the phase velocities for two slightly different structures and then took the numerical difference. The variational approach for both Love and Rayleigh waves was suggested by Jeffreys (1961). A practical scheme for obtaining partial derivatives of group velocity is given by Rodi et al. (1975).

where (7.76) was used in simplifying the denominator on the right-hand side. The weight function for $\delta\rho/\rho$ is again the kinetic energy, and that for $\delta\lambda$ is proportional to the dilatational strain energy. By rewriting (7.78) in terms of $\delta(\lambda + \frac{2}{3}\mu)$ and $\delta\mu$, we would find weight functions proportional to dilatational- and shear-strain energies (see Problem 2.6).

7.3.3 Rayleigh-Ritz method

The variational principles, such as given in (7.67) and (7.73) for Love and Rayleigh waves, are used in the Rayleigh-Ritz method to find the eigenvalues and eigenfunctions. In this approach, we approximate the eigenfunction by a linear combination of basis functions $\phi_i(z)$, which are chosen usually to satisfy boundary conditions at $z = 0$ and $z = \infty$:

$$l_1(z) = \sum_{i=1}^{n} c_i \phi_i(z). \tag{7.79}$$

If we substitute this form into the energy integrals in (7.66), we find

$$I_1 = \mathbf{c}^T \mathbf{A}_1 \mathbf{c}, \qquad I_2 = \mathbf{c}^T \mathbf{A}_2 \mathbf{c}, \qquad I_3 = \mathbf{c}^T \mathbf{A}_3 \mathbf{c}, \tag{7.80}$$

where $\mathbf{c} = (c_1, c_2, \ldots, c_n)^T$ and the matrices $\mathbf{A}_1$, $\mathbf{A}_2$, and $\mathbf{A}_3$ have elements

$$\begin{aligned} A_{1ij} &= \tfrac{1}{2} \int_0^\infty \rho \phi_i \phi_j \, dz, \\ A_{2ij} &= \tfrac{1}{2} \int_0^\infty \mu \phi_i \phi_j \, dz, \\ A_{3ij} &= \tfrac{1}{2} \int_0^\infty \mu \frac{\partial \phi_i}{\partial z} \frac{\partial \phi_j}{\partial z} \, dz. \end{aligned} \tag{7.81}$$

The coefficient vector $\mathbf{c}$, which minimizes the Lagrangian integral $\omega^2 I_1 - k^2 I_2 - I_3$, is obtained by setting the derivative of the integral with respect to c_i $(i = 1, 2, \ldots, n)$ equal to zero. The result is given by

$$(\omega^2 \mathbf{A}_1 - k^2 \mathbf{A}_2 - \mathbf{A}_3)\mathbf{c} = \mathbf{0}. \tag{7.82}$$

This equation defines an eigenvalue-eigenvector problem for $\mathbf{c}$. Nontrivial solutions $\mathbf{c}$ exist only when k is an eigenvalue that satisfies

$$|\omega^2 \mathbf{A}_1 - k^2 \mathbf{A}_2 - \mathbf{A}_3| = 0 \tag{7.83}$$

for a given ω. Once an eigenvalue is determined, the corresponding eigenvector is obtained by solving (7.82).

Since the eigenvalue is stationary with respect to any perturbation in the eigenfunction, the approximate eigenvector determined by solving (7.82) can be used to find a more accurate eigenvalue using (7.69).

Takeuchi and Kobayashi (1959) used exponentially decaying functions as the basis functions and determined the Love-wave dispersion in a half-space in which rigidity increases linearly with depth. They demonstrated the power of Rayleigh-Ritz methods by obtaining essentially the same results as the numerical integration method of Satô (1959): the former took about the same time on a hand-calculator as the latter on an IBM650. More recently, Wiggins (1976b) used cubic polynomials, as shown in Figure 7.6, for the basis functions.

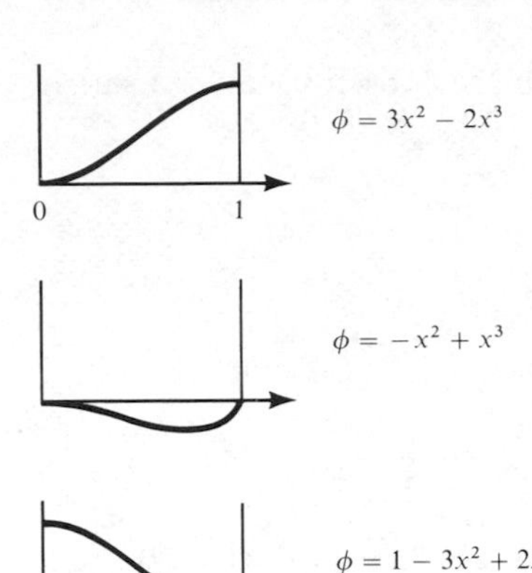

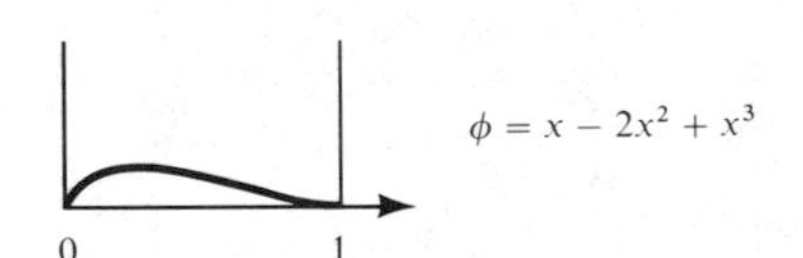

$\phi = x - 2x^2 + x^3$

0 1

FIGURE **7.6**

Basis functions used by Wiggins (1976b) to find eigenfunctions via the Rayleigh-Ritz method. Many inhomogeneous layers may be present in the medium; shown here are the four normalized functions used for each layer ($x = 0$ and $x = 1$ correspond to layer boundaries). For each of the four functions shown, ϕ or $d\phi/dx$ is zero or one at $x = 0$ and 1.

Their values and derivatives are unity or vanish at the end points, which is convenient for constructing a trial eigenfunction satisfying prescribed boundary conditions. Wiggins concluded that the Rayleigh-Ritz method using these basis functions is competitive with the most efficient matrix method. A similar approach was taken also by Buland and Gilbert (1976).

7.3.4 *Attenuation of surface waves*

In Chapter 5, we introduced spatial $Q(\omega)$ to describe the attenuation of plane body-waves with the travel distance and discussed the associated dispersion; we have seen that the propagation factor $\exp[i(kx - \omega t)] = \exp[i\omega(x/c - t)]$ for elastic media is to be replaced by

$$\exp[i(Kx - \omega t)] = \exp\left[\frac{(-\omega x)}{2c(\omega)Q(\omega)}\right]\exp\left\{i\omega\left[\frac{x}{c(\omega)} - t\right]\right\}.$$

When a solution of an elastic-wave problem is given in terms of $c = c_e$, then the corresponding anelastic solution is given by replacing c_e via the rule

$$\frac{1}{c_e} \to \frac{1}{c(\omega)}\left(1 + \frac{i}{2Q(\omega)}\right). \qquad \text{(5.87 again)}$$

In particular, if Q is effectively constant, we found that dispersion is logarithmic, and is given by

$$\frac{c(\omega_1)}{c(\omega_2)} = 1 + \frac{1}{\pi Q}\ln\left(\frac{\omega_1}{\omega_2}\right).$$

For a lossy medium with high Q, the rule (5.87) can be considered to introduce a small perturbation to the velocity c_e (Anderson and Archambeau, 1964; Anderson et al., 1965). The change δc, which is made at fixed frequency, will have real and imaginary components:

$$\left(\frac{\delta c}{c}\right)_\omega = \left(\frac{c}{c_e} - 1\right) - \frac{i}{2Q^{\text{spatial}}}. \tag{7.84}$$

Let us first consider the change in phase velocity of Love waves due to perturbations in S-wave speeds at each depth. The fractional change $\delta c/c$ in phase velocity due to $\delta\mu/\mu$ and $\delta\rho/\rho$ is already given in (7.71). Since

$$\delta\beta/\beta = \tfrac{1}{2}(\delta\mu/\mu - \delta\rho/\rho),$$

we have from (7.71), neglecting density perturbation,

$$\left(\frac{\delta c}{c}\right)_\omega = \frac{\displaystyle\int_0^\infty \left[k^2 l_1^2 + \left(\frac{dl_1}{dz}\right)^2\right] 2\mu \frac{\delta\beta}{\beta}\,dz}{\displaystyle 2k^2 \int_0^\infty \mu l_1^2\,dz}. \tag{7.85}$$

Substituting the form (7.84) for $\delta\beta/\beta$ in the above equation, we obtain $(\delta c/c)$, with real and imaginary parts. The real part will introduce material dispersion for Love waves due to anelasticity in addition to geometric dispersion due to vertical heterogeneity of the medium. The imaginary part will cause attenuation of Love-wave amplitudes, and from (7.84) we define the spatial Q_L of Love waves by

$$\textit{spatial } Q_L^{-1} = -2\,\mathrm{Im}(\delta c/c)_\omega. \tag{7.86}$$

Putting $\mathrm{Im}(\delta\beta/\beta) = -1/(2Q_\beta)$ in (7.85), we obtain

$$\textit{spatial } Q_L^{-1} = \frac{\displaystyle\int_0^\infty \left[k^2 l_1^2 + \left(\frac{dl_1}{dz}\right)^2\right] \mu Q_\beta^{-1}\,dz}{\displaystyle k^2 \int_0^\infty \mu l_1^2\,dz}. \tag{7.87}$$

This formula shows that Q_L^{-1} for Love waves is expressed as an integral of Q_β^{-1} for shear waves with weight function proportional to the strain-energy density of the particular mode.

For Rayleigh waves, the perturbations in λ and μ are replaced by those in α and β, and recognizing the term $\delta(\lambda + 2\mu)/(\lambda + 2\mu) = 2\delta\alpha/\alpha + \delta\rho/\rho$ in (7.78), we obtain

$$\left(\frac{\delta c}{c}\right)_\omega = \frac{1}{4UcI_1}\left\{\int_0^\infty \left(kr_1 + \frac{dr_2}{dz}\right)^2 (\lambda + 2\mu)2\left(\frac{\delta\alpha}{\alpha}\right) dz \right.$$
$$\left. + \int_0^\infty \left[\left(kr_2 - \frac{dr_1}{dz}\right)^2 - 4kr_1\frac{dr_2}{dz}\right]2\left(\frac{\delta\beta}{\beta}\right) dz\right\}. \tag{7.88}$$

Defining the spatial Q_R of Rayleigh waves in a manner similar to that of (7.86), and putting $\text{Im}(\delta\alpha/\alpha) = -1/(2Q_\alpha)$, we get from (7.88),

$$\text{spatial } Q_R^{-1} = \frac{1}{4UcI_1}\left\{\int_0^\infty \left(kr_1 + \frac{dr_2}{dz}\right)^2 (\lambda + 2\mu)2Q_\alpha^{-1}\, dz \right.$$
$$\left. + \int_0^\infty \left[\left(kr_2 - \frac{dr_1}{dz}\right)^2 - 4kr_1\frac{dr_2}{dz}\right]2\mu Q_\beta^{-1}\, dz\right\}. \tag{7.89}$$

Again, Q_R^{-1} for Rayleigh waves is expressed as an integral of Q^{-1} for P- and S-waves with appropriate weight functions.

So far, we have considered the spatial Q defined for the attenuation with travel distance (Box 5.7). This definition is suitable for measurement on propagating waves. However, there are several cases of seismic measurements in which the temporal Q is directly observed. The most important case is the attenuation of a particular free oscillation of the whole Earth. In this case, k is fixed in contrast to a propagating-wave experiment in which ω is fixed. The change of phase velocity due to model-parameter perturbation is different between experiments with fixed ω and fixed k. Thus the phase function $\exp[i(kx - \omega t)]$ is replaced by

$$\exp[i(kx - \Omega t)] = \exp\left[-\frac{\omega t}{2Q(\omega)}\right]\exp[ik(x - ct)].$$

The replacement of an elastic velocity c_e is now made via the rule

$$c_e \to c\left(1 - \frac{i}{2Q^{\text{temporal}}}\right);$$

that is,

$$\left(\frac{\delta c}{c}\right)_k = \left(\frac{c}{c_e} - 1\right) - \frac{i}{2Q^{\text{temporal}}}. \tag{7.90}$$

Since both $(\omega, k + \delta k)$ and $(\omega + \delta\omega, k)$ are pairs of eigenvalues for the perturbed medium, the slope of the line connecting the two points in an ω-k diagram will give group velocity U for the perturbed medium:

$$U = \frac{\omega - (\omega + \delta\omega)}{(k + \delta k) - k} = -\frac{\delta\omega}{\delta k}. \tag{7.91}$$

But in the one case we have $k + \delta k = K = k(1 + i/2Q^{\text{spatial}})$, and in the other we have $\omega + \delta\omega = \Omega = \omega(1 - i/2Q^{\text{temporal}})$. From these we construct $-\delta\omega/\delta k$

BOX **7.7**

Some effects of anisotropy

The general theory of surface-wave propagation in anisotropic media with vertical inhomogeneity is conceptually a simple extension of the propagator matrix techniques given in Section 7.2. The displacement-stress vector (see (5.101)) is now of length 6, and in general we cannot speak of Love and Rayleigh waves as separate families of surface waves. Rather, there exists one generalized family of surface waves, which, in weakly anisotropic media, can be broken down into quasi-Love and quasi-Rayleigh motions.

If anisotropy is weak, then the phase velocity c of surface waves can be shown to have dependence on the azimuthal direction ϕ of propagation via

$$c(\omega, \phi) = A(\omega) + B(\omega)\cos 2\phi + C(\omega)\sin 2\phi + E(\omega)\cos 4\phi + F(\omega)\sin 4\phi \tag{1}$$

(Smith and Dahlen, 1973). This formula is approximate, and similar to the body-wave result (5.102). Forsyth (1975) has applied (1) to surface-wave data for the Pacific, though he did not include the 4ϕ terms because in practice (with present data) their effect is minor.

For an anisotropic half-space composed of a welded stack of homogeneous layers, a detailed discussion of surface-wave dispersion has been given by Crampin (1970, 1971, 1975, 1977). If anisotropy is present in the Earth because of some preferential alignment of anisotropic crystals (e.g., of olivine) that originally solidified near a spreading center such as a mid-oceanic ridge, then horizontal planes through the anisotropy are likely to be planes of symmetry. Taking x_3 as the depth direction, Love (1944) showed that there is symmetry about horizontal planes if and only if $c_{ijkl} = 0$ when either one or three of the subscripts is the number 3. Crampin (1975) described a surface-wave particle motion characteristic of this type of anisotropy. Called *Inclined-Rayleigh* motion, it consists of elliptical particle-motion with vertical and horizontal axes, but the plane of the ellipse is inclined to the sagittal plane (see Section 5.6). Thus particle motion has a component transverse to the slowness direction in which the wave is propagating. The horizontal components are in phase, but are $\pm\pi/2$ out of phase with the vertical component. Crampin and King (1977) have described several observations of this particle motion at an array of seismometers in Norway (NORSAR), for surface waves that have travelled across Eurasia.

and obtain the general rule

$$\textit{temporal } Q^{-1} = \frac{U}{c} \times \textit{spatial } Q^{-1}. \tag{7.92}$$

These two kinds of Q were first distinguished by Brune (1962) in a study of surface waves. The above formula is important when we compare Q of dispersive waves determined by a standing-wave experiment and by a propagating-wave experiment. For nondispersive waves, $c = U$ and there is no distinction between the two Q's.

Equation (7.92) can be easily understood if we measure the attenuation of dispersed waves using the stationary-phase approximation (7.18). At a given x and t, the frequency ω given by $x/t = U(\omega)$ dominates the record. Since the wave with frequency ω has existed in the medium over the time period $t = x/(U(\omega))$, it must have been attenuated by a factor

$$\exp\left[\frac{-\omega t}{2 \textit{ temporal } Q(\omega)}\right] = \exp\left[\frac{-\omega x}{2U(\omega) \textit{ temporal } Q(\omega)}\right]. \tag{7.93}$$

Since, by definition, this is equal to

$$\exp\left[\frac{-\omega x}{2c(\omega) \textit{ spatial } Q(\omega)}\right],$$

we obtain (7.92).

7.4 Surface-wave Terms of Green's Function for a Vertically Heterogeneous Medium

In this section, we shall obtain a simple compact solution for surface waves generated by a point force with time dependence $e^{-i\omega t}$ buried in a vertically heterogeneous medium. We shall first express the general solution for the equations of motion in cylindrical coordinates (r, ϕ, z), using the motion-stress vector introduced in previous sections and a set of vectors that are functions of r, ϕ alone to describe the horizontal propagation. A stress discontinuity equivalent to a point source is expanded in a series of these horizontally-varying vector functions. We then find a solution with the prescribed discontinuity that also satisfies the boundary conditions at free surface and infinite depth. This approach was used by Haskell (1964), Harkrider (1964), Ben-Menahem et al. (1970), and Hudson (1969). A compact result is obtained by applying the variational principle to the residue evaluated at poles in the k-plane. This last step was taken by Keilis-Borok and Yanovskaya (1962), Harkrider and Anderson (1966), Vlaar (1966) and Saito (1967).

We shall first describe a simple method that works for two-dimensional cases, using a reciprocal theorem of Herrera (1964) that, unfortunately, does not apply in three dimensions. We shall then follow Saito (1967) to find the solution in three dimensions.

7.4.1 *Two-dimensional case*

Let us first find orthogonality relations for eigenfunctions of Love and Rayleigh waves by the use of our reciprocal theorem (2.35). Neglecting body forces, and writing the Fourier transforms of $\mathbf{v}(\mathbf{x}, t)$ and $\mathbf{u}(\mathbf{x}, t)$ as $\mathbf{v}(\mathbf{x}, \omega)$ and $\mathbf{u}(\mathbf{x}, \omega)$, respectively, the transform of (2.35) is rewritten as

$$0 = \iint_S \{\mathbf{v}(\mathbf{x}, \omega) \cdot \mathbf{T}(\mathbf{u}(\mathbf{x}, \omega), \mathbf{n}) - \mathbf{u}(\mathbf{x}, \omega) \cdot \mathbf{T}(\mathbf{v}(\mathbf{x}, \omega), \mathbf{n})\} \, dS. \tag{7.94}$$

(*Note:* The convolution in (2.35) has here become a product of transforms.) This equation should hold for any pair of solutions $\mathbf{u}$ and $\mathbf{v}$ of the equation of motion for a given medium without body forces. We take two different modes of Love waves propagating in the x-direction as $\mathbf{u}$ and $\mathbf{v}$ and define S as the surface enclosing the shaded slab with corners at $(a, 0)$, $(b, 0)$, (a, D), and (b, D), as shown in Figure 7.7.

For Love waves, the traction $\mathbf{T}$ vanishes at the free surface $z = 0$ and the motion vanishes at $z = \infty$. The displacement and traction components along the *vertical* plane are

$$\begin{aligned} u_y &= l_1(\omega, k_n, z) \exp(ik_n x) \\ T_y &= \mu \frac{\partial u_y}{\partial x} = ik_n \mu l_1 \exp(ik_n x) \end{aligned} \tag{7.95}$$

for the first solution and

$$\begin{aligned} v_y &= l_1(\omega, k_m, z) \exp(ik_m x) \\ T_y &= ik_m \mu l_1 \exp(ik_m x) \end{aligned} \tag{7.96}$$

for the second. Other components do not contribute to the integral (7.94).

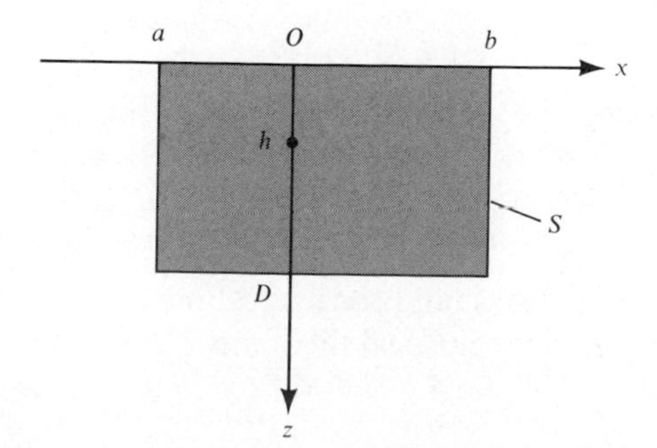

FIGURE **7.7**
Integration in (7.94) is taken over the surface S enclosing the shaded slab.

Substituting (7.95) and (7.96) in (7.94), we find as $D \to \infty$,

$$\int_0^\infty i(k_n - k_m)\mu(z)l_1(\omega, k_m, z)l_1(\omega, k_n, z)\,dz \exp[i(k_n + k_m)a]$$
$$= \int_0^\infty i(k_n - k_m)\mu(z)l_1(\omega, k_m, z)l_1(\omega, k_n, z)\,dz \exp[i(k_n + k_m)b]$$

This equality holds for arbitrary a and b if and only if

$$\int_0^\infty \mu(z)l_1(\omega, k_m, z)l_1(\omega, k_n, z)\,dz = 0 \qquad \text{for } k_n \neq k_m. \tag{7.97}$$

The eigenfunctions of different modes of Love waves, for fixed ω, are therefore orthogonal to each other with the rigidity distribution $\mu(z)$ as a weight function.

For Rayleigh waves, we similarly take the nth mode as **u**, and displacement and traction components along the vertical plane are then

$$\begin{aligned} u_x &= r_1(\omega, k_n, z)\exp(ik_nx), \\ u_z &= ir_2(\omega, k_n, z)\exp(ik_nx), \\ \tau_{xx} &= iT_1(\omega, k_n, z)\exp(ik_nx) = i\left\{\lambda\frac{dr_2}{dz} + (\lambda + 2\mu)k_nr_1\right\}\exp(ik_nx), \\ \tau_{zx} &= T_2(\omega, k_n, z)\exp(ik_nx) = \left(\mu\frac{dr_1}{dz} - \mu k_nr_2\right)\exp(ik_nx). \end{aligned} \tag{7.98}$$

Those for **v** are obtained by replacing k_n by k_m. The orthogonality relation for Rayleigh waves is then obtained as

$$\sum_{i=1}^{2}\int_0^\infty \{r_i(\omega, k_n, z)T_i(\omega, k_m, z) - r_i(\omega, k_m, z)T_i(\omega, k_n, z)\}\,dz = 0$$

or

$$\begin{aligned} \int_0^\infty \Bigg\{&(\lambda + 2\mu)(k_m - k_n)r_1(\omega, k_n, z)r_1(\omega, k_m, z) - \mu(k_m - k_n)r_2(\omega, k_n, z)r_2(\omega, k_m, z) \\ &+ \lambda\left[r_1(\omega, k_n, z)\frac{dr_2(\omega, k_m, z)}{dz} - r_1(\omega, k_m, z)\frac{dr_2(\omega, k_n, z)}{dz}\right] \\ &+ \mu\left[r_2(\omega, k_n, z)\frac{dr_1(\omega, k_m, z)}{dz} - r_2(\omega, k_m, z)\frac{dr_1(\omega, k_n, z)}{dz}\right]\Bigg\}\,dz = 0. \end{aligned} \tag{7.99}$$

With these orthogonality relations in hand, we shall proceed to find surface-wave terms of Green's function, using again the reciprocal theorem. We use the same solution for **u** as defined in (7.95) and (7.98) for Love and Rayleigh waves, respectively, but for $\mathbf{v}$ we use the two-dimensional Green function $\mathbf{G}_k \exp(-i\omega t)$, i.e., the response to a line force with time dependence $\exp(-i\omega t)$ located at $x = 0$, $z = h$. The line is parallel to the y-axis (i.e., x_2-axis), and the force acts in the x_k-direction (i.e., x or y or z), so that

$$-\rho\omega^2 G_{ik}(x, z; 0, h; \omega) = \frac{\partial}{\partial x_j}\tau_{ij}(\mathbf{G}_k) + \delta_{ik}\,\delta(x)\,\delta(z - h). \tag{7.100}$$

Following the same steps as taken from (2.35) to the representation theorem (2.41), we find that

$$u_k(0, h) = \int_S \{G_{ik}(x, z; 0, h; \omega)T_i(\mathbf{u}, \mathbf{n}) - u_i(x, z)T_i(\mathbf{G}_k, \mathbf{n})\}\, dS. \tag{7.101}$$

Now we choose again the same integration surface S as shown in Figure 7.11, but this time make not only $D \to \infty$, but also $a \to -\infty$ and $b \to +\infty$. Since surface waves from a line source suffer no geometric spreading but body waves do, we expect that the Green function at a large distance will consist entirely of surface waves propagating outward from the source. Thus for a large x, the y-component of Green's function is given by a sum of Love waves

$$G_{yy}(x, z; 0, h; \omega) \sim \begin{cases} \sum_m b_{m2}^{+}(h) l_1(\omega, k_m, z)\exp(ik_m x) & x \gg 0 \\ \sum_m b_{m2}^{-}(h) l_1(\omega, k_m, z)\exp(-ik_m x) & x \ll 0. \end{cases} \tag{7.102}$$

The x- and z-components are composed of Rayleigh waves,

$$G_{xk}(x, z; 0, h; \omega) \sim \begin{cases} \sum_m a_{mk}^{+}(h) r_1(\omega, k_m, z)\exp(ik_m x) & x \gg 0 \\ \sum_m a_{mk}^{-}(h) r_1(\omega, k_m, z)\exp(-ik_m x) & x \ll 0, \end{cases}$$

$$G_{zk}(x, z; 0, h; \omega) \sim \begin{cases} \sum_m i a_{mk}^{+}(h) r_2(\omega, k_m, z)\exp(ik_m x) & x \gg 0 \\ -\sum_m i a_{mk}^{-}(h) r_2(\omega, k_m, z)\exp(-ik_m x) & x \ll 0. \end{cases} \tag{7.103}$$

Substituting (7.102) and (7.95) into (7.101), we find the y-component of displacement is given by

$$u_y(0, h) = \sum_m b_{m2}^{+}(h) \int_0^\infty i(k_n - k_m)\mu(z) l_1(\omega, k_n, z) l_1(\omega, k_m, z)\exp[i(k_n + k_m)b]\, dz$$

$$- \sum_m b_{m2}^{-}(h) \int_0^\infty i(k_n + k_m)\mu(z) l_1(\omega, k_n, z) l_1(\omega, k_m, z)\exp[i(k_m - k_n)a]\, dz.$$

Using the orthogonality relation (7.97), this is reduced to

$$\begin{aligned} u_y(0, h) &= l_1(\omega, k_n, h) \\ &= -2ik_n b_{n2}^-(h) \int_0^\infty \mu(z) l_1^2(\omega, k_n, z)\, dz \end{aligned}$$

or

$$b_{n2}^-(h) = \frac{l_1(\omega, k_n, h)}{-4ik_n I_2}, \tag{7.104}$$

where I_2 was defined in (7.66). Similarly, using $u_y = l_1 \exp(-ik_n x)$ instead of (7.95), we obtain

$$b_{n2}^+(h) = b_{n2}^-(h). \tag{7.105}$$

Putting (7.104) and (7.105) into (7.102), we obtain the Love-wave terms of Green's function due to a line force oriented in the y-direction,

$$G_{yy}(x, z; 0, h; \omega) \sim \sum_n \frac{l_1(\omega, k_n, h) l_1(\omega, k_n, z)}{4k_n I_2} \exp\left(ik_n x + i\frac{\pi}{2}\right) \qquad x \gg 0, \tag{7.106}$$

or using (7.70),

$$G_{yy}(x, z; 0, h; \omega) \sim \sum_n \frac{l_1(\omega, k_n, h) l_1(\omega, k_n, z)}{4k_n c U I_1} \exp\left(ik_n x + i\frac{\pi}{2}\right) \qquad x \gg 0,$$

where I_1 and I_2 are given in (7.66).

The above equation shows the remarkable simplicity of Love-wave terms of Green's function. The effect of source depth h, receiver depth z, travel distance x, and medium properties expressed by c, U, and I_1 are separated for each mode. Once the eigenvalue-eigenfunction problem is solved, the calculation of the Love-wave amplitude generated by a line force is extremely simple. We shall show later, that the same formula with a minor modification applies to Love waves from a point force.

The Rayleigh-wave terms of Green's function can be obtained similarly by substituting (7.103) and (7.98) into (7.101) and applying the orthogonal relation (7.99). For example, putting subscript $k = 1$ (the x-direction) in (7.101), we obtain

$$\begin{aligned} u_x(0, h) &= r_1(\omega, k_n, h) \\ &= -ia_{n1}^-(h) \int_0^\infty \Bigg\{ 2k_n[(\lambda + 2\mu) r_1^2(\omega, k_n, z) + \mu r_2^2(\omega, k_n, z)] \\ &\quad + 2\left[\lambda r_1(\omega, k_n, z) \frac{dr_2(\omega, k_n, z)}{dz} - \mu r_2(\omega, k_n, z) \frac{dr_1(\omega, k_n, z)}{dz}\right]\Bigg\} dz \end{aligned}$$

or

$$a_{n1}^{-}(h) = \frac{r_1(\omega, k_n, h)}{-i4k_n\left(I_2 + \dfrac{I_3}{2k_n}\right)} = \frac{r_1(\omega, k_n, h)}{-i4k_n cUI_1} \tag{7.107}$$

where I_1, I_2, I_3 are given in (7.74). Again, we have $a_{n1}^{+}(h) = a_{n1}^{-}(h)$. Substituting them into (7.103), we obtain Rayleigh-wave terms of Green's function due to a line force acting in the x-direction:

$$G_{xx}(x, z; 0, h; \omega) \sim \sum_n \frac{r_1(\omega, k_n, h)r_1(\omega, k_n, z)}{4k_n cUI_1} \exp\left(ik_n x + i\frac{\pi}{2}\right) \qquad x \gg 0,$$

$$G_{zx}(x, z; 0, h; \omega) \sim \sum_n \frac{r_1(\omega, k_n, h)r_2(\omega, k_n, z)}{4k_n cUI_1} \exp(ik_n x + i\pi) \qquad x \gg 0, \tag{7.108}$$

Again, this is an extremely simple form that separates the source, receiver, and path effects. Green's function for a line force oriented in the z-direction can be obtained likewise as

$$G_{xz}(x, z; 0, h; \omega) \sim \sum_n \frac{r_2(\omega, k_n, h)r_1(\omega, k_n, z)}{4k_n cUI_1} \exp(ik_n x) \qquad x \gg 0, \tag{7.109}$$

$$G_{zz}(x, z; 0, h; \omega) \sim \sum_n \frac{r_2(\omega, k_n, h)r_2(\omega, k_n z)}{4k_n cUI_1} \exp\left(ik_n x + i\frac{\pi}{2}\right) \qquad x \gg 0, \tag{7.110}$$

Note that the phase shifts appearing in G_{xx}, G_{yy}, and G_{zz} are all $\frac{1}{2}\pi$ phase delay.

BOX **7.8**

Sign convention on vertical motion

Since x_3, or the z-axis, is directed downward, the positive direction in the vertical displacement is here chosen downward. This is opposite to the customary practice. Seismologists have naturally considered the depth to be positive downward, but have then inconsistently chosen the vertical component of seismic motion to be positive upward. In this book, we have chosen that the z-axis be positive downward for both displacement and depth wherever we develop results with (x, y, z) or (r, ϕ, z) coordinates. Of course, in spherical polars, with $r = 0$ at the center of the Earth and r as a vertical coordinate, the vertically upward direction is naturally positive, as we shall assume in Chapter 8 and relevant parts of Chapter 9.

A cursory glance at the form of these solutions, with k_n in the denominator, suggests ω^{-1} dependence of the amplitude spectrum. However, I_1 (with normalized r_1 and r_2) depends on ω in such a way that $k_n I_1$ is only a weakly varying function of ω. See Problem 7.7.

7.4.2 *Three-dimensional case*

A natural frame for Green's function for a point source is the cylindrical system (r, ϕ, z) defined in Figure 7.8. Let us go back to Box 6.5 and start with the three scalar potentials ϕ, ψ, χ representing *P*-, *SV*-, and *SH*-waves, respectively. We showed that the elastic displacement in a homogeneous body can be expressed as

$$\mathbf{u} = \nabla\phi + \nabla \times \nabla \times (0, 0, \psi) + \nabla \times (0, 0, \chi), \tag{7.111}$$

the potentials here satisfying wave equations (3) of Box 6.5. It follows by the method of separation of variables that general solutions can be obtained by a superposition of the basic solutions

$$\begin{aligned} \phi(\mathbf{x}, \omega) &= J_m(kr)e^{im\phi}(Ae^{-\gamma z} + Be^{\gamma z}) \exp(-i\omega t), \\ \psi(\mathbf{x}, \omega) &= J_m(kr)e^{im\phi}(Ce^{-\nu z} + De^{\nu z}) \exp(-i\omega t), \\ \chi(\mathbf{x}, \omega) &= J_m(kr)e^{im\phi}(Ee^{-\nu z} + Fe^{\nu z}) \exp(-i\omega t), \end{aligned} \tag{7.112}$$

where $J_m(kr)$ is the mth-order Bessel function; m is an integer; A, B, C, D, E, F are constants; and $\gamma = (k^2 - \omega^2/\alpha^2)^{1/2}$, $\nu = (k^2 - \omega^2/\beta^2)^{1/2}$. (Compare (7.112) with the axisymmetric solutions in Box 6.6.) We shall find it useful to lump together the r, ϕ dependence as

$$Y_k^m(r, \phi) = J_m(kr)e^{im\phi}, \tag{7.113}$$

an expression that can naturally be called the basic *horizontal wavefunction*, since it alone characterizes the horizontal propagation for potentials.

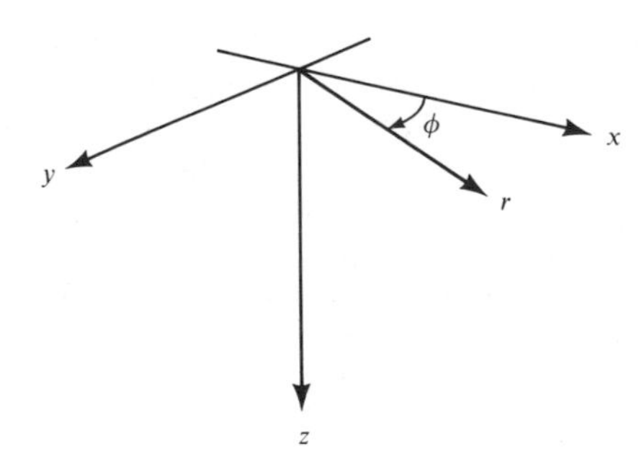

FIGURE **7.8**
Orientations of Cartesian and cylindrical polar coordinates used to analyze waves from a point source in a vertically heterogeneous medium.

An apparent barrier to our continuing with the method of potentials lies in the fact that our present interest is in vertically heterogeneous media, whereas the solutions (7.112) require homogeneity. Potentials can, however, help us to sort out the different horizontal wavefunctions that are appropriate for different physical variables, since this aspect of the problem is unchanged by vertical heterogeneity. Continuing, then, with a homogeneous medium, we can construct (u_r, u_ϕ, u_z) from potentials via (7.111). The P-wave component, grad ϕ, is easily obtained, and for SV and SH we have

$$\mathbf{u}^{SV} = \left(\frac{\partial^2\psi}{\partial r\,\partial z}, \frac{1}{r}\frac{\partial^2\psi}{\partial z\,\partial\phi}, -\frac{1}{r}\frac{\partial}{\partial r}\left(r\frac{\partial\psi}{\partial r}\right) - \frac{1}{r^2}\frac{\partial^2\psi}{\partial\phi^2}\right),$$
$$\mathbf{u}^{SH} = \left(\frac{1}{r}\frac{\partial\chi}{\partial\phi}, -\frac{\partial\chi}{\partial r}, 0\right). \tag{7.114}$$

In Section 2.6 we found the stress-displacement relations for general orthogonal curvilinear coordinates, and from (2.50) and (2.45) we can give the traction acting on horizontal planes in terms of displacement components:

$$\tau_{rz} = \mu\left(\frac{\partial u_z}{\partial r} + \frac{\partial u_r}{\partial z}\right),$$
$$\tau_{z\phi} = \mu\left(\frac{1}{r}\frac{\partial u_z}{\partial\phi} + \frac{\partial u_\phi}{\partial z}\right), \tag{7.115}$$
$$\tau_{zz} = \lambda\,\mathrm{div}\,\mathbf{u} + 2\mu\frac{\partial u_z}{\partial z}.$$

If we substitute potentials of the form (7.112) into (7.115), we find that the displacement $\mathbf{u}$ and the traction $\mathbf{T}$ acting on the horizontal plane at depth z have the following form

$$\mathbf{u} = [l_1(\omega, k, z)\mathbf{T}_k^m(r, \phi) + r_1(\omega, k, z)\mathbf{S}_k^m(r, \phi) + r_2(\omega, k, z)\mathbf{R}_k^m(r, \phi)]\exp(-i\omega t),$$
$$\mathbf{T} = [l_2(\omega, k, z)\mathbf{T}_k^m(r, \phi) + r_3(\omega, k, z)\mathbf{S}_k^m(r, \phi) + r_4(\omega, k, z)\mathbf{R}_k^m(r, \phi)]\exp(-i\omega t). \tag{7.116}$$

The r, ϕ dependence of $\mathbf{u}$ and $\mathbf{T}$ is described by three orthogonal vector functions defined by

$$\mathbf{T}_k^m(r, \phi) = \frac{1}{kr}\frac{\partial Y_k^m}{\partial\phi}\hat{\mathbf{r}} - \frac{1}{k}\frac{\partial Y_k^m}{\partial r}\hat{\boldsymbol{\phi}},$$
$$\mathbf{S}_k^m(r, \phi) = \frac{1}{k}\frac{\partial Y_k^m}{\partial r}\hat{\mathbf{r}} + \frac{1}{kr}\frac{\partial Y_k^m}{\partial\phi}\hat{\boldsymbol{\phi}}, \tag{7.117}$$
$$\mathbf{R}_k^m(r, \phi) = -Y_k^m(r, \phi)\hat{\mathbf{z}},$$

where $\hat{\mathbf{r}}$, $\hat{\boldsymbol{\phi}}$, and $\hat{\mathbf{z}}$ are unit vectors in directions r, ϕ, z. The z-dependence is described by six scalar functions l_1, l_2, r_1, r_2, r_3, and r_4. It can be verified that these functions (derived here from z-dependent terms in the potentials (7.112)) do satisfy the familiar equation $\partial\mathbf{f}/\partial z = \mathbf{Af}$ discussed in Section 7.2, first with $\mathbf{f} = (l_1, l_2)^T$ to give (7.24), and then $\mathbf{f} = (r_1, r_2, r_3, r_4)^T$ to give (7.28).

This result for homogeneous media motivates us to *start* with the form (7.116), even in vertically *heterogeneous* media, substituting it into the equations of motion and the stress-strain relation. By this route we arrive *directly* at the first-order simultaneous differential equations (7.24) and (7.28), finding that the z-dependence for the motion-stress vector in cylindrical waves is exactly the same as the z-dependence in plane-wave problems.

We have just given a precise reason why cylindrical waves and plane waves share a common z-dependence. The result becomes intuitively reasonable if we make r large in the vector functions (7.117). First, we replace $J_m(kr)$ by $[H_m^{(1)}(kr) + H_m^{(2)}(kr)]/2$, as we did in (6.15), and use the asymptotic expansion for outgoing waves

$$H_m^{(1)}(kr) \sim \sqrt{\frac{2}{\pi kr}} \exp\left[i\left(kr - \frac{2m+1}{4}\pi\right)\right].$$

We also neglect terms attenuating with distance more rapidly than $1/\sqrt{r}$. Then

$$\begin{aligned}
\text{outgoing } \mathbf{T}_k^m(r, \phi) &\sim -i\sqrt{\frac{1}{2\pi kr}} \exp\left(ikr - i\frac{2m+1}{4}\pi + im\phi\right)\hat{\boldsymbol{\phi}}, \\
\text{outgoing } \mathbf{S}_k^m(r, \phi) &\sim i\sqrt{\frac{1}{2\pi kr}} \exp\left(ikr - i\frac{2m+1}{4}\pi + im\phi\right)\hat{\mathbf{r}}, \qquad (7.118) \\
\text{outgoing } \mathbf{R}_k^m(r, \phi) &\sim -\sqrt{\frac{1}{2\pi kr}} \exp\left(ikr - i\frac{2m+1}{4}\pi + im\phi\right)\hat{\mathbf{z}},
\end{aligned}$$

showing that at a large distance, these waves do behave locally like plane Love and Rayleigh waves with appropriate polarization. Therefore, their z-dependence can be obtained by studying plane waves, and we can use all the results obtained on the motion-stress vectors for plane waves in the cylindrical wave problem.

Let us now consider surface waves generated by a point force $\mathbf{F}\exp(-i\omega t)$ in a vertically heterogeneous medium. Taking the point of application as $r = 0$, $z = h$, the corresponding body force (force per unit volume) is $\mathbf{F}\exp(-i\omega t)[\delta(r)/(2\pi r)]\,\delta(z - h)$. In the following we shall make frequent use of Cartesian coordinates (x, y, z), related to (r, ϕ, z) via $x = r\cos\phi$, $y = r\sin\phi$, and note then that $\delta(r)/(2\pi r) = \delta(x)\,\delta(y)$ (since each side of this equality has the same effect when integrated over horizontal planes). By equation (3.4), the applied

BOX **7.9**

On horizontal wave functions

The horizontal wavefunction for plane waves described in Chapters 6 and 7 is e^{ikx}. It is a unique feature of plane waves that this same phase factor is common to all the physical field variables of interest (such as displacement, dilatation, strain, and stress). For non-Cartesian coordinates, there is first the problem that separable solutions may not even exist. However, separable solutions *do* exist for cylindrical and spherical polars, which are the most important cases to examine for wave propagation in geophysics.

The problem next to be faced is the fact that horizontal wavefunctions in (for example) cylindrical coordinates may differ from one physical variable to another. We saw a simple example of this in Box 6.6, where for u_z and τ_{zz} it is J_1. To handle the general axisymmetric case, there is a dependence on azimuthal order number m as well as on k, and in (7.116) and (7.117) we found it convenient to work with three different horizontal wavefunctions, each of which is a vector.

With the plane-wave horizontal wavefunction, it is possible to synthesize functions of the horizontal spatial variable x by the Fourier inverse transform,

$$f(x) = \frac{1}{2\pi}\int_{-\infty}^{\infty} f(k)e^{+ikx}\,dk. \tag{1}$$

In Box 6.10 we developed similar results (the inverse Hankel transform) for the horizontal phase function J_m. It is then possible to combine these with the Fourier series

$$\begin{aligned} f(\phi) &= \frac{1}{2\pi}\sum_{m=-\infty}^{\infty} f(m)e^{im\phi} \\ f(m) &= \int_0^{2\pi} f(\phi)e^{-im\phi}\,d\phi \end{aligned} \tag{2}$$

to obtain an expansion for vector functions of (r, ϕ). This expansion is

$$\mathbf{f}(r,\phi) = \frac{1}{2\pi}\sum_{m=-\infty}^{\infty}\int_0^{\infty}[f_T(k,m)\mathbf{T}_k^m(r,\phi) + f_S(k,m)\mathbf{S}_k^m(r,\phi) + f_R(k,m)\mathbf{R}_k^m(r,\phi)]k\,dk, \tag{3}$$

in which $\mathbf{T}_k^m$, $\mathbf{S}_k^m$, $\mathbf{R}_k^m$ are the horizontal wavefunctions given in (7.117) and the coefficients in (3) are given by

$$\begin{aligned} f_T(k,m) &= \int_0^{2\pi}\int_0^{\infty} \mathbf{f}(r,\phi)\cdot[\mathbf{T}_k^m(r,\phi)]^* r\,dr\,d\phi, \\ f_S(k,m) &= \int_0^{2\pi}\int_0^{\infty} \mathbf{f}(r,\phi)\cdot[\mathbf{S}_k^m(r,\phi)]^* r\,dr\,d\phi, \\ f_R(k,m) &= \int_0^{2\pi}\int_0^{\infty} \mathbf{f}(r,\phi)\cdot[\mathbf{R}_k^m(r,\phi)]^* r\,dr\,d\phi. \end{aligned} \tag{4}$$

For the Cartesian case, the horizontal wavefunctions have an orthogonality expressed as $\delta(k - k') = 1/(2\pi)\int_{-\infty}^{\infty} \exp[i(k - k')x]\,dx$. An equivalent result for our vector functions is stated in (7.121).

body force is equivalent to a discontinuity in traction on horizontal planes at $z = h$, given by

$$\mathbf{T}(h + 0) - \mathbf{T}(h - 0) = -\mathbf{F}\exp(-i\omega t)\,\delta(x)\,\delta(y). \tag{7.119}$$

Our method of solution will be (i) to decompose this discontinuity into its (k, m) components; (ii) to solve the equation $\partial\mathbf{f}/\partial z = \mathbf{A}\mathbf{f}$ for each (k, m), where $\mathbf{f}$ is the z-dependent motion-stress vector with known discontinuity across $z = h$; and then (iii) to construct the solution as a function of (r, ϕ, z) by superposition of its (k, m) components.

We seek to find the coefficients in

$$-\mathbf{F}\exp(-i\omega t)\,\delta(x)\,\delta(y) = \frac{\exp(-i\omega t)}{2\pi}\sum_m \int_0^\infty k[f_T(k, m)\mathbf{T}_k^m + f_S(k, m)\mathbf{S}_k^m + f_R(k, m)\mathbf{R}_k^m]\,dk. \tag{7.120}$$

Fortunately, $\mathbf{T}_k^m$, $\mathbf{S}_k^m$, and $\mathbf{R}_k^m$ are orthogonal to each other, and each of these three types of horizontal wavefunction satisfies an orthogonality relation (for different k and m) in the form

$$\int_0^{2\pi}\int_0^\infty \mathbf{T}_k^m(r, \phi)\cdot[\mathbf{T}_{k'}^{m'}(r, \phi)]^* r\,dr\,d\phi = \frac{2\pi\delta_{mm'}\,\delta(k - k')}{\sqrt{kk'}}, \tag{7.121}$$

where * indicates the complex conjugate. It follows that the expansion coefficients in (7.120) are given by

$$f_T(k, m) = -\int_0^{2\pi}\int_0^\infty [\mathbf{T}_k^m(r, \phi)]^*\cdot\mathbf{F}\,\delta(x)\,\delta(y) r\,dr\,d\phi, \tag{7.122}$$

with similar results for f_S and f_R.

Evaluation of the double integral in (7.122) is simplified by using Cartesians, recognizing $\int_0^{2\pi}\int_0^\infty (\quad) r\,dr\,d\phi = \int_{-\infty}^\infty\int_{-\infty}^\infty (\quad)\,dx\,dy$ and the important property

$$\mathbf{T}_k^m(r, \phi) = k^{-1}\nabla\times(0, 0, Y_k^m). \tag{7.123}$$

Interpreting (7.123) in Cartesians, we find from (7.122) that

$$\begin{aligned} f_T(k, m) &= -\int_{-\infty}^\infty\int_{-\infty}^\infty \frac{1}{k}\left(F_x\frac{\partial Y_k^{-m}}{\partial y} - F_y\frac{\partial Y_k^{-m}}{\partial x}\right)\delta(x)\,\delta(y)\,dx\,dy \\ &= -F_x e^{-im\pi/2}\frac{\partial}{\partial(ky)}J_m(ky)\bigg|_{y=0} + F_y\frac{\partial}{\partial(kx)}J_m(kx)\bigg|_{x=0}. \end{aligned} \tag{7.124}$$

Since J_m has a zero derivative at the origin except for $m = \pm 1$, $f_T(k, m) = 0$ unless $m = \pm 1$. Then $(\partial/\partial\zeta)J_{\pm 1}(\zeta) = \pm\frac{1}{2}$ at $\zeta = 0$, so that

$$f_T(k, 1) = \tfrac{1}{2}(F_y + iF_x),$$
$$f_T(k, -1) = \tfrac{1}{2}(-F_y + iF_x). \tag{7.125}$$

A similar method works for $f_S(k, m)$, using

$$\mathbf{S}_k^m(r, \phi) = k^{-1}\nabla Y_k^m(r, \phi) \tag{7.126}$$

and working in Cartesians to find $f_S(k, m) = 0$ unless $m = \pm 1$, and then

$$f_S(k, 1) = \tfrac{1}{2}(-F_x + iF_y),$$
$$f_S(k, -1) = \tfrac{1}{2}(F_x + iF_y). \tag{7.127}$$

For $f_R(k, m)$ we work directly in cylindrical coordinates:

$$f_R(k, m) = \int_0^{2\pi}\int_0^{\infty} F_z \frac{\delta(r)}{2\pi r} e^{-im\phi} J_m(kr) r\, dr\, d\phi,$$

so that $f_R(k, m) = 0$ unless $m = 0$, and then

$$f_R(k, 0) = J_0(0)F_z = F_z. \tag{7.128}$$

All the expansion coefficients for our traction discontinuity have now been found, the only contributing terms being those with $m = 0$, $m = \pm 1$.

Our next step is to find the motion-stress vector $(l_1, l_2, r_1, r_2, r_3, r_4)$ that has prescribed discontinuities f_T, f_S, f_R in traction components l_2, r_3, r_4 (respectively). Of course, our solution must satisfy the equations of motion and the free surface condition

$$l_2 = r_3 = r_4 = 0 \qquad \text{on } z = 0. \tag{7.129}$$

It must also satisfy the radiation condition, requiring that l_1, r_1, and r_2 must contain only downward-travelling waves as $z \to \infty$, or, if the wavenumber is large enough so that $\gamma = (k^2 - \omega^2/\alpha^2)^{1/2}$ and $\nu = (k^2 - \omega^2/\beta^2)^{1/2}$ become real (positive), then all of

$$l_1, r_1, r_2 \to 0 \qquad \text{as } z \to \infty. \tag{7.130}$$

When the surface-wave eigenvalue problem was introduced in Section 7.2, we found that a continuous solution of the equations of motion did not satisfy the homogeneous boundary conditions (7.129) and (7.130) unless, for given ω, k

took on special discrete values. It is therefore interesting to find that when a traction discontinuity at the source depth is added to the problem, it becomes possible to find a solution for *every* value of k. We give this solution below, noting here that once the motion-stress vector problem is solved as a function of (k, m, z, ω) it becomes possible to write the steady-state displacement due to point force $\mathbf{F}\exp(-i\omega t)$ in the form

$$\mathbf{u}(r, \phi, z, t) = \exp(-i\omega t)\frac{1}{2\pi}\sum_{m=-\infty}^{\infty}\int_0^\infty k[l_1(k, m, z, \omega)\mathbf{T}_k^m(r, \phi) + r_1(k, m, z, \omega)\mathbf{S}_k^m(r, \phi) + r_2(k, m, z, \omega)\mathbf{R}_k^m(r, \phi)]\,dk. \quad (7.131)$$

Let us first find the solution $l_1(k, m, z, \omega)$, recapitulating that

$$\mathbf{l} = \begin{pmatrix} l_1 \\ l_2 \end{pmatrix}$$

satisfies

$$\frac{\partial \mathbf{l}}{\partial z} = \begin{pmatrix} 0 & \mu^{-1} \\ -\omega^2\rho + k^2\mu & 0 \end{pmatrix}\mathbf{l}, \quad \text{(7.24 again)}$$

with $l_2 = 0$ on $z = 0$. As $z \to \infty$,

$$\begin{aligned} &\textit{either } \mathbf{l} \text{ becomes a downgoing wave (a homogeneous body wave)} \\ &\textit{or } \mathbf{l} \to \mathbf{0} \text{ (an inhomogeneous wave),} \end{aligned} \quad (7.132)$$

depending on the value of k/ω (the horizontal slowness). Finally

$$\mathbf{l}\bigg|_{z=h^+} - \mathbf{l}\bigg|_{z=h^-} = \begin{pmatrix} 0 \\ f_T(k, m) \end{pmatrix}. \quad (7.133)$$

We shall construct a discontinuous solution $\mathbf{l}'$ that satisfies (7.133) and then construct a continuous solution $\mathbf{l}''$ such that a linear combination

$$\mathbf{l} = \mathbf{l}' + \frac{\mathbf{l}''}{\Delta(k)} \quad (7.134)$$

satisfies *all* the required conditions. Both $\mathbf{l}'$ and $\mathbf{l}''$ solve the equation of motion, and boundary conditions are

$$\begin{aligned} l_1' &= 0 \quad \text{and} \quad l_2' = 0 \qquad \text{for all } z > h, \\ l_1' &= 0 \quad \text{and} \quad l_2' = -f_T \quad \text{at } z = h - 0, \end{aligned} \quad (7.135)$$

$$\begin{aligned} &l_1'' \to \text{downgoing} \quad \text{or} \quad l_1'' \to 0 \qquad \text{as } z \to \infty, \\ &l_2'' = -\Delta(k) l_2' \qquad \text{at } z = 0. \end{aligned} \tag{7.136}$$

It can readily be verified that $\mathbf{l}$ given by (7.134), with $\mathbf{l}'$ and $\mathbf{l}''$ as defined above, does satisfy all the required conditions (7.132) and (7.133). The radiation condition for l_1'' in (7.136), as $z \to \infty$, takes two forms, depending on the value of the horizontal slowness k/ω. If this value is small enough, then l_1'' at great depth is a homogeneous downward-traveling body wave. But if $(k/\omega) > \lim_{z\to\infty} (1/\beta(z))$, then l_1'' is an inhomogeneous wave tending to zero as $z \to \infty$. The function $\Delta(k)$ is defined so that the surface shear stress in the discontinuous solution $\mathbf{l}'$, when multiplied by $-\Delta(k)$, exactly equals this stress in the continuous solution. But if k happens to be a surface-wave eigenvalue, the surface stress vanishes in the continuous solution $\mathbf{l}''$. It follows that $\Delta(k)$ must be zero when k is an eigenvalue and $\mathbf{l}''$ is a surface-wave eigenfunction. No downgoing wave (as $z \to \infty$) is allowed in this case, since $\mathbf{l}''$ satisfies a wave equation with homogeneous boundary conditions and no source term. Eigenvalues of k must therefore be large enough so that $l_1'' \to 0$ as $z \to \infty$, to prevent the downward loss of energy that would be carried by a body wave.

In Figure 7.9 we give more detail on obtaining $\mathbf{l}'$ and $\mathbf{l}''$. Constructing $\mathbf{l}$ via (7.134), we can now use (7.131) to synthesize displacement as

$$\mathbf{u}(r, \phi, z, t) = \exp(-i\omega t) \frac{1}{2\pi} \sum_m \int_0^\infty k \left[l_1' + \frac{l_1''}{\Delta(k)} \right] \mathbf{T}_k^m(r, \phi)\, dk. \tag{7.137}$$

Our next goal is to identify the surface-wave contributions in this displacement field. Recall from (7.124) that contributions come only from $m = \pm 1$ if the source is a point force.

The integrand in (7.137) has poles at $\Delta = 0$. Since we have chosen for Δ a function that vanishes when k is an eigenvalue, the contribution from these poles will give the normal modes, or Love waves in this case. However, (7.137) also contains all the body-wave and leaky-mode contributions that may be present, and our evaluation of the integral will follow some of the steps taken in the discussion of (6.11)–(6.15) and (6.30). Thus we replace $J_m(kr)$ by $[H_m^{(1)}(kr) + H_m^{(2)}(kr)]/2$, converting the integral of $H_m^{(2)}$ over positive k to an integral of $H_m^{(1)}$ over negative k, so that (7.137) becomes an integral over the whole real k-axis. We write $\mathbf{T}_k^m$ with J_m replaced by $\frac{1}{2}H_m^{(1)}$ as $\mathbf{T}_k^{m(1)}$, and then the pole contributions will be

$$\mathbf{u}^{\text{LOVE}} = \exp(-i\omega t) \sum_m \sum_n ik_n \frac{l_1''(k_n, m, z, \omega)}{\left(\dfrac{\partial \Delta}{\partial k}\right)_{k=k_n}} \mathbf{T}_{k_n}^{m(1)}(r, \phi), \tag{7.138}$$

where k_n ($n = 0, 1, 2, \ldots$) are the positive real roots of $\Delta(k) = 0$. (Negative real roots do not contribute, by an argument based on the effect of slight

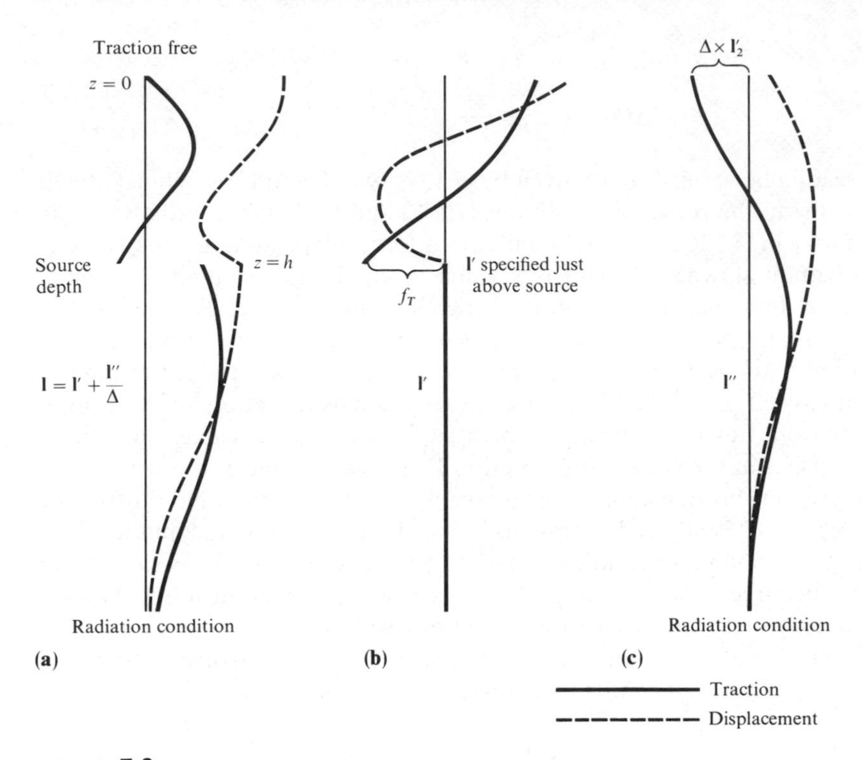

FIGURE **7.9**

(**a**) Diagramatic representation of properties of the *SH*-wave with a traction discontinuity at $z = h$. This wave is $\mathbf{l}$, satisfying homogeneous conditions at $z = 0$ and $z = \infty$.
(**b**) The solution $\mathbf{l}'$ is zero below $z = h$. For given k, $\mathbf{l}'$ can be obtained above the source by taking initial values

$$\begin{pmatrix} 0 \\ -f_T \end{pmatrix}$$

for $\mathbf{l}'$ on $z = h^-$ and integrating $d\mathbf{l}'/dz = \mathbf{A}\mathbf{l}'$ upward to $z = 0$. For surface-wave excitation, we shall find that $\mathbf{l}'$ does not actually have to be evaluated. (**c**) The solution $\mathbf{l}''$ is continuous. It can be found (for given k) by taking a solution that satisfies the radiation condition at great depth and integrating the equation of motion upward to $z = 0$. This solution is multiplied (at all depths) by a scalar constant, chosen so that the surface traction becomes $-\Delta \times$ surface traction in $\mathbf{l}'$.

Note that $\mathbf{l} = \mathbf{l}' + \Delta^{-1}\mathbf{l}''$ is defined for all choices of k. For small enough wavenumbers, downward radiation of energy is possible (body waves). With large enough wavenumbers, however, there is exponential decay of $\mathbf{l}$ (and $\mathbf{l}''$) with depth. It is this situation that is shown in the figure.

anelasticity. Poles on the negative real axis move down into the third quadrant, and are outside a closed contour formed in the upper half plane, whereas poles on the positive real k-axis move up into the first quadrant, and their residues *are* picked up. We shall discuss these integration paths further, and associated branch cuts and body waves, in Sections 7.6 and 9.3.)

The evaluation of $(\partial\Delta/\partial k)_{k=k_n}$ can be done very simply by using the variational principles of Section 7.3. From (7.68), we find for a continuous solution $\mathbf{l}''$ of the equations of motion that

$$\frac{1}{2}[l_1'' \, l_2'']_0^\infty = -\omega^2 I_1 + k^2 I_2 + I_3.$$

Then, from boundary conditions (7.136),

$$\frac{1}{2}\Delta(k)(l_2' l_1'')_{z=0} = -\omega^2 I_1 + k^2 I_2 + I_3.$$

If k is nearly an eigenvalue, then $\mathbf{l}''$ is nearly an eigenfunction. Because of the stationarity of $(-\omega^2 I_1 + k^2 I_2 + I_3)$ for slight departures of $\mathbf{l}''$ from a true eigenfunction (see (7.67)), it follows that the change in Δ due to a perturbation in k can be written as

$$\left(\frac{\partial\Delta}{\partial k}\right)_{k=k_n} (l_2' l_1'')_{z=0} = 4k_n I_2 = 4k_n c U I_1, \tag{7.139}$$

where c and U are the phase and group velocity corresponding to ω and k_n (see (7.70)). We can even eliminate l_2' at $z = 0$ from the above equation, because it is easy to show from the equations of motion that $(d/dz)(l_1' l_2'' - l_1'' l_2') = 0$. Evaluating the constant $l_1' l_2'' - l_1'' l_2'$ at $z = 0$ and $z = h - 0$, and using (7.135) and (7.136), we find

$$l_1'' l_2' \Big|_{z=0} = -f_T(k_n, m) l_1'' \Big|_{z=h}. \tag{7.140}$$

Assembling (7.138)–(7.140), we now find

$$\mathbf{u}^{\text{LOVE}} = -i \exp(-i\omega t) \sum_m \sum_n \frac{f_T(k_n, m) l_1''(h) l_1''(z)}{4cUI_1} \mathbf{T}_{k_n}^{m(1)}. \tag{7.141}$$

The sum over m is easy to carry out, because $f_T = 0$ unless $m = \pm 1$ (see (7.125)). Since l_1'' in (7.141) is merely a particular Love-wave mode, we shall drop the primes and use l_1; i.e., we revert to our original notation for individual modes. We also replace $\mathbf{T}_k^{m(1)}$ by the asymptotic form given in (7.118). Our final result, giving the excitation of Love modes by a point force $\mathbf{F}\exp(-i\omega t)$ applied at $r = 0$, $z = h$, is then

$$\mathbf{u}^{\text{LOVE}} = \exp(-i\omega t)\sum_n \frac{(F_y\cos\phi - F_x\sin\phi)l_1(k_n, h, \omega)}{8cUI_1} \times \sqrt{\frac{2}{\pi k_n r}}\,[l_1(k_n, z, \omega)\hat{\boldsymbol{\phi}}]\exp\left[i\left(k_n r + \frac{\pi}{4}\right)\right], \tag{7.142}$$

where $I_1 = \frac{1}{2}\int_0^\infty \rho[l_1(k_n, z, \omega)]^2\,dz$ and $l_1(k_n, z, \omega)$ is a continuous eigenfunction.

For Rayleigh waves, following similar steps, the final result is

$$\mathbf{u}^{\text{RAYLEIGH}} = \exp(-i\omega t)\sum_n \frac{F_z r_2(k_n, h, \omega) + i(F_x\cos\phi + F_y\sin\phi)r_1(k_n, h, \omega)}{8cUI_1} \times \sqrt{\frac{2}{\pi k_n r}}\,[r_1(k_n, z, \omega)e^{-i\pi/4}\hat{\mathbf{r}} + r_2(k_n, z, \omega)e^{i\pi/4}\hat{\mathbf{z}}]\exp(ik_n r), \tag{7.143}$$

where $I_1 = \frac{1}{2}\int_0^\infty \rho(r_1^2 + r_2^2)\,dz$.

The compact result, i.e., the pole residue in terms of $1/(cUI_1)$ in equations (7.142) and (7.143), was given for Love waves by Keilis-Borok and Yanovskaya (1962). The corresponding result for Rayleigh waves was guessed at by Harkrider and Anderson (1966) and verified by them numerically. Saito (1967) was the first to obtain the same compact form for the excitation of free oscillations of the whole Earth by a point source (a subject we take up in the next chapter).

Comparison with the line-source solutions given in (7.106), (7.108), and (7.109) shows a remarkable similarity between the two- and three-dimensional solutions. The solution for a point source is advanced in phase by $\pi/4$ and contains more high frequencies by a factor proportional to $\omega^{1/2}$, as compared to the line source. This is easily understood because the latter may be considered as a superposition or spatial smoothing of the former. The basic simplicity of the solutions, i.e., the separation of source, medium, and receiver factors, applies to both cases (see also Box 6.1).

To coordinate these results for steady-state displacements with our notation for the Green function, note that we can drop an $\exp(-i\omega t)$ factor in (7.142) and (7.143), and then regard the left-hand side as displacement in the frequency domain due to a point force $\mathbf{F}(\omega)$ acting at $\boldsymbol{\xi} = (0, 0, h)$. It follows that

$$u_i(\mathbf{x}, \omega) = F_p(\omega)G_{ip}(\mathbf{x}; \boldsymbol{\xi}; \omega). \tag{7.144}$$

By $\mathbf{G}(\mathbf{x}; \boldsymbol{\xi}; \omega)$ we mean the Fourier transform of the t-dependent $\mathbf{G}(\mathbf{x}, t; \boldsymbol{\xi}, \tau)$, with τ fixed at zero. Convolutions equivalent to (7.144) appeared frequently in Chapters 3 and 4. Comparisons with (7.142) and (7.143) now show that surface-wave excitation is described by Love and Rayleigh wave terms in the Green function. Specifically, these are

$$\mathbf{G}^{\text{LOVE}} = \sum_n \frac{l_1(z)l_1(h)}{8cUI_1} \begin{pmatrix} \sin^2\phi & -\sin\phi\cos\phi & 0 \\ -\sin\phi\cos\phi & \cos^2\phi & 0 \\ 0 & 0 & 0 \end{pmatrix} \left(\frac{2}{\pi k_n r}\right)^{1/2} \exp\left[i\left(k_n r + \frac{\pi}{4}\right)\right] \tag{7.145}$$

and

$$\mathbf{G}^{\text{RAYLEIGH}} = \sum_n \frac{1}{8cUI_1} \begin{pmatrix} r_1(z)r_1(h)\cos^2\phi & r_1(z)r_1(h)\cos\phi\sin\phi & -ir_1(z)r_2(h)\cos\phi \\ r_1(z)r_1(h)\sin\phi\cos\phi & r_1(z)r_1(h)\sin^2\phi & -ir_1(z)r_2(h)\sin\phi \\ ir_2(z)r_1(h)\cos\phi & ir_2(z)r_1(h)\sin\phi & r_2(z)r_2(h) \end{pmatrix}$$

$$\times \left(\frac{2}{\pi k_n r}\right)^{1/2} \exp\left[i\left(k_n r + \frac{\pi}{4}\right)\right]. \tag{7.146}$$

7.5 Love and Rayleigh Waves from a Point Source with Arbitrary Seismic Moment

If surface waves are excited by a point source described only by its moment tensor $\mathbf{M}$, then one way to study the excitation is in terms of the equivalent body force, just as we did in the previous section. This is Saito's (1967) approach, and the (k, m) expansion of the traction discontinuity (7.120) is now found to have nonzero terms from $m = \pm 2$. However, we have already obtained the Love and Rayleigh components of Green's function, so that a quicker route is simply to use the formula

$$u_i(\mathbf{x}, \omega) = M_{pq}(\omega) \frac{\partial}{\partial \xi_q} G_{ip}(\mathbf{x}; \boldsymbol{\xi}; \omega), \tag{3.22 again}$$

In differentiating $\mathbf{G}^{\text{LOVE}}$ and $\mathbf{G}^{\text{RAYLEIGH}}$, we shall retain only the largest terms, namely, those that involve depth derivatives of the vertical eigenfunctions or horizontal derivatives of $\exp(ik_n r)$. Using $\partial r/\partial \xi_1 = -\cos\phi$, $\partial r/\partial \xi_2 = -\sin\phi$, we find, for example, that

$$u_x^{\text{LOVE}}(\mathbf{x}, \omega) = -\sum_n \sin\phi \frac{l_1(z)}{8cUI_1} \left(\frac{2}{\pi k_n r}\right)^{1/2} \exp\left[i\left(k_n r + \frac{\pi}{4}\right)\right]$$

$$\times \left\{ ik_n l_1(h)[M_{xx}\sin\phi\cos\phi - M_{yx}\cos^2\phi + M_{xy}\sin^2\phi\right.$$

$$\left. - M_{yy}\sin\phi\cos\phi] - \left.\frac{dl_1}{dz}\right|_h [M_{xz}\sin\phi - M_{yz}\cos\phi] \right\}. \tag{7.147}$$

The derivation of (7.147) is similar to our derivation of (4.29) via (4.28), although we are now keeping only the far-field terms. A similar expression can be found for u_y^{LOVE}, and it is soon found that $u_z^{\text{LOVE}} = 0$, so that a vector formula can be given as

$$\mathbf{u}^{\text{LOVE}}(\mathbf{x}, \omega) = \sum_n \frac{l_1(z)}{8cUI_1}\left(\frac{2}{\pi k_n r}\right)^{1/2} \exp\left[i\left(k_n r + \frac{\pi}{4}\right)\right]\{\quad\}\hat{\boldsymbol{\phi}}, \tag{7.148}$$

where $\{\quad\}$ is the same as the bracket $\{\quad\}$ of (7.147).

For Rayleigh waves, we find from the last row of components in (7.146) that

$$\begin{aligned} u_z^{\text{RAYLEIGH}} = \sum_n \frac{r_2(z)}{8cUI_1}\left(\frac{2}{\pi k_n r}\right)^{1/2} \exp\left[i\left(k_n r + \frac{\pi}{4}\right)\right] \times \Big\{ k_n r_1(h)[M_{xx}\cos^2\phi \\ + (M_{xy} + M_{yx})\sin\phi\cos\phi + M_{yy}\sin^2\phi] + i\left.\frac{dr_1}{dz}\right|_h [M_{xz}\cos\phi \\ + M_{yz}\sin\phi] - ik_n r_2(h)[M_{zx}\cos\phi + M_{zy}\sin\phi] + \left.\frac{dr_2}{dz}\right|_h M_{zz}\Big\}. \end{aligned} \tag{7.149}$$

For the radial (i.e., horizontal) component of Rayleigh-wave motion, one finds

$$u_r^{\text{RAYLEIGH}} = \sum_n \frac{r_1(z)}{8cUI_1}\left(\frac{2}{\pi k_n r}\right)^{1/2} \exp\left[i\left(k_n r - \frac{\pi}{4}\right)\right]\{\quad\}, \tag{7.150}$$

where $\{\quad\}$ is the same as the bracket $\{\quad\}$ of (7.149).

Note that these formulas for surface-wave excitation in terms of the moment tensor are appropriate not only for small dislocation sources, but also for small volume sources. To see this, consider a volume source with moment density $m_{pq} = \partial M_{pq}/\partial V$. This is the moment density per unit volume described in Section 3.4, and we start with an equation equivalent to (3.30) for the x_i-component of displacement:

$$u_i(\mathbf{x}, \omega) = \int_V m_{pq}(\boldsymbol{\eta}, \omega)\frac{\partial}{\partial \eta_q} G_{ip}(\mathbf{x}; \boldsymbol{\eta}; \omega)\, dV(\boldsymbol{\eta}).$$

If the moment density is concentrated at a point $\boldsymbol{\xi}$, then

$$m_{pq}(\boldsymbol{\eta}, \omega) = M_{pq}(\omega)\,\delta(\boldsymbol{\eta} - \boldsymbol{\xi})$$

as $\boldsymbol{\eta}$ varies within V, and the above representation does indeed give $u_i = M_{pq}G_{ip,q}$ again.

As an example, let us first consider the case of an explosive source. Regarded as a point source, the moment tensor is a diagonal matrix with equal elements $M_{xx} = M_{yy} = M_{zz} = M_0$. In this case it is apparent that Love waves are not excited. Rayleigh waves are azimuthally isotropic, and the vertical displacement is given by (7.149) as

$$u_z(\mathbf{x}, \omega) = \sum_n \frac{r_2(z)}{8cUI_1}\left(\frac{2}{\pi k_n r}\right)^{1/2}\left(\frac{-1}{i\omega}\right)\exp\left[i\left(k_n r + \frac{\pi}{4}\right)\right]\left\{k_n r_1(h) + \left.\frac{dr_2}{dz}\right|_h\right\}, \tag{7.151}$$

where we have assumed that $M_0(t)$ is a step with amplitude 1 dyne-cm, so that $M_0(\omega) = -1/i\omega$.

Using the phase velocity, group velocity, and eigenfunctions obtained earlier for Gutenberg's continental Earth model (see Box 7.5), it is a simple matter to compute the amplitude of fundamental Rayleigh waves from an explosive source buried at various depths. Only the first term (i.e., $n = 0$) is taken in (7.151), which has units of cm-sec. Numerical results at an epicentral distance of 2000 km are shown in Figure 7.10, and it is clear that Rayleigh-wave excitation from an explosive source decreases smoothly with depth.

On the other hand, an earthquake source with the moment tensor of double-couple symmetry (3.19) generates a Rayleigh-wave spectrum with complicated

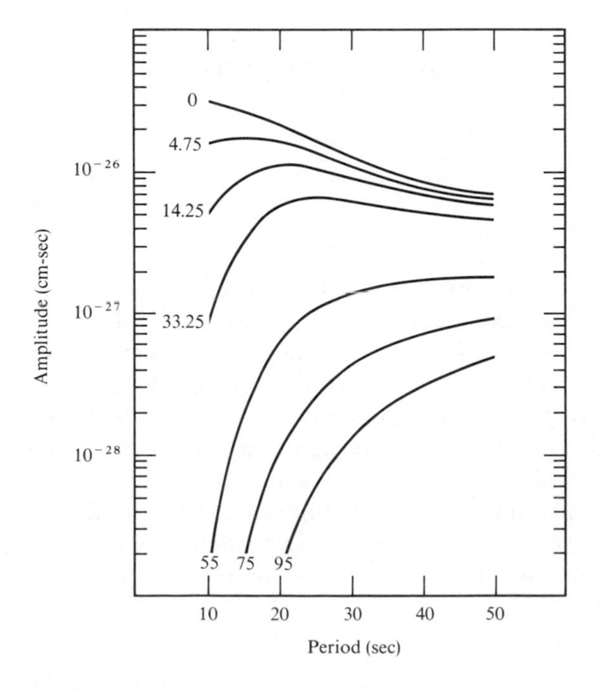

FIGURE **7.10**

Amplitude spectra of Rayleigh waves (vertical displacement) at $\Delta = 2000$ km for an underground explosion with unit seismic moment (diagonal element). The number attached to each curve represents the source depth in kilometers. [Reproduced from Tsai and Aki (1971); copyrighted by the American Geophysical Union.]

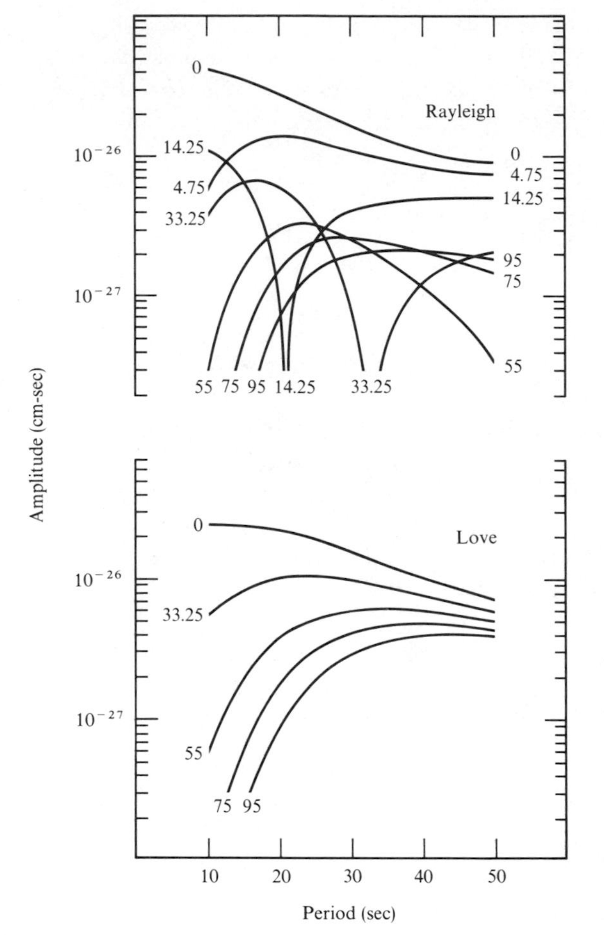

FIGURE **7.11**

Amplitude spectra of Rayleigh (vertical) and Love waves at $\Delta = 2000$ km and $\phi = 30^\circ$ from the fault strike for a vertical strike-slip earthquake with unit step-function seismic moment. The number attached to each curve represents the source depth in kilometers. [Reproduced from Tsai and Aki (1971); copyrighted by the American Geophysical Union.]

dependences on source mechanism and focal depth. The case of pure strike-slip motion on a vertical fault and the case of a pure dip-slip motion along a fault plane with dip angle 45° are shown in Figures 7.11 and 7.12, respectively. In both cases, the receiver is located at an epicentral distance of 2000 km and azimuth 30° from the fault strike, and a step-like change of double-couple moment by 1 dyne-cm is assumed for both Rayleigh and Love waves. The Cartesian components of **M**, needed in (7.148)–(7.149), are obtained from Box 4.4. Although the Love-wave spectrum decreases with focal depth smoothly in both cases, the Rayleigh-wave spectrum shows strong dependence on both the source mechanism and the focal depth, offering a powerful method for determining these source parameters.

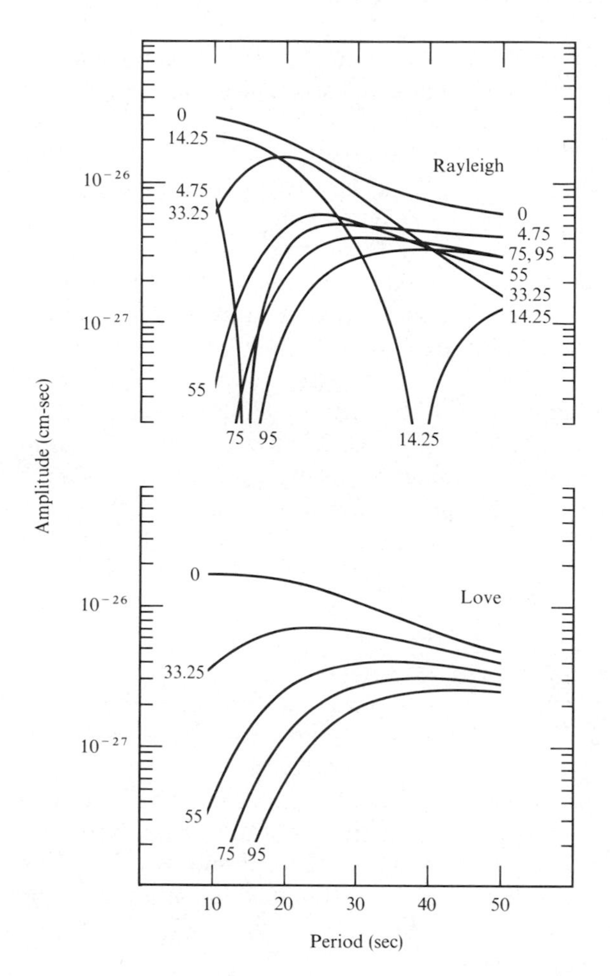

FIGURE **7.12**

Amplitude spectra of Rayleigh (vertical) and Love waves at $\Delta = 2000$ km and $\phi = 30°$ from the fault strike for a dip-slip fault with dip angle 45° and unit step-function seismic moment. The number by each curve gives the source depth in kilometers. [Reproduced from Tsai and Aki (1971); copyrighted by the American Geophysical Union.]

7.6 Leaky Modes

So far in this chapter, we have considered only surface waves or normal modes that arise as residue contributions at poles in the wavenumber plane. The solution for the normal mode vanishes at and beyond the cutoff frequency, because the poles move from the "top" Riemann sheet (where the imaginary part of vertical wavenumber is chosen positive to assure the radiation condition at an infinite depth) to lower sheets. A complete solution requires evaluation of additional integrals around branch cuts, such as those shown in Figure 6.12 for a half-space problem. In this section, we shall show that the branch-cut integrals may be transformed into a sum of residue contributions from poles in the lower sheets, which, unlike the normal modes, attenuate exponentially with time because of leakage to the underlying half-space. For this reason, they are called leaky modes.

We shall study the leaky modes for the simplest layered medium; i.e., a liquid layer overlying a liquid half-space. Wave generation and propagation in this medium were extensively studied by Pekeris (1948) in order to explain observations on ocean acoustics obtained by Ewing and Worzel (1948). Here we shall follow Rosenbaum (1960) in deriving a complete representation of the total wave field as a sum of normal modes and leaky modes. The result is useful to obtain simple approximate formulas valid for a large lapse of time.

Consider a liquid layer with thickness H, density ρ_1, and acoustic wave velocity α_1 overlying a liquid half-space with density ρ_2 and velocity α_2 (Fig. 7.13). We shall assume a spherical pressure source located at a depth h within the layer. We shall use the same coordinates and start with the same source expressions as in the problem of two liquid half-spaces in contact, studied in Section 6.2. We shall first write the primary pressure waves in the form (6.11),

$$P^{\text{inc}} = \frac{A}{R} \exp\left[i\omega\left(\frac{R}{\alpha_1} - t\right)\right], \tag{7.152}$$

where R is the distance between the source point Q and the receiver point P. We use a trial form of solution similar to (6.12) and (6.13), with additional terms representing waves going both ways in the layer, and determine the unknown factor in integrands to meet the boundary conditions. The boundary conditions are (i) vanishing pressure at $z = 0$, (ii) continuity of pressure and vertical displacement across the interface at $z = H$, and (iii) no waves coming from $z = +\infty$. Then (for $0 \leq z \leq h$) the total pressure field in the layer can be written as

$$2 \exp(-i\omega t) \int_0^\infty J_0(kr)k \frac{\sinh \gamma_1 z}{\gamma_1} F_1(\omega, k)\, dk, \tag{7.153}$$

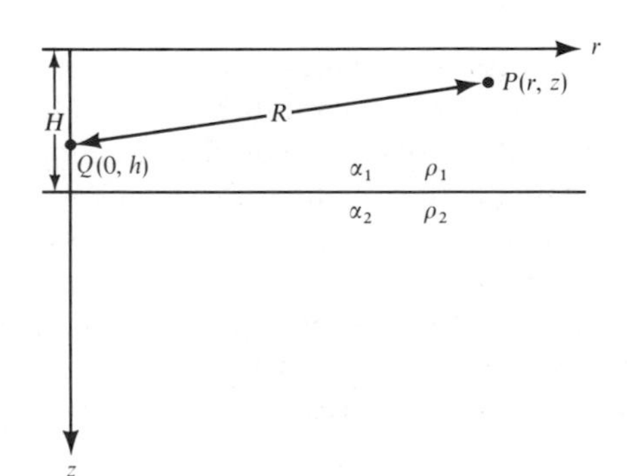

FIGURE **7.13**

Parameters for a point source at Q in a fluid layer over a fluid half-space.

where

$$F_1(\omega, k) = A\frac{\rho_2\gamma_1 \cosh\{\gamma_1(H-h)\} + \rho_1\gamma_2 \sinh\{\gamma_1(H-h)\}}{\rho_2\gamma_1 \cosh(\gamma_1 H) + \rho_1\gamma_2 \sinh(\gamma_1 H)}. \quad (7.154)$$

Here $\gamma_i = (k^2 - \omega^2/\alpha_i^2)^{1/2}$, and the signs of γ_1 and γ_2 are chosen so that $\operatorname{Re}\gamma_1 \geq 0$, $\operatorname{Re}\gamma_2 \geq 0$. A convenient rule to remember is that $i\gamma_2$ has real and imaginary parts that are positive or zero if the radiation condition is satisfied.

In order to derive the modal solution for frequencies below the cutoff frequency, it is necessary to consider a transient source with a continuous spectrum. Let us consider primary pressure waves of the form

$$P^{\text{inc}}(t) = \frac{A\exp\left[\sigma\left(\dfrac{R}{\alpha_1} - t\right)\right]}{R}, \qquad t > \frac{R}{\alpha_1},$$

$$= 0, \qquad t < \frac{R}{\alpha_1}. \quad (7.155)$$

Since the above source can be expressed as a Fourier inverse integral via

$$P^{\text{inc}}(t) = \frac{A}{2\pi R}\int_{-\infty}^{\infty} \frac{\exp\left[-i\omega\left(t - \dfrac{R}{\alpha_1}\right)\right]}{\sigma - i\omega}\, d\omega, \quad (7.156)$$

the corresponding total pressure field can be obtained by an integral of (7.153) with respect to ω:

$$P^{\text{total}}(t) = \frac{1}{\pi}\int_{-\infty}^{\infty} \frac{\exp(-i\omega t)}{\sigma - i\omega}\, d\omega \int_0^{\infty} J_0(kr)k\,\frac{\sinh\gamma_1 z}{\gamma_1}\,F_1(\omega, k)\, dk. \quad (7.157)$$

Since $P^{\text{total}}(t)$ is real, its Fourier transforms for $\pm\omega$ are complex conjugate, and we can rewrite (7.157) as

$$P^{\text{total}}(t) = \frac{2}{\pi}\operatorname{Re}\int_0^{\infty} \frac{\exp(-i\omega t)}{\sigma - i\omega}\, d\omega \int_0^{\infty} J_0(kr)k\,\frac{\sinh\gamma_1 z}{\gamma_1}\,F_1(\omega, k)\, dk. \quad (7.158)$$

Now let us change the order of integration and first evaluate the integral with respect to ω by deforming the path of integration in the complex ω-plane. Since the integrand is an even function of γ_1, $\omega = \pm\alpha_1 k$ is not a branch point, and the only branch cuts in the problem are those associated with γ_2. The branch cuts around $\omega = \pm\alpha_2 k$ are made along $\operatorname{Re}\gamma_2 = 0$, which corresponds to that part of the real ω-axis for which $|\operatorname{Re}\omega| > \alpha_2 k$, as shown in Figure 7.14.

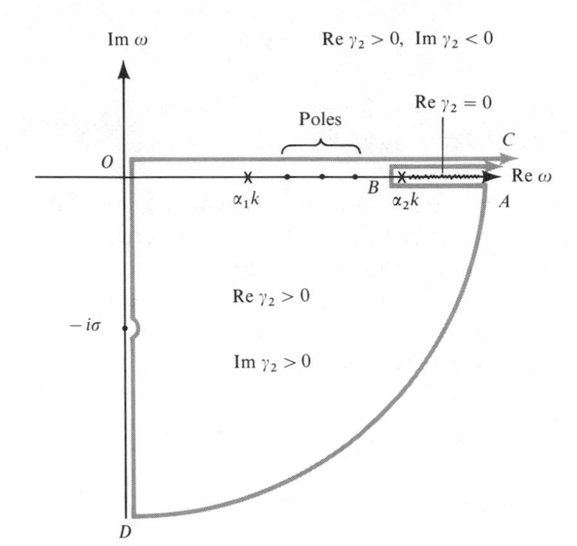

FIGURE **7.14**
Poles (closed circles) and branch points (crosses) in the ω-plane.

Because of the radiation condition, our original integration path lies on the top sheet, for which Re $\gamma_2 > 0$. We next show that only a finite number of real roots of

$$f_1(\omega, k) \equiv \rho_2\gamma_1 \cosh \gamma_1 H + \rho_1\gamma_2 \sinh \gamma_1 H = 0 \tag{7.159}$$

exist on the top sheet. Putting

$$i\gamma_1 H = p_1 + iq_1, \qquad i\gamma_2 H = p_2 + iq_2,$$

where p_1, p_2, q_1, q_2 are real, substituting them into $f_1(\omega, k)$, and taking its imaginary part, we find that

$$\frac{\sinh q_1 \cosh q_1}{\cos^2 p_1 \cosh^2 q_1 + \sin^2 p_1 \sinh^2 q_1} + \frac{\rho_2}{\rho_1}\frac{p_1p_2 + q_1q_2}{p_2^2 + q_2^2} = 0 \tag{7.160}$$

if (7.159) holds. Since $\gamma_1^2 - \gamma_2^2 = \omega^2/\alpha_2^2 - \omega^2/\alpha_1^2$ is real, $p_1q_1 = p_2q_2$. Therefore,

$$\frac{p_1p_2 + q_1q_2}{p_2^2 + q_2^2} = \frac{q_1}{q_2}\cdot\frac{p_1^2 + q_2^2}{p_2^2 + q_2^2}.$$

Substituting this into (7.160), we see that the first and second terms of (7.160) have the same sign if $q_2 > 0$. Thus the root of $f_1(\omega, k)$ cannot exist on the top sheet, where $q_2 = \text{Re}\,\gamma_2 > 0$, except when $q_1 = \text{Re}\,\gamma_1 = 0$. When Re $\gamma_1 = 0$, the roots are real and the corresponding phase velocity c is given by the formula

$$\tan \omega H \sqrt{\frac{1}{\alpha_1^2} - \frac{1}{c^2}} = -\frac{\rho_2 \sqrt{\frac{1}{\alpha_1^2} - \frac{1}{c^2}}}{\rho_1 \sqrt{\frac{1}{c^2} - \frac{1}{\alpha_2^2}}}. \tag{7.161}$$

This is similar to the period equation (7.6) for Love waves but with an important difference. Because of the negative sign on the right side, the zeroth-order tangent curve does not contribute a root (see Fig. 7.2). Thus *all* the modes in a layered acoustic medium have finite cutoff frequencies, and we shall take $n = 1$ as the largest mode (rather than $n = 0$). The cutoff frequency ω_{cn} for the nth mode is given by setting $c = \alpha_2$ in (7.161):

$$\omega_{cn} = \frac{\pi\,(n - \frac{1}{2})}{H \sqrt{\frac{1}{\alpha_1^2} - \frac{1}{\alpha_2^2}}}. \tag{7.162}$$

The corresponding cutoff wavenumber is given by

$$k_{cn} = \frac{\omega_{cn}}{\alpha_2}. \tag{7.163}$$

For a given k, there are only a finite number of roots with $k_{cn} \leq k$ that appear on the top sheet, as shown in Figure 7.14. In addition to these poles on the real axis, another pole due to the source-time function exists at $-i\sigma$ on the negative imaginary axis.

Now let us deform the initial integration path OC into $ODABC$, as shown in Figure. 7.14, in such a way that

$$\int_{OC} = \int_{OD} + \int_{DA} + \int_{ABC} - 2\pi i \Sigma \quad \text{residues at poles.}$$

For a large t, because of the factor $\exp(-i\omega t)$, the integrand of (7.158) vanishes on DA as the radius of the arc approaches infinity. On the other hand, along the path OD, the integrand is pure imaginary except along the small semicircle around $-i\sigma$, and its only contribution is minus half the residue at $-i\sigma$ times $2\pi i$. We shall write this as

$$P_L = \int_{OD} = 2e^{-\sigma t} \int_0^\infty J_0(kr)k \frac{\sinh \gamma_1(\sigma)z}{\gamma_1(\sigma)} F_1(-i\sigma, k)\, dk \tag{7.164}$$

where $\gamma_1(\sigma) = (\sigma^2/\alpha_1^2 + k^2)^{1/2}$. This represents nonoscillatory motion with time dependence identical to the source function. Although this expression violates the acoustic causality (motion starts before the arrival of primary waves), it does not affect the result for a large t, in which we are primarily interested.

The normal-mode solution is a sum of residues from poles on the real axis, given by

$$P_N = -2\pi i \Sigma \quad \text{residues}$$

$$= 4 \operatorname{Re} \sum_n \int_{k_{cn}}^{\infty} J_0(kr)k \sinh(\gamma_1 z) \frac{\exp(-i\omega_n t)}{\omega_n + i\sigma} F_2(\omega_n, k)\, dk, \qquad (7.165)$$

where

$$F_2(\omega_n, k) = \frac{A\rho_2 \sinh(\gamma_1 h)}{\sinh(\gamma_1 H)\left[\dfrac{\partial}{\partial \omega} f_1(\omega, k)\right]_{\omega=\omega_n}}. \qquad (7.166)$$

Here ω_n are the roots of $f_1(\omega_n, k) = 0$, and k_{cn} is the cutoff wavenumber defined in (7.163). The integration limit is imposed because, for $k < k_{cn}$, the nth pole does not exist on the top sheet.

We are left now, with the integral along the path ABC. If the integrand is continuous between AB and BC, this integral of course vanishes. It does not vanish because γ_2 has opposite signs across the branch cut. In fact, $\operatorname{Im} \gamma_2$ is negative along BC and positive along AB. To show this, using $r_1, r_2, \theta_1, \theta_2$ defined in Figure 7.15, we write

$$\alpha_2\gamma_2 = -i(\omega^2 - \alpha_2^2 k^2)^{1/2} = -i\sqrt{r_1 r_2}\left(\cos\frac{\theta_1 + \theta_2}{2} + i \sin\frac{\theta_1 + \theta_2}{2}\right).$$

The first quadrant may be defined by $0 < \theta_1 + \theta_2 < \pi$. Thus, on the top sheet, where $\operatorname{Re} \gamma_2 > 0$, $\operatorname{Im} \gamma_2$ is negative. Likewise, we can show that $\operatorname{Im} \gamma_2$ is positive

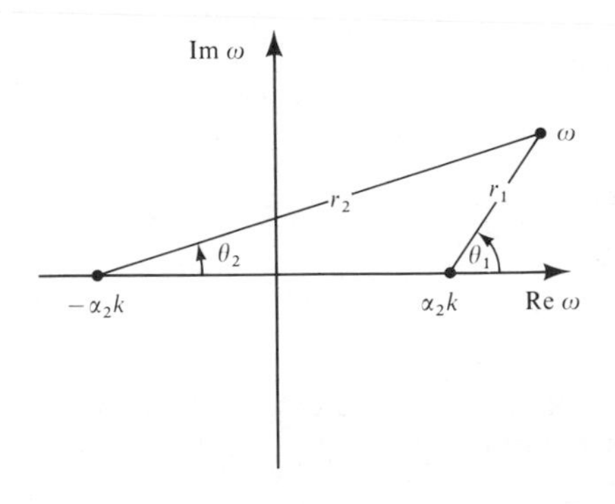

FIGURE **7.15**
Definition of the root chosen for γ_2 in Figure 7.14. Use $0 \le \theta_1 < 2\pi$, $-\pi < \theta_2 \le \pi$, and

$$\alpha_2\gamma_2 = (r_1 r_2)^{1/2} \exp[i(\theta_1 + \theta_2 - \pi)/2].$$

in the fourth quadrant of the top sheet of the ω-plane. (On the bottom sheet, the first quadrant is defined by $2\pi < \theta_1 + \theta_2 < 3\pi$.)

With the sign of γ_2 clearly defined, we can rewrite the integral along ABC as

$$\int_{ABC} = \int_{BC} - \int_{BA}$$

$$= \frac{4}{\pi} \operatorname{Im} \int_0^\infty J_0(kr)k\,dk \int_{\alpha_2 k}^\infty \frac{\exp(-i\omega t)}{\omega + i\sigma} \sinh \gamma_1 z F_3(\omega, k)\,d\omega, \tag{7.167}$$

where

$$F_3(\omega, k) = \frac{A\rho_1\rho_2\gamma_2 \sinh \gamma_1 h}{(\rho_2\gamma_1 \cosh \gamma_1 H + \rho_1\gamma_2 \sinh \gamma_1 H)(\rho_2\gamma_1 \cosh \gamma_1 H - \rho_1\gamma_2 \sinh \gamma_1 H)}. \tag{7.168}$$

To evaluate the above integral, we further deform the integration path from BC to BC', as shown in Figure 7.16. This time we shall keep the integrand of (7.167) analytic. It follows that we are forced to take the root $\operatorname{Re} \gamma_2 < 0$ in the fourth quadrant; i.e., we descend to a lower Riemann sheet. The integral along a quarter-circle CC' vanishes as the radius tends to infinity. This deformation replaces the branch-line integral partially by the residue contribution from poles located in the area swept by the deformation. Thus

$$\int_{ABC} = P_{B1} + P_{B2}$$

where

$$P_{B1} = \frac{4}{\pi} \operatorname{Im} \int_0^\infty J_0(kr)k\,dk \int_{\alpha_2 k}^{\alpha_2 k - i\infty} \frac{\exp(-i\omega t)}{\omega + i\sigma} \sinh \gamma_1 z F_3(\omega, k)\,d\omega \tag{7.169}$$

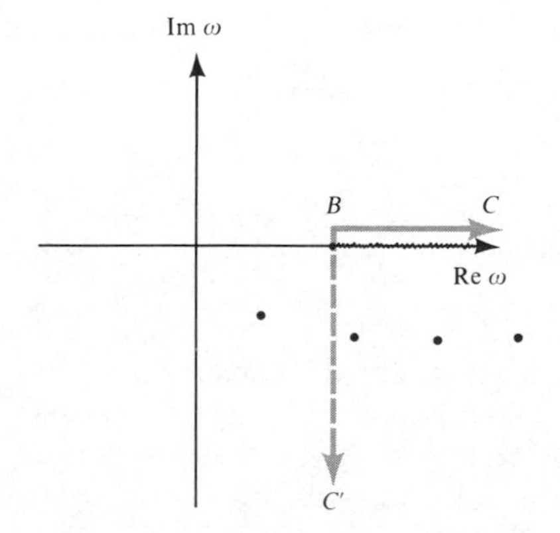

FIGURE **7.16**
Poles on the lower Riemann sheet ($\operatorname{Re} \gamma_2 < 0$) of the complex ω-plane.

and

$$P_{B2} = -2\pi i \sum \text{residues}$$

$$= 4 \operatorname{Re} \sum_n \int_0^{k_{on}} J_0(kr)k\, dk \frac{\exp(-i\omega_n t)}{\omega_n + i\sigma} \sinh \gamma_1 z F_2(\omega_n, k)\, dk. \quad (7.170)$$

In evaluating the residues, note that all the poles of F_3 encountered in the lower sheet are the roots of

$$\rho_2\gamma_1 \cosh \gamma_1 H + \rho_1\gamma_2 \sinh \gamma_1 H = 0,$$

i.e., of $f_1(\omega, k) = 0$. This follows from our discussion of (7.159), which implies here that $\rho_2\gamma_1 \cosh \gamma_1 H - \rho_1\gamma_2 \sinh \gamma_1 H$ cannot have complex roots if $\operatorname{Re} \gamma_2 < 0$. The expression (7.170) is identical to the normal-mode solution (7.165) except for the integration limits. These limits arise because the poles exist in the area on the right of BC' only for $0 < k < k_{on}$. This requires an explanation. We shall trace the movement of poles on the ω-plane as k increases from zero to infinity.

Let us pretend that k represents time. At $k = 0$, $i\gamma_1 = \omega/\alpha_1$, $i\gamma_2 = \omega/\alpha_2$, and the roots of $f_1(\omega, 0) = 0$ are given by

$$\omega_n = \frac{\alpha_1(n - \frac{1}{2})\pi}{H} - i\frac{\alpha_1}{2H} \ln \frac{\alpha_2\rho_2 - \alpha_1\rho_1}{\alpha_2\rho_2 + \alpha_1\rho_1}. \quad (7.171)$$

The poles are located in the lower sheet of the ω-plane, as shown in Figure 7.16 (we assume $\rho_2\alpha_2 > \alpha_1\rho_1$). At $k = 0$, initial velocity and acceleration in the ω-plane are, respectively,

$$\frac{d\omega_n}{dk} = 0,$$

$$\frac{d^2\omega_n}{dk^2} = \frac{\alpha_1^2}{\omega_n}\left[1 + i\frac{\rho_1\alpha_2}{\rho_2\omega_n H}\left(1 - \frac{\alpha_1^2}{\alpha_2^2}\right)\Big/\left(1 - \frac{\rho_1^2\alpha_1^2}{\rho_2^2\alpha_2^2}\right)\right]. \quad (7.172)$$

Examining the real and imaginary parts of acceleration, we find that the poles are accelerated in the direction shown by arrows in Figure 7.17. The lowest-order pole is accelerated most.

At $k = 0$, on the other hand, the integration path BC' is located on the negative imaginary axis. The path BC' moves to the right at a constant velocity α_2, which is higher than that of the poles in the initial stage. Thus BC' passes poles one by one as k increases. The integration limit k_{on} introduced in (7.170) is the value of k at the instant when the nth pole is caught up by BC'. For

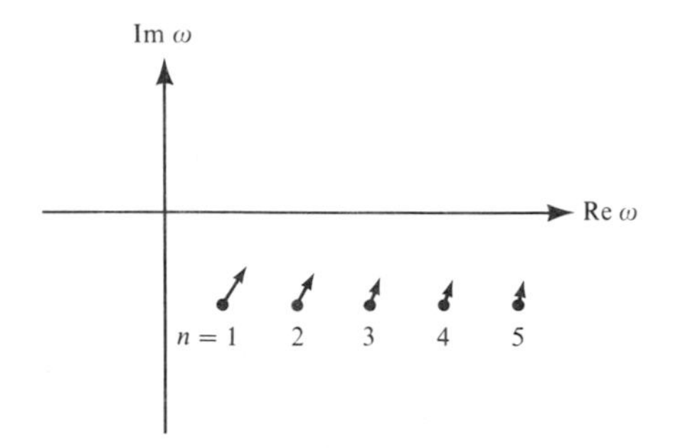

FIGURE **7.17**
Initial position (i.e., at $k = 0$) of poles on the lower Riemann sheet. Arrows show the direction and relative magnitude of "acceleration" of pole positions.

$k < k_{on}$, the nth pole is on the right of BC', hence the residue contribution. The location ω_{on} of the nth pole at $k = k_{on}$ satisfies the following equation:

$$\text{Re } \omega_{on} = \alpha_2 k_{on}. \tag{7.173}$$

After being passed by BC', a pole proceeds to the right and upward to the real ω-axis. At the real axis, it meets the other pole coming from the first quadrant (double roots). Then one of the roots moves to the left, the other to the right, both along the real axis. The root that moves to the right is now accelerated and catches up with BC' again at $k = k_{cn}$, where the pole jumps up to the top sheet at the cutoff frequency $\omega_{cn} = \alpha_2 k_{cn}$. Thereafter, it slows down to reach $\alpha_1 k$ as k goes to infinity.

We must now evaluate the integral P_{B1} along path BC' given in (7.169). To do this, we shall change the order of integration again and return to the k-plane. At the same time, we change the variable ω to q by

$$q = i(\omega - \alpha_2 k). \tag{7.174}$$

Rewriting (7.169), we have

$$P_{B1} = \frac{4}{\pi} \text{Im} \int_0^\infty e^{-qt}\, dq \int_0^\infty J_0(kr)k \frac{e^{-i\alpha_2 kt}}{q - \sigma + i\alpha_2 k} \sinh \gamma_1 z F_3(\omega, k)\, dk. \tag{7.175}$$

Since $\gamma_2 = (iq/\alpha_2)^{1/2}(2k - iq/\alpha_2)^{1/2}$, the branch point is located at $k = i(q/2\alpha_2)$, as shown in Figure 7.18. The cut is made from the branch point to $-i\infty$. This choice of branch cut and the original integration path BC' lying on the lower sheet (Fig. 7.16) give the signs of real and imaginary parts of γ_2, as indicated in Figure 7.18.

We change the integration path from the positive real k-axis to the negative imaginary axis so that the factor $e^{-i\alpha_2 kt}$ in the integrand will attenuate rapidly along the latter path. By this deformation, we pick up contributions from the complex poles in the fourth quadrant. Since Re $\gamma_2 < 0$ there, $\rho_2\gamma_1 \cosh \gamma_1 H - \rho_1\gamma_2 \sinh \gamma_1 H = 0$ cannot have complex roots. Therefore, the poles are the

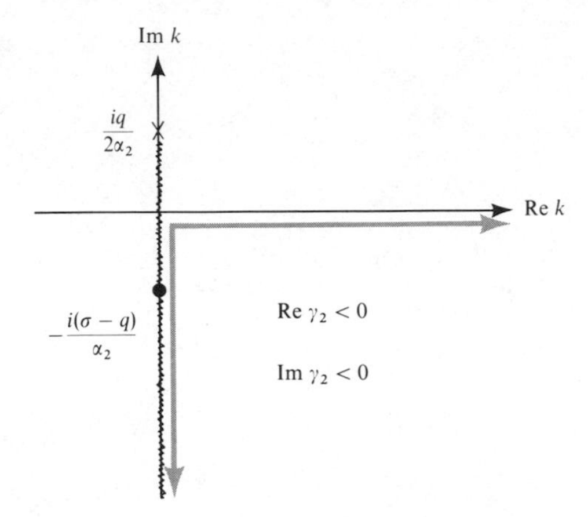

FIGURE **7.18**
A pole (closed circle) and a branch point (cross) in the k-plane.

roots of $\rho_2\gamma_1 \cosh \gamma_1 H + \rho_1\gamma_2 \sinh \gamma_1 H = 0$. Thus we have

$$P_{B1} = P_{BL} + P_{B3}$$

$$P_{BL} = \frac{4}{\pi} \operatorname{Im} \int_0^\infty e^{-qt}\, dq \int_0^{-i\infty} J_0(kr)k \frac{e^{-i\alpha_2 kt}}{q - \sigma + i\alpha_2 k} \sinh \gamma_1 z F_3(\omega, k)\, dk \tag{7.176}$$

and

$$P_{B3} = -2\pi i \sum \text{residues}$$

$$= 4 \operatorname{Re} \sum_n \int_0^{q_{\text{on}}} e^{-qt} J_0(k_n r) k_n \frac{\exp(-i\alpha_2 k_n t)}{q - \sigma + i\alpha_2 k_n} \sinh \gamma_1 F_4(q, k)\, dq, \tag{7.177}$$

where

$$F_4(q, k) = \frac{A\rho_2 \sinh \gamma_1 z}{\sinh \gamma_1 H \left[\dfrac{\partial}{\partial k} (\rho_2\gamma_1 \cosh \gamma_1 H + \rho_1\gamma_2 \sinh \gamma_1 H) \right]_{k = k_n}}. \tag{7.178}$$

The integral with respect to k in (7.176) vanishes if $\sigma < q$, because the pole $i(q - \sigma)/\alpha_2$ will not be on the integration path. Therefore, the integration range for q is $0 < q < \sigma$. Hence P_{BL} vanishes if $\sigma = 0$ (i.e., a step-function source), and it cannot be an important term to us.

The integration limit q_{on} for P_{B3} requires an explanation. As mentioned before, for real positive k there are only two values of k, namely k_{on} and k_{cn}, at which the real part of ω becomes equal to $\alpha_2 k$. In other words, for real positive k and real positive q, there are two roots of $f_1(\omega, k) = 0$ that have the form

$$\omega = \alpha_2 k - iq. \tag{7.179}$$

One root corresponds to k_{cn}, at which $q = 0$, and the other root corresponds to k_{on} at which $q = q_{\text{on}}$. For real positive q, there are no other roots of $f_1(\omega, k) = 0$ on the real positive k-axis.

Now, near $q = 0$, it can be shown that $\text{Im}(dk/dq)_{f_1=0} < 0$. Therefore, a pole moves to the fourth quadrant as q increases from zero, as shown in Figure 7.19. Since the pole crosses the real k-axis only once more at $q = q_{\text{on}}$, it will not exist in the fourth quadrant for $q > q_{\text{on}}$. For this reason, the integration limits for P_{B3} are $0 < q < q_{\text{on}}$.

Finally, we change the integration variable from q to k, by the following relation:

$$f_1(q, k) = \rho_2 \gamma_1 \cosh \gamma_1 H + \rho_1 \gamma_2 \sinh \gamma_1 H = 0.$$

Then

$$\frac{dq}{dk} = -\frac{(\partial f_1/\partial k)_{q=\text{const.}}}{(\partial f_1/\partial q)_{k=\text{const.}}} = \frac{(\partial f_1/\partial k)_{q=\text{const.}}}{i(\partial f_1/\partial \omega)_{k=\text{const.}}}$$

Substituting this into (7.177), we have

$$P_{B3} = 4 \text{ Re} \sum_n \int_{k_{\text{on}}}^{k_{cn}} \frac{J_0(kr)k \exp(-i\omega_n t)}{\omega_n + i\sigma} \sinh \gamma_1 z F_2(\omega_n, k)\, dk, \tag{7.180}$$

which has the integrand identical to P_N and P_{B2} in (7.165) and (7.170). Putting all the terms together, we finally have

$$p^{\text{total}} = 4 \text{ Re} \sum_n \int_0^{\infty} \frac{J_0(kr)k \exp(-i\omega_n t)}{\omega_n + i\sigma} \sinh \gamma_1 z F_2(\omega_n, k)\, dk, \tag{7.181}$$

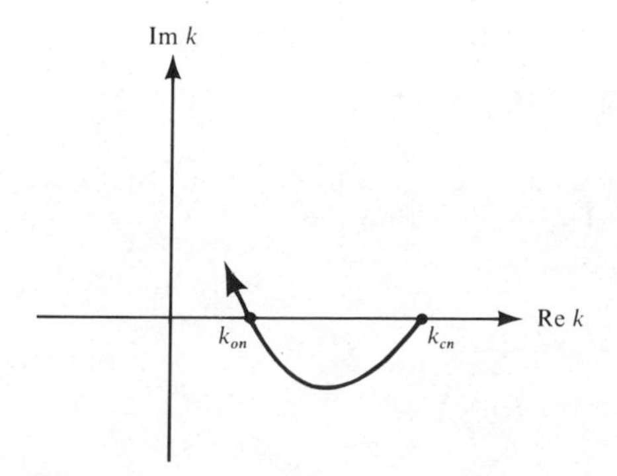

FIGURE **7.19**
Path of a pole in the k-plane.

where nonoscillatory terms P_L and P_{BL} are neglected. The integration path deviates from the real k-axis between k_{on} and k_{cn}, as shown in Figure 7.19.

7.6.1 *Organ-pipe mode*

We first consider the case in which the receiver is located above the source, i.e., $r = 0$, as may be encountered in a reflection survey. The integral (7.181) then takes a form considered in Box 6.3, with the exponent $-i\omega_n t$, and the integral may be evaluated by the steepest-descents method for a large t. Since, as shown in (7.172), $d\omega_n/dk = 0$ at $k = 0$, the saddle point is located at $k = 0$ in this case. Assuming the main contribution to come from the vicinity of $k = 0$, we expand the integrand phase factor $\exp(-i\omega_n t)$ as a power series in k, keeping only the first two terms. Values of ω_n and $d^2\omega_n/dk^2$ at $k = 0$ are given in (7.171) and (7.172), respectively. The steepest-descents path is determined by

$$-\frac{it}{2}\,\omega_n''(0)k^2 = -i\frac{t}{2}\,k^2\left[\frac{d^2\omega_n}{dk^2}\right]_{k=0} = -x^2,$$

where x is real. Changing the variable from k to x by the above equation, we obtain

$$p^{\text{total}} = 4\,\text{Re}\sum_n \frac{F_2(\omega_n, 0)}{\omega_n(0) + i\sigma}\exp(-i\omega_n(0)t)\,\frac{2}{i\omega_n''(0)t}\int_0^\infty xe^{-x^2}\,dx.$$

Substituting $F_2(\omega_n, 0)$, $\omega_n(0)$ from (7.166) and (7.171), and recognizing that $\int_0^\infty xe^{-x^2}\,dx = \frac{1}{2}$, the above formula can be rewritten as

$$p^{\text{total}} = 4A\,\frac{\alpha_1^2}{Ht}\sum_n \left|\frac{\sin(\omega_n h/\alpha_1)\sin(\omega_n z/\alpha_1)}{\omega_n(-i\omega_n + \sigma)\omega_n}\right|_{k=0}$$
$$\times\left(\frac{\alpha_2\rho_2 - \alpha_1\rho_1}{\alpha_2\rho_2 + \alpha_1\rho_1}\right)^{-\alpha_1 t/2H}\cos\left[\frac{\alpha_1}{H}\left(n - \tfrac{1}{2}\right)\pi t + \psi\right], \tag{7.182}$$

where ψ is the phase angle of the quantity inside the absolute value symbol.

The above formula shows a damped oscillation, with period $4H/[(2n - 1)\alpha_1]$. The lowest-order mode ($n = 1$) has wavelength equal to four times the layer thickness H: a quarter-wavelength oscillator. The damping is such that the corresponding temporal Q (Box 5.7) is given by

$$Q = \frac{\left(n - \frac{1}{2}\right)\pi}{\ln\left(\dfrac{\alpha_2\rho_2 + \alpha_1\rho_1}{\alpha_2\rho_2 - \alpha_1\rho_1}\right)}. \tag{7.183}$$

If the impedance contrast is large, or $\alpha_2\rho_2 \gg \alpha_1\rho_1$, the logarithm approaches 0 and Q becomes large, resulting in a long-lasting reverberation. This type of reverberation is called an organ-pipe mode, because it is essentially a one-dimensional oscillation in the vertical direction.

7.6.2 *Phase velocity and attenuation*

Let us consider the case in which the receiver is located far from the source, so that we can use the asymptotic expansion of the Bessel function $J_0(kr)$. Substituting (6.16) into (7.181) and retaining only the leading term, we have

$$p^{\text{total}} \sim \frac{4}{\pi r} \operatorname{Re} \sum_n \int_{-\infty}^{\infty} (k)^{1/2} \frac{\exp\left(-i\omega_n t + ikr - i\frac{\pi}{4}\right)}{\omega_n + i\sigma} F_2(\omega_n, k)\, dk. \tag{7.184}$$

In this case, the saddle point is determined by

$$\frac{d(-i\omega_n t + ikr)}{dk} = 0$$

or

$$\frac{d\omega_n}{dk} = \frac{r}{t}. \tag{7.185}$$

Since r/t is real, we see that the group velocity $d\omega_n/dk$ is real at the saddle point. To locate the saddle point, one must find the (ω_n, k_n) that satisfies (7.185) and $f_1(\omega_n, k_n) = 0$ (7.159) simultaneously. Once we find the saddle point, say k_{sn}, and the corresponding ω_{sn}, the steepest-descents path is determined by

$$\begin{aligned} -i\omega_n t + ikr &\sim -i\omega_{ns} t + ik_{sn} r - \frac{i}{2}\omega_{sn}''(k - k_{sn})^2 t \\ &= -i\omega_{sn} t + ik_{sn} r - x^2 \end{aligned}$$

with real x. Carrying out the integration along the steepest-descents path, we obtain

$$p^{\text{total}} \sim \frac{4}{\sqrt{rt}} \sum_n |Q_n| \exp(-L_n t) \cos(\operatorname{Re}\omega_{sn} t - \operatorname{Re} k_{sn} r + \psi), \tag{7.186}$$

where $L_n = -\operatorname{Im}\omega_{sn} + \operatorname{Im} k_{sn} r/t$,

$$Q_n = \frac{\sqrt{k_{sn}} F_2(\omega_{sn}, k_{sn})}{(i\omega_{sn} + \sigma)\sqrt{\omega_{sn}''}},$$

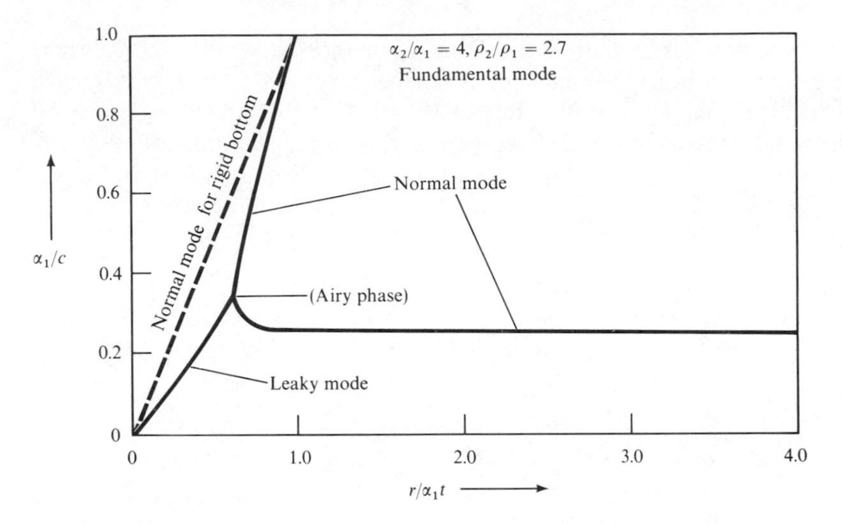

FIGURE **7.20**

Relation between group velocity and phase velocity for the fundamental normal mode and leaky mode in the case $\alpha_2/\alpha_1 = 4$ and $\rho_2/\rho_1 = 2.7$. The relation for the leaky mode is similar to that for the normal mode when the half-space is rigid. [From Rosenbaum (1960); copyrighted by the American Geophysical Union.]

and ψ is the phase angle of Q_n. These formulas give the phase velocity and attenuation associated with each leaky mode.

Rosenbaum (1960) gives some numerical results. For example, Figure 7.20 shows the relation between group velocity $d\omega_n/dk = r/t$ and phase velocity $\text{Re}(\omega_{sn})/\text{Re}(k_{sn}) = c$ for the lowest mode ($n = 1$) in the case $\alpha_2/\alpha_1 = 4$ and $\rho_2/\rho_1 = 2.7$. Figure 7.21 shows the corresponding attenuation factor L. The acoustic leaking mode has lower group velocity for higher phase velocity, and

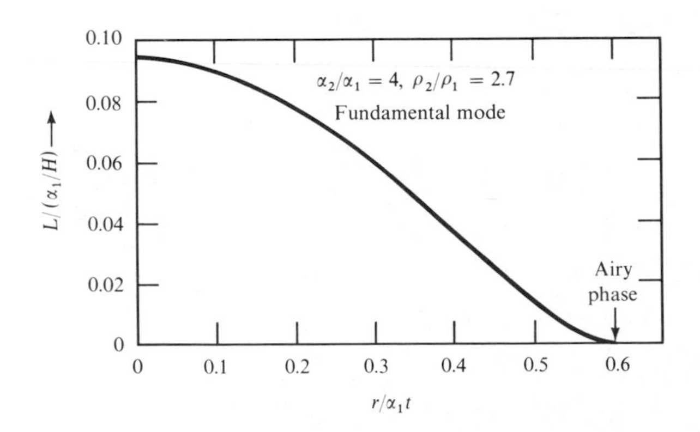

FIGURE **7.21**

The attenuation factor of a leaky mode. [From Rosenbaum (1960) copyrighted by the American Geophysical Union.]

its group velocity is always lower than that of the normal mode. The relation between group and phase velocity for a normal mode in a layer overlying a rigid surface is shown by a broken line in Figure 7.20. For an elastic body, on the other hand, leaky modes arriving earlier than normal modes do exist. A decaying oscillatory long-period motion following P-waves, called the PL-wave, is often observed at short distances. Such motions are interpreted as leaky modes by Oliver and Major (1960).

SUGGESTIONS FOR FURTHER READING

Ewing, M., J. L. Worzel, and C. L. Pekeris. *Propagation of Sound in the Ocean*. Geological Society of America, Memoir 27, 1948.

Ewing, W. M., W. S. Jardetzky, and F. Press. *Elastic Waves in Layered Media*. New York: McGraw-Hill, 1957.

Gantmacher, F. R., *The Theory of Matrices* (2 vols.). New York: Chelsea Publishing Co., 1959.

Gilbert, F., and G. E. Backus. Propagator matrices in elastic wave and vibration problems. *Geophysics*, **31**, 326–332, 1966.

Lighthill, M. J., Group velocity. *Journal of the Institute of Mathematics and Its Applications*, **1**, 1–28, 1965.

Brune, J. N., J. E. Nafe, and J. E. Oliver. A simplified method for the analysis and synthesis of dispersed wave trains. *Journal of Geophysical Research*, **65**, 287–304, 1960.

Schwab, F. A., and L. Knopoff. Fast surface wave and free mode computations. *In* B. A. Bolt (editor), *Seismology: Surface Waves and Earth Oscillations* (Methods in Computational Physics, Vol. 11). New York: Academic Press, 1972.

Schultz, M. H. *Spline Analysis*. Englewood Cliffs, New Jersey, Prentice-Hall, 1973.

Oliver, J., and M. Major. Leaking modes and the PL phase. *Bulletin of the Seismological Society of America*, **50**, 165–180, 1960.

Bamford, D., S. Crampin, and others. Special edition on anisotropy. *Geophysical Journal of the Royal Astronomical Society*, **49**, 1–243, 1977.

PROBLEMS

7.1 We develop here an alternative view of Love waves in a single layer over a half-space. A physical mechanism for such a surface wave (with frequency ω and phase velocity $c(\omega)$) is provided by body waves that are reflected supercritically within the upper layer ($\beta_1 < \beta_2$). For two points A and B at the same depth and taken distance X apart along the direction of propagation of the surface wave, the phase at B must be shifted from that at A by $X\omega/c$.

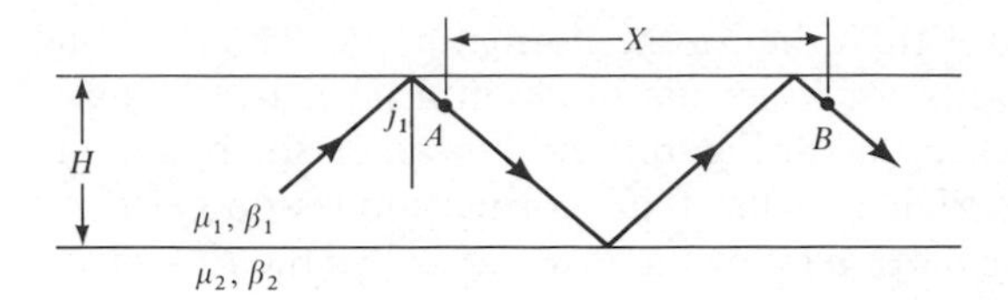

a) For the body wave between A and B, show that this phase shift is

$$\psi = \frac{2H\omega}{\beta_1} \sec j_1 - 2 \tan^{-1} \left[\frac{\mu_2 \left(\frac{1}{c^2} - \frac{1}{\beta_2^2} \right)^{1/2}}{\mu_1 \left(\frac{1}{\beta_1^2} - \frac{1}{c^2} \right)^{1/2}} \right].$$

(*Hint:* Equation (5.32) is useful here.)

b) Show that ψ and $X\omega/c$ can differ only by an integer multiple of 2π for a surface wave to exist. From this requirement, obtain the dispersion relation (7.6).

c) Why cannot Love waves exist if $\beta_1 > \beta_2$?

7.2 Show that the propagator matrix for SH-waves in a homogeneous layer is given by a matrix multiplication of (7.50) (evaluated at $z = z_0$) substituted in (7.49). Obtain the corresponding result for P-SV waves from (7.55) and (7.56). (In essence, this is the way in which Thomson and Haskell originally obtained the propagator, rather than from (7.41).)

7.3 Generalize the idea expressed in the previous question by relating both $\mathbf{f}(z_l)$ and $\mathbf{f}(z_{l-1})$ to $\mathbf{w}_l$ and then showing that

$$\mathbf{f}(z_l) = \mathbf{F}_l(z_l)\mathbf{F}_l^{-1}(z_{l-1})\mathbf{f}(z_{l-1}).$$

Hence show that the propagator from z_0 to z_k is

$$\mathbf{P}(z_k, z_0) = [\mathbf{F}_k(z_k)\mathbf{F}_k^{-1}(z_{k-1})][\mathbf{F}_{k-1}(z_{k-1})\mathbf{F}_{k-1}^{-1}(z_{k-2})] \cdots [\mathbf{F}_1(z_1)\mathbf{F}_1^{-1}(z_0)]$$

(Note that these results are still true if the medium consists of a stack of *inhomogeneous* layers, provided $\mathbf{F}_l(z)$ is a matrix whose columns are linearly independent solutions of $\partial\mathbf{f}/\partial z = \mathbf{A}(z)\mathbf{f}$ in the lth layer.)

7.4 In Sections 5.4 and 7.2, we showed that $\mathbf{Fw}$ can be thought of as a sum of all the possible wave types that solve $\partial\mathbf{f}/\partial z = \mathbf{Af}$; that each of the columns of $\mathbf{F}$ is separately a basic solution of $\partial\mathbf{f}/\partial z = \mathbf{Af}$; and that $\mathbf{w}$ is a vector of constants that give the weight of each basic solution present in the sum $\mathbf{Fw}$. Consider the first column of $\mathbf{F}$ in (7.55) when $k > \omega/\alpha$, and show that the corresponding wave in the sum $\mathbf{Fw}$ of (7.54) is an inhomogeneous P-wave with displacement amplitude

$$\acute{P}e^{-\gamma z}[\alpha^2k^2/\omega^2 - \sin^2(kx - \omega t)]^{1/2}.$$

7.5 Show by redefining the origin that one way to construct $G_{np}(0, 0, h; x, y, z; \omega)$ is by making the switches $(r \to r;\ \phi \to \phi + \pi;\ z \to h;\ h \to z)$ in formulas for $G_{np}(x, y, z; 0, 0, h; \omega)$. Use this approach to verify that the reciprocity

$$G_{np}(0, 0, h; x, y, z; \omega) = G_{pn}(x, y, z; 0, 0, h; \omega)$$

is satisfied for surface-wave components (7.145) and (7.146).

7.6 Show that the change in phase velocity of Love waves at fixed wavenumber, which will result from a perturbation ($\delta\rho$, $\delta\mu$) in the structure, is

$$\left(\frac{\delta c}{c}\right)_k = \frac{\int_0^\infty \left[k^2 l_1^2 + \left(\frac{dl_1}{dz}\right)^2\right] \delta\mu \, dz - \int_0^\infty \omega^2 l_1^2 \, \delta\rho \, dz}{2\omega^2 \int_0^\infty \rho l_1^2 \, dz}.$$

7.7 Show that the Rayleigh-wave eigenfunction for a half-space with Poisson's ratio 0.25 is given by

$$r_1 = e^{-0.8475kz} - 0.5773e^{-0.3933kz},$$

$$r_2 = 0.8475e^{-0.8475kz} - 1.4679e^{-0.3933kz},$$

and that the energy integral I_1 is equal to 0.6205 ρ/k (k is the horizontal wavenumber and ρ is the density). Then, using (7.149), obtain an explicit formula for Rayleigh waves due to a point source with arbitrary moment tensor located at depth h.

7.8 The waveform of a dispersed wave train given by

$$f(t) = \frac{1}{2\pi} \int_{-\infty}^{\infty} |F(\omega)| e^{-i\omega t + i\phi(\omega)} \, d\omega$$

$$= \frac{1}{\pi} \int_0^\infty |F(\omega)| \cos[\omega t - \phi(\omega)] \, d\omega$$

can be computed easily and accurately if the group delay time $t_g = d\phi/d\omega$ and $d|F(\omega)|/d\omega$ are known at discrete frequencies ω_i. Show that

$$f(t) = \sum \frac{1}{\pi} \int_{\omega_i - \Delta\omega_i/2}^{\omega_i + \Delta\omega_i/2} |F(\omega)| \cos(\omega t - \phi(\omega)) \, d\phi$$

$$\sim \sum_i \frac{\Delta\omega_i}{\pi} \left[|F(\omega_i)| \cos \omega_i(t - t_{pi}) \frac{\sin\{(\Delta\omega_i/2)(t - t_{gi})\}}{(\Delta\omega_i/2)(t - t_{gi})} - \frac{d|F(\omega_i)|}{d\omega} \frac{\sin \omega_i(t - t_{pi})}{t - t_{gi}} \left\{ \frac{\sin[(\Delta\omega_i/2)(t - t_{gi})]}{(\Delta\omega_i/2)(t - t_{gi})} - \cos\left[\frac{\Delta\omega_i}{2}(t - t_{gi})\right] \right\} \right],$$

where $t_{pi} = \phi(\omega_i)/\omega_i$ is the phase delay time at ω_i.

7.9 Determine the phase velocity as a function of frequency for surface waves that are observed with the following properties:

a) their group velocity is 4.4 km/sec independent of frequency;

b) their wave shape changes with the travel distance, but comes back to the same shape every 8800 km;

c) their phase velocity increases with period. (Long-period Love waves in the period range 40–200 sec roughly show the above properties. They show an impulsive form because of the frequency-independent group velocity, and are sometimes called G waves.)

CHAPTER 8

Free Oscillations of the Earth

Long-period surface waves generated by a large earthquake may be recorded several times at a given station as they continue to travel around the world. The flat Earth model studied in Chapter 7 is obviously inadequate for analyzing these long-period waves. Even for periods as short as 20 sec, the effect of the Earth's curvature on the surface-wave dispersion cannot be ignored, especially when a pronounced low-velocity layer in the upper mantle traps a large part of the energy and effectively reduces the travel distance. Furthermore, the dispersion can be so strong that long-period surface waves that have traveled $N + 1$ times around the Earth can interfere with slower, short-period waves that have traveled N times around and with waves that have traveled around in the opposite direction. It is therefore important, just to complete the surface-wave analysis, that we study the spherical Earth model. But our main goal in this chapter is to understand that surface waves and body waves (and leaking modes) are phenomena that can be expressed in terms of more basic motions, the Earth's free oscillations.

Because the Earth is a body of finite size, it can resonate as a whole only at certain discrete frequencies. Let us first understand why this must be so.

In the previous chapter, wave systems were characterized by their horizontal wavenumber (k) and frequency (ω). Because these waves were propagating in media that extend indefinitely in the lateral (horizontal) direction, k in that

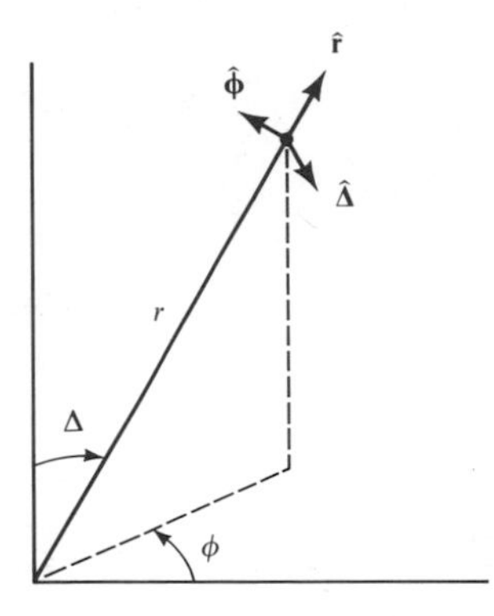

FIGURE **8.1**
Spherical polar coordinates (r, Δ, ϕ), with origin at the center of the Earth. The line $\Delta = 0$ is often taken to pass through a seismic source, in which case (r, Δ, ϕ) are known as *epicentral coordinates.*

case was a continuous variable. We emphasized an approach in which, for each ω, only certain wavenumbers $k = k_n(\omega)$ $(n = 0, 1, 2, \ldots)$ are possible. But we could instead have taken k as the independent variable, finding that only a certain number of eigenfrequencies $\omega = {}_n\omega(k)$ are possible for surface-wave modes (though, for each n, ${}_n\omega$ varies continuously). It is this latter form of the dispersion relation that connects naturally to a discussion of free oscillations in the spherical Earth. This follows because in spherical geometry the "horizontal wavenumber" is fixed at certain discrete values by the finite lateral extent of the medium ($0 \leq \Delta < 2\pi$, where Δ is shown in Fig. 8.1). In fact, we shall work with horizontal *angular* wavenumber l, instead of k, and find that l must be zero or a positive integer. Rather than ${}_n\omega(l)$, we write ${}_n\omega_l$, and one of our first goals must be to obtain the possible eigenfrequencies ${}_n\omega_l$ $(n = 0, 1, 2, \ldots)$ for given l.

We shall start this analysis with a discussion of the "dispersion relation" $\omega = {}_n\omega_l$ for a homogeneous liquid sphere—that is, we need to know how ω depends on l for the fundamental $(n = 0)$ and overtones $(n > 0)$. We shall neglect self-gravitation, so that we shall be working with an unrealistic but simple model that is useful for introducing many of the wavefunctions needed in the analysis of any spherical model. We shall then describe the excitation of free oscillations by a point source with emphasis on the effect of the Earth's sphericity on the propagation of surface waves. For waves with periods longer than about 500 sec, the effect of gravity becomes important, and we shall show how self-gravitation can be taken into account. Finally, we shall discuss the splitting of spectral lines due to the rotation and (in Chapter 13) the splitting due to lateral variation of structure in the Earth.

8.1 Free Oscillations of a Homogeneous Liquid Sphere

Perhaps the simplest of all free-oscillation systems that shares some major aspects with free oscillations of the Earth is the solution of the "violin-string" problem. The transverse motion of the string, as a function of position x and time t, satisfies a one-dimensional wave equation $c^2\partial^2 y/\partial x^2 = \partial^2 y/\partial t^2$, and $y(x, t) = 0$ at $x = 0$ and $x = L$. We shall expect the reader to have some familiarity with the associated Sturm-Liouville theory. The most important result is that *any* solution $y(x, t)$ satisfying the above wave equation and boundary conditions can be expanded as

$$y = \sum_n a_n y_n(x, t),$$

where the a_n are constants and the y_n are free oscillations; $y_n = \sin({}_n\omega x/c)\cos({}_n\omega t)$ with eigenfrequencies ${}_n\omega = (n + 1)\pi c/L$.

To develop some familiarity with the special properties of a *spherical* medium, we shall imagine a homogeneous compressible fluid sphere of radius r_0, with elastic properties specified by bulk modulus κ and density ρ. No body forces will be allowed to act within this medium, and we shall examine the free oscillations of small pressure perturbations P about the equilibrium pressure field. The usual equation of motion (2.17) becomes

$$\rho\ddot{\mathbf{u}} = -\nabla P, \tag{8.1}$$

because stress τ_{ij} is merely $-P\delta_{ij}$; and Hooke's law (2.18) reduces to

$$P = -\kappa\nabla\cdot\mathbf{u}. \tag{8.2}$$

Therefore, since ρ and κ are constants,

$$c^2\nabla^2 P = \partial^2 P/\partial t^2, \tag{8.3}$$

where $c^2 = \kappa/\rho$.

We shall assume a "free" boundary condition—namely, that $P(\mathbf{x}, t) = 0$ on the surface with radius r_0. Choosing a spherical polar coordinate system (r, Δ, ϕ), as in Figure 8.1, it is natural to try to find special solutions in which the dependence upon (r, Δ, ϕ, t) is separated into four different factors: $P = R(r)\Theta(\Delta)\Phi(\phi)T(t)$. Our discussion of (5.8) indicates that $T = \exp(\pm i\omega t)$ for some constant ω, and in order to separate spatial dependences we need to use the explicit form of ∇^2 in spherical polars, which is

$$\nabla^2 P = \frac{1}{r^2}\frac{\partial}{\partial r}\left(r^2\frac{\partial P}{\partial r}\right) + \frac{1}{r^2 \sin\Delta}\frac{\partial}{\partial\Delta}\left(\sin\Delta\frac{\partial P}{\partial\Delta}\right) + \frac{1}{r^2\sin^2\Delta}\frac{\partial^2 P}{\partial\phi^2}. \tag{8.4}$$

In Box 8.1, we find that

$$\Theta(\Delta)\Phi(\phi) = Y_l^m(\Delta, \phi) \equiv (-1)^m\left[\frac{2l+1}{4\pi}\frac{(l-m)!}{(l+m)!}\right]^{1/2} P_l^m(\cos\Delta)e^{im\phi}, \tag{8.5}$$

where l and m are integers, $-l \le m \le l$, and $P_l^m(\cos\Delta)$ is the associated Legendre function. Although the Δ and ϕ dependences have been separated, it is common practice to write $Y_l^m(\Delta, \phi)$ for the product $\Theta\Phi$ in (8.5).

BOX **8.1**

Spherical surface harmonics

A long list of important properties can be derived for the special functions $\Theta(\Delta)\Phi(\phi)$ that separate the horizontal variation of solutions to $c^2\nabla^2 P = \partial^2 P/\partial t^2$ in spherical geometry. We here outline the formal derivation of some of these properties, which are needed frequently in geophysics because of the need to define continuous bounded functions over spherical surfaces within the Earth.

Trying a solution $P(\mathbf{x}, t) = R(r)\Theta(\Delta)\Phi(\phi)\exp(-i\omega t)$, we find from (8.3) and (8.4) that

$$\frac{\sin^2 \Delta}{R}\frac{d}{dr}\left(r^2\frac{dR}{dr}\right) + \frac{\sin \Delta}{\Theta}\frac{d}{d\Delta}\left(\sin \Delta\frac{d\Theta}{d\Delta}\right) + \frac{\omega^2 r^2}{c^2}\sin^2 \Delta = -\frac{1}{\Phi}\frac{d^2\Phi}{d\phi^2}.$$

The left-hand side is independent of ϕ, hence $(1/\Phi)(d^2\Phi/d\phi^2)$ is a constant. Solving for Φ and noting that $\Phi(\phi)$ must be periodic with period 2π if $P(\mathbf{x}, t)$ is to be a single-valued function of position, we find the eigenfunctions

$$\Phi = e^{im\phi} \qquad m = 0, \pm 1, \pm 2, \pm 3, \ldots. \tag{1}$$

The equation in (r, Δ) for R and Θ is now

$$\frac{1}{R}\frac{d}{dr}\left(r^2\frac{dR}{dr}\right) + \frac{\omega^2 r^2}{c^2} = \frac{m^2}{\sin^2 \Delta} - \frac{1}{\sin \Delta\, \Theta}\frac{d}{d\Delta}\left(\sin \Delta\frac{d\Theta}{d\Delta}\right),$$

where it has been arranged that the left-hand side depends only on r and the right-hand side only on Δ. The equation can thus be satisfied for all (r, Δ) only if there is some constant K for which

$$\frac{d}{dr}\left(r^2\frac{dR}{dr}\right) + \left(\frac{\omega^2 r^2}{c^2} - K\right)R = 0 \tag{2}$$

and

$$\frac{d}{d\Delta}\left(\sin \Delta\frac{d\Theta}{d\Delta}\right) = \left(\frac{m^2}{\sin^2 \Delta} - K\right)\sin \Delta\, \Theta \tag{3}$$

We continue with an analysis of the Θ-equation, beginning with:

THE CASE $m = 0$

The function $\Phi(\phi)$ is constant, and the solution $P(\mathbf{x}, t)$ has axial symmetry. Θ satisfies $d/d\Delta(\sin \Delta\, d\Theta/d\Delta) = -K \sin \Delta\, \Theta$, and it is convenient to get away from the angle Δ and use instead the variable $x = \cos \Delta$, since then the trigonometric terms in the Θ-equation are suppressed. We find

$$(1 - x^2)\frac{d^2\Theta}{dx^2} - 2x\frac{d\Theta}{dx} + K\Theta = 0, \tag{4}$$

known as the *Legendre equation.* For general values of the constant K, the solutions have singularities at the end-points of the range $-1 \leq x \leq 1$. This is the range corresponding to

$0 \le \Delta \le \pi$, which is needed to describe position in the Earth. But for certain special values of K, there are nonsingular solutions Θ that turn out to be polynomials in x.

To prove these statements, one assumes a solution exists in the form

$$\Theta(x) = x^k \sum_{i=0}^{\infty} b_i x^i \qquad (b_0 \neq 0). \tag{5}$$

Substituting (5) into (4) and equating the coefficient of each power of x to zero, we find

$$b_0 k(k - 1) = 0, \tag{6}$$

$$b_1(k + 1)k = 0, \tag{7}$$

and, in general,

$$b_{i+2} = b_i\left[\frac{(k + i)(k + i + 1) - K}{(k + i + 1)(k + i + 2)}\right]. \tag{8}$$

From (6), we know $k = 0$ or $k = 1$. Without loss of generality, we shall assume $k = 0$ (since $k = 1$ implies from (7) and (8) that all the b_i with odd subscripts are zero, and thus that b_0 alone determines all the nonzero b_i: this solution is included in the case $k = 0$).

From (8) we see in general that $|b_{i+2}/b_i| \to 1$ as $i \to \infty$. Thus, by comparison with a geometric series, there is convergence of (5) provided $-1 < x < 1$. But what happens at $x = \pm 1$ ($\Delta = 0$ or π)? It can be shown that, in this case, the infinite series for $\Theta(x)$ will *diverge*, unless by some chance one of the even-suffix b_i is zero and one of the odd-suffix b_i is zero. (For then all further b_i are zero, so that the infinite series is reduced to a polynomial, which clearly does "converge" for the special values $x = \pm 1$.) Looking at (8) with $k = 0$, we see that the only way the coefficient b_{i+2} can be zero (if $b_i \neq 0$) is if $K = i(i + 1)$.

We have then the important result that the constant K, which was introduced to separate the radial equation from the Θ-equation, must be the product of two successive integers. Otherwise, the Θ-equation does not have a solution valid throughout $0 \le \Delta \le \pi$.

Furthermore, if $K = l(l + 1)$ and l is even, then it is the even-suffix b_i that make up the solution. The odd-suffix series must be stopped by requiring $b_1 = 0$. Similarly, if $K = l(l + 1)$ and l is odd, then $b_0 = 0$. In either case, the solution for Θ is a polynomial of order l. The customary choice for b_0 or b_1 is made by requiring

$$\Theta(x) - 1 \qquad \text{for } x = 1. \tag{9}$$

The polynomials that result are the *Legendre polynomials*. Writing them out as a sum of descending powers, a great deal of manipulation gives, for l either even or odd, the expression

$$\Theta = P_l(x) = \frac{(2l)!}{2^l(l!)^2}\left[x^l - \frac{l(l-1)x^{l-2}}{2(2l-1)} + \frac{l(l-1)(l-2)(l-3)x^{l-4}}{2 \cdot 4 \cdot (2l-1)(2l-3)} - \cdots\right], \tag{10}$$

stopping at either x or 1 (times a constant) as the last term. The first few Legendre polynomials are

$$P_0(x) = 1, \qquad P_1(x) = x, \qquad P_2 = \tfrac{1}{2}(3x^2 - 1),$$

$$P_3(x) = \tfrac{1}{2}(5x^3 - 3x), \qquad P_4(x) = \tfrac{1}{8}(35x^4 - 30x^2 + 3),$$

and, in general,

$$P_l(x) = \frac{1}{2^l l!} \frac{d^l}{dx^l} (x^2 - 1)^l,$$

known as *Rodrigues' formula.*

THE CASE $m \neq 0$

For m a nonzero integer, we shall initially assume it to be positive. Then with $x = \cos \Delta$ in (3), we find

$$\frac{d}{dx}\left[(1 - x^2)\frac{d\Theta}{dx}\right] = \frac{m^2}{1 - x^2} - K\Theta. \tag{11}$$

We might attempt a power-series solution like (5). This method bogs down, however, because one finds that the formula for b_{i+2} involves not just b_i (as it did before for Θ with $m = 0$), but also b_{i+1}, and the general solution of such a three-term recursion relation is laborious. To guess at what might be done, we recall that for $m = 0$ the properties of Θ near $x = \pm 1$ are important. They may also be expected to be important for $m > 0$, by inspection of the coefficients in (11). We thus turn to a brief examination of Θ near $x = \pm 1$. With $\varepsilon = x \pm 1$ and ε small, (11) is approximately

$$\varepsilon \frac{d^2\Theta}{d\varepsilon^2} + \frac{d\Theta}{d\varepsilon} - \frac{m^2\Theta}{4\varepsilon} = 0,$$

which has solutions $\Theta = \varepsilon^{m/2}$ and $\varepsilon^{-m/2}$. The second solution is not well-behaved at $\varepsilon = 0$, and can be rejected. It seems then that Θ should have zeros of order $m/2$ at $x = \pm 1$. They can both be factored out by writing

$$\Theta(x) = (1 - x^2)^{m/2} A(x),$$

and we can hope to study Θ by studying $A(x)$.

This method turns out to be fruitful, because A satisfies the ordinary differential equation

$$(1 - x^2)\frac{d^2A}{dx^2} - 2(m + 1)\frac{dA}{dx} + [K - m(m + 1)]A = 0, \tag{12}$$

and this *does* have just a two-term recursive formula for the coefficients in an expansion of the form $A(x) = x^k \sum_{i=0}^{\infty} c_i x^i$. The recursive formula turns out to be

$$c_{i+2} = \frac{[(i + m)(i + m + 1) - K]}{(i + 1)(i + 2)} c_i.$$

In general, this formula will generate two series-solutions for $A(x)$ (one of even powers of x, with factor c_0, and one of odd powers, with factor c_1). If these series were not terminated at some power x^r, they would behave like $(1 - x^2)^{-m}$. The requirement that Θ have no singularities in $-1 \leq x \leq 1$ $(0 \leq \Delta \leq \pi)$ thus leads to the result $c_{r+2} = 0$, also with $c_1 = 0$

if r is even, and $c_0 = 0$ if r is odd. Thus

$$(r + m)(r + m + 1) = K,$$

and K has eigenvalues that again (i.e., as for $m = 0$) are the product of consecutive integers; $r \geq 0$, $m \geq 0$, hence we take $K = l(l + 1)$ for some integer $l \geq 0$. Since $r \geq 0$, we find also the important result $m \leq l$.

Note that, since K takes the same eigenvalues if $m = 0$ or $m > 0$, the radial function $R(r)$ is unchanged by dropping the requirement of axial symmetry.

We have shown that $\Theta(x) = (1 - x^2)^{m/2} A(x)$, where A is now a polynomial in x. There is no particular difficulty in finding the coefficients of this polynomial. However, a quick way to get an explicit formula for A is available, since, if the equation satisfied by the Legendre polynomial P_l (see (4)) is differentiated m times, there results

$$(1 - x^2)\frac{d^{m+2}}{dx^{m+2}} P_l - 2(m + 1)x \frac{d^{m+1}}{dx^{m+1}} P_l + [l(l + 1) - m(m + 1)] \frac{d^m}{dx^m} P_l = 0.$$

Comparing this with the equation (12) satisfied by $A(x)$, we see that a solution for A is $A(x) = d^m P_l(x)/dx^m$. Since $P_l(x)$ is a polynomial involving nonnegative powers of x, there is no danger of $A(x)$ blowing up anywhere in $-1 \leq x \leq 1$.

The product $(1 - x^2)^{m/2}\, d^m P_l(x)/dx^m$ is therefore a solution for the angular function $\Theta(X)$. It is called the *associated* Legendre function, $P_l^m(x)$.

The equation (11) for Θ depends upon m only via m^2. Therefore, if $m < 0$, the nonsingular solution must be proportional to $P_l^{|m|}(\cos \Delta)$. We adopt the convention

$$P_l^{-m}(x) = (-1)^m \frac{(l - m)!}{(l + m)!} P_l^m(x), \tag{13}$$

in which the constant of proportionality has been chosen so that

$$P_l^m(x) = \frac{(1 - x^2)^{m/2}}{2^l l!} \frac{d^{l+m}}{dx^{l+m}} (x^2 - 1)^l \tag{14}$$

applies for all l, m such that $-l \leq m \leq l$.

Several books have been written on the properties of P_l and P_l^m (Robin, 1957; Hobson, 1955), and Wiggins and Saito (1971) showed how they may be computed efficiently. Summarizing the most important formulas, it is known that

$$\frac{1}{(1 + r^2 - 2r \cos \Delta)^{1/2}} = \sum_{l=0}^{\infty} r^l P_l(\cos \Delta) \qquad 0 < r < 1 \tag{15}$$

$$(l - m + 1)P_{l+1}^m(x) - (2l + 1)xP_l^m(x) + (l + m)P_{l-1}^m(x) = 0 \tag{16}$$

$$(1 - x^2)\frac{d}{dx} P_l^m(x) = (l + 1)xP_l^m(x) - (l - m + 1)P_{l+1}^m(x). \tag{17}$$

It is convenient to define fully normalized surface harmonics

$$Y_l^m(\Delta, \phi) = (-1)^m \left[\frac{2l + 1}{4\pi} \frac{(l - m)!}{(l + m)!}\right]^{1/2} P_l^m(\cos \Delta) e^{im\phi} \tag{18}$$

for integers $l \geq 0$ and integers m such that $-l \leq m \leq l$. Then

$$\int_0^{2\pi} d\phi \int_0^{\pi} [Y_l^m(\Delta, \phi)]^* Y_{l'}^{m'}(\Delta, \phi) \sin \Delta \, d\Delta = \delta_{ll'}\delta_{mm'}, \tag{19}$$

and

$$Y_l^{-m}(\Delta, \phi) = (-1)^m [Y_l^m(\Delta, \phi)]^*.$$

In the theory for excitation of normal modes by a point source, we need values of Y_l^m and some of its derivatives at $\Delta = 0$. A key result is

$$P_l^m(\cos \Delta) \to \frac{1}{2^m m!} \frac{(l+m)!}{(l-m)!} \Delta^m \qquad \text{as } \Delta \to 0, \quad \text{for } m \geq 0.$$

Next we must examine the radial wavefunction $R = R(r)$, and this must satisfy

$$\frac{1}{r^2} \frac{d}{dr}\left(r^2 \frac{dR}{dr}\right) + \left[\frac{\omega^2}{c^2} - \frac{l(l+1)}{r^2}\right] R = 0. \tag{8.6}$$

Since c here is constant, (8.6) is a standard equation that is known to have spherical Hankel functions as its solutions. Furthermore, since we are interested in solutions that have no singularity in pressure anywhere in the range $0 \leq r \leq r_0$, the solutions of (8.6) must be $R(r) \propto j_l(\omega r/c)$, where j_l is the spherical Bessel function of order l. Because l is an integer, we can take advantage of the relation

$$j_l(x) = x^l \left(-\frac{1}{x}\frac{d}{dx}\right)^l \left(\frac{\sin x}{x}\right) \tag{8.7}$$

so that, for $l = 0$,

$$rR(r) \propto \sin(\omega r/c), \tag{8.8}$$

and for $l = 2$,

$$R(r) \propto \left(\frac{3c^5}{\omega^5 r^5} - \frac{c^3}{\omega^3 r^3}\right) \sin(\omega r/c) - \frac{3c^4}{\omega^4 r^4} \cos(\omega r/c). \tag{8.9}$$

It remains to take account of the "free" boundary condition on $r = r_0$, which requires $R(r_0) = 0$. It is this condition that gives the permissible eigenfrequencies ${}_n\omega_l$. Thus, from (8.8), we find for $l = 0$ that $\sin(\omega r_0/c) = 0$. It follows that

$$ {}_n\omega_0 = \frac{(n+1)\pi c}{r_0} \qquad (n = 0, 1, 2, \ldots). \tag{8.10} $$

$n = 0$ gives the *fundamental* mode, and $n \geq 1$ constitute the *overtones*, or higher modes. Motions with $l = 0$ are purely *radial modes*, because the associated u_Δ and u_ϕ are zero. In the case that $r_0 = 6000$ km and $c = 5$ km/sec, the periods (in seconds) corresponding to (8.10) are ${}_nT_0 = (2400)/(n + 1)$, with the fundamental ${}_0T_0$ being 40 min. For $l = 2$, the first zero of $j_l(x)$ occurs where $x \sim 1.8\,\pi$, hence ${}_0T_2 \sim 22$ min. In Figure 8.2 we show the eigenfunctions ${}_nR_0 \propto j_0({}_n\omega_0 r/c)$ for $n = 0, 1, 2$; and ${}_nR_2 \propto j_2({}_n\omega_2 r/c)$ for $n = 0, 1, 2$.

Important properties of the free oscillations of our homogeneous fluid sphere are their *orthogonality*, and their *degeneracy* with respect to azimuthal order number m. Thus, for the vector space of steady-state $[\exp(-i\omega t)]$ solutions to $c^2\nabla^2 P = \partial^2 P/\partial t^2$ within $|\mathbf{x}| < r_0$, and satisfying $P(\mathbf{x}, t) = 0$ for $|\mathbf{x}| = r_0$, we define an inner product. If f and g are two members of this vector space,

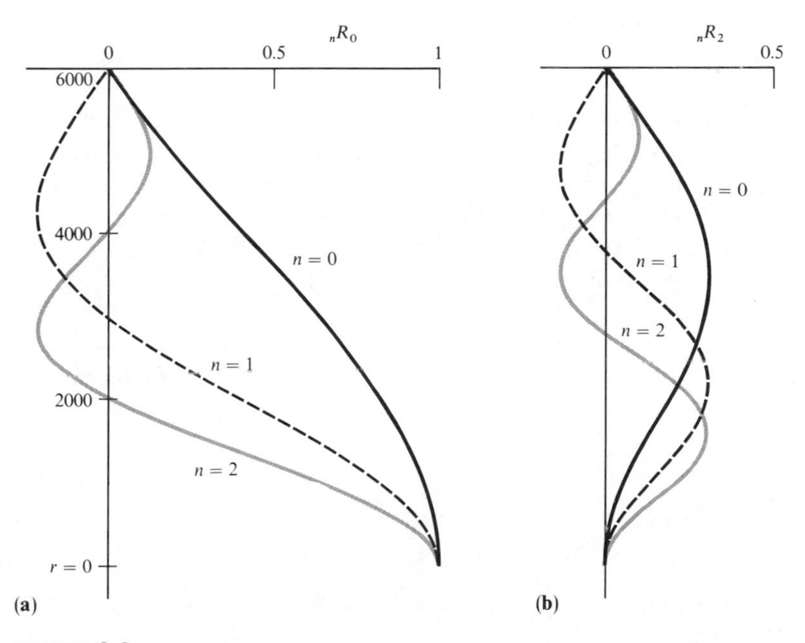

FIGURE **8.2**

Eigenfunctions ${}_nR_l$ of pressure within a fluid sphere of radius 6000 km are shown as a function of radius. **(a)** ${}_nR_0(r)$ for $n = 0, 1, 2$. **(b)** ${}_nR_2(r)$ for $n = 0, 1, 2$.

we define

$$\{f, g\} = \iiint\limits_{|\mathbf{X}| \le r_0} \rho f^* g \, dV. \tag{8.11}$$

Suppose further that f and g are normal modes, with eigenfrequencies ω_f and ω_g, respectively. Then some algebra yields

$$[\omega_g^2 - (\omega_f^*)^2]\{f, g\} = 0.$$

Putting $f = g$ here, it follows that eigenfrequencies must be real. With $f \neq g$ and $\omega_f \neq \omega_g$, it also follows that f and g are *orthogonal* in the sense that $\{f, g\} = 0$. Thus, in our present problem,

$$\{ j_l({}_n\omega_l r/c) Y_l^m(\Delta, \phi), j_{l'}({}_{n'}\omega_{l'} r/c) Y_{l'}^{m'}(\Delta, \phi)\} = 0 \tag{8.12}$$

unless $l = l'$ and $m = m'$ and $n = n'$.

For each choice of integers $l \ge 0$, $n \ge 0$, there are $(2l + 1)$ free oscillations sharing the same eigenfrequency ${}_n\omega_l$ and the same radial eigenfunction $j_l({}_n\omega_l r/c)$. This degeneracy arises because the one-dimensional wave equation (8.6) is independent of m, and each m in the range $-l \le m \le l$ provides a different normal mode, via pressure $= {}_nP_l^m(\mathbf{x}, t) = j_l({}_n\omega_l r/c) Y_l^m(\Delta, \phi) \exp(-i{}_n\omega_l t)$.

It is possible to extend the analysis of free oscillations to spherical media that are considerably better models of the Earth than our homogeneous fluid sphere, and still retain the same features of orthogonality and degeneracy and the same numbering scheme for the modes in terms of (l, m, n). A significant improvement over the fluid model is the homogeneous elastic solid sphere, which in 1882 was thoroughly studied by Horace Lamb. This is the simplest compressible medium that exhibits two independent families of modes, spheroidal and toroidal, which differ in their properties precisely in the way that we have found *P-SV* motions to differ from *SH* in a flat Earth model (Chapter 7). Just as we introduced vector surface harmonics for cylindrical geometry (7.117), so must we now introduce such vectors for spherical geometry. We define

$$\begin{aligned}
\mathbf{R}_l^m(\Delta, \phi) &= Y_l^m \hat{\mathbf{r}}, \\
\mathbf{S}_l^m(\Delta, \phi) &= \frac{1}{[l(l+1)]^{1/2}} \left(\frac{\partial Y_l^m}{\partial \Delta} \hat{\boldsymbol{\Delta}} + \frac{1}{\sin \Delta} \frac{\partial Y_l^m}{\partial \phi} \hat{\boldsymbol{\phi}} \right), \\
\mathbf{T}_l^m(\Delta, \phi) &= \frac{1}{[l(l+1)]^{1/2}} \left(\frac{1}{\sin \Delta} \frac{\partial Y_l^m}{\partial \phi} \hat{\boldsymbol{\Delta}} - \frac{\partial Y_l^m \hat{\boldsymbol{\phi}}}{\partial \Delta} \right),
\end{aligned} \tag{8.13}$$

where $Y_l^m(\Delta, \phi)$ is the fully normalized surface harmonic defined in (8.5) and $\hat{\mathbf{r}}$, $\hat{\boldsymbol{\Delta}}$, $\hat{\boldsymbol{\phi}}$ are unit vectors in directions r, Δ, ϕ, respectively (Fig. 8.1). It is easy to see that **R**, **S**, and **T** are perpendicular to each other. Because $Y_l^m(\Delta, \phi)$ is fully normalized, the vector functions satisfy the following orthogonal relations:

$$\int_0^\pi \sin \Delta \, d\Delta \int_0^{2\pi} d\phi \mathbf{R}_l^{m*} \cdot \mathbf{R}_{l'}^{m'} = \delta_{mm'}\delta_{ll'},$$

$$\int_0^\pi \sin \Delta \, d\Delta \int_0^{2\pi} d\phi \mathbf{S}_l^{m*} \cdot \mathbf{S}_{l'}^{m'} = \delta_{mm'}\delta_{ll'},$$

$$\int_0^\pi \sin \Delta \, d\Delta \int_0^{2\pi} d\phi \mathbf{T}_l^{m*} \cdot \mathbf{T}_{l'}^{m'} = \delta_{mm'}\delta_{ll'}.$$

A *spheroidal* motion is one for which the radial component of $\nabla \times \mathbf{u}$ is zero; and a *toroidal* motion has both $u_r = 0$ and $\nabla \cdot \mathbf{u} = 0$. The homogeneous fluid sphere clearly can support only spheroidal motions. If a normal mode in this medium has the pressure field proportional to $j_l({}_n\omega_l r/c) Y_l^m(\Delta, \phi) \exp(-i_n\omega_l t)$, then, from (8.1), the corresponding displacement field is proportional to

$$\left\{ \left(\frac{dj_l}{dr} \right) \mathbf{R}_l^m(\Delta, \phi) + \frac{[l(l+1)]^{1/2}}{r} j_l \mathbf{S}_l^m(\Delta, \phi) \right\} \frac{\exp(-i_n\omega_l t)}{\rho({}_n\omega_l)^2}. \tag{8.14}$$

It is apparent from this result, and directly from (8.13), that $\mathbf{R}_l^m$ and $\mathbf{S}_l^m$ are required to describe vector fields associated with spheroidal motion, and $\mathbf{T}_l^m$ for toroidal motion.

Provided we consider Earth models that are spherically symmetric and do not rotate, we shall find it possible to describe spheroidal modes by generalizing the radial functions appearing in (8.14). But the modes will still display the same horizontal wavefunctions, the same degeneracy, and an orthogonality similar to that of (8.12). The symbol ${}_nS_l$ is often used to identify a spheroidal mode, and ${}_nT_l$ a toroidal mode.

In a sense, the spherical Earth model is easier to investigate than a flat Earth model, for any motion within a sphere can be expressed by a superposition of normal modes. The leaky mode, for which we allocated considerable space in the previous chapter, does not exist for a sphere. As shown in the next section, the formulas for excitation of free oscillations by a point source can be derived more simply than those for surface waves.

8.2 Excitation of Free Oscillations by a Point Source

Since a sphere is a finite body, any disturbance can be expressed as a superposition of normal modes. Following Gilbert (1971), we shall go back to the nineteenth century and start with the vibration of a system of N particles studied by Rayleigh and Routh.

Consider N particles initially in a state of equilibrium. A set of external forces is then applied with $\mathbf{f}_\alpha(t)$ acting on the αth particle ($\alpha = 1, \ldots, N$), putting all the particles in motion. Let the mass of the αth particle be m_α and the displacement of the αth particle from its equilibrium position be $\mathbf{u}_\alpha$. For small displacements, we may assume that the change in internal force between particles is a linear sum of displacements. Then the equation of motion is written as

$$m_\alpha \frac{d^2\mathbf{u}_\alpha(t)}{dt^2} + \sum_{\beta=1}^{N} c_{\alpha\beta}\mathbf{u}_\beta(t) = \mathbf{f}_\alpha(t) \qquad t > 0\ (\alpha = 1, 2, \ldots, N), \tag{8.15}$$

where the initial conditions are

$$\mathbf{u}_\alpha(0) = \mathbf{0} \quad \text{and} \quad \frac{d}{dt}\mathbf{u}_\alpha(0) = \mathbf{0}$$

and $\mathbf{c}$ is symmetric and positive definite (cf. (2.32)).

The normal modes of the system are given by the solution of (8.15) for $\mathbf{f}_\alpha = \mathbf{0}$. Since there are three equations for each particle, there are $3N$ eigenfrequencies ω_i and $3N$ eigenvectors (normal modes) for the total system. The normal modes are denoted by ${}_i\mathbf{u}_\alpha \exp(-i\omega_i t)$, this being the displacement of the αth particle in the ith normal mode. For each i, the motion of all the particles ($\alpha = 1, \ldots, N$) is needed to describe the mode. The motion of the αth particle in the ith normal mode is obtained by solving

$$-\omega_i^2 m_\alpha({}_i\mathbf{u}_\alpha) + \sum_\beta c_{\alpha\beta}({}_i\mathbf{u}_\beta) = \mathbf{0} \qquad (\alpha = 1, \ldots, N). \tag{8.16}$$

The eigenvectors ${}_i\mathbf{u}_\alpha$ are orthogonal, in the sense that

$$\sum_\alpha m_\alpha({}_j\mathbf{u}_\alpha^*) \cdot ({}_i\mathbf{u}_\alpha) = 0$$

unless $i = j$, but their amplitudes are not determined by (8.16). We choose to normalize them by

$$\sum_\alpha m_\alpha({}_j\mathbf{u}_\alpha^*) \cdot ({}_i\mathbf{u}_\alpha) = \delta_{ij}. \tag{8.17}$$

Taking the scalar product of (8.16) with ${}_j\mathbf{u}_\alpha^*$, summing over α, and using (8.17), we obtain

$$\sum_{\alpha\beta} c_{\alpha\beta}({}_j\mathbf{u}_\alpha^*) \cdot ({}_i\mathbf{u}_\beta) = \omega_i^2\delta_{ij}. \tag{8.18}$$

To find the solution of (8.15) as a superposition of normal modes, we need to find the coefficients a_i in the expansion

$$\mathbf{u}_\alpha(t) = \sum_i a_i({}_i\mathbf{u}_\alpha) \exp(-i\omega_i t) \qquad (t > 0). \tag{8.19}$$

(Of course, it is the real part of this summation that we use for real $\mathbf{u}_\alpha(t)$.) We work with the Laplace transform, $\mathbf{u}_\alpha(t) \to \mathbf{u}_\alpha(s)$, finding from (8.19) that

$$\mathbf{u}_\alpha(s) = \sum_i \frac{a_i}{s + i\omega_i} \, {}_i\mathbf{u}_\alpha. \tag{8.20}$$

To find the coefficients a_i, we substitute from (8.20) into the Laplace transform of (8.15):

$$m_\alpha s^2 \sum_i \left(\frac{a_i}{s + i\omega_i}\right) {}_i\mathbf{u}_\alpha + \sum_\beta c_{\alpha\beta} \sum_i \left(\frac{a_i}{s + i\omega_i}\right) {}_i\mathbf{u}_\beta = \mathbf{f}_\alpha(s). \tag{8.21}$$

From (8.17), (8.18), and (8.21), it is then easy to show that

$$\frac{a_j}{s + i\omega_j} = \frac{\sum_\alpha {}_j\mathbf{u}_\alpha^* \cdot \mathbf{f}_\alpha(s)}{s^2 + \omega_j^2},$$

so that from (8.20) we have

$$\mathbf{u}_\alpha(s) = \sum_i \left[\frac{\sum_\beta {}_i\mathbf{u}_\beta^* \cdot \mathbf{f}_\beta(s)}{s^2 + \omega_i^2}\right] {}_i\mathbf{u}_\alpha, \tag{8.22}$$

which is simple to invert to the time domain. Assuming that the forces vary as a step function in time, $\mathbf{f}_\alpha(t) = \mathbf{F}_\alpha H(t)$, we find that $\mathbf{f}_\beta(s) = s^{-1}\mathbf{F}_\beta$ in (8.22). The Laplace inverse of $[s(s^2 + \omega_i^2)]^{-1}$ is $H(t)(1 - \cos \omega_i t)/\omega_i^2$, so that for $t > 0$ we obtain

$$\mathbf{u}_\alpha(t) = \sum_i \left(\sum_\beta {}_i\mathbf{u}_\beta^* \cdot \mathbf{F}_\beta\right) {}_i\mathbf{u}_\alpha \frac{1 - \cos \omega_i t}{\omega_i^2}. \tag{8.23}$$

When the medium is dissipative, we find, using the temporal Q (Box 5.7), that

$$\mathbf{u}_\alpha(t) = \sum_i \left(\sum_\beta {}_i\mathbf{u}_\beta^* \cdot \mathbf{F}_\beta\right) {}_i\mathbf{u}_\alpha \frac{1 - \exp[-(\omega_i t/2Q_i)] \cos \omega_i t}{\omega_i^2}. \tag{8.24}$$

The structure of this solution has some very interesting properties. First, we note that the static displacement $\mathbf{u}_\alpha(t \to \infty)$ is easily obtained from (8.24) as a sum of the normal modes. In fact, (8.24) indicates that the motion in each mode can be thought of as a decaying oscillation about a new reference level, this new level being initiated at $t = 0$, as shown in Figure 8.3. Second, the form of the normal-mode summation (8.24) tells us what can be learned about the source from observations of the displacement of just one particle. Suppose we know all the normal modes of the system. If we observe the αth particle and

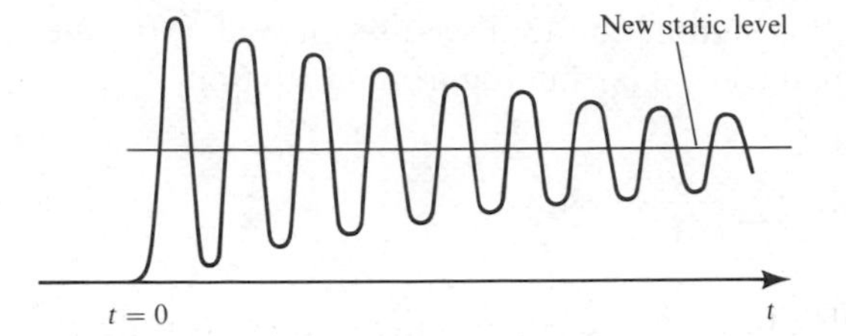

FIGURE **8.3**
Record of ground motion in a given normal mode due to a source that acts as a step function in time at $t = 0$.

its displacement $\mathbf{u}_\alpha(t)$, we can hope to infer the excitation coefficients

$$\sum_\beta {}_i\mathbf{u}_\beta^* \cdot \mathbf{F}_\beta$$

for $i = 1, 2, \ldots, 3N$. (This is not possible for those modes that have a node at the αth particle. In these cases, ${}_i\mathbf{u}_\alpha = \mathbf{0}$, and the excitation coefficient of the ith mode cannot be determined from knowledge of $\mathbf{u}_\alpha(t)$ because the mode is not observed there. The process of obtaining the excitation coefficients involves looking at the Fourier spectrum of $\mathbf{u}_\alpha(t)$ and measuring the height of spectral peaks centered on each ω_i. Again, there is difficulty in determining excitation coefficients for two modes that have nearly the same frequency.) It is interesting that, by observing just one particle and obtaining the excitation coefficients, it is then possible to use (8.24) with different α and predict the motion of all other particles in the system. Even if some of the excitation coefficients are not determined (for reasons noted above), it may still be possible to predict major parts of the spectrum of $\mathbf{u}_\alpha(t)$ for all α.

We have made these comments on the solution (8.24) because it can easily be modified to apply to the Earth. We consider increasing the number of particles so that they approach a continuum. In the limit, a sum over particles such as $\sum_{\beta=1}^N {}_i\mathbf{u}_\beta^* \cdot \mathbf{F}_\beta$ is replaced by a volume integral such as $\int {}_i\mathbf{u}^*(\boldsymbol{\xi}) \cdot \mathbf{f}(\boldsymbol{\xi})\, dV(\boldsymbol{\xi})$, where $\mathbf{f}(\boldsymbol{\xi})$ is now the body force per unit volume. We shall continue to assume that this body force acts as a step function in time. From (8.24) we immediately obtain

$$\mathbf{u}(\mathbf{x}, t) = \sum_i \left(\int_V {}_i\mathbf{u}^*(\boldsymbol{\xi}) \cdot \mathbf{f}(\boldsymbol{\xi})\, dV \right) {}_i\mathbf{u}(\mathbf{x}) \frac{1 - \exp[-(\omega_i t/2Q_i)] \cos \omega_i t}{\omega_i^2}. \tag{8.25}$$

Our use of i here denotes the ith normal mode of the whole Earth. That is, each i corresponds to some value for the triple of integers (l, m, n) that we found in Section 8.1 were necessary for characterizing individual modes. The sum in (8.25) is thus an infinite sum, but, as shown by Rayleigh (1945, paragraph 101), it does converge because of the factor ω_i^{-2}. The normal modes in (8.25) have been normalized (cf. (8.17)) by

$$\int_V \rho(\boldsymbol{\xi})\, {}_j\mathbf{u}^*(\boldsymbol{\xi}) \cdot {}_i\mathbf{u}(\boldsymbol{\xi})\, dV = \delta_{ij}, \tag{8.26}$$

where $\rho(\xi)$ is the density, and the volume integrals above are taken over the whole Earth.

We shall now find the vibration of a spherical Earth model due to a point source that is specified by a moment tensor. Using a result that was previously given as an exercise (Problem 3.4), the body force becomes

$$f_p(\xi, t) = -M_{pq}(t) \frac{\partial}{\partial \xi_q} \delta(\xi - \mathbf{x}_s). \tag{8.27}$$

We shall assume that $\mathbf{M}$ acts as a step function in time at $\mathbf{x}_s$, so that the body force is also a step function, and (8.25) is directly applicable. The ith excitation coefficient is now

$$\begin{aligned} \int_V {}_i\mathbf{u}^*(\xi) \cdot \mathbf{f}(\xi)\, dV &= -M_{pq} \int_V {}_i\mathbf{u}_p^*(\xi) \frac{\partial}{\partial \xi_q} \delta(\xi - \mathbf{x}_s)\, dV(\xi) \\ &= {}_iu^*_{p,q}(\mathbf{x}_s) M_{pq} = {}_ie^*_{pq}(\mathbf{x}_s) M_{pq}, \end{aligned} \tag{8.28}$$

where ${}_ie_{pq}$ is the (pq) strain component in the ith normal mode. To obtain the last equality in (8.28), we used the symmetry $M_{pq} = M_{qp}$. Putting (8.28) into (8.25), we finally obtain the displacement for an arbitrary point source $\mathbf{M}H(t)$ acting at $\mathbf{x}_s$:

$$\mathbf{u}(\mathbf{x}, t) = \sum_i ({}_ie^*_{pq}(\mathbf{x}_s) M_{pq})\, {}_i\mathbf{u}(\mathbf{x}) \frac{1 - \exp(-\omega_i t/2Q_i) \cos \omega_i t}{\omega_i^2}. \tag{8.29}$$

Thus, once the normal modes ${}_i\mathbf{u}$ of the Earth are known, it is a simple matter to calculate the response of the Earth to a point source with an arbitrary moment tensor.

To find explicit forms for the normal modes, we must be more specific about the Earth model. We shall consider here a nonrotating spherically symmetric Earth in which the density $\rho(r)$ and Lamé parameters $\lambda(r)$ and $\mu(r)$ depend only on the distance r from the center of symmetry. The equations of motion (2.47)–(2.50) for this model can fruitfully be studied by the motion-stress vector approach that we adopted in Chapter 7. In spherical polar coordinates, the appropriate ansatz for displacement in the mode (l, m, n) is

$$[{}_nU_l(r)\mathbf{R}_l^m(\Delta, \phi) + {}_nV_l(r)\mathbf{S}_l^m(\Delta, \phi) + {}_nW_l(r)\mathbf{T}_l^m(\Delta, \phi)] \exp(-i_n\omega_l t). \tag{8.30}$$

The associated traction working on spherical surfaces r = constant is

$$[{}_nR_l(r)\mathbf{R}_l^m(\Delta, \phi) + {}_nS_l(r)\mathbf{S}_l^m(\Delta, \phi) + {}_nT_l(r)\mathbf{T}_l^m(\Delta, \phi)] \exp(-i_n\omega_l t) \tag{8.31}$$

and we can write the equations for the radial functions in the following two separate forms:

$$\frac{d}{dr}\begin{pmatrix} V \\ U \\ S \\ R \end{pmatrix} = \begin{pmatrix} \frac{1}{r} & -\frac{[l(l+1)]^{1/2}}{r} & \frac{1}{\mu} & 0 \\ \frac{\lambda[l(l+1)]^{1/2}}{r(\lambda+2\mu)} & -\frac{2\lambda}{r(\lambda+2\mu)} & 0 & \frac{1}{\lambda+2\mu} \\ \frac{4l(l+1)\mu(\lambda+\mu)}{r^2(\lambda+2\mu)} - \rho\omega^2 - \frac{2\mu}{r^2} & -\frac{2\mu(3\lambda+2\mu)[l(l+1)]^{1/2}}{r^2(\lambda+2\mu)} & -\frac{3}{r} & -\frac{\lambda[l(l+1)]^{1/2}}{r(\lambda+2\mu)} \\ \frac{2\mu(3\lambda+2\mu)[l(l+1)]^{1/2}}{r^2(\lambda+2\mu)} & -\rho\omega^2 + \frac{4\mu(3\lambda+2\mu)}{r^2(\lambda+2\mu)} & \frac{[l(l+1)]^{1/2}}{r} & \frac{-4\pi}{r(\lambda+2\mu)} \end{pmatrix}\begin{pmatrix} V \\ U \\ S \\ R \end{pmatrix} \tag{8.32}$$

and

$$\frac{d}{dr}\begin{pmatrix} W \\ T \end{pmatrix} = \begin{pmatrix} \frac{1}{r} & \frac{1}{\mu} \\ \frac{\mu(l-1)(l+2)}{r^2} - \rho\omega^2 & -\frac{3}{r} \end{pmatrix}\begin{pmatrix} W \\ T \end{pmatrix}. \tag{8.33}$$

(We have dropped subscripts l and n from the dependent variables and from ω. Note that m does not enter the matrix equations.)

Thus the vibrations of a spherically symmetric Earth without rotation can be separated into two types of modes. One is the spheroidal mode with horizontal wavefunctions $\mathbf{R}_l^m$ and $\mathbf{S}_l^m$ and radial wavefunctions determined by (8.32). The other is the toroidal or torsional mode with horizontal wavefunction $\mathbf{T}_l^m$ and radial wavefunction determined by (8.33). It is clear from a comparison of matrices in (8.32) and (7.28) that the spheroidal modes include Rayleigh waves. Comparing (8.33) and (7.24), we see that the toroidal modes include Love waves. Such comparisons require that the horizontal wavenumber k of surface waves be identified with $[l(l+1)]^{1/2}/r$ for free oscillations. We shall present a more detailed comparison of surface waves and free oscillations in the next section.

To find the normal modes, we must solve the eigenvalue-eigenvector problems (8.32) and (8.33) under the boundary conditions that the solutions are regular at $r = 0$ and the tractions vanish at the Earth's surface. The numerical method and Rayleigh-Ritz method described in Chapter 7 can be adapted to solve these problems. One method of handling the condition at $r = 0$ (Takeuchi and Saito, 1972) is to assume that the Earth is uniform in $r < r_1$ and solve the differential equations in powers of r. The power series are then evaluated at $r = r_1$, and numerical integration is initiated from these values and taken upward. For each integer l, there are eigenvalues ${}_n\omega_l (n = 0, 1, 2, \ldots)$, and for each ${}_n\omega_l$ there is an eigenfunction for the motion-stress vector. Again, we note a degeneracy, in that eigenfrequency and radial eigenfunction are independent of m in the range $-l \leq m \leq l$.

The normalization formula (8.26) for normal modes together with that for horizontal wavefunctions leads to the following normalization for radial functions:

$$\int_0^{r_\oplus} \rho(r)\{[{}_nU_l(r)]^2 + [{}_nV_l(r)]^2\}r^2\,dr = 1 \tag{8.34}$$

and

$$\int_0^{r_\oplus} \rho(r)[{}_nW_l(r)]^2 r^2\,dr = 1, \tag{8.35}$$

where $r_\oplus$ is the radius of the Earth.

With the normal-mode solution thus completely defined, our next step toward computing the point-source response is to evaluate the strain tensor for the normal-mode displacement at the source point, since this is needed in (8.29). For the spherical coordinates (r, Δ, ϕ), we obtain the strain components from (2.45) by putting $h^r = 1$, $h^\Delta = r$, and $h^\phi = r \sin \Delta$:

$$\begin{aligned}
e_{rr} &= \frac{\partial u_r}{\partial r},\\
e_{\Delta\Delta} &= \frac{1}{r}\frac{\partial u_\Delta}{\partial \Delta} + \frac{u_r}{r},\\
e_{\phi\phi} &= \frac{1}{r \sin \Delta}\frac{\partial u_\phi}{\partial \phi} + \frac{u_\Delta}{r}\cot \Delta + \frac{u_r}{r},\\
e_{\Delta\phi} = e_{\phi\Delta} &= \frac{1}{2}\left[\frac{1}{r}\left(\frac{\partial u_\phi}{\partial \Delta} - u_\phi \cot \Delta\right) + \frac{1}{r \sin \Delta}\frac{\partial u_\Delta}{\partial \phi}\right],\\
e_{\phi r} = e_{r\phi} &= \frac{1}{2}\left[\frac{1}{r \sin \Delta}\frac{\partial u_r}{\partial \phi} + \frac{\partial u_\phi}{\partial r} - \frac{u_\phi}{r}\right],\\
e_{r\Delta} = e_{\Delta r} &= \frac{1}{2}\left[\frac{\partial u_\Delta}{\partial r} - \frac{u_\Delta}{r} + \frac{1}{r}\frac{\partial u_r}{\partial \Delta}\right],
\end{aligned} \tag{8.36}$$

where u_r, u_Δ, and u_ϕ are the r, Δ, ϕ components of displacement given in (8.30).

Following Gilbert and Dziewonski (1975), we shall put the source point at the pole ($\Delta = 0$) and evaluate normal-mode strain components as $\lim_{\Delta \to 0} \mathbf{e}(\Delta)$. We then find that all the components vanish for $|m| > 2$. The strain components for a spheroidal mode are shown in Table 8.1. Corresponding results for a

TABLE **8.1**
Strain components for a spheroidal mode.

	$m = 0$	$m = \pm 1$	$m = \pm 2$
e_{rr}	$b_0 \dfrac{dU}{dr}$	0	0
$e_{\Delta\Delta}$	$\dfrac{b_0}{r}\left[U - \frac{1}{2}\sqrt{l(l+1)}V\right]$	0	$\dfrac{b_0\sqrt{(l+2)(l-1)}}{4}\dfrac{V}{r}$
$e_{\phi\phi}$	$e_{\Delta\Delta}$	0	$-e_{\Delta\Delta}$
$2e_{r\Delta}$	0	$\dfrac{-b_0}{2}\left[\dfrac{\sqrt{l(l+1)}}{r}U + \dfrac{dV}{dr} - \dfrac{V}{r}\right]$	0
$2e_{r\phi}$	0	$2ime_{r\Delta}$	0
$2e_{\Delta\phi}$	0	0	$ime_{\Delta\Delta}$

toroidal mode are shown in Table 8.2. The constant b_0 appearing in these tables is $[(2l + 1)/4\pi]^{1/2}$. Evaluating the strain components at the source level $r = r_s$ and substituting them into (8.29), we find the response of the Earth to a point source with an arbitrary moment tensor that varies as a step function in time. For general time-dependence we can use $M_{pq}(t) = \int_{-\infty}^{\infty} \dot{M}_{pq}(\tau)H(t - \tau)\, d\tau$. Interpreting this integrand as a step-function source acting at $t = \tau$, we find from (8.29) that

$$\mathbf{u}(\mathbf{x}, t) = \sum_i \left[{}_i e^*_{pq}(\mathbf{x}_s)\dot{M}_{pq}(t)\right] * {}_i\mathbf{u}(\mathbf{x})\left(\frac{1 - \exp(-\omega_i t/2Q_i)\cos\omega_i t}{\omega_i^2}\right). \tag{8.37}$$

TABLE **8.2**
Strain components for a toroidal mode.

	$m = 0$	$m = \pm 1$	$m = \pm 2$
e_{rr}	0	0	0
$e_{\Delta\Delta}$	0	0	$\dfrac{imb_0}{8}\sqrt{(l+2)(l-1)}\dfrac{W}{r}$
$e_{\phi\phi}$	0	0	$-e_{\Delta\Delta}$
$2e_{r\Delta}$	0	$\dfrac{-imb_0}{2}\left[\dfrac{dW}{dr} - \dfrac{W}{r}\right]$	0
$2e_{r\phi}$	0	$\dfrac{b_0}{2}\left[\dfrac{dW}{dr} - \dfrac{W}{r}\right]$	0
$2e_{\Delta\phi}$	0	0	$\dfrac{-b_0\sqrt{(l+2)(l-1)}}{2}\dfrac{W}{r}.$

The asterisk here following $\dot{M}_{pq}$ indicates convolution, and we have written out (8.37) to indicate that the point source is naturally characterized by its moment-rate tensor, $\dot{\mathbf{M}}(t)$.

In this section we have followed the simple and straightforward steps due to Gilbert (1971) and Gilbert and Dziewonski (1975) in deriving the formula for excitation of free oscillations. Earlier, Saito (1967) solved the same problem using a method similar to the one we described for surface-wave excitation in Chapter 7, and he obtained a formula equivalent to (8.29). Saito's results were used by Mendiguren (1973) in devising a stacking technique for high-resolution identification of spectral peaks, as described in Chapter 11. Figure 8.4 shows a

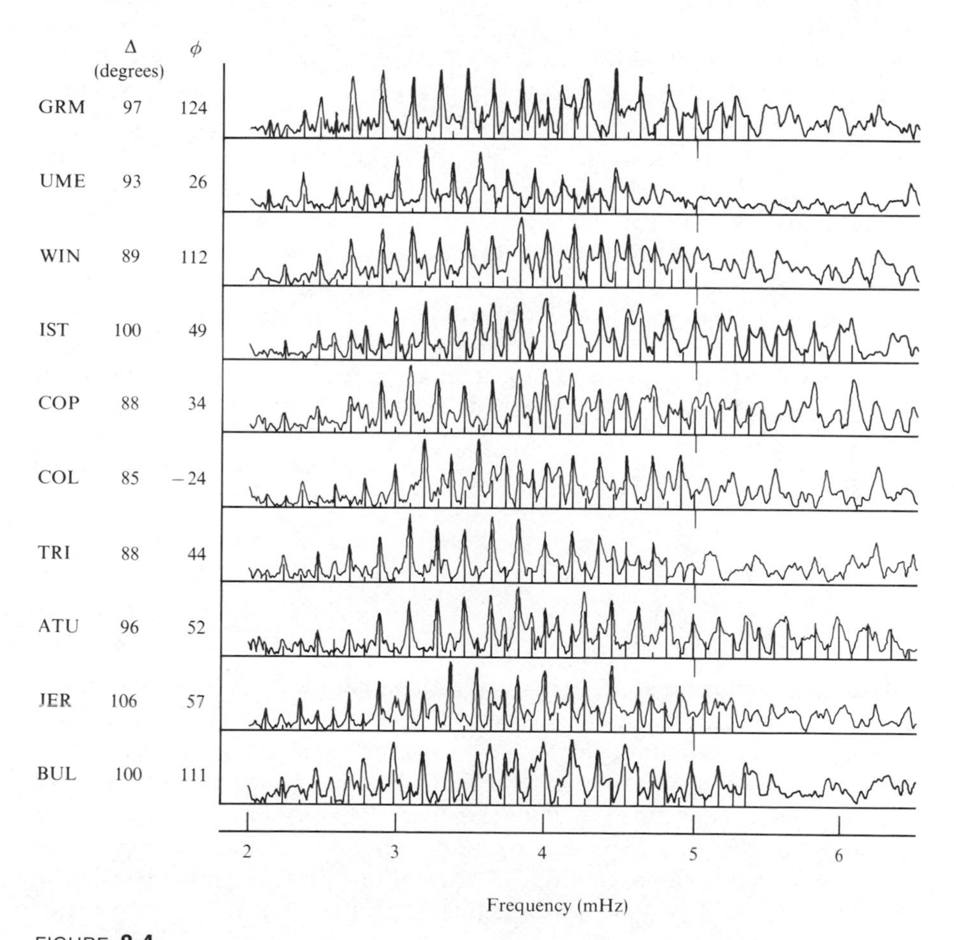

FIGURE **8.4**

Continuous lines indicate the observed radial displacement spectrum. Vertical bars are the theoretical spectral lines for the fundamental spheroidal mode. Δ is the epicentral distance. ϕ is the azimuth at the epicenter measured clockwise from North. [From Mendiguren, 1973.]

comparison between observed and calculated spectral peaks at several WWSSN stations for a large deep earthquake in Colombia. The continuous lines indicate the observed radial displacement spectrum, and the vertical bars show the theoretical amplitudes of free oscillation calculated by Mendiguren (1973) using Saito's formula and a focal mechanism determined from the observed P-wave first-motion pattern.

8.3 Surface Waves on the Spherical Earth

In the preceding section, we found that the coefficient matrix for free oscillations ((8.32) or (8.33)) approaches that for surface waves as r tends to infinity provided that the wavenumber k of surface waves is replaced by $[l(l+1)]^{1/2}/r$ for free oscillations. For most purposes here, $[l(l+1)]^{1/2} \sim l + \frac{1}{2}$, and we can identify surface wave k as $(l + \frac{1}{2})/r$. (In fact, $k \leftrightarrow (l + \frac{1}{2})/r$ is more appropriate than $k \leftrightarrow [l(l+1)]^{1/2}/r$, as shown in Box 9.3 after equation (9).) Since the wavelength λ is $2\pi/k$, this means that λ is equal to $2\pi r/(l + \frac{1}{2})$, so that the circumference is not exactly an integer multiple of wavelength; there is half a wavelength extra. In other words, the distance between neighboring nodes of a free oscillation cannot be equal to half a wavelength everywhere. To examine this more closely, we shall consider the simple case of spheroidal modes generated by an explosive point source $M_{rr} = M_{\Delta\Delta} = M_{\phi\phi} = M_0$, with $M_{\Delta\phi} = M_{\phi r} = M_{r\Delta} = 0$. Substituting from Table 8.1 into (8.29), we find that the r-component of displacement is the spheroidal mode sum

$$u_r(\mathbf{x}, t) = \sum_n \sum_l {}_nA_l \cdot P_l(\cos\Delta) \cdot [1 - \exp(-{}_n\omega_l t/2_n Q_l) \cos {}_n\omega_l t], \quad (8.38)$$

where

$${}_nA_l = \frac{2l+1}{4\pi({}_n\omega_l)^2}\left[\frac{d}{dr}{}_nU_l + \frac{2_nU_l}{r} - \frac{(l(l+1))^{1/2}}{r}{}_nV_l\right]\Bigg|_{r=r_s} {}_nU_l(r), \quad (8.39)$$

and we have assumed that $M_0(t)$ is a step with amplitude 1 dyne-cm. In (8.38) we are using $\mathbf{x} = (r, \Delta, \phi)$, and the sum over all modes i has been written as a sum over the fundamental and overtones n and order numbers l. Only $m = 0$ contributes. Thus, (8.38) shows a superposition of standing-wave patterns called *zonal harmonics*, determined by $P_l(\cos\Delta)$. Since $P_l(\cos\Delta)$ has exactly l nodes in the interval $0 < \Delta < \pi$, there are l cycles of oscillation around the great circle. On the other hand, the asymptotic expansion of $P_l(\cos\Delta)$, which is valid for large l except near $\Delta = 0$ or $\Delta = \pi$, is (see Boxes 8.3 and 9.3)

$$P_l(\cos\Delta) \sim \left(\frac{2}{l\pi\sin\Delta}\right)^{1/2} \cos[(l + \tfrac{1}{2})\Delta - \pi/4]. \quad (8.40)$$

This shows again that the wavelength is approximately $2\pi r/(l + \frac{1}{2})$ except near $\Delta = 0$ or $\Delta = \pi$. Taking l cycles of such waves, we get $2\pi rl/(l + \frac{1}{2})$ instead of

$2\pi r$. This means that the distance between neighboring nodes in the vicinity of $\Delta = 0$ or π is longer than elsewhere, and therefore that the apparent phase velocity is faster in these special regions.

In measuring surface-wave phase velocity, the above effect causes an apparent phase advance amounting to $\pi/2$ at each polar passage, which must be allowed for if the path contains the epicenter or its antipode. (The phase shift of $\pi/4$ in (8.40) is doubled for entrance to and exit from the pole.) This is known as the *polar phase shift*, introduced in modern seismology by Brune *et al.* (1961), who

BOX **8.2**

An example of the Poisson sum formula

Given a function $g = g(v)$, we shall first prove that

$$\sum_{l=-\infty}^{\infty} g(l + \tfrac{1}{2}) = \sum_{s=-\infty}^{\infty} (-1)^s \int_{-\infty}^{\infty} g(v) e^{2i\pi s v}\, dv. \qquad (1)$$

Then, applying (1) to the function $f(v)$, where $f = g$ for $v \geq 0$ and $f = 0$ for $v < 0$, it follows that

$$\sum_{l=0}^{\infty} g(l + \tfrac{1}{2}) = \sum_{s=-\infty}^{\infty} (-1)^s \int_{0}^{\infty} g(v) e^{2i\pi s v}\, dv. \qquad (2)$$

To prove (1), define $S = S(x)$ by $S(x) = \sum_{l=-\infty}^{\infty} g(l + \frac{1}{2} + x)$. Then S is periodic with period 1. We can write out the Fourier series expansion for $S(x)$ as

$$S(x) = \sum_{s=-\infty}^{\infty} S_s e^{-2i\pi s x}, \qquad \text{where } S_s = \int_0^1 S(y) e^{2i\pi s y}\, dy.$$

Taking $x = 0$ and substituting for S_s and $S(y)$, we get

$$S(0) = \sum_{s=-\infty}^{\infty} \int_0^1 \sum_{l=-\infty}^{\infty} g(l + \tfrac{1}{2} + y) e^{2i\pi s y}\, dy,$$

and note that $S(0)$ equals the left-hand side of (1). Moreover,

$$e^{2i\pi s y} = (-1)^s e^{2i\pi s(l + 1/2 + y)}$$

because s, l, and sl, are integers. Thus

$$\sum_{l=-\infty}^{\infty} g(l + \tfrac{1}{2}) = \sum_{s=-\infty}^{\infty} (-1)^s \int_0^1 \sum_{l=-\infty}^{\infty} g(l + \tfrac{1}{2} + y) e^{2i\pi s(l + 1/2 + y)}\, dy$$

$$= \sum_{s=-\infty}^{\infty} (-1)^s \sum_{l=-\infty}^{\infty} \int_{l+1/2}^{l+3/2} g(v) e^{2i\pi s v}\, dv,$$

which equals the right-hand side of (1).

showed that it resolved what previously had been inconsistent results for the phase velocities measured over minor arcs, major arcs, and full great circles.

For a more quantitative comparison of free oscillations and surface waves, we shall use the Poisson sum formula for each overtone in (8.38). From equation (2) of Box 8.2, we then obtain

$$u_r(\mathbf{x}, t) = \sum_n \int_0^\infty d\nu \sum_{s=-\infty}^{\infty} (-1)^s ({}_nA_{\nu-1/2}) P_{\nu-1/2}(\cos\Delta) C(\nu, t) e^{2i\pi s\nu}, \quad (8.41)$$

where

$$C(\nu, t) = H(t)[1 - \cos({}_n\omega_{\nu-1/2}t) \cdot \exp(-{}_n\alpha_{\nu-1/2}t)] \quad (8.42)$$

and α is related to the temporal Q via $\alpha = \omega/2Q$.

Following Gilbert (1976), we rewrite the sum over s in (8.41) using Legendre functions of the second kind, denoted by $Q_l(\cos\Delta)$ (see Box 8.3.) Then

$$u_r(\mathbf{x}, t) = \sum_n \int_0^\infty d\nu\, {}_nA_{\nu-1/2} \sum_{N=1}^{\infty} R_N(\nu, \Delta) C(\nu, t), \quad (8.43)$$

where, for N odd,

$$R_N = (-1)^{(N-1)/2}\Big\{P_{\nu-1/2}(\cos\Delta)\cos[(N-1)\pi\nu] + \frac{2}{\pi} Q_{\nu-1/2}(\cos\Delta)\sin[(N-1)\pi\nu]\Big\}, \quad (8.44)$$

and for N even,

$$R_N = (-1)^{N/2}\left\{P_{\nu-1/2}(\cos\Delta)\cos(N\pi\nu) - \frac{2}{\pi} Q_{\nu-1/2}(\cos\Delta)\sin(N\pi\nu)\right\}. \quad (8.45)$$

The equivalence of (8.41) and (8.43) follows from

$$\begin{aligned} R_1 + R_2 + R_3 + R_4 + \cdots \\ &= P_{\nu-1/2}(1 - \cos 2\pi\nu - \cos 2\pi\nu + \cos 4\pi\nu + \cos 4\pi\nu - \cos 6\pi\nu - \cdots) \\ &\quad + \frac{2}{\pi} Q_{\nu-1/2}(0 + \sin 2\pi\nu - \sin 2\pi\nu - \sin 4\pi\nu + \sin 4\pi\nu \\ &\quad + \sin 6\pi\nu - \cdots) \\ &= \sum_{s=-\infty}^{\infty} (-1)^s P_{\nu-1/2} e^{2i\pi\nu s}. \end{aligned}$$

BOX **8.3**

Different Legendre functions and their asymptotic approximations

In Box 8.1 we examined Legendre polynomials $P_l(\cos\Delta)$. Although these are the only solutions that are physically allowed near $\Delta = 0$ and $\Delta = 180°$, it is often worthwhile to work with different solutions to the Legendre equation, since these may locally be suitable for representing seismic motions at distances away from $\Delta = 0$ and $\Delta = 180°$.

Quoting from Nussenzveig (1965),

$$P_{\nu-1/2}(\cos\Delta) = \left(\frac{2}{\pi\nu\sin\Delta}\right)^{1/2}\left[\cos\left(\nu\Delta - \frac{\pi}{4}\right) + \frac{\cot\Delta}{8\nu}\sin\left(\lambda\Delta - \frac{\pi}{4}\right) + O\left(\frac{1}{\nu^2}\right)\right] \quad (1)$$

for $0 < \varepsilon \le \Delta \le \pi - \varepsilon$, $|\nu| \gg 1$, and $|\nu|\varepsilon \gg 1$. This solution, behaving like

$$(\sin\Delta)^{-1/2}\cos(\nu\Delta - \tfrac{1}{4}\pi),$$

suggests that a linearly independent solution might exist, with asymptotic behavior like $(\sin\Delta)^{-1/2}\sin(\nu\Delta - \frac{1}{4}\pi)$. By summing and subtracting these solutions, one can thus obtain traveling waves. In fact, there are Legendre solutions $Q^{(i)}$ $(i = 1, 2)$ with the following properties:

$$Q^{(1)(2)}_{\nu-1/2}(\cos\Delta) = \frac{e^{\mp i(\nu\Delta - \pi/4)}}{(2\pi\nu\sin\Delta)^{1/2}}\left[1 \pm i\frac{\cot\Delta}{8\nu} + O\left(\frac{1}{\nu^2}\right)\right] \quad (2)$$

for $0 < \varepsilon \le \Delta \le \pi - \varepsilon$, $|\nu| \gg 1$, $|\nu|\varepsilon \gg 1$, and $(\nu - \frac{1}{2})$ not near the negative integers, where $Q^{(i)}_{\nu-1/2}$ has poles. These poles cancel in the sum

$$P_l = Q_l^{(1)} + Q_l^{(2)}.$$

The formal definition of $Q_l^{(i)}$ is

$$Q_l^{(1)(2)} = \frac{1}{2}\left(P_l \pm \frac{2i}{\pi}Q_l\right), \quad (3)$$

where Q_l is the Legendre function of the second kind, but the major properties of $Q_l^{(i)}$ are those that follow from (2).

Two further solutions of Legendre's equation are given by the combinations (8.44) and (8.45). Writing these in terms of $Q^{(1)}$ and $Q^{(2)}$, one obtains, for N odd,

$$\begin{aligned} R_N &= (-1)^{(N-1)/2}[Q^{(1)}_{\nu-1/2}(\cos\Delta)e^{-i(N-1)\pi\nu} + Q^{(2)}_{\nu-1/2}(\cos\Delta)e^{i(N-1)\pi\nu}] \\ &\sim (-1)^{(N-1)/2}\left(\frac{2}{\pi\nu\sin\Delta}\right)^{1/2}\cos\left\{\nu[(N-1)\pi + \Delta] - \frac{\pi}{4}\right\}, \end{aligned} \quad (4)$$

and for N even,

$$\begin{aligned} R_N &= (-1)^{N/2}[Q^{(2)}_{\nu-1/2}(\cos\Delta)e^{-iN\pi\nu} + Q^{(1)}_{\nu-1/2}(\cos\Delta)e^{iN\pi\nu}] \\ &\sim (-1)^{N/2}\left(\frac{2}{\pi\nu\sin\Delta}\right)^{1/2}\cos\left\{\nu[N\pi - \Delta] + \frac{\pi}{4}\right\}. \end{aligned} \quad (5)$$

From (4) and (5), it is clear that each R_N is a standing-wave pattern.

A uniformly asymptotic approximation for $P_{\nu-1/2}$, which works even near $\Delta = 0$, is

$$P_{\nu-1/2}(\cos\Delta) = \left(\frac{\Delta}{\sin\Delta}\right)^{1/2}\left[J_0(\nu\Delta) + \frac{1}{8}(\Delta\cot\Delta - 1)\frac{J_1(\nu\Delta)}{\nu\Delta} + O\left(\frac{1}{\nu^2}\right)\right].$$

The effort to identify body waves and surface waves, in some expression that includes these as well as other waves, is primarily an exercise in manipulating the six Legendre functions, P, Q, $Q^{(1)}$, $Q^{(2)}$, R_N (N even), R_N (N odd). This problem will recur in Chapter 9, where we choose to use a Watson transformation rather than a Poisson sum.

The reason for introducing the R_N is that they are the appropriate horizontal wavefunctions for the Nth orbit of waves around the Earth. The orbit numbering is shown in Figure 8.5. Note, however, that each R_N as defined in (8.44)–(8.45) is still a standing wave, rather than a traveling wave. This is demonstrated in Box 8.3, and our point here is that for purposes of comparison with surface waves we should expect, for each N, to have horizontal phase functions that, for positive wavenumber, are standing waves. This follows because in Chapters 6 and 7 we frequently found solutions

$$u(X, \omega) = \int_0^\infty f(k, \omega) J_0(kX)\, dk \tag{8.46}$$

for flat Earth models. Here we have introduced X as the horizontal distance (the range); k is horizontal wavenumber, and J_0 is the zero-order Bessel function. The integrand of (8.46) is a standing wave, and to obtain a traveling-wave representation we used $f(-k, \omega) = -f(k, \omega)$ and

$$u(X, \omega) = \tfrac{1}{2}\int_{-\infty}^\infty f(k, \omega) H_0^{(1)}(kX)\, dk \tag{8.47}$$

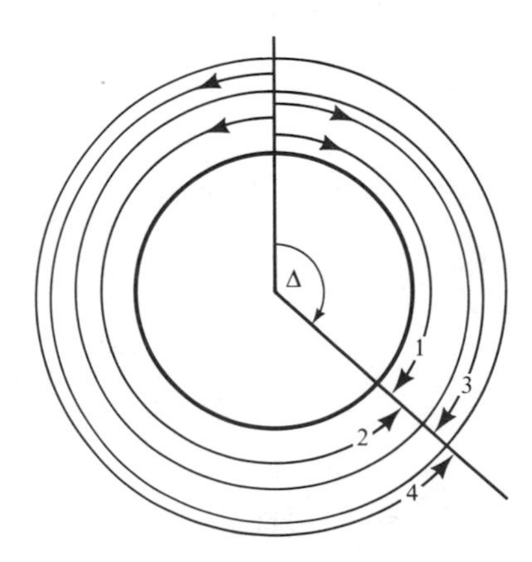

FIGURE **8.5**

Orbit numbering for waves traveling around the Earth. $N = 1$ is the minor arc; $N = 2$ is the major arc; $N \geq 3$ involves at least one great-circle path.

(see, e.g., (6.15)). It is interesting that the traveling wavefunctions $H_0^{(1)}$ and $H_0^{(2)}$ for a flat Earth have a branch cut on the negative real-wavenumber axis ($k < 0$), and that the analogue wavefunctions for a spherical Earth, $Q_{\nu-1/2}^{(2)}$ and $Q_{\nu-1/2}^{(1)}$, have poles on the negative real-wavenumber axis ($\nu < 0$). In Chapter 9 we shall find several further examples of strings of poles that, in many ways, behave like branch cuts.

Working only with the first-arriving waves from (8.43), $N = 1$, we write $l + \frac{1}{2} = \nu = kr$, where $r = |\mathbf{x}|$ is the radial position of the receiver (usually $r = r_\oplus$, the Earth's radius). Then

$$u_r(\mathbf{x}, t) = \left(\frac{\Delta}{\sin \Delta}\right)^{1/2} \sum_n \int_0^\infty r\, dk\, {}_nA(k) J_0(kX) C(kr, t), \tag{8.48}$$

where we have used

$$R_1(\nu, \Delta) = P_{\nu-1/2}(\cos \Delta) \sim \left(\frac{\Delta}{\sin \Delta}\right)^{1/2} J_0(kX). \tag{8.49}$$

X is the horizontal range $r\Delta$, so that $\nu\Delta = kX$; and ${}_nA(k) = {}_nA_{\nu-1/2}$.

Ignoring the static offset term in $C(kr, t)$, we note that the Fourier transform $u_r(\mathbf{x}, \omega)$ is given by substituting

$$\frac{1}{2}\left[\frac{1}{i[\omega + {}_n\omega(k)] - {}_n\alpha(k)} + \frac{1}{i[\omega - {}_n\omega(k)] - {}_n\alpha(k)}\right]$$

for C in (8.48). The coefficient of $J_0(kX)$ in (8.48) is odd in k, hence

$$u_r(\mathbf{x}, \omega) = \frac{1}{4}\left(\frac{\Delta}{\sin \Delta}\right)^{1/2} \sum_n \int_{-\infty}^\infty r\, dk\, {}_nA(k) H_0^{(1)}(kX) \frac{1}{i[\omega - {}_n\omega(k)] - {}_n\alpha(k)}. \tag{8.50}$$

The other term, involving $\omega + {}_n\omega(k)$, has been dropped, since it generates a pole in the lower half of the complex k-plane, and our intention is to complete a closed path in the upper half-plane. To locate the pole of the above integrand and evaluate its residue, we can obtain the Taylor series for ${}_n\omega(k)$ as k varies near that value $k_n(\omega)$ which is the eigenvalue for the nth surface-wave mode. Thus

$${}_n\omega(k) = \omega + (k - k_n)U_n(\omega) + \cdots,$$

where $U_n(\omega) = d\omega/dk =$ the group velocity. Almost always the group velocity is positive, so that the pole of (8.50) lies in the first quadrant of the complex k-plane, at $k = k_n(\omega) + i\alpha_n/U_n(\omega)$. Here α_n is based on a temporal Q, so that if we shift to a description of attenuation based on spatial Q we must replace $\alpha_n/U_n = \omega/2Q_n^{\text{temporal}}\, U_n$ by $\omega/2Q_n{}^{\text{spatial}}\, c_n = k_n/2Q_n^{\text{spatial}}$ (see (7.92)). Using the integration path shown in Figure 8.6, with the far-field approximation to $H_0^{(1)}$, we can

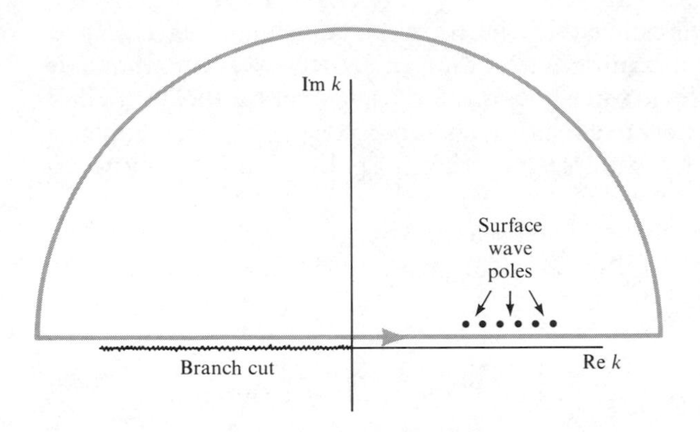

FIGURE **8.6**
Features in the complex wavenumber plane, used to evaluate the surface waves present in equation (8.50).

evaluate the residue from the poles in the first quadrant (one pole for each n; $n = 0, 1, 2, \ldots$).

$$u_r(\mathbf{x}, \omega) = i\frac{\pi}{2}\left(\frac{\Delta}{\sin\Delta}\right)^{1/2} \sum_n \frac{rA(k_n)}{U_n}\left(\frac{2}{\pi k_n X}\right)^{1/2} \exp\left[i\left(k_n X + \frac{\pi}{4}\right)\right] \exp(-k_n X/2Q_n). \tag{8.51}$$

Here we are writing $A(k_n)$ for ${}_nA(k)$ evaluated at $k = k_n(\omega)$, and we have assumed that factors in (8.50) are slowly varying functions of k near each pole, except for the phase function e^{ikX}. Equation (8.51) gives a representation of surface waves due to an explosive unit step-function source in a spherical Earth.

In order to compare (8.51) with the corresponding formula (7.151) for Rayleigh waves in a plane-stratified medium, we must allow for the difference in normalization of radial/vertical eigenfunctions. In fact, $r_n U_l(r)$ corresponds to $r_2(z)/\sqrt{2I_1}$; $rd({}_nU_l(r))/dr$ to $-(dr_2(z)/dz)/\sqrt{2I_1}$; and $r_n V_l(r)$ to $r_1(z)/2I_1$. Thus $rA(k_n)$ in (8.51) is replaced to first order by

$$-\frac{k_n}{2\pi\omega^2}\left[\frac{dr_2}{dz} + k_n r_1\right]\Bigg|_{\substack{\text{source}\\ \text{level}}} \frac{r_2(z)}{2I_1}.$$

Noting that (8.51) is for *upward* motion, but (7.151) is for the *downward* component, it follows that the two formulas are virtually identical. The only differences lie in the spatial attenuation factor, which we have explicitly included in (8.51), and in the factor $(\Delta/\sin\Delta)^{1/2}$, which accounts for the azimuthal geometrical spreading over the surface of a sphere as opposed to the spreading over a flat surface.

In summary, the effect of the Earth's sphericity can be taken into account by the use of a geometrical spreading factor proportional to $(\sin \Delta)^{-1/2}$ and by appropriate eigenfunctions obtained by solving (8.32) and/or (8.33) for the motion-stress vector in the spherical Earth model.

8.4 Free Oscillations of a Self-gravitating Earth

Gravity does not affect the torsional modes of free oscillation, because these have no radial component of displacement and are divergence-free, so that the density distribution is unchanged. But density perturbations do occur in spheroidal motions. Since gravity is effectively a body force, its time-dependence will influence the overall motion. Self-gravitation cannot be neglected for spheroidal modes with periods longer than about 500 sec, and before giving the necessary theory it is interesting to look at the historical background.

From the theory of free oscillations of a homogeneous solid elastic sphere, published by Lamb in 1882, it is apparent that the mode with the greatest period is ${}_0S_2$. Bromwich (1898) studied the period equation numerically, and was able to incorporate the effect of self-gravitation in the case of an incompressible medium. He showed that for an incompressible sphere having the same size and mass as the Earth, and with rigidity that of steel, the period of ${}_0S_2$ is decreased from about 65 minutes to about 55 minutes by including self-gravitation. Love (1911) was able to allow for self-gravitation even in a compressible medium, and by taking Poisson's ratio equal to $\frac{1}{4}$ in an Earth model that was otherwise the same as that considered by Bromwich, he found the period of ${}_0S_2$ would be almost exactly one hour. Clearly, if this mode were to be observed, instruments would be needed that had sensitivity at periods very much longer than the periods of seismic motion then being routinely recorded.

In the 1940's and 1950's, it became apparent that observed seismic surface waves had properties that to a large extent could be understood in terms of the theory for surface waves within a stack of homogeneous flat-lying plates welded together at the interfaces. The theory for such a medium was extensively studied (Haskell, 1953; Ewing et al., 1957), and in particular the dispersion for such an elastic medium could be calculated. There was clearly a need to design seismometers that were useful at longer and longer periods, because the observed surface-wave dispersion could then be used to infer Earth structure deeper within the mantle. Hugo Benioff, at the California Institute of Technology, was a leader in instrument design, and for the Kamchatka earthquake of 1952 November 4 his quartz-rod strain seismometer produced the record shown in Figure 8.7. As reported by Benioff et al. (1954), the 57-min oscillation "may represent free oscillations of the Earth as a whole." Many different individuals were stimulated by this result to develop better seismometers for very long periods, and also to undertake the theoretical and numerical efforts needed

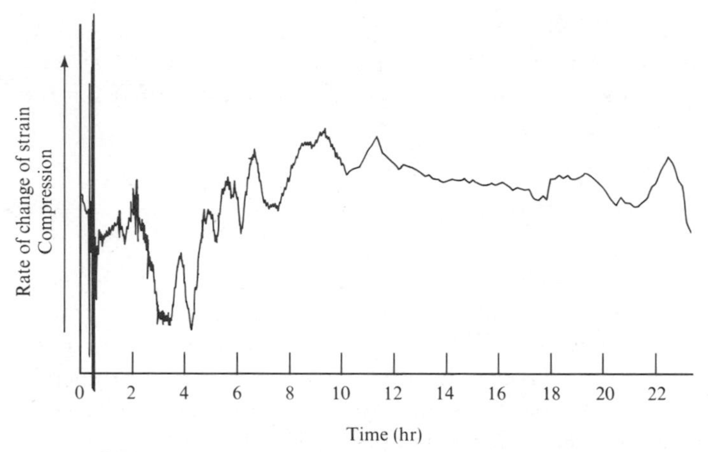

FIGURE **8.7**

Seismogram of the Kamchatka earthquake of 1952 November 4, recorded by the Benioff strain seismograph at Pasadena, California. Beginning approximately $3\frac{1}{2}$ hours after the time of the earthquake, there appears an oscillation, or wave train, of four cycles, originally reported as having a period of 57 min. [Reproduced from Benioff, 1958; copyrighted by the American Geophysical Union.]

to predict the free oscillations of realistic Earth models. Thus Alterman et al. (1959) published the theoretical value for the ${}_0S_2$-period in Earth models that had earlier been proposed by Bullen and Bullard: in different models, with and without an inner solid core, the values were always around 53.5–53.7 min. Concurrent improvements in instrumentation had been made, but observational confirmation had to await a large enough earthquake. In 1960 May 22, one of the greatest earthquakes of modern times occurred in Chile. (The magnitude was greater than 8, but this event was so big that it saturated conventional magnitude scales. See Kanamori, 1977.) Amid great excitment, three different groups of investigators reported at the IASPEI (International Association of Seismology and Physics of the Earth's Interior) meeting, held in the summer of 1960 in Helsinki, that they had observed a wide range of normal modes of the whole Earth (Benioff et al., 1961; Ness et al., 1961; Alsop et al., 1961a). Clearly, a whole new branch of seismology had been opened.

In this section we give the theory for spheroidal modes in an Earth model that is spherically symmetric, nonrotating, self-gravitating, elastic, and having an isotropic stress-strain relation and an isotropic initial stress.

The equations of elastic motion that we developed in Chapter 2 were for the most part based upon a Lagrangian formulation in which the particle $\mathbf{x}$ at time t_0 was assumed to move to $\mathbf{x} + \mathbf{u} = \mathbf{X}$ at time t, and we studied $\mathbf{u}$ as the function $\mathbf{u} = \mathbf{u}(\mathbf{x}, t)$. Unfortunately, the Lagrangian formulation does not easily handle variations of density and the consequent effects of a fluctuating

BOX **8.4**

A fortunate accident or a unique record of true ground motions?

Although the seismogram shown in Figure 8.7 was a great stimulus to the study of normal modes, it became apparent in the early 1960's that the long-period oscillations recorded at Pasadena for this event are highly anomalous. (i) The reported period of 57 min is longer than current estimates for ${}_0S_2$ ($\sim$54 min). (ii) The 57-min period is so dominant for a part of the record that it is hard to imagine why other free oscillations are so weak. (iii) Only about four cycles have periods of 57 min, and it is hard to see why they should cease. For several years it was therefore thought that some instrumental defect had produced these long-period cycles. However, Kanamori (1976) has shown (i) that the period is not 57 min, but (after careful Fourier analysis) about 54 min. He has speculated (ii) that possibly some post-seismic slow deformation of the Kamchatka source region occurred, with a time scale so long that ${}_0S_2$ was favorably excited; and (iii) that the cessation after four cycles was due to destructive interference from the different members of ${}_0S_2^m$ ($-2 \leq m \leq 2$), which are significantly different due to the Earth's rotation, or that indeed there was an instrumental defect beginning at about seven hours into the record.

gravity field, so we shall here take an Eulerian approach: **x**, **X**, and **u** are defined as above, but we study quantities such as **u** and ρ as functions of ($\mathbf{X}$, t).

Let us then consider a spherical Earth model that is in hydrostatic equilibrium under the self-gravitation. The equilibrium state is described by its density $\rho_0(r)$, pressure $P_0(r)$, and gravitational potential $V_0(r)$, all of which are functions only of r. Since V_0 is due to the density distribution ρ_0, we can apply Poisson's equation

$$\nabla^2 V_0 = -4\pi\gamma\rho_0, \tag{8.52}$$

where γ is the gravitational constant ($\sim 6.670 \times 10^{-8}$ dyne cm^2/g^2). The equation of equilibrium is given by matching the body force to the pressure gradient, i.e.,

$$-\rho_0 g_0 \hat{\mathbf{r}} = \rho_0 \nabla V_0 = \nabla P_0, \tag{8.53}$$

where $g_0 = g_0(r)$, which is positive, is the magnitude of the unperturbed gravitational force per unit volume.

Superimposed on this initial state, we consider a small displacement perturbation $\mathbf{u}(\mathbf{X}) \exp(-i\omega t)$. We shall describe the density and gravitational potential in the perturbed state by $\rho(\mathbf{X}, t)$ and $V_0 + K(\mathbf{X}) \exp(-i\omega t)$, respectively, taking here an Eulerian approach.

From the conservation of mass, the increase in mass within a volume V (fixed in space) is equal to the influx of mass through the surface S enclosing V, so that

$$\int_V \rho\, dV = \int_V \rho_0\, dV - \int_S \rho_0 \mathbf{u} \exp(-i\omega t) \cdot \mathbf{n}\, dS, \tag{8.54}$$

where $\mathbf{n}$ is the outward unit normal for the surface element dS. (Equation (8.54) is an approximation, in that the density of material crossing S is changing with time. We have taken it as ρ_0, which is adequate for a first-order approximation because $\mathbf{u}$ is small.) Changing the surface integral to a volume integral by Gauss's theorem, we obtain

$$\rho(\mathbf{X}, t) = \rho_0 - \nabla \cdot (\rho_0 \mathbf{u} \exp(-i\omega t)).$$

Since ρ_0 is a function only of r, we can rewrite the above equation as

$$\rho = \rho_0 - \left(u_r \frac{d\rho_0}{dr} + \rho_0 \nabla \cdot \mathbf{u} \right) \exp(-i\omega t). \tag{8.55}$$

The gravitational potential in the perturbed state satisfies the Poisson equation

$$\nabla^2 (V_0 + K \exp(-i\omega t)) = -4\pi\gamma\rho,$$

which with (8.52) and (8.55) can be used to show that

$$\nabla^2 K = 4\pi\gamma \left(u_r \frac{d\rho_0}{dr} + \rho_0 \nabla \cdot \mathbf{u} \right). \tag{8.56}$$

To obtain the equation of motion, we use Cartesian coordinates (X_1, X_2, X_3) and begin with the relation between rate of change of momentum, body force (gravity), and stress gradient:

$$-\rho\omega^2 u_i \exp(-i\omega t) = \rho(V_0 + K \exp(-i\omega t))_{,i} + \sigma_{ij,j}. \tag{8.57}$$

(To be exact, the left-hand side here should contain the material acceleration $D^2\mathbf{u} \exp(-i\omega t)/Dt^2$. But $\mathbf{u}$ and the particle velocity are small, so that the acceleration is $-\omega^2 \mathbf{u} \exp(-i\omega t)$ to first order.) The stress $\sigma_{ij}(\mathbf{X}, t)$ is evaluated in the perturbed state. Since the particle at $\mathbf{X}$ at time t was initially at $\mathbf{x} = \mathbf{X} - \mathbf{u} \exp(-i\omega t)$, the stress at $\mathbf{X}$ is the initial stress at $\mathbf{X} - \mathbf{u} \exp(-i\omega t)$ together with an additional effect due to the distortion of the region around the $\mathbf{x}$-particle, i.e., due to the strains associated with $\mathbf{u}$. For an isotropic body with Lamé's constants λ and μ, we therefore have

$$\begin{aligned} \sigma_{ij}(\mathbf{X}, t) &= -P_0 \delta_{ij}\Big|_{\mathbf{X} - \mathbf{u} \exp(-i\omega t)} + \text{stress due to } \mathbf{u} \\ &= -P_0(r)\delta_{ij} + \left[u_r \frac{dP_0}{dr} \delta_{ij} + \lambda \nabla \cdot \mathbf{u} \delta_{ij} + \mu(u_{i,j} + u_{j,i}) \right] \exp(-i\omega t). \end{aligned} \tag{8.58}$$

Substituting the above form into (8.57) and retaining first-order terms, it follows that

$$-\rho_0\omega^2\mathbf{u} = \left(u_r\frac{d\rho_0}{dr} + \rho_0\nabla\cdot\mathbf{u}\right)g_0\hat{\mathbf{r}} + \rho_0\,\nabla K - \nabla(\rho_0 u_r g_0)$$
$$+\left[(\lambda + 2\mu)\nabla\nabla\cdot\mathbf{u} - \mu\nabla\times(\nabla\times\mathbf{u}) + \frac{d\lambda}{dr}(\nabla\cdot\mathbf{u})\hat{\mathbf{r}} + 2\frac{d\mu}{dr}\frac{\partial\mathbf{u}}{dr}\right.$$
$$\left. + \frac{d\mu}{dr}\hat{\mathbf{r}}\times(\nabla\times\mathbf{u})\right]. \tag{8.59}$$

The term [ ] here is merely the vector having its ith Cartesian component given by $\{\lambda\nabla\cdot\mathbf{u}\delta_{ij} + \mu(u_{i,j} + u_{j,i})\}_{,j}$, assuming that λ and μ depend only on radius r. The remaining terms on the right-hand side of (8.59) quantify the effect of self-gravitation. When $\rho_0(r)$, $\lambda(r)$, and $\mu(r)$ are known, it is possible to find $g_0(r)$ and then solve (8.56) and (8.59) for K and $\mathbf{u}$ under appropriate initial and boundary conditions. Note that (8.59) is valid also for toroidal modes, although most of the terms in this equation are then zero.

We have previously found that equations of motion are conveniently handled if they are transformed into $d\mathbf{f}/dr = \mathbf{Af}$, with $\mathbf{f}$ continuous across discontinuities in the coefficient matrix $\mathbf{A} = \mathbf{A}(r, \omega)$. This result motivates the following discussion.

When there is a discontinuous jump in some medium property (e.g., ρ_0, λ, or μ) at a certain radius r_d, then displacement and traction components are continuous across r_d, and so is the gravitational potential perturbation K. But dK/dr is discontinuous if there is a density jump. We therefore seek a quantity related to dK/dr, which is continuous.

Rewriting (8.56) as $\nabla\cdot(\nabla K - 4\pi\gamma\rho_0\mathbf{u}) = 0$, we apply Gauss's theorem to any volume V with surface S and find $\int_S(\nabla K - 4\pi\gamma\rho_0\mathbf{u})\cdot\mathbf{n}\,dS = 0$. Choosing V as the thin disc with one flat face just above the interface r_d and one just below, it follows that $dK/dr - 4\pi\gamma\rho_0 u_r$ must be the same on either side of the interface. Thus we might choose the continuous quantity $dK/dr - 4\pi\gamma\rho_0 u_r$ as one of the dependent variables.

For a spheroidal mode, we have already seen that the displacement and traction in the mode (l, m, n) are given by

$$u_r = {}_nU_l(r)Y_l^m(\Delta,\phi),\qquad u_\Delta = \frac{{}_nV_l(r)}{[l(l+1)]^{1/2}}\frac{\partial Y_l^m}{\partial\Delta},\qquad u_\phi = \frac{{}_nV_l(r)}{[l(l+1)]^{1/2}}\frac{1}{\sin\Delta}\frac{\partial Y_l^m}{\partial\phi},$$
$$\tau_{rr} = {}_nR_l(r)Y_l^m,\qquad \tau_{r\Delta} = \frac{{}_nS_l(r)}{[l(l+1)]^{1/2}}\frac{\partial Y_l^m}{\partial\Delta},\qquad \tau_{r\phi} = \frac{{}_nS_l(r)}{[l(l+1)]^{1/2}}\frac{1}{\sin\Delta}\frac{\partial Y_l^m}{\partial\phi}, \tag{8.60}$$

where $(\tau_{rr}, \tau_{r\Delta}, \tau_{r\phi})$ is the traction derived from $\mathbf{u}$. The gravitational potential can also be separated as

$$K = {}_nK_l(r)Y_l^m(\Delta,\phi), \tag{8.61}$$

and following Takeuchi and Saito (1972) we introduce

$$_nG_l(r) = \frac{d_nK_l}{dr} - 4\pi\gamma\rho_o({}_nU_l) + \frac{l+1}{r}\,{}_nK_l. \tag{8.62}$$

We have already pointed out that $(dK/dr - 4\pi\gamma\rho_0 u_r)$ and K are continuous across density jumps, so that $_nG_l$ is continuous. The additional virtue of $_nG_l$ is that $_nG_l(r_\oplus) = 0$ (i.e., at the Earth's surface), a result that follows from K satisfying Laplace's equation in $r > r_\oplus$ and hence $_nK_l(r) = {}_nK_l(r_\oplus)(r_\oplus/r)^{l+1}$ outside the Earth.

Substituting (8.60)–(8.62) into (8.56) and (8.59), we arrive at a set of six first-order linear differential equations for the radial functions:

$$\begin{aligned}
\frac{dU}{dr} &= \frac{1}{\lambda+2\mu}\left\{R - \frac{\lambda}{r}[2U - (l(l+1))^{1/2}V]\right\},\\
\frac{dR}{dr} &= -\omega^2\rho_0 U + \frac{2}{r}\left(\lambda\frac{dU}{dr} - R\right) + \frac{1}{r}\left[\frac{2(\lambda+\mu)}{r} - \rho_0 g_0\right][2U - (l(l+1))^{1/2}V]\\
&\quad + \frac{(l(l+1))^{1/2}}{r}S - \rho_0\left(G - \frac{l+1}{r}K + \frac{2g_0}{r}U\right),\\
\frac{dV}{dr} &= \frac{1}{\mu}S + \frac{1}{r}[V - (l(l+1))^{1/2}U],\\
\frac{dS}{dr} &= -\omega^2\rho_0 V - \frac{\lambda}{r}(l(l+1))^{1/2}\frac{dU}{dr} - \frac{\lambda+2\mu}{r^2}[2(l(l+1))^{1/2}U - l(l+1)V]\\
&\quad + 2\frac{\mu}{r^2}[(l(l+1))^{1/2}U - V] - \frac{3}{r}S - \frac{\rho_0}{r}(l(l+1))^{1/2}(K - g_0U),\\
\frac{dK}{dr} &= G + 4\pi\gamma\rho_0 U - \frac{l+1}{r}K,\\
\frac{dG}{dr} &= \frac{l-1}{r}(G + 4\pi\gamma\rho_0 U) + \frac{4\pi\gamma\rho_0}{r}[2U - (l(l+1))^{1/2}V],
\end{aligned} \tag{8.63}$$

where we have dropped the subscripts n and l. The equations (8.63) can be written as $d\mathbf{f}/dr = \mathbf{Af}$ with $\mathbf{f}$ continuous and $\mathbf{A} = \mathbf{A}(r, \omega)$. The existence of a solution for $\mathbf{f}$ that is regular at $r = 0$, and for which $_nR_l(r_\oplus)$, $_nS_l(r_\oplus)$ and $_nG_l(r_\oplus)$ are zero, requires that ω be an eigenvalue. Because the numerical effort to solve (8.63) for the eigenvalues $_n\omega_l$ and eigenvectors $\mathbf{f} = ({}_nV_l, {}_nU_l, {}_nS_l, {}_nR_l, {}_nK_l, {}_nG_l)^T$ is very similar to the effort described in Section 7.2 for surface waves, we shall not give further details. For toroidal modes, (8.59) gives the coupled system (8.33), since gravity has no direct influence.

Of course, the variational methods of Section 7.3 also apply to normal modes with only notational changes. Perturbation in phase velocity (due to small changes in structure) is replaced by perturbation in eigenfrequency. For example, for a toroidal mode (see (8.33) and (8.35)),

$$\left(\frac{\delta\omega}{\omega}\right)_l = \int_0^{r_\oplus}\left[\frac{\rho}{\omega}\left(\frac{\partial\omega}{\partial\rho}\right)_{l,\beta}\frac{\delta\rho}{\rho} + \frac{\beta}{\omega}\left(\frac{\partial\omega}{\partial\beta}\right)_{l,\rho}\frac{\delta\beta}{\beta}\right]dr, \tag{8.64}$$

where

$$\left(\frac{\rho}{\omega}\frac{\partial\omega}{\partial\rho}\right)_{l,\beta} = \frac{1}{2\omega^2 l(l+1)}\left[-\omega^2\rho r^2 W^2 + \frac{r^2T^2}{\mu} + (l-1)(l+2)\mu W^2\right]$$

and

$$\left(\frac{\beta}{\omega}\frac{\partial\omega}{\partial\beta}\right)_{l,\rho} = \frac{1}{\omega^2 l(l+1)}\left[\frac{r^2T^2}{\mu} + (l-1)(l+2)\mu W^2\right]. \tag{8.65}$$

Partial derivatives for toroidal and spheroidal modes in a transversely isotropic Earth model are given by Takeuchi and Saito (1972), and Woodhouse (1976) has shown how the perturbation of internal boundaries affects the eigenfrequencies.

We have now given a fairly complete account of normal-mode theory in a spherically symmetric nonrotating Earth model. About 40 different eigenfrequencies were distinguished in 1960 following the Chilean earthquake, and the initial efforts of various investigators was simply to identify these modes of oscillation correctly. The data provided by eigenfrequencies were quite clearly of a different nature from travel-time data of body waves, since it was known how to construct Earth models having a given travel-time curve, but it was not clear how normal-mode eigenfrequencies could be used to infer Earth structure other than by trial and error. The importance of this problem was a stimulus to develop new methods of inverting geophysical data, and the work of Backus and Gilbert (1967, 1968, 1970) on inversion theory was carried out using normal-mode data as an example.

With modern instrumentation it is possible to observe and identify normal modes from earthquakes with magnitude (M_s) as low as 6.5 (Block et al., 1970). To determine which normal modes will be favorably excited by a given earthquake, we see from (8.37) that the crucial quantity is the strain in the mode, evaluated at the source. Thus shallow earthquakes favorably excite the fundamental modes (${}_0S_l$ and ${}_0T_l$ for different l), whereas deep earthquakes would be best for observing overtones. Using the spectra of 211 WWSSN seismograms

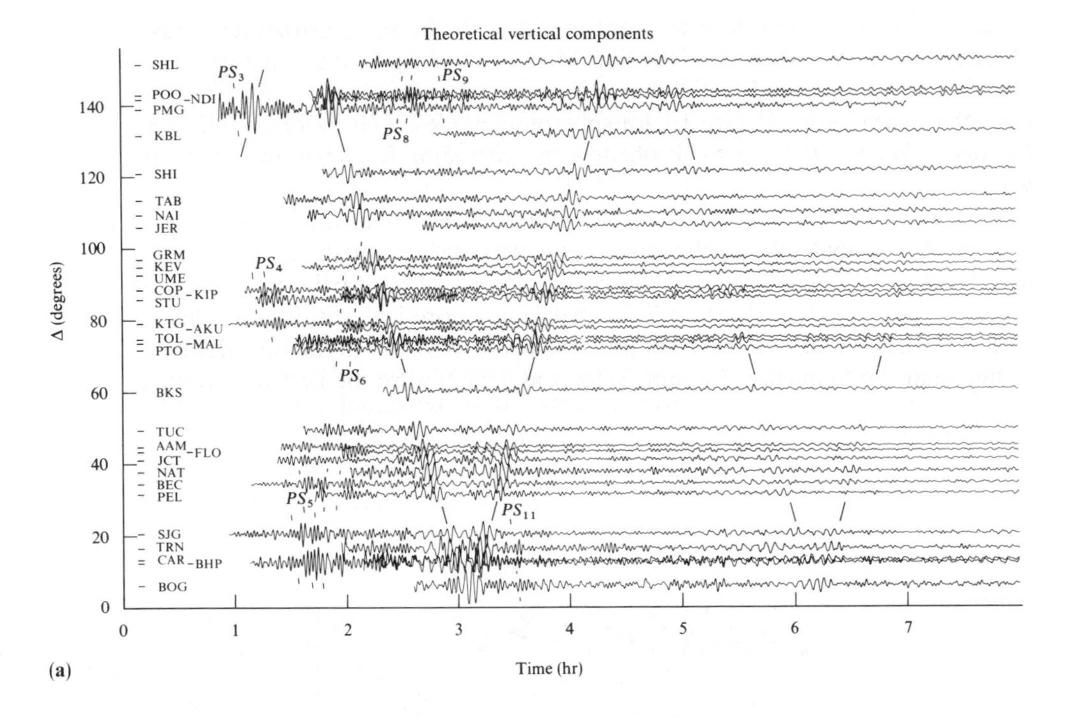

Theoretical vertical components
SHL
POO
NDI
PMG
KBL
SHI
TAB
NAI
JER
GRM
KEV
UME
COP
KIP
STU
KTG
AKU
TOL
MAL
PTO
BKS
TUC
AAM
FLO
JCT
NAT
BEC
PEL
SJG
TRN
CAR
BHP
BOG
PS_3
PS_9
PS_8
PS_4
PS_6
PS_5
PS_{11}
Δ (degrees)
Time (hr)
0
20
40
60
80
100
120
140
0
1
2
3
4
5
6
7

(a)

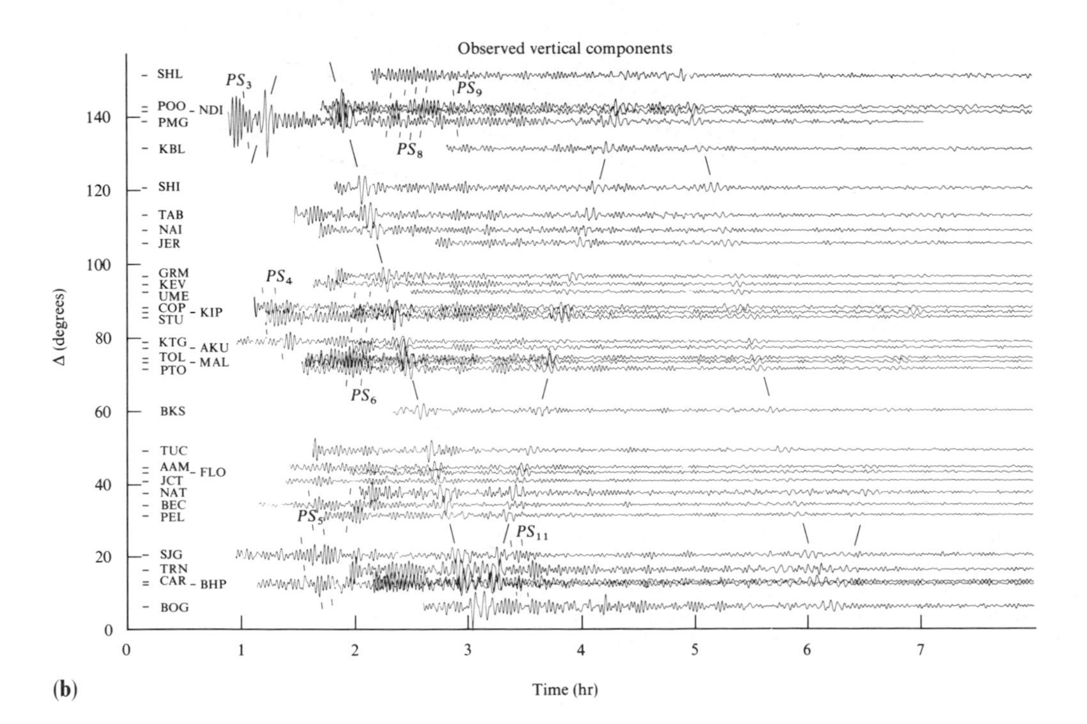

Observed vertical components
SHL
POO
NDI
PMG
KBL
SHI
TAB
NAI
JER
GRM
KEV
UME
COP
KIP
STU
KTG
AKU
TOL
MAL
PTO
BKS
TUC
AAM
FLO
JCT
NAT
BEC
PEL
SJG
TRN
CAR
BHP
BOG
PS_3
PS_9
PS_8
PS_4
PS_6
PS_5
PS_{11}
Δ (degrees)
Time (hr)
0
20
40
60
80
100
120
140
0
1
2
3
4
5
6
7

(b)

seismograms by summing normal modes. A summary of early work in this field is given by Landisman et al. (1970), who compared various features of the normal-mode summation (e.g., the amplitude of a particular body wave) with corresponding features in seismic data. The first comparisons in the time domain–whole wave trains (including body waves) from normal-mode summation, and observed seismograms—were done by Luh and Dziewonski (1975) for the deep Colombian earthquake of 1970. Using a previously determined moment rate tensor for this event and a particular Earth model, these authors computed 75 seismograms by superimposing more than a thousand modes in the period range 100–1000 sec. Their results are shown in Figure 8.10. Straight lines across the record sections indicate Rayleigh waves, R_1, R_2, etc., which are prominent at these periods, even for such a deep event (depth assigned as 651 km). Prominent also are body-wave phases in which *P*- and *S*-waves traversing the mantle have multiple legs, e.g., *PSPSPS*, the conversion between *P* and *S* taking place on reflection at the Earth's surface. The example we have just given is denoted PS_3, and several PS_n are marked on the figure. General characteristics of the data are reproduced by the theoretical record section, and where there is disagreement, Luh and Dziewonski suggest that the inadequacy can be traced to the assumed moment-rate tensor and/or the *Q*-model. Clearly, this method of interpreting seismograms has great potential. In execution, the principal difficulties lie in adequate computation and storage of the large number of "short-period" normal modes (periods down to about 5 sec) that are observed but are not separately identifiable on "long-period" WWSSN seismograms. For toroidal modes, a transformation is known (Box 9.9) that converts the spherical geometry of the Earth to a plane-layered problem. In this case, highly efficient programs have been written that permit modal computations down to periods of about 10 sec (Schwab et al., 1976; Nakanishi et al., 1977; Mantovani et al., 1977). For spheroidal modes, Rayleigh-Ritz methods have been implemented that allow computation of all modes down to periods of about 45 sec (Buland and Gilbert, 1976, 1980). At even shorter periods, an accurate asymptotic theory has been developed for each mode (Woodhouse, 1978). Therefore, it is possible that (8.37) may be used to generate comparisons like those shown in Figure 8.10a,b, but without the drastic filtering of the data.

FIGURE **8.10**

Vertical component for the deep Colombian earthquake of 31 July 1970. (**a**) Theoretical records for 34 WWSSN stations, with each seismogram plotted at its correct distance Δ from the epicenter. (This method of display is known as a *record section.*) Computation is via (8.37), using modes in the period range 100–1000 sec and the source model and Earth model 1066A of Gilbert and Dziewonski (1975). (**b**) Observed vertical ground motion for the same stations and the same earthquake as (a). (These seismograms have been corrected for instrument response and filtered to pass the same frequency range as that used in computing (a).) [From Luh and Dziewonski (1975).]

BOX **8.5**

Consideration of initial stress

Stresses within the Earth can reach values of the order of megabars. Since the elastic constants in Hooke's law are also of this order, it follows that strains of order one may be present in the Earth, even in the simple situation where there is a reference state in which strain and stress are both zero, such that the Earth's present configuration can be obtained from this state by linear stress-strain theory. In fact, we have no reason to suppose there is such a reference state; even if there were, it would not be a suitable reference for studying seismic waves, as wave-propagation theory for finite strains is nonlinear. Instead, we work with a theory for small incremental stresses, using as reference an initial state in which the stress may be large but strains are zero.

In seismology, it is natural to take as the initial state the Earth's configuration just before an earthquake. Because the initial stress is predominantly isotropic (being due at any given depth to the gravity field acting on the overburden of material), it turns out that there are virtually no complications introduced in all the formulas we have so far obtained, which altogether ignore the initial stress field. However, the formal development of the theory of elasticity is very much more complicated. The work necessary to check whether formulas based on classical elasticity need modification for Earth models with high initial stress has largely been carried out by Dahlen (1972, 1973, 1976a, 1977).

Some of the complications of working with a general initial stress lie in the fact that incremental stress at a material particle consists of two parts: one depending on local strains, and familiar from the theory given in Chapter 2; the other depending on local rotations, which act to rotate the initial stress field. Thus total stresses in general depend upon $e_{ij} = \frac{1}{2}(u_{i,j} + u_{j,i})$ and $\omega_{ij} = \frac{1}{2}(u_{i,j} - u_{j,i})$. In order to describe the stress and strain fields, Dahlen (1973) advocates a Lagrangian approach. Furthermore, in order to refer boundary conditions (such as a description of faulting) back to the initial configuration, the Cauchy stress tensor used in Chapter 2 is inappropriate. The area elements (in magnitude and orientation) across which traction acts are in the deformed state for definition of the Cauchy stress. Instead, Dahlen (1972, 1973) recommends the use of area elements in the initial state to describe subsequent tractions. Suppose an element of material area centered on $\mathbf{x}$ has, in the initial state, magnitude $dS(\mathbf{x})$ and orientation $\mathbf{n}(\mathbf{x})$. At some later time t, the particle originally at $\mathbf{x}$ has moved to $\mathbf{X} = \mathbf{x} + \mathbf{u}$ and the area element of material has magnitude $dS(\mathbf{X}, t)$ and orientation $\mathbf{n}(\mathbf{X}, t)$. The Cauchy stress tensor $\boldsymbol{\sigma}^C$ has Cartesian components such that traction across the material element has jth component $n_i(\mathbf{X}, t)\sigma_{ij}^C(\mathbf{X}, t)\, dS(\mathbf{X}. t)$. This traction is equated to $n_i(\mathbf{x})\sigma_{ij}^{PK}(\mathbf{x}, t)\, dS(\mathbf{x})$ to define Cartesian components of the *Piola-Kirchhoff stress tensor*, a tensor that can thus quantify time-dependent tractions per unit original area of a deforming surface. Accounting for the initial stress $\boldsymbol{\sigma}^0$, the incremental Piola-Kirchhoff stress tensor $\tilde{\boldsymbol{\tau}}$ and the incremental Cauchy stress $\boldsymbol{\tau}$ are given by

$$\boldsymbol{\sigma}^{PK} = \boldsymbol{\sigma}^0 + \tilde{\boldsymbol{\tau}} \qquad \boldsymbol{\sigma}^C = \boldsymbol{\sigma}^0 + \boldsymbol{\tau}. \tag{1}$$

Dahlen (1972) and Malvern (1969) point out that the Piola-Kirchhoff stress tensor is not symmetric. The relation between incremental stress tensors is (to first order in initial stresses and subsequent strains)

$$\tilde{\tau}_{ij} = \tau_{ij} + \sigma^0_{ij}\frac{\partial u_k}{\partial x_k} - \sigma^0_{jk}\frac{\partial u_i}{\partial x_k},$$

reducing to

$$\tilde{\tau}_{ij} = \tau_{ij} - p_0\,\delta_{ij}\frac{\partial u_k}{\partial x_k} + p_0\frac{\partial u_i}{\partial x_j}$$

for an isotropic initial stress, $\sigma^0_{ij} = -p_0\delta_{ij}$.

For an Earth model that is rotating, self-gravitating, and elastic for stresses and strains about an isotropically stressed initial state, Dahlen (1972) showed that the body-force equivalent for a shear dislocation (which we derived in Chapters 2 and 3) is still a distribution of double couples. If the initial state has shearing stresses too, then the body-force equivalent for shearing consists of double couples plus extra terms, of order [(initial shearing stresses) $\div$ (*in situ* elastic constants)]. These extra terms are very small, since rocks typically cannot support shear stresses greater than one or two kilobars. For shear faulting in the x_3-plane, Dahlen (1976a) found still the continuity $[\tau_{31}] = [\tau_{32}] = [\tau_{33}] = 0$ for the incremental Cauchy stress tensor, and still the formula

$$\Delta E = -\frac{1}{2}\int_\Sigma [u_i(\mathbf{x}, \infty)](\sigma^0_{ij} + \sigma^1_{ij})\nu_j\, dS \tag{2}$$

(see Box 3.4) for the change in strain energy throughout a medium due to faulting across Σ. Formulas for frictional work on Σ, and work needed to create new fault surface, are also unchanged by the presence of an isotropic initial stress. Even if changes in length of day and changes in the gravitational field may occur, Dahlen (1977) showed that (2) still expresses the energy of faulting in a prestressed Earth model (i.e., the sum of changes in rotational energy, gravitational energy, and internal shear-strain energy, although individual terms in this sum can range up to four orders of magnitude greater than the right-hand side of (2)).

8.5 Splitting of Normal Modes Due to the Earth's Rotation

The worst features of the Earth models we discussed in Section 8.4 are that they did not rotate and were spherically symmetric. We consider here the effect of rotation, and defer a discussion of lateral heterogeneity until Chapter 13.

The Earth's daily rotation can be expected to produce quantitative effects upon normal modes that are roughly in the dimensionless ratio (normal mode period)/(24 hours). Clearly, this can amount to a few percent for low-order fundamental modes, and should be observable. Observations of the 1960 Chilean earthquake indeed indicated that modes such as ${}_0S_2$ did not have a single spectral peak, but were composed of at least two lines with periods of

54.7 and 53.1 minutes (Benioff et al., 1961). The effect was recognized to be mathematically similar to the Zeeman effect (the splitting of degenerate energy levels of a hydrogen atom in a magnetic field). For each l, we found previously that there exist $2l + 1$ modes, i.e., $-l \leq m \leq l$, with the same eigenfrequency and radial eigenfunction. But rotation removes this degeneracy, and instead of a single eigenfrequency ${}_n\omega_l$, we must speak of the multiplet ${}_n\omega_l^m(-l \leq m \leq l)$, each having a different radial eigenfunction. Before giving the relevant theory, it is useful to note some of the practical problems that the Earth's rotation imposes on normal-mode observations.

First, there is the problem of determining the temporal Q of a given mode. A standard method for measuring the decay rate of a given frequency component in a time series is to compute the spectrum and measure the width of the spectral peak at its half-power level. Thus, for a time function $f(t) = H(t)[e^{-\omega_0 t/2Q} \cos \omega_0 t]$, the amplitude spectrum near ω_0 is

$$|f(\omega)| \sim \frac{1}{2}\left[(\omega - \omega_0)^2 + \left(\frac{\omega_0}{2Q}\right)^2\right]^{-1/2}. \tag{8.66}$$

When $\omega = \omega_0 \pm \omega_0/2Q$, the amplitude is down by a factor of $1/\sqrt{2}$ from its peak value, and the associated power spectrum $|f(\omega)|^2$ is down by a factor of $\frac{1}{2}$. Measuring the width $\Delta\omega$ of the peak at its half-power level, it follows that $Q^{-1} = \Delta\omega/\omega_0$. Clearly, this standard method of finding Q will not work, if spectral peaks within a given multiplet overlap. Then individual modes are not resolved, and (8.66) is an inadequate model of the spectrum near ω_0. Alsop et al. (1961b) found that another common method of measuring Q could still be made to work. Their approach was to interpret the relative amplitude of spectra determined from carefully chosen time windows (of equal length) along the original seismogram, one spectrum for each window.

The second problem associated with multiplet-splitting by rotation is that it obscures the multiplet-splitting caused by the Earth's departure from spherical symmetry. Dahlen (1968, 1969) showed that rotation effects to both first and second order must be removed if the lateral heterogeneity of the Earth is to be studied by its splitting effect on normal modes.

With these preliminaries set forth, we give now the first-order perturbation theory for rotational splitting.

Consider a toroidal mode of oscillation, which from (8.59) satisfies

$$\rho_0 \frac{\partial^2 u_i}{\partial t^2} = [\mu(u_{i,j} + u_{j,i})]_{,j}. \tag{8.67}$$

This describes small motions away from a nonrotating equilibrium state in which a fixed inertial coordinate system can be identified. But the Earth has a daily rotation, and it is therefore appropriate to modify (8.67) so that **u** describes small motions away from a rotating equilibrium state. To do this, we use a coordinate system rotating with constant angular velocity $\mathbf{\Omega} = \Omega\hat{\mathbf{z}}$, where Ω is 2π per 24 hours (actually, per sidereal day) and $\hat{\mathbf{z}}$ is a unit vector along the Earth's rotation axis. In this noninertial system, particle velocity is given by $\partial\mathbf{u}/\partial t + \mathbf{\Omega} \times \mathbf{u}$ and particle acceleration by $\partial^2\mathbf{u}/\partial t^2 + 2\mathbf{\Omega} \times \partial\mathbf{u}/\partial t + \mathbf{\Omega} \times (\mathbf{\Omega} \times \mathbf{u})$. Since Ω is so small (relative to normal-mode frequencies), we shall neglect terms of order Ω^2, and (8.67) becomes

$$\rho_0 \frac{\partial^2 \mathbf{u}}{\partial t^2} + 2\rho_0 \mathbf{\Omega} \times \frac{\partial \mathbf{u}}{\partial t} = \mathbf{X}, \tag{8.68}$$

where $X_i = [\mu(u_{i,j} + u_{j,i})]_{,j}$.

We shall assume that (8.68) without the Coriolis acceleration $2\mathbf{\Omega} \times \partial\mathbf{u}/\partial t$ has a degenerate eigenvalue

$$\omega_0 = {}_n\omega_l. \tag{8.69}$$

The solution $\mathbf{u}$ of (8.68) that we wish to investigate has spatial dependence dominated by ${}_nW_l(r)\mathbf{T}_l^m(\Delta, \phi)$ and an eigenfrequency $\omega = {}_n\omega_l^m$ that is slightly perturbed from (8.69). It can be shown (Backus and Gilbert, 1961) that $\mathbf{u}$ is analytic in Ω for small Ω, so that we may use power series in the form

$$\begin{aligned}
\frac{\omega}{\omega_0} &= 1 + \sigma_1\left(\frac{\Omega}{\omega_0}\right) + \sigma_2\left(\frac{\Omega}{\omega_0}\right)^2 + \cdots, \\
\mathbf{u} &= \mathbf{u}_0 + \mathbf{u}_1\left(\frac{\Omega}{\omega_0}\right) + \cdots, \\
\mathbf{X} &= \mathbf{X}_0 + \mathbf{X}_1\left(\frac{\Omega}{\omega_0}\right) + \cdots,
\end{aligned} \tag{8.70}$$

where ω is the eigenvalue of (8.68) (i.e., $\mathbf{u} \propto \exp(-i\omega t)$). Substituting $\mathbf{\Omega} = \Omega\hat{\mathbf{z}}$ and (8.70) into (8.68) and equating powers of (Ω/ω_0), we find

$$-\rho_0\omega_0^2\mathbf{u}_0 = \mathbf{X}_0 \tag{8.71}$$

$$-\rho_0\omega_0^2\mathbf{u}_1 - 2\rho_0\omega_0^2[\sigma_1\mathbf{u}_0 + i\hat{\mathbf{z}} \times \mathbf{u}_0] = \mathbf{X}_1. \tag{8.72}$$

Our goal now is to obtain the first-order perturbation σ_1. Clearly, $\mathbf{u}_1$ must be linearly independent of $\mathbf{u}_0$, so that by redefining $\mathbf{u}_0$ if necessary, we can assume $\mathbf{u}_1$ and $\mathbf{u}_0$ are orthogonal. Pekeris et al. (1961) have also shown that $\mathbf{X}_1$ is orthogonal to $\mathbf{u}_0$, hence from (8.72),

$$\int \rho_0[\sigma_1\mathbf{u}_0^* \cdot \mathbf{u}_0 + i\mathbf{u}_0^* \cdot (\hat{\mathbf{z}} \times \mathbf{u}_0)]\, dV = 0.$$

This is an equation for σ_1, and can be rewritten (using (8.35)) as

$$\sigma_1 = -i\hat{\mathbf{z}} \cdot \int \rho_0 \mathbf{u}_0 \times \mathbf{u}_0^* \, dV. \tag{8.73}$$

Since $\mathbf{u}_0 = {}_nW_l(r)\mathbf{T}_l^m(\Delta, \phi) \exp[-i_n\omega_l t]$, the integral in (8.73) gives

$$\sigma_1 = \frac{-m}{l(l+1)}. \tag{8.74}$$

This is somewhat disappointing from a geophysical viewpoint, for it tells us that the first-order splitting of toroidal modes is independent of Earth structure. The perturbed eigenfrequencies are, from (8.70),

$${}_n\omega_l^m = {}_n\omega_l - \frac{m}{l(l+1)}\Omega \qquad \text{for } -l \le m \le l, \tag{8.75}$$

and the degeneracy has been removed.

Remarkably, equation (8.73) still holds for spheroidal modes, where $\mathbf{u}_0 = [{}_nU_l\mathbf{R}_l^m + {}_nV_l\mathbf{S}_l^m] \exp[-i_n\omega_l t]$, and the normalization (8.34) is enforced. Again, σ_1 can be evaluated, giving

$${}_n\omega_l^m = {}_n\omega_l - m\Omega_n\beta_l, \tag{8.76}$$

where

$${}_n\beta_l = \int_0^{r_\oplus} \frac{\rho_0({}_nV_l)^2}{l(l+1)} r^2 \, dV + 2\int_0^{r_\oplus} \frac{\rho_0({}_nU_l)({}_nV_l)}{[l(l+1)]^{1/2}} r^2 \, dV.$$

The values of ${}_n\beta_l(\Omega/{}_n\omega_l)$ have been computed by Dahlen (1968), and are largest for the modes ${}_1S_1$ and ${}_0S_2$, amounting to about 0.015.

Perhaps the main use of the theory of rotational splitting lies in modeling part of an observed spectral peak in terms of closely spaced components of a split multiplet so that a method of data analysis can be designed to estimate Q. Stein and Geller (1977) advocate a synthesis of ground motion in the time domain—that is, obtaining the excitation for each mode within the multiplet $-l \le m \le l$ associated with a given (n, l). They sum the modes within the multiplet and compare this synthetic record with the data. (Both synthetics and data are narrow-band filtered to isolate the multiplet of interest.) Figure 8.11 gives an example for ${}_0S_2$ observed after the great Chilean earthquake of 1960, and the 150-hour length of the record shows interference between different modes within the multiplet.

We have in this section outlined only the simplest aspects of rotational splitting. Dahlen and Smith (1975) have given an extended treatment, showing that normal modes for a rotating Earth are no longer orthogonal. When part of the Earth model is fluid (e.g., the outer core), a whole new range of theoretical

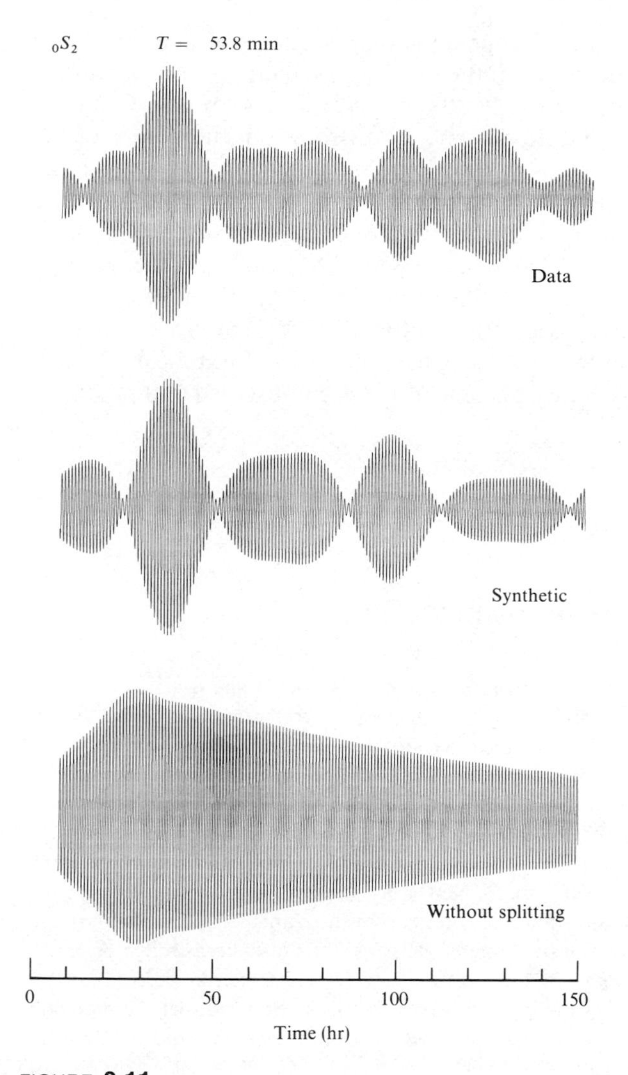

FIGURE **8.11**

Comparison between observations of $_0S_2$ in the time domain and a synthesis of this record using methods that include and exclude the effect of rotational splitting. [From Stein and Geller (1977).]

problems is opened up, even without the complications of rotation, because suites of normal modes with very long periods can exist. They can have periods of many hours, and have been called *undertones* (Pekeris and Accad, 1972).

There are many influences acting upon the Earth with time scales that are long compared to the period of $_0S_2$. It might be hoped that a quasistatic treatment of deformation could be given, but unfortunately this is not simple, due to the presence of the fluid core. Longman (1963) pointed out that equations (8.63) become mutually inconsistent in the limit as $\mu \to 0$ and $\omega \to 0$, and many authors have suggested ways to resolve the paradox (Crossley and Gubbins, 1975). In the context of normal modes and Earth rotation, the theory of the Chandler wobble is complicated by Longman's paradox. Smith (1977) has avoided a quasistatic treatment, and showed how the dynamic equations for a rotating Earth (with fluid core but without oceans) do have the Chandler wobble as a normal mode. He calculated the period to be 405.19 sidereal days for an Earth model with a stably stratified fluid core. The observed period is 434.14 ± 1.02 sidereal days (Currie, 1974), but Dahlen (1976b) showed that the world's oceans would act to increase the Chandler period for an Earth model such as Smith's, giving satisfactory agreement between observed and predicted Chandler periods.

SUGGESTIONS FOR FURTHER READING

Biot, M. A. *Mechanics of Incremental Deformations*, New York: Wiley, 1965.

Brune, J. N., and F. Gilbert. Torsional overtone dispersion from correlations of S waves to SS waves. *Bulletin of the Seismological Society of America*, **64**, 313–320, 1974.

Gilbert, F., and A. M. Dziewonski. An application of normal mode theory to the retrieval of structural parameters and source mechanisms from seismic spectra. *Philosophical Transactions of the Royal Society* (Lond.), **A278**, 187–269, 1975.

Love, A. E. H. *Some Problems of Geodynamics*. New York: Dover Publications, 1967.

Rayleigh, Baron. *Theory of Sound* (Chap. 5). New York: Dover Publications, 1945.

Satô, Y., T. Usami, and M. Landisman. Theoretical seismograms of spheroidal type on the surface of a gravitating elastic sphere—II. Case of Gutenberg-Bullen A′ earth models. *Bulletin of the Earthquake Research Institute, Tokyo Univ.*, **45**, 601–624, 1967.

Takeuchi, H., and M. Saito. Seismic surface waves. In B. A. Bolt (editor), *Seismology: Surface Waves and Earth Oscillations* (Methods in Computational Physics, Vol. 11). New York: Academic Press, 1972.

PROBLEMS

8.1 Give the normal-mode theory for torsional oscillations of a homogeneous elastic solid sphere, and find the two longest periods. What generally is the most favorable depth for exciting ${}_0T_2$ by a single horizontal force? By a couple with moment about a vertical axis?

8.2 Show that the temporal Q of a torsional mode is related to the $Q_\beta(r)$ of body waves at radius r by

$$Q^{-1} = \int_0^{r_\oplus} \frac{\beta}{\omega} \left[\frac{\partial \omega}{\partial \beta}\right] Q_\beta^{-1}(r)\, dr,$$

the partial derivative here being given in (8.65).

8.3 What is wrong with the following argument? The Earth's daily rotation is not excited by external forces, and is therefore a free oscillation of the Earth. It must be a torsional oscillation, and corresponds to $l = 1$. For an Earth that is spherically symmetric, it follows that rotation can have no effect on any of the other modes (toroidal and spheroidal), since modes in such a model do not interact.

8.4 Consider a sequence of G-waves, $G1$, $G2$, . . . , Gn, arriving at a given station after a large earthquake (see Appendix 1). Taking into account the polar phase shift, derive a formula for determining the phase velocity over the great circle from the Fourier transforms of Gn and Gm.

The formula does not include the initial phase due to source effect if n and m differ by an even number. Otherwise, it requires the knowledge of difference between initial phases at opposite radiation azimuths. Is there any simple rule relating the initial phases at opposite azimuths from a point source (7.148) that can be used to simplify the formula for the latter case?

Consider the same problem for mantle Rayleigh waves using (7.149).

CHAPTER 9

Body Waves in Media with Depth-dependent Properties

The most successful single method for using seismic waves to investigate the Earth's internal structure has historically been the application of ray-theory formulas (Section 4.4) to interpret short-period body waves. In fact, just from observations of the travel-time function $T(\mathbf{x}, \xi)$ for varying receiver position $\mathbf{x}$ and varying source position ξ, Beno Gutenberg in 1913 accurately estimated the depth to the core-mantle boundary as 2900 km. Accurate estimates of crustal thickness began with Mohorovičić in 1909, and the existence of an inner core was recognized in 1936 by Inge Lehmann. By 1939, the independent efforts of Gutenberg and Harold Jeffreys in applying ray theory had led to velocity models of the whole Earth which differ from each other by only a few percent. In the depth range 800–2800 km, there is less than 1% difference between the P-wave velocity in Jeffreys' and Gutenberg's models, and less than 1% too between Jeffreys' model and the modern model 1066B of Gilbert and Dziewonski (1975). However, these early results of Gutenberg and Jeffreys, achieved with ray theory, are not enough to resolve many questions about the composition and state of materials at great depth within the Earth. For example, in the upper mantle there are two depth ranges (around 400 km and 650 km) within which the velocity gradient ($d\alpha/dz$) is anomalously large. The cause is likely to be either a compositional gradient or a transition between different solid phases of the same mineral (or possibly both), and any generally acceptable resolution of this difficult question will depend heavily on seismologists being able to give very accurate estimates of $d\alpha/dz$ and $d\beta/dz$ with depth. Similarly, to understand the energetics of a dynamo in the Earth's fluid core, it is important to know if the velocity gradient there has values that will either inhibit or promote convection.

It has therefore turned out that one of the most classical aspects of seismology is still an active research field: namely, the deduction of the Earth's internal structure, but now with emphasis on relatively fine detail, such as velocity gradients within restricted depth ranges and the estimation of jumps in properties across recognized discontinuities such as the Moho.

Body waves have here an important role to play, since their relatively short period, and the possibility of isolating them in time from other waves on a seismogram, permits their use in studying highly localized regions within the Earth. This fact has long been recognized, and the problem has largely been to find a way that significantly improves upon the classical method of working with travel-time data. In Chapter 11, we describe the ability of seismic arrays to measure slowness directly (i.e., the gradient of $T(\mathbf{x}, \xi)$ as $\mathbf{x}$ varies within the array), and in Chapter 12 we show how this is a significant improvement, because it is the slowness that is needed in inverse calculations. Furthermore, since T is often a multiple-valued function, arrays can help to identify later arrivals by measuring their different slowness. However, these are improvements that, for the most part, are still based on timing the arrivals and working with the ray theory of Section 4.4.

Attempts to use ray theory for the *amplitude* of body-wave arrivals have been summarized by Julian and Anderson (1968), Wesson (1970), Shimshoni and Ben-Menahem (1970), and Chapman (1971). These papers show that the geometrical spreading function ($\mathscr{R}(\mathbf{x}, \xi)$ in the notation of Chapter 4) can be an exceedingly sensitive function of velocity gradients, which is a desirable feature for purposes of inverting for Earth structure. However, it is at this stage that one can clearly recognize the limitations in applying classical ray theory to the interpretation of data. The problem is a breakdown of the theory itself, for in body-wave data can be seen a variety of frequency-dependent effects that ray theory cannot quantify. Body waves are seen in "shadow zones" (which could not be penetrated according to Snell's law); they are seen near caustics (a surface that is a weak type of focus, on which $\mathscr{R}(\mathbf{x}, \xi) = 0$) with frequency-dependent amplitudes that certainly do not have the singularity predicted by the geometrical-spreading function; and they are seen in situations where a variety of rays between source and receiver all have similar arrival times, so that complicated interference effects are observed. The wave theory needed to understand these observations is basically an extension of the methods used in Chapter 6 to solve Lamb's problem, plus the systematic procedures described in Chapter 7 for handling waves in layered media. In fact, we shall distinguish three rather different theoretical methods for successfully studying body waves. At present, although the methods of calculation are substantially different, each theory is now used in the same way—namely, to generate synthetic seismograms in some given Earth model with a given seismic source. A comparison is then made between theory and observations. Because whole waveforms present in the data are used to judge whether particular models of the Earth and source are successful or not, synthetic seismograms are a major advance over the use merely of ray-theoretical travel-time curves and polarity of first motions.

The first of the theoretical methods we shall describe is an extension of Cagniard's approach (Section 6.5), used in practice to handle the case of a medium composed of very many (possibly several hundred) homogeneous plane layers. As shown by Helmberger (1968), the key here is a numerical approach for obtaining the Cagniard path in the complex ray-parameter plane (rather than an evaluation of simple hyperbolas, such as that given following (6.91)). The second theoretical approach, originated by Fuchs (1968, 1970, 1971) and often called the "reflectivity method," again is directed toward the computation of waves within a medium composed of very many homogeneous layers. Based on the matrix methods developed in seismology by Haskell (see Chapter 7), Fuchs' method involves numerical integration over a limited range of real ray parameters, followed by an inverse Fourier transform over frequency to obtain the synthetic seismogram.

The Earth, of course, is not composed of a stack of homogeneous plane layers; therefore, to apply the above two methods to the interpretation of seismograms, it is necessary to note two important results. One, due to a theorem given by Volterra (see Gilbert and Backus, 1966), is that the wave systems in a medium with *continuous* spatial variations of density and wave speeds can indeed be studied by solving for the waves in a *discrete* medium composed of many homogeneous parts. In fact, taking the wave solution for the discrete medium (in the limit as the number of homogeneous regions tends to infinity in such a way as to give the density and wave speeds of the continuous medium) this solution does tend to the wave solution for the continuous medium. In practice, for a medium in which ρ, α, β depend only on depth (i.e., on the Cartesian coordinate z), it is therefore permissible to model a continuous profile with a large number of homogeneous plane layers. Such an approach has been widely adopted, though Chapman (1976a) has extended the Cagniard method to allow a solution directly, for continously varying $\rho(z)$, $\alpha(z)$, $\beta(z)$.

The second result needed to apply Cagniard's and Fuchs' methods in seismology involves a consideration of the Earth's spherical geometry. (Throughout this chapter, we shall be concerned only with media in which ρ, α, β are functions of r in a spherical polar coordinate system or functions of depth z with z belonging to either a Cartesian system or a cylindrical system with vertical axis of symmetry.) The question, then, is how to transform a wave-propagation problem posed for a medium with spherical symmetry (i.e., radial heterogeneity) into a problem posed for a plane-stratified medium, in which our first two methods of solution are directly applicable. By using such an "Earth-flattening transformation," it has been shown by Andrianova *et al.* (1967) and by Biswas and Knopoff (1970) that, for SH-waves, it is possible to obtain *exact* wave solutions in spherical media from the corresponding solution in plane media. Unfortunately, Chapman (1973) showed that an exact transform for *P-SV* problems does not appear possible, and that certainly the Earth-flattening transformations currently used for *P-SV* problems are only approximate.

The third method of solution we shall describe is one in which the Earth model is composed of radially inhomogeneous layers. The seismic motions (as a function of time, at different positions in the Earth) are broken up into signals arriving within different time windows, each signal corresponding basically to some particular type of body-wave ray path. Instead of geometrical ray theory, one uses a more accurate solution for the signal of interest by considering an integral that is conducted over both frequency and ray parameter. Within this general procedure, several different methods have been proposed for executing the double integration. Either the frequency integral or the ray-parameter integral can be done first. The ray-parameter integral can be carried out over real values or over complex values. Many different scientists have contributed toward the underlying theory. A major contribution was made by Scholte (1956), who used well-established methods for summing partial wave expansions in spherical geometry and showed in some detail how the solution in the frequency domain for specific seismic rays could be identified. The solution involves an integral in the complex plane using a variable that (essentially) is the ray parameter, and a major contribution was made by Phinney and Alexander (1966) and Phinney and Cathles (1969) when they showed that this integral can easily be obtained numerically. Taking the approach of complex integration over ray parameter, followed by a real integration over frequency, we shall show some practical applications to several seismic waves that have interacted with the Earth's solid and fluid cores. Chapman (1978a,b) has described several alternative approaches. One that is often very effective entails first an integration over real frequencies (in many cases, this is possible analytically rather than numerically), followed by a fairly simple inversion of the ray parameter integral.

Because the Earth is such a complicated medium (even when assumed to be spherically symmetric), it is necessary in practice to make approximations in each of the different solution methods we shall describe. It is therefore fortunate that there *are* several methods, since it often appears that the accuracy of any one method (and, more important, the geophysical conclusions that may be drawn from its application) can be verified only by a comparison against other methods. In fact, these comparisons have only recently begun to be made, and, although results are incomplete, it appears that the different methods have some unique and some overlapping areas of applicability.

9.1 Cagniard's Method for a Medium with Many Plane Layers: Analysis of a Generalized Ray

This section is a natural extension of Section 6.5, in which we showed how to evaluate the generalized reflection from the interface between two half-spaces. Supposing that the reader is familiar with a derivation of (6.91), and the steps detailed in (6.92)–(6.95), we follow Helmberger (1968) in describing how to evaluate the generalized primary reflection shown in Figure 9.1. Helmberger's

ρ_0 α_0 β_0

z_0

Th_1 ρ_1 α_1 β_1

z_1

Th_2 ρ_2 α_2 β_2

z_2

z_{n-1}

Th_n ρ_n α_n β_n

z_n

ρ_{n+1} α_{n+1} β_{n+1}

(a)

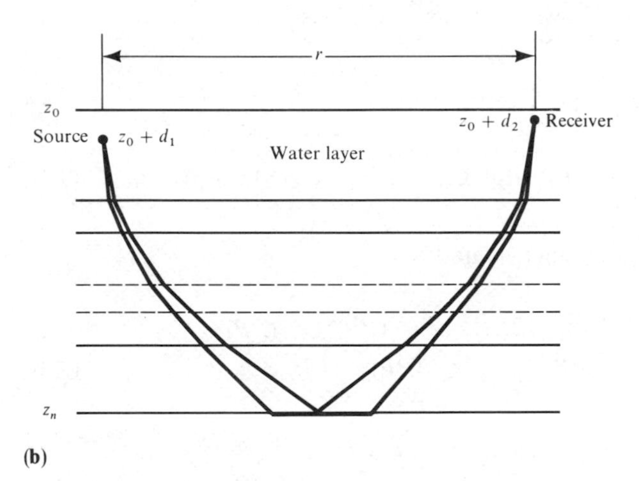

(b)

FIGURE **9.1**

(a) Notation for the density (ρ) and two wave speeds (α, β) in a stack of homogeneous layers. The boundary between layers n and $n + 1$ is at depth z_n, and the thickness of the nth layer is $Th_n = z_n - z_{n-1}$. **(b)** Ray interpretation for the two main contributions to the generalized P-wave reflection associated with the nth boundary; source and receiver are in layer 1. We have assumed $\alpha_{n+1} > \alpha_n$, so that a head wave (involving horizontal propagation at the top of the $(n + 1)$-th layer) can exist, together with a wide-angle reflection, as shown. Because this generalized reflection is associated with only one interface, it is known as a *primary* reflection. No mode conversions (from P to SV) are shown. In practice, for a P-wave source, it is often true that the total P-wave response at the receiver is given quite accurately by summing such primary reflections, one for each interface (i.e., by ignoring multiple internal reflections and conversions from P to SV and back to P).

original application of this method was to the interpretation of hydrophone records due to a point source of pressure, a charge of TNT, fired at depth d_1 below the surface of the Bering Sea. With a receiver at depth d_2, a major goal was to interpret the response at ranges r of about 30–70 km in terms of the head wave and reflection at the Moho. Therefore, the theory required extensions to handle possible layering in the crust.

Taking z_0 as the ocean-air interface, and layer 1 as the ocean itself, the source is specified by an incident pressure field in layer 1:

$$P^{\text{inc}}(r, z, t) = \frac{R_S}{[r^2 + (z - z_0 - d_1)^2]^{1/2}} P_0\left(t - \frac{[r^2 + (z - z_0 - d_1)^2]^{1/2}}{\alpha_1}\right). \tag{9.1}$$

That is, the initial pressure pulse has shape $P_0(t - R_S/\alpha_1)$ at a standard (fixed) distance R_S from the source. It follows from our analysis of pressure sources in Chapter 6 that we can immediately write down the Laplace transform for the generalized P-wave primary shown in Figure 9.1b. It is

$$P(r, z, s) = \frac{2s}{\pi} R_S P_0(s) \operatorname{Im} \int_0^{i\infty} \frac{p}{\xi_1} \times K_0(spr)[\text{PRODUCT}(p)] \exp\{-s[\text{SUM}(p)]\}\, dp, \tag{9.2}$$

where K_0 is a modified Bessel function and

$$\text{PRODUCT}(p) = (\grave{P}\grave{P})_1 \cdot (\grave{P}\grave{P})_2 \cdot \cdots \cdot (\grave{P}\grave{P})_{n-1} \cdot (\grave{P}\acute{P})_n \cdot (\acute{P}\acute{P})_{n-1} \cdot \cdots \cdot (\acute{P}\acute{P})_2 \cdot (\acute{P}\acute{P})_1, \tag{9.3}$$

$$\text{SUM}(p) = (Th_1 - d_1)\xi_1 + Th_2\xi_2 + \cdots + Th_n\xi_n + Th_n\xi_n + \cdots + Th_2\xi_2 + (Th_1 - d_2)\xi_1. \tag{9.4}$$

Here, $\xi_i = (\alpha_i^{-2} - p^2)^{1/2}$ with branch cuts chosen by $\operatorname{Re} \xi_i \geq 0$, and $(\grave{P}\grave{P})_{n-1}$ (etc.) is a transmission coefficient for the $(n - 1)$th boundary. The product in (9.3), involving plane-wave transmission and reflection coefficients, is easily written down with an eye on Figure 9.1b, following the generalized ray across interfaces. (Although our physical variable is *pressure*, we have used *displacement* coefficients in this case because source and receiver are in the same layer, and pressure coefficients would give the same product.) Similarly, the sum in (9.4) is the accumulated vertical phase delay along the generalized ray.

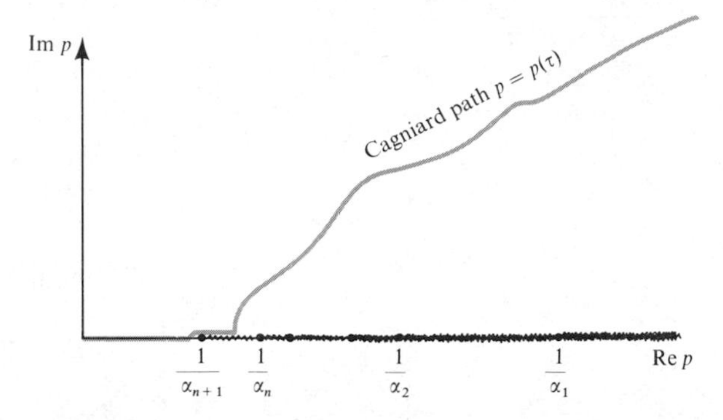

FIGURE **9.2**
The first quadrant of the complex p-plane, showing many branch points on the real axis (with cuts extending to the right), and a Cagniard path solving (9.5) (i.e., for the generalized ray of Figure 9.1). The head-wave contribution arises near $p = 1/\alpha_{n+1}$, and the reflection arises near the departure of the Cagniard path from the real p-axis.

The Cagniard path for inversion of (9.2) is now the solution $p = p(\tau)$ of

$$\tau = pr + \mathrm{SUM}(p), \tag{9.5}$$

where τ is real and positive. Although this step must be accomplished numerically, a solution can readily be found that has many of the features described for the generalized reflection in Figure 6.15. We show the solution for a multilayered case in Figure 9.2, assuming that P-wave speeds $\alpha_1, \alpha_2, \ldots, \alpha_{n+1}$ increase with depth and that r is great enough so that a head wave from the $(n + 1)$th layer is indeed present.

If the source pulse $P_0(t)$ is a unit step at distance R_S, then the time-domain response for the generalized primary is exactly

$$\frac{2}{\pi} R_S \,\mathrm{Im} \int_0^t \frac{p}{\xi_1}\left(\frac{dp}{d\tau}\right) \frac{\mathrm{PRODUCT}(p)\, d\tau}{(t - \tau)^{1/2}(t - \tau + 2pr)^{1/2}} \tag{9.6}$$

(compare with (6.92)). In practice, it is adequate to replace $(t - \tau + 2pr)^{1/2}$ in this denominator by $(2pr)^{1/2}$, so that the step response (9.6) becomes

$$\frac{2}{\pi}\psi(t) * \frac{1}{t^{1/2}}, \qquad \text{where } \psi(t) = R_S \,\mathrm{Im}\left\{\frac{p^{1/2}}{\xi_1}\frac{dp}{d\tau}\frac{\mathrm{PRODUCT}(p)}{(2r)^{1/2}}\right\}. \tag{9.7}$$

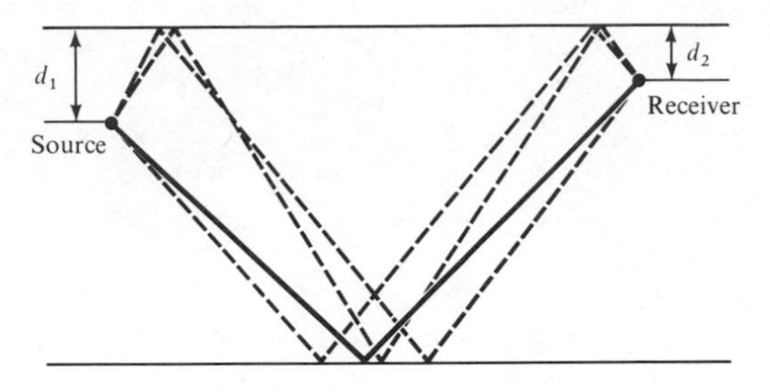

FIGURE **9.3**

The effect of surface reflections, for near-surface source and receiver, is to add three extra generalized rays to the original ray shown in Figure 9.1b. Thus, for the near-source reflection, a time delay $2d_1/\alpha_1$ is introduced (the ocean sound speed being so much slower than α_{n+1} that, for Moho refraction, ray paths in the ocean are nearly vertical). Since $d_1 \ll Th_1$ and $d_1 \ll r$, this time delay is the only important effect (apart from a sign change on reflection), the remainder of the ray being essentially the same as for the original generalized ray.

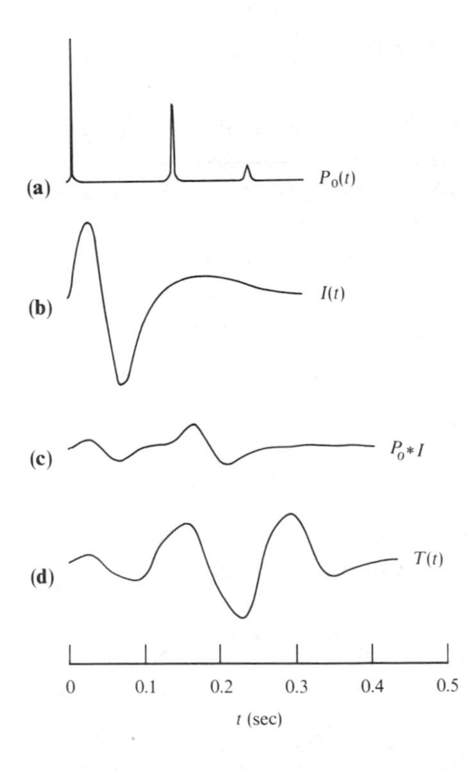

FIGURE **9.4**

Construction of the transfer function.
(a) Typical source function $P_0(t)$ for a charge of TNT fired at a depth of 160 feet. An oscillating gas bubble is responsible for the later pulses. **(b)** The instrument function (hydrophone plus recording system), being the response that would be recorded for a delta-function pressure wave passing the receiver. **(c)** Convolution of (a) and (b).
(d) The transfer function itself, after a typical convolution to describe near-source and near-receiver reflections (see (9.8)). [After Fig. 6 of Helmberger, 1968.]

Before one can compare this response with the data, it is of course necessary to carry out convolutions with a realistic source pulse $P_0(t)$, with the instrument response $I(t)$, and to make some allowance for near-source and near-receiver reflections. These effects together are summarized by a transfer function $T(t)$, and

$$T(t) = P_0(t) * I(t) * \left[\delta(t) - \delta\left(t - \frac{2d_1}{\alpha_1}\right) - \delta\left(t - \frac{2d_2}{\alpha_1}\right) + \delta\left(t - \frac{2d_1}{\alpha_1} - \frac{2d_2}{\alpha_1}\right)\right]. \tag{9.8}$$

The assumptions behind this choice of three extra delta functions to represent surface reflections are described in Figure 9.3, and Figure 9.4 shows the way in which Helmberger (1968) developed $T(t)$ from its components. It then follows that the synthetic response is

$$T(t) * \frac{d}{dt}[\text{step-function response given in (9.6)}] = \frac{dT}{dt} * \frac{2}{\pi}\psi(t) * \frac{1}{t^{1/2}}. \tag{9.9}$$

Recall that only the generalized P-wave primary reflection from the interface at z_n has been considered here. For the oceanic crustal model shown in Figure 9.5, there are P-wave primaries to be considered from four other interfaces above

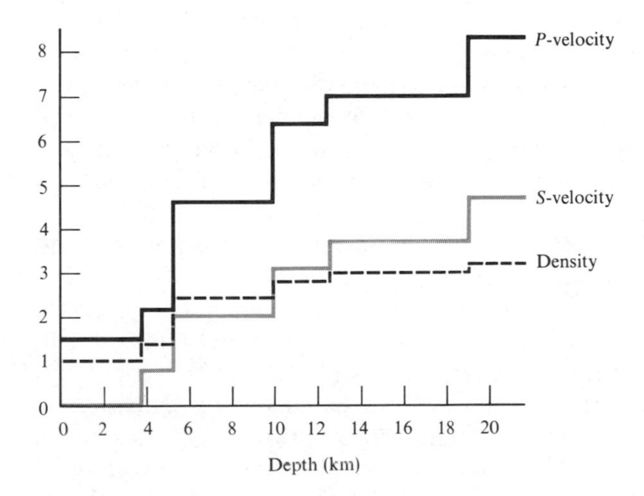

FIGURE **9.5**

A model of the oceanic crust beneath the Northern Aleutian Basin. Moho at depth 19 km, water layer 3.75 km. [After Fig. 10 of Helmberger, 1968.]

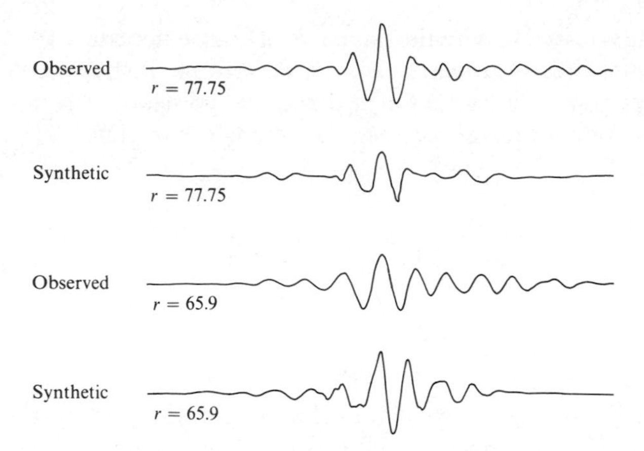

FIGURE **9.6**

Comparison of observed and synthetic hydrophone records for the head wave and wide-angle reflection from the Moho beneath the Northern Aleutian Basin. The reflection is relatively quite strong. Synthetics were computed from the generalized *P*-wave primary reflections in the model of Figure 9.5. The observations have been attenuated by a factor $10^{1/2}$, whereas the synthetic waveforms are reduced by 2. [After Fig. 11 of Helmberger, 1968.]

the Moho at depth 19 km. However, these can easily be included by summing the appropriate ψ-functions defined in (9.7), each with a different Cagniard path, and then carrying out the two convolutions expressed in (9.9). The result is shown in Figure 9.6, a comparison of synthetic and observed waveforms given by Helmberger (1968) for a refraction experiment conducted in the Bering Sea, about 500 km north of Adak. The figure does show remarkably good agreement between synthetic and observed pulse shapes, although some discrepancies in relative amplitude remain, due possibly to errors in calibration of the source. Further details of the Cagniard path, and the wavefront expansion for head wave and reflection, are described in Problems 9.1 and 9.2.

The ease with which the contribution from each generalized ray can be computed makes it feasible to consider cases in which several hundred generalized rays are summed together. A large number of interesting geophysical problems have therefore yielded to an analysis based on this extension to Cagniard's theory. Thus Mellman and Helmberger (1974) have shown how a thin high-velocity layer can cause attenuation of high-frequency waves; Vered and Ben-Menahem (1974) and Langston and Helmberger (1975) have shown how to analyze shear dislocation sources; Helmberger and Malone (1975) have matched synthetics and observations of *SH* waves to show how crustal layering can severely distort the waveforms due to local earthquakes; and Spudich and

Helmberger (1979) have examined the sedimentary structures that may be expected at the ocean floor, finding the practical effect of leaking modes like the $\bar{P}$ wave we have described in Chapter 6 for a half-space.

Often, however, the number of generalized rays that must be considered is prohibitively large. This will occur for structures having strong velocity gradients and/or synthetics that are to be evaluated for extended times. Because of the importance of such problems, we turn now to an alternative method of computation, in which *all* the multiples are retained.

9.2 The Reflectivity Method for a Medium with Many Plane Layers

In Chapters 6 and 7, we showed how to obtain doubly transformed solutions as a function of (k, z, ω), where k is the horizontal wavenumber. These chapters emphasized manipulations in the complex k-plane (or p-plane, with $k = \omega p$), in order to evaluate an inverse transform giving the solution as a function of (r, z, ω). Fuchs' approach is to evaluate this inverse transform numerically, by integrating over real values of k, and then also to integrate over real values of ω to obtain the desired solution as a function of (r, z, t). For example, in describing the excitation of Love waves (Section 7.4) we obtained a formula for the multi-transformed solution. This can be computed by using the known form of the propagator for SH-waves in a stack of homogeneous layers. The integral over real k is carried out at a sequence of different frequencies, and then a discrete form of $(1/\pi)\,\mathrm{Re}\int_0^\infty s(r, z, \omega)\exp(-i\omega t)\,d\omega$ is used to obtain the synthetic seismogram.

A difficulty in execution of the above program is caused by poles in the multi-transformed solution that correspond to surface-wave modes. These lie on the real k-axis itself, and hence are on the path of integration. It is likely that the difficulty can be removed by adding a small complex part to the velocity in each layer, accounting for anelasticity (see (5.88)), which moves surface-wave poles off the positive real k-axis and up into the first quadrant of the k-plane. However, Fuchs' approach has principally been used in body-wave studies to get the response at the free surface of a stack of homogeneous plane layers within which a point source is active. If interest is restricted to body waves, then integration over k can be limited to a part of the real k-axis that has no surface-wave poles. We now discuss a particular example of this numerical approach, which has come to be known as the *reflectivity method*.

Referring to Figure 9.7, suppose a point source at depth h generates waves that, at the receiver, appear to be reflected from some structure between depths z_m and z_n. For example, in the Earth this might be a region of high velocity gradient, modeled here by many thin homogeneous layers.

Using the motion-stress vector approach, we have seen that the source is handled by a discontinuity in $\mathbf{f}(k, z, \omega)$ at $z = h$, where components of $\mathbf{f}$ are multi-transformed components of displacement and traction. If the source has

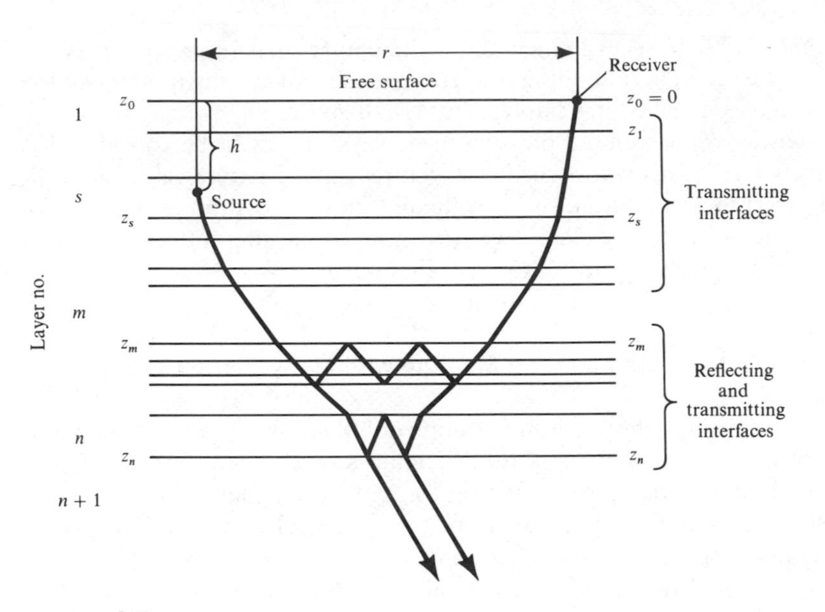

FIGURE **9.7**

The reflectivity method. A point source at depth h is located in layer s. It is desired to synthesize the displacement at range r on the free surface for body waves that have interacted with structure in the medium between depths z_m and z_n. Since the propagator matrix is used between these two depths, all multiples occurring within the structure of interest are retained. For example, a P head-wave and S-waves are shown in layer n. Only those waves that depart downward from the source as P-waves and up to the receiver as P-waves are described in the text. The effect of structure above the source is conceptually simple to include.

azimuthal (ϕ) dependence, then we have seen that a transform from ϕ to m is necessary, that each m component is solved, and that $\mathbf{f}(k, \phi, z, \omega)$ contains a sum over m. However, only a few terms are necessary in practice ($-2 \leq m \leq 2$ for double-couple sources). The discontinuity in $\mathbf{f}$, for a shear dislocation of general orientation, has been described by Hudson (1969).

If the source were in an infinite homogeneous whole space with the same elastic parameters as layer s, it would not be difficult to obtain the downgoing motion-stress vector below the source. We shall consider only a P-SV problem, with an interest in the P-wave mode of propagation in the region above the reflecting structure. Recall in our notation $\mathbf{f} = \mathbf{Fw}$ that $\mathbf{F} = \mathbf{F}(z)$ is the *layer matrix*, defined in each layer (see (5.65)) and that $\mathbf{w}$ is a vector that weights the columns of $\mathbf{F}$. The four components of $\mathbf{w}$ are constant in general in each layer, and give the amplitude of downgoing P, downgoing SV, upgoing P, upgoing SV (respectively) in the layer. The source layer differs from all others in that $\mathbf{w}_s$ takes two values, one appropriate for $z < h$ (there are extra upgoing waves from the source), and one appropriate for $z > h$ (with extra downgoing waves). It follows from (5.65) that the motion-stress vector for downgoing P-waves from the source alone is given by the first column of the matrix $\mathbf{E}$, times phase factor $\exp[i\omega\xi_s(z - h)]$, times a scalar P^{inc} that is related to the discontinuity in $\mathbf{f}$ at h. It is this vector, $\mathbf{f}^{\text{inc}\,P}$, that we need to start the propagation part of the problem.

$$\mathbf{f}^{\text{inc}\,P} = \mathbf{f}^{\text{inc}\,P}(k, m, z, \omega) = P^{\text{inc}} \begin{pmatrix} \alpha_s p \\ \alpha_s \xi_s \\ 2i\omega\rho_s\alpha_s\beta_s^2 p\xi_s \\ i\omega\rho_s\alpha_s(1 - 2\beta_s^2 p^2) \end{pmatrix} \exp[i\omega\xi_s(z - h)]. \tag{9.10}$$

The source spectrum, and any azimuthal effects, are contained within the scalar P^{inc}. Since our main interest is in displacement, we shall not continue to write out the last two components of the motion-stress vector.

Carrying the wave on to a downgoing P-wave at the bottom of layer m, we multiply the downgoing wave by transmission coefficients and vertical phase factors, obtaining

$$\begin{pmatrix} \alpha_m p \\ \alpha_m \xi_m \end{pmatrix} \left(\prod_{j=s}^{m-1} \right) \grave{P}\grave{P}_j \exp\left[i\omega \left(\sum_{j=s+1}^{m} \xi_j Th_j \right) \right] P^{\text{inc}} \exp[i\omega\xi_s(z_s - h)]$$

for the motion that is incident on the stack of layers between z_m and z_n in which we are interested.

Let the P-P reflection coefficient for this stack of layers be denoted by $\{\grave{P}\acute{P}\}_n^m$, with similar notation for converted ($P \to S$) and transmitted phases. To obtain this coefficient, we work with the propagator matrix $\mathbf{P}(z_m, z_n)$ which is known explicitly (see Box 9.1) for homogeneous layers. Recalling the property

$$\mathbf{f}(z_m) = \mathbf{P}(z_m, z_n)\mathbf{f}(z_n), \tag{9.11}$$

we substitute

$$\mathbf{f}(z_m) = \mathbf{E}_m \begin{pmatrix} 1 \\ 0 \\ \{\grave{P}\acute{P}\}_n^m \\ \{\grave{P}\acute{S}\}_n^m \end{pmatrix} \quad \text{and} \quad \mathbf{f}(z_n) = \mathbf{E}_{n+1} \begin{pmatrix} \{\grave{P}\grave{P}\}_n^m \\ \{\grave{P}\grave{S}\}_n^m \\ 0 \\ 0 \end{pmatrix}. \tag{9.12}$$

The reference level for phase is here taken as z_m for reflections and z_n for transmission coefficients, so it is the matrices $\mathbf{E}$ that appear in (9.12) (because $\mathbf{E}_m = \mathbf{F}_m(z_m)$ and $\mathbf{E}_{n+1} = \mathbf{F}_{n+1}(z_n)$, from (5.65)). The first of equations (9.12) contains our assumption that the only wave incident on the stack is downgoing P, and the second of (9.12) has no upgoing waves in the half-space below z_n. Since these forms for the motion-stress vector, substituted in (9.11), give four scalar equations in four unknowns, we can solve for the particular unknown of interest, namely $\{\grave{P}\acute{P}\}_n^m$. Detailed discussion is given by Červený (1974) and Kind (1976) (see also Box 9.6). The upgoing P-wave at the bottom of layer m is

$$\begin{pmatrix} \alpha_m p \\ -\alpha_m \xi_m \end{pmatrix} \left(\prod_{j=s}^{m-1} \grave{P}\grave{P}_j \right) \exp\left[i\omega \left(\sum_{j=s+1}^{m} \xi_j Th_j \right) \right] P^{\text{inc}} \{\grave{P}\acute{P}\}_n^m \exp[i\omega \xi_s (z_s - h)].$$

Next, we carry this motion up to the free surface as a P-wave, obtaining

$$\begin{pmatrix} \alpha_1 p \\ -\alpha_1 \xi_1 \end{pmatrix} (\text{PRODUCT}) \times \exp[i\omega(\text{SUM})] \{\grave{P}\acute{P}\}_n^m P^{\text{inc}}, \tag{9.13}$$

where

$$\text{PRODUCT} = \left(\prod_{j=s}^{m-1} \grave{P}\grave{P}_j \right) \times \left(\prod_{j=1}^{m-1} \acute{P}\acute{P}_j \right),$$
$$\text{SUM} = \left(\sum_{j=s+1}^{m} \xi_j Th_j \right) + \left(\sum_{j=1}^{m} \xi_j Th_j \right) + \xi_s (z_s - h). \tag{9.14}$$

The displacement (9.13) must be corrected for the effect of a free surface by adding the downward-reflected P- and S-waves at $z = 0$ to the upcoming P-wave. The correction is given in Problem 5.6, and at last we can state the

BOX **9.1**

Propagator matrices for SH and for P-SV problems

We introduced propagator matrices in the context of Love and Rayleigh waves (Section 7.2). Particle motion, we noted, decays with depth, hence it was convenient to work with $e^{\pm \gamma z}$, $e^{\pm \nu z}$, in which (for the most part) γ and ν were real. Moreover, for P-SV, a $\pi/2$ phase shift was explicitly recognized between vertical and horizontal components of displacement and traction (see (7.25)–(7.27)), so that r_1, r_2, r_3, r_4 were real. Although the resulting equations are correct for body waves too, the notation γ, ν is inconvenient because γ and ν become negative imaginary. Note also that (for homogeneous waves) there is no phase shift between displacement components or between traction components, so that r_1, r_2, r_3, r_4 cannot all be real. In short, it is better to work with the motion-stress vectors introduced (for plane waves) in Section 5.4 and to work with slownesses p, ξ, η rather than horizontal wavenumber k and decay constants γ and ν.

For SH-waves in the far field, we use

$$\mathbf{f} = \mathbf{f}(k, m, z, \omega) = \mathbf{f}(\omega p, m, z, \omega) = \begin{pmatrix} u_\phi(\omega p, m, z, \omega) \\ \tau_{\phi z}(\omega p, m, z, \omega) \end{pmatrix} \tag{1a}$$

for cylindrical coordinates (r, ϕ, z). The underlying horizontal wavefunction is $e^{im\phi}\, d[J_m(kr)]/dr$. This representation is an approximation, since it neglects a horizontal component of motion perpendicular to the ϕ-direction. An exact approach is based on displacement $\mathbf{u} = U(k, m, z, \omega)\mathbf{T}_k^m(r, \phi)$ and traction $= T(k, m, z, \omega)\mathbf{T}_k^m(r, \phi)$, where $\mathbf{T}_k^m$ is the vector surface harmonic defined in (7.117). In the exact representation, valid in the near field and for long times, we take

$$\mathbf{f} = \begin{pmatrix} U(\omega p, m, z, \omega) \\ T(\omega p, m, z, \omega) \end{pmatrix}. \tag{1b}$$

In either (1a) or (1b), the associated equations $\mathbf{f} = \mathbf{Fw}$, $\mathbf{F} = \mathbf{E\Lambda}$ are described in Section 5.4, and layer matrix $\mathbf{F}$ and its inverse are given explicitly in (5.63)–(5.64). The propagator is

$$\mathbf{P}(z, z_0) = \mathbf{F}(z)\mathbf{F}^{-1}(z_0) = \begin{pmatrix} \cos[\omega\eta(z - z_0)] & \dfrac{1}{\omega\mu\eta} \sin[\omega\eta(z - z_0)] \\ -\omega\mu\eta \sin[\omega\eta(z - z_0)] & \cos[\omega\eta(z - z_0)] \end{pmatrix}, \tag{2}$$

provided z and z_0 are in the same (homogeneous) layer. If these depths are in different layers of a structure composed of homogeneous welded layers, then use (7.44).

For *P-SV* in the far field, we use u_r, u_z, u_{rz}, u_{zz} for the components of $\mathbf{f}$. Then $\mathbf{F}$ is found from (5.65) and its inverse from (5.67). The propagator (for z and z_0 in the same homogeneous layer) is then

$$\mathbf{P}(z, z_0) = \begin{pmatrix} 2\beta^2p^2C_\xi + (1 - 2\beta^2p^2)C_\eta & \dfrac{ip}{\xi}(1 - 2\beta^2p^2)S_\xi - 2i\beta^2p\eta S_\eta \\ 2i\beta^2p\xi S_\xi - \dfrac{ip}{\eta}(1 - 2\beta^2p^2)S_\eta & (1 - 2\beta^2p^2)C_\xi + 2\beta^2p^2C_\eta \\ -4\omega\rho\beta^4p^2\xi S_\xi - \dfrac{\omega\rho}{\eta}(1 - 2\beta^2p^2)^2S_\eta & 2i\omega\rho\beta^2p(1 - 2\beta^2p^2)[C_\xi - C_\eta] \\ 2i\omega\rho\beta^2p(1 - 2\beta^2p^2)[C_\xi - C_\eta] & -\dfrac{\omega\rho}{\xi}(1 - 2\beta^2p^2)^2S_\xi - 4\omega\rho\beta^4p^2\eta S_\eta \end{pmatrix}$$

$$\begin{pmatrix} \dfrac{p^2}{\omega\rho\xi}S_\xi + \dfrac{\eta}{\omega\rho}S_\eta & -\dfrac{ip}{\omega\rho}[C_\xi - C_\eta] \\ -\dfrac{ip}{\omega\rho}[C_\xi - C_\eta] & \dfrac{\xi}{\omega\rho}S_\xi + \dfrac{p^2}{\omega\rho\eta}S_\eta \\ 2\beta^2p^2C_\xi + (1 - 2\beta^2p^2)C_\eta & 2i\beta^2p\xi S_\xi - \dfrac{ip}{\eta}(1 - 2\beta^2p^2)S_\eta \\ \dfrac{ip}{\xi}(1 - 2\beta^2p^2)S_\xi - 2i\beta^2p\eta S_\eta & (1 - 2\beta^2p^2)C_\xi + 2\beta^2p^2C_\eta \end{pmatrix} \tag{3}$$

where columns 1 and 2 are written above columns 3 and 4; $C_\xi = \cos[\omega\xi(z - z_0)]$, $C_\eta = \cos[\omega\eta(z - z_0)]$, $S_\xi = \sin[\omega\xi(z - z_0)]$, and $S_\eta = \sin[\omega\eta(z - z_0)]$.

Note here the checkerboard pattern of real and imaginary entries in (3). A product of such matrices has the same pattern, so that it is easy to program the required manipulations of the propagator (e.g., (7.44)) using only real variables if p is real. Some of the entries in (2) and (3) are $O(\omega)$, and some are $O(\omega^{-1})$. This frequency dependence can be removed by working with particle velocities and tractions (rather than particle displacements and tractions).

displacement components of the motion-stress vector at the receiver. They are

$$\begin{pmatrix} u_r(k, m, 0, \omega) \\ u_z(k, m, 0, \omega) \end{pmatrix} = \begin{pmatrix} 4\alpha_1 \beta_1^2 p \xi_1 \eta_1 \\ -2\alpha_1 \xi_1 (1 - 2\beta_1^2 p^2) \end{pmatrix}$$
$$\times \frac{\text{PRODUCT} \times \exp[i\omega \, \text{SUM}] \times \{\grave{P}\acute{P}\}_n^m \times P^{\text{inc}}}{(1 - 2\beta_1^2 p^2)^2 + 4\beta_1^4 p^2 \xi_1 \eta_1}. \tag{9.15}$$

This is the multi-transformed solution, which is to be summed over m (P^{inc} is m-dependent, in general) and integrated over k and ω. Thus, working from (7.131), we find that in the frequency domain the solution is

$$\mathbf{u}(r, \phi, 0, \omega) = \frac{1}{2\pi} \sum_{m=-\infty}^{\infty} \int_0^\infty k[u_r(k, m, 0, \omega)\mathbf{S}_k^m - iu_z(k, m, 0, \omega)\mathbf{R}_k^m] \, dk.$$

At body-wave values of k, and with some horizontal separation between source and receiver, the $\hat{\boldsymbol{\phi}}$ component of $\mathbf{S}_k^m$ (see (7.117)) is negligible, hence (9.15) is integrated via

$$\mathbf{u}(r, \phi, 0, \omega) = \frac{1}{2\pi} \sum_{m=-\infty}^{\infty} \int_0^\infty \omega^2 p \left[u_r \frac{d}{d(\omega pr)} J_m(\omega pr)\hat{\mathbf{r}} + iu_z J_m(\omega pr)\hat{\mathbf{z}} \right] e^{im\phi} \, dp, \tag{9.16}$$

followed by an integral over ω to recover $\mathbf{u}(r, \phi, 0, t)$.

In practice, only a finite range of p-values is used in (9.16), and Fuchs and Müller (1971) recommend $0 \le p \le 1/\alpha_{\max}$, where $\alpha_{\max}$ is the fastest P-wave speed in the layers above the reflecting region. It follows that SUM is always real, and inhomogeneous waves are avoided above the reflecting region.

The reflecting region itself appears in the computation only via $\{\grave{P}\acute{P}\}_n^m$, and we shall shortly look at a simple example that illustrates that all multi-reflected rays within the range z_m to z_n are automatically accounted for. First, however, we show a record section of synthetic seismograms obtained by the reflectivity method for an explosive source acting at the surface of a particular crustal structure. The structure is shown in Figure 9.8a, and Fuchs and Müller (1971) have obtained the synthetics shown in Figure 9.8c, using as reflector the lower boundary of the low-velocity channel and the layers below this channel. A guide in the interpretation is given by the *reduced travel-time curve*, Figure 9.8b. (If t were plotted against r instead of against the reduced time $t - r/6$, the vertical scale would spread over about 26 sec instead of the 6 sec shown, and much detail of the temporal separation between arrivals could be lost.) This curve is also a guide to the range of p-values needed in the integration (9.16), since the ray parameter at which prominent waves will arrive at the receiver can be obtained from the travel-time curve. In fact, it is the gradient of the travel-time curve (see Problem 9.2(i)). Since this gradient ranges from about 1/8.15 to $\frac{1}{6}$, the range of p-values used in the integration for Figure 9.8c was $1/8.15 \le p \le \frac{1}{6}$. The most time-consuming part of the computation is the evaluation of $\{\grave{P}\acute{P}\}_n^m$ at different values of p and ω. However, it is a step that need be done only

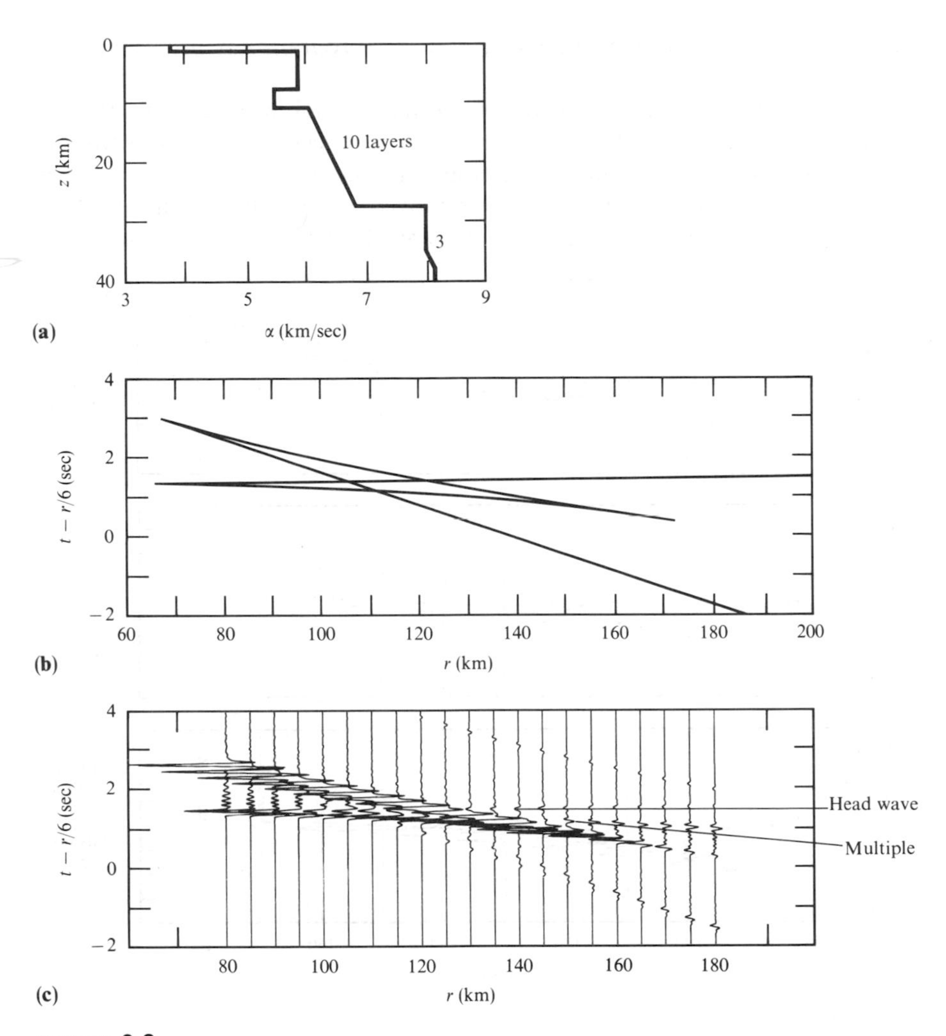

FIGURE **9.8**

An application of the reflectivity method. (**a**) A model of crustal structure. Only the P-wave speed is shown. The linear increase in the lower crust (down to the Moho at depth 27 km) is in fact modeled by 10 homogeneous layers, and a thin linear region below the Moho by 3 layers. (**b**) A reduced travel-time curve for P-waves in (a) that have slowness less than 1/6 sec/km. (**c**) Synthetic seismograms (vertical component of displacement) corresponding to (b), obtained via the reflectivity method. Most prominent are the wave that has been turned around once within the velocity gradient in the lowermost crust and the wide-angle reflection at the Moho. P_n is small, but its early arrival beyond 120 km would permit its identification in the Earth if this crustal structure is accurate. In the range 150–180 km is seen a prominent multiple at reduced travel time 1 sec that may be identified as a wave turned around twice within the lowermost crust, having also been reflected once from below the low-velocity zone. The "head wave" from just below the lower boundary of the low-velocity zone is very weak beyond about 140 km because of the positive velocity gradient. Latest of all to arrive is a phase with slowness equal to 1/8.15 (like the head wave). This is an artifact of truncating the integration in (9.16) to a finite range of p. [After Fig. 7 of Fuchs and Müller 1971.]

and equations corresponding to (9.12) for this SH-problem are

$$\mathbf{f}(z_1) = \begin{pmatrix} 1 & 1 \\ i\omega\mu_1\eta_1 & -i\omega\mu_1\eta_1 \end{pmatrix}\begin{pmatrix} 1 \\ \{\grave{S}\acute{S}\}_2^1 \end{pmatrix}, \quad \mathbf{f}(z_2) = \begin{pmatrix} 1 & 1 \\ i\omega\mu_3\eta_3 & -i\omega\mu_3\eta_3 \end{pmatrix}\begin{pmatrix} \{\grave{S}\grave{S}\}_2^1 \\ 0 \end{pmatrix}.$$

Substituting these into $\mathbf{f}(z_1) = \mathbf{P}(z_1, z_2)\mathbf{f}(z_2)$ and using the SH propagator matrix (Box 9.1, equation (2)), we obtain two equations for the two unknowns $\{\grave{S}\acute{S}\}_2^1$ and $\{\grave{S}\grave{S}\}_2^1$. The reflection solution is

$$\{\grave{S}\acute{S}\}_2^1 = \frac{\left(C - \dfrac{i\mu_3\eta_3 S}{\mu_2\eta_2}\right) - \left(\dfrac{\mu_3\eta_3}{\mu_1\eta_1} C - \dfrac{i\mu_2\eta_2 S}{\mu_1\eta_1}\right)}{\left(C - \dfrac{i\mu_3\eta_3 S}{\mu_2\eta_2}\right) + \left(\dfrac{\mu_3\eta_3}{\mu_1\eta_1} C - \dfrac{i\mu_2\eta_2 S}{\mu_1\eta_1}\right)}, \tag{9.17}$$

where $C = \cos(\omega\eta_2 Th_2)$, $S = \sin(\omega\eta_2 Th_2)$.

Shown in Figure 9.10b is the system of multiples that may be expected from a causal approach to the problem. That is, a reflection $\grave{S}\acute{S}_1$ is expected from the top of the layer, followed by a wave once-reflected from the bottom of the layer, $\grave{S}\grave{S}_1 \cdot \grave{S}\acute{S}_2 \cdot \acute{S}\acute{S}_1 e^{2i\omega\eta_2 Th_2}$, etc. The phase factor allows for vertical propagation, and the total of reflections should be given as

$$\{\grave{S}\acute{S}\}_2^1 = \grave{S}\acute{S}_1 + \grave{S}\grave{S}_1\left[\sum_{j=1}^{\infty} (\grave{S}\acute{S}_2 e^{2i\omega\eta_2 Th_2})^j (\acute{S}\grave{S}_1)^{j-1}\right]\acute{S}\acute{S}_1 \tag{9.18}$$

$$= \grave{S}\acute{S}_1 + \frac{\grave{S}\grave{S}_1 \cdot \grave{S}\acute{S}_2 \cdot \acute{S}\acute{S}_1 \cdot e^{2i\omega\eta_2 Th_2}}{1 - \grave{S}\acute{S}_2 \cdot \acute{S}\grave{S}_1 \cdot e^{2i\omega n_2 Th_2}}. \tag{9.19}$$

This is the key stage. An infinite geometrical series has been replaced by a single exact expression for the sum. Individual terms of (9.18) correspond to particular rays, and the whole infinite family of multiples is clearly contained in (9.19). To evaluate (9.19), we can use (5.32) for individual reflections/transmissions, finding

$$\{\grave{S}\acute{S}\}_2^1 = \left(\frac{\mu_1\eta_1 - \mu_2\eta_2}{\mu_1\eta_1 + \mu_2\eta_2}\right) + \frac{\dfrac{2\mu_1\eta_2}{\mu_1\eta_1 + \mu_2\eta_2} \cdot \dfrac{\mu_2\eta_2 - \mu_3\eta_3}{\mu_2\eta_2 + \mu_3\eta_3} \cdot \dfrac{2\mu_2\eta_2}{\mu_1\eta_1 + \mu_2\eta_2} \cdot e^{2i\omega\eta_2 Th_2}}{1 - \dfrac{\mu_2\eta_2 - \mu_3\eta_3}{\mu_2\eta_2 + \mu_3\eta_3} \dfrac{\mu_2\eta_2 - \mu_1\eta_1}{\mu_1\eta_1 + \mu_2\eta_2} \cdot e^{2i\omega\eta_2 Th_2}},$$

which does reduce (after some algebraic manipulation) exactly to (9.17). In this way we have shown that the propagator-derived reflection for the double interface can be written as an infinite set of generalized rays.

If more than two interfaces are present, and with the complication of conversion between P and SV, it is clear that an immensely involved system of generalized rays may be present in the propagator-derived reflection, e.g., in $\{\grave{P}\acute{P}\}_n^m$. Spencer (1960, 1965) pointed out the main virtue of working with generalized rays—namely, that each term in the infinite series has an extremely simple dependence on frequency or on the Laplace transform variable s. For example, in (9.18) we note that $\omega\eta_2 = \omega(1/\beta_2^2 - p^2)^{1/2} = is(1/\beta_2^2 - p^2)^{1/2}$, and s enters each generalized ray only via the exponent function. A Cagniard method can therefore be used to get the contribution (in the time domain) of each generalized ray. The ω-dependence of (9.15) and (9.17) is far more complicated, and Cagniard methods do not apply directly to the sum of the whole system of multiples. Hron (1971, 1972) showed how subfamilies of generalized rays could be found that all used the same Cagniard path, and Cisternas et al. (1973) gave a systematic method for finding out the generalized rays contained within the exact (i.e., propagator-derived) solution.

The reflectivity method was developed by Fuchs and his colleagues for crustal studies. Orcutt et al. (1976) have applied the method to study of the structure of the East Pacific Rise. By using a flat Earth model that has the same travel times as a spherical Earth model (Section 9.6), Müller (1973) has used the reflectivity method to interpret long-period body waves that pass through the Earth's core, and Müller and Kind (1976) have analyzed long-period body-wave seismograms for the whole Earth.

Our description of wave propagation in a stack of homogeneous layers has led us so far to two different numerical procedures. However, whether Cagniard's method is adopted (for many generalized rays) or Fuchs' method (with its double integration), there is one skill that turns out again and again to be useful in acquiring some insight into the waves that propagate within a given structure—namely, the ability to recognize the way in which particular waves in the elastic medium are related to properties of certain wavefunctions in the complex ray-parameter plane. For example, we have examined saddle points, branch cuts, and poles in the p-plane and have related them to ray paths, head waves, and surface waves. For a stack of homogeneous layers, the analysis of a particular generalized ray is fairly simple in the p-plane. This is largely because the horizontal distance to which a particular ray can propagate is always an *increasing* function of ray parameter (see Problem 9.2). However, in the remainder of this chapter, we shall be looking at waves in media for which there can be a continuous change in seismic velocities and density within each layer. This can lead, first, to the phenomenon of a ray being completely turned around *within* a given layer. An example is shown in Figure 9.11. Second, we shall find that the horizontal distance reached by a particular ray (propagating in a stack of inhomogeneous layers) is very often a *decreasing* function of ray parameter. The distance can also increase with p, and (for a given distance Δ) there can exist rays of both types (i.e., with $d\Delta/dp$ less than and greater than zero).

These complications will lead us, in Section 9.4, to call upon several mathematical methods that we have so far not needed to analyze waves in

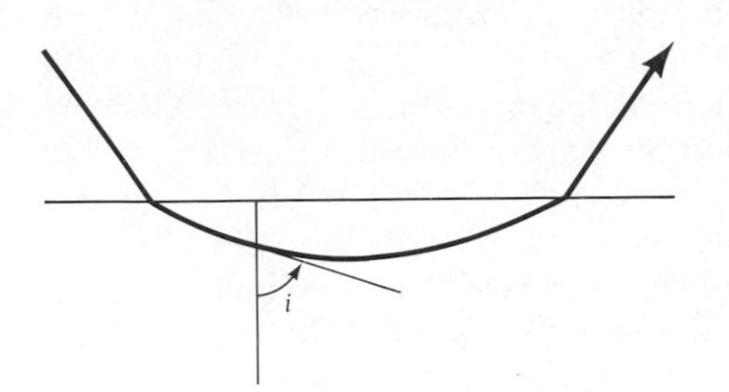

FIGURE **9.11**
A ray is shown that turns around within an inhomogeneous layer. The point at which the ray is traveling horizontally is said to be the *turning point*. Snell's law, $(\sin i)/v = p$ (a constant), applies along the ray. Since $i = 90^\circ$ for a turning point, the equation $1/v(z) = p$ is satisfied if z is the depth of a turning point.

homogeneous layers. However, properties of wavefunctions in the complex ray-parameter plane will still be of paramount importance. We prepare for this material by reviewing some elementary ray theory.

So far, we have emphasized plane-stratified media in which the ray parameter p is $\sin i/\alpha$ (both i and velocity α being depth dependent). There is frequently a need to express travel time and distance as a function of p (see Problem 9.2). The results for P-waves are

$$\begin{aligned} \mathrm{TIME}(p) &= \int \frac{dz}{\alpha(z)\cos i(z)} = \int (1 - p^2\alpha^2)^{-1/2}\frac{dz}{\alpha} = \int \frac{dz}{\alpha^2\xi}, \\ \mathrm{DISTANCE}(p) &= \int \tan i \, dz = \int \frac{p}{\xi}\, dz, \end{aligned} \tag{9.20}$$

in which

$$\xi = \xi(p, z) = \left(\frac{1}{\alpha^2(z)} - p^2\right)^{1/2} = \frac{\cos i}{\alpha}$$

and the integration is carried out over the range of depths traversed by the ray. However, the next section is devoted to spherically stratified media, in which the ray parameter p is $(r \sin i)/\alpha$, since it is for such media that ray theory in seismology is most extensively applied.

9.3 Classical Ray Theory in Seismology

In Chapter 4 we derived the ray-theory solutions for displacement for high-frequency waves radiated into the far field from a point source. The key concepts of geometrical spreading and travel time along a ray path were well understood by Christiaan Huygens and Pierre de Fermat about 300 years ago. Our interest is in the particular properties of the rays, as receiver position varies, for a medium that is laterally homogeneous. For example, if travel time T is plotted against distance for a particular observed body wave, how might we conclude that the wave is a reflection from some internal discontinuity within the Earth? Or how might we find evidence for a low-velocity zone?

We begin in Figure 9.12 with a look at the S-wave rays that are present for a surface source in a model of the upper mantle. Clearly, several rays might arrive at a given receiver, and the travel-time function (Figure 9.12c) is multivalued for a certain range of distances. However, each point along the travel-time curve has a unique slope, the value decreasing from A to B, B to C, etc. This suggests a useful independent variable. It follows from Figure 5.2 that this slope, $dT/d\Delta$, which is the horizontal slowness for an obliquely propagating wave, is nothing but the ray parameter p. (Horizontal distance is Δ in dimensionless units and is $r\Delta$ in units of length.)

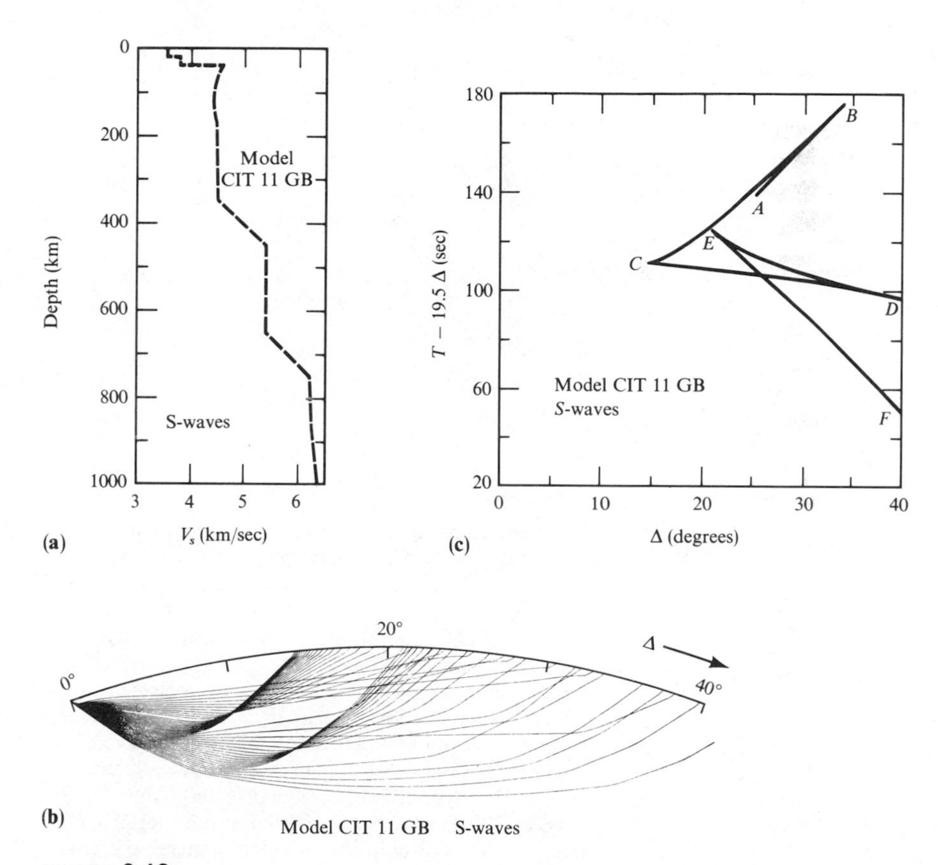

FIGURE **9.12**

(a) The S-wave velocity for the upper mantle, taken from model CIT 11 GB. **(b)** Corresponding S-wave rays for a point source at the surface, calculated for take-off angles increasing from 28° to 50° in $\frac{1}{2}$° increments. Note that distance between source and receiver in the Earth is measured by the angle Δ subtended at the Earth's center. **(c)** Corresponding reduced travel-time curve. Point C is clearly identified with strong focusing of rays in (b) at Δ near 14°, and amplitudes there will be large. Lines AB, BC, and CD together constitute a *triplication*, and each of the two triplications shown is associated with a major velocity increase (with depth) in the Earth model. [After Julian and Anderson, 1968.]

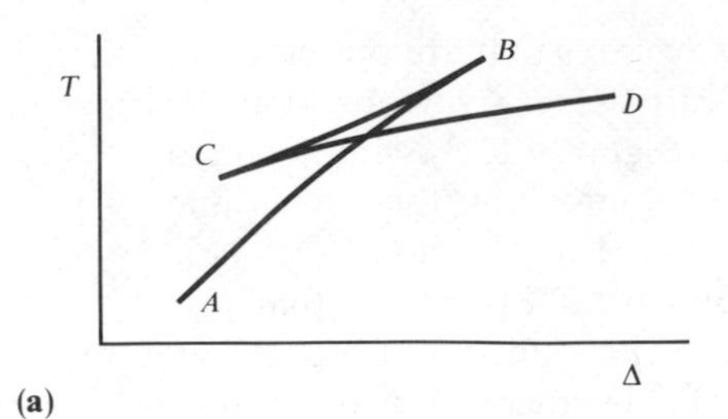

(a)

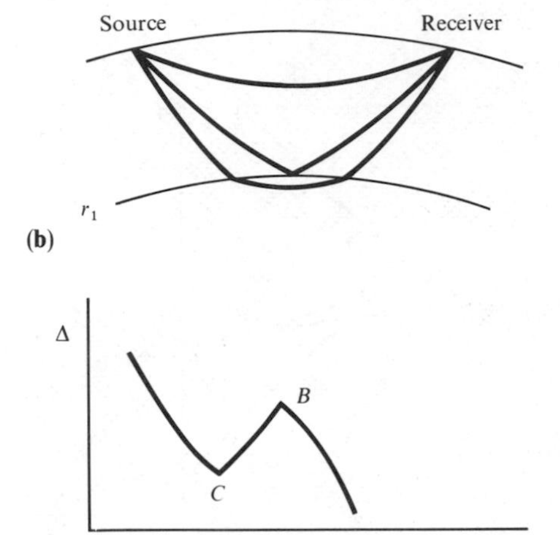

(b)

(c)

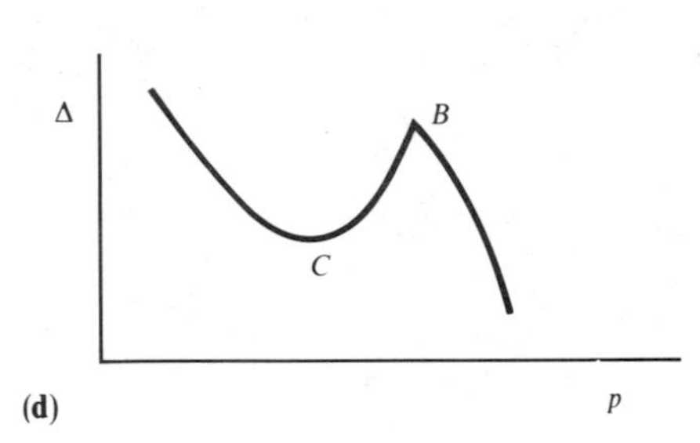

(d)

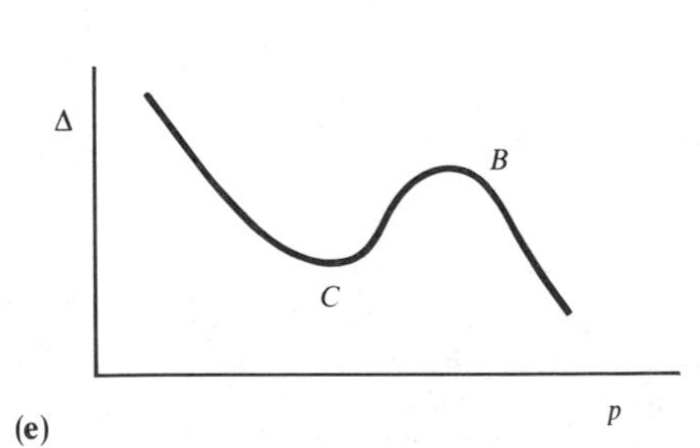

(e)

FIGURE **9.13**

The behavior of T, Δ, and p for a velocity increase with depth. (**a**) Nomenclature for a typical triplication. (**b**) Three rays are shown, all having the same Δ. The velocity is assumed to be increasing with depth, and at the level r_1 an anomalous increase of velocity with depth also occurs. The uppermost ray here may be called the direct ray (branch AB); then there is a reflection-like ray (branch BC); and finally a head-wave-like ray (branch CD). The degree of continuity in the velocity profile across r_1 determines whether rays actually change direction when they reach this level or have high curvature. (**c**) $\Delta(p)$ is shown for a discontinuous velocity increase across r_1. Note that $d\Delta/dp$ is discontinuous at B and C, and may never be small. (The quantity $(d\Delta/dp)^{-1/2}$ is a controlling factor in the wave amplitude predicted by ray theory.) (**d**) $\Delta(p)$ is shown for a continuous velocity profile in which the velocity gradient is discontinuous across r_1. Then $d\Delta/dp = 0$ at C, giving a caustic. (**e**) $\Delta(p)$ is shown when both velocity and velocity gradient are continuous. Both B and C are caustics.

Thus consider how the distance function depends on p. From Figure 9.12c it is clear that distance and slope decrease together for the branch BC, but along AB and CD distance is increasing while slope decreases. This is shown explicitly in Figure 9.13, and some special significance is attached to the points B and C. Note that $d\Delta/dp$ changes sign at these points, and it can happen that $d\Delta/dp$ remains continuous, so that $d\Delta/dp = 0$ at C or at B and C. Since the geometrical spreading function $\mathscr{R}^{-1}$ is proportional to $(d\Delta/dp)^{-1/2}$ (see Problem 4.3), ray theory predicts a singular amplitude for the displacement. This is the phenomenon of a *caustic*, and an example is shown in Figure 9.12 at the distance 14°. A caustic is the envelope of a system of rays, and for the source and Earth model of Figure 9.12 the envelope in three dimensions is a surface intersecting the Earth's surface on a circle. Of course, the prediction of ray theory here is incorrect: amplitudes may be large in the vicinity of a caustic (as shown by the focusing of rays in Figure 9.12b), but there is no singularity at finite frequencies.

In fact, the sensitivity of amplitudes (calculated by ray theory) to the quantity $d\Delta/dp$ leads to some difficulties in computation. The problem is that Earth models are ordinarily specified by giving the values of density (ρ) and P- and S-wave speeds (α and β) at several different radii. But different methods of interpolation between such discrete values can often lead to completely different estimates for the geometrical spreading. Chapman (1971) recommends a cubic spline for the velocity interpolation, since the second derivative of velocity with respect to depth can then be made continuous with depth, and this degree of continuity is needed if $d\Delta/dp$ is to be continuous. Bullen (1960) gives some examples of how $d\Delta/dp$ can change character for different kinds of velocity increase with depth, as shown here in Figure 9.13. The effect of caustics is widely observed in seismic waves, and wave-theoretical methods of calculation are described in Section 9.4.

The specific formulas for travel time T and distance Δ, as a function of p, are easily derived if we note that an element of ray length ds is related to a change dr in radius by $\cos i\, ds = dr$, where i is the angle between local vertical and the ray direction. Thus, for the P-wave velocity $\alpha = \alpha(r)$,

$$\begin{aligned} T(p) &= \int \frac{ds}{\alpha} = \int \frac{dr}{\alpha(r)\cos i(r)} = \int \left(1 - \frac{p^2\alpha^2}{r^2}\right)^{-1/2} \frac{dr}{\alpha} = \int \frac{dr}{\alpha^2 \xi}, \\ \Delta(p) &= \int \sin i \frac{ds}{r} = \int \frac{p}{r^2}\left(1 - \frac{p^2\alpha^2}{r^2}\right)^{-1/2} dr = \int \frac{p\, dr}{r^2 \xi}, \end{aligned} \tag{9.21}$$

where we have used $p = (r \sin i)/\alpha$ (from (4.45b)), and $\xi = \xi(r) = (1/\alpha^2 - p^2/r^2)^{1/2} = \cos i/\alpha$ for vertical slowness. The integration is carried out over that range of radii traversed by the ray, and thus it will often occur that a turning point appears as one of the limits in the integration. Using r_p as the turning-point radius corresponding to ray parameter p, it follows that $p = r_p/\alpha(r_p)$, and hence that the integrands in (9.21) can have singularities.

A quantity formed from T and Δ that does not have a singular integrand at a turning point is

$$\tau = \tau(p) \equiv T - p\Delta = \int \frac{\cos i}{\alpha}\, dr = \int \xi\, dr. \tag{9.22}$$

We shall find that $\tau(p)$ has a surprising number of uses. It appears in inverse theories (Johnson and Gilbert, 1972; Bessonova et al., 1976) and in theories of wave propagation. However, even in the restricted context of ray theory it has remarkable properties due to the relation

$$\frac{d\tau}{dp} = -\Delta(p) \tag{9.23}$$

(which is proved by differentiating $\int \xi\, dr$ and comparing with Δ in (9.21)). It follows that τ is a monotonically decreasing function of p, unlike $\Delta(p)$, so that it is an even better function than $\Delta(p)$ for unfolding triplications in travel-time curves. [Previously, we have often used the symbol τ for dummy times in writing out convolutions explicitly: $f * g = \int_0^t f(\tau)g(t - \tau)\, d\tau$. It was therefore natural to develop Cagniard paths, for problems with a point source, as the solution $p = p(\tau)$ of $\tau = pr + \text{SUM}(p)$ (see (9.5)). However, $\text{SUM}(p)$ here is effectively the new variable $\tau(p)$ that we have just defined in (9.22). This fact is brought out in Problem 9.2(b). To avoid confusion, a different symbol is necessary for times along the Cagniard path, in problems where τ is reserved for integration over vertical slowness.]

We have seen in Figures 9.12 and 9.13 that triplications are associated with velocity increasing with depth. If velocity *decreases* sufficiently rapidly with depth (i.e., $d\alpha/dr$ positive and large enough), then another characteristic feature appears in the travel-time curve. This is the phenomenon of a *shadow zone*, illustrated in Figure 9.14. Within the low-velocity zone, there is a range of depths in which no turning point is present. When a turning point is present at the bottom of a ray, it is achieved by the angle i (between ray and vertical) being an increasing function of decreasing radius along the ray. Thus a shadow zone occurs whenever the velocity gradient in the low-velocity zone is such as to make $di/dr > 0$, for then i can never increase to the turning-point value (90°) as r is decreased. Since $(r \sin i)/\alpha$ is constant, we find that

$$r\frac{di}{dr} = -[1 - \zeta(r)] \tan i, \tag{9.24}$$

where ζ is the normalized velocity gradient $(r/\alpha)(d\alpha/dr)$. The condition $di/dr > 0$, under which a shadow is generated, therefore translates to $\zeta(r) > 1$, or $d\alpha/dr > \alpha/r$. Comparing Figures 9.14c and 9.14d, note that there are two or one or no

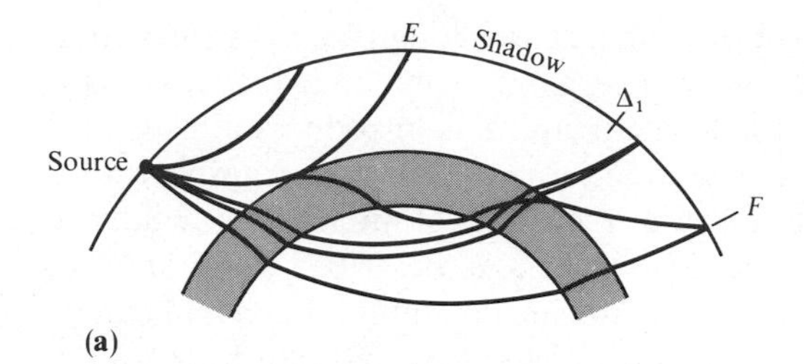

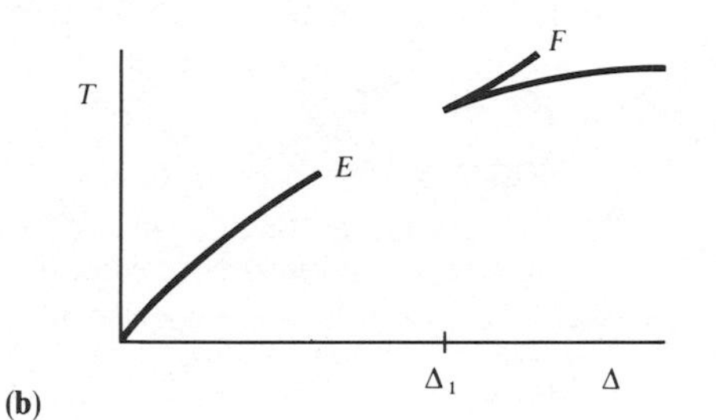

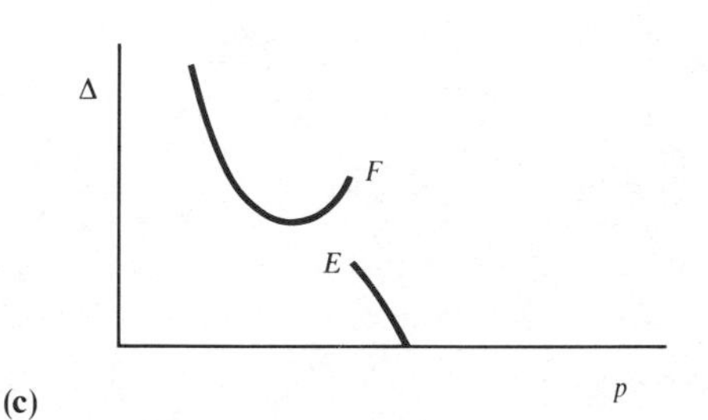

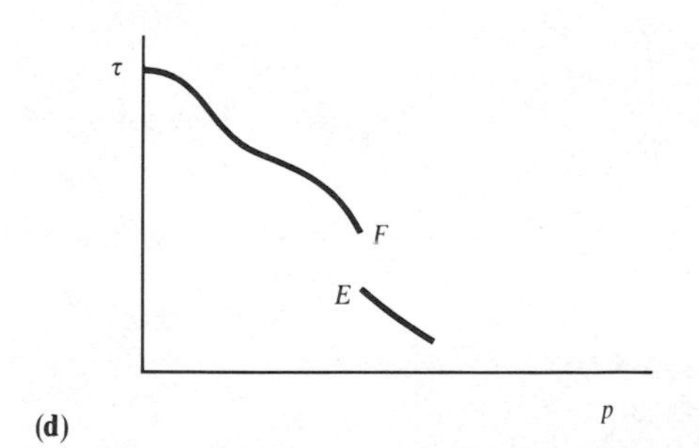

FIGURE **9.14**

The behavior of T, Δ, and p for a velocity decrease with depth. (**a**) A low-velocity zone (within which $d\alpha/dr > \alpha/r$ and there are no turning points) is shown shaded, and a shadow within which no rays are received is observed at the surface. (**b**) The travel-time curve. The upper boundary of the low-velocity zone is the turning point for the ray emerging at point E. Point F has the same ray parameter, but lies on a ray going through the low-velocity zone itself. As ray parameter decreases slightly from its value at F, distance Δ *decreases* until a caustic is reached at Δ_1. (**c**) The values of $\Delta = \Delta(p)$. These show that the further boundary of the shadow is in fact a caustic. (**d**) Upper and lower boundaries of the low-velocity zone are turning points for rays that differ infinitesimally in their ray parameter. The turning-point radius is therefore a discontinuous function of p. This is also a discontinuity in $\tau(p) = T - p\Delta$ and in the gradient $d\tau/dp = -\Delta(p)$.

values of Δ corresponding to a given value of p, whereas $T - p\Delta$ has just one value almost everywhere.

To conclude this brief review of ray theory, a few remarks are needed to augment our Chapter 4 formulas for pulse shapes. Recall that we found (e.g., (4.57)) that amplitudes at $\mathbf{x}$ are controlled by the factor $1/\mathscr{R}(\mathbf{x}, \boldsymbol{\xi})$ for a point source at $\boldsymbol{\xi}$. However, in the course of propagation along the ray from $\boldsymbol{\xi}$ to $\mathbf{x}$, there are several effects that can introduce a frequency-independent phase advance to the waves received in the far field, and this introduces a distortion into the pulse shape that can often be seen in seismic data. One example is that of a wave incident upon a discontinuity in the Earth at an angle of incidence great enough to excite an inhomogeneous wave. As described in Chapter 5, the resulting phase shift leads to a change in the incident pulse shape that can be calculated via the Hilbert transform. Another example is that of a ray between $\boldsymbol{\xi}$ and $\mathbf{x}$ that, though stationary, is not a minimum time path. Some of the rays most important to seismology in which this arises are: direct *P*- and *S*-waves

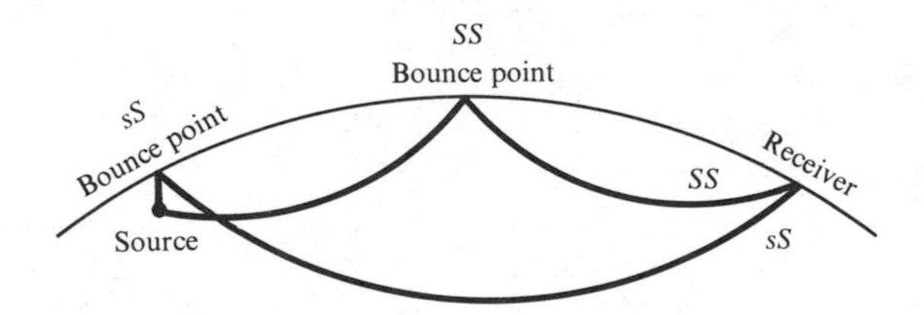

FIGURE **9.15**
An *S*-wave departing downward from the source and once-reflected at the Earth's surface between source and receiver, is known as *SS*, whereas *sS* departs upward from the source and is reflected near the source.

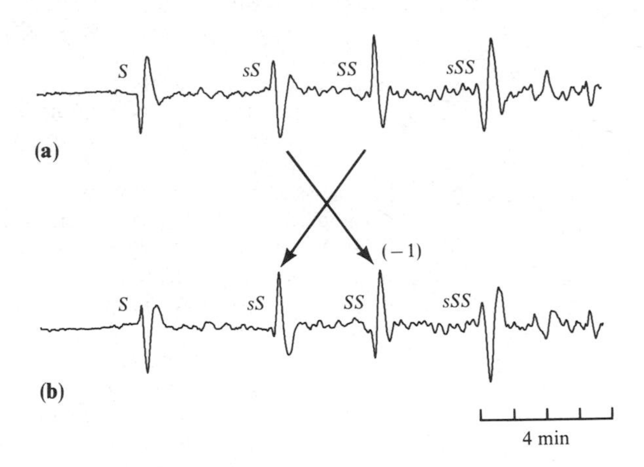

FIGURE **9.16**
(a) Original EW seismogram (orientation is for *SH*-waves) of a deep earthquake (608 km) below the Sea of Japan on 8 October 1960. Recorded at PAL, $\Delta = 96.3°$. **(b)** The Hilbert transform of (a). For convenience, the four main phases are still labeled with the original names. One arrow indicates that *SS* in (a) is the same shape as the Hilbert transform of *sS*, shown in (b). The second arrow indicates that *sS* in (a) is the same shape (after a polarity reversal) as the Hilbert transform of *SS*, shown in (b). [After Choy and Richards, 1975.]

that depart downward from the source and are received on a branch of the travel-time curve for which $d^2T/d\Delta^2 > 0$; *PP* and *SS*; and *SKKS*. Thus *SS* has a minimax time path (see Fig. 9.15), because a perturbation of the "bounce point" to points on the Earth's surface in the same plane as the source and receiver will actually decrease the travel time (and perturbation perpendicular to this direction will increase the travel time). On the other hand, *sS* is a true minimum time path, and Choy and Richards (1975) showed observations of *SS* that were quite accurately the Hilbert transform of *sS* (see Fig. 9.16).

Although we have emphasized spherically symmetric media in the above review, all the ideas and formulas have an analogue in plane-stratified media. This we shall call upon, when developing the Earth-flattening approximation to apply Cagniard's and Fuchs' methods to teleseismic waves in the Earth.

9.4 Wave Propagation in Media Having Smoothly Varying Depth-dependent Velocity Profiles within Which Turning Points Are Present

When a ray departs downward from some point source within the Earth, it can be observed again at a surface receiver only after it is reflected back upward from some internal discontinuity, or after it gradually changes from "downgoing" to "upgoing" simply by passing through a turning point. We have mentioned turning points in the previous section, in the context of ray theory, but our present aim is to understand turning points in the context of a frequency-dependent wave theory for inhomogeneous media. The importance of this is that it avoids the artificiality of using large numbers of layers to model the inhomogeneity.

The key elements of elastic-wave theory in homogeneous media include Lamé's theorem, the analysis of plane waves, the Sommerfeld integral, and ways to evaluate such integrals by manipulations in the complex ray-parameter plane. Each of these elements is generalized in this section, and we shall find that WKBJ theory (reviewed in Box 9.3) has great utility. In fact, in situations where WKBJ approximations are accurate, our computation of body waves will merge classical ray theory with some of the methods used to solve Lamb's problem. But where these approximations are inaccurate (this is the case if a turning point is present at a depth close to some discontinuity within the Earth), then distinctly new elements emerge.

In studying homogeneous elastic media, we found it convenient to move from a study of potentials to a method based on the motion-stress vector. A similar development is convenient in inhomogeneous media, because potentials display the properties of scalar wave solutions in their simplest form, yet the motion-stress vector approach enables one more easily to do the necessary bookkeeping for physical quantities. Beginning, then, with potentials, we find in Box 9.2 that displacement **u** in a spherically symmetric medium can be written as a sum of three vector terms

$$\mathbf{u} = \rho^{-1/2}[\nabla P + \nabla \times \nabla \times (rV, 0, 0)] + \mu^{-1/2}\nabla \times (rH, 0, 0). \quad (9.25a)$$

Here, the three vectors on the right-hand side are, respectively, the *P*-, *SV*-, and *SH*-components of displacement, with scalar potentials P, V, H satisfying (in the frequency domain) the decoupled wave equations

$$\nabla^2 P + \frac{\omega^2}{\alpha^2} P = 0, \qquad \nabla^2 V + \frac{\omega^2}{\beta^2} V = 0, \qquad \nabla^2 H + \frac{\omega^2}{\beta^2} H = 0. \quad (9.25b)$$

BOX **9.2**

Scalar potentials for P-, SV-, and SH-waves in spherically symmetric media

We now use spherical coordinates (r, Δ, ϕ) in which $r = 0$ is the center of symmetry in the medium and the direction $\Delta = 0$ is taken to lie on some point source. A second-order vector wave equation for displacement $\mathbf{u} = (u_r, u_\Delta, u_\phi)$ can then be obtained from relations between acceleration and stress (2.47), stress and strain (2.50), and strain and displacement (2.45). The equation is

$$\begin{aligned}\rho\omega^2\mathbf{u} = &-(\lambda + 2\mu)\nabla\nabla \cdot \mathbf{u} + \mu\nabla \times \nabla \times \mathbf{u} - \lambda'(\nabla \cdot \mathbf{u}, 0, 0)\\ &- 2\mu' \frac{\partial \mathbf{u}}{\partial r} + \mu'(0, \mathrm{curl}_\phi\, \mathbf{u}, -\mathrm{curl}_\Delta\, \mathbf{u}) - \mathbf{f}.\end{aligned} \quad (1)$$

Here, and below, a prime denotes radial differentiation, and $\mathrm{curl}_\alpha\, \mathbf{u}$ is the α-coordinate of $\nabla \times \mathbf{u}$ $(\alpha = r, \Delta, \phi)$. We shall here consider free solutions, i.e., taking the body force $\mathbf{f}$ equal to $\mathbf{0}$.

In deriving potentials for $\mathbf{u}$, we shall follow in outline the same method used in Section 4.1. (Further details for the present problem are given by Richards, 1974.) Thus we seek to turn (1) directly into the required representation by giving a specific construction for the potentials.

This construction is

$$\begin{aligned}P &\equiv \frac{-(\lambda + 2\mu)}{\rho^{1/2}\omega^2}\left[\nabla \cdot \mathbf{u} + \frac{\mu}{\lambda + \mu}\left(\frac{2\mu'}{\mu} - \frac{\rho'}{\rho}\right)u_r\right],\\ \mathbf{S} &\equiv \frac{\mu}{\rho^{1/2}\omega^2}\left[\nabla \times \mathbf{u} + \frac{\lambda + 2\mu}{\lambda + \mu}\left(\frac{\rho'}{\rho} - \frac{2\mu'}{\lambda + 2\mu}\right)(0, -u_\phi, u_\Delta)\right].\end{aligned} \quad (2)$$

After dividing (1) through by $\rho\omega^2$, the equation of motion can be written in the form

$$\mathbf{u} = \rho^{-1/2}(\nabla P + \nabla \times \mathbf{S}) + \text{terms of order } |\mathbf{u}|/\omega. \quad (3)$$

For the body waves we shall be studying, these last terms are negligible. They are typically of order $(\beta'/\omega)|\mathbf{u}|$, which reaches about 2% of $|\mathbf{u}|$ for 20-sec waves in the upper mantle of Earth models with a fairly high shear-velocity gradient (0.005 km/sec per km).

The definitions (2) are now to be turned into wave equations, substituting for **u** from (3). In this way, it is found that

$$\nabla^2 P + \frac{\rho\omega^2}{\lambda + 2\mu} P = \text{terms of order } \frac{|\mathbf{u}|}{\omega},$$
$$\nabla^2 \mathbf{S} + \frac{\rho\omega^2}{\mu} \mathbf{S} = \text{terms of order } \frac{|\mathbf{u}|}{\omega}. \tag{4}$$

But each separate term on the left-hand side in (4) is of order $\omega|\mathbf{u}|$, so that terms on the right-hand side are negligible for most purposes, being two orders lower in frequency.

It remains only to find scalar potentials separately for SV and SH. This entails an inversion of the Beltrami operator B, where

$$Bf \equiv \frac{1}{\sin\Delta}\frac{\partial}{\partial\Delta}\left(\sin\Delta\frac{\partial f}{\partial\Delta}\right) + \frac{1}{\sin^2\Delta}\frac{\partial^2 f}{\partial\phi^2}.$$

Note that B contains the horizontal derivatives present in the Laplacian operator, so that $BY_l^m = -l(l+1)Y_l^m$ for any surface harmonic Y_l^m. The inverse B^{-1} of B has been discussed by Backus (1958): if $g = g(r, \Delta, \phi)$ can be expressed as a sum $g = \sum_{l=1}^{\infty}\sum_{m=-l}^{l} g_l^m(r) Y_l^m$, then we invert $Bf = g$ to obtain $f = B^{-1}g = -\sum_{l=1}^{\infty}\sum_{m=-l}^{l} [l(l+1)]^{-1} g_l^m Y_l^m$.

SV-POTENTIAL

For spheroidal motion, $\text{curl}_r\, \mathbf{u} = 0$, and then $S_r = 0$ and $\nabla \cdot \mathbf{S} = 0$ both follow from (2). Thus $\partial(\sin\Delta S_\Delta)/\partial\Delta + \partial S_\phi/\partial\phi = 0$, which is a condition that there exists a function V such that $S_\Delta = (1/\sin\Delta)\,\partial V/\partial\phi$, $S_\phi = -\partial V/\partial\Delta$. Hence

$$\mathbf{S} = \nabla \times (rV, 0, 0) \qquad \text{for } SV \text{ waves.} \tag{5}$$

To construct V, we form $\text{curl}_r\, \mathbf{S}$ and note that $-r\, \text{curl}_r\, \mathbf{S} = BV$, an operator we have already found how to invert. Richards (1974) shows that

$$\nabla^2 V + \frac{\rho\omega^2}{\mu} V = \text{terms of order } \frac{|\mathbf{u}|}{\omega^2}. \tag{6}$$

SH-POTENTIAL

For toroidal motion, $u_r = 0$ and $\nabla \cdot \mathbf{u} = 0$. We can therefore follow the same stages as discussed above for **S**, introducing a potential H_0 via $H_0 = -B^{-1}(r\, \text{curl}_r\, \mathbf{u})$. To get the canonical form of the wave equation for SH-potential, it is convenient to work with $H = \mu^{1/2} H_0$. Then

$$\mathbf{u} = \mu^{-1/2}\nabla \times (rH, 0, 0) \qquad \text{for } SH\text{-waves,}$$

and

$$\nabla^2 H + \frac{\rho\omega^2}{\mu} H = \text{terms of order } \frac{|\mathbf{u}|}{\omega}. \tag{7}$$

Equations (9.24) and (9.25) are obtained by ignoring small terms in (3)–(7). (All these terms are zero in homogeneous media.)

In (9.25a) and (9.25b), the quantities ρ, μ, α, and β are each functions only of radius r in the spherical polar system (r, Δ, ϕ). Thus each of the wave equations has the form of a Helmholtz equation with radially varying wavenumber.

Suppose next that a point source is introduced into the medium. We first obtain a representation of the source that plays a role similar to the Sommerfeld integral for waves in homogeneous media. The simplest source to consider is a delta function introduced into the wave equation for $P(\mathbf{r}, \omega)$:

$$\nabla^2 P + \frac{\omega^2}{\alpha^2(r)} P = K_s\, \delta(\mathbf{r} - \mathbf{r}_s), \tag{9.26}$$

where the source is at $\mathbf{r}_s$ and K_s is a constant (subscript s refers throughout to constants determined by the source).

The solution of (9.26) in an infinite homogeneous medium is

$$P(\mathbf{r}, \omega) = -\frac{K_s \exp(i\omega|\mathbf{r} - \mathbf{r}_s|/\alpha_s)}{4\pi|\mathbf{r} - \mathbf{r}_s|}, \tag{9.27}$$

which has a well-known partial wave expansion in terms of spherical Hankel functions and Legendre polynomials:

$$P(\mathbf{r}, \omega) = -\frac{iK_s\omega}{4\pi\alpha_s} \sum_{l=0}^{\infty} (l + \tfrac{1}{2}) h_l^{(1)}(\omega r/\alpha_s)\left[h_l^{(1)}(\omega r_s/\alpha_s) + h_l^{(2)}(\omega r_s/\alpha_s)\right] P_l(\cos \Delta) \tag{9.28}$$

(for $r > r_s$), where α_s is the (constant) P-wave velocity throughout.

This last expansion must be generalized if we are to progress in our study of *inhomogeneous* media. Fortunately, the method of separation of variables shows that horizontal wave functions are still Legendre polynomials, and all the complications are confined to the vertical (now, radial) wavefunction. Thus we shall try to solve (9.26) by a sum of separated solutions in the form

$$P(\mathbf{r}, \omega) = \sum_{l=0}^{\infty} a(r, l) P_l(\cos \Delta). \tag{9.29}$$

Then $a(r, l) = (l + \frac{1}{2}) \int_0^\pi P(\mathbf{r}, \omega) P_l(\cos \Delta) \sin \Delta \, d\Delta$, and substitution of (9.29) in the wave equation gives

$$\frac{d^2}{dr^2}[ra(r, l)] + \left[\frac{\omega^2}{\alpha^2(r)} - \frac{l(l + 1)}{r^2}\right] ra(r, l) = K_s\, \delta(r - r_s) \frac{(2l + 1)}{4\pi r}. \tag{9.30}$$

Following Seckler and Keller (1959) and Friedman (1951), we introduce three particular solutions of the homogeneous equation related to (9.30). Let $f_l(r)$ be that solution for $a(r, l)$ which is regular at the central point $r = 0$. For very large values of r, the wave equation becomes roughly $d^2a/dr^2 = -\omega^2 a/\alpha^2$. By analogy with the solutions $e^{\pm i\omega r/\alpha}$ when α is constant, we expect that two independent solutions of our wave equation can be chosen, one with a phase that increases with r, and another with a phase that decreases. We label these solutions $g_l^{(1)}(r)$ and $g_l^{(2)}(r)$, respectively. In association with the factor $\exp(-i\omega t)$, $g^{(1)}$ is an outgoing wave and $g^{(2)}$ is ingoing. Apart from a normalization, the three solutions are completely defined and we may take

$$a(r, l) = \begin{cases} c_1 g_l^{(1)}(r) & r_s \le r \\ c_2 f_l(r) & \text{for } 0 \le r \le r_s. \end{cases} \tag{9.31}$$

The constants c_1 and c_2 are determined by noting that (9.30) implies both continuity of $a(r, l)$ as r increases through r_s and a step of height $K_s(2l + 1)/(4\pi r_s^2)$ in $d(a(r, l))/dr$. It follows that

$$c_1 = -\frac{K_s(2l + 1)}{4\pi r_s^2}\frac{f_l(r_s)}{W(r_s)}, \qquad c_2 = -\frac{K_s(2l + 1)}{4\pi r_s^2}\frac{g_l^{(1)}(r_s)}{W(r_s)}, \tag{9.32}$$

where $W(r) = g_l^{(1)}(r)df_l/dr - f_l(r)\,dg_l^{(1)}/dr$ is the Wronskian of $g_l^{(1)}$ and f_l. Fortunately, we can show $W(r)$ is proportional to $1/r^2$, since the wave equations satisfied by $g_l^{(1)}$ and f_l can be used to find $d(r^2W)/dr = 0$. The constant of proportionality depends on the normalization of our wave functions, and at this stage we appeal to WKBJ theory. From Box 9.3 (equation 11), it follows for large ω that

$$rg_l^{\substack{(1)\\(2)}}(r) \sim \frac{K^{\substack{(1)\\(2)}}}{\left[\dfrac{1}{\alpha^2(r)} - \left(\dfrac{l+\frac{1}{2}}{\omega r}\right)^2\right]^{1/4}} \exp\left\{\pm i\omega \int_{r_p}^{r}\left[\frac{1}{\alpha^2(r')} - \left(\frac{l+\frac{1}{2}}{\omega r'}\right)^2\right]^{1/2} dr'\right\}, \tag{9.33}$$

where $K^{(1)}$ and $K^{(2)}$ are constants and r_p is that radius at which the integrand vanishes.

Defining

$$p \equiv \frac{l + \frac{1}{2}}{\omega}, \text{ we have } p = \frac{r_p}{\alpha(r_p)}. \tag{9.34}$$

Since the ray parameter in spherical geometry is $(r \sin i)/\alpha$, it follows that r_p has a physical interpretation as the radius to the turning point along the ray with ray parameter p.

BOX **9.3**

WKBJ theory

We are concerned here with finding approximate solutions to the second-order equation

$$\frac{d^2\phi}{dx^2} + \omega^2 s^2 \phi = 0, \tag{1}$$

where ω is large and positive and $s = s(x)$ is such that s^2 is a monotonically increasing function of x. The method we are about to describe has great generality, because any linear homogeneous equation of second order can be transformed into (1), often with the special properties of ω and s^2 mentioned above. Many authors have rediscovered the method (including Wentzel, Kramers, Brillouin, and Jeffreys, hence the name). The main ideas go back certainly to Green and Liouville, and were used by Rayleigh (1912).

We shall suppose that s^2 has a zero (called a *turning point*) at $x = x_p$, so that $s^2 \gtrless 0$ according as $x \gtrless x_p$. It follows from (1) that the curvature of a solution $\phi = \phi(x)$ has the opposite or the same sign as ϕ, according as $x \gtrless x_p$. This fact alone suggests the main character of the solutions, which is oscillatory above the turning point x_p, and exponential below.

Where $s(x)$ is constant, solutions are $\phi = e^{\pm i\omega s x}$, and this suggests that we try the form $\phi = e^{i\omega\tau(x)}$. From (1), then,

$$i\omega\tau'' - \omega^2(\tau')^2 + \omega^2 s^2 = 0,$$

where a prime denotes d/dx. As a first approximation, we neglect $\omega\tau''$ here, giving $\tau' \sim \pm s(x)$ and $\tau(x) \sim \pm\int s(x)\,dx$. This would give $\tau'' \sim \pm s'$, so that the second approximation for τ satisfies $(\tau')^2 = s^2 \pm is'/\omega$, i.e., $\tau' = \pm s + is'/2s\omega$, and then $\tau(x) = \pm\int s(x)\,dx + (i/2\omega)\ln s$. Corresponding solutions for ϕ are

$$\phi(x) \sim \frac{A}{s^{1/2}(x)}\exp\left(i\omega\int_{x_p}^{x} s\,dx\right) + \frac{B}{s^{1/2}(x)}\exp\left(-i\omega\int_{x_p}^{x} s\,dx\right) \tag{2}$$

(A and B are constants). This is valid provided $|s'/\omega| \ll |s^2|$. Since, very crudely, $\omega s \times$ wavelength $= 2\pi$, the validity condition amounts to $|s' \times \text{wavelength}| \ll 2\pi|s|$, and the change in $s(x)$ in one wavelength must be much less than s itself. Clearly, (2) will be invalid near $x = x_p$, where s is zero.

Note that in the exponential region, $x < x_p$, s is imaginary, so that an appropriate notation for the general solution is

$$\phi(x) \sim \frac{C}{(-s^2)^{1/4}}\exp\left[\omega\int_{x}^{x_p}(-s^2)^{1/2}\,dx\right] + \frac{D}{(-s^2)^{1/4}}\exp\left[-\omega\int_{x}^{x_p}(-s^2)^{1/2}\,dx\right]. \tag{3}$$

If C is nonzero, then there is an exponentially growing component in the solution below x_p, and it will dominate in this region. Interest then centers on the solution $\phi(x)$ below x_p

for which the WKBJ approximation (3) has $C = 0$. Given a value of D, this special solution is defined by (1) in all ranges of x, and we seek to find its asymptotic expression in the form (2) for the region above the turning point. One cannot simply carry (3) up to the turning point, equate it there to (2), and then continue upward, because neither of these formulas is valid near the turning point (and they have singularities there that are absent from the actual solution $\phi(x)$). We speak of "connecting" the exponentially decaying solution to the two propagating solutions summed in (2), and the method suggested by Rayleigh (1912) is to use Airy functions.

We note that $Ai(-y)$ is a solution of $dw/dy + yw = 0$, which is a special example of our equation (1). The following results are known (Abramowitz and Stegun, 1964, formulas 10.4.60-61):

$$\begin{aligned} Ai(-y) &\sim \pi^{-1/2} y^{-1/4} \cos[\tfrac{2}{3} y^{3/2} - \pi/4] \qquad \text{as } y \to \infty, \\ Ai(-y) &\sim \tfrac{1}{2}\pi^{-1/2}(-y)^{-1/4} \exp[-\tfrac{2}{3}(-y)^{3/2}] \qquad \text{as } y \to -\infty. \end{aligned} \tag{4}$$

Thus $Ai(-y)$ is the special solution that has exponential decay below the turning point.

Near $x = x_p$, we approximate $s^2(x)$ by $s^2 = \lambda(x - x_p)$, where $\lambda = ds^2/dx$ at $x = x_p$. Then the decaying solution in (3) is

$$\frac{D}{\lambda^{1/4}(x_p - x)^{1/4}} \exp[-\tfrac{2}{3}\omega\lambda^{1/2}(x_p - x)^{3/2}]$$

for very large ω and x just below x_p. Comparing with the second of (4), this is

$$2D\pi^{1/2}(\omega/\lambda)^{1/6}\, Ai(-\omega^{2/3}\lambda^{1/3}(x - x_p)). \tag{5}$$

But now we can use the *first* of (4) to find how this solution behaves just above the turning point. It must be like

$$\frac{2D}{\lambda^{1/4}(x - x_p)^{1/4}} \cos[\tfrac{2}{3}\omega\lambda^{1/2}(x - x_p)^{3/2} - \pi/4]. \tag{6}$$

However, in this region, it is given by (2) as

$$\frac{1}{\lambda^{1/4}(x - x_p)^{1/4}} \{A \exp[i\tfrac{2}{3}\omega\lambda^{1/2}(x - x_p)^{3/2}] + B \exp[-i\tfrac{2}{3}\omega\lambda^{1/2}(x - x_p)^{3/2}]\}. \tag{7}$$

Comparing (6) and (7), we conclude that

$$A = De^{-i\pi/4}, \qquad B = De^{+i\pi/4}, \tag{8}$$

and we have finally found the asymptotic behavior above the turning point for the solution that decays below.

Actually, the equation to which we wish to apply this method is

$$\frac{d^2}{dr^2}(ra) + \left(\frac{\omega^2}{\alpha^2} - \frac{l(l+1)}{r^2}\right) ra = 0. \tag{9) (cf. 9.30}$$

There is a singularity at $r = 0$ that we remove by working with x via $r = r_0 e^{x/x_0}$; r_0 and x_0 are constants. Then $r^{1/2}a(r, l) = \phi$ satisfies

$$\frac{d^2\phi}{dx^2} + \frac{r_0^2}{x_0^2} e^{2x/x_0} \left[\frac{\omega^2}{\alpha^2} - \left(\frac{l + \frac{1}{2}}{r_0 e^{x/x_0}} \right)^2 \right] \phi = 0,$$

which *is* in the form (1). Terms like $\omega \int_{x_p}^{x} s\, dx$ become

$$\omega \int_{r_p}^{r} \left[\frac{1}{\alpha^2} - \left(\frac{l + \frac{1}{2}}{\omega r} \right)^2 \right]^{1/2} dr,$$

and the exponentially decaying solution for $a(r, l)$ is approximately given by

$$ra(r, l) = r^{1/2}\phi \sim \frac{D}{\left[\left(\frac{l + \frac{1}{2}}{\omega r} \right)^2 - \frac{1}{\alpha^2} \right]^{1/4}} \exp\left\{ -\omega \int_{r}^{r_p} \left[\left(\frac{l + \frac{1}{2}}{\omega r'} \right)^2 - \frac{1}{\alpha^2} \right]^{1/2} dr' \right\} \tag{10}$$

in the region $r < r_p$. This same solution $a(r, l)$ is given, above the turning point ($r > r_p$), by

$$ra(r, l) = r^{1/2}\phi \sim \frac{D}{\left[\frac{1}{\alpha^2} - \left(\frac{l + \frac{1}{2}}{\omega r} \right)^2 \right]^{1/4}} \left[\exp\left(i\omega\tau - \frac{\pi}{4} \right) + \exp\left(-i\omega\tau + \frac{\pi}{4} \right) \right],$$

where

$$\tau = \int_{r_p}^{r} \left[\frac{1}{\alpha^2} - \left(\frac{l + \frac{1}{2}}{\omega r'} \right)^2 \right]^{1/2} dr'. \tag{11}$$

Several of the special functions used in applied mathematics have widely used approximations that are particular examples of (2) and (3). An example has already been given in (4) for the Airy function. Others include spherical Hankel functions $h_l^{(1)}(\omega r/\alpha)$ and $h_l^{(2)}(\omega r/\alpha)$ (in which α is a constant) and Legendre functions $P_l(\cos \Delta)$. Spherical Hankel functions satisfy (9) (if α is constant), and the corresponding WKBJ formula (10) (known in this case as the Debye approximation to $h_l^{(1)} + h_l^{(2)}$) has the normalization $D = \alpha^{1/2}/\omega$. Since α in this case is constant, τ in (11) can be integrated to

$$\left[\frac{r^2}{\alpha^2} - \left(\frac{l + \frac{1}{2}}{\omega} \right)^2 \right]^{1/2} + (l + \tfrac{1}{2}) \sin^{-1}\left[\frac{(l + \frac{1}{2})\alpha}{\omega r} \right].$$

The spherical Bessel function $j_l(\omega r/\alpha)$ is approximated by (10), with $D = \alpha^{1/2}/2\omega$ when $r < r_p$.

The WKBJ approximation to Legendre's equation gives

$$P_l(\cos \Delta) \sim \left(\frac{1}{2\pi l \sin \Delta} \right)^{1/2} \left\{ \exp\left[-i(l + \tfrac{1}{2})\Delta + \frac{i\pi}{4} \right] + \exp\left[+i(l + \tfrac{1}{2})\Delta - \frac{i\pi}{4} \right] \right\}, \tag{12}$$

and in this case the turning point (where (12) breaks down) occurs for $\sin \Delta = 0$, e.g., $\Delta = 0$ and π. Note that (12) is a standing wave, proportional to $\cos[(l + \frac{1}{2})\Delta - \pi/4]$, although it has been written in traveling-wave components (cf. (9.39)).

Our normalization for $g_l^{(j)}$ is made after comparison with the Debye approximation to spherical Hankel functions (see discussion in Box 9.3). This is appropriate, for we shall assume $g_l^{(j)}$ actually equals $h_l^{(j)}(\omega r/\alpha_s)$ if the medium is homogeneous with constant velocity α_s. $K^{(1)}$ and $K^{(2)}$ are now determined, and

$$g_l^{\substack{(1)\\(2)}}(r) \sim e^{\mp i\pi/4} \frac{\alpha_s^{1/2}}{\omega r} \cdot \frac{1}{\left[\dfrac{1}{\alpha^2(r)} - \dfrac{p^2}{r^2}\right]^{1/4}} \cdot \exp\left\{\pm i\omega \int_{r_p}^{r} \left[\frac{1}{\alpha^2(r')} - \frac{p^2}{r^2}\right]^{1/2} dr'\right\}$$

in the region for which $r/\alpha(r) > p$ (i.e., above the depth of deepest penetration of the ray with parameter p). Pursuing this link with ray theory even further, we recognize $(1/\alpha^2 - p^2/r^2)^{1/2}$ as $\alpha^{-1} \cos i$, which is the P-wave vertical slowness we have previously labeled as ξ. Thus

$$g_l^{\substack{(1)\\(2)}}(r) \sim \frac{e^{\mp i\pi/4}\alpha_s^{1/2}}{\omega r \xi^{1/2}(r)} \cdot \exp\left[\pm i\omega \int_{r_p}^{r} \xi(r')\, dr'\right]. \tag{9.35}$$

Similarly, from WKBJ theory, we have

$$f_l(r) \sim \frac{\alpha_s^{1/2}}{2\omega r} \cdot \frac{1}{\left[\dfrac{p^2}{r^2} - \dfrac{1}{\alpha^2}\right]^{1/4}} \cdot \exp\left\{-\omega \int_{r}^{r_p} \left[\frac{p^2}{(r')^2} - \frac{1}{\alpha^2(r')}\right]^{1/2} dr'\right\} \tag{9.36}$$

in the region for which $r/\alpha(r) < p$ (i.e., $r < r_p$), so that $f_l(r)$ is exponentially decaying with depth below the turning point. The normalization implied by (9.36) is such as to make $f_l(r) = j_l(\omega r/\alpha_s)$ (the spherical Bessel function) in a homogeneous medium.

The three functions f_l, $g_l^{(1)}$, $g_l^{(2)}$ are solutions in all ranges of radius, and it is important to obtain and to understand the connection formula between them. From Box 9.3, we find it is the same relation as that for spherical Bessel/Hankel functions, i.e., the connection formula has the simple form

$$f_l(r) = \tfrac{1}{2}[g_l^{(1)}(r) + g_l^{(2)}(r)]. \tag{9.37}$$

The physical interpretation is that amplitudes decay below the turning point. But above, there is a standing wave composed of downgoing and upgoing components, just as the ray first goes down to the turning point and then up. Below the turning point, each of $g_l^{(1)}$ and $g_l^{(2)}$ will grow exponentially with depth.

With the normalization above, it is now possible to evaluate the Wronskian. We find $W(r) = -i\alpha_s/(\omega r^2)$, exactly. Thus, from (9.29), (9.31), (9.32), and (9.37), we finally obtain

$$P(\mathbf{r}, \omega) = -\frac{iK_s\omega}{4\pi\alpha_s} \sum_{l=0}^{\infty} (l + \tfrac{1}{2}) g_l^{(1)}(r_>)[g_l^{(1)}(r_<) + g_l^{(2)}(r_<)] P_l(\cos\Delta) \tag{9.38}$$

where $r_>$ = greater of (r, r_s); $r_<$ = lesser of (r, r_s).

It is often useful to split the Legendre function into its traveling-wave components,

$$P_l = Q_l^{(1)} + Q_l^{(2)}, \qquad \text{where } Q_l^{\binom{(1)}{(2)}}(\cos\Delta) \sim \frac{\exp\{\mp i[(l+\frac{1}{2})\Delta - \pi/4]\}}{(2\pi l \sin\Delta)^{1/2}} \tag{9.39}$$

for large l (provided that l is not near a negative integer and that Δ is not near 0 or 180°). Recall for Cartesian coordinates that the horizontal wavefunction is $e^{i\omega px}$, in which p is the horizontal slowness/ray parameter. Similarly, from (9.39), we can recognize $(l + \frac{1}{2})/\omega$ as the ray parameter in spherical geometry, as indicated already in (9.34).

The result (9.38) is conceptually similar to the Sommerfeld integral, in the sense that it shows how the waves from a simple point source can be expanded in vertical and horizontal wavefunctions appropriate to the coordinate system in which boundary conditions can easily be analyzed. However, although the partial wave expansion (9.38) is a convergent series, the individual terms do not significantly decrease until l is large enough to make $g_l^{(1)}(r_<) + g_l^{(2)}(r_<)$ exponentially decaying. This requires l greater than $\omega r_< \div \alpha(r_<)$. Since $r_<$ is usually several thousand km and $\alpha(r_<)$ is about 10 km/sec, several thousand terms in (9.38) would have to be summed to study short-period (1- or 2-sec) body waves. Such numerical difficulty is effectively avoided by converting the summation over l into an integral (the Watson or Poisson transform), and then direct integration is simple to carry out after a suitable contour is found in the (complex) order plane. A rigorous derivation of the suitable contour is not easy (Nussenzveig, 1965; Ansell, 1978), since (9.38) is essentially a sum of four different types of waves. Each is of type $g_l^{(j)}(r_<)Q_l^{(k)}$, with $j = 1$ or 2, $k = 1$ or 2, and can be recognized in terms of downward or upward departure from $r_<$ to $r_>$ and of arrival toward or away from the source. But if the series (9.38) is separated into four series, one for each wave type, each series *diverges*. Quite intricate manipulations are needed to avoid the divergence, and in Box 9.4 we give an example for the most important case, that illustrated in Figure 9.17, in which a turning point is present.

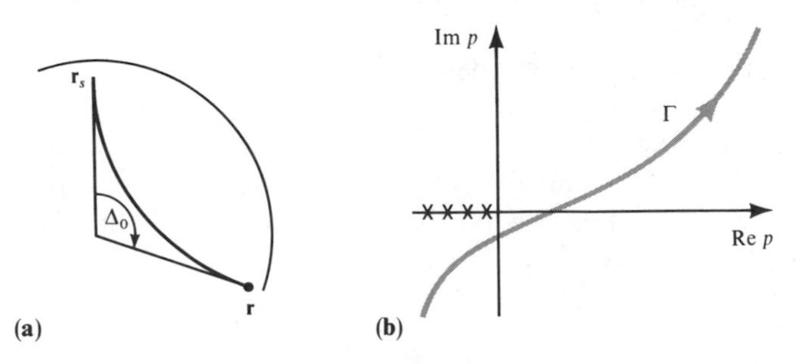

FIGURE **9.17**

(**a**) Parameters for a ray departing downward from source to receiver. A turning point is present. (**b**) Associated integration path Γ. See equation (9.40). [From Richards, 1973b.]

BOX **9.4**

A sample application of the Watson transform

This is a device for converting sums to integrals. Using poles of sec $\nu\pi$ at $\nu = \frac{1}{2}, 1\frac{1}{2}, 2\frac{1}{2}, \ldots$, the transform is

$$\sum_{l=0}^{\infty} f(l + \tfrac{1}{2}) = \tfrac{1}{2} \int_C f(\nu) e^{-i\nu\pi} \sec \nu\pi \, d\nu, \tag{1}$$

where C is taken around the positive real ν-axis as shown in the figure.

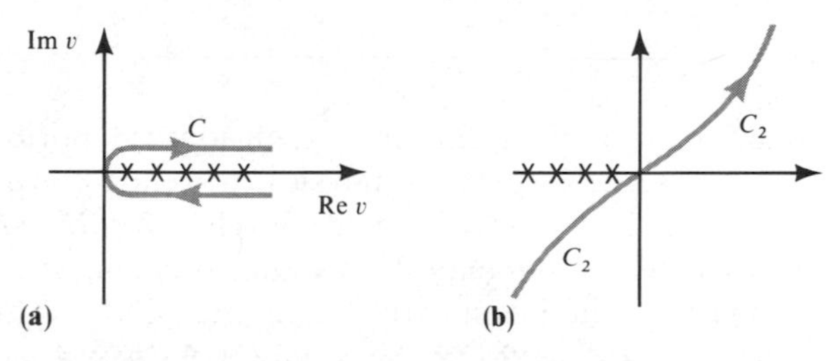

(**a**) Path for Watson transformation. Poles occur all along the real axis and are shown for positive ν at $\nu = l + \frac{1}{2}$; $l = 0, 1, 2, \ldots$. (**b**) Deformation of C for direct ray with a turning point, after the $Q^{(2)}_{-\nu-\frac{1}{2}}$ integration is reflected in the origin.

We shall apply (1) to the partial wave expansion (9.38) in the case that the ray from source $\mathbf{r}_s$ to receiver $\mathbf{r}$ has a turning point. This configuration is shown in Figure 9.17a, and using $P_l(\cos\Delta) = e^{il\pi} P_l(-\cos\Delta)$ we find

$$P(\mathbf{r}, \omega) = -\frac{K_s \omega}{4\pi\alpha_s} \int_C \nu g^{(1)}_{\nu-\frac{1}{2}}(r) f_{\nu-\frac{1}{2}}(r_s) P_{\nu-\frac{1}{2}}(-\cos\Delta) \sec \nu\pi \, d\nu.$$

To manipulate this integrand and integration path into a more useful form, many special properties of the vertical and horizontal wavefunctions must be deployed to avoid divergent integrals at intermediate steps in the development. Thus it follows from Nussenzveig (1965) that C may be deformed across the fourth quadrant, and parts of the first and third, into path C_2, which is symmetric about the origin (see figure). At this stage, the integrand has poles on the positive real ν-axis. Writing

$$P_{\nu-\frac{1}{2}}(-\cos\Delta) \sec \nu\pi = \left[-e^{-i\nu\pi} Q^{(2)}_{\nu-\frac{1}{2}}(\cos\Delta) + e^{i\nu\pi} Q^{(2)}_{\nu-\frac{1}{2}}(\cos\Delta)\right] \csc \nu\pi$$

allows the integral C_2 to be split into two terms. Replacing ν by $-\nu$ in the integral containing $Q^{(2)}_{-\nu-\frac{1}{2}}$, and using $g^{(1)}_{-\nu-\frac{1}{2}}(r_s) e^{-i\nu\pi} = g^{(1)}_{\nu-\frac{1}{2}}(r_s)$ and

$$f_{-\nu-\frac{1}{2}}(r_s) = f_{\nu-\frac{1}{2}}(r_s) e^{-i\nu\pi} + i \sin \nu\pi \; g^{(1)}_{-\nu-\frac{1}{2}}(r_s),$$

we get

$$P(\mathbf{r}, \omega) = -\frac{iK_s\omega}{4\pi\alpha_s} \int_{C_2} \nu g_{\nu-\frac{1}{2}}^{(1)}(r) g_{\nu-\frac{1}{2}}^{(1)}(r_s) Q_{\nu-\frac{1}{2}}^{(2)}(\cos\Delta)\, d\nu. \tag{2}$$

We note that: (i) the final integral here can be identified with just one of the four traveling-wave terms present in the partial wave expansion (9.38); (ii) the factor $Q_{\nu-\frac{1}{2}}^{(2)}(\cos\Delta)$ in (2) has poles only on the negative real axis, so that C_2 can now be distorted to cross the positive real ν-axis wherever convenient; (iii) if there is no turning point present between $\mathbf{r}_s$ and $\mathbf{r}$, e.g., if the ray is everywhere going upward, then the expected traveling-wave term in (9.38) is $g_l^{(1)}(r) g_l^{(2)}(r_s) Q_l^{(2)}(\cos\Delta)$. Ansell (1978) has given a detailed analysis, pointing out divergent integrals in Richards' (1973b) discussion.

Although the rigorous manipulation of partial wave expansions is laborious, the results make a great deal of sense in that the physical traveling wave of interest can still be identified in the partial wave expansion. The effect of using complex integration paths is twofold: to eliminate all unwanted traveling waves; and to obtain a computationally attractive integral for the remainder. Thus, for source and receiver as shown in Figure 9.17a, the result (2) of Box 9.4 identifies precisely the traveling wave one should expect. We obtain

$$P(\mathbf{r}, \omega) = \frac{-iK_s\omega^3}{4\pi\alpha_s} \int_{\Gamma} p g_{\omega p-\frac{1}{2}}^{(1)}(r) g_{\omega p-\frac{1}{2}}^{(1)}(r_s) Q_{\omega p-\frac{1}{2}}^{(2)}(\cos\Delta)\, dp, \tag{9.40}$$

in which p is ray parameter ($\nu = l + \frac{1}{2} = \omega p$) and Γ, shown in Figure 9.17b, can be deformed to cross any convenient point or points of the positive real p-axis.

The effect of discontinuities in the Earth, such as the core-mantle boundary, is examined in the next section. We shall find it necessary to include reflection and transmission coefficients in the integrand of (9.40). However, the effect of multi-pathing between source and receiver, of the type shown in Figure 9.12b,c, is already included in (9.40). Implicit in our definition of $g_l^{(j)}$ is the assumption that just one turning point r_p solves $p = r/\alpha(r)$, hence low-velocity zones have been excluded. (WKBJ expressions may be modified to permit a relaxation of this assumption.) Because the T-Δ curve has no gaps, the branches of the travel-time curve are such as to give one ray, three rays, or, in general, some odd number of rays between source and receiver.

The uses made of (9.40) have included computation of the vertical wave functions directly (by integration of the equations of motion), followed by integration as in (9.40) over a path Γ chosen near a steepest-descents path (Chapman and Phinney, 1972). However, approximations for the wavefunctions are often adequate, and this permits much more rapid computation. We shall first discuss WKBJ approximations, applied to the radial displacement derived from (9.40) for a receiver at distance Δ_0 from the source. Then, from (9.25a), (9.35), and (9.39),

$$u_r(r, \Delta_0, \omega) = [\rho(r)]^{-1/2}\partial P/\partial r = \frac{K_s e^{-3i\pi/4}\omega^{3/2}}{[2\pi\rho(r)\sin\Delta_0]^{1/2}}\frac{1}{4\pi r r_s}\int_\Gamma \left(\frac{p\xi(r)}{\xi(r_s)}\right)^{1/2} e^{i\omega J}\,dp, \tag{9.41}$$

where the phase-delay integral is

$$J = J(p) = \int_{r_p}^{r_s} \xi\,dr + \int_{r_p}^{r} \xi\,dr + p\Delta_0.$$

As pointed out in (9.22), $\int\xi\,dr$ is related to the travel-time and distance integrals, so that

$$J = T(p) - p\,\Delta(p) + p\Delta_0 = \tau(p) + p\Delta_0, \tag{9.42}$$

where $T(p)$ and $\Delta(p)$ are the time and distance at which the ray with ray parameter p arrives at radius r from the source at level r_s. Since $\partial J/\partial p = \Delta_0 - \Delta(p)$ (see (9.23)), it follows that the integrand for u_r has saddle points at values of p such that $\Delta(p) = \Delta_0$, i.e., at just the ray parameters for which there is a ray between source and receiver. Near such a saddle point p_0 (say), Taylor series expansion gives

$$J(p) = T(p_0) + \tfrac{1}{2}(p - p_0)^2(-\partial\Delta/\partial p). \tag{9.43}$$

If there is just one real ray between $\mathbf{r}$ and $\mathbf{r}_s$, then $\partial\Delta/\partial p$ is negative (the AB branch in Fig. 9.13), and the saddle is oriented favorably for Γ to be taken as the path of steepest descent, leading to the approximation of (9.41) by

$$u_r(r, \Delta_0, \omega) \sim \frac{-iK_s\omega}{4\pi}\left(\frac{\alpha_s}{\rho\alpha}\right)^{1/2}\frac{e^{i\omega T}}{r r_s}\left[\frac{\cos i_s \sin\Delta_0}{p_0\cos i}\left(-\frac{\partial\Delta}{\partial p}\right)\right]^{-1/2}.$$

Identifying the geometrical spreading via

$$\alpha_s\mathcal{R}(\mathbf{r}, \mathbf{r}_s) = r r_s[\cos i \cos i_s \sin\Delta_0|\partial\Delta/\partial p|/p_0]^{1/2} \tag{9.44}$$

(see Problem 4.3), we have

$$u_r(r, \Delta_0, \omega) \sim \frac{-iK_s\omega}{4\pi(\rho\alpha\alpha_s)^{1/2}}\frac{e^{i\omega T}}{\mathcal{R}(\mathbf{r}, \mathbf{r}_s)}\cos i(r). \tag{9.45}$$

We have described the source strength by K_s. If the P-wave source is described instead by its moment tensor $M_{ij}(t) = M_0(t)\delta_{ij}$, then the body force is $f_i(\mathbf{x}, t) = -M_0(t)\,\partial\delta(\mathbf{x} - \mathbf{r}_s)/\partial x_i$. To relate K_s and M_0 when gradients of material properties in the source region are small, note that dominant terms in the vector wave equation (Box 9.2, equation (1)) become $\rho\omega^2\mathbf{u} + \nabla(\rho\alpha^2\nabla\cdot\mathbf{u}) = \nabla[M_0(\omega)\delta(\mathbf{x} - \mathbf{r}_s)]$, and with $\mathbf{u} = \rho^{-1/2}\,\nabla P$ this gives (comparing with (9.26)) $K_s = M_0(\omega)/\rho_s^{1/2}\alpha_s^2$. In the time domain we therefore find from (9.45) that

$$u_r(r, \Delta_0, t) \sim \frac{\cos i(r)\dot{M}_0(t - T)}{4\pi[\rho(r)\rho_s\alpha(r)\alpha_s^5]^{1/2}\,\mathscr{R}(\mathbf{r}, \mathbf{r}_s)}, \tag{9.46}$$

which is exactly what geometrical ray theory would predict.

A more complex configuration is shown in Figure 9.18, where for the distance Δ_0 there are five ray parameters that solve the equation $\Delta(p) = \Delta_0$. The second derivative of the phase integral, $-\omega\,\partial\Delta/\partial p$, is alternately positive and negative, giving saddles with the orientation shown in Figure 9.18b. Γ may be chosen to cross each saddle by the steepest-descents path to give approximations of the type (9.45) for the first, third, and fifth saddles. For the second and fourth, the approximation is of type (9.45) times $(-i)$, i.e., with a $\pi/2$ phase advance. Inverting to the time domain, arrivals corresponding to even-numbered saddles have pulse shapes that are the Hilbert transform of those for odd numbers. The phenomenon of a caustic occurs when $\Delta_c = \Delta(p)$ has a double root (at p_c, say), so that $\partial\Delta/\partial p$ also is zero at $p = p_c$. Figure 9.19 describes wavefunction properties in this case, and the point to emphasize is that (9.41) is still quite easily evaluated numerically, after generalizing programs to compute T and Δ for complex ray parameter. In this way, Richards (1973b) has described the effect of a caustic in $PKKP$ on amplitudes near 240°.

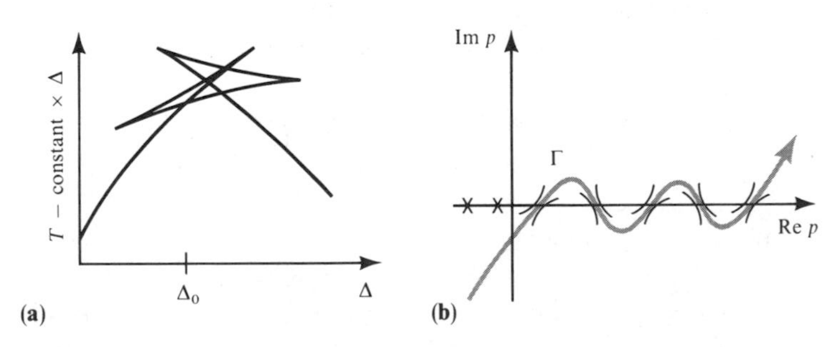

FIGURE **9.18**

(a) A reduced travel-time curve, with two overlapping triplications.
(b) Integration path in the complex ray-parameter plane crossing five saddles, each at a value of p satisfying $\Delta(r, p) = \Delta_0$. [From Richards, 1973b.]

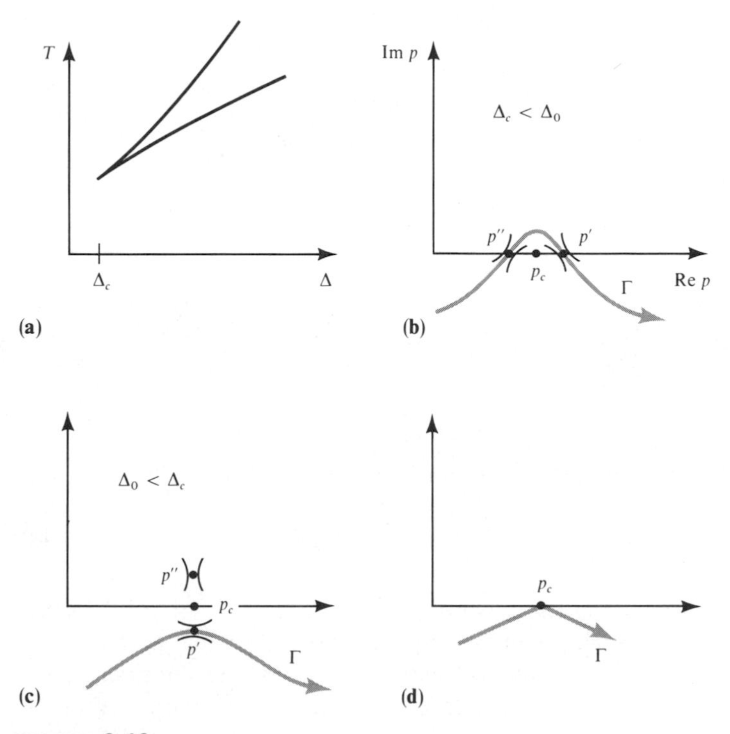

FIGURE **9.19**

Behavior of saddle points near a caustic. (**a**) Two branches of $T = T(\Delta)$ touch at the distance Δ_c, where there is a caustic. (**b**) For distances Δ_0 just greater than Δ_c, two real solutions of $\Delta(p) = \Delta_0$ are present, each giving a saddle point on the real p-axis. As Δ_0 moves through Δ_c, these two saddles merge at p_c, and then (**c**) for distances Δ_0 just less than Δ_c, two complex conjugate solutions of $\Delta(p) = \Delta_0$ are present, one parallel to and below the real p-axis, and this now is the steepest-descents path. (**d**) Two straight-line segments are shown, which are adequate for numerical evaluation of (9.41) in both cases $\Delta_0 \gtrless \Delta_c$. [After Richards, 1973b.]

Chapman (1976b) has described how representations like (9.41) can be written in the time domain as a sum of convolutions. The result is important, because it leads to perhaps the simplest useful method of synthesizing seismograms when multiple arrivals and caustics can be present. Remarkably, the method has close links to the Cagniard technique, although it works even when turning points are present. We begin with (9.41), but using an isotropic moment tensor $M_{ij} = M_0(\omega)\delta_{ij}$ so that

$$u_r(r, \Delta_0, \omega) = \frac{-i\omega M_0(\omega)\, e^{-i\pi/4}\omega^{1/2}}{(2\pi\rho\rho_s \sin\Delta_0)^{1/2} 4\pi r r_s \alpha_s^2} \int_\Gamma B(p) e^{i\omega J(p)}\, dp, \qquad (9.47)$$

where

$$B(p) = \left(\frac{p\xi(r)}{\xi_s}\right)^{1/2} = p^{1/2}\left(\frac{1}{\alpha^2} - \frac{p^2}{r^2}\right)^{1/4} \bigg/ \left(\frac{1}{\alpha_s^2} - \frac{p^2}{r_s^2}\right)^{1/4}.$$

In the time domain, this becomes (see Box 9.5)

$$u_r(r, \Delta_0, t) = \frac{\dfrac{dM_0}{dt}}{(2\pi\rho\rho_s \sin\Delta_0)^{1/2} 4\pi r r_s \alpha_s^2} * \frac{d}{dt}\frac{H(t)}{(\pi t)^{1/2}} * \frac{1}{\pi}\,\mathrm{Re}\int_0^\infty \int_\Gamma B(p) e^{i\omega(J-t)} dp\, d\omega. \tag{9.48}$$

But the last of the three functions of time convolved here is $f(t)$ with

$$f(t) = \mathrm{Re}\int_\Gamma B(p)\,\delta[J(p) - t]\,dp. \tag{9.49}$$

Therefore, the integrand makes a contribution precisely from p-values for which $J(p)$ is equal to t. Such p-values are parameterized by real t, and are, in part, identical to values on a Cagniard path. To see this, we consider first a simple case where just one ray parameter p_0 solves $\Delta(p) = \Delta_0$, with travel time

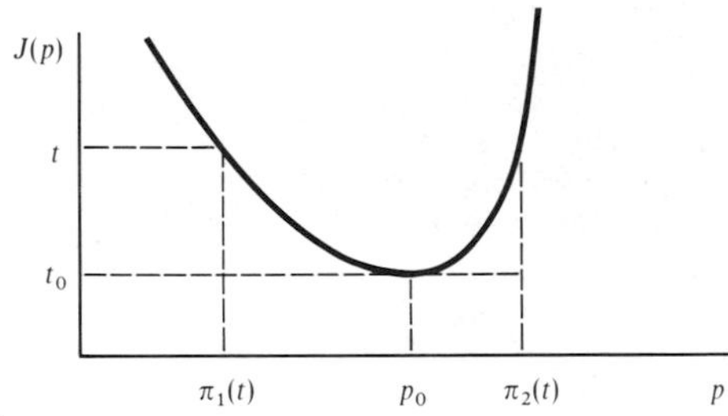

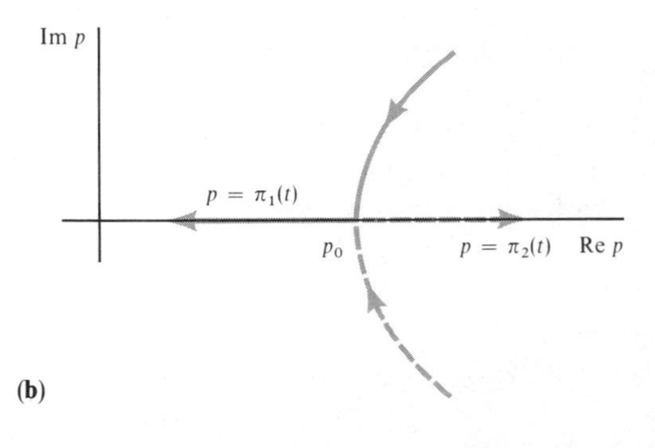

FIGURE **9.20**

(**a**) The function $J(p)$ is obtained from $\tau(p)$ (Fig. 9.14d) by adding a linear term $p\Delta_0$. A minimum at p_0 corresponds to time $t_0 = J(p_0)$. (**b**) The paths $p = \pi_j(t)$ of solutions to $t = J(p)$, *which lie on the real* p-axis for $t > t_0$.

t_0 between $\mathbf{r}_s$ and $\mathbf{r}$ (Fig. 9.17). The quantity $J(p) = T(p) - p\Delta(p) + p\Delta_0$ is shown in Figure 9.20a, and its slope $\Delta_0 - \Delta(p)$ is zero at p_0, where $J(p) = t_0$. For times greater than t_0, the equation

$$t = J(p) \tag{9.50}$$

has two roots, which Chapman (1976b) labels $\pi_1(t)$ and $\pi_2(t)$. In fact, equation (9.50) is just like an equation for a Cagniard path. Previously, we have had an interest in just one root, and we have followed it off the real p-axis for times later than some ray-arrival time. Here we shall keep both roots and hold Γ only to the real p-axis (in the vicinity of p_0) so that complex ray-parameter values are avoided. The paths $\pi_1(t)$ and $\pi_2(t)$, shown in Figure 9.20b, have real values only for $t \geq t_0$. Note that the direction taken along $\pi_1(t)$ as t increases is opposite to the direction familiar from Chapter 6 and Section 9.1. The reason is that $\partial\Delta/\partial p < 0$ at p_0, and values along $\pi_j(t)$ near p_0 are given via the Taylor expansion

$$t = t_0 + \tfrac{1}{2}(\pi_j - p_0)^2 \frac{\partial^2 J}{\partial p^2} = t_0 - \tfrac{1}{2}(\partial\Delta/\partial p)\big|_{p_0} (\pi_j - p_0)^2. \tag{9.51}$$

Once the real solutions $\pi_j = \pi_j(t)$ of (9.50) are known, we can approximate the delta-function argument in (9.49) by

$$J(p) - t = (p - \pi_j) \left.\frac{\partial J}{\partial p}\right|_{\pi_j} = (p - \pi_j)[\Delta_0 - \Delta(\pi_j)]$$

near each $\pi_j(t)$. In fact, it is invalid to include higher-order terms here, since the expression being approximated is itself the outcome of a WKBJ approximation. Thus (9.48) becomes

$$u_r(r, \Delta_0, t) = \frac{\dfrac{dM_0}{dt}}{(2\rho\rho_s \sin\Delta_0)^{1/2} 4\pi^2 r r_s \alpha_s^2} * \frac{d}{dt}\frac{H(t)}{t^{1/2}} * \sum_j \frac{B(\pi_j)}{|\Delta_0 - \Delta(\pi_j)|}. \tag{9.52}$$

To work with, this result is similar to our earlier Cagniard-derived solutions, but without the complex arithmetic. To be reassured that geometrical ray theory is contained in (9.52), we expand $\Delta_0 - \Delta(\pi_j)$ near p_0 as $(-\partial\Delta/\partial p)_{p_0}(\pi_j - p_0)$ and then use (9.51) to write

$$\Delta_0 - \Delta(\pi_j) \sim [-2\,\partial\Delta/\partial p]_{p_0}^{1/2}(t - t_0)^{1/2}. \tag{9.53}$$

From convolutions in (9.52) and Box 9.5, it follows that each of the two terms ($j = 1, 2$) contributes just half the geometrical approximation (9.46).

BOX **9.5**

Useful transform pairs

Because our time series are real, their Fourier transforms have real parts that are even in frequency and imaginary parts that are odd. With this understood, we write formulas only for positive real frequencies: $f(t) = (1/\pi) \operatorname{Re} \int_0^\infty f(\omega) \exp(-i\omega t)\, d\omega$. The connections between convolutions with $t^{-1/2}$ and multiplication by $\omega^{-1/2}$ are then based on the following transform pairs:

Frequency	*Time*
$e^{i\pi/4}\left(\dfrac{\pi}{\omega}\right)^{1/2}$	$\dfrac{H(t)}{t^{1/2}}$
$e^{-i\pi/4}\left(\dfrac{\pi}{\omega}\right)^{1/2}$	$\dfrac{H(-t)}{(-t)^{1/2}}$
$e^{-i\pi/4}(\pi\omega)^{1/2}$	$\dfrac{d}{dt}\dfrac{H(t)}{t^{1/2}}$
$e^{i\pi/4}(\pi\omega)^{1/2}$	$-\dfrac{d}{dt}\dfrac{H(-t)}{(-t)^{1/2}}$

The Hilbert transform of $H(t)/t^{1/2}$ is $H(-t)/(-t)^{1/2}$, and a basic transform to remember is

$$\frac{H(t)}{t^{1/2}} * \frac{H(t-T)}{(t-T)^{1/2}} = \pi H(t-T).$$

Convolutions that contain the Hilbert transform operator and the operator d/dt can be written in many different forms by using commutative properties and associating these operators with different functions. Thus

$$\frac{d\psi}{dt} * \frac{H(t)}{t^{1/2}} * \frac{H(t-T)}{(t-T)^{1/2}} = \psi * \frac{d}{dt}[\pi H(t-T)] = \pi\psi(t-T),$$

and

$$\psi(t) * \left[-\frac{d}{dt}\frac{H(-t)}{(-t)^{1/2}} * \frac{H(T-t)}{(T-t)^{1/2}}\right] = \left(-\frac{d\psi}{dt}\right) * (-\pi H(t-T)) = \pi\psi(t-T).$$

To get geometrical ray theory out of (9.52)–(9.53), we use

$$\frac{d}{dt}\frac{H(t)}{t^{1/2}} * \frac{H(t)}{(t-t_0)^{1/2}} = \pi\delta(t-t_0).$$

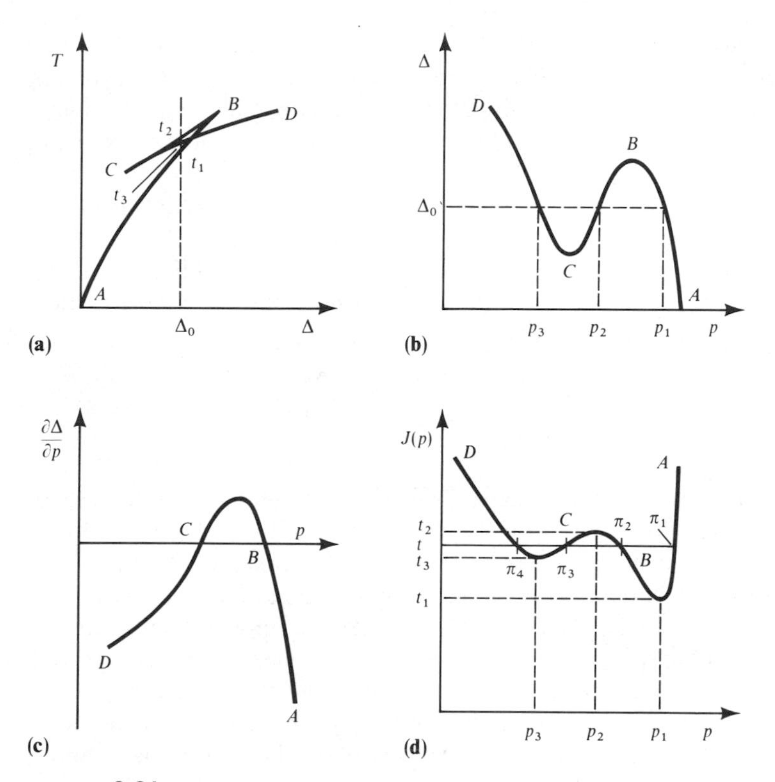

FIGURE **9.21**

A travel-time curve with caustics at B and C, and derived curves. Geometrical ray quantities are indicated for a distance Δ_0 in the triplication. (**a**) Travel-time curve with arrivals at t_1, t_3, t_2, (**b**) $\Delta = \Delta(p)$, with three solutions p_1, p_2, p_3 to $\Delta(p) = \Delta_0$. (**c**) Zeros in $\partial\Delta/\partial p$ occur at B and C. (**d**) The construction of solutions $\pi_j(t)$ to $t = J(p)$ is shown for a particular distance Δ_0. [After Chapman, 1976b; copyrighted by the American Geophysical Union.]

The main virtue of (9.52) is that multi-pathing and caustics are handled straightforwardly. Thus, in Figure 9.21 is shown a more complicated travel-time curve, with three solutions p_1, p_2, p_3 to $\Delta(p) = \Delta_0$. There are four solutions to $t = J(p)$, and it is fairly clear how to apply (9.52). The caustics at B and C are inflections in $J = J(p)$, and if Δ_0 is a caustic then the corresponding inflection in Figure 9.21d is horizontal. One finds $\Delta_0 - \Delta(\pi_j) \propto (t - t_0)^{2/3}$, which gives strong amplitudes via (9.52).

It is interesting that (9.52) was discovered by methods quite different from the one we have followed. Thus, Chapman (1976a) investigated Cagniard's method for large numbers of homogeneous plane layers, and was able to obtain an expression for the generalized primary reflection (see (9.6)). He summed over all interfaces and took the limit as the layer thicknesses all shrank to zero. Near wavefronts, the expression was $\pi/3$ times that given in (9.52) (but adapted for plane layering), which, for turning-point rays, implies the surprising result that fine layering and large numbers of primary reflections do not, in the limit, give the geometrical ray theory formula at wavefronts. Rather, the total effect of primaries is about 5% too large. However, tertiary reflections (triply reflected waves) tend in the limit to $-\pi^3/648$ times (9.52) near wavefronts, and Chapman showed from $(2l + 1)$th order reflections that the wavefront approximation was $2(-1)^l(\pi/6)^{2l+1}/(2l + 1)!$ times (9.52). Adding successive orders of reflection does then give a rapidly converging series, which does converge to (9.52) because

$$\frac{\pi}{3} - \frac{\pi^3}{648} + \cdots = 2 \sin \frac{\pi}{6} = 1.$$

Wiggins (1976a) formulated another approach, identifying $J(p)$ as the arrival time at Δ_0 of a wavefront (or "disc") with slowness p. By picking equal time increments along the $J(p)$ curve, he obtains a discrete version of (9.52) and calls this *disc-ray theory*. Chapman (1978a) has suggested that an appropriate name for the solution (9.52) is the *WKBJ seismogram*. This is appropriate, because it is an expression in the space/time domain that is equivalent to WKBJ approximations in the ray-parameter/frequency domain. The WKBJ seismogram is cheaply computed, and in many cases it is probably the method that should be preferred for investigating waveforms. We have obtained it as a special example of what Chapman (1978a,b) has called the *slowness method*, i.e., intermediate results (after the frequency integration) are obtained in the slowness (ray parameter) domain. However, in the next section we shall find that there are important examples in which WKBJ approximations are inaccurate, so that alternatives to (9.52) must be explored.

Throughout this section we have assumed a smoothly varying velocity profile. If discontinuities are to be allowed for, such as the Earth's free surface, then the associated reflection and transmission coefficients must be introduced. For the WKBJ seismogram, it turns out that these are simply plane-wave coefficients. Going beyond WKBJ theory, the principal techniques are well known (Scholte, 1956) for simple Earth models consisting of a few homogeneous shells, since in these the vertical wavefunctions are made up from spherical Hankel functions. It is found in such models that the wave path of interest (for example, *SKKP*) can be analyzed by isolating its partial wave series in an expression for the total response. This is conceptually the same exercise as identifying a particular generalized ray within an expression that quantifies an infinite family of rays (see Fig. 9.10). A Watson transform can still be used, but now any path deformation in the complex p-plane must take account of poles in the reflection/transmission coefficients for the ray path of interest. Most of these poles are located by studying properties of the WKBJ approximation to spherical Hankel functions, hence in more realistic Earth models, with radial variation of elastic properties between depths of discontinuity, we may still expect to locate relevant poles by using WKBJ formulas (9.35)–(9.36) for the generalized vertical wavefunctions $g^{(j)}_{\omega p - \frac{1}{2}}$. The rigorous analysis of path deformations needed to change the Watson path into paths such as Γ (Fig. 9.17) is a formidable project for realistic Earch models. However, several different canonical problems in wave propagation can be identified, and these are now well understood. In what follows, we shall emphasize *spectral methods*, in which intermediate results are obtained in the frequency domain (after an integration over complex ray parameter).

9.5 Body-wave Problems for Earth Models in Which Discontinuities Are Present

To handle systematically the reflections and transmissions at discontinuities within the Earth, it is important to develop an approach based on the motion-stress vector. Without this device, an *ad hoc* approach based (usually) on potentials must be used, and detailed algebraic manipulation must be carried out for each particular application. In this section, we shall develop the motion-stress vector for *P-SV* motion in spherically symmetric media, using a notation similar to that we have developed for homogeneous plane layers. The scattering matrix is easy to write down, and we shall use it to study several different body waves.

The appropriate horizontal wavefunctions to use are the vector surface harmonics mentioned in Chapter 8. Our notation is based on the fully normalized surface harmonic $Y_l^m(\Delta, \phi)$ (see (8.5)), and

$$\begin{aligned}
\mathbf{R}_l^m(\Delta, \phi) &= Y_l^m \hat{\mathbf{r}}, \\
\mathbf{S}_l^m(\Delta, \phi) &= \frac{1}{[l(l+1)]^{1/2}} \left(\frac{\partial Y_l^m}{\partial \Delta} \hat{\mathbf{\Delta}} + \frac{1}{\sin \Delta} \frac{\partial Y_l^m}{\partial \phi} \hat{\boldsymbol{\phi}} \right), \qquad (9.54) \\
\mathbf{T}_l^m(\Delta, \phi) &= \frac{1}{[l(l+1)]^{1/2}} \left(\frac{1}{\sin \Delta} \frac{\partial Y_l^m}{\partial \phi} \hat{\mathbf{\Delta}} - \frac{\partial Y_l^m}{\partial \Delta} \hat{\boldsymbol{\phi}} \right).
\end{aligned}$$

With these we make the expansions

$$\mathbf{u} = \mathbf{u}(r, \Delta, \phi, \omega) = (u_r, u_\Delta, u_\phi) = \sum_{l=0}^{\infty} \sum_{m=-l}^{l} [U_l^m \mathbf{R}_l^m + V_l^m \mathbf{S}_l^m + W_l^m \mathbf{T}_l^m], \qquad (9.55)$$

and

$$\text{traction} = (\tau_{rr}, \tau_{r\Delta}, \tau_{r\phi}) = \sum_{l=0}^{\infty} \sum_{m=-l}^{l} [R_l^m \mathbf{R}_l^m + S_l^m \mathbf{S}_l^m + T_l^m \mathbf{T}_l^m].$$

It is the six quantities U_l^m, V_l^m, W_l^m, R_l^m, S_l^m, T_l^m for which we wish to solve. Each is a function of (r, l, m, ω), and in this sense is the triple transform of a solution in the domain (r, Δ, ϕ, t), in which we are interested. Only radial derivatives remain in the multi-transformed equations of motion (2.47)–(2.50), and these separate into 2 groups:

$$\frac{d}{dr}\begin{pmatrix} V \\ U \\ S \\ R \end{pmatrix} = \begin{pmatrix} \dfrac{1}{r} & -\dfrac{[l(l+1)]^{1/2}}{r} & \dfrac{1}{\mu} & 0 \\ \dfrac{\lambda[l(l+1)]^{1/2}}{r(\lambda+2\mu)} & -\dfrac{2\lambda}{r(\lambda+2\mu)} & 0 & \dfrac{1}{\lambda+2\mu} \\ \dfrac{4\omega^2 p^2 \mu(\lambda+\mu)}{r^2(\lambda+2\mu)} - \rho\omega^2 - \dfrac{2\mu}{r^2} & -\dfrac{2\mu(3\lambda+2\mu)[l(l+1)]^{1/2}}{r^2(\lambda+2\mu)} & -\dfrac{3}{r} & -\dfrac{\lambda[l(l+1)]^{1/2}}{r(\lambda+2\mu)} \\ -\dfrac{2\mu(3\lambda+2\mu)[l(l+1)]^{1/2}}{r^2(\lambda+2\mu)} & -\rho\omega^2 + \dfrac{4\mu(3\lambda+2\mu)}{r^2(\lambda+2\mu)} & \dfrac{[l(l+1)]^{1/2}}{r} & \dfrac{-4\mu}{r(\lambda+2\mu)} \end{pmatrix}\begin{pmatrix} V \\ U \\ S \\ R \end{pmatrix} \tag{9.56}$$

and

$$\frac{d}{dr}\begin{pmatrix} W \\ T \end{pmatrix} = \begin{pmatrix} \dfrac{1}{r} & \dfrac{1}{\mu} \\ \dfrac{\mu(l-1)(l+2)}{r^2} - \rho\omega^2 & -\dfrac{3}{r} \end{pmatrix}\begin{pmatrix} W \\ T \end{pmatrix}. \tag{9.57}$$

We have dropped subscript l and superscript m from the dependent variables. Note that m does not even enter the equations, but we must remember that an m-dependence is present, in general, because of the m-dependence of the source expansion. In Chapter 8, we also associated an overtone number, n, with the radial functions. Overtones do not arise in our present approach, because we do not now seek the special solutions subject to regularity at $r = 0$ and zero stresses at $r = r_\oplus$ (the Earth's surface). Rather, we shall work with a complete set of linearly independent solutions without (yet) imposing boundary conditions.

Clearly, (9.56) describes *P-SV* motion (cf. (5.60)), and (9.57) describes *SH* (cf. (5.58)), and these two groups of equations are in the standard form $d\mathbf{f}/dr = \mathbf{A}(r)\mathbf{f}$. Our next goal is to find how $\mathbf{f}$ can be written as $\mathbf{f} = \mathbf{F}\mathbf{w} = \mathbf{E}\boldsymbol{\Lambda}\mathbf{w}$, where the layer matrix $\mathbf{F}$ has been factored into a term $\mathbf{E}$ describing the amplitude of displacement and stress components in each layer and a term $\boldsymbol{\Lambda}$ (a diagonal matrix) giving vertical wavefunctions for downgoing and upgoing P and SV. The simplest way to generate $\mathbf{F}$ and its factors $\mathbf{E}$ and $\boldsymbol{\Lambda}$ is via the potentials (9.25a) appropriate to spherical geometry, since each column of $\mathbf{F}$ is derivable from an identifiable vertical wavefunction (Problem 5.8). Another way is to invert transformations given by Chapman (1973, 1974). The final result is quite remarkably similar to plane-wave theory (5.65). For *P-SV* problems, it is

$$\mathbf{f}(r) = \begin{pmatrix} V \\ U \\ S \\ R \end{pmatrix} = \mathbf{E\Lambda w}, \qquad \text{where } \mathbf{\Lambda} = \begin{pmatrix} \pi^{(2)}(r) & 0 & 0 & 0 \\ 0 & \sigma^{(2)}(r) & 0 & 0 \\ 0 & 0 & \pi^{(1)}(r) & 0 \\ 0 & 0 & 0 & \sigma^{(1)}(r) \end{pmatrix} \tag{9.58}$$

and

$$\mathbf{E} = \begin{pmatrix} -\frac{i\alpha p}{r} & -i\beta\grave{\eta} & -\frac{i\alpha p}{r} & -i\beta\acute{\eta} \\ -\alpha\grave{\xi} & \frac{\beta p}{r} & \alpha\acute{\xi} & -\frac{\beta p}{r} \\ -2\omega\rho\alpha\beta^2\frac{p}{r}\grave{\xi} & -\omega\rho\beta\left(1-2\beta^2\frac{p^2}{r^2}\right) & 2\omega\rho\alpha\beta^2\frac{p}{r}\acute{\xi} & \omega\rho\beta\left(1-2\beta^2\frac{p^2}{r^2}\right) \\ i\omega\rho\alpha\left(1-2\beta^2\frac{p^2}{r^2}\right) & -2i\omega\rho\beta^3\frac{p}{r}\grave{\eta} & i\omega\rho\alpha\left(1-2\beta^2\frac{p^2}{r^2}\right) & -2i\omega\rho\beta^3\frac{p}{r}\acute{\eta} \end{pmatrix}. \tag{9.59}$$

The wavefunctions $\pi^{(j)}$ are for *P*-waves, and are related to our previous $g_l^{(j)}$ via

$$\pi^{\binom{(1)}{(2)}}(r, p, \omega) \equiv \left(\frac{\rho_s}{\rho}\right)^{1/2}\frac{\alpha_s}{\alpha}\,g_l^{\binom{(1)}{(2)}} \sim e^{\mp i\pi/4}\left(\frac{\rho_s\alpha_s}{\rho\alpha\cos i}\right)^{1/2}\left(\frac{\alpha_s}{\omega r}\right)\exp\left[\pm i\omega\int_{r_p}^{r}\xi(r')\,dr'\right]. \tag{9.60}$$

Wavefunctions $\sigma^{(j)}$ for *SV* are defined from the shear-velocity profile $\beta(r)$, and similarly

$$\sigma^{\binom{(1)}{(2)}}(r, p, \omega) \sim e^{\mp i\pi/4}\left(\frac{\rho_s\beta_s}{\rho\beta\cos j}\right)^{1/2}\left(\frac{\beta_s}{\omega r}\right)\exp\left[\pm i\omega\int_{r_p}^{r}\eta(r')\,dr'\right] \tag{9.61}$$

(r_p here is the *S*-wave turning point, where $\eta = (1/\beta^2 - p^2/r^2)^{1/2} = \beta^{-1}\cos j$ is zero). Accented vertical slownesses in (9.59) are defined by

$$\grave{\xi} \equiv \frac{1}{-i\omega\pi^{(2)}}\frac{d\pi^{(2)}}{dr}, \quad \acute{\xi} \equiv \frac{1}{i\omega\pi^{(1)}}\frac{d\pi^{(1)}}{dr}, \quad \grave{\eta} \equiv \frac{1}{-i\omega\sigma^{(2)}}\frac{d\sigma^{(2)}}{dr}, \quad \acute{\eta} \equiv \frac{1}{i\omega\sigma^{(1)}}\frac{d\sigma^{(1)}}{dr}, \tag{9.62}$$

and to first order it follows from WKBJ theory that, above the turning points, $\grave{\xi} \sim \acute{\xi} \sim \alpha^{-1}\cos i$ and $\grave{\eta} \sim \acute{\eta} \sim \beta^{-1}\cos j$. However, below the turning points, (compare with (5.34)–(5.37) for one interface between two homogeneous half-spaces. The left-hand side of (4) represents the scattered waves, and **S** operates on the incident waves.) Entries in **S** now refer to generalized scattering coefficients, so that, for example, $\grave{P}\acute{S}$ denotes the phase and amplitude of an *SV*-wave transmitted downward in inhomogeneous layer number M due to downward P in layer $N + 1$. Thus $\grave{P}\acute{S}$ contains all the multiples within the stack of layers between interfaces M and N. To obtain **S** explicitly in terms of entries of **H**, manipulate (2) into the form (4) to find

$$\mathbf{S} = -\begin{pmatrix} H_{13} & H_{14} & -1 & 0 \\ H_{23} & H_{24} & 0 & -1 \\ H_{33} & H_{34} & 0 & 0 \\ H_{43} & H_{44} & 0 & 0 \end{pmatrix}^{-1} \begin{pmatrix} H_{11} & H_{12} & 0 & 0 \\ H_{21} & H_{22} & 0 & 0 \\ H_{31} & H_{32} & 1 & 0 \\ H_{41} & H_{42} & 0 & 1 \end{pmatrix}$$

$$= \frac{1}{\mathsf{D}} \begin{pmatrix} 0 & 0 & -H_{44} & H_{34} \\ 0 & 0 & H_{43} & -H_{33} \\ \mathsf{D} & 0 & H_{14}H_{43} - H_{13}H_{44} & H_{13}H_{34} - H_{14}H_{33} \\ 0 & \mathsf{D} & H_{24}H_{43} - H_{23}H_{44} & H_{23}H_{34} - H_{24}H_{33} \end{pmatrix} \begin{pmatrix} H_{11} & H_{12} & 0 & 0 \\ H_{21} & H_{22} & 0 & 0 \\ H_{31} & H_{32} & 1 & 0 \\ H_{41} & H_{42} & 0 & 1 \end{pmatrix}, \quad (6)$$

where $\mathsf{D} = H_{33}H_{44} - H_{34}H_{43}$.

Multiplying out the two matrices in (6), one can find any desired entry in **S**. The method we have given here is due to Červený (1974), who also showed how to avoid numerical difficulties by working with the 6×6 matrix of 2×2 minors of **H** (the so-called *delta matrix*). The delta matrix method, reduced to manipulations with 5×5 matrices, has been described in detail by Kind (1976) for homogeneous layers.

It is often important to obtain **S** in the special case of a single interface between two inhomogeneous layers. Matrix **H** is then obtained from (1). The method in (5.34)–(5.36) leads us to define

$$a = \rho_2(1 - 2\beta_2^2 p^2/r_1^2) - \rho_1(1 - 2\beta_1^2 p^2/r_1^2), \qquad b = \rho_2(1 - 2\beta_2^2 p^2/r_1^2) + 2\rho_1\beta_1^2 p^2/r_1^2,$$

$$c = \rho_1(1 - 2\beta_1^2 p^2/r_1^2) + 2\rho_2\beta_2^2 p^2/r_1^2, \qquad d = 2(\rho_2\beta_2^2 - \rho_1\beta_1^2),$$

$$E = b\grave{\xi}_1 + c\acute{\xi}_2, \qquad F = b\grave{\eta}_1 + c\acute{\eta}_2, \qquad G = a - d\grave{\xi}_1\acute{\eta}_2,$$

$$H = a - d\acute{\xi}_2\grave{\eta}_1, \qquad \mathsf{D} = EF + GHp^2/r_1^2,$$

where subscript 1 is for the lower layer and subscript 2 for the upper. The interface itself is at r_1, and the *P-SV* coefficients themselves are

$$\grave{P}\acute{P} = -\{[(b\grave{\xi}_1 - c\grave{\xi}_2)F + (a + d\grave{\xi}_2\grave{\eta}_1)Gp^2/r_1^2]/\mathsf{D}\}(\pi_2^{(2)}/\pi_2^{(1)}),$$

$$\grave{P}\acute{S} = \{(\acute{\xi}_2 + \grave{\xi}_2)(ac + bd\grave{\xi}_1\grave{\eta}_1)p\alpha_2/(r_1\beta_2\mathsf{D})\}(\pi_2^{(2)}/\sigma_2^{(1)}),$$

$$\grave{P}\grave{P} = \{\rho_2(\acute{\xi}_2 + \grave{\xi}_2)F\alpha_2/(\alpha_1\mathsf{D})\}(\pi_2^{(2)}/\pi_1^{(2)}),$$

$$\grave{P}\grave{S} = -\{\rho_2(\acute{\xi}_2 + \grave{\xi}_2)Gp\alpha_2/(r_1\beta_1\mathsf{D})\}(\pi_2^{(2)}/\sigma_1^{(2)}),$$

$$\grave{S}\acute{P} = \{(\acute{\eta}_2 + \grave{\eta}_2)(ac + bd\grave{\xi}_1\grave{\eta}_1)p\beta_2/(r_1\alpha_2\mathsf{D})\}(\sigma_2^{(2)}/\pi_2^{(1)}),$$

$$\grave{S}\acute{S} = \{[b\grave{\eta}_1 - c\grave{\eta}_2)E + (a + d\grave{\xi}_1\grave{\eta}_2)Hp^2/r_1^2]/\mathsf{D}\}(\sigma_2^{(2)}/\sigma_2^{(1)}),$$

$$\grave{S}\grave{P} = \{\rho_2(\acute{\eta}_2 + \grave{\eta}_2)Hp\beta_2/(r_1\alpha_1\mathsf{D})\}(\sigma_2^{(2)}/\pi_1^{(2)}),$$

$$\grave{S}\grave{S} = \{\rho_2(\acute{\eta}_2 + \grave{\eta}_2)E\beta_2/(\beta_1\mathsf{D})\}(\sigma_2^{(2)}/\sigma_1^{(2)}),$$

$$\acute{P}\acute{P} = \{\rho_1(\grave{\xi}_1 + \acute{\xi}_1)F\alpha_1/(\alpha_2\mathsf{D})\}(\pi_1^{(1)}/\pi_2^{(1)}),$$

$$\acute{P}\acute{S} = \{\rho_1(\grave{\xi}_1 + \acute{\xi}_1)Hp\alpha_1/(r_1\beta_2\mathsf{D})\}(\pi_1^{(1)}/\sigma_2^{(1)}),$$

$$\acute{P}\grave{P} = \{[b\acute{\xi}_1 - c\acute{\xi}_2)F - (a + d\acute{\xi}_1\acute{\eta}_2)Hp^2/r_1^2]/\mathsf{D}\}(\pi_1^{(1)}/\pi_1^{(2)}),$$

$$\acute{P}\grave{S} = -\{(\grave{\xi}_1 + \acute{\xi}_1)(ab + cd\acute{\xi}_2\acute{\eta}_2)p\alpha_1/(r_1\beta_1\mathsf{D})\}(\pi_1^{(1)}/\sigma_1^{(2)}),$$

$$\acute{S}\acute{P} = -\{\rho_1(\grave{\eta}_1 + \acute{\eta}_1)Gp\beta_1/(r_1\alpha_2\mathsf{D})\}(\sigma_1^{(1)}/\pi_2^{(1)}),$$

$$\acute{S}\acute{S} = \{\rho_1(\grave{\eta}_1 + \acute{\eta}_1)E\beta_1/(\beta_2\mathsf{D})\}(\sigma_1^{(1)}/\sigma_2^{(1)}),$$

$$\acute{S}\grave{P} = -\{(\grave{\eta}_1 + \acute{\eta}_1)(ab + cd\acute{\xi}_2\acute{\eta}_2)p\beta_1/(r_1\alpha_1\mathsf{D})\}(\sigma_1^{(1)}/\pi_1^{(2)}),$$

$$\acute{S}\grave{S} = -\{[(b\acute{\eta}_1 - c\acute{\eta}_2)E - (a + d\acute{\xi}_2\acute{\eta}_1)Gp^2/r_1^2]/\mathsf{D}\}(\sigma_1^{(1)}/\sigma_1^{(2)}).$$

The last factor in each coefficient, a ratio of wavefunctions, is evaluated at r_1.

Our sign convention here is geared to the dependent variables in (9.58). For example, if

$$\mathbf{w}_2 = \begin{pmatrix} 0 \\ 1 \\ \grave{S}\acute{P} \\ \grave{S}\acute{S} \end{pmatrix} \quad \text{then} \quad \mathbf{w}_1 = \begin{pmatrix} \grave{S}\grave{P} \\ \grave{S}\grave{S} \\ 0 \\ 0 \end{pmatrix}.$$

If one of the two layers is a fluid, then only nine of the sixteen coefficients have meaning. They can easily be found from those given here by a limiting procedure. For example, if layer 1 is the fluid core, use $\beta_1 \to 0$, $\eta_1 \to \infty$, and $bc - adp^2/r_1^2 = \rho_1\rho_2$. In this way, coefficients $\grave{P}\grave{P}$, $\acute{P}\grave{P}$, $\acute{P}\acute{S}$ defined above are used for the core-mantle boundary to derive $\grave{P}\grave{K}$, $\acute{K}\grave{K}$, $\acute{K}\acute{S}$, respectively.

Corresponding results for SH and a single interface are

$$\grave{S}\acute{S} = [(\mu_2\grave{\eta}_2 - \mu_1\grave{\eta}_1)/\mathsf{D}]\sigma_2^{(2)}/\sigma_2^{(1)},$$

$$\grave{S}\grave{S} = [\mu_2(\grave{\eta}_2 + \acute{\eta}_2)/\mathsf{D}]\sigma_2^{(2)}/\sigma_1^{(2)},$$

$$\acute{S}\acute{S} = [\mu_1(\acute{\eta}_1 + \grave{\eta}_1)/\mathsf{D}]\sigma_1^{(1)}/\sigma_2^{(1)},$$

$$\acute{S}\grave{S} = [(\mu_1\acute{\eta}_1 - \mu_2\acute{\eta}_2)/\mathsf{D}]\sigma_1^{(1)}/\sigma_1^{(2)},$$

where $\mathsf{D} = \mu_1\grave{\eta}_1 + \mu_2\acute{\eta}_2$ and μ_1, μ_2 are, respectively, the rigidities below and above the interface.

theory in the range $p < p_d$, the *amplitude* of $\grave{P}\acute{P}$ turns out to be equal to the plane-wave reflection coefficient (still a function of p). Going on to make the saddle point approximation for the integral over this *PcP* saddle (Box 6.3), we should merely recover classical ray theory for this phase (i.e., geometrical spreading together with a plane-wave reflection coefficient to describe the scattering at the core-mantle boundary).

This preliminary discussion of *PcP* brings out a special example of a general rule: if a real ray is present between source and receiver, then a combination of two approximations (WKBJ theory and the saddle point approximation) applied to the integral representation for the generalized ray (cf. (9.64)) will merely yield geometrical ray theory. We first obtained a result of this nature in our derivation of (6.19) for homogeneous media. Seckler and Keller (1959) gave many examples for inhomogeneous media. Of course, our interest is in improving upon classical ray theory. We shall give several examples of the improvement that results when neither the WKBJ approximations nor the saddle-point approximation is used.

Returning to Figure 9.23, observe that, as Δ_0 increases, rays for direct *P* and *PcP* move closer together until they merge at the special distance Δ_d. Taking a phrase from optics, it is natural to call Δ_d the *shadow boundary*, because it marks the edge of a region $\Delta_d < \Delta_0$ within which neither direct *P* nor *PcP* can arrive. As shown in Figure 9.23b, rays arriving within the shadow must creep around the base of the mantle. This arrival is widely observed in seismology, and is known as the *diffracted P-wave* (written P_{diff}). It is similar in geometry to the diffraction of radio waves over the Earth's curved surface, a problem for which Watson originally developed the transformation that now bears his name. At $\Delta_0 = \Delta_d$, the two saddles have merged at $p = p_d$. Within the shadow, the line integral (9.64) can be evaluated as a rapidly converging series of residues from a string of poles (zeros of the denominator of $\grave{P}\acute{P}$, described in Box 9.8) that stretches up into the first quadrant from the first pole just above p_d. If the receiver is well into the shadow, so that Δ_0 is considerably greater than Δ_d, only the first residue is significant. It is located at p_{diff}, close to p_d, in that $p_{\text{diff}} = p_d + \lambda e^{i\pi/3}\omega^{-2/3}$, where λ is a positive constant having a magnitude of the same order as $p_d^{1/3}$. Well into the shadow, it follows that the amplitude of the diffracted wave P_{diff} is proportional to

$$\exp[-\omega^{1/3}\lambda\sin(\pi/3)(\Delta_0 - \Delta_d)].$$

Amplitudes in P_{diff} are thus more attenuated at higher frequencies and at greater distances into the shadow.

All these properties of (9.64) were described by Scholte (1956). Phinney and Cathles (1969) pointed out that a numerical evaluation of the line integral (9.64) has many advantages. A suitable path is the two broken line segments shown in Figure 9.23c, and the result of an integration at two different frequencies is shown in Figure 9.24.

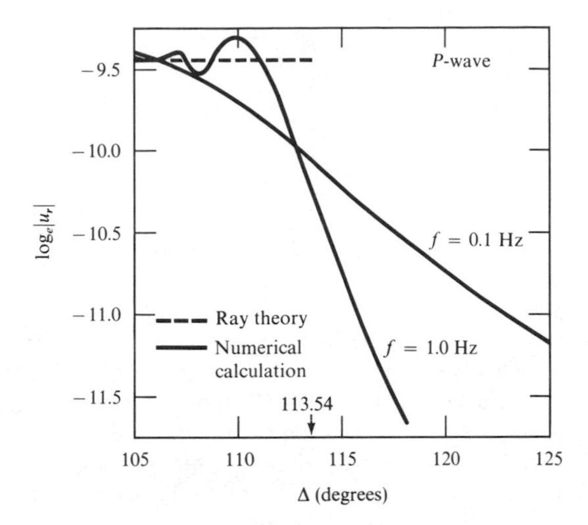

FIGURE **9.24**
Vertical component of P-wave displacement amplitude due to an explosive source for which $M_0(t)$ is a step function. Computation is via (9.63) at two different frequencies, and the path of integration (line segments) is shown in the previous figure. The shadow boundary Δ_d is at 113.54° in this crude Earth model. For $\Delta_0 < \Delta_d$ there is interference of P and PcP. The broken line gives ray theory for direct P only. Although there is no sharp shadow boundary, the amplitudes at 1 Hz do decay quite rapidly with distance in the vicinity of Δ_d. The longer-period waves would be observable far into the shadow.

In the remainder of this section, we shall be describing the results of numerical integration along paths consisting of straight-line segments in the complex ray-parameter plane. But one new element is needed in the theory. We find there is frequently the need to evaluate wavefunctions like $\pi_m^{(1)}(r_{\text{CMB}}, p, \omega)$ as p varies near values for which the radial argument is a turning point. Such is the case, for example, near p_d in Figure 9.23c, and note that almost all the energy in P_{diff} has this ray parameter. The problem is that WKBJ approximations break down in precisely this turning-point region, as described in Box 9.3. (The computation in Figure 9.24 presented no problem, because "mantle" and "core" were homogeneous, so that vertical wavefunctions $\pi^{(j)}$ and $\sigma^{(j)}$ were exactly given by spherical Hankel functions.) Fortunately, for many cases of radial inhomogeneity, there *is* a uniformly asymptotic approximation for $\pi^{(j)}$ and $\sigma^{(j)}$ that is satisfactory. A brief review is given in Box 9.7, and we state results here as the Langer approximation

$$\pi^{\substack{(1)\\(2)}}(r, p, \omega) \sim \left(\frac{\pi \rho_s \alpha_s}{2\rho\alpha}\right)^{1/2} \frac{\alpha_s e^{\pm i\pi/6}}{\omega r} \left(\frac{\omega\tau}{\alpha\xi}\right)^{1/2} H_{1/3}^{\substack{(1)\\(2)}}(\omega\tau), \tag{9.65}$$

where $\tau = T - p\Delta = \int_{r_p}^{r} \xi(r')\,dr'$ and $\xi = \xi(r) = (1/\alpha^2(r) - p^2/r^2)^{1/2}$. The associated vertical slownesses are

$$\grave{\xi} \sim \left(\frac{1}{\alpha^2} - \frac{p^2}{r^2}\right)^{1/2} e^{-i\pi/6} \frac{H_{2/3}^{(2)}(\omega\tau)}{H_{1/3}^{(2)}(\omega\tau)} \quad \text{and} \quad \acute{\xi} \sim \left(\frac{1}{\alpha^2} - \frac{p^2}{r^2}\right)^{1/2} e^{i\pi/6} \frac{H_{2/3}^{(1)}(\omega\tau)}{H_{1/3}^{(1)}(\omega\tau)}. \tag{9.66}$$

Analogous expressions for $\sigma^{(j)}$, $\grave{\eta}$, and $\acute{\eta}$ are based on $\beta(r)$ and the shear-wave turning point.

BOX **9.7**

A uniformly asymptotic approximation for vertical wavefunctions

In our early discussions of vertical wavefunctions ($g^{(1)}$, $g^{(2)}$, f, $\pi^{(1)}$, $\pi^{(2)}$, $\sigma^{(1)}$, $\sigma^{(2)}$), the main emphasis was on the equations of motion that these functions satisfy, hence we regarded them principally as functions of r. But to evaluate the waves set up by a point source, we usually work with vertical wavefunctions at fixed r (a source or receiver position or a level on which some boundary condition is imposed) and varying p and ω.

After an appropriate normalization has been determined, the WKBJ approximation (9.35)–(9.36), (9.60) is often suitable, evaluating wavefunctions as p and ω vary for large ω. However, this is a nonuniform approximation, being poor or grossly wrong for p-values near that particular p-value p_r which makes the radial argument r a turning point. To obtain a uniform approximation, we shall again revert to the equation of motion.

In what follows, we shall study $g^{(1)}$ and $g^{(2)}$, introduced as the outgoing and ingoing solutions of

$$\frac{d^2}{dr^2}[ra(r, l)] + \omega^2\left[\frac{1}{\alpha^2(r)} - \frac{l(l+1)}{\omega^2 r^2}\right]ra(r, l) = 0. \tag{1}$$

Early attempts to solve (1) as r varies near r_p hinged on developing a Taylor series expansion for the coefficient of the term $ra(r, l)$, i.e., expanding this term as a series in powers of $r - r_p$. In the turning-point region itself, this leads to Airy-function solutions with argument proportional to $(r - r_p)$ (e.g., see the derivation of equation (5), Box 9.3). However, this again is a nonuniform approximation, because it works *only* in the turning-point region.

The uniformly asymptotic solution we shall use is based on work of Langer (1951) and Olver (1954a,b). Effectively, the method is to use a new depth variable, derived from r and $\alpha(r)$, in terms of which the equation (1) is close to being an Airy equation at *all* depths, including the turning-point level. We shall merely state the results, referring the reader to Richards (1976b) for more details.

Thus, working from $\tau = \tau(r, p) = \int_{r_p}^{r} \xi \, dr$ as defined for the WKBJ approximation, we introduce the new depth variable $\zeta = (\frac{3}{2}\omega\tau)^{2/3}$. It is assumed here that the velocity profile is an analytic function of radius, so that r_p is an analytic function of p. It follows that ξ, τ, and ζ are analytic functions of p, and so are $g^{(1)}$ and $g^{(2)}$. If the velocity profile is *not* analytic (e.g., it might be made up from many Mohorovičić layers, in each of which the velocity is given by ar^b with a and b constant in that layer), then r_p is not an analytic function of p, and neither are $g^{(1)}$ and $g^{(2)}$. Even so, $g^{(1)}$ and $g^{(2)}$ may vary sufficiently smoothly with p so as to be treated for our purposes as analytic functions in the complex plane.

In terms of ζ, (1) can be written in the form

$$\frac{d^2}{d\zeta^2}\left[r\left(\frac{d\zeta}{dr}\right)^{1/2} a(r, l)\right] + \zeta\left[r\left(\frac{d\zeta}{dr}\right)^{1/2} a(r, l)\right] = \text{terms two orders lower in frequency.} \tag{2}$$

For large ω, the right-hand side is uniformly negligible (i.e., for *all* depths). It follows that $r(d\zeta/dr)^{1/2}a(r, l)$ has solutions made up from a linear combination of Airy functions

$$\mathrm{Ai}(-\zeta),\ \mathrm{Ai}(-e^{2i\pi/3}\zeta),\ \mathrm{Ai}(-e^{-2i\pi/3}\zeta). \tag{3}$$

We have now obtained the desired uniformly asymptotic solutions for $a(r, l)$, and it remains only to identify the particular linear combination of (3) for each of the special solutions $g^{(1)}$ (outgoing) and $g^{(2)}$ (ingoing). This is most easily done by comparing the WKBJ approximations for the three functions in (3) with the WKBJ approximations for $g^{(1)}$ and $g^{(2)}$. The comparison tells us that

$$g^{\substack{(1)\\(2)}} \sim \frac{2}{\omega r}\left(\frac{\pi\alpha_s}{\xi}\right)^{1/2} \zeta^{1/4}\,\mathrm{Ai}(-e^{\pm 2i\pi/3}\zeta). \tag{4}$$

WKBJ theory was used only to obtain the *normalization* above, and (4) gives the uniformly asymptotic approximations we require. They are little more difficult to evaluate than WKBJ approximations. Both require $\tau = \tau(r, p)$, and then (4) works with Airy functions, whereas WKBJ approximations use exponentials.

The Langer approximations quoted in the text (9.65) are based upon (4) above, (9.60), and standard formulas relating Airy functions to Hankel functions of order $\frac{1}{3}$. Various branch cuts needed for individual factors in (4) and (9.65) are chosen so that the product of factors, the Langer approximation itself, is an entire function (single valued and analytic), bounded everywhere in the p-plane for finite p. Zeros of $g^{(1)}$ lie in the first and third quadrants of this plane, and zeros of $g^{(2)}$ lie in the second and fourth quadrants (see Box 9.8).

In the remainder of this section, we shall draw upon examples of body waves in seismology having properties that are readily quantified by using line integrals like (9.64) in the complex ray-parameter plane, but with the Langer approximation for the vertical wavefunctions.

Thus we look now at P-waves that have been transmitted into the Earth's fluid core, these being an excellent example of what can happen when waves interact with a discontinuity in which the slower medium (in this case, the fluid core) lies below the interface. Specializing to the example of $P4KP$ (see Fig. 9.25a), which has been observed after large nuclear explosions (Adams, 1972, Buchbinder, 1972), our theory allows us to write down the radial displacement at distance Δ_0 as

$$u_r^{P4KP}(r, \Delta_0, \omega) = \frac{-i\omega M_0(\omega)i\omega}{4\pi\rho_s\alpha_s^4}$$

$$\times \int_\Gamma \omega^2 p\pi^{(1)}(r_s)\grave{P}\grave{K}\cdot(\acute{K}\grave{K})^3\cdot(\acute{K}\acute{P})[\alpha(r)\xi(r)\pi^{(1)}(r)]Q^{(2)}_{\omega p-\frac{1}{2}}(\cos\Delta_0)\,dp. \tag{9.67}$$

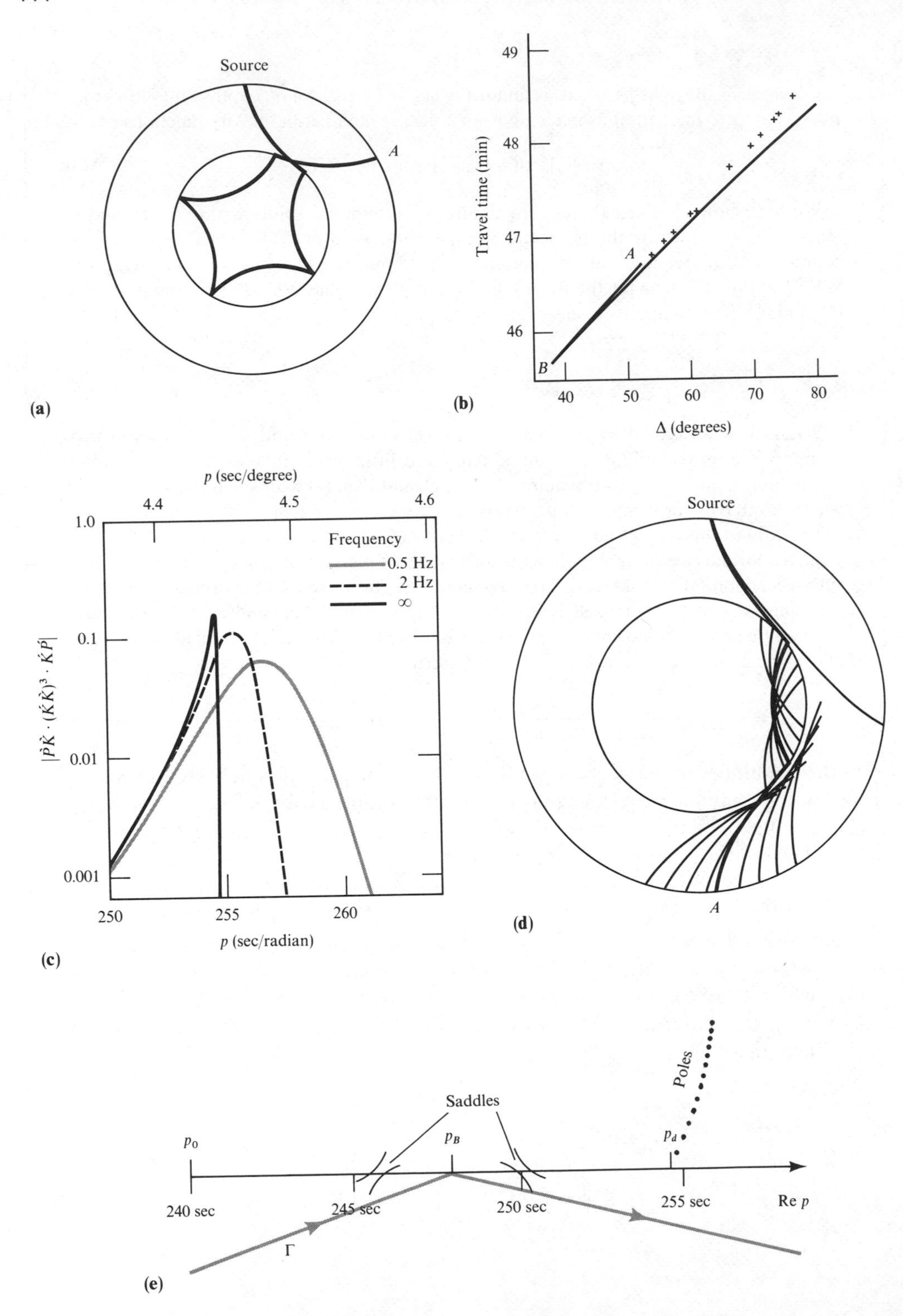
Source
A
(a)
49
48
47
46
B
Travel time (min)
40
50
60
70
80
Δ (degrees)
(b)
p (sec/degree)
4.4
4.5
4.6
1.0
0.1
0.01
0.001
Frequency
0.5 Hz
2 Hz
∞
250
255
260
p (sec/radian)
(c)
(d)
Poles
Saddles
p_0
p_B
p_d
240 sec
245 sec
250 sec
255 sec
Re p
Γ
(e)

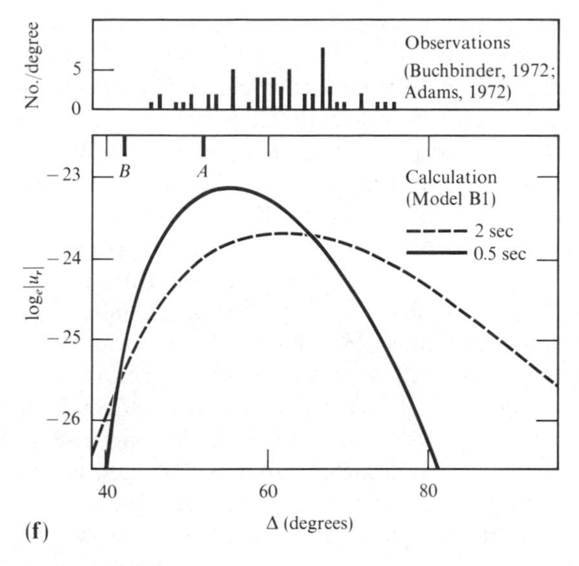

(f)

FIGURE **9.25**

Waves that travel as P in the mantle, and which have been internally reflected below the Earth's core-mantle boundary, are known as $PmKP$, with m an integer specifying the number of legs within the fluid core. (**a**) The ray path for $P4KP$. The ray parameter shown is exactly the core-grazing ray parameter p_d. For this special ray parameter, the point of emergence of $P4KP$ at the Earth's surface is designated as A. On the basis of classical ray theory, it may be expected that $P4KP$ is observed only for ray parameters less than or equal to p_d. Such, however, is not the case. (**b**) The travel-time curve for $P4KP$ in the Jeffreys-Bullen Earth model is shown as a solid line. From the maximum ray parameter at A, where the curve begins, there is a decrease to a minimum (in $\Delta = \Delta(p)$) at B, where there is a caustic. Continued decrease of ray parameter is then on a branch for which $\partial\Delta/\partial p < 0$. Shown as crosses are some travel times observed for presumed nuclear explosions in Novaya Zemlya, recorded at North American stations. (**c**) The product of reflection and transmission coefficients relevant to $P4KP$ is shown against ray parameter for two finite frequencies. The limiting case, as frequency tends to infinity, is also shown (it is zero for $p_d < p$, and is merely a product of plane-wave coefficients (5.39) for $p < p_d$). [After Richards, 1973b.] (**d**) The ray drawn heaviest is the core-grazing ray, emerging at A. There is exponential decay of energy below a turning point. Thus, for a ray with ray parameter slightly greater than p_d, the core-mantle boundary has energy leaking down from above. Below the boundary, energy can propagate again along real rays. The high-velocity region just below the turning point therefore acts as a barrier, through which, at finite frequencies, some energy can pass into the low-velocity zone (the core in this case). For clarity, we here show only PKP, but this tunneling phenomenon occurs for all $PmKP$. Energy can also tunnel back out into the mantle. Tunneling will spread energy over a range of distances, and this range extends well beyond the point A. (**e**) Features in the complex p-plane, for the integrand in (9.67). Saddles on either side of $p = p_B$ behave similarly to those shown in Figure 9.19b,c. At $p = p_0$, there is a zero of $\grave{K}\grave{K}$. (**f**) The amplitudes of $P4KP$ at two frequencies are shown against distance, as computed from (9.67) using the Earth model B1 of Jordan and Anderson (1974). Points A and B for this model are marked, and note that the maximum amplitudes occur beyond the point A. $P4KP$ is observed typically with period about 1 sec, and in the top half of the figure is shown a histogram based on observations reported by Buchbinder (1972) and Adams (1972). Maxima in the computed curves do occur at distances where observations are relatively plentiful.

We are here using K as the P-wave in the fluid core, so that $\grave{P}\grave{K}$ is a transmission coefficient (readily obtained from the general $\grave{P}\grave{P}$ given in Box 9.6), $\acute{K}\grave{K}$ is the internal reflection, and $\acute{K}\acute{P}$ is another transmission coefficient. The ray-theoretical travel-time curve for $P4KP$ is shown in Figure 9.25b, and it is curious that observations of the phase (which are plentiful on short-period WWSSN instruments, so that the frequency content is high enough for one to expect classical ray theory to apply) are associated only with the later branch. Even more curious, the observations are predominantly at distances beyond the "end" of this later branch, i.e., beyond the cut-off point A, which marks the limit out to which geometrical rays can reach. The explanation of these phenomena can best be appreciated by examining $\grave{P}\grave{K} \cdot (\acute{K}\grave{K})^3 \cdot \acute{K}\acute{P}$ at values of p in the vicinity of saddle points corresponding to the two ray arrivals (i.e., travel-time branches) shown in Figure 9.25b. This product of coefficients is plotted in Figure 9.25c for frequencies $\frac{1}{2}$ Hz, 2 Hz, and for the limiting case as frequency tends to infinity. A critical ray-parameter value at 254.6 sec is apparent, beyond which the amplitude is zero in the limiting case. The critical value is the ray parameter p_d for a mantle ray that just grazes the core. However, at finite frequencies, energy *can* be transmitted along this generalized ray, even for $p > p_d$. This phenomenon is known as *tunneling* (after a phenomenon in quantum mechanics for which there is a similar theory), and is described further in Figure 9.25d. Numerically, $\grave{P}\grave{K}$ becomes small as p increases beyond p_d because of the factor $(\acute{\xi}_m + \grave{\xi}_m)\pi_m^{(2)}(r_{\text{CMB}})$. Using (9.62) and the Wronskian for $(\pi^{(1)}, \pi^{(2)})$, this factor is of order $[\pi_m^{(1)}(r_{\text{CMB}})]^{-1}$. Both $\pi^{(1)}$ and $\pi^{(2)}$ *grow* exponentially below the turning point, so that the factor we are examining must decay exponentially. Numerically, $\acute{K}\acute{P}$ also behaves like $[\pi_m^{(1)}(r_{\text{CMB}})]^{-1}$. However, $\acute{K}\grave{K}$ is of order 1. It just happens that, for this case of the core-mantle boundary, $\acute{K}\grave{K}$ has a zero very near to $p = 240$ sec (i.e., to the left of p_d), and $\acute{K}\grave{K}$ grows rapidly from zero as p increases near p_d. It is the cube of this reflection coefficient that we must evaluate, and the rapid growth of $(\acute{K}\grave{K})^3$ outweighs the decay in $\grave{P}\grave{K} \cdot \acute{K}\acute{P}$ for a short part of the range $p_d < p$, giving the broken and gray curves in Figure 9.25c, which actually have their maxima in the tunneling region. The choice of Γ for a computation of (9.67) can be taken as shown in Figure 9.25e. More simply, it can be taken as the real axis from $p \sim 240$ sec to $p \sim 265$ sec (because the zero in $\acute{K}\grave{K}$ gives a natural tapering on the left, and there is exponential tapering to the right, as shown in Figure 9.25c). Finally, shown in Figure 9.25f is the result of evaluating our line integral (9.67) for $P4KP$ at different distances. Indeed, this does show that the strongest arrivals, in the period range $\frac{1}{2}$–2 sec, do lie beyond the ray-theoretical cutoff point A. The effect is not small, because arrivals are seen, and are explained by the theory, at distances 20° (i.e., around 2000 km) into the "forbidden" region. Fuchs and Schulz (1976) have shown that tunneling also can be significant for body waves in crustal structures within which there is a high-velocity layer.

We next examine the wave *SKS*, and related multiples *SKKS*, *SKKKS*, etc., denoted in brief as *SmKS* for various integers m. These *SV*-waves are an excellent example of what can happen when waves interact with a discontinuity in which the faster medium lies below the interface. (The *S*-wave speed at the base of the mantle is less than the *P*-wave speed at the top of the core.) Choy (1977) has used the Langer approximation for these phases, integrating over complex p and also over real ω to give synthetic seismograms in the time domain.

For *SV*-waves, we must first look at a source other than the isotropic *P*-wave source we have so far considered in this chapter. A useful source for which to obtain explicit results is the shear dislocation with strike ϕ_s, rake λ, dip δ, and strength (i.e., moment) $\mu A\bar{u} = M_0(\omega)$. We obtained the radiation pattern for such a source, as well as the geometrical spreading formula, in (4.85) and (4.89). Our present goal is therefore to improve upon that approximation, for which the horizontal component is

$$u_\Delta^{SV}(r, \Delta_0, \phi, \omega) \sim \frac{\mathscr{F}^{SV}(-i\omega M_0(\omega))(-\cos j)\exp(i\omega T^S)}{4\pi[\rho_s\rho(r)\beta_s^5\beta(r)]^{1/2}\mathscr{R}^s(\mathbf{r}, \mathbf{r}_s)}. \tag{9.68}$$

(A factor $-\cos j(r)$ appears here, because (4.89) gives amplitude in the $\hat{\mathbf{p}}$-direction and we want the horizontal component in the direction of Δ increasing.) From our previous experience with *P*-waves, we know that

$$\frac{i\omega}{\alpha_s^4}\int_\Gamma \omega^2 p\pi^{(1)}(r_s)[\alpha(r)\xi(r)\pi^{(1)}(r)]Q^{(2)}_{\omega p-\frac{1}{2}}(\cos\Delta_0)\,dp$$

$$\sim \left[\frac{\rho_s}{\rho(r)\alpha_s^5(r)}\right]^{1/2}\frac{\cos i}{\mathscr{R}^P(\mathbf{r}, \mathbf{r}_s)}\exp(i\omega T^P).$$

Therefore, the corresponding result for *SV* allows us to write

$$u_\Delta^{\mathrm{inc}\,SV}(r, \Delta_0, \phi, \omega) = \frac{i\omega M_0(\omega)i\omega}{4\pi\rho_s\beta_s^4}$$

$$\times\int_\Gamma \omega^2 p\mathscr{F}^{SV}\sigma^{(1)}(r_s)[\beta(r)\acute{\eta}(r)\sigma^{(1)}(r)]Q^{(2)}_{\omega p-\frac{1}{2}}(\cos\Delta_0)\,dp. \tag{9.69}$$

This represents an *SV*-wave that is downgoing at the source and is received at (r, Δ_0) as an upgoing *SV*-wave. From this description alone, we could have written down most of the integrand factors in (9.69). The remaining factors simply give the normalization, and for this it is perfectly appropriate to use classical ray theory. Thus $\mathscr{F}^{SV}$ in (9.69) is given as a function of (p, ϕ) from (4.85):

$$\mathscr{F}^{SV}(p, \phi) \sim [\sin\lambda \cos 2\delta \sin(\phi - \phi_s) - \cos\lambda \cos\delta \cos(\phi - \phi_s)](1 - 2\beta_s^2 p^2/r_s^2) + [\cos\lambda \sin\delta \sin 2(\phi - \phi_s) - \sin\lambda \sin 2\delta(1 + \sin^2(\phi - \phi_s))] \frac{\beta_s^2 p}{r_s}\left(\frac{1}{\beta_s^2} - \frac{p^2}{r_s^2}\right)^{1/2}. \tag{9.70}$$

Since (9.70) originates from ray theory, it is only approximate. However, it is highly accurate for body waves when the turning point for the ray is well below the source level, and this of course is always the case for core phases. Although (9.70) substituted into (9.69) gives two different integrands, $\mathscr{F}^{SV}(p, \phi)$ may vary so slowly with p along those parts of Γ from which there is a numerically significant integrand that $\mathscr{F}^{SV}$ is effectively constant and can be taken outside the integration (see also Box 9.10).

If the wavelength of waves radiated from the source is not very much greater than the spatial dimensions of the rupture surface of the fault, then the finite extent of faulting can have a significant effect on seismic waves in the far-field. This subject is covered in Chapter 14, and the necessary correction to our present theory is given in Box 14.1.

In (9.69), we have obtained the representation of the "incident wave" from which *SmKS* waves are derived. Thus, for *SKS*,

$$u_\Delta^{SKS}(r, \Delta_0, \phi, \omega) = \frac{i\omega M_0(\omega) i\omega}{4\pi\rho_s\beta_s^4} \times \int_\Gamma \omega^2 p \mathscr{F}^{SV} \sigma^{(1)}(r_s) \grave{S}\grave{K} \cdot \acute{K}\acute{S}[\beta(r)\acute{\eta}(r)\sigma^{(1)}(r)] Q^{(2)}_{\omega p - \frac{1}{2}}(\cos\Delta_0)\, dp, \tag{9.71}$$

where Γ is shown in Figure 9.26b. Choy (1977) has studied (9.71) using the Earth model 1066B of Gilbert and Dziewonski (1975). For large enough Δ_0, the saddle point on the real p-axis, corresponding to *SKS*, can move to the left of the line of diffraction poles emanating from $p = p_d$. The effect of integrating along Γ is then to pick up the contribution from diffraction poles, as well as from the saddle. Thus (9.71) automatically contains the phase $SP_{\text{diff}}KS$ described in Figure 9.26a. Choy has carried out also the ω-integration of (9.71), and some of his results are shown in Figure 9.26c. Clearly, the diffraction as *P* along the base of the mantle can be significant, even after convolution with a long-period WWSSN seismometer.

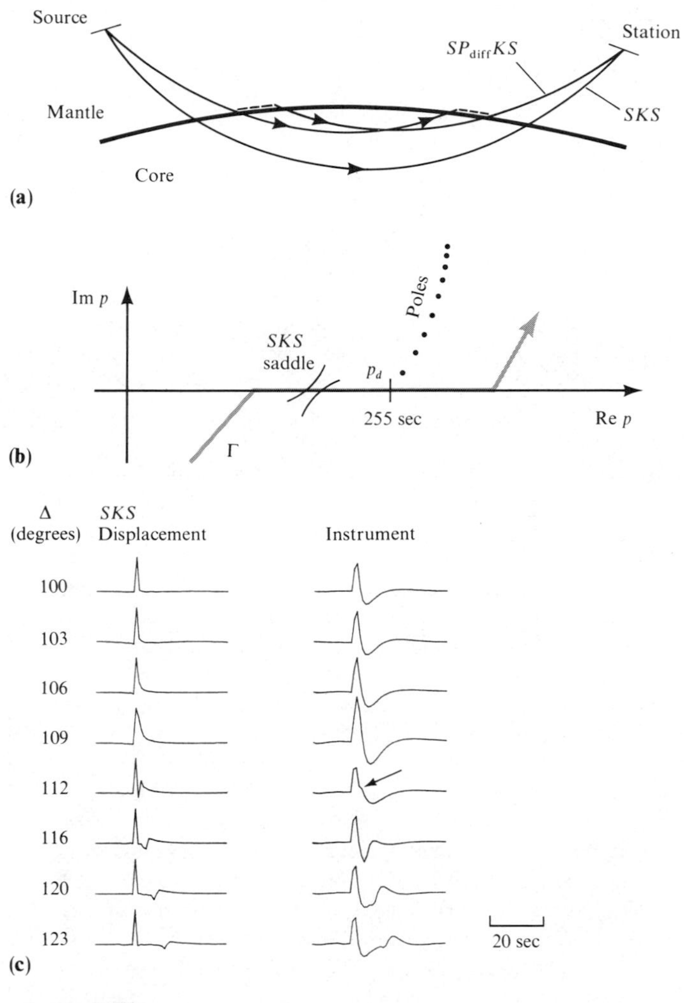

FIGURE **9.26**

If SKS is observed at distances beyond 100° then the ray is incident upon the core-mantle boundary with a slowness that allows some coupling to a P-wave diffracted around the base of the mantle. We label this phase as $SP_{\text{diff}}KS$, shown in (**a**). The diffracted part of the path can occur at either end of the K-leg. In (**b**) is shown the path of integration for (9.71), and the effect of $SP_{\text{diff}}KS$ will be apparent whenever the saddle for SKS lies to the left of diffraction poles emanating from p_d. (**c**) *Left.* The impulse response computed from (9.71) in Earth model 1066B of Gilbert and Dziewonski (1975). *Right.* The impulse response, after convolution with a long-period WWSSN instrument (15–100) (T_s = 15 sec, T_g = 100 sec; see Section 10.3.3). An arrow marks the emergence of $SP_{\text{diff}}KS$, and a strong effect of this phase can be seen in the coda of SKS beyond 112°. [After Choy, 1977.]

BOX **9.8**

Poles of scattering coefficients

The sixteen *P-SV* coefficients obtained in Box 9.6 all share the same factor (1/D), where

$$\mathsf{D} = (b\grave{\xi}_1 + c\acute{\xi}_2)(b\grave{\eta}_1 + c\acute{\eta}_2) + (a - d\grave{\xi}_1\acute{\eta}_2)(a - d\acute{\xi}_2\grave{\eta}_1)p^2/r_1^2.$$

Zeros of D (as p varies) are then poles of the scattering coefficients, and must receive special consideration when paths Γ are deformed in the complex p-plane.

In fact, D^{-1} has Stoneley poles on the real p-axis, just as we found for plane-wave theory (see discussion following (5.55)). More important, D^{-1} has strings of poles that have properties similar to the branch cuts we encountered in solving Lamb's problem. Fortunately, the locations of pole strings for D^{-1} are about the same as the locations of pole strings for the four quantities $\grave{\xi}_1, \acute{\xi}_2, \grave{\eta}_1, \acute{\eta}_2$ that appear in D. Our notation has been chosen to emphasize the similarity of these four quantities to the vertical slownesses $\xi = \alpha^{-1}\cos i$ and $\eta = \beta^{-1}\cos j$ that appear in plane-wave theory for homogeneous media. (The appearance of $\alpha^{-1}\cos i$ and $\beta^{-1}\cos i$ in plane-wave scattering coefficients (5.39) was due to vertical differentiation of wave functions $e^{\pm i\omega\xi z}$ and $e^{\pm i\omega\eta z}$. In (9.62) we have generalized this for waves with turning points, finding that downgoing and upgoing vertical slownesses must be distinguished.)

For real values of the ray parameter, $\grave{\xi}$ and $\acute{\xi}$ are complex conjugates. If $p < r/\alpha$, then WKBJ theory applied to (9.62) gives $\grave{\xi} \sim (1/\alpha^2 - p^2/r^2)^{1/2} \sim \acute{\xi}$. As p approaches r/α, Langer approximation (9.66) for $\grave{\xi}$ and $\acute{\xi}$ is necessary. For $r/\alpha < p$, $\pi^{(1)}$ and $\pi^{(2)}$ are exponentially large but $\pi^{(1)} + \pi^{(2)}$ is exponentially small. It follows that $\pi^{(1)} \sim -\pi^{(2)}$ and $\grave{\xi} \sim -\acute{\xi} \sim -i(p^2/r^2 - 1/\alpha^2)^{1/2}$.

More generally, it may be shown that $\grave{\xi}$ and $\acute{\xi}$ are analytic functions of p, with singularities located as shown in the first two figures. The strings depart from the real p-axis near $r/\alpha(r)$. The poles of $\acute{\xi}$ are the zeros of $H^{(1)}_{1/3}(\omega\tau)$ (see (9.66)), and they lie near lines (called Stokes' lines) given by requiring $\tau = \tau(r, p)$ to be real. Away from the singularities, $\acute{\xi} \sim (1/\alpha^2 - p^2/r^2)^{1/2}$. For example, this approximation would be accurate for lines such as AB shown in the figure below. As a line of singularities is approached, the approximation fails—and fails

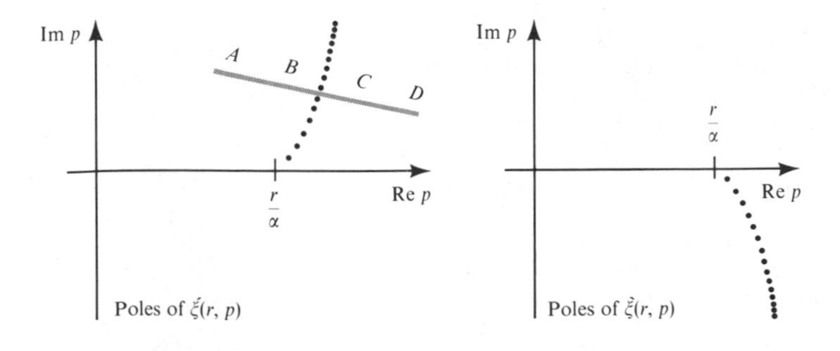

Cormier and Richards (1977) have shown that a good way to avoid the difficulty is to work directly with an expression that accounts for *all* the multiples taken together. The geophysical problem they studied was actually that of *P*-waves that have traversed both the outer fluid core and the inner solid core of the Earth, and the interference head wave here arises at the inner-core/outer-core interface (having radius r_1). In Figure 9.30a is shown the travel-time curve for *PKP*, also with branch *CD* for *PKiKP* (the reflection from r_1); branch *DF* for *PKIKP* (with a single transmission leg below r_1); and a branch for *PKIIKP*. A natural notation for the reflection coefficient in *PKiKP* is $\grave{K}\acute{K}$, with similar notation for transmissions across r_1. The whole family of rays shown in Figure 9.29 is therefore described by

$$\grave{K}\acute{K} + \grave{K}\grave{I} \cdot \acute{I}\acute{K} + \grave{K}\grave{I} \cdot \acute{I}\grave{I} \cdot \acute{I}\acute{K} + \cdots = \grave{K}\acute{K} + \frac{\grave{K}\grave{I} \cdot \acute{I}\acute{K}}{1 - \acute{I}\grave{I}}. \tag{9.72}$$

The summation of this geometric series is similar to a problem of multiples we described earlier in plane-wave theory, (9.18)–(9.19).

In the derivation of each of $\grave{K}\acute{K}$, $\grave{K}\grave{I}$, $\acute{I}\acute{K}$, $\acute{I}\grave{I}$, it has been assumed that the energy scattered into the lower medium is only downgoing energy. In fact, from Box 9.6 with a liquid layer 2 above solid layer 1,

$$\grave{K}\acute{K} = \frac{[\rho_1(1 - 2\beta_1^2 p^2/r_1^2)^2 \xi_2 - \rho_2 \grave{\xi}_1 + 4\rho_1\beta_1^4(p^2/r_1^2)\grave{\xi}_2\hat{\eta}_1\grave{\xi}_1]\pi_2^{(2)}}{[\rho_1(1 - 2\beta_1^2 p^2/r_1^2)^2 \acute{\xi}_2 + \rho_2 \grave{\xi}_1 + 4\rho_1\beta_1^4(p^2/r_1^2)\acute{\xi}_2\hat{\eta}_1\grave{\xi}_1]\pi_2^{(1)}}. \tag{9.73}$$

The *P*-wave wavefunctions for the solid core appear here only via $\grave{\xi}_1$, and indeed this *is* the appropriate vertical slowness for a downgoing wave. But, as

FIGURE **9.30**

Generalized *PKP*. (**a**) The travel-time curve for *PKP* and associated reflections and transmissions at the inner-core/outer-core boundary. The numerical integration in (9.77) is handled by separating the *PKiKP* contribution from a single term describing the interference wave. Thus $\grave{K}(\check{I})\acute{K}$ in (9.77) is broken into two terms via (9.76). (**b**) With $\grave{K}\acute{K}$ in (9.77), the integration is conducted over the path Γ_1. (**c**) With $(\grave{K}\grave{I} \cdot \acute{I}\acute{K})/(1 - \acute{I}\grave{I})$ in (9.77), the integration is conducted over Γ_2. In each case, there are poles trending down into the fourth quadrant (p_c is the critical ray parameter, corresponding to rays with turning point at the top of the solid core; p_g is the ray parameter for the grazing ray, with turning point at the bottom of the fluid core). (**d**) For the remainder of the path, the two integrands are combined together into one term, proportional to $\grave{K}(\check{I})\acute{K}$. Poles associated with $\hat{\xi}_1$ have been replaced by poles associated with $\check{\xi}_1$. To the right of p_c, $\grave{K}(\check{I})\acute{K}$ tends to $+1$, hence (9.77) quantifies the *AB* and *BC* branches of *PKP*.

can be seen in Figure 9.29, this downgoing wave ($\pi_1^{(2)}$) is soon turned around to an upgoing wave ($\pi_1^{(1)}$), because the turning point is not far below the interface. Therefore, for a P-wave incident downward from the fluid core, it is appropriate instead to use $\pi_1^{(1)} + \pi_1^{(2)}$ for the wavefunction in the lower medium, since this is the wavefunction that correctly describes amplitudes throughout the solid core. The reflection coefficient $\grave{K}\acute{K}$ is modified if this new wavefunction in the solid core is used, and a natural notation is $\grave{K}(\check{I})\acute{K}$. Satisfying boundary conditions at the interface, we find

$$\grave{K}(\check{I})\acute{K} = \frac{[\rho_1(1 - 2\beta_1^2 p^2/r_1^2)^2 \grave{\xi}_2 - \rho_2 \check{\xi}_1 + 4\rho_1 \beta_1^4 (p^2/r_1^2) \grave{\xi}_2 \grave{\eta}_1 \check{\xi}_1]\pi_2^{(2)}}{[\rho_1(1 - 2\beta_1^2 p^2/r_1^2)^2 \acute{\xi}_2 + \rho_2 \check{\xi}_1 + 4\rho_1 \beta_1^4 (p^2/r_1^2) \acute{\xi}_2 \grave{\eta}_1 \check{\xi}_1]\pi_2^{(1)}}, \quad (9.74)$$

where

$$\check{\xi}_1 \equiv \frac{1}{-i\omega(\pi_1^{(1)} + \pi_1^{(2)})} \frac{d}{dr} (\pi_1^{(1)} + \pi_1^{(2)}). \quad (9.75)$$

Our new reflection coefficient, $\grave{K}(\check{I})\acute{K}$, has two important properties. First, it can be shown after some algebra that

$$\grave{K}(\check{I})\acute{K} = \grave{K}\acute{K} + \frac{\grave{K}\grave{I} \cdot \acute{I}\acute{K}}{1 - \acute{I}\grave{I}}, \quad (9.76)$$

and hence this single coefficient accounts for all the rays present in Figure 9.29. The expansion of $\grave{K}(\check{I})\acute{K}$ into the left-hand terms of (9.72) is an example of the *rainbow expansion* (sometimes called the *Debye expansion*), so named because of a similar expansion arising in the theory of the rainbow. (In fact, the refractive index of water in air is similar to that of the core relative to the mantle, and the seismological analogue of the primary rainbow is the caustic in *PKKP*.) It is usually appropriate to make the rainbow expansion when the medium below the interface is slower than that above, because the multiples then set up in the lower medium will not interfere with each other. An example is that of *PcP* + *PKP* + *PKKP* + $\cdots$, and each of the transmitted waves can be studied separately. However, it is often inappropriate to make the expansion when the medium below is faster.

The second important property of $\grave{K}(\check{I})\acute{K}$ concerns the position of poles in the complex p-plane. Previously, for $\grave{K}\acute{K}$, there was a system of poles associated with $\grave{\xi}_1$, and therefore trending down into the fourth quadrant (Box 9.8). But the new ratio defining $\check{\xi}_1$ has poles and zeros stringing just above the real p-axis, between 0 and $r_1/\alpha_1(r_1)$, and it is here that poles of $\grave{K}(\check{I})\acute{K}$ are found.

The representation of P-waves originating from a point shear dislocation with moment $M_0(\omega)$ and transmitted through the inner core can now be written as

$$u_r^{PKP}(r, \Delta_0, \phi, \omega) = \frac{-i\omega M_0(\omega) i\omega}{4\pi\rho_s\alpha_s^4}$$

$$\times \int_\Gamma \omega^2 p \mathscr{F}^P \pi^{(1)}(r_s) \grave{P}\grave{K} \cdot \grave{K}(\check{I})\acute{K} \cdot \acute{K}\acute{P}[\alpha(r)\xi(r)\pi^{(1)}(r)] Q^{(2)}_{\omega p - \frac{1}{2}}(\cos \Delta_0)\, dp. \quad (9.77)$$

The radiation pattern here is derived from (4.84), and $\mathscr{F}^P$ depends upon the strike ϕ_s, dip δ, and rake λ via

$$\begin{aligned}\mathscr{F}^P &= \sin\lambda \sin 2\delta + [\cos\lambda \sin\delta \sin 2(\phi - \phi_s) \\ &\quad - \sin\lambda \sin 2\delta(1 + \sin^2(\phi - \phi_s))]\alpha_s^2 p^2/r_s^2 + 2[\sin\lambda \cos 2\delta \sin(\phi - \phi_s) \\ &\quad - \cos\lambda \cos\delta \cos(\phi - \phi_s)] \frac{\alpha_s^2 p}{r_s}\left(\frac{1}{\alpha_s^2} - \frac{p^2}{r_s^2}\right)^{1/2}. \end{aligned} \quad (9.78)$$

It is correct to designate (9.77) as the generalized PKP-wave, because it accounts for *all* the compressional phases generated by a P-wave transmitted downward once and upward once at the core-mantle interface. For p such that the wave has a turning point well above the inner core, the factor $\grave{K}(\check{I})\acute{K}$ in (9.77) is approximately $+1$, and the effect of the inner core is negligible. Even the caustic in PKP near 142° is accounted for. At lower values of p, the interference head wave is correctly quantified, provided the integrand is broken up into two parts and integrated along separate paths, as shown in Figure 9.30b,c,d.

Using a representation equivalent to (9.77), Cormier and Richards (1977) have computed PKP in the parameterized Earth model of Dziewonski *et al.* (1975). Their results are shown in Figure 9.31 and compared with observations in Figure 9.32.

It is clear that integrands like (9.77) can be built up for more and more complicated body waves. Even the effect of crustal and upper mantle layering can be accounted for, because the propagator matrix corresponding to (9.58)–(9.59) can be written in closed form. In this way, it is possible to compute leaking modes, together with other phases that arrive within the same time window. Consider, for example, what happens when an SV-wave is incident from below upon the crust-mantle interface at a ray parameter in the vinicity of 770 sec/radian. This is near the critical value at which P is propagating horizontally just beneath the Moho, and a variety of possible ray arrivals for a receiver on

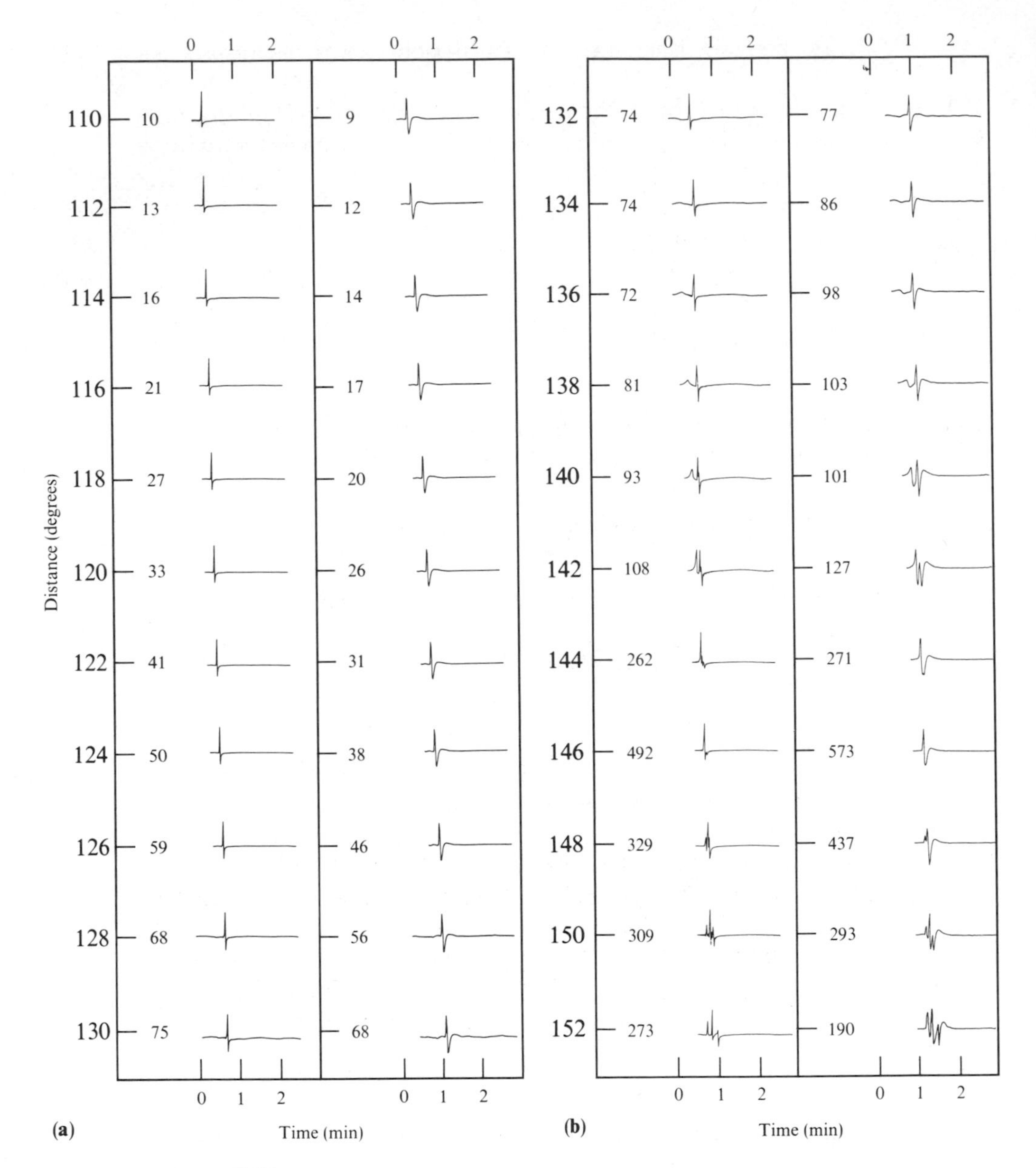

FIGURE **9.31**

Left in (**a**) and (**b**). Impulse response for *PKP*, computed via (9.77) in the parameterized Earth model (PEM) of Dziewonski *et al.* (1975). *Right* in (**a**) and (**b**). Impulse response convolved with the response of a 15–100 (long-period) WWSSN vertical instrument. Numbers at the left of responses indicate relative amplitude in a sequence of increasing distances. Note that the waveform does not change across the critical point near 120°. The position of the critical point merely affects amplitudes. A long-period precursive phase is apparent in 134°–140°, and is associated with the caustic in *PKP* (point *B*, Fig. 9.30a) at 143°. The complicated pulses in 138°–152° are a result of interference between all the arrivals shown in Figure 9.30a. [After Cormier and Richards, 1977.]

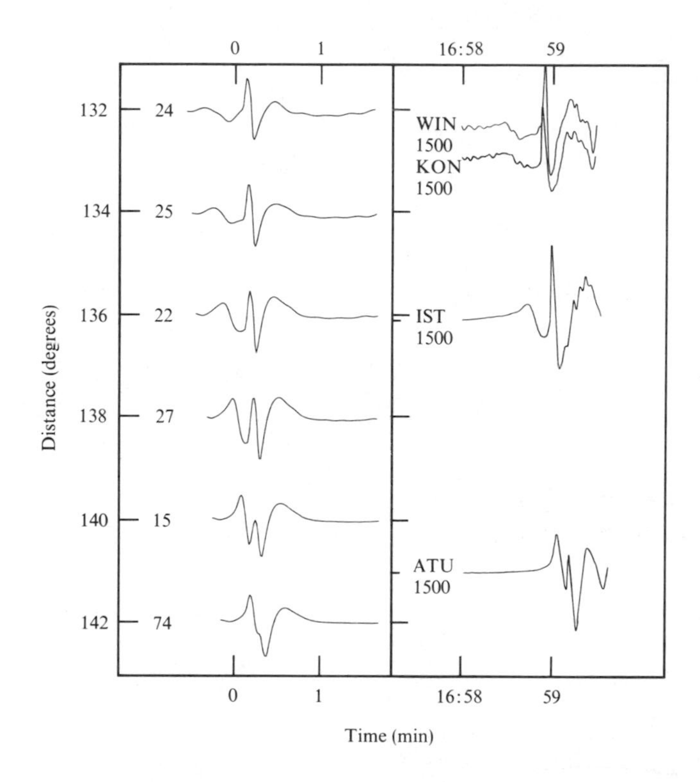

FIGURE **9.32**
Comparison between synthetic seismograms (shown at left), and observed (shown at right) for an earthquake on 9 July 1964. [After Cormier and Richards, 1977.]

the surface are shown in Figure 9.33, including a whispering gallery effect. It is also known that the crust can support leaking modes, called *PL*-modes, in which the *P*-wave energy below the Moho decays exponentially with depth but *SV*-wave energy is radiated downward. The corresponding phase slowness (ray parameter) of the fundamental is only slightly greater than that of SP_{diff} (see Fig. 9.33e), so that this leaking mode can also be excited by the incident *SV*-wave of Figure 9.33. It has been labelled as shear-coupled *PL* by Oliver (1961), and prominent examples of this phase, together with a discussion of how it may efficiently be excited, are given by Chander *et al.* (1968) and Poupinet and Wright (1972). Frazer (1977) has shown how a single integrand can be developed to describe all the phases of Figure 9.33, plus multiples in the crust and shear-coupled *PL*-waves, and he has integrated over real p and real ω to obtain synthetic record sections, such as the one given in Figure 9.34. To handle the waves scattered into the mantle from a boundary condition imposed at the Moho, he used the wavefunction $\sigma^{(2)}$ for downgoing *SV*, but $\pi^{(1)} + \pi^{(2)}$ for *P*. This is similar to the way in which a whispering gallery phenomenon was handled in Figures 9.29–9.32. He allowed for reverberations in the crust by using the propagator matrix between Moho and surface, and imposed free surface boundary conditions. Frazer was able to show that prominent shear-coupled *PL*-waves require a favorably oriented source and a mantle with

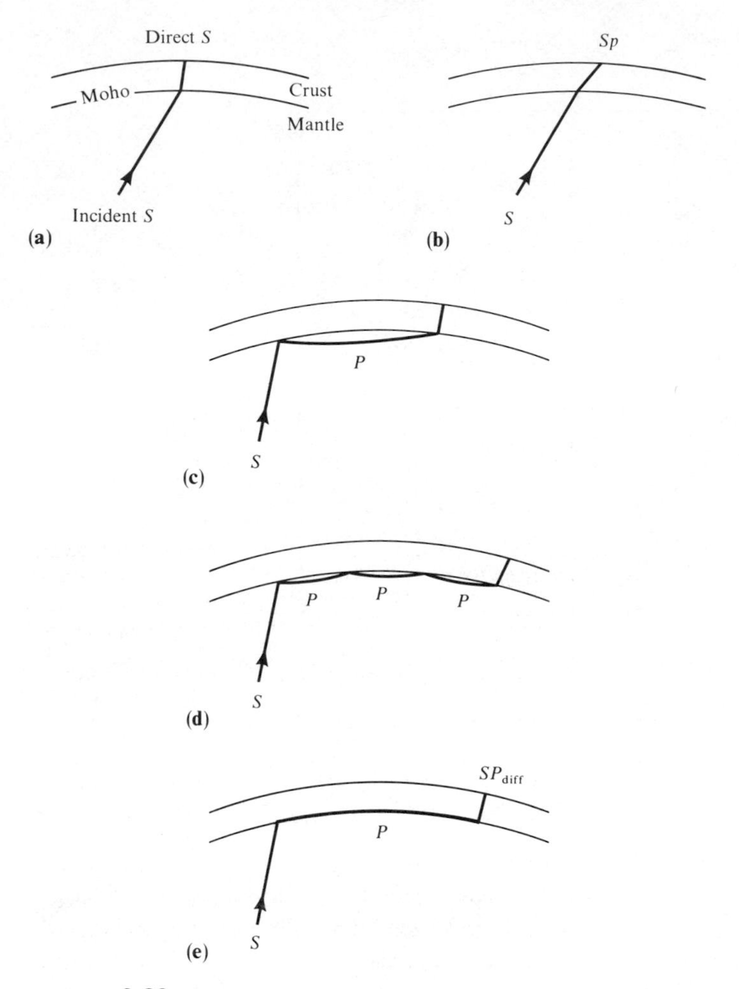

FIGURE **9.33**

Ray paths associated with SV incident upon the Moho at an angle close to that which generates horizontally propagating P at the top of the mantle. (**a**) The direct S-wave (SV in the crust). (**b**) Traveling as P in the crust, this arrival is denoted by Sp. It arrives a little earlier than the direct S-wave, and has been described in detail by Jordan and Frazer (1975). (**c**) A single leg as P in the mantle is included (the crustal path could be P or S). (**d**) Three mantle P-wave legs. (**e**) The number of mantle P-wave legs can be increased indefinitely, until in the limit (at a critical value of ray parameter) the ray path is horizontal. Frazer (1977) labels the associated arrival at the surface SP_{diff}. It contains superposed P- and S-rays through the crust, and, in practice, includes the whole family of rays such as (c) and (d), forming an interference head wave. For a given point source and a given receiver, all the ray paths (a)–(e) may be present together with reverberations in the crust, though each particular ray will in general have a slightly different ray parameter.

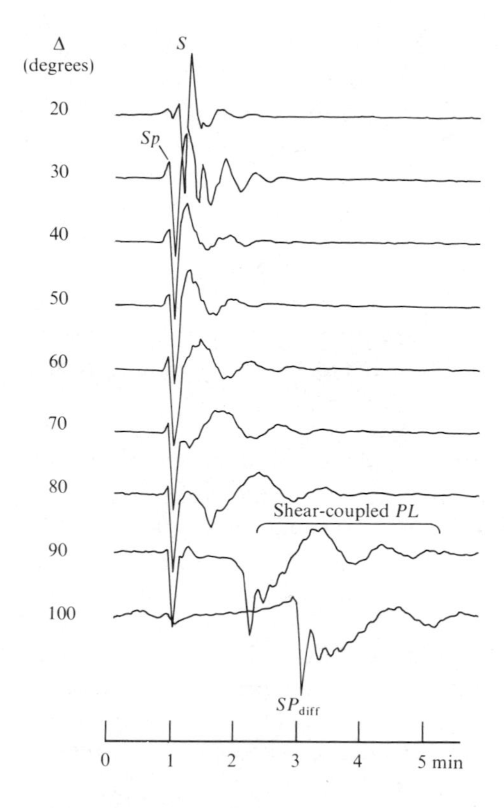

FIGURE **9.34**

A record section showing component phases that make up the complex SV-arrival for a receiver on the free upper surface of a crustal layer. The main SV-path in the mantle has a single turning point between source and receiver. The P-velocity is taken as constant just below the Moho, generating a very effective whispering gallery, as seen by the large SP_{diff} arrival. It is the horizontal component of motion that is computed here, and the traces at different distances Δ have been time-shifted to align the first arrival. [After Frazer, 1977.]

anomalously high shear-velocity gradient near 650 km depth. Because seismograms in practice do not contain SP_{diff} phases, such as those in the synthetic records of Figure 9.34, Frazer concluded that the P-velocity must decrease with depth below the Moho. If it decreases sufficiently rapidly ($d\alpha/dr > \alpha/r$), then a horizontally traveling ray is refracted downward: the whispering gallery of Figure 9.33 is absent, and $\pi^{(2)}$ alone is needed for the scattered P-wave below the Moho.

9.6 Comparison Between Different Methods

Although different parts of the theory we have presented here have been known for some time, it is only in the past few years that the effort to compute accurate body-wave pulse shapes in realistic structures has begun to be more successful than classical ray theory. We have found it convenient to distinguish three computational methods: (i) Cagniard inversion of generalized rays that do not have turning points, (ii) double integration (over real slowness and real frequency) retaining all multiples, and (iii) double integration (over real or complex slowness and real frequency) for generalized rays (possibly with turning points) in inhomogeneous layers. From the third method we have derived the WKBJ seismogram, and in practice this solution shares some of the features of Cagniard inversion. Although these are probably the main methods now being used in seismology, they may in some applications be superseded either by the direct sum of normal modes (as described in Chapter 8) or by the method of Alekseyev and Mikhaylenko (1976). In the latter method, a spatial horizontal transform is applied to the equations of motion, leaving a partial differential equation in depth and time alone. This is solved by a finite-difference scheme for many values of horizontal wavenumber, and then the inverse spatial transform (a numerical integral over horizontal wavenumber) is computed to give synthetic record sections.

It should be recognized that the theory of wave propagation in vertically inhomogeneous media is still a research field, and it is not yet clear which of the main methods we have emphasized is appropriate for a given problem. Furthermore, in each of the methods, the user must make some choices in the details of how a particular method is executed. In the remainder of this chapter, we shall briefly state some advantages and disadvantages of each method and the choices that a user must make in working with them.

First, we note that the methods of Cagniard and Fuchs have been developed for plane homogeneous layers. Their application in a spherical Earth requires some preliminary discussion, given in Box 9.9. For each of these two methods, a major choice must be made in the way a given inhomogeneous Earth model is approximated by a layered model. Coarser layering can substantially decrease computation costs, but at the expense of accuracy. The decision here, of what layer thickness is appropriate, will of course depend upon the frequency range for which accurate results are required.

For broad-band pulse shapes, Cagniard's method and the WKBJ seismogram have inherent advantages because the response (apart from convolutions) is calculated in the time domain. When the pulse amplitude is rapidly changing with time, details of this change can easily be studied by sampling along the Cagniard path at points that are closely spaced.

If pulse shapes are to be computed at many distances, then Fuch's and Scholte's methods have an inherent advantage, since the same path of integration can serve for many distances, whereas the Cagniard path (and the equivalent path for the WKBJ seismogram) must be newly determined for each distance. (It should be recognized that, by "Cagniard's method" we are referring to a technique that has benefitted from the contributions of many individuals. Computational details are given, for example, by Wiggins and Helmberger (1974). Fuch's reflectivity method also has been extended from its original formulation. For example, Kind (1978) modified the method to allow it to handle reverberations in layers both above and below the source. He was also able to incorporate the effects of attenuation, and surface waves and leaking

BOX **9.9**

Earth-flattening transformations and approximations

In Sections 7.2 and 9.2, we showed how to obtain and use the motion-stress vector

$$\begin{pmatrix} u_\phi \\ \tau_{\phi z} \end{pmatrix}$$

that solves

$$\frac{d}{dz}\begin{pmatrix} u_\phi \\ \tau_{\phi z} \end{pmatrix} = \begin{pmatrix} 0 & 1/\mu(z) \\ k^2\mu(z) - \omega^2\rho(z) & 0 \end{pmatrix}\begin{pmatrix} u_\phi \\ \tau_{\phi z} \end{pmatrix}. \tag{1}$$

This is the SH-wave equation, and matrix methods for its solution are highly developed, particularly when the medium is approximated by large numbers of homogeneous plane layers. For example, the reflectivity method makes this approximation.

Our present interest, however, lies in

$$\frac{d}{dr}\begin{pmatrix} W \\ \\ T \end{pmatrix} = \begin{pmatrix} \dfrac{1}{r} & \dfrac{1}{\mu} \\ \\ \dfrac{\mu(l-1)(l+2)}{r^2} - \omega^2\rho & -\dfrac{3}{r} \end{pmatrix}\begin{pmatrix} W \\ \\ T \end{pmatrix}, \tag{2}$$

which is the equivalent of (1) in spherical geometry. We are also interested in the corresponding P-SV equations. Fortunately, Andrianova et al. (1967) and Biswas and Knopoff (1970) have found a transformation of (2) that gives, exactly, the equation (1). Therefore, the reflectivity method can be applied in spherical geometry with no approximations (other than the replacement of depth-dependent $\rho(z)$, $\mu(z)$ by many thin layers and the truncation of integration ranges $\int_0^\infty dk \int_0^\infty d\omega$). For P-SV problems, however, the transformation between the spherical wave equation and the Cartesian wave equation cannot be made exact, and new approximations are introduced.

First we shall describe the SH transformation. It is convenient to use a subscript s for quantities relevant to the spherical medium, such as μ_s, ρ_s, β_s, and a subscript f for quantities in the flat-layered (i.e., z-dependent) medium. Given profiles μ_s, ρ_s, and an order number l, we want to transform the equation (2) for

$$\begin{pmatrix} W \\ T \end{pmatrix}$$

into the equation (1) for

$$\begin{pmatrix} u_\phi \\ \tau_{\phi z} \end{pmatrix}.$$

The new depth variable z is defined by

$$e^{-z/a} = \frac{r}{a}, \tag{3}$$

modes. By using the solution (9.58) and Box 9.6, it is possible to extend Fuch's method further to handle all multiples within *inhomogeneous* layers. By "Scholte's method" we are loosely referring to the whole range of methods summarized in Section 9.5, based on *complex* ray-parameter and real-frequency integration. Where this method improves on WKBJ approximations, it is sometimes referred to as a "full-wave theory.")

If attenuation is strong, in the sense that $Q \gg 1$ but waves are attenuated by propagating over many wavelengths, then the theory for Fuch's and Scholte's methods is virtually unchanged. The velocity profile itself is defined to be complex (see (5.88)). However, Cagniard's method and the WKBJ seismogram cannot readily accommodate causal attenuation, except by *ad hoc* methods based on classical ray theory.

The various methods differ most strongly in their treatment of problems in which the wave to be synthesized has been influenced by Earth structure in the vicinity of a turning point. Just this kind of body wave is likely to be of great geophysical interest, because anomalous features in a velocity profile are likely to have their strongest effect on body waves with ray paths that bottom near the anomalous structure. To put this another way: the waves that contain maximum information about some velocity anomaly are the waves that spend the longest time in the vicinity of the anomaly. They will therefore be the waves that travel horizontally at the depth of the anomaly (i.e., have their turning point there). Cagniard's and Fuchs' methods differ for these problems, because the first is based on generalized rays (and it is impractical to retain more than a few hundred), whereas the second contains all multiples within a prescribed range of depths and within a prescribed range of slowness ($dT/d\Delta$). These methods are compared in Figure 9.9.

With Scholte's method, turning-point computations are treated no differently from problems of narrow-angle reflection. However, it is important to be able to compute vertical wavefunctions accurately. It is this fact that renders the WKBJ seismogram inaccurate for problems on grazing incidence. Whether the Langer approximation is used or a numerical integration of the equations of motion, the precise way in which velocities are interpolated (in the Earth model) must have some influence on the computed pulse shapes. Earth models are most commonly tabulated as values of density and P- and S-wave speeds at different depths, and geophysicists have no convention about how to obtain values at intermediate levels. Linear interpolation is common, but so is the law ar^b (with a and b being constants in a given layer). The Earth-flattening transformation, together with the stack of homogeneous plane layers used in Cagniard's and Fuchs' methods, is another form of interpolation (see the figure in Box 9.9). It will probably be impossible to decide which of the methods of computing body-wave seismograms is most appropriate for a particular problem until the effect of different interpolation laws in the underlying Earth model is thoroughly quantified.

BOX **9.10**

The moment tensor and generalized rays

For a seismic point source described objectively by a second-order moment tensor, we have given specific results about how the surface waves and normal modes of the Earth will be excited. We give here the corresponding excitation formulas for body waves.

We made a preliminary attempt at finding the body-wave excitation when we obtained the radiation patterns in (4.91) based on geometrical ray theory. In terms of the take-off angle i_ξ (for P-waves) or j_ξ (for S-waves), which describes the direction of departure of the ray from the hypocenter, the key factors for P-, SV-, and SH-waves can be written out as

$$(P)\qquad \boldsymbol{\gamma}\cdot\dot{\mathbf{M}}\cdot\boldsymbol{\gamma} = \sin^2 i_\xi[\cos^2\phi\dot{M}_{xx} + \sin 2\phi\dot{M}_{xy} + \sin^2\phi\dot{M}_{yy} - \dot{M}_{zz}] + 2\sin i_\xi\cos i_\xi[\cos\phi\dot{M}_{xz} + \sin\phi\dot{M}_{yz}] + [\dot{M}_{zz}], \tag{1}$$

$$(SV)\qquad \hat{\mathbf{p}}\cdot\dot{\mathbf{M}}\cdot\boldsymbol{\gamma} = \sin j_\xi\cos j_\xi\,[\cos^2\phi\dot{M}_{xx} + \sin 2\phi\dot{M}_{xy} + \sin^2\phi\dot{M}_{yy} - \dot{M}_{zz}] + (1 - 2\sin^2 j_\xi)[\cos\phi\dot{M}_{xz} + \sin\phi\dot{M}_{yz}], \tag{2}$$

$$(SH)\qquad \hat{\boldsymbol{\phi}}\cdot\dot{\mathbf{M}}\cdot\boldsymbol{\gamma} = \sin j_\xi[\tfrac{1}{2}\sin 2\phi(\dot{M}_{yy} - \dot{M}_{xx}) + \cos 2\phi\dot{M}_{xy}] + \cos j_\xi[\cos\phi\dot{M}_{yz} - \sin\phi\dot{M}_{xz}]. \tag{3}$$

These formulas use the notation of Chapter 4, and they have been arranged here to display the dependence on certain combinations of moment-tensor elements. When setting up a theory for generalized rays in which the body-wave solution associated with a particular ray path is given by an integral over ray parameter, then one of the expressions (1), (2), or (3) will enter as a factor in the integrand. This will be true for the Cagniard method, for the reflectivity method as given in Section 9.2, for the WKBJ seismogram, and also (with possible modifications, described below) for the full-wave-theory methods described in Section 9.6. Common to each of these methods is a decision about whether the ray path of interest is one that departs upward or downward from the source. To illustrate the procedure for general **M**, we shall use a full-wave theory for SH and look at $S + ScS$ waves (becoming S_{diff} in the shadow of the Earth's core). Thus, in the frequency domain, we can write

$$u_\phi^{S+ScS}(r, \Delta_0, \phi, \omega) = \frac{-i\omega\cdot i\omega}{4\pi\rho_s\beta_s^4}\int_\Gamma \omega^2 p(\hat{\boldsymbol{\phi}}\cdot\mathbf{M}\cdot\boldsymbol{\gamma})\sigma^{(1)}(r_s)\grave{S}\acute{S}\sigma^{(1)}(r)Q^{(2)}_{\omega p-\frac{1}{2}}(\cos\Delta_0)\,dp. \tag{4}$$

This expression can be written down with an eye on (9.64) or (9.69). $\grave{S}\acute{S}$ is the ScS reflection coefficient (for SH), and Γ is a path very similar to that shown in Figure 9.23c. Our interest here is in the source-dependent factor in (4), and to obtain its dependence on ray parameter we note that (1)–(3) can be written as

$$(P)\qquad \boldsymbol{\gamma}\cdot\dot{\mathbf{M}}\cdot\boldsymbol{\gamma} = (p\alpha_s/r_s)^2[\cos^2\phi\dot{M}_{xx} + \sin 2\phi\dot{M}_{xy} + \sin^2\phi\dot{M}_{yy} - \dot{M}_{zz}] + 2(p\alpha_s^2/r_s)\varepsilon\xi_s[\cos\phi\dot{M}_{xz} + \sin\phi\dot{M}_{yz}] + [\dot{M}_{zz}], \tag{5}$$

$$(SV)\qquad \hat{\mathbf{p}}\cdot\dot{\mathbf{M}}\cdot\boldsymbol{\gamma} = (p\beta_s^2/r_s)\varepsilon\eta_s[\cos^2\phi\dot{M}_{xx} + \sin 2\phi\dot{M}_{xy} + \sin^2\phi\dot{M}_{yy} - \dot{M}_{zz}] + (1 - 2\beta_s^2p^2/r_s^2)[\cos\phi\dot{M}_{xz} + \sin\phi\dot{M}_{yz}], \tag{6}$$

$$(SH)\qquad \hat{\boldsymbol{\phi}}\cdot\dot{\mathbf{M}}\cdot\boldsymbol{\gamma} = (p\beta_s/r_s)[\tfrac{1}{2}\sin 2\phi(\dot{M}_{yy} - \dot{M}_{xx}) + \cos 2\phi\dot{M}_{xy}] + \beta_s\varepsilon\eta_s[\cos\phi\dot{M}_{yz} - \sin\phi\dot{M}_{xz}]. \tag{7}$$

The symbols ξ_s, η_s denote vertical slowness (e.g., $\xi = \alpha^{-1} \cos i = (\alpha^{-2} - p^2/r^2)^{1/2}$) evaluated at the source, and ε is $+1$ or -1 according as the ray is downgoing or upgoing at the source. Substituting from (7) into (4), we find that the following scheme is convenient for computation.

Compute and store two fundamental solutions,

$$\begin{aligned} SH_1(\Delta_0, \omega) &= \int_\Gamma \sin j_\xi f(p, \omega)\, dp, \\ SH_2(\Delta_0, \omega) &= \int_\Gamma \cos j_\xi f(p, \omega)\, dp, \end{aligned} \tag{8}$$

where for our present problem $f(p, \omega) = i\omega(4\pi\rho_s\beta_s^4)^{-1}\omega^2 p\sigma^{(1)}(r_s)\grave{S}\acute{S}\sigma^{(1)}(r)Q^{(2)}_{\omega p - \frac{1}{2}}(\cos \Delta_0)$, $\sin j_\xi = p\beta_s/r_s$, and $\cos j_\xi = \beta_s\eta_s$. To generate the record section, given any moment rate tensor $\dot{\mathbf{M}}$, we can multiply the spectrum of $[\frac{1}{2} \sin 2\phi(\dot{M}_{yy} - \dot{M}_{xx}) + \cos 2\phi\dot{M}_{xy}]$ by $SH_1(\Delta_0, \omega)$, and the spectrum of $[\cos \phi\dot{M}_{yz} - \sin \phi\dot{M}_{xz}]$ by $SH_2(\Delta_0, \omega)$, add, and invert to the time domain by integrating over real frequencies. This exercise can be done very quickly. We can set up an even simpler numerical procedure if we decide in advance that $\dot{\mathbf{M}}$ is appropriate for a shear dislocation with a certain strike, dip, and rake, and a certain time function for $M_0(t) = \mu A\bar{u}$. In this case we form $\dot{M}_0(t) * SH_1(\Delta_0, t)$ and $\dot{M}_0(t) * SH_2(\Delta_0, t)$. These two record sections are shown on the opposite page after a recursive filter has been applied to give the effect of a typical long-period seismometer. The virtue of this approach is that the seismogram for a given distance (Δ_0) is given in terms merely of a *linear sum* of the two seismograms shown at that distance in the figure. Coefficients in the linear sum are a function of strike, dip, rake, and azimuth (ϕ) from source to receiver, so that it is easy to test for the effect of a variety of fault-plane solutions. The two coefficients are given in detail by $[\frac{1}{2}\sin 2\phi(M_{yy} - M_{xx}) + \cos 2\phi M_{xy}]/M_0$ and $[\cos \phi M_{yz} - \sin \phi M_{xz}]/M_0$, with components for $\mathbf{M}$ as in Box 4.4.

If, instead of SH, we consider a P-wave departing from the source, then it is necessary to store *three* fundamental types of solution:

$$\begin{aligned} P_1(\Delta_0, \omega) &= \int_\Gamma \sin^2 i_\xi f(p, \omega)\, dp, \\ P_2(\Delta_0, \omega) &= \int_\Gamma 2 \sin i_\xi \cos i_\xi f(p, \omega)\, dp, \\ P_3(\Delta_0, \omega) &= \int_\Gamma f(p, \omega)\, dp. \end{aligned}$$

If the source is an explosion, only the P_3 solution is needed. For an SV-wave leaving the source, we must in general work with *two* fundamental solutions,

$$\begin{aligned} SV_1(\Delta_0, \omega) &= \int_\Gamma \sin j_\xi \cos j_\xi f(p, \omega)\, dp, \\ SV_2(\Delta_0, \omega) &= \int_\Gamma (1 - 2 \sin^2 j_\xi) f(p, \omega)\, dp. \end{aligned}$$

Underlying the discussion we have just given is the implicit assumption that WKBJ approximations are accurate for vertical wavefunctions evaluated at the source level, r_s. This assumption would have been plain if we had given a formal derivation of, for example,

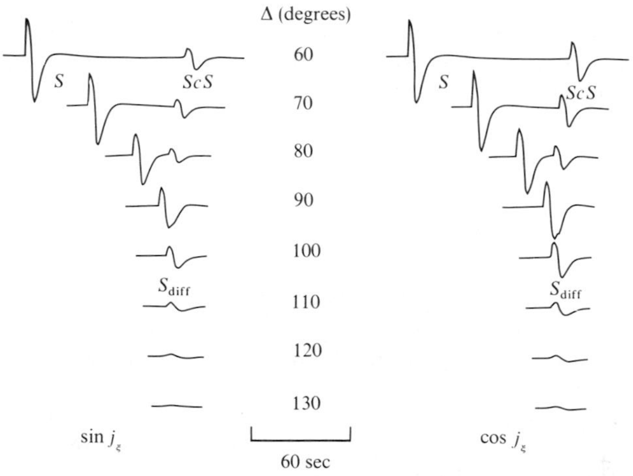

Two record sections for $S + ScS$ waves (polarized as SH). Computation is for the PEM-C Earth model (Dziewsonski et al., 1975), with allowance for a 15–100 long-period seismometer and a moment $M_0(t)$ acting as a linear ramp of 4-sec duration. The record section at the left uses a dependence on $\sin j_\xi$ for the integrand; on the right the dependence is on $\cos j_\xi$. Note that the ScS/S amplitude ratio is larger for $\cos j_\xi$ dependence than for $\sin j_\xi$. The reason for this is that the saddle for ScS occurs at a significantly lower ray parameter than does the saddle for S. There is a corresponding effect on $\sin j_\xi$, but $\cos j_\xi$ is relatively insensitive to saddle-point position.

(4). This would entail the usual procedure of writing the source equivalent to $\mathbf{M}$ in terms of a discontinuity in displacement and traction components (Hudson, 1969) and evaluating the excitation of radiated waves via transforms over frequency, horizontal distance, and azimuth (Section 7.4). Where vertical differentiation is necessary at r_s, the result in the transform plane will (if WKBJ methods apply) be like multiplication by plus or minus $i\omega\xi_s$ for P and $i\omega\eta_s$ for S. This arises because vertical wavefunctions have such phase factors as $\exp(\pm i\omega\int\xi\, dr)$. Clearly, then, we took a major short cut when we avoided the rigorous derivation of (4) and instead wrote this expression using the form that WKBJ theory would give us if it also were to reproduce the radiation patterns we derived in Chapter 4.

Although WKBJ theory will be accurate for waves departing steeply downward or upward from the source, it is necessary to use a more accurate method for rays departing nearly horizontally. For such rays, two different problems are present together. First, the WKBJ approximation for $\varepsilon\xi_s$ (or $\varepsilon\eta_s$) in (5)–(7) may be inaccurate. Second, the choice of

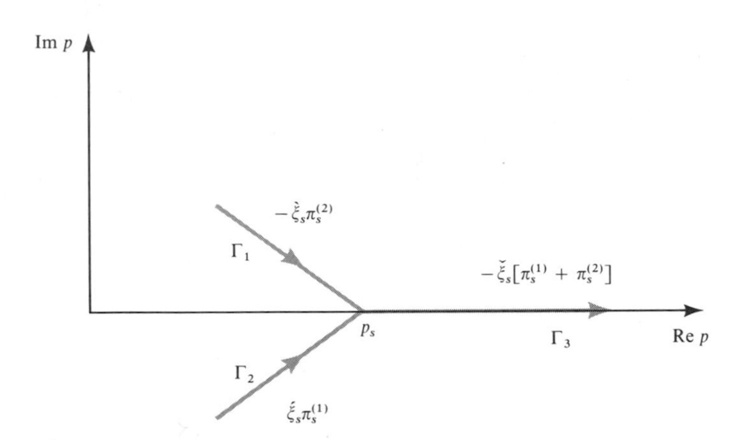

The complex ray-parameter plane, with path Γ_1 for the ray departing upward from the source, path Γ_2 for the ray departing downward, and path Γ_3 for the combined integrand, quantifying tunneling upward from the source depth. Associated with each path is shown the appropriate wavefunction and generalized vertical slowness for the source level r_s.

integration path in the complex p-plane is complicated by the need to consider wavefunctions for both upward- and downward-departing rays. The first problem is easily solved by using a uniformly asymptotic approximation such as (9.66). (Note that downgoing rays require $\varepsilon\xi_s = \acute{\xi}_s$, since $\acute{\xi}_s$ has a phase that increases if r_s is increased above the turning point, and upgoing rays require $\varepsilon\xi_s = -\grave{\xi}_s$.) For the second problem, the solution is to break up the computation into three different line integrals in a way that is very similar to that used in solving the whispering-gallery problem (see Fig. 9.30b,c,d). The critical ray parameter, at which the source level itself is the turning point, is $p_s = r_s/\alpha_s$. To the left of this point in the complex ray-parameter plane, as shown in the figure above, we compute downward-departing waves via a path Γ_2 with $\xi_s = \acute{\xi}_s\pi_s^{(1)}$ in the integrand, and use Γ_1 (with $\xi_s = -\grave{\xi}_s\pi_s^{(2)}$) for upward-departing waves. Both integrands are combined at a point near p_s. Then, to the right, on Γ_3, the integrand decays exponentially, quantifying the effect of energy tunneling upward from the source level r_s to the turning-point level in the source layer, from which level it can further propagate upward, but with relatively small loss.

Finally, we note that just five different linear combinations of $\dot{M}_{ij}$ components are needed in the representation of body waves set up by a system of couples. (From (1) and (2), we see that just three different combinations of moment-tensor elements are needed for P and SV, and two further combinations are apparent in (3) for SH.) The far-field surface-wave formulas we obtained in Chapter 7, (7.147)–(7.150), can also be written in terms of these same five linear combinations of $\dot{M}_{ij}$. Therefore, at each azimuth from the source, one can hope at best to obtain (from far-field data) only five linear constraints on the six independent components of $\mathbf{M}$.

SUGGESTIONS FOR FURTHER READING

Abramovici, F. Numerical seismograms for a layered elastic solid. *Bulletin of the Seismological Society of America*, **60**, 1861–1876, 1970.

Budden, K. G. *Radio Waves in the Ionosphere*. Cambridge University Press, 1961.

Bullen, K. E. *An Introduction to the Theory of Seismology* (3rd ed., Chaps. 7 and 8 for ray theory). Cambridge University Press, 1963.

Chapman, C. H., and R. A. Phinney. Diffracted seismic signals and their numerical solution. *In* B. A. Bolt (editor) *Seismology: Body Waves and Sources* (Methods of Computational Physics, Vol. 12). New York: Academic Press, 1972.

Coddington, E. A., and N. Levinson. *Theory of Ordinary Differential Equations* (Chap. 6 for asymptotic solutions). New York: McGraw-Hill, 1955.

Dey-Sarkar, S. K., and C. H. Chapman. A simple method for the computation of body-wave seismograms. *Bulletin of the Seismological Society of America*, **68**, 1577–1593, 1978.

Felsen, L. B., and N. Marcuvitz. *Radiation and Scattering of Waves*. Engelwood Cliffs, New Jersey: Prentice-Hall, 1973.

Fuchs, K. The reflection of spherical waves from transition zones with arbitrary depth-dependent elastic moduli and density. *Journal of the Physics of the Earth*, **16**, special issue, 27–41, 1968.

Fuchs, K., and G. Müller. Computation of synthetic seismograms with the reflectivity method and comparison of observations. *Geophysical Journal of the Royal Astronomical Society*, **23**, 417–433, 1971.

Heading, J. *An Introduction to Phase-Integral Methods*. London: Methuen and Co., 1962.

Helmberger, D. V., and S. D. Malone. Modeling local earthquakes as shear dislocations in a layered half-space. *Journal of Geophysical Research*, **80**, 4881–4888, 1975.

Nussenzveig, H. M. High-frequency scattering by an impenetrable sphere. *Annals of Physics (New York)*, **34**, 23–95, 1965.

Wait, J. R. *Electromagnetic Waves in Stratified Media*. New York: MacMillan, 1962.

Wasow, W. *Asymptotic Expansions for Ordinary Differential Equations* (Pure and Applied Mathematics, Vol. 14). New York: Interscience, 1965.

Wiggins, R. A., and D. V. Helmberger. Synthetic seismogram computation by expansion in generalized rays. *Geophysical Journal of the Royal Astronomical Society*, **37**, 73–90, 1974.

PROBLEMS

9.1 a) Given the generalized ray described in Figures 9.1 and 9.2 and the Cagniard path $p = p(\tau)$ that solves (9.5), show that the distance L traveled by the head wave in

It follows from (f) that $1/t^{1/2} * dp/dt$, for times near t_{refl}, is

$$\left(\frac{1}{2\dfrac{d}{dp}\text{DISTANCE}}\right)^{1/2}\left[i\pi H(t - t_{\text{refl}}) - \ln\left(\frac{|t - t_{\text{refl}}|}{t_{\text{refl}}}\right)\right].$$

Use this result to obtain the step-function response for the wide-angle reflection as

$$R_S\left[\frac{1}{\xi_1}\left(\frac{p}{r\dfrac{d}{dp}\text{DISTANCE}}\right)^{1/2}(\grave{P}\grave{P}_1)\cdots(\grave{P}\grave{P})_{n-1}(\acute{P}\acute{P})_{n-1}\cdots(\acute{P}\acute{P})_1\right]_{p=p_0}$$

$$\times\left[\text{Re}(\grave{P}\acute{P}_n)H(t - t_{\text{refl}}) - \frac{\text{Im}(\grave{P}\acute{P}_n)}{\pi}\ln\left(\frac{|t - t_{\text{refl}}|}{t_{\text{refl}}}\right)\right].$$

(Recall that R_S is the distance at which the incident wave from the source is a unit step in pressure. It is presumed that R_S is not large enough to place the calibrating receiver outside layer one.)

h) The geometrical ray theory of Chapter 5 would indicate that the unit step-function response is given by

$$\frac{R_S}{\mathscr{R}(\mathbf{x},\boldsymbol{\xi})}H(t - T(\mathbf{x}, \boldsymbol{\xi})) \times \text{factor to account for loss of energy at interfaces,}$$

where $T(\mathbf{x}, \boldsymbol{\xi})$ is the travel time between source at $\boldsymbol{\xi}$ and receiver at $\mathbf{x}$ and $\mathscr{R}(\mathbf{x}, \boldsymbol{\xi})$ is the geometrical spreading function. Clearly, $T(\mathbf{x}, \boldsymbol{\xi}) = t_{\text{refl}}$. Use the method of Problem 4.3 to show that

$$\mathscr{R}(\mathbf{x}, \boldsymbol{\xi}) = \xi_1\left(\frac{r}{p}\frac{d}{dp}\text{DISTANCE}\right)^{1/2}.$$

(*Note*: Source and receiver are both in layer one)

i) Since transmission coefficients for our generalized ray are all real, and only $\grave{P}\acute{P}_n$ is complex, use the results of (h) and of Problem 5.14 to give an alternative derivation of the approximate step-function response, which we obtained in (g) by approximation of Cagniard's method.

9.3 Show that the entries in the propagator matrix $\mathbf{P}(z_m, z_n)$ (for *SH*-waves, or for *P-SV*, in a medium composed of homogeneous plane layers, with z_m and z_n possibly being separated by several different layers) do not have branch cuts in ξ_m, η_m, or in any other vertical slowness.

Outline, in the form (9.16), an exact expression for the Fourier transform of the total pressure at point $(r, z_0 + d_2)$ shown in Figure 9.1b (that is, with *all* multiples included and with the source (9.1)). Show that the only branch cuts of the integrand in the p-plane are those associated with ξ_{n+1}, η_{n+1}, and $J_0(\omega pr)$. (This result greatly facilitates a discussion of path deformation. Note that the corresponding integrand for the generalized primary Moho reflection alone, given by (9.2) for the Laplace transform, has many more branch cuts.)

9.4 a) For a deep source in a spherically symmetric medium, show that the ray departing horizontally is the ray with maximum ray parameter. Show that this ray corresponds to an inflection on the travel-time curve $T = T(\Delta)$, and hence corresponds to a zero value for $dp/d\Delta$, so that the usual expression for geometrical spreading (9.44) cannot be used.

b) Consider a point source S at radius r_s in a spherically symmetric Earth model with velocity profile $\alpha(r)$. Suppose that a ray departing horizontally from S is received at P (radius r, distance Δ from S). Show in this case that the geometrical spreading at P is given by

$$\mathscr{R}(P, S) = r\left(\frac{\sin \Delta \cos i(r)}{[1 - \zeta(r_s)]}\right)^{1/2},$$

where $\zeta(r)$ is the normalized velocity gradient, $r/\alpha(r)\,(d\alpha/dr)$. (*Hints*: What is $\mathscr{R}^2(S, P)$ in terms of $\delta r_s/\delta i(r)$? How can $r_s/\alpha(r_s) = p$ give $\delta r_s/\delta i(r)$?)

9.5 The phase of a vertical wavefunction is given, in WKBJ theory, by identifying $\omega\alpha^{-1} \cos i$ as the vertical wavenumber and integrating this quantity over the vertical distance traveled from some reference depth. In this context, interpret the sum of waves $g_n^{(1)}(r) + g_n^{(2)}(r)$ at a level $r > r_p$ in terms of an incident wave $g_n^{(2)}$ and a reflected wave $g_n^{(1)}$, and show that a $\pi/2$ phase advance occurs in the "reflection" from the turning point.

9.6 If WKBJ approximations are used for

$$P(\mathbf{r}, \omega) = g^{(1)}_{\omega p - \frac{1}{2}}(r) Q^{(2)}_{\omega p - \frac{1}{2}}(\cos \Delta)$$

in the depth range $r > r_p$, show that the phase of P varies most rapidly along paths on which $r(\sin i)/\alpha(r) = p$.

9.7 In association with the WKBJ approximation for $\mathbf{R}_l^0$ and $\mathbf{S}_l^0$, show that $\pi^{(j)}$ and $\sigma^{(j)}$ approximated by (9.60) and (9.61) have the property of conserving the flux rate of energy across horizontal levels in cones with solid angle $\sin \Delta\, \delta\Delta\, \delta\phi$ at the center of the Earth.

9.8 Show that the group velocity for body waves with travel time T at distance Δ on a spherical Earth is Δ/T. Show that, as $\Delta \to 0$ for the body-wave reflection ScS, the group velocity tends to zero but the phase velocity $(d\Delta/dT)$ tends to infinity.

9.9 In general terms, what are likely to be the typical frequency-dependent and distance-dependent properties of seismic waves near a caustic?

the experimental aspects of seismometry distinctly different from those of most other branches of physical science.

There are two basic methods in use today for measuring seismic motion. The one most commonly used is based on the principle of inertia, and the *pendulum seismometer* is the inertial sensor. The other is based on the deformation of a small part of the Earth, and the *strainmeter*, or *strain seismometer*, is the sensor.

It seems possible that a method based on the conservation of angular momentum could be developed in the future. A *gyroscope* has the potential for measuring the rigid-body rotation of a point on the Earth. Suppose that the high-speed wheel of the gyroscope were suspended in a balanced, frictionless gimbal system, so that it is perfectly free to move about its center of gravity. In the absence of external torque, the direction of the angular momentum of the gyroscope would remain invariable in the inertial reference frame, allowing the measurement of rotation of the Earth-fixed gimbal system. Or, another useful property of the gyroscope, the precession under the action of torque, could be used for detecting rotation. Such devices are commonly used in inertial navigation, and should be used in seismology for more complete descriptions of Earth motion. The state-of-the-art sensitivity and stability are, however, not yet good enough to give a useful geophysical measurement.

10.1.1 Pendulum seismometer

A pendulum seismometer consists of a mass M attached to a point of the Earth through a parallel arrangement of a spring and a dashpot, as shown in Figure 10.1. Assuming that all motion is restricted to the x-direction, we denote the motion of the Earth in the inertial reference frame as $u(t)$ and the motion of mass M relative to the Earth as $\xi(t)$. The spring will exert a force proportional to its elongation $\xi - \xi_0$, and the dashpot will exert a force proportional to the relative velocity $\dot{\xi}(t)$ between the mass and the Earth. Representing these constants of proportionality as k and D, the equation of motion is given by

$$M \frac{d^2}{dt^2}[\xi(t) + u(t)] + D\frac{d\xi(t)}{dt} + k[\xi(t) - \xi_0] = 0. \tag{10.1}$$

Rewriting the displacement $\xi(t) - \xi_0$ relative to the equilibrium position ξ_0 as $\xi(t)$, we have

$$\ddot{\xi} + 2\varepsilon\dot{\xi} + \omega_s^2\xi = -\ddot{u} \tag{10.2}$$

where $2\varepsilon = D/M$ and $\omega_s^2 = k/M$.

The above equation shows that by a linear combination of $\xi(t)$ and its time derivatives (which are measurable), we can reproduce the acceleration $\ddot{u}$ of the Earth motion. For very rapid Earth motion, the first term of the left-hand side of (10.2) dominates, and $\ddot{\xi}$ becomes nearly equal to $-\ddot{u}$. In other words,

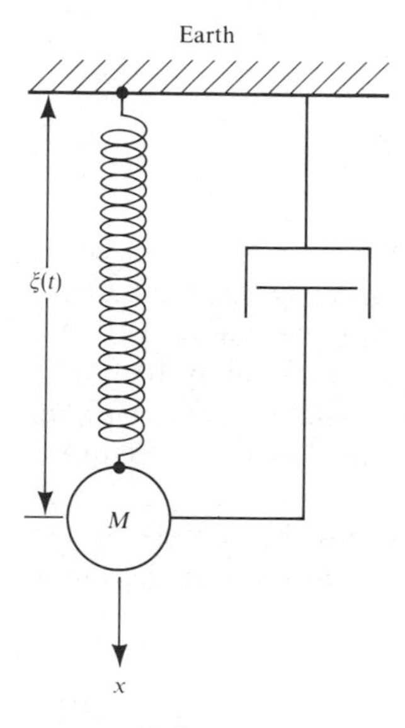

FIGURE **10.1**

A pendulum seismograph is a mass suspended by a parallel arrangement of spring and dashpot.

the record $\xi(t)$ is reproducing the Earth's displacement $-u(t)$ for rapid movement. For very slow motion, in which the third term dominates, the term $\omega_s^2\xi$ becomes nearly equal to $-\ddot{u}$. Thus the record $\xi(t)$ is reproducing the acceleration $\ddot{u}(t)$ for slow motion. The sensitivity is proportional to T_s^2, where $T_s = 2\pi/\omega_s$ is the undamped pendulum period and ω_s the corresponding undamped angular frequency.

The response of a pendulum seismometer to sinusoidal ground displacement $\exp(-i\omega t)$ can be written as $X(\omega)\exp(-i\omega t)$. $X(\omega)$ is called the *frequency response function*, and can completely describe the performance of the pendulum. For $u(t) = \exp(-i\omega t)$ and $\xi(t) = X(\omega)\exp(-i\omega t)$, equation (10.2) gives

$$X(\omega) = \frac{-\omega^2}{\omega^2 + 2i\varepsilon\omega - \omega_s^2}.$$

Defining the *amplitude response* $|X(\omega)|$ and the *phase delay* $\phi(\omega)$ by

$$X(\omega) = |X(\omega)|e^{i\phi(\omega)},$$

we find

$$|X(\omega)| = \frac{\omega^2}{\sqrt{(\omega^2 - \omega_s^2)^2 + 4\varepsilon^2\omega^2}}$$

and

$$\phi(\omega) = -\tan^{-1}\frac{2\varepsilon\omega}{\omega^2 - \omega_s^2} + \pi. \tag{10.3}$$

For $\omega \gg \omega_s$, $|X(\omega)| \to 1$ and $\phi(\omega) \to \pi$. In other words, the pendulum records the ground displacement faithfully, but with reversed sign. The sign difference is usually eliminated by indicating the direction of ground motion properly on the record. Figure 10.2 shows $|X(\omega)|$ and $\phi(\omega)$ without the π term in (10.3). The curves are shown with $h = \varepsilon/\omega_s$ as a parameter; h is the *damping constant*, equal to half the reciprocal of the Q-value (the quality factor of a damped oscillator).

The performance of a pendulum seismometer can also be completely described by its response $f(t)$ to an impulsive acceleration $\ddot{u}(t) = \delta(t)$, the Dirac δ-function. From equation (10.2), $f(t)$ satisfies

$$\ddot{f} + 2\varepsilon\dot{f} + \omega_s^2 f = -\delta(t). \tag{10.4}$$

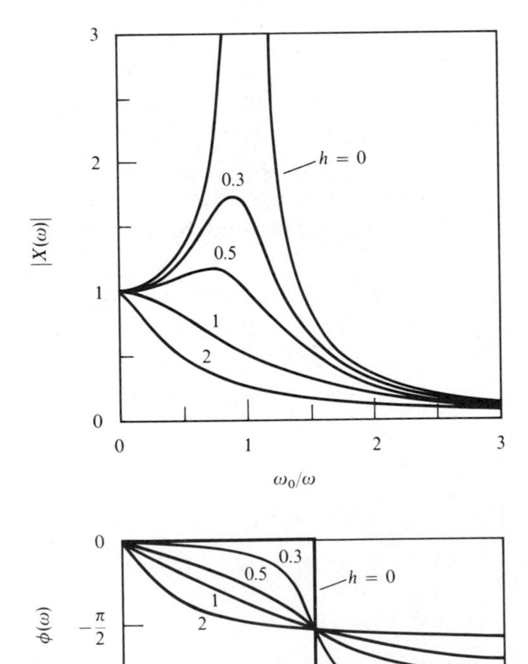

FIGURE **10.2**

Amplitude response $|X(\omega)|$ and phase delay $\phi(\omega)$ of a pendulum seismometer according to (10.3).

Taking the Fourier transform of both sides of (10.4) and putting

$$\int_{-\infty}^{\infty} f(t) \exp(i\omega t)\, dt = f(\omega), \tag{10.5}$$

we have

$$f(\omega) = \frac{-1}{-\omega^2 - 2i\varepsilon\omega + \omega_s^2}.$$

Since $f(\omega) = X(\omega)/\omega^2$, the frequency-response function can be obtained from the impulse-response function by the Fourier transform, and vice versa. From the inverse transform of (10.5),

$$f(t) = \frac{1}{2\pi}\int_{-\infty}^{\infty} \frac{\exp(-i\omega t)\, d\omega}{\omega^2 + 2i\varepsilon\omega - \omega_s^2} = \frac{1}{2\pi}\int_{-\infty}^{\infty} \frac{\exp(-i\omega t)\, d\omega}{(\omega - \omega_1 + i\varepsilon)(\omega + \omega_1 + i\varepsilon)}, \tag{10.6}$$

where $\omega_1^2 = \omega_s^2 - \varepsilon^2$. If we extend ω to the complex ω-plane, we find that poles of the integrand always lie in the lower half-plane because $\varepsilon > 0$. Then, for $t < 0$, the integral along the real axis can be replaced by one along the infinite semicircle in the upper half-plane, which vanishes because of the factor $|\exp(-i\omega t)| = \exp(+\mathrm{Im}\{\omega t\})$. Thus

$$f(t) = 0, \qquad t < 0. \tag{10.7}$$

For $t > 0$, the residue contribution from the poles gives, in the case $\omega_s > \varepsilon$,

$$f(t) = \frac{-1}{\omega_1} e^{-\varepsilon t} \sin \omega_1 t, \tag{10.8}$$

and in the case $\omega_s < \varepsilon$,

$$f(t) = \frac{-1}{\sqrt{2\varepsilon^2 - \omega_s^2}} \{\exp[-(\varepsilon - \sqrt{\varepsilon^2 - \omega_s^2})t] - \exp[-(\varepsilon + \sqrt{\varepsilon^2 + \omega_s^2})t]\}. \tag{10.9}$$

In the limiting case of $\omega_s = \varepsilon$, both formulas give

$$f(t) = -te^{-\varepsilon t}, \qquad t > 0. \tag{10.10}$$

If the seismometer is underdamped ($\omega_s \gg \varepsilon$), the impulse response will show undesirable ringing. On the other hand, if it is overdamped ($\omega_s \ll \varepsilon$), one of the exponential decay constants in equation (10.9), $\varepsilon - \sqrt{\varepsilon^2 - \omega_s^2}$, is nearly $\frac{1}{2}(\omega_s^2/\varepsilon)$, and the response will be delayed longer for greater damping. For these reasons, the best result is obtained near the critical damping $\omega_s = \varepsilon$.

The response $\xi(t)$ of the seismometer to an arbitrary ground acceleration $\ddot{u}(t)$ can be obtained by a convolution with the impulse response $f(t)$:

$$\xi(t) = \int_0^\infty \ddot{u}(t - \tau) f(\tau)\, d\tau, \tag{10.11}$$

and this is the general solution of (10.2).

10.1.2 Stable long-period pendulums

Since the sensitivity of a pendulum to acceleration at low frequency is proportional to the square of its free period, finding a stable, long-period pendulum was the most important problem of instrumental seismology for many years, until LaCoste invented the "zero-initial-length" spring in 1935. LaCoste's spring can theoretically achieve an infinite period without instability.

Usually, when a spring is stretched, the applied load f is proportional, via a modulus k, to the elongation $l - l_0$, where l is the actual length and l_0 is known as the initial length. Thus a zero-initial-length spring is one in which the tension is proportional to the actual length. The zero-initial-length spring must be wound with a twist applied to the wire as it is coiled, and it has a residual tension even when no load is applied because of the finite size of the coils. When loaded slightly, the residual tension will keep the spring length unchanged until the load increases to a certain limit. Then the coils begin to separate and the spring length l will increase proportionally with the load, $f = kl$.

An easy but accurate test for checking whether a given spring has zero initial length is as follows: suspend the spring vertically with a load that extends it to some new equilibrium length l. Let it oscillate vertically about this equilibrium position, and measure the period T. For any spring that obeys Hooke's law, it is simple to show that $T = 2\pi[(l - l_0)/g]^{1/2}$, and from this formula one can obtain $l - l_0$. Then, if $(l - l_0)$ is equal to the directly measured value of l, the spring has zero initial length.

Figure 10.3 illustrates how the spring is used to measure the vertical component of ground motion. Mass M is fixed to a boom that rotates around a hinge point B. One end of the zero-initial-length spring is connected to a point A through a short flexible wire, and the other end is similarly connected to a point on the boom near mass M. Point A is vertically above point B, and both are fixed to the Earth.

Let us consider the equilibrium of the pendulum under no Earth movement. The torque around point B due to the gravitational force exerted on M is equal to $Mgh_1 \sin \phi$, where h_1 is the boom length and ϕ is the angle between the

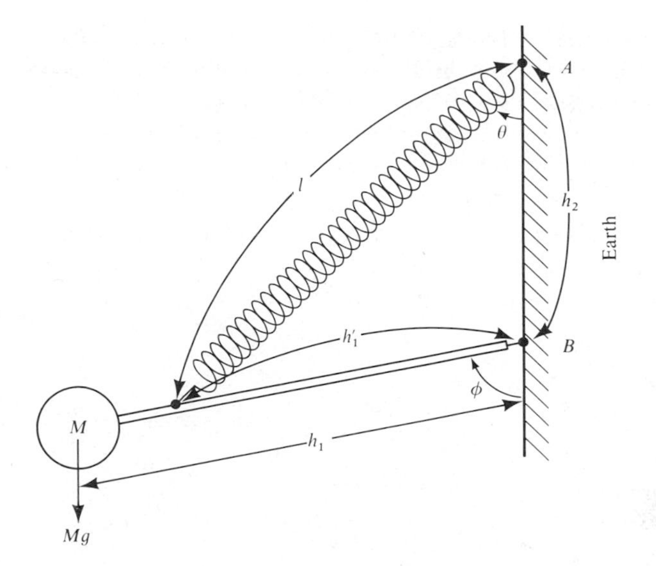

FIGURE **10.3**
Vertical (LaCoste) pendulum, with a zero-initial-length spring. Since the boom will also respond to a change in gravity, this instrument is occasionally referred to as a *gravimeter*.

boom and the vertical. Because of the zero initial length, the torque due to the spring is equal to $klh_2 \sin \theta$, where l is the spring length, k is the spring constant, h_2 is the distance between A and B, and θ is the angle between the spring and vertical. At the equilibrium position, the two torques balance each other:

$$Mgh_1 \sin \phi = klh_2 \sin \theta.$$

On the other hand, from geometry, we find that

$$l \sin \theta = h_1' \sin \phi, \tag{10.12}$$

where h_1' is the distance along the boom from B to the point of spring attachment. Therefore, the equilibrium condition is given by

$$Mgh_1 \sin \phi = kh_1'h_2 \sin \phi$$

or

$$Mgh_1 = kh_1'h_2. \tag{10.13}$$

Surprisingly, the condition is independent of ϕ. In other words, once the lengths and mass are properly chosen, there is no restoring force, and any mass position is in equilibrium. Therefore, we see that the pendulum has an infinite period.

There are many ways of suspending mass and spring other than the one shown in Figure 10.3 to achieve the same infinite period. In practice, the infinite period is unattainable because of inexact spring length and hinge positions; finite restoring force of the hinge, as well as variations in spring constant k due to temperature change; variations in gravity, and other disturbances. The maximum period maintained in a routine observation was 80 sec by Francis Lehner of the Seismological Laboratory of the California Institute of Technology. Standard long-period seismometers are usually operated at periods of 15 to 30 sec.

As mentioned before, the sensitivity of a pendulum to long-period acceleration is proportional to the square of pendulum period T_s. From (10.2), for a gradual change in acceleration by an amount $-\Delta a$, we have

$$\xi = \frac{\Delta a}{\omega_s^2} = \frac{\Delta a}{(2\pi)^2} T_s^2.$$

For example, if $T_s \sim 60$ sec, $\xi = \Delta a \cdot 10^2$ cm. The pendulum mass will move by 1 mm when the acceleration changes by 1 milligal ($=10^{-3}$ cm/sec^2).

Stable long-period pendulums for a horizontal-component seismometer have been known for many years. The long-period horizontal pendulum of the present standard seismograph is shown schematically in Figure 10.4a. The top

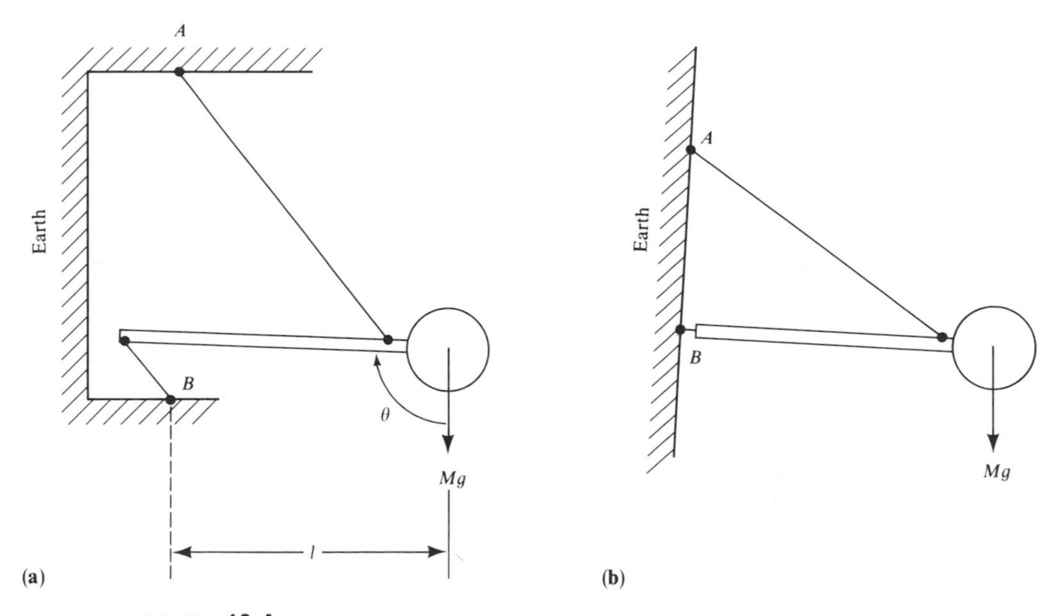

FIGURE **10.4**

Horizontal pendulums. (**a**) The Zöllner suspension. Both wires are kept in tension to avoid buckling instability. (**b**) Often called the "garden gate" suspension. The line AB, fixed to the Earth, is tipped slightly away from the vertical in both (a) and (b). In practice, the pivot at B in (**b**) and in Figure 10.3 is provided with a hook, so that the pivot is under tension.

and bottom wires supporting the mass and boom are kept in tension by the weight. This suspension, originally due to Zollner (1869), eliminated the friction at pivots of other earlier horizontal pendulums, which were based on the design shown in Figure 10.4b. For small-amplitude oscillations, the pendulum moves in a plane inclined at an angle θ from the vertical plane. The component of gravity in this plane is $g \cos \theta$. Therefore, the pendulum period is $2\pi\sqrt{l/(g \cos \theta)}$, where l is the boom length. As the boom direction approaches horizontal ($\theta \to \pi/2$), the oscillation period approaches infinity and becomes unstable. The maximum stable period achieved by a standard instrument is around 30 sec.

10.1.3 Measurement of horizontal acceleration

For almost all designs of pendulum seismometer, it is not possible to distinguish between a horizontal acceleration of the ground and a contribution from gravity due to tilt. Figure 10.5 illustrates this situation using a simple pendulum. When the seismometer frame is tilted by an amount $\delta\Psi$, a torque will be exerted around the hinge and will cause pendulum motion relative to the frame. The effect of this torque is the same as that produced by a horizontal acceleration of magnitude $g \sin \delta\Psi$.

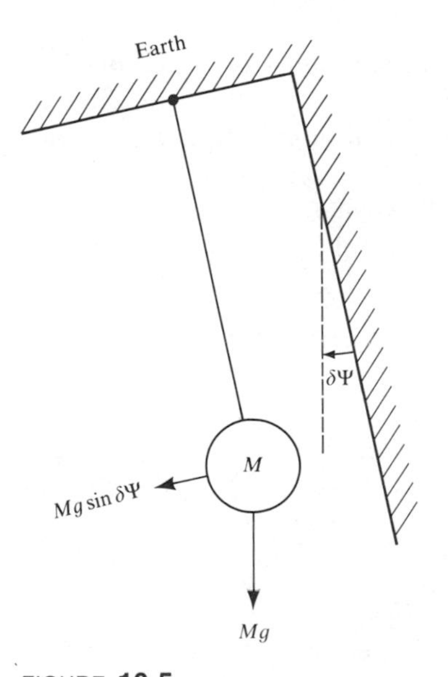

FIGURE **10.5**
Tilt causes apparent horizontal acceleration of a suspended mass.

A preliminary attempt to separate the effects of tilt and horizontal acceleration was made by Farrell (1969), using gyros, but so far, most seismologists have been content with the rather arbitrary assumption that either acceleration or tilt dominates a particular signal. The response of the horizontal pendulum (Fig. 10.4) has been analyzed in detail by Rodgers (1968), and, apart from the ambiguity between tilt and horizontal acceleration, he describes another problem with this design. Although the instrument is sensitive primarily to horizontal accelerations (and their tilt equivalent) perpendicular to the boom, its response is affected also by along-the-boom accelerations and by tilt components that change the angle θ (see Fig. 10.4). Rodgers calls these "parametric effects," because they act to change the free period of the instrument.

It has been customary to consider the motion of a seismometer pendulum in Cartesian coordinates, where the gravitational force is directed parallel to the z-axis. Actually, the gravitational force is directed to a point, the center of the Earth. A basic design that takes this into account is the Schuler pendulum. It is much used in inertial navigation, and may also be important for seismology in the future. This design is shown in Figure 10.6; the pendulum is hinged at P, and for simplicity in the derivation below, we shall consider a motion restricted to the xy-plane. The equation of pendulum motion is

$$J_p \frac{d^2\theta}{dt^2} = M(\mathbf{h} \times \mathbf{g})_z - M\left(\mathbf{h} \times \frac{d^2\mathbf{r}}{dt^2}\right)_z, \tag{10.14}$$

where J_p is the moment of inertia of the pendulum around P; θ is the rotation of the pendulum measured from the x-axis; M is the mass of the pendulum; $\mathbf{h}$ is the position of the mass relative to P; $\mathbf{g}$ is the gravitational acceleration; and $\mathbf{r}$ is the position of P relative to the center of the Earth.

Let us consider the motion of the pendulum when P moves horizontally, i.e., perpendicular to the direction toward the center of the Earth. In this case,

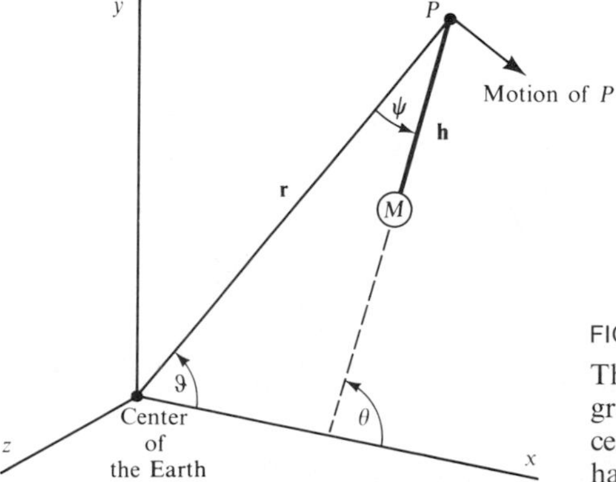

FIGURE **10.6**

The motion of a pendulum subjected to a gravitational force directed toward the center of the Earth. Cartesian axes here have directions fixed in inertial space.

the components of the acceleration vector $d^2\mathbf{r}/dt^2$ are $(-r\ddot{\vartheta}\sin\theta, r\ddot{\vartheta}\cos\theta, 0)$, where ϑ is the angle between $\mathbf{r}$ and the x-axis. The components of other vectors are $\mathbf{h} = (-h\cos\theta, -h\sin\theta, 0)$ and $\mathbf{g} = (-g\cos\vartheta, -g\sin\vartheta, 0)$. Equation (10.14) can now be written as

$$J_p \frac{d^2\theta}{dt^2} = Mhg\sin(\vartheta - \theta) + Mhr\ddot{\vartheta}. \tag{10.15}$$

Since the angle ψ of deflection of the pendulum from vertical is equal to $\theta - \vartheta$, we can rewrite the above equation as

$$J_p\left(\frac{d^2\psi}{dt^2} + \frac{d^2\vartheta}{dt^2}\right) + Mgh\sin\psi = Mhr\ddot{\vartheta} \tag{10.16}$$

or

$$J_p\left(\frac{d^2\psi}{dt^2}\right) + Mhg\sin\psi = -(J_p - Mhr)\ddot{\vartheta}. \tag{10.17}$$

The pendulum period T_0 for small ψ, for which $\sin\psi \sim \psi$, is given by

$$\frac{2\pi}{T_0} = \left(\frac{Mgh}{J_p}\right)^{1/2}. \tag{10.18}$$

The force due to horizontal acceleration vanishes when $J_p = Mhr$. This condition is equivalent to

$$\frac{2\pi}{T_0} = \left(\frac{Mgh}{Mhr}\right)^{1/2} = \left(\frac{g}{r}\right)^{1/2}, \tag{10.19}$$

which is the period of a simple pendulum with length equal to the Earth's radius. The corresponding period is 84 min. A pendulum with this period is called a *Schuler pendulum.*

If one carries a Schuler pendulum by airplane it will always point to the center of the Earth, and will be unaffected by horizontal acceleration. If we mount an accelerometer on the Schuler pendulum, measure the horizontal acceleration during a flight from one point to another on the Earth, and then integrate that acceleration twice with respect to time, we can measure the distance between the two points.

Instruments developed for inertial navigation thus have intrinsic potential for more exact description of Earth movements, such as distinguishing the acceleration from tilt. The sensitivities of the state-of-the-art devices are, however, not yet enough for useful geophysical application.

10.1.4 Measurement of strain and rotation

Consider a strain seismometer (or *strainmeter*) that measures the relative displacement of two nearby points in the Earth, in a manner sketched in Figure 10.7a. The movement at P_1 is transmitted to P'_1 by a bar fixed at P_1, and the relative position of P'_1 and P_2 is measured. Since the ground motion to be measured has, in practice, periods very much longer than the natural period of waves in the bar, the strains in the bar itself are negligible and it can be considered rigid.

Initially, P_1 is located at $\mathbf{x}$ and P_2 at $\mathbf{x} + \delta\mathbf{x}$. These points subsequently undergo displacements $\mathbf{u}(\mathbf{x})$ and $\mathbf{u}(\mathbf{x} + \delta\mathbf{x})$, respectively, so that P_2 moves to $\mathbf{u}(\mathbf{x}) + (\delta\mathbf{x} \cdot \nabla)\mathbf{u}$. (We are using a Lagrangian description of the motion: see Section 2.1.)

The displacement of P'_1 is that of P_1 plus an additional term due to rotation $\frac{1}{2}$ curl $\mathbf{u}$ (see equation (2.2)). To first order, the length and the orientation of the bar are given by the vector $\delta\mathbf{x}$, so that the relative displacement of P_2 from

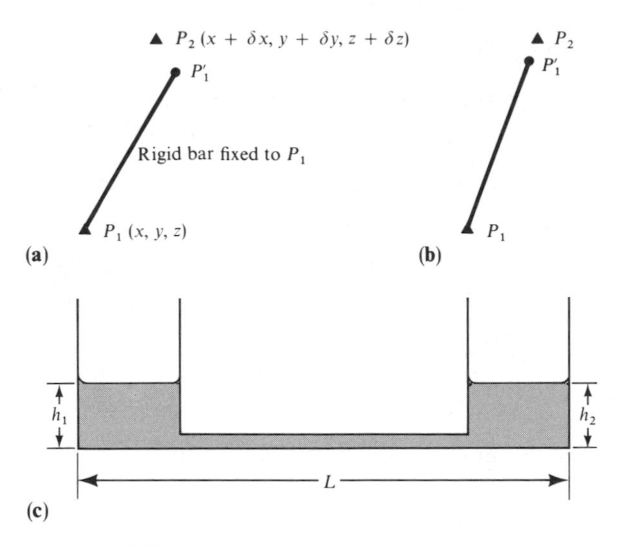

FIGURE **10.7**

Measurement of strain and tilt. (**a**) The bar $P_1P'_1$ is anchored at P_1 and rotates with P_1. (**b**) The bar $P_1P'_1$ is fixed at P_1, but P_1, P'_1, P_2 are effectively collinear. This arrangement is achieved either by pivoting the bar at P_1 or by having the sensor of P'_1P_2 be a capacitance detector that is insensitive to slight misalignments of P_1, P'_1, P_2. (**c**) A tiltmeter consisting of two fluid reservoirs, connected by a tube. Fluid levels h_1 and h_2 are monitored, and temporal changes in the quantity $(h_2 - h_1)/L$ are interpreted as tilt. See Beavan and Bilham (1977) for discussion of temperature effects.

P_1' is $\mathbf{u} + (\delta\mathbf{x} \cdot \nabla)\mathbf{u} - \mathbf{u} - \frac{1}{2}(\text{curl } \mathbf{u}) \times \delta\mathbf{x}$. The ith component of this vector is

$$\delta x_j \frac{\partial u_i}{\partial x_j} - \varepsilon_{ijk}\left(\frac{1}{2}\varepsilon_{jlm}\frac{\partial u_m}{\partial x_l}\right)\delta x_k,$$

which can be rewritten as $e_{ij}\,\delta x_j$, with e_{ij} as the Cartesian components of the strain tensor:

$$e_{ij} = \frac{1}{2}\left(\frac{\partial u_i}{\partial x_j} + \frac{\partial u_j}{\partial x_i}\right).$$

This result shows that the strainmeter is indeed sensitive to strain components, but not to the rotation.

In practice, it is not the vector $e_{ij}\,\delta x_j$ that is measured, but the scalar difference between P_2P_1 and $P_1'P_1$, so that P_1 need provide only a pivot for the bar and the "bar" can in fact be a wire. Rotation is not considered, and P_2P_1 and $P_1'P_1$ are aligned as in Figure 10.7b. As shown following equation (2.3), the length $|\delta\mathbf{x} + \delta\mathbf{u}| - |\delta\mathbf{x}|$ of P_2P_1' in this case is $e_{ij}v_iv_j|\delta\mathbf{x}|$, where $\boldsymbol{v}$ is a unit vector along the bar. In order to determine six components of strain, we need in general six different strainmeters, each oriented along a different direction $\boldsymbol{v}$. For a measurement made near the Earth's free surface, the condition of zero traction imposes three linear constraints on the stress components, so that only three different strainmeters are needed to determine the strain tensor completely. A variety of site effects can contaminate the local (measured) strain field, making it different from the strain field of geophysical interest. In particular, there can be a significant effect from the cavity within which the instrument is emplaced (Harrison, 1976a) and from the topography in the region of the instrument site.

An array of pendulum seismographs spread over a certain area of the Earth's surface may be used to calculate the strain as well as the rotation, but, in practice, performance is limited by shallow heterogeneities in the Earth's structure (Saito, 1968).

A rotation sensor may be constructed by suspending a mass at its center of mass and coupling it to the Earth through a spring and a dashpot. This approach has been attempted but has never succeeded in producing a useful result.

Another approach is the use of a gyroscope. From the law of conservation of angular momentum, a spinning body suspended at its center of mass without external torque will keep its axis of revolution fixed in inertial space. This will allow the measurement of the Earth's motion relative to an inertial frame. The possibility of measuring Chandler wobbles and movement of lithospheric plates with the gyroscope may be approaching. However, the state-of-the-art sensitivity of the general rotation-sensor is not yet enough for a useful geophysical application.

As a special case, rotation about a horizontal axis is known in geophysics as *tilt*. In the sense that small rotations can be represented vectorially with three components, tilt is concerned with two horizontal components. We have remarked already that the principles upon which tiltmeters and horizontal accelerometers are presently designed do not in any way permit a separation of the two effects. Pendulum tiltmeters have been used extensively for measuring solid Earth tides, as described in detail by Melchior (1966). Apart from pendulums, most common in geophysics is the use of two fluid-filled reservoirs connected by a tube (Fig. 10.7c); measurement is made of the differential level in the reservoirs. Spatial fluctuations in temperature within the working fluid can give a spurious signal, which is reduced if the connection between the reservoirs is an open half-filled channel rather than a closed tube. Using a tiltmeter of this design, with 500 feet between reservoirs and with interferometers to measure water height, Michelson and Gale in 1919 were able to report on tidal tilting.

More recently, it has become popular to use an air bubble that is free to move within the working fluid trapped under a slightly curved horizontal plate. Such instruments measure tilt over a relatively short base (~ 20 cm). The bubble position is monitored continuously, so that, if tilt occurs, a signal is generated that is proportional to the bubble displacement. Instruments of this design are operated in arrays in Central California to monitor ground tilts that might be associated with activity on the San Andreas fault (McHugh and Johnston, 1976). Harrison (1976b) has described the operation of a similar instrument, using a bubble trapped under an optical flat. If tilt occurs, the plate is re-leveled to maintain the bubble position. Using this feedback as the output signal, Harrison was able to measure the M_2 tide to about 0.5% after 3 months of observation. However, short-base tiltmeters are inherently susceptible to very local site effects, which it is not yet possible to quantify. If nearby rock units are heterogeneous, then an applied overall strain can locally cause tilting, and this partly explains why closely spaced short-baseline tiltmeters do not always give the same signal.

10.2 Frequency and Dynamic Range of Seismic Signals

Most of the signals studied in seismology are transient, i.e., they have in practice a finite length, such as seismograms from explosions and earthquakes. For such a transient signal $f(t)$, the Fourier transform $f(\omega)$ exists, where

$$f(\omega) = \int_{-\infty}^{\infty} f(t) \exp(i\omega t)\, dt, \tag{10.20}$$

and

$$f(t) = \int_{-\infty}^{\infty} f(\omega) \exp(-i\omega t) \frac{d\omega}{2\pi}, \tag{10.21}$$

where t is the time and ω is the angular frequency. We shall define the *amplitude spectral density* as the absolute value of $f(\omega)$, and the *phase-delay spectrum* $\phi(\omega)$ by the following equation:

$$f(\omega) = |f(\omega)|e^{i\phi(\omega)}. \tag{10.22}$$

(Our reason for calling ϕ the phase delay is given in Box 5.5.) Since $f(t)$ is real, the following relations are valid:

$$\begin{aligned} f^*(\omega) &= f(-\omega), \\ |f(\omega)| &= |f(-\omega)|, \\ \phi(\omega) &= -\phi(-\omega), \end{aligned} \tag{10.23}$$

where the asterisk indicates the complex conjugate.

The unit of $|f(\omega)|$ is the unit of $f(t)$ divided by the unit of frequency $(\omega/2\pi)$. For example, if $f(t)$ represents the ground displacement in cm, then the unit of $|f(\omega)|$ is cm per Hz. The most commonly used units of $\phi(\omega)$ are 1 radian or 1 circle (i.e., 2π radians).

There are three other distinct types of signals, for which the ordinary Fourier transform does not exist. One is the superposition of sinusoidal oscillations with frequencies ω_n, such as the tidal Earth-strain caused by the gravitational attraction of the Sun and the Moon. For this, we define amplitude A_n and phase delay ϕ_n in the following manner:

$$f(t) = \sum_n A_n \exp(-i\omega_n t + i\phi_n), \tag{10.24}$$

where A_n has the same physical dimension as $f(t)$.

Another type of signal we shall consider is the stationary stochastic process, such as ambient seismic ground noise caused by the atmosphere, the oceans, some volcanic processes, industrial activities, and traffic. These signals cannot be expressed either by (10.21) or by (10.24). We must introduce the *power spectral density* $P(\omega)$, which is the Fourier transform of the *autocorrelation function* $P(\tau)$, defined as

$$P(\tau) = \langle f(t)f(t + \tau)\rangle$$

where the symbols $\langle\ \rangle$ indicate averaging over time t.

$$P(\omega) = \int_{-\infty}^{\infty} P(\tau)e^{i\omega\tau}\, d\tau. \tag{10.25}$$

The unit of $P(\omega)$ is the square of the unit of $f(t)$, divided by the unit of frequency $(\omega/2\pi)$. For displacement $f(t)$ measured in cm, the unit of $P(\omega)$ is cm^2/Hz. $P(\omega)$

does not contain information about the phase. This type of signal may be expressed (Chapter 11) as a convolution of white noise $w(t)$ with a weight function $g(t)$:

$$f(t) = \int_{-\infty}^{t} w(t')g(t - t')\, dt', \tag{10.26}$$

where $g(t)$ has a Fourier transform $g(\omega)$ that is related to $P(\omega)$ by

$$P(\omega) = g^*(\omega)g(\omega) = |g(\omega)|^2. \tag{10.27}$$

The phase-delay spectrum of $g(\omega)$ may be determined from $|g(\omega)|$ if the logarithm of $g(\omega)$ is analytic in the upper half of the ω-plane.

The final type of signal we must consider is nonstationary, nonsinusoidal, and has unknown behavior outside the finite length of our record. We shall refer to signals of this type as "drift." Approximating them as some overall time-dependent change, these signals may be characterized by the total change and the rise time. They may be approximated by a linear trend, in which case the rate of change will be the parameter.

When a signal has an approximately sinusoidal form with a certain frequency, the amplitude may be measured from the record. If the bandwidth of the signal is known, then one can estimate the amplitude or power spectral density. Likewise, if the spectral densities and the signal bandwidth are known, one can estimate the signal amplitude.

For a rough approximation, the amplitude of a wavelet is the product of amplitude spectral density and bandwidth of the wavelet. For example, if $|f(\omega)| = F$ (a constant) for $-\omega_0 < \omega < \omega_0$, $|f(\omega)| = 0$ otherwise, and $\phi(\omega) = 0$ for all ω, then the corresponding signal $f(t)$ is written as

$$f(t) = \frac{1}{2\pi}\int_{-\omega_0}^{\omega_0} F \exp(-i\omega t)\, d\omega = 2 \cdot Ff_0 \cdot \frac{\sin \omega_0 t}{\omega_0 t},$$

where $f_0 = \omega_0/2\pi$. The maximum amplitude is at $t = 0$, and is equal to $F \cdot 2f_0$ because $x^{-1} \sin x = 1$ at $x = 0$. Likewise, for a band-passed signal with $|f(\omega)| = F$ for $\omega_0 < |\omega| < \omega_1$, $|f(\omega)| = 0$ otherwise, and $\phi(\omega) = 0$, we obtain

$$f(t) = 2Ff_1 \frac{\sin \omega_1 t}{\omega_1 t} - 2Ff_0 \frac{\sin \omega_0 t}{\omega_0 t},$$

where $f_1 = \omega_1/2\pi$. The maximum amplitude occurs at $t = 0$, and

$$f(t)_{t=0} = F \cdot 2(f_1 - f_0), \tag{10.28}$$

which again is the product of the amplitude spectral density and the bandwidth.

For the power spectral density $P(\omega)$ defined for noise, we obtain

$$P(\tau) = \frac{1}{2\pi} \int_{-\infty}^{\infty} P(\omega) e^{-i\omega\tau} \, d\omega, \tag{10.29}$$

which is the inverse transform of (10.25). Thus

$$P(\tau)\Big|_{\tau=0} = \frac{1}{2\pi} \int_{-\infty}^{\infty} P(\omega) \, d\omega.$$

On the other hand, by definition,

$$P(\tau)\Big|_{\tau=0} = \langle f^2(t) \rangle.$$

Therefore, we have

$$\frac{1}{2\pi} \int_{-\infty}^{\infty} P(\omega) \, d\omega = \langle f^2(t) \rangle. \tag{10.30}$$

For example, if $P(\omega) = P$ for $\omega_0 < |\omega| < \omega_1$, and $P(\omega) = 0$ otherwise, then we have

$$\langle f^2(t) \rangle = P \cdot 2(f_1 - f_0). \tag{10.31}$$

Thus, for a rough approximation, the mean square amplitude of noise in the time domain is equal to the product of the power spectral density and the bandwidth.

By the use of (10.28) and (10.31), and knowing the bandwidth of the signal, we can approximately relate the amplitude spectral density to the amplitude, and the power spectral density to the mean square amplitude.

Let us now make a survey of seismic signals and explore the frequency and dynamic range covered by various observations. In each case, we shall characterize the observations in terms of one of the four types of signal we have considered above.

10.2.1 Surface waves with periods around 20 seconds

The most prominent signals recorded by a standard long-period seismograph, especially classical mechanical ones, for a distant shallow earthquake are surface waves with a period around 20 sec. Body waves show smaller amplitude because of stronger geometrical spreading than surface waves. The surface waves with periods less than about 10 or 15 sec suffer from scattering due to shallow heterogeneities. The attenuation of surface waves is minimal at about 20 sec, because those with periods longer than 25 sec begin to lose energy into the asthenosphere. Besides, the magnification of standard seismographs decreases with periods of more than 20 sec. It was quite natural, therefore, that a magnitude scale M_s was introduced by Gutenberg and Richter (1936) based on the surface-wave amplitude at a period of 20 sec. (See Appendix 2, p. 533.)

Figure 10.8 shows the amplitude of surface waves from a shallow earthquake with magnitude $M_s = 3$ as a function of epicentral distance. The curve is obtained from the table given by Richter (1958, p. 346). The most sensitive long-period seismometer can detect a distant earthquake with $M_s \sim 3$, which will show an amplitude of 100 mμ (10^{-5} cm) at $\Delta = 20°$ and 10 mμ at $\Delta = 80°$. On the other hand, the largest earthquake ($M_s = 8\frac{1}{2}$) will show an amplitude of several cm at $\Delta = 20°$ and several mm at 80°. This large signal dynamic range (10^{-6} to 1 cm) imposes heavy demands on seismic instrumentation and recording media (see Box 11.1).

As described in Section 10.1, there are three basic sensors that can be used in seismology: the inertial sensor, which is sensitive to acceleration; the strainmeter, which measures the strain in the Earth beneath it; and the gyro, which has the potential for measuring rotation. It is instructive to give the value of acceleration α, strain ε, and rotation θ associated with the 20-sec surface waves from a small earthquake, say $M_s \sim 3$. For plane Rayleigh waves with period 20 sec, phase velocity 3.5 km/sec, and displacement amplitude 0.1 μ, we have

$$\alpha \sim \left(\frac{2\pi}{20}\right)^2 \cdot 10^{-5}\ \text{cm/sec}^2 \sim 10^{-6}\ \text{gal} = 10^{-9}\ \text{g},$$

$$\varepsilon \sim \frac{2\pi}{3.5 \times 10^5 \times 20} \cdot 10^{-5} \sim 10^{-11},$$

$$\theta \sim \varepsilon = 10^{-11} \sim 2 \times 10^{-6}\ \text{arc-sec}.$$

Large explosions, such as an underground nuclear test, can also generate 20-sec surface waves with amplitude approximately proportional to the yield. A 1-megaton shot in hard rock roughly corresponds to $M_s \sim 5\frac{3}{4}$, and thus the most sensitive long-period seismometer will detect a distant shot of several kilotons (corresponding to $M_s \sim 3$).

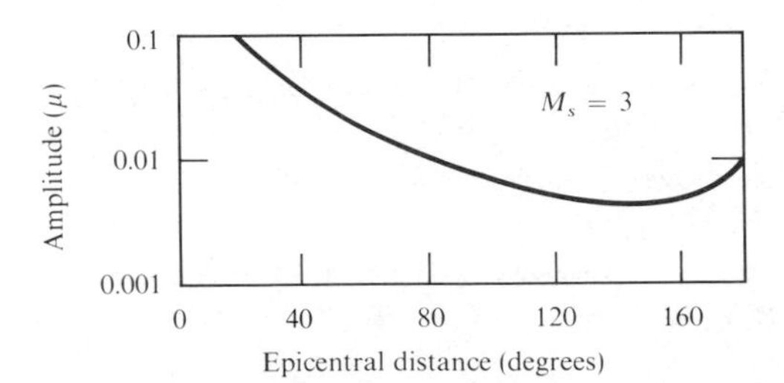

FIGURE **10.8**

Amplitude of surface waves from a shallow earthquake with magnitude $M_s = 3$ as a function of epicentral distance.

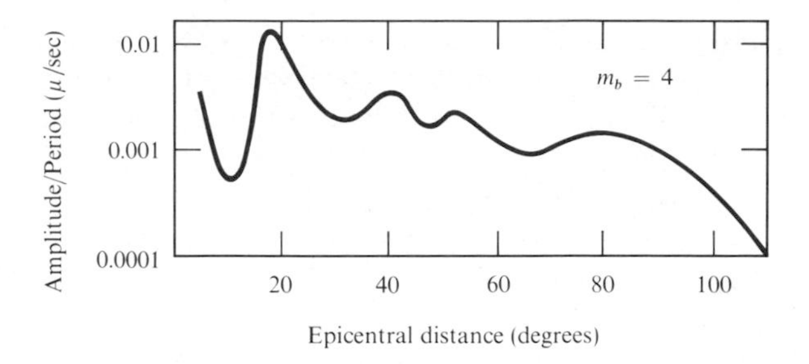

FIGURE **10.9**
The value of A/T (A = amplitude, T = period) for a shallow earthquake with $m_b = 4$ as a function of epicentral distance.

10.2.2 P-waves for 5° < Δ < 110°

The signal level of P-waves from a distant earthquake may be found from Gutenberg's calibration curve (see Richter, 1958, p. 688) for determining the body-wave magnitude m_b. Figure 10.9 shows the value of A/T as a function of epicentral distance, where A is the amplitude in microns and T is the period in sec for a shallow earthquake with $m_b = 4$. This curve can be used to find m_b for any shallow earthquake, as

$$m_b = \log(A/T)_{\text{obs}} - \log(A/T)_{m_b=4} + 4,$$

where $(A/T)_{\text{obs}}$ is the observed value of A/T at a certain epicentral distance (which must be known), and $(A/T)_{m_b=4}$ is the value obtained from Figure 10.9 for the distance. For P-waves recorded by standard seismographs, T is usually around 1 sec, and the amplitude is about 10 mμ at $\Delta = 20°$ and 1 mμ at $\Delta = 90°$ for $m_b = 4$. These signals may be detected by the most sensitive short-period seismometers. The greatest earthquake ($m_b \sim 8$) will show A/T of 1 mm/sec at $\Delta = 20°$. For such large earthquakes, T may be about 10 sec, and the amplitude on the order of 1 cm. Again, we see a requirement for large dynamic range from 10^{-7} to 1 cm.

The 1 mμ (millimicron) displacement at $T = 1$ sec corresponds to an acceleration of 4×10^{-10} g, and to rotations and strains of around 10^{-12}.

10.2.3 Range of amplitude spectral densities for surface waves and P-waves

In Figures 10.8 and 10.9, we showed the signal amplitudes as a function of distance. Figure 10.10 shows the amplitude spectral density defined in Section 10.2 for both surface waves and P-waves at $\Delta = 90°$ as a function of frequency.

The solid lines with specified magnitudes represent typical spectral densities for the smallest detectable events and the largest events. They are obtained by schematizing the empirical data with the aid of theoretical spectra calculated for a realistic Earth model (Chapter 7) and using a scaling law for earthquake sources (Chapter 14, Fig. 14.14).

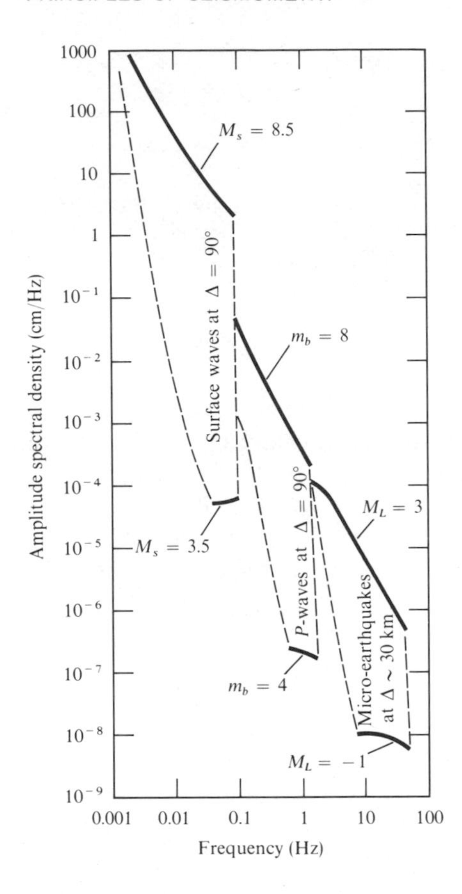

FIGURE **10.10**

The ranges of amplitude spectral density for surface waves at the epicentral distance 90° corresponding to the range of M_s from $3\frac{1}{2}$ to $8\frac{1}{2}$; for *P*-waves at 90° corresponding to the range of m_b from 4 to 8; and for *S*-waves from microearthquakes at distance 30 km corresponding to magnitude M_L from -1 to 3.

The broken lines indicate the rough limits of the spectral range usually studied for each type of wave. For surface waves studied teleseismically, the high-frequency limit corresponds to a period of about 10 sec. Shorter waves are either hidden in the high ambient seismic noise shown later in Figure 10.11 or scattered by the strong lateral heterogeneity of the Earth near the surface. The effect is path-dependent: paths in an ocean basin wipe out Rayleigh waves with periods shorter than about 15 sec, whereas paths in a shield area transmit short-period surface waves over long distances. The low-frequency limit is determined by the sensitivity of conventional seismometers.

For *P*-waves, the high-frequency limit is due primarily to attenuation. The low-frequency limit for small events is due to the characteristic of standard short-period seismographs, which are designed to suppress the frequency range of high ambient noise, discussed below. For larger events, it is possible to extend the spectral range to longer periods.

The spectral range for microearthquake signals is discussed in the next section.

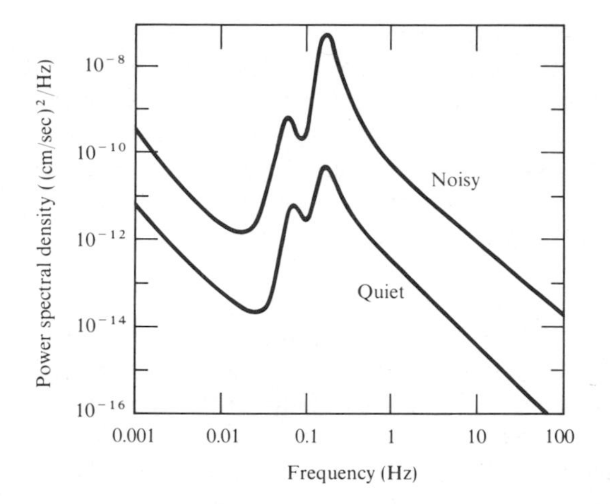

FIGURE **10.11**
Power spectra of ambient seismic noise at noisy and quiet conditions for a typical station on hard basement rock.

10.2.4 Microearthquake waves at short distance

Figure 10.10 also shows the frequency and dynamic range of S-waves from microearthquakes observed at $\Delta \sim 30$ km. The microearthquake is usually defined as an earthquake with $M_L < 3$, where M_L is the Richter magnitude originally defined for local earthquakes in southern California. The lower-magnitude limit $M_L \sim -1$ is typical and is determined by the ambient noise. The displacement amplitude for $M_L \sim -1$ at $\Delta = 30$ km is on the order of 10^{-7} cm, with frequencies up to several tens of Hz. The corresponding acceleration ($\sim 10^{-6}$ g) is, however, considerably higher than the thresholds for teleseismic P- and S-waves, making the instrumentation easier. The corresponding strain and rotation are on the order of 10^{-9}.

10.2.5 Ambient seismic noise

Figure 10.11 shows two representative power spectra of seismic noise, at quiet and noisy conditions, for a typical station on hard basement rock remote from cultural activities when there is no strong wind in the vicinity of the station. The figure was obtained by modifying the curve published by Haubrich (1967) using more recent observations. A significant modification was necessary at periods longer than about 30 sec as a result of the work of Savino et al. (1972), who discovered a noise power minimum between 30 and 40 sec using the records of high-gain long-period (HGLP) seismographs developed by Pomeroy et al. (1969) and installed at eight stations around the world.

The main features of the noise spectrum are the two peaks at about 0.14 Hz and about 0.07 Hz. Both peaks are due to ocean waves. The smaller peak occurs at the primary frequency at which most ocean waves are observed and is considered to be due to the action of ocean waves on coasts, as proposed first by Wiechert in 1904. The main peak was explained by Longuet-Higgins in 1950 as due to the pressure from standing ocean waves, which may be formed by waves traveling in opposite directions in the source region of a storm or near the coast. This mechanism generates seismic waves with a frequency twice that of ocean waves. Hasselman (1963) showed that both theories quantitatively explain seismic and ocean-wave observations.

The level of noise recorded on vertical-component seismographs housed in an air-tight casing on competent rock is independent of local meteorological conditions and of the depth of overburden. On the other hand, the horizontal-component long-period seismographs, such as HGLP seismographs, are affected by the noise due to local meteorological conditions if the vault is shallow. This effect, however, attenuates quickly with depth. The noise level in power can be reduced to 10% of the level at the surface by placing the seismograph at a depth of 100 meters. This substantial decay was correctly predicted by Sorrells (1971), who modeled the noise source by a wind-induced pressure field propagating as a plane wave with ground-level wind velocities (see Problem 10.3).

The power spectrum in Figure 10.11 is given for the ground particle velocity in units of $(\text{cm/sec})^2/\text{Hz}$. Comparing this spectrum with the range of signal level given in Figure 10.10, we see that the gap between the range for surface waves and that for *P*-waves is caused by the largest peak of seismic noise. The peak power density is 10^{-7} $(\text{cm/sec})^2/\text{Hz}$ at frequency 0.15 Hz. Putting the bandwidth as half the frequency, the mean square particle velocity is obtained as 1.5×10^{-8} $(\text{cm/sec})^2$ from (10.31). Then the RMS (root mean square) particle velocity will be about 10^{-4} cm/sec and the corresponding RMS displacement noise level will be about 10^{-4} cm. On the other hand, the amplitude spectral density for a signal with amplitude 10^{-4} cm and bandwidth 0.075 Hz will be about 7×10^{-4} cm/Hz. This explains why there is a notch-shaped gap in Figure 10.10 between the range for surface waves and that for *P*-waves; the tip of the notch is near the point (0.1 Hz, 10^{-3} cm/Hz).

Similarly, the displacement noise level at frequency 0.01 Hz will be 10^{-5} cm and the corresponding amplitude spectral density for the threshold signal will be 10^{-3} cm/Hz. This explains the low-frequency limit of the signal range for surface waves shown in Figure 10.10.

10.2.6 Amplitude of free oscillations

The amplitude of free oscillations excited by an earthquake attenuates with time. The initial amplitude corrected for attenuation was obtained for the great Chilean earthquake of 1960 by Benioff et al. (1961). This measurement was made by a strainmeter, and the initial strain amplitudes for the fundamental-mode spheroidal free oscillations, ${}_0S_2$ (T = 54 min), ${}_0S_5$ (20 min), and ${}_0S_{19}$ (6 min) were 2×10^{-11}, 8×10^{-11}, and 2×10^{-9} respectively.

The free oscillations from the same earthquake were also observed by the LaCoste-Romberg tidal gravimeter. Power spectral analysis by Ness et al. (1961) revealed spectral peaks with amplitude about $0.1-1 \times (10^{-9}\ \mathrm{g})^2$/cph (cycles per hour) and bandwidth a small fraction of 1 cph. The vertical acceleration amplitude of these oscillations, therefore, was a fraction of 10^{-9} g.

An estimate for the initial displacement amplitude was given by Abe (1970) for the Kurile Islands earthquake of 1963 October 13 ($M_s = 8\frac{1}{4}$, seismic moment $= 7.5 \times 10^{28}$ dyne-cm) and for the great Alaska earthquake of 1964 March 28 ($M_s = 8.3$, seismic moment $= 7.6 \times 10^{29}$ dyne-cm). The initial amplitudes were about 10^{-2} cm for ${}_0S_{10}$ through ${}_0S_{14}$ for the Alaskan earthquake.

One of the most sensitive acceleration sensors (Block and Moore, 1970) can detect free-oscillation peaks of higher-order numbers from an earthquake as small as $M_s = 6.5$. The signal level of these peaks in the frequency range 10 to 20 cph is about $10^{-21}\ \mathrm{g}^2$/cph.

10.2.7 Amplitudes of solid Earth tide, Chandler wobble, plate rotation, and moonquakes

The semi-diurnal and diurnal Earth tides are the largest signals in the frequency range lower than 1 cph. Their acceleration peak-to-peak amplitude is about 3×10^{-7} g, and their strain amplitude is about 10^{-7}.

The variation of latitude due to free nutation was predicted by Euler in 1765, and was named after Chandler, a merchant in Cambridge, Massachusetts, who discovered the wobble period of 428 days. The amplitude of the wobble is about 0.2 arc-sec, or 10^{-6} radian.

The fastest relative rotation of lithospheric plates is that of the Pacific plate relative to the Antarctic plate at the rate of 10^{-6} degree/year (Le Pichon, 1968) or 2×10^{-8} radian/year.

The natural moonquakes (category A, Latham et al., 1971) are recorded with amplitudes in the range 10^{-8}–10^{-7} cm with periods around 1 sec. The noise spectra (Fig. 10.11) for the Earth show that they could not be detected teleseismically on the Earth.

10.2.8 Seismic motion in the epicentral area

Figure 10.12 shows the peak horizontal ground acceleration (in units of gravitational acceleration) as a function of distance to the closest surface trace of fault slip for three ranges of earthquake magnitude summarized by Page et al. (1975).

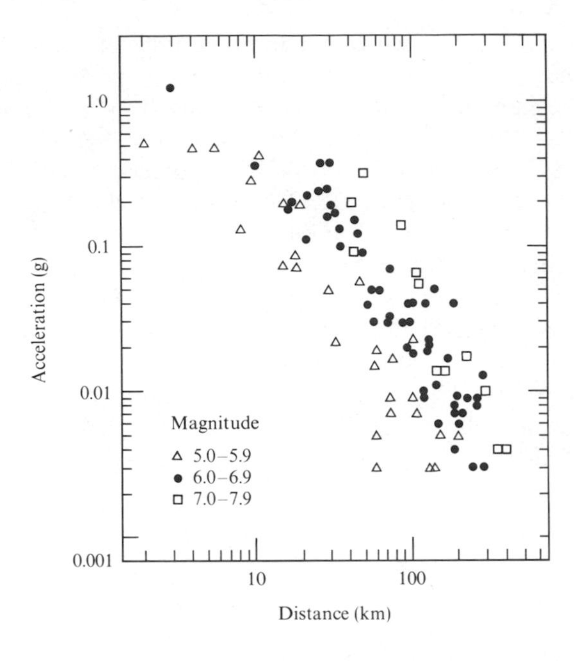

FIGURE **10.12**

Peak horizontal ground acceleration (in units of gravitational acceleration) as a function of distance to the closest surface trace of fault slip. Different symbols are used for different magnitude ranges. [After Page et al., 1975; copyrighted by the American Association for the Advancement of Science.]

The largest value is 1.25 g recorded on a rock ridge 3 km from the fault plane during the 1971 San Fernando earthquake (magnitude 6.6).

A curve for the typical acceleration spectral density is given in Figure 10.13 for a station at a distance of 8 km from the fault break of a medium-sized earthquake (Temblor station record during the Parkfield, California, earthquake 1966 June 28; $M_s = 6.3$, $M_L = 5.5$, seismic moment $= 1.4 \times 10^{25}$ dyne-cm). The curve was obtained from the actual record for periods shorter than 1 sec and by an extrapolation for periods longer than 1 sec using an appropriate dislocation model.

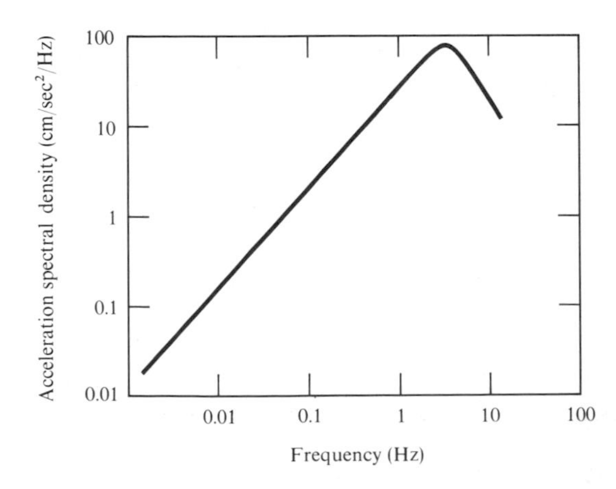

FIGURE **10.13**

An example of acceleration spectral density in the epicentral area of a medium-sized earthquake.

For a complete description of the source mechanism of an earthquake, it is necessary to obtain the wide spectral range, such as that shown in Figure 10.13. This requirement poses a difficult problem in seismic instrumentation. As Figure 10.13 shows, the main signal has its peak acceleration spectral density of 100 (cm/sec^2)/Hz at 3 Hz, which corresponds to the acceleration amplitudes of 0.3 g. On the other hand, the acceleration amplitude of the signal contained in the spectral band around 0.01 Hz is about 10^{-6} g. The problem is to isolate low-frequency signal with amplitude one millionth of that of the coexisting high-frequency signal. Therefore, linearity is an important requirement for the instrument in wide-band near-field seismic experiments. Since no physical sensor is perfectly linear, a small fraction of high frequency waves will be rectified and show up as a spurious slow motion that will contaminate our long-period signals.

10.2.9 Strain steps

A sensitive strainmeter often records a step-like change in strain during a distant earthquake (Press, 1965). Figure 10.14 shows the amount of strain step as a function of earthquake magnitude and epicentral distance (Wideman and Major, 1967). The cause of these steps is not yet understood. Some attribute them to the global static deformation directly related to the earthquake source; others seek an explanation in the instrumental hysteresis or in a minute change in properties of the rock mass to which the strainmeter is attached. The passage of seismic waves may cause small permanent displacement along joints in the rock mass, and rainfall has also been held responsible for some observed strain steps. The Japanese network of crustal movement observatories (1970) has observed strain steps in association with the Gifu earthquake of September 1969. Unlike the lines shown in Figure 10.14, the general trend in these data shows the reciprocal cubic distance dependence as predicted for a dislocation in an unbounded medium (differentiating equation (4.34) to get strain).

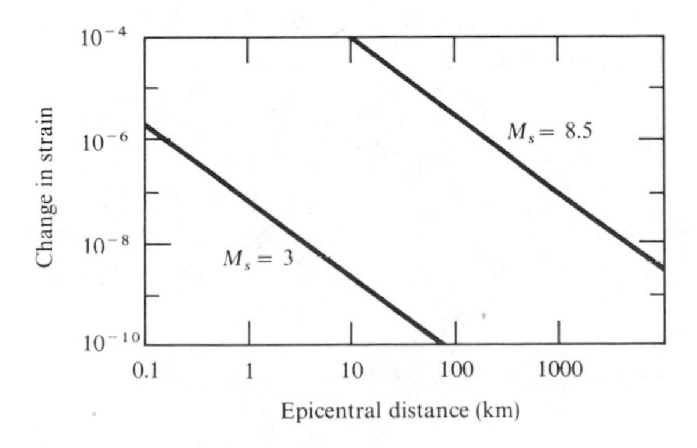

FIGURE **10.14**
Strain steps as a function of epicentral distance for magnitudes 3 and 8.5. [After Wideman and Major, 1967.]

10.3 Detection of Signal

Now that we know the frequency and dynamic range of seismic signals, let us consider more practical aspects of the seismic sensors. First, of course, the seismometer must be sensitive enough to detect the signal. The sensitivity of a sensor is ultimately limited by the thermal noise generated in its dissipative elements.

10.3.1 Brownian motion of a seismometer pendulum

A pendulum with one degree of freedom (movement restricted to one direction) has mean kinetic energy given by $\frac{1}{2}kT$ (k = Boltzman's constant) when it is in thermal equilibrium with the surrounding air at the absolute temperature T (degrees Kelvin). If it had more energy, it would warm the air by accelerating the motion of air molecules. If it had less energy, the collision of air molecules would accelerate the pendulum motion.

Referring back to Figure 10.1 and equations (10.1) and (10.2), the motion of the pendulum due to the force $e(t)$ exerted by the collision of air molecules is the solution of the equation

$$M\ddot{\xi} + D\dot{\xi} + k\xi = e(t). \tag{10.32}$$

The solution is given by the convolution of $e(t)$ with the impulse response $f(t)$ obtained earlier in equation (10.8):

$$\xi(t) = \frac{1}{M}\int_{-\infty}^{t} e(t')f(t - t')\,dt'. \tag{10.33}$$

Since the motions of individual molecules are independent and uncorrelated, the autocorrelation of $e(t)$ should have the form

$$\langle e(t)e(t')\rangle = \sigma^2\delta(t - t'), \tag{10.34}$$

where σ^2 is determined by the condition that the mean kinetic energy of the pendulum is equal to $\frac{1}{2}kT$. From (10.33), we have

$$\begin{aligned}
\frac{1}{2}M\langle\dot{\xi}^2(t)\rangle &= \frac{1}{2M}\int_{-\infty}^{t}\dot{f}(t - t')\int_{-\infty}^{t}\langle e(t')e(t'')\rangle\dot{f}(t - t'')dt'' \\
&= \frac{\sigma^2}{2M}\int_{-\infty}^{t}[\dot{f}(t - t')]^2\,dt' \\
&= \frac{\sigma^2}{2M}\int_{0}^{\infty}[\dot{f}(\tau)]^2\,d\tau.
\end{aligned}$$

Then, using equation (10.7) and (10.8), we have

$$\frac{1}{2} M\langle \dot{\xi}^2(t) \rangle = \frac{\sigma^2}{2M} \cdot \frac{1}{4\varepsilon}$$

$$= \frac{1}{2} kT \tag{10.35}$$

or

$$\sigma^2 = 4M\varepsilon kT.$$

Since the pendulum mass is moving with velocity $\dot{\xi}(t)$ under the force $e(t)$, the power received is

$$\langle \dot{\xi}(t) e(t) \rangle = \frac{1}{M} \int_{-\infty}^{t} \langle e(t) e(t') \rangle \dot{f}(t - t')\, dt'$$

$$= \frac{\sigma^2}{M} \dot{f}(0)$$

$$= 4\varepsilon kT. \tag{10.36}$$

This shows that there is no thermal noise power if $\varepsilon = 0$. In other words, if there is no lossy element, energy cannot flow into the system, because the outflow and inflow must balance at equilibrium. Only the dashpot can convert its kinetic energy into heat; the mass and spring cannot.

In order to see the noise generation of the dashpot more directly, we shall remove the spring and compute the power:

$$\langle \dot{\xi}(t) e(t) \rangle = \frac{\langle e(t) e(t) \rangle}{2\varepsilon M}$$

$$= \frac{\sigma^2\, \delta(0)}{2\varepsilon M}. \tag{10.37}$$

The power is infinite! But the power spectral density defined in (10.25) is finite. Since $\delta(t) = 1/(2\pi) \int_{-\infty}^{\infty} \exp(-i\omega t)\, d\omega$, we can write

$$\langle \dot{\xi}(t) e(t) \rangle = \frac{\sigma^2\, \delta(0)}{2\varepsilon M}$$

$$= \frac{1}{2\pi} \cdot \frac{\sigma^2}{2\varepsilon M} \int_{-\infty}^{\infty} d\omega$$

$$= \int_{0}^{\infty} 4kT\, df, \tag{10.38}$$

where $f = \omega/(2\pi)$. In other words, the available thermal power of the dashpot in the frequency band of width Δf is $4kT\,\Delta f$. This form of thermal power is independent of ε, and is applicable not only to any mechanical dissipative system but also to an electrical dissipative system. For example, the bandwidth of our pendulum system is about ε, and the total noise power is $4\varepsilon kT$ according to equation (10.36), giving the power density $4kT$. The noise power from an electric resistance R may be expressed as $\langle V^2 \rangle / R$, where V is voltage. Equating this power to $4kT\,\Delta f$, we obtain the well-known formula for Johnson noise:

$$\langle V^2 \rangle = 4RkT\,\Delta f. \tag{10.39}$$

Let us now consider the ground acceleration $\alpha(t)$ that would generate pendulum motion equal to its Brownian motion. Since the acceleration produces the force $M\alpha(t)$, the available noise power of the dashpot is

$$\begin{aligned} M\langle \dot{\xi}(t)\alpha(t)\rangle &= \frac{M}{2\varepsilon}\langle \alpha(t)\alpha(t)\rangle \\ &= 4kT\,\Delta f \end{aligned}$$

or

$$\langle \alpha^2(t) \rangle = \frac{8\varepsilon kT\,\Delta f}{M}. \tag{10.40}$$

This gives a ground acceleration equivalent to the Brownian motion of the pendulum. Using $Q^{-1} = 2\varepsilon/\omega_s$, where $\omega_s/2\pi$ is the undamped pendulum period, (10.40) is rewritten as

$$\frac{\langle \alpha^2(t) \rangle}{\Delta f} = \frac{\omega_s}{MQ} \cdot 4kT. \tag{10.41}$$

Thus the acceleration-noise power-density is inversely proportional to the product of mass, Q, and the pendulum period.

Most seismographs are set at critical damping, which corresponds to $Q = \frac{1}{2}$. In that case,

$$\frac{\langle \alpha^2(t) \rangle}{\Delta f} = \frac{\omega_s}{M} \cdot 8kT. \tag{10.42}$$

For comparison with the ground-noise spectra given in Figure 10.11, we shall calculate the spectral density for ground velocity v equivalent to thermal noise.

$$\frac{\langle v^2 \rangle}{\Delta f} = \frac{8\omega_s kT}{(2\pi f)^2 M}. \tag{10.43}$$

It is very interesting to find from this figure that the "observed" ground-noise spectra show roughly the same frequency dependence f^{-2}, separately, for periods longer than about 30 sec, and shorter than about 3 sec (note the asymptotic behavior as $\omega \to 0$ and $\omega \to \infty$). For the "quiet" curve, those parts with slope f^{-2} correspond to

$$\frac{8\omega_s kT}{(2\pi)^2 M} = 10^{-17}\ (\text{cm/sec})^2\text{-Hz} \qquad \text{for long periods,}$$

$$= 10^{-13}\ (\text{cm/sec})^2\text{-Hz} \qquad \text{for short periods.} \tag{10.44}$$

Putting the pendulum period $T_s = 2\pi/\omega_s$ and $kT = 4 \times 10^{-14}$ erg, we find that the requirement for critically damped pendulum noise to be below the ground noise is given by

$$MT_s > 5 \times 10^3 \text{ gram-sec} \qquad \text{for long periods}$$

and

$$MT_s > 5 \times 10^{-1} \text{ gram-sec} \qquad \text{for short periods.}$$

The standard long-period seismograph has about a 10-kg mass and a period of 15–30 sec, satisfying the above requirement. The Block-Moore (1970) accelerometer, housed in a hard vacuum (10^{-7} mm Hg) at a tightly controlled temperature, has an unusually small mass of 10 grams and a relatively short period of 1 sec, and does not satisfy the requirement. Low thermal noise in this instrument is achieved by making the pendulum Q high (200) and using capacitance sensing and electrostatic feedback (electronic refrigeration).

Even with this high Q-value, however, we notice that the thermal power density of the Block-Moore accelerometer is slightly higher than the seismic noise level at quiet periods determined by Savino et al. (1972) using the records of instruments developed by Pomeroy et al. (1969), which are equipped with a pendulum of mass 10 kg and period 30 sec. The low noise of the latter instrument was accomplished by a rigid environmental control, putting the system in an airtight chamber.

Short-period seismometers in common use have pendulum periods of 0.1 to 1 sec. The required mass to overcome the thermal noise is then only 0.5 to 5 grams. The pendulum mass commonly used in a short-period seismometer is much larger than that. The sensing device of these instruments (a moving coil in a magnet gap) requires a larger pendulum mass for greater signal power, as shown in the next section.

Detailed discussions of the thermal noise produced when the pendulum is coupled to the electromagnetic transducer, amplifier, and galvanometer are given by Byrne (1961).

10.3.2 Electromagnetic velocity sensor

The motion of a pendulum relative to the seismometer frame is measured most commonly by the electromagnetic velocity sensor shown schematically in Figure 10.15. Voltage $V(t)$ is generated across the terminals of a coil that is fixed to the mass M and is moving through a magnetic field with velocity $\dot{\xi}(t)$. The terminal is shunted by resistance R. Let l represent the length of coil wire within the magnetic field of flux density B, and assume that the directions of coil movement, magnetic field, and electric current in the coil are perpendicular to each other.

Consider first the mechanical work done by the mass moving through the magnetic field. The force F encountered is, by Biot-Savart's law,

$$F = IlB, \tag{10.45}$$

where I is the current in the coil. The mechanical power consumed is

$$F\dot{\xi} = IlB\dot{\xi}. \tag{10.46}$$

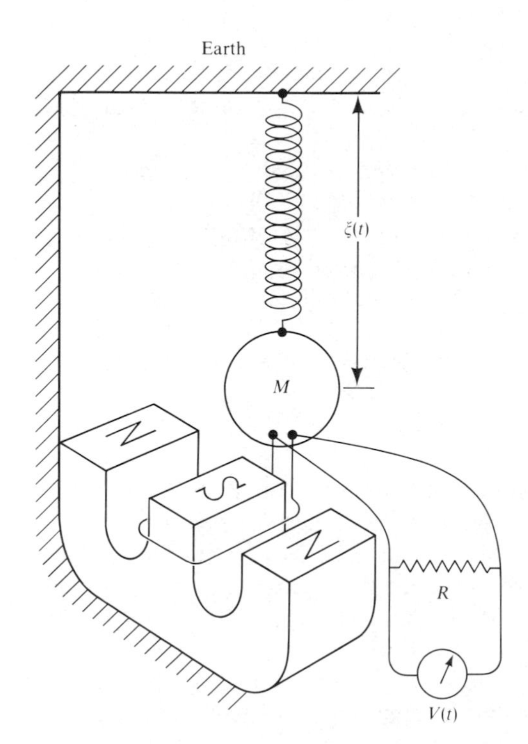

FIGURE **10.15**
When a coil attached to the mass moves through the magnetic field, the voltage across the coil terminals is proportional to the relative velocity between mass and magnet.

This power must be consumed by the resistance $R + R_0$ (R for the shunt and R_0 for the coil), since these are the only dissipative elements of the system. Thus we have

$$VI = IlB\dot{\xi}$$

or

$$V = lB\dot{\xi}. \tag{10.47}$$

If we write G for lB, we find the interesting result that $V = G\dot{\xi}$ and $F = GI$. This is a special case of Onsager's reciprocal theorem on irreversible thermodynamics.

It follows that

$$I = \frac{G\dot{\xi}}{R_0 + R} \tag{10.48}$$

and

$$F = \frac{G^2\dot{\xi}}{R_0 + R}.$$

With this force F acting on the mass, the equation of motion of the pendulum is

$$\ddot{\xi} + \omega_s^2\xi = -\ddot{u} - \frac{G^2}{R_0 + R}\frac{\dot{\xi}}{M}. \tag{10.49}$$

Comparing (10.49) with the equation for a pendulum with dashpot (10.2), we find

$$2\varepsilon = \frac{G^2}{(R_0 + R)M}. \tag{10.50}$$

In general, if the pendulum's mechanical attenuation is not zero but ε_0, the total attenuation is given by

$$\varepsilon = \varepsilon_0 + \frac{G^2}{2(R_0 + R)M}. \tag{10.51}$$

The electric power generated in the shunt resistance by the pendulum motion

$\dot{\xi}(t)$ is, from (10.48),

$$\langle VI \rangle = R\langle I^2 \rangle = \frac{RG^2}{(R_0 + R)^2} \langle \dot{\xi}^2(t) \rangle. \tag{10.52}$$

In order to make the sensor powerful, $G^2/(R_0 + R)$ must be large. Then, from (10.51), we find that the attenuation ε becomes large, making the pendulum response insensitive to ground motion. In order to keep ε small, we have to increase M. This is why a large mass is needed for a sensitive instrument without an electronic amplifier.

Neglecting ε_0, critical damping is achieved for

$$\frac{\varepsilon}{\omega_s} = \frac{G^2}{2(R_0 + R)M\omega_s} = 1 \tag{10.53}$$

or

$$\frac{G^2}{R_0 + R} = 2M\omega_s. \tag{10.54}$$

Putting (10.54) into (10.52), the power is given by

$$\langle VI \rangle = \frac{R}{R_0 + R} \cdot 2M\omega_s \langle \dot{\xi}^2 \rangle. \tag{10.55}$$

The power is proportional to mass and inversely proportional to pendulum period. The internal resistance of the coil cannot be much larger than the shunt resistance for effective operation. For a typical seismograph with mass 5 kg and period 1 sec, (10.55) gives

$$\langle VI \rangle \sim 5 \times 10^3 \times 2\pi \times \langle \dot{\xi}^2 \rangle.$$

The highest ground-noise level, such as caused by traffic in a city built on soft sediment, may amount to $\sqrt{\langle \dot{\xi}^2 \rangle} \simeq 10^{-2}$ cm/sec. Then,

$$\begin{aligned} \langle VI \rangle &\sim 1 \text{ erg/sec} \\ &= 10^{-7} \text{ watt.} \end{aligned}$$

This shows that the idea of generating electricity using seismograph motion due to traffic noise is not very practical.

The electromagnetic sensor was introduced to seismology by Galitzin in 1914. He treated seismometry with the exact methods of experimental physics, and cleared up the question of identification of P- and S-waves once and for all for teleseismic events.

Galitzin used a galvanometer to measure the voltage generated by the electromagnetic sensor. A galvanometer is a coil suspended by a thin fiber at its center of mass in a magnetic field, so that electric current passing through the coil will exert a torque around the fiber. A mirror is attached to the coil, and the deflection of the mirror is optically recorded. This is still one of the most common seismograph systems currently used in observatory seismology.

The power sensitivity of this system is proportional to the pendulum mass, as shown by (10.55). If the mass is large, it is difficult to damp its oscillation electromagnetically as shown in equation (10.50). We need a high G-value to obtain a high-gain seismometer. Before 1939, the quality of magnets was poor and the available magnetic flux B was limited; the only way to increase G was to increase the length l of the coil. Benioff's (1932) variable reluctance seismograph achieved this by using a magnet as the pendulum mass, which moves between two armatures fixed to the seismometer frame and wound with a long wire. He was able to damp critically a pendulum mass of 100 kg at periods of about 1 sec. His seismograph supplied valuable data on short-period body waves at teleseismic distances.

Rihn (1969) gives an estimate for the volume V of magnet necessary to damp critically a pendulum with mass M and period T: for the ALNICO V magnet, $V\,(\text{cm}^3) = 50\,M\,(\text{kg})/T\,(\text{sec})$.

With the improved quality of magnets, we no longer need a 100-kg mass to gain the required sensitivity at short periods. Willmore (1960) summarizes the later development of electromagnetic sensors and concludes that a pendulum mass of 5 kg is just right for a system directly coupled to a galvanometer. He also showed that, for pendulum periods longer than 1 sec, one can minimize the total instrument weight by using the magnet as the pendulum mass.

One can further reduce the mass by using an amplifier or by using a displacement transducer. The Block-Moore accelerometer with a mass of 10 grams can sense a displacement of 1 mμ at 5-sec periods. The classic pendulum seismograph of Wiechert had to overcome the friction of the pen running over smoked paper. The largest mass of a Wiechert seismograph was 17 tons and achieved a magnification of 10^4, which corresponds to a minimum detectable displacement at a 5-sec period of more than 10 mμ.

10.3.3 The response characteristics of standard observatory seismographs

In the usual observatory seismographs, such as those at the stations of the World-Wide Standard Seismograph Network, the output current of the electromagnetic transducer coil is fed into a sensitive galvanometer through an attenuating circuit, such as that shown in Figure 10.16. V_1 is the electromotive force induced in the transducer coil, and V_2 is that in the galvanometer coil. The currents I_1 and I_2 may be expressed in terms of the motions of the two coils.

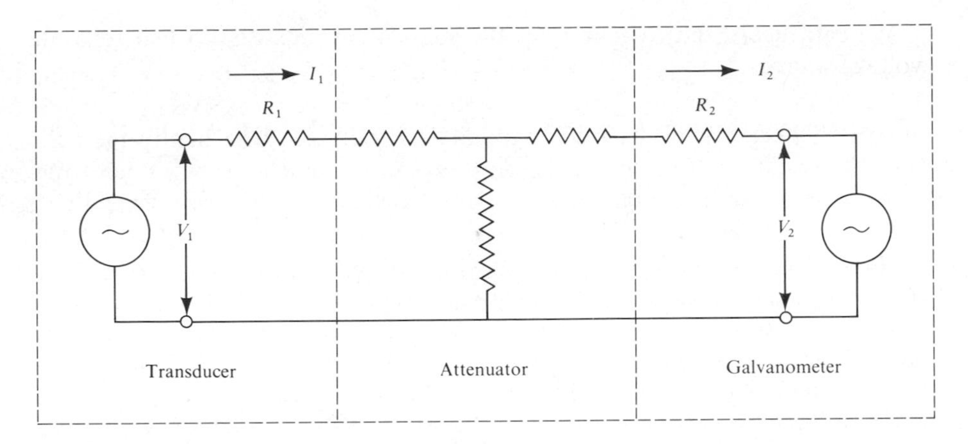

FIGURE **10.16**
Transducer-recorder circuit of a standard observatory seismograph.

The equation of motion for the coils, on the other hand, includes the force terms due to these currents. The complete description of the galvanometer deflection corresponding to a given ground motion requires the solution of a fourth-order differential equation. A historical summary of the analysis of this problem can be found in Eaton (1957). Here we shall follow Hagiwara (1958), who gave a simple and practical description of the solution.

The seismometer motion ξ is affected by a restoring force $-G_1 I_1$, where G_1 is the electrodynamical constant of the seismometer (coil and magnet), so that the seismometer equation of motion is

$$\ddot{\xi} + 2\varepsilon_{0s}\dot{\xi} + \omega_s^2 \xi = -\ddot{u} - \frac{G_1 I_1}{M} \tag{10.56}$$

Here, ε_{0s} is the mechanical damping of the seismometer, ω_s is the undamped resonant frequency, u is the ground motion, and M is the moving mass of (10.2).

The current I_2 in the galvanometer leads to a couple, $G_2 I_2$, which acts to rotate the galvanometer mirror through an angle Φ. Here, G_2 is the electrodynamical constant of the galvanometer, and

$$\ddot{\Phi} + 2\varepsilon_{0g}\dot{\Phi} + \omega_g^2 \Phi = \frac{G_2 I_2}{K}, \tag{10.57}$$

where ε_{0g} is the mechanical damping of the galvanometer, ω_g is its undamped resonant frequency, and K is the moment of inertia of the galvanometer coil and mirror.

We can regard the current I_1 as the sum of currents due to two separate voltage sources:

$$I_1 = (\text{current through } R_1 \text{ with } V_2 = 0) + (\text{current through } R_1 \text{ with } V_1 = 0)$$
$$= I_1\Big|_{V_2=0} \qquad\qquad + I_1\Big|_{V_1=0} \tag{10.58}$$

R_1 here is the resistance of the seismometer coil. V_2 can be maintained at zero if the galvanometer coil is physically restrained from moving; such a coil is said to be *clamped*. Similarly, $I_1|_{V_1=0}$ is given by clamping the seismometer coil. From (10.48) we have

$$I_1\Big|_{V_2=0} = \frac{G_1}{Z_{11}}\dot{\xi},$$

with Z_{11} here being the ratio of V_1 to I_1 when the galvanometer is clamped. Similarly,

$$I_2\Big|_{V_1=0} = -\frac{G_2}{Z_{22}}\frac{d\Phi}{dt},$$

with Z_{22} as the impedance V_2/I_2 with a clamped seismometer. This last current flows partly through the attenuator circuit and partly through the seismometer coil, so that

$$I_1\Big|_{V_1=0} = -\frac{\mu_2 G_2}{Z_{22}}\frac{d\Phi}{dt},$$

where $\mu_2 = (I_1/I_2)|_{V_1=0}$ is an attenuation factor. From (10.58) we can then conclude that

$$I_1 = \frac{G_1}{Z_{11}}\dot{\xi} - \frac{\mu_2 G_2}{Z_{22}}\frac{d\Phi}{dt}, \tag{10.59}$$

and similarly

$$I_2 = \frac{\mu_1 G_1}{Z_{11}}\dot{\xi} - \frac{G_2}{Z_{22}}\frac{d\Phi}{dt} \qquad \text{with } \mu_1 = \left(\frac{I_2}{I_1}\right)_{V_2=0} \tag{10.60}$$

When (10.59) is substituted into the seismometer equation (10.56), we see that an extra damping term is present, and

$$\ddot{\xi} + 2\varepsilon_s\dot{\xi} + \omega_s^2\xi = -\ddot{u} + \frac{\mu_2 G_1 G_2}{Z_{22}M}\frac{d\Phi}{dt}. \tag{10.61}$$

ε_s now is the total seismometer damping, $\varepsilon_{0s} + \frac{1}{2}G_1^2/(Z_{11}M)$ (cf. (10.51)). The reaction of the galvanometer on the seismometer motions is quantified by the last term in (10.61).

When (10.60) is substituted into the galvanometer equation (10.57), we find

$$\ddot{\Phi} + 2\varepsilon_g\dot{\Phi} + \omega_g^2\Phi = \frac{\mu_1 G_1 G_2}{Z_{11}K}\dot{\xi}, \tag{10.62}$$

in which ε_g is the total galvanometer damping, $\varepsilon_{0g} + \frac{1}{2}G_2^2/(Z_{22}K)$.

The seismograph response $\Phi(t)$, for a given input $u(t)$, can be obtained by solving the fourth-order equation that results when ξ is eliminated between (10.61) and (10.62). We find that

$$\ddot{\ddot{\Phi}} + 2(\varepsilon_s + \varepsilon_g)\dddot{\Phi} + [\omega_s^2 + \omega_g^2 + 4\varepsilon_s\varepsilon_g(1 - \sigma^2)]\ddot{\Phi}$$

$$+ 2(\varepsilon_s\omega_g^2 + \varepsilon_g\omega_s^2)\dot{\Phi} + \omega_s^2\omega_g^2\Phi = -\frac{\mu_1 G_1 G_2}{Z_{11}K}\dddot{u}, \tag{10.63}$$

where

$$\sigma^2 = \frac{\mu_1\mu_2 G_1^2 G_2^2}{4Z_{11}Z_{22}MK\varepsilon_s\varepsilon_g} = \frac{\mu_1\mu_1(\varepsilon_s - \varepsilon_{0s})(\varepsilon_g - \varepsilon_{0g})}{\varepsilon_s\varepsilon_g}.$$

The quantity σ^2 is a dimensionless measure of the *coupling* between seismometer and galvanometer. The coupling is small when the attenuation is strong, and the electromagnetic damping is small relative to the mechanical damping. The effect of coupling on the shape of the seismograph response is strongest when the periods of pendulum and galvanometer are equal. Even then, according to Hagiwara (1958), if we take $\sigma^2 < \frac{1}{4}$, the maximum deviation from the zero coupling response is about 20% in amplitude and about 15° in phase. If the coupling is neglected ($\sigma^2 = 0$), the reaction term $\propto d\Phi/dt$) in (10.61) is taken as zero, and the response function of the seismograph can be easily calculated. Taking $u = U\exp(-i\omega t)$, the solution of (10.61) is $\xi = \omega^2 U\exp(-i\omega t)/[-\omega^2 - 2i\varepsilon_s\omega + \omega_s^2]$. From this as input to (10.62) we find

$$\Phi = \frac{\mu_1 G_1 G_2(-i\omega)}{Z_{11}K[-\omega^2 - 2i\varepsilon_g\omega + \omega_g^2]}\frac{\omega^2 U\exp(-i\omega t)}{[-\omega^2 - 2i\varepsilon_s\omega + \omega_s^2]}. \tag{10.64}$$

For large ω, we have

$$\Phi \sim \frac{\mu_1 G_1 G_2}{Z_{11}K}\left(\frac{1}{i\omega}\right)U\exp(-i\omega t). \tag{10.65}$$

For small ω, we have

$$\Phi \sim \frac{\mu_1 G_1 G_2}{Z_{11} K} \left(\frac{\omega^3}{i\omega_s^2 \omega_g^2} \right) U \exp(-i\omega t). \qquad (10.66)$$

The above equations show that the amplitude response peaks at frequencies between ω_s and ω_g, decreases with increasing frequency as ω^{-1} for higher frequencies, and decreases with increasing period as T^{-3} for longer periods.

The absolute value of seismograph sensitivity is determined by the product $G_1 G_2 \mu_1 / Z_{11} K$, which is easily measured. For example, if we put a constant current I_0 through the galvanometer coil, the deflection Φ is given by (10.57) as

$$\Phi = \frac{G_2}{K\omega_g^2} I_0,$$

from which G_2/K can be calculated. Other quantities are discussed earlier.

The phase response can also be calculated easily using (10.64). Putting $\Phi = |\Phi| \exp[-i\omega t + i\phi(\omega)]$, $\phi(\omega)$ represents the instrumental phase-delay. A small problem here is the choice of sign in (10.64) or a phase uncertainty by π. The choice depends on how the instrument is calibrated. Suppose we are calibrating a vertical-component seismograph and we apply a downward impulse on the pendulum mass. The galvanometer deflection trace will swing to one direction, which will be marked as "up" because the downward mass movement corresponds to an upward ground movement. For an impulsive movement, the seismograph will respond according to the high-frequency asymptotic characteristics (10.65). With the choice of sign given in (10.64)–(10.66), the response would be like a *negative* step, because this is the time-domain signal corresponding to $(1/i\omega)$ as $\omega \to \infty$. But since we now designate the direction of galvanometer swing as "up" when the ground moves impulsively "up," we change the sign of (10.65) so that

$$\Phi \sim \frac{\mu_1 G_1 G_2}{Z_{11} K} U \cdot \frac{1}{\omega} \exp(-i\omega t + i\pi/2) \qquad \text{as } \omega \to \infty.$$

The signs for (10.64) and (10.66) must also be reversed. The phase delay is $+\pi/2$ for infinite frequency, decreases monotonically with decreasing frequency, and reaches $-3\pi/2$ at zero frequency (see Box 10.1).

The phase values shown in Hagiwara's original figures must be corrected by adding π in order to be consistent with this conventional method of marking the direction of Earth movement on the record. Note also that we obtain phase *delay* from his figures, since he used $\exp(+i\omega t)$. Corrected phase delays as well

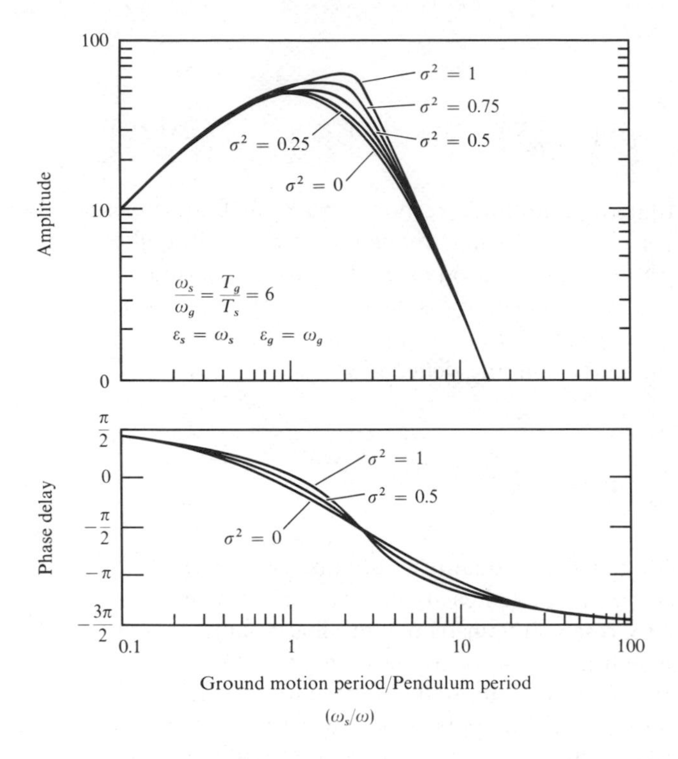

FIGURE **10.17**
Amplitude and phase-delay response for various coupling constants σ^2. [After Hagiwara, 1968.]

as the amplitude response curves are shown in Figure 10.17 for various coupling constants σ^2, in the case of $\omega_s/\omega_g = 6$ and $\varepsilon_s/\omega_s = \varepsilon_g/\omega_g = 1$, which roughly apply to the long-period seismographs of WWSSN stations. (Such seismographs are often referred to as "15–100 instruments." The numbers here are, respectively, the seismometer free period, and the galvanometer free period, both in seconds. Also common are 30–100 instruments. A typical short-period seismograph has the configuration 1–0.75.)

10.3.4 High sensitivity at long periods

The frequency response of the standard velocity-transducer-galvanometer system attenuates inversely proportional to the cube of period at long periods, as shown in (10.66). By increasing the pendulum period, the highly attenuating range can be pushed further toward longer periods. In Section 10.1 we have mentioned LaCoste's pendulum, which operates stably at periods of several tens of seconds. Another more recent effort is Benioff's mercury tube tiltmeter; it makes use of the inertia of mercury, which oscillates with a long period between two containers connected by a tube.

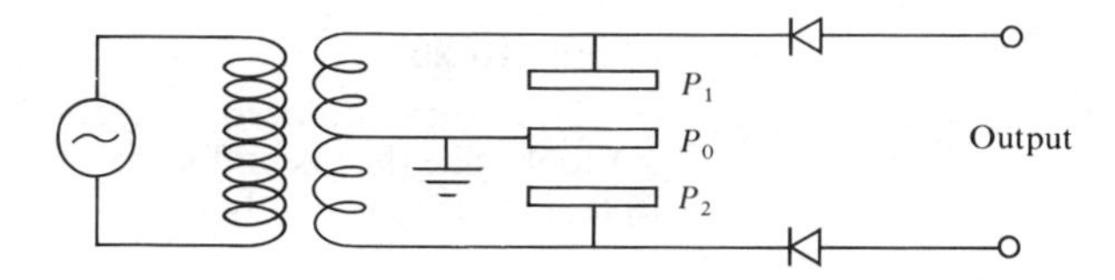

FIGURE **10.18**

Displacement transducer used in Benioff's strainmeter.

Two basically different ways of improving high sensitivity at long periods are the use of (1) a displacement transducer instead of a velocity transducer and (2) a strain sensor instead of an inertial sensor. Either way, the low-frequency response is improved proportionately with the period. Both improvements were combined in Benioff's extensometer (1959), which successfully recorded the Earth's free oscillation during the great Chilean earthquake of 1960.

The displacement transducer used by Benioff consists of a capacitor plate P_0 attached to the moving part of the seismometer sandwiched between two plates P_1 and P_2 fixed to the frame (Fig. 10.18). The direction of movement of P_0 is perpendicular to the plane of the plate. A fixed-frequency oscillator (5 MHz in Benioff's, and 1.5 MHz in that of Major et al., 1964) drives two tuned, parallel, inductance-capacitance circuits at a frequency close to, but not equal to, their common resonant frequency. When the plate P_0 is in the midposition between P_1 and P_2, the averaged voltage across the output terminal is zero. Movement of P_0 makes one circuit more nearly resonant and the other less so. The resulting imbalance in magnitudes of voltage is nearly proportional to the displacement of P_0. The relation between the output voltage and displacement is schematically shown in Figure 10.19. Because of the pronounced nonlinearity for large displacement, more recent displacement transducers use a different circuitry.

The new circuit shown in Figure 10.20 works as follows. A fixed-frequency oscillator (16 kHz in Block and Moore, 1966; 3 kHz in Stacey et al., 1969; and

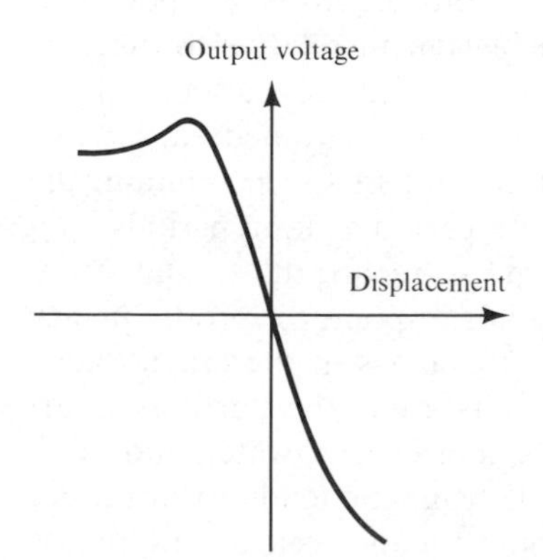

FIGURE **10.19**

Response of displacement transducer shown in Figure 10.18.

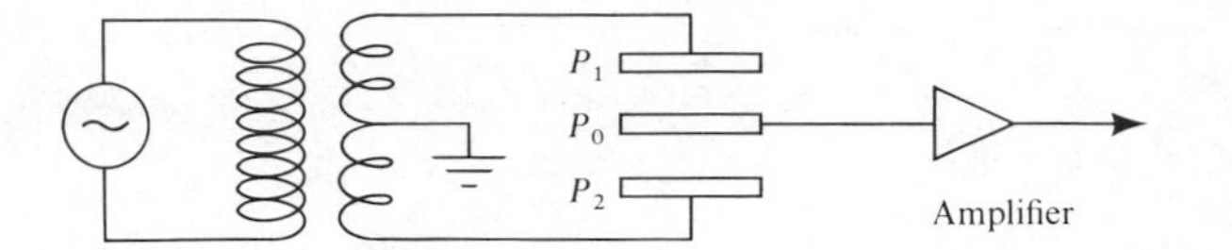

FIGURE **10.20**
Displacement transducer. Although this looks like the one in Figure 10.18, it works differently.

500 kHz in Buck et al., 1971) supplies equal and opposite voltage on the capacitor plates P_1 and P_2 by use of a transformer with a split secondary. The center tap of the secondary is grounded. When the plate P_0 is in the midposition, the voltage of P_0 relative to the ground is zero. Since the capacity formed by parallel plates is inversely proportional to the distance between the plates, the voltage of P_0 is directly proportional to the displacement of P_0 from the midpoint. This voltage can be measured very precisely by the tuned amplifier (lock-in amplifier or synchronous detector) because the signal frequency is precisely known. This device has been successfully used by Block and Moore in their accelerometer, and by Stacey et al. and Buck et al. in their mercury-tube tiltmeters.

The strain seismometer has a certain advantage over the inertial seismometer as a sensor of long-period waves. Consider a simple harmonic wave with velocity c propagating in the x-direction. Let the displacement u in the x-direction be

$$u = U \exp[-i\omega(t - x/c)].$$

Then the extensional strain in the x-direction is

$$\frac{\partial u}{\partial x} = i\frac{\omega}{c} U \exp[-i\omega(t - x/c)].$$

If a displacement transducer is used, the output voltage will be proportional to $\omega U/c$. On the other hand, the pendulum displacement relative to the seismometer framework, at periods longer than the pendulum period, will be proportional to $\omega^2 U$ (see (13.2)). In other words, as mentioned earlier, the strain sensor's frequency response is better than that of the pendulum, the improvement increasing proportionately with the period at long periods.

Another advantage of the strainmeter is that it can record the secular strain change, whereas the pendulum seismometer cannot respond to zero frequency signals except for those produced by tilt and by changes in the gravity field.

A disadvantage of the strainmeter, however, is its use of the Earth as a part of the instrument. The presence of cracks, joints, loose rocks, water, and other weak material having unpredictable mechanical behavior tends to introduce undesirable noise and makes the system more nonlinear and even nonstationary.

BOX **10.1**

General features of instrument response

We have seen how an instrument that records seismic motions can usefully be thought of in terms of several components. It is informative to look at the frequency response of each component, using an approximate theory if necessary to get a simple result. We shall find responses that are proportional to different powers of frequency, so that graphical display is best done with log-log plots.

As an example of the basic inertial sensor, we have considered both a mass on a spring and pendulums. For these devices, the output ξ (Fig. 10.1) has an amplitude response $|X(\omega)|$ given by (10.3). $|X(\omega)|$ is just the ratio between output and input amplitude spectra, $|\xi(\omega)|/|u(\omega)|$. It tends to a constant at high frequencies, and is proportional to ω^2 at low frequencies, so that the behavior is roughly as shown in part (a) of the figure below.

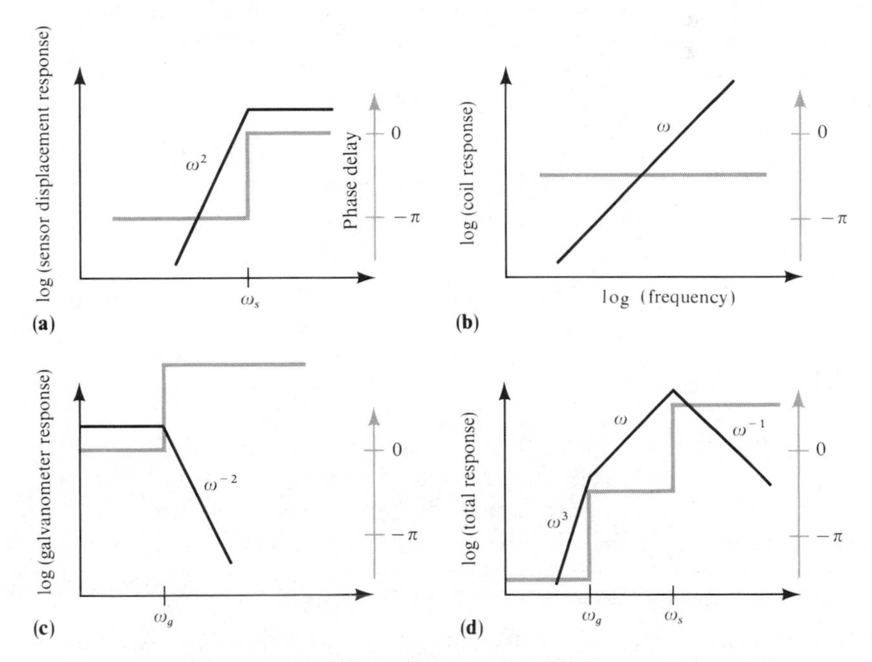

(**a**) The main features of the sensor's displacement response are its constant amplitude at frequencies above ω_s and its rapid fall-off (12 db per octave) below ω_s. (**b**) The coil response is proportional to frequency at all frequencies. (**c**) The galvanometer response is constant below its natural frequency (ω_g) and falls off rapidly at higher frequencies. (**d**) Summing the first three log (response) curves, in the case $\omega_g < \omega_s$, we see an extremely rapid fall-off at low frequencies. If, instead, $\omega_s < \omega_g$, the shape of the total response is unchanged. Note that double differentiation to get acceleration response (i.e., response to ground acceleration) corresponds to multiplying (a) by ω^{-2}, so that it will be flat below ω_s and decay according to ω^{-2} at lower frequencies.

The next component may be a coil, for which $\xi(\omega)$ is the input and current $I(\omega)$ is the output. From (10.48), we see that the amplitude-response spectrum of this component is proportional to ω (part (b) of figure), because $\dot{\xi}(\omega) = -i\omega\xi(\omega)$.

The next component may be a galvanometer, for which $I(\omega)$ is the input and the mirror angle $\Phi(\omega)$, perhaps multiplied by some frequency-independent constant, is the output. We have made the simplifying assumptions that the current in the galvanometer is directly related (proportional) to the current in the seismometer coil and that neither current has a feedback effect (coupling) on the pendulum motion. From (10.57), it follows that the galvanometer response, $\Phi(\omega)/I(\omega)$, has an amplitude that is constant at low frequencies and behaves like ω^{-2} at high frequencies (part (c) of figure).

The total response is a product of the component responses, becoming a summation when studied with log (response) plots. Typically, one finds for the so-called "velocity pickup" (i.e., the coil), that the total response is as shown in part (d) of the figure. If the inertial sensor has its displacement output (instead of its velocity output) directly converted to an electrical signal, then the final response is merely the sum of parts (a) and (c). In modern instruments, the seismometer output (inertial sensor plus coil) may become the input to a variety of electronic amplifiers with different frequency characteristics, in which case curves in parts (a) and (b) must be added to the amplifier response.

The phase-delay spectrum is approximated, for the inertial sensor, by a step jump from $-\pi$ to 0 as ω increases through ω_s. (It is assumed that the direction of ground motion has been properly marked, as discussed following equation (10.3).) For the coil, the phase delay is constant at $-\pi/2$, and for the galvanometer the phase delay is a step jump from 0 to π as ω increases through ω_g. These delays are shown in gray in the figure.

10.3.5 The nonlinearity of the seismic sensor

The response of any physical system is nonlinear unless the magnitude of the input is very small. The seismograph is no exception. Let us start with the nonlinearity of the standard seismograph described in Section 10.3.3. Berckhemer and Schneider (1964) made a careful study of ground displacement by solving the fourth-order differential equation of the Galitzin-Wilip seismograph (T_s (seismometer pendulum period) = 12 sec, T_g (galvanometer period) = 12 and 50 sec) and the Press-Ewing seismograph (T_s = 30 sec, T_g = 100 sec), on an analogue computer at the Stuttgart Institute of Technology.

The records of earthquakes at long distances indicated the ground displacement coming back to the initial position after the passage of seismic waves. On the other hand, the waveforms from nearby earthquakes indicated a residual displacement, or more precisely a parabolic increase in displacement corresponding to a permanent change in acceleration. A typical record of such a long-period transient waveform is shown in Figure 10.21. In order to explain these records, it was required that an earthquake of magnitude 5 produce a permanent change in acceleration of the order of 10^{-7} to 10^{-8} g at a distance

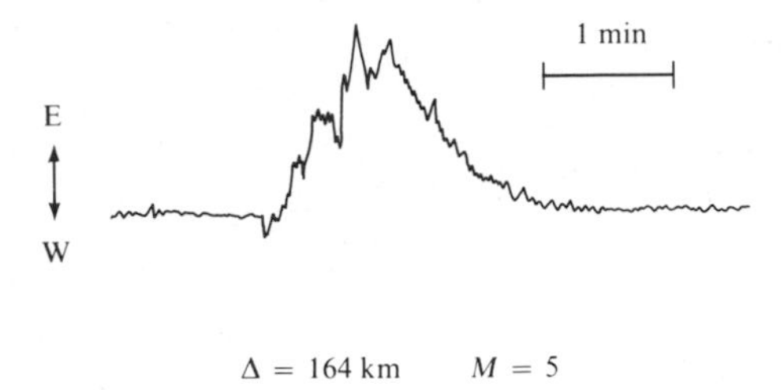

FIGURE **10.21**
Typical record of a near earthquake by a standard long-period seismograph.

of about 100 km. This means a change of gravity by 0.01–0.1 milligal for the vertical component, and a tilt of 10^{-7}–10^{-8} radian for the horizontal component. Press (1965) suspected instrumental hysteresis as the cause, inasmuch as the observed tilt was three to four orders of magnitude larger than the value he calculated for that size earthquake. Berckhemer and Schneider, however, considered that the tilt was a real phenomenon, but rejected the reality of gravity change. They attributed it to the nonlinearity of the vertical-component seismograph.

The nonlinearity considered by them is due to the bowstringing effect of the helical spring. A strong earthquake at a short distance will shake the seismograph strongly and may cause an oscillation of the spring, as depicted in Figure 10.22. Assuming a sinusoidal fundamental mode of oscillation with the amplitude a at the center, the length of spring will be

$$l \sim l_0\left[1 + \frac{1}{4}\left(\frac{a\pi}{l_0}\right)^2\right].$$

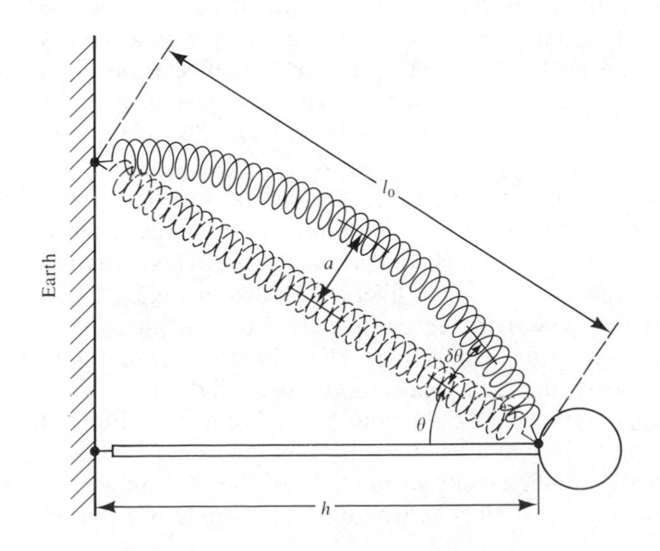

FIGURE **10.22**
Bow-stringing of a helical spring.

On the other hand, the angle $\delta\theta$ will be

$$\delta\theta \sim \frac{a\pi}{l_0}.$$

The torque exerted by the spring is

$$(T + \delta T)h \sin(\theta + \delta\theta) \sim Th \sin\theta + \delta Th \sin\theta + Th \cos\theta\, \delta\theta - Th \sin\theta \frac{(\delta\theta)^2}{2},$$

where T is the tensional force exerted by the spring to maintain itself at length l_0, and δT is the change in T due to the bow-stringing of the spring.

For the zero-initial-length spring, another mysterious effect is found. Putting the spring constant at k, we have, for the zero-initial-length spring,

$$T = kl_0$$

and

$$\delta T = k\frac{l_0}{4}\left(\frac{a\pi}{l_0}\right)^2.$$

Then

$$(T + \delta T)h \sin(\theta + \delta\theta) = Th \sin\theta\left[1 - \frac{1}{4}\left(\frac{a\pi}{l_0}\right)^2\right] + Th \cos\theta\, \delta\theta.$$

Since $\delta\theta$ is a rapidly oscillating term due to bowstringing, it vanishes on the average. Thus the effect of bowstringing averaged over time would always reduce the torque in the spring and lower the mass position. This effect transforms a high-frequency disturbance into a low-frequency spurious signal. For the LcCoste-type vertical seismometer, this effect would always look like a step increase of gravity.

An advantage of the Block-Moore accelerometer over the LaCoste gravimeter is its simple pendulum structure, which minimizes the nonlinear effect. Their pendulum is simply a brass plate attached to a single piece of fused quartz, parts of which have been drawn out into two fibers as shown in Figure 10.23. This structure eliminates the bow-stringing completely, but can have other high-frequency modes of oscillation, which may give rise to spurious low-frequency signals superimposed upon the main mode of oscillation.

Block and Moore discuss several possible causes of nonlinearity. First of all, they force the pendulum to stay at a fixed position by the use of feedback. This eliminates the effect of inhomogeneity in the field of the sensing device, as well as the approximation that $\sin\theta \sim \theta$, where θ is the angle of rotation

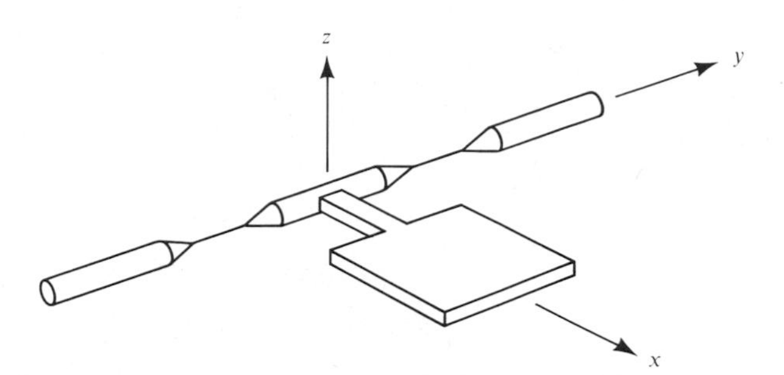

FIGURE **10.23**

Metal plate attached to a short bar of quartz, two parts of which have been drawn out into fibers. The plate is small ($\sim 10\ \text{cm}^2$), and is held horizontally by torsion in the fibers. It acts as a vertical pendulum in the Block-Moore accelerometer. [After Block and Moore (1970); copyrighted by American Geophysical Union.]

of the pendulum around the pivotal point. Then they discuss the effect of cross-coupling from the twisting mode (Fig. 10.24) and the swinging mode (Fig. 10.25), both of which can generate a spurious long-period signal.

Nonlinearity is also a serious problem for the strain seismometer. Sacks et al. (1971) specifically designed a strainmeter free from the effects of nonlinearity and hysteresis, and installed three of them at Matsushiro, Japan, close to a quartz-tube strainmeter. Their instrument consists of a liquid-filled resilient tube, which is buried in a borehole and held in tight contact with the rock wall. The coupling between the wall and the tube is made by the use of expanding cement. The tube is supposed to follow faithfully the minute distortion of the borehole down to 10^{-6} microns. The instrument measures only the dilatational strain. Their principal observation is that a number of strain changes associated with near earthquakes have been recorded by the quartz extensometer, but not by the new strainmeter. The sensitivity of the latter has been tested by comparing the amplitude of *P*-waves and Rayleigh waves with the standard

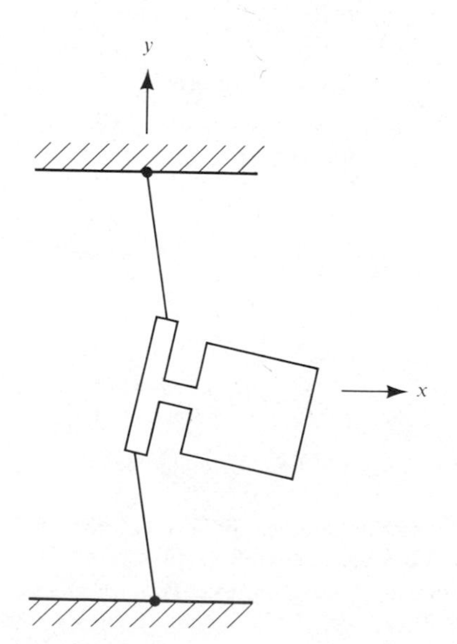

FIGURE **10.24**

Twisting Mode. [After Block and Moore (1970); copyrighted by the American Geophysical Union.]

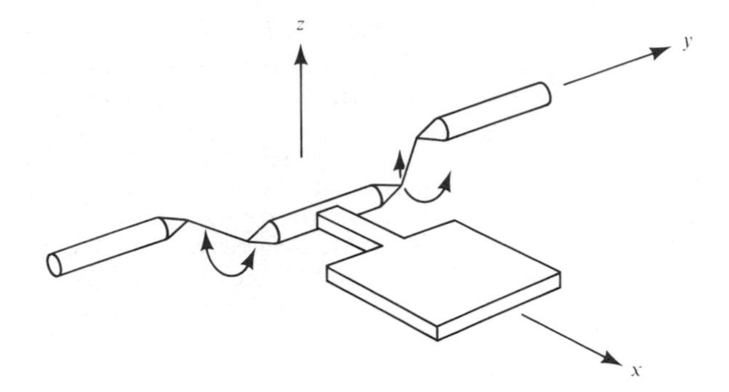

FIGURE **10.25**
Swinging mode. [After Block and Moore (1970); copyrighted] by the American Geophysical Union.

pendulum seismograph. This result suggests that many strain changes observed by the quartz extensometer may be spurious.

Currently, a program is underway to upgrade the World-Wide Standard Seismograph Network by installing a new set of instruments at selected sites designated as Seismic Research Observatories (SRO's) (Peterson et al., 1976). To reduce the noise on horizontal components (Section 10.2.5), the seismometers are placed in a 7-inch-diameter borehole at a depth of about 100 m. The mechanical configuration of the sensors is of more conventional design than that of Block and Moore. A LaCoste pendulum is used for the vertical component and "garden-gate" suspensions (Fig. 10.4b) for the horizontal components. To gain in linearity and stability, the sensor is force-balanced through a capacitance transducer. The sensors are sealed in containers, then baked and evacuated to drive out moisture and air, thus avoiding internal convections. Another new program, known as IDA (International Deployment of Accelerometers) and operated by the University of California at San Diego, is planned to incorporate 20 stations around the world (Agnew et al., 1976). Each station is to have a digitally recorded LaCoste-Romberg gravimeter. Dynamic range of the recording is 120 db. The signal is filtered prior to recording, with one channel intended for studying Earth tides, and another amplifying the band between 1 and 30 cph for studying normal modes.

SUGGESTIONS FOR FURTHER READING

Beauchamp, K. G. (editor). *Exploitation of Seismograph Networks*. Leiden, Noordhoff: International Publishing, 1975.

Berger, J. Application of laser techniques to geodesy and geophysics. *In* H. E. Landsberg (editor), *Advances in Geophysics* (Vol. 16). New York: Academic Press, 1973.

Hagiwara, T. A note on the theory of the electromagnetic seismograph. *Bulletin of the Earthquake Research Institute*, **36**, 139–164, 1958.

Kosminskaya, I. P., N. N. Ruzyrev, and A. K. Alekseev. Explosion seismology, its past, present, and future. *Tectonophysics*, **13**, 309–323, 1972.

Murphy, A., J. Savino, J. Rynn, G. Choy, and K. McCamy. Observations of long-period (10s–100s) seismic noise at several worldwide locations. *Journal of Geophysical Research*, **77**, 5042–5049, 1972.

Rodgers, P. W. The response of the horizontal pendulum seismometer to Rayleigh and Love waves, tilt, and free oscillations of the earth. *Bulletin of the Seismological Society of America*, **58**, 1384–1406, 1968.

Sax, L. R. Stationarity of seismic noise. *Geophysics*, **33**, 668–674, 1968.

Shapiro, I. I., and C. A. Knight. Geophysical applications of long-baseline radio interferometry. *In* L. Mansinha, D. E. Smylie, and A. E. Beck (editors), *Earthquake Displacement Fields and the Rotation of the Earth*. New York: Springer-Verlag, 284–301, 1970.

Willmore, P. L. The detection of earth movements. *In* S. K. Runcorn (editor), *Methods and Techniques in Geophysics* (Vol. 1). New York: Interscience, 1960.

PROBLEMS

10.1 Show that the convolution (10.11) can also be written in the form

$$\xi(t) = \int_0^\infty u(t - \tau)\ddot{f}(\tau)\, d\tau.$$

10.2 Show that the period T of a mass suspended vertically by a spring is equal to $T = 2\pi[(l - l_0)/g]^{1/2}$, where l is the equilibrium length of the spring (under its gravitational load) and l_0 is the initial length (under no tension).

10.3 This problem is motivated by the need to know how much ground motion is caused by atmospheric disturbances, and how much this motion (which is noise, for most studies) can be reduced by burying seismometers at some depth (Sorrells, 1971).

a) Consider a homogeneous half-space with its surface at $z = 0$. Suppose that a moving pressure is applied to the surface, and is modeled by boundary conditions $\tau_{zx} = \tau_{yz} = 0$, $\tau_{zz} = -P\exp[i\omega(x/c - t)]$.

b) Show that *P-SV* motion in the half-space is given by the motion-stress vector $\mathbf{f} = \mathbf{F}\mathbf{w}$, where $\mathbf{f}$ and $\mathbf{F}$ are given by (5.60) and (5.65), and

$$\mathbf{w} = \frac{iP}{\omega\rho\beta^2 \mathsf{R}(p)} \begin{pmatrix} \left(\dfrac{1}{\beta^2} - 2p^2\right)\dfrac{1}{\alpha} \\ \dfrac{-2p\xi}{\beta} \\ 0 \\ 0 \end{pmatrix}$$

where $p = c^{-1}$, $\mathsf{R}(p)$ is the Rayleigh function (5.54), and $\xi = (1/\alpha^2 - p^2)^{1/2}$.

c) If $c \ll \beta$ and $\omega zc/2\beta^2 \ll 1$ show that the horizontal and vertical components of displacement are, respectively,

$$u_x = \frac{icP}{\omega\rho\beta^2}\left[\frac{\beta^2}{\alpha^2 - \beta^2} - \frac{\omega z}{c}\right]e^{-\omega z/c} \exp\left[i\omega\left(\frac{x}{c} - t\right)\right]$$

and

$$u_z = \frac{cP}{\omega\rho\beta^2}\left[\frac{\alpha^2}{\alpha^2 - \beta^2} + \frac{\omega z}{c}\right]e^{-\omega z/c} \exp\left[i\omega\left(\frac{x}{c} - t\right)\right].$$

10.4 From the outputs of a pendulum and a strainmeter at the same site, how could you obtain the surface-wave dispersion? (That is, how could you study the frequency dependence of phase velocity $c(\omega)$, where c is given in terms of frequency and horizontal wavenumber by $c = \omega/k$?)

APPENDIX 1

Glossary of Waves

Air waves Audible sounds are sometimes generated by earthquakes; a local earthquake may sound like distant thunder. Instrumental measurements show that these sounds arrive simultaneously with the first *P*-waves (Hill et al., 1976). Long-period (minutes to hours) acoustic-gravity waves are also excited by great earthquakes, as well as by volcanic explosions, meteorite falls, and nuclear blasts in the atmosphere (Harkrider, 1964).

Air-coupled surface waves Despite the great density contrast between air and ground, atmospheric pressure disturbances traveling over the Earth's surface can amplify surface waves if the phase velocity is equal to the acoustic velocity in the air. Such a coupling with air has been observed for flexural waves in ice sheets floating on the ocean and for Rayleigh waves in ground with low-velocity surface layers (Press et al., 1951). A simultaneous arrival of air waves and tidal disturbances was observed after the famous explosion of the volcano Krakatoa in 1883.

Airy phase Portions of dispersed wavetrains associated with the maxima or minima of the group velocity (as a function of frequency). The stationary phase approximation blows up for them, but the method we describe in Problem 7.8 is still accurate. Alternatively, an Airy function can be used for an approximate calculation of waveform (Savage, 1969b). Examples are continental Rayleigh waves at periods around 15 sec and mantle Rayleigh waves at periods of 200 to 250 sec.

Body waves Waves that propagate through an unbounded continuum are called body waves, as opposed to surface waves, which propagate along the boundary surface.

c This symbol is used to indicate the reflection at the core-mantle boundary for waves incident from the mantle. For example, incident *S*-waves are designated *ScS*. A major study of *PcP* is described by Frasier and Chowdury (1974).

Coda waves of a local earthquake The seismograms of a local earthquake usually show some vibrations long after the passage of body waves and surface waves. This portion of the seismogram to its end is called the coda. They are believed to be back-scattering waves due to lateral inhomogeneity distributed throughout the Earth's crust and upper mantle (Aki and Chouet, 1975).

Converted waves Conversion of *P* to *S* and *S* to *P* occurs at a discontinuity for non-normal incidence. These converted waves sometimes show distinct arrivals on the seismogram between the *P* and *S* arrivals, and may be used to determine the location of the discontinuity.

Crary waves Crary waves are a train of sinusoidal waves with nearly constant frequency observed on a floating ice sheet. They are multi-reflected *SV*-waves with horizontal phase velocity near the speed of compressional waves in ice.

Depth phases (*pP*, *pS*, *sP*, *sS*) The symbol *pP* has been used for *P*-waves propagated upward from the hypocenter, turned into downward propagating *P*-waves by reflection at the free surface, and observed at teleseismic distances. *sS*, *sP*, and *pS* have analogous meanings. For example, *sP* corresponds to a phase that ascends from the focus to the surface as *S*-waves and then, after reflection, travels as *P*-waves to the recording station. These phases are useful for an accurate determination of focal depth. See Figure 9.15 for *sS*.

Diffracted *P* The *P*-wave ray path from a surface focus that grazes the Earth's core emerges at an epicentral distance of about 103°. Although geometrical optics predicts no direct arrivals of *P*-waves in the shadow zone beyond this distance, we continue to observe *P*-waves, especially of long period, up to distances of at least 130°. They are diffracted around the core boundary. See Figure 9.23b.

Flexural waves A normal mode in an infinite plate in vacuum with motion antisymmetric with respect to the median plane of the plate. Examples in nature are the waves in floating ice. (For short waves, the period equation for a normal mode reduces to that for flexural waves in a plate modified slightly by the presence of water. For long waves, however, the gravity term in the period equation dominates, and the mode approaches that of gravity waves in water.)

Frozen waves In the epicentral area of a great earthquake, walls, embankments, and the like are sometimes left in the form of a wave. These "frozen waves" are attributed to cracking open of the ground at the crests of the waves, sometimes with the emission of sand and water.

Gravity waves Normal modes in a surface layer with very low shear velocity, such as unconsolidated sediments, may be affected significantly by gravity at long periods (Gilbert, 1967). Waves similar to the gravity waves in a fluid layer are possible in addition to the shortening of wavelength of normal modes by gravity. So-called "visible waves" with large amplitude and relatively long periods observed in the epicentral area of a great earthquake have been suggested to be gravity waves (Lomnitz, 1970).

Ground roll A term used in exploration seismology to refer to surface waves generated from explosions. They are characterized by low velocity, low frequency, and high amplitude, and are observed in regions where the near-surface layering consists of poorly consolidated, low-velocity sediments overlying more competent beds with higher velocity. Thus ground roll usually consists of Rayleigh waves.

Guided waves Guided waves are trapped in a wave-guide by total reflections or bending of rays at the top and bottom boundaries. An outstanding example is the acoustic waves in the SOFAR channel, a low-velocity channel in the ocean. Since the absorption coefficient for sound in seawater is quite small for frequencies on the order of a few hundred cycles per second, transoceanic transmission is easily achieved. If we consider the Earth's surface as the top of a wave-guide, surface waves, such as Rayleigh, Love, and their higher modes, are guided waves. The waves associated with a low-velocity channel in the crust or mantle may be interpreted as normal modes with concentration of energy in the channel. Where they can exist, guided waves may propagate to considerable distances, because they are effectively spreading in only two spatial dimensions.

***G*-waves (*Gn*)** Another name for long-period Love waves. Because the group velocity of Love waves in the Earth is nearly constant (4.4 km/sec) over the period range from about 40 to 300 sec, their waveform is rather impulsive, and they have received this additional name. They are called *G*-waves after Gutenberg. It takes about $2\frac{1}{2}$ hours for *G*-waves to make a round trip of the Earth. After a large earthquake, we observe a sequence of *G*-waves named *G*1, *G*2, . . . , *Gn*, according to the arrival time. The odd numbers refer to *G*-waves traveling in the direction from epicenter to station, and the even numbers to those leaving the epicenter in the opposite direction and approaching the station from the antipode of the epicenter. See Problem 7.9 and Figure 8.5.

Head waves Head waves are observed in a half-space that is in welded contact with another half-space with higher velocity when the seismic source is located in the lower-velocity medium. The ray path of head waves is along the interface, and the wavefront in the lower-velocity half-space is a part of the surface of an expanding cone. For this reason, head waves are also called "conical waves." An example of head waves is P_n with a ray path along the top of the mantle.

I*, *i The symbol *I* is used to indicate that a part of the ray path lies inside the Earth's inner core as *P*-waves. For example, *PKIKP* refers to *P*-waves that have penetrated to the interior of the inner core and returned to the surface without conversion to *S*-waves throughout the entire path. On the other hand, *i* is used to indicate reflection at the boundary between outer and inner core in the same manner that *c* is used for reflection at the core-mantle boundary.

Inhomogenous plane waves Plane waves with amplitudes varying in a direction different from the direction of propagation. The velocity of propagation is lower than that of the regular plane waves. They are also called "evanescent waves."

J The symbol *J* is used to indicate that a part of the ray path lies inside the Earth's inner core as *S*-waves.

K *P*-waves in the outer core are designated as *K*. For example, *S*-waves incident from the mantle, converted to *P*-waves at the core boundary, propagated through the outer core as *P*-waves, and converted back to *S*-waves at the re-entry to the mantle are designated as *SKS*. Just as *PP*, *PPP*, etc. are used to designate surface reflections, *KK*, *KKK*, etc. are used for *P*-waves in the core reflected at the core-mantle interface from below.

***L* (*LQ*, *LR*)** The symbol *L* is used to designate long-period surface waves. When the type of surface wave is known, *LQ* and *LR* are used for Love and Rayleigh waves, respectively.

Leaking modes Normal modes in a layered half-space, in general, have cutoff frequencies below which the phase velocity exceeds the *P*- and/or *S*-velocities of the half-space, and the energy leaks through the half-space as body waves. Because of the leakage, the amplitude of leaky modes attenuates exponentially with distance.

***Lg*-waves** Short-period (1–6 sec) large-amplitude arrivals with predominantly transverse motion (Ewing et al., 1957). *Lg*-waves propagate along the surface with velocities close to the average shear velocity in the upper part of the continental crust (Herrin and Richmond, 1960). The waves are observed only when the wave path is entirely continental. As little as 2° of intervening ocean is sufficient to eliminate the waves. When *Lg*-waves arrive in two distinct groups, they are called *Lg*1 and *Lg*2.

***Li*-waves** These are similar to *Lg*-waves, but their existence is not as widely accepted as that of *Lg*. The velocity of *Li*-waves is 3.8 km/sec (as compared to 3.5 km/sec for *Lg*) and may be associated with the lower continental crust (Båth, 1954, 1957).

Longitudinal waves Displacement associated with far-field *P*-waves in a homogeneous isotropic solid is parallel to the direction of propagation. For this reason, *P*-waves are also called "longitudinal waves."

Love waves *SH*-waves having their largest amplitudes confined near the surface of an elastic body. Their existence was first predicted (by A. E. H. Love) for a homogeneous layer overlying a homogeneous half-space with an *S*-wave velocity greater than that of the layer. They can exist, in general, in a vertically heterogeneous medium. They cannot, however, exist in a homogeneous half-space as long as the surface is plane.

M This symbol is used to designate wave groups with the maximum amplitude in a seismogram.

***M*1 and *M*2 waves** *M*1 and *M*2 are symmetric and antisymmetric normal modes in an infinite plate in a vacuum. Each has a sequence of fundamental and higher modes, designated as *M*11, *M*12, . . . , or *M*21, *M*22, ... A similar classification has been attempted for the fundamental and higher modes of Rayleigh waves in a layer over a half-space. However, the distinction between *M*1 and *M*2 (sometimes called Sezawa waves) for a layered half-space is not as clear as for a plate.

Mantle Rayleigh waves Just as long-period Love waves are given another name, "*G*-waves," long-period Rayleigh waves are sometimes called mantle Rayleigh waves.

Microseisms Continuous ground motion constituting background noise for any seismic experiment. Microseisms with frequencies higher than about 1 Hz are usually caused by artificial sources, such as traffic and machinery, and are sometimes called microtremors, to be distinguished from longer-period microseisms due to natural disturbances. At a typical station in the interior of a continent, the microseisms have predominant periods of about 6 sec. They are caused by the pressure from standing ocean waves, which may be formed by waves traveling in opposite directions in the source region of a storm or near the coast (Longuet-Higgins, 1950).

Normal modes Normal modes were originally defined as free vibrations of a system with a finite number of degrees of freedom, such as a finite number of particles connected by a massless spring. Each mode is a simple harmonic vibration at a certain frequency called an eigenfrequency. There are as many independent modes as the number of degrees of freedom. An arbitrary motion of the system can be expressed as a superposition of normal modes. Free vibrations of a finite continuum body, such as the Earth, are also called normal modes. In this case, there are an infinite number of normal modes, and an arbitraty motion of the body can be expressed by their superposition. The concept of normal modes has been extended to wave-guides in which free waves with a certain phase velocity can exist without external force. Examples are Rayleigh waves in a half-space and Love waves in a layered half-space. In these cases, however, one cannot express an arbitrary motion by superposing normal modes.

P*   Designates *P*-waves refracted through an intermediate layer in the Earth's crust with a velocity near 6.5 km/sec. The upper boundary of this layer has been called the Conrad discontinuity. See Box 6.4, page 212.

P′   Another symbol for *PKP*.

***P* coda**   The portion of *P*-waves after the arrival of the primary waves. They may be due to *P* to *S* conversions at interfaces or to multiple reflections in layers or to scattering by three-dimensional inhomogeneities.

PdP   This wave is like the surface reflection *PP*, except that the reflection occurs at an interface at depth d (expressed in kilometers, e.g., $P_{600}P$) instead of at the surface.

P_d, P_u, P_r   The travel time for *P*-waves near $\Delta = 20°$ shows a triplication due to a sharp velocity increase in the upper mantle below the low-velocity layer. Three branches are designated in the order of decreasing $dt/d\Delta$ as P_d (direct), P_u (upper), and P_r (refracted).

Plate waves   The period equation for normal modes in an infinite plate in a vacuum can be split into two. One of the equations governs the mode with motion symmetric with respect to the median plane of the plate, and the other governs the mode with antisymmetric motion. The former is sometimes called the $M1$-wave, and the latter $M2$. For example, $M11$ and $M12$ are the fundamental and first higher modes of the $M1$ wave, respectively. For very short waves, both $M11$ and $M21$ approach Rayleigh waves in an elastic half-space made of the plate material. For wavelengths that are long compared with plate thickness, $M21$ are called flexural waves. They are dispersive, with phase velocities decreasing to zero with increasing wavelength (Satô, 1951).

***PL*-waves**   A train of long-period waves (30 to 50 sec) observed in the interval between *P*- and *S*-waves for distances less than about 30°. They show dispersion, with longer periods arriving earlier. They are explained as a leaking mode of the crust-mantle wave-guide (Oliver and Major, 1960; Su and Dorman, 1965).

P_n   Beyond a certain critical distance, generally in the range from 100 to 200 km, the first arrival from seismic sources in the crust corresponds to waves refracted from the top of the mantle. Called P_n, these waves are relatively small, with long-period motion followed by larger and sharper waves of shorter period called $\bar{P}$, which are propagated through the crust. The P_n-wave has long been interpreted as a head (conical) wave along the interface of two homogeneous media—namely, crust and mantle. The observed amplitude, however, is usually greater than that predicted for head waves, implying that the velocity change is not exactly step-like but has a finite gradient at or below the transition zone. The designation P_n has in recent years been applied to short-period *P*-waves that propagate over considerable distances (even up to 20°) with horizontal phase velocities in the range 7.8–8.3 km/sec. Thus

Herrin (1969) defines P_n as "the first arrival of seismic energy in the range from a few degrees to a distance where the travel-time function begins to show appreciable curvature." An interpretation in terms of head waves at the Moho is here unsatisfactory (although the horizontal velocity and travel times would be explained), because head waves must decay rapidly with distance. More likely is an explanation in terms of guided waves, within a high-Q layer several tens of kilometers in thickness at the top of the mantle. See description of the related wave S_n.

$\bar{P}$ (Pg) Travel-time curves at short distances (up to a few hundred km) for seismic sources in the Earth's crust usually consist of two intersecting straight lines; one with velocity about 6 km/sec at shorter distances and the other about 8 km/sec at greater distances. The former is attributed to direct P-waves propagating through the crust and is designated as $\bar{P}$ or Pg, which stands for granitic layer.

***P*-waves** Compressional elastic waves are called P-waves in seismology, P standing for "primary." In a homogeneous isotropic body, the velocity of P-waves is equal to $\sqrt{\kappa + \frac{4}{3}\mu)/\rho}$, where κ, μ, and ρ are bulk modulus, rigidity, and density, respectively. The particle displacement associated with P-waves is often parallel to the direction of wave propagation. For this reason, P-waves are sometimes called longitudinal waves.

Rayleigh waves The most fundamental of the surface waves, with strongest amplitudes in the neighborhood of the plane free surface of an elastic body. For the case of a homogeneous body, the velocity of propagation is 0.88 to 0.95 times the shear velocity, depending on Poisson's ratio. Rayleigh waves in a vertically heterogeneous half-space have frequency-dependent phase and group velocities. Higher modes can exist in the vertically heterogeneous half-space.

Rg Short-period, fundamental-mode Rayleigh waves (in the range 8 to 12 sec) observed for continental paths are sometimes designated as Rg (Ewing et al., 1957).

***S*-waves** Elastic shear waves are called S-waves in seismology, S standing for "secondary." In a homogeneous isotropic body, the velocity of S-waves is equal to $\sqrt{\mu/\rho}$, where μ and ρ are rigidity and density, respectively. The particle displacement associated with S-waves is often perpendicular to the direction of wave propagation. For this reason, S-waves are sometimes called transverse waves.

$\bar{S}$ (Sg) S-waves propagating through the crust like $\bar{P}$ or Pg.

***Sa*-waves** Sa-waves typically have periods of 10 to 30 sec and a group velocity of 4.4 to 4.5 km/sec measured along the surface. They can have both SV- and SH-components of motion. Their waveforms are usually complex and vary from station to station in an irregular manner (Brune, 1965).

Seiche A free oscillation of the surface of an enclosed body of water, such as a lake, pond, or bay with a narrow entrance. They are sometimes excited by earthquakes and by tsunamis. The period of oscillation ranges from a few minutes to a few hours, and the oscillation may last for several hours to one or two days.

Shear-coupled *PL*-waves This is a long-period wavetrain that follows S for distances up to about 80°. It has been explained as being due to the coupling of S-waves with a leaking mode of the crust-mantle wave-guide, i.e., PL-waves. The coupling of PL-waves with SS and SSS has also been observed (Chander et al., 1968; Poupinet and Wright, 1972).

***SH*-wave** *S*-waves with displacement only in the horizontal direction. For a vertically heterogeneous medium, *SH*-waves do not interact with *P*-waves and are simpler than *SV*-waves.

S_n Early use of the designation S_n was in reference to short-period *S*-waves that were presumed to propagate as head waves along the top of the mantle. Quite commonly, the term is also now applied to a prominent arrival of short-period shear waves that may be observed (with a straight-line travel-time curve) at epicentral distances as great as 40° (Molnar and Oliver, 1969). Stephens and Isacks (1977) suggest that these waves travel in a wave-guide at the top of the mantle. Propagation for shorter periods is more efficient, because longer-period waves have a substantial fraction of their energy within lower-*Q* material, at greater depth, and hence are filtered out. Examples of S_n at great distances are given by Walker et al. (1978).

Stoneley waves These are interface waves with the largest amplitudes confined to the neighborhood of a plane interface of two elastic media. They are always possible at a solid-fluid interface, but can exist at a solid-solid interface only under the stringent condition that shear-wave velocities in the two media are nearly equal.

Surface *P*-waves The ray path of surface *P* consists of two segments: an *S*-wave path from the source to the free surface with an apparent horizontal velocity equal to the *P*-wave velocity, and a *P*-wave path along the free surface to the receiver. The surface *P*-waves appear at the critical distance and can be a sharper arrival than the direct *S*-waves, although they attenuate very rapidly with distance.

Surface reflections (*PP*, *SS*, *SP*, *PS*, *PPP*, *SSS*) *P*-waves that have undergone one reflection at the surface before arriving at the recording station are denoted as *PP* if the wave initially left the hypocenter in the downward direction (in contrast to *pP*, which leaves in an upward direction). Those reflected twice at the surface are denoted as *PPP*. Likewise, *PS* is a once-reflected wave arriving at the station as an *S*-wave after conversion by reflection from *P*-waves.

***SV*-wave** For an isotropic homogeneous body, the displacement associated with far-field *S*-waves is restricted to a plane perpendicular to the propagation direction. The particle motion in the plane can be described by two orthogonal vectors, one in the horizontal direction and the other perpendicular to it. The latter component is the *SV*-wave. *SV*-waves interact with *P*-waves in a vertically heterogeneous medium.

***T*-phase** *T*, standing for tertiary waves (after *P* and *S*), has been used to designate late-arriving phases with period less than 1 sec, observed at stations of island or coastal regions for earthquakes in which the path of propagation is mostly oceanic. They travel the oceanic part of the path with the velocity of sound in water, probably through the SOFAR channel (Lineham, 1940; Ewing et al., 1952).

Tsunami, or "tidal wave" Gravity waves set up on the surface of the sea by disturbances in the sea bed. This disturbance may be an upheaval or subsidence due to a submarine earthquake, submarine landslide, or volcanic explosion.

Tube waves in a bore hole In an empty cylindrical hole, a kind of surface wave can propagate along the axis of the hole with energy confined to the vicinity of the hole. They exhibit dispersion with phase velocity increasing with the wavelength. At wavelengths much shorter than the hole radius, they approach Rayleigh waves. The phase velocity reaches the shear velocity at wavelengths of about three times the radius. Beyond this cutoff wavelength, they attenuate quickly by radiating S-waves. In a fluid-filled cylindrical hole, in addition to a series of multi-reflected conical waves propagating in the fluid, tube waves exist without a cutoff for the entire period range. At short wavelengths, they approach Stoneley waves for the plane liquid-solid interface. For wavelengths longer than about 10 times the hole radius, the velocity of tube waves becomes constant, given by the bulk modulus κ of the fluid and the rigidity μ of the solid, as $v = c(1 + \kappa/\mu)^{-1/2}$, where c is the acoustic velocity in the fluid (Biot, 1952; White, 1965).

Visible earthquake waves Slow waves with long period and short wavelengths reported by eyewitnesses in the epicentral area of a great earthquake (Lomnitz, 1970).

Volcanic tremors The seismic signals generated by volcanic activities are quite variable in character, ranging from those indistinguishable from tectonic earthquakes to continuous vibrations with sharply peaked spectra. The continuous vibrations are known as volcanic, or harmonic, tremors (Minakami, 1974; Aki et al., 1977).

APPENDIX 2

Definition of Magnitudes

An earthquake magnitude was originally defined by Richter (1935) as the logarithm of maximum amplitude measured in microns on the record of a standard torsion seismograph with a pendulum period of 0.8 sec, magnification of 2800, and damping factor 0.8, located at a distance of 100 km from the epicenter. This standard instrument, known after its designers as the Wood-Anderson seismometer, consists of a small copper cylinder (with mass less than 1 gram and with its axis vertical) attached to a vertical metal fiber. The fiber is under tension, so that the restoring force is supplied by tension. The instrument as a whole is sensitive to horizontal motions, which are detected via light reflected from a small mirror in the cylinder. A calibration curve was constructed to reduce the amplitude observed at an arbitrary epicentral distance to that expected at 100 km. This magnitude scale is now referred to as local magnitude M_L.

A magnitude scale based on teleseismic surface waves was described by Gutenberg and Richter (1936) and developed more extensively by Gutenberg (1945). For shallow earthquakes at distances $15^\circ < \Delta < 130^\circ$, he found the formula

$$M_s = \log A + 1.656 \log \Delta + 1.818, \tag{A.1}$$

where A is the horizontal component of the maximum ground displacement (in microns) due to surface waves with periods of 20 sec.

Many formulas for M_s have been proposed since that of Gutenberg. They were summarized by Vaněk et al. (1962), who proposed the formula

$$M_s = \log(A/T)_{\max} + 1.66 \log \Delta + 3.3, \tag{A.2}$$

which has been adopted officially by the IASPEI (International Association for Seismology and Physics of the Earth's Interior). In equation (A.2), $(A/T)_{\max}$ is the maximum of all A/T (amplitude/period) values of the wave groups on a record. For $T = 20$ sec, equation (A.2) becomes nearly identical to (A.1).

Another important magnitude scale is the one based on the amplitude of teleseismic body waves. It is defined by the formula

$$m_b = \log(A/T) + Q, \tag{A.3}$$

where Q is a function of epicentral distance and focal depth empirically determined by Gutenberg and Richter (1956) for eliminating the path effect from observed amplitude (see Fig. 10.9). A/T is the maximum in the wave group of either P, PP, or SH, with separate tables and charts of Q for each phase. Vaněk et al. (1962) also summarized later formulas for m_b and proposed a revised calibration function for Q.

The practice of m_b determination used by the U.S. Geological Survey for current routine reporting is, however, significantly different from what was used by Gutenberg and Richter. The most important difference is in the instrument characteristics used for the determination. Gutenberg and Richter used broad-band instruments that register relatively long-period P-waves (4–10 sec) for major events. On the other hand, the U.S.G.S. measurements are currently made using the short period WWSSN instruments, which show P-waves with period nearly always about 1 sec. Furthermore, the U.S.G.S. requires that (A/T) must be measured in the first 5 sec of the record. In earlier practice, the maximum was sought over a signal duration of up to about 10 sec to allow for an earthquake with a gradual onset.

Bibliography

Abe, K.

1970 Determination of seismic moment and energy from the Earth's free oscillation. *Physics of the Earth and Planetary Interiors* 4:49–61.

Abramowitz, M., and I. A. Stegun

1964 *Handbook of Mathematical Functions.* U.S. National Bureau of Standards.

Adams, R. D.

1972 Multiple inner core reflections from a Novaya Zemlya explosion. *Bulletin of the Seismological Society of America* 62:1063–1071.

Agnew, D., J. Berger, R. Buland, W. Farrell, and F. Gilbert

1976 International deployment of accelerometers: A network for very long period seismology. *EOS, Transactions of the American Geophysical Union* 57:180–188.

Aki, K.

1966 Generation and propagation of *G* waves from the Niigata earthquake of June 16, 1964. 2. Estimation of earthquake movement, released energy, and stress-strain drop from *G* wave spectrum. *Bulletin of the Earthquake Research Institute* 44:23–88.

Aki, K., and L. B. Chouet

1975 Origin and coda waves: Source, attenuation, and scattering effects. *Journal of Geophysical Research* 80:3322–3342.

Aki, K., M. Fehler, and S. Das

1977 Source mechanism of volcanic tremor; fluid-driven crack models and their application to the 1973 Kilauea eruption. *Journal of Volcanology and Geothermal Research* 2:259–287.

Alekseyev, A. S., and B. G. Mikhaylenko

1976 Solution of Lamb's problem for a vertically inhomogeneous elastic halfspace. *Izvestiya, Physics of the Solid Earth*. December: pp. 11–25.

Alsop, L. E., G. H. Sutton, and M. Ewing

1961a Free oscillations of the Earth observed on strain and pendulum seismographs. *Journal of Geophysical Research* 66:631–641.

1961b Measurement of Q for very long period free oscillations. *Journal of Geophysical Research* 66:2911–2915.

Alterman, Z., H. Jarosch, and C. L, Pekeris

1959 Oscillations of the Earth. *Proceedings of the Royal Society of London* A252:80–95.

Anderson, D. L., and C B. Archambeau

1964 The anelasticity of the Earth. *Journal of Geophysical Research* 69:2071–2084.

Anderson, D. L., A. Ben-Menahem, and C. B. Archambeau

1965 Attenuation of seismic energy in the upper mantle. *Journal of Geophysical Research* 70:1441–1448.

Andrianova, Z. S., V. I. Keilis-Borok, A. L. Levshin, and M. G. Neiganz

1967 *Seismic Love Waves*. New York: Consultants Bureau.

Ansell, J. H.

1978 On the scattering of *SH* waves from a point source by a sphere. *Geophysical Journal of the Royal Astronomical Society* 54:359–387.

Archambeau, C. B., E. A. Flinn, and D. G. Lambert

1969 Fine structure of the upper mantle. *Journal of Geophysical Research* 74: 5825–5865.

Arons, A. B., and D. R. Yennie

1950 Phase distortion of acoustic pulses obliquely reflected from a medium of higher sound velocity. *Journal of the Acoustical Society of America* 22:231–237.

Azimi, Sh. A., A. V. Kalinin, V. V. Kalinin, and B. L. Pivovarov

1968 Impulse and transient characteristics of media with linear and quadratic absorption laws. *Izvestiya, Physics of the Solid Earth*. February: pp. 88–93.

Backus, G.

1958 A class of self-sustaining dissipative spherical dynamos. *Annals of Physics* (N.Y.) 4:372–447.

1965 Possible forms of seismic anisotropy in the uppermost mantle under oceans. *Journal of Geophysical Research* 70:3429–3439.

Backus, G., and F. Gilbert

1961 The rotational splitting of the free oscillations of the earth. *Proceedings of the National Academy of Science* 47:362–371.

1967 Numerical applications of a formalism for geophysical inverse problems. *Geophysical Journal of the Royal Astronomical Society* 13:247–276.

1968 The resolving power of gross Earth data. *Geophysical Journal of the Royal Astronomical Society* 16:169–205.

1970 Uniqueness in the inversion of inaccurate gross Earth data. *Philosophical Transactions of the Royal Society of London* A266:123–192.

Backus, G., and M. Mulcahy

1976 Moment tensors and other phenomenological descriptions of seismic sources. I. Continuous displacements. *Geophysical Journal of the Royal Astronomical Society* 46:341–361.

1976 Moment tensors and other phenomenological descriptions of seismic sources. II. Discontinuous displacements. *Geophysical Journal of the Royal Astronomical Society* 47:301–329.

Båth M.

1954 The elastic waves *Lg* and *Rg* along Eurasiatic paths. *Arkiv Geophysik* **2**:295–342.

1957 A continental channel wave guided by the intermediate layer in the crust. *Geofisica pura e applicata* 38 : 19–31.

Beavan, J., and R. Bilham

1977 Thermally induced errors in fluid tube tiltmeters. *Journal of Geophysical Research* 82:5699–6704.

Benioff, H.

1932 A new vertical seismograph. *Bulletin of the Seismological Society of America* 22:155–169.

1958 Long waves observed in the Kamchatka earthquake of November 4, 1952. *Journal of Geophysical Research* 63:589–593.

1959 Fused-quartz extensometer for secular, tidal, and seismic strain. *Bulletin of the Geological Society of America* 70:1019–1032.

Benioff, H., B. Gutenberg, and C. F. Richter

1954 Progress report. *Transactions of the American Geophysical Union* 35:979–987.

Benioff, H., F. Press, and S. W. Smith

1961 Excitation of the free oscillations of the Earth by earthquakes. *Journal of Geophysical Research* 66:605–620.

Ben-Menahem, A., S. W. Smith, and T.-L. Teng

1965 A procedure for source studies from spectrums of long-period seismic body waves. *Bulletin of the Seismological Society of America* 55:203–235.

Ben-Menahem, A., M. Rosenman, and D. G. Harkrider

1970 Fast evaluation of source parameters from isolated surface-wave signals. I. Universal tables. *Bulletin of the Seismological Society of America* 60:1337–1387.

Berckhemer, H., and G. Schneider

1964 Near earthquakes recorded with long-period seismographs. *Bulletin of the Seismological Society of America* 54:973–985.

Berry, M. J., and G. F. West

1966 Reflected and head wave amplitudes in a medium of several layers. In *The Earth Beneath Continents* (Geophysical Monograph 10). American Geophysical Union.

Bessonova, E. N., V. M. Fishman, M. G. Shnirman, G. A. Sitnikova, and L. R. Johnson

1976 The tau method for inversion of travel times. II. Earthquake data. *Geophysical Journal of the Royal Astronomical Society* 46:87–108.

Biot, M. A.
1952 Propagation of elastic waves in a cylindrical bore containing a fluid. *Journal Applied Physics* 23:997–1005.

Biswas, N. N., and L. Knopoff
1970 Exact earth-flattening calculation for Love waves. *Bulletin of the Seismological Society of America* 60:1123–1137.

Block, B., J. Dratler, and R. D. Moore
1970 Earth and normal modes from a 6.5 magnitude earthquake. *Nature* 226: 343–344.

Block, B., and R. D. Moore
1966 Measurements in the earth mode frequency range by an electrostatic sensing and feedback gravimeter. *Journal of Geophysical Research* 71:4361–4375.
1970 Tidal to seismic frequency investigations with a quartz accelerometer of new geometry. *Journal of Geophysical Research* 75:1493–1506.

Bolt, B. A.
1962 Gutenberg's early *PKP* observations. *Nature* 196:122–124.
1964 The velocity of seismic waves near the Earth's center. *Bulletin of the Seismological Society of America* 54:191–208.

Borcherdt, R. D.
1973 Energy and plane waves in linear viscoelastic media. *Journal of Geophysical Research* 78:2442–2453.
1977 Reflection and refraction of type. II. *S* waves in elastic and anelastic media. *Bulletin of the Seismological Society of America* 67:43–67.

Bortfeld, R.
1961 Approximation to the reflection and transmission coefficients of plane longitudinal and transverse waves. *Geophysical Prospecting* 9:485–502.

Bracewell, R. B.
1965 *The Fourier Transform and Its Applications.* New York: McGraw-Hill.

Brekhovskikh, L. M.
1960 *Waves in Layered Media* (Chapter 4). New York: Academic Press.

Bromwich, T. J. I'A.
1898 On the influence of gravity on elastic waves, and, in particular, on the vibrations of an elastic globe. *Proceedings of the London Mathematical Society* Vol. 30.

Brune, J. N.
1962 Attenuation of dispersed wave trains. *Bulletin of the Seismological Society of America* 52:109–112.
1965 The *Sa* phase from the Hindu Kush earthquake of July 6, 1962. *Pure and Applied Geophysics* 62:81–95.
1968 Seismic moment, seismicity, and rate of slip along major fault zones. *Journal of Geophysical Research* 73:777–784.

Brune, J. N., and J. Dorman
1963 Seismic waves and Earth structure in the Canadian Shield. *Bulletin of the Seismological Society of America* 53:167–210.

Brune, J. N., J. E. Nafe, and L. E. Alsop
1961 The polar phase shift of surface waves on a sphere. *Bulletin of the Seismological Society of America* 51:247–257.

Brune, J. N., J. E. Nafe, and J. Oliver
1960 A simplified method for the analysis and synthesis of dispersed wave trains. *Journal of Geophysical Research* 65:287–304.

Buchbinder, G. G. R.
1972 Travel times and velocities in the outer core from *PmKP*. *Earth and Planetary Science Letters* 14:161–168.

Buchen, P. W.
1971 Plane waves in linear viscoelastic media. *Geophysical Journal of the Royal Astronomical Society* 23:531–542.

Buck, S. W., F. Press, D. Shepard, M. N. Toksöz, and H. Trantham
1971 Development of a mercury tiltmeter for seismic recording. Final Technical Report to the Advanced Research Projects Agency on Contract No. F44620-69-C-0126, Massachusetts Institute of Technology.

Buland, R. P., and F. Gilbert
1976 The theoretical basis for the rapid and accurate computation of normal mode eigenfrequencies and eigenfunctions. In *Retrieving the Seismic Moment Tensor*, Ph.D. thesis of R. P. Buland, University of California, San Diego.
1980 The rapid and accurate computation of normal mode eigenfrequencies and eigenfunctions. *Journal of Computational Physics* (in press).

Bullen, K. E.
1937 The ellipticity correction to travel-times of *P* and *S* earthquake waves. *Monthly Notices of the Royal Astronomical Society, Geophysical Supplement* 4:143–157.
1960 Notes on cusps in seismic travel-times. *Geophysical Journal of the Royal Astronomical Society* 3:354–359.

Burridge, R., and L. Knopoff
1964 Body force equivalents for seismic dislocations. *Bulletin of the Seismological Society of America* 54:1875–1888.

Byrne, C. J.
1961 Instrument noise in seismometers. *Bulletin of the Seismological Society of America* 51:69–84.

Cagniard, L.
1962 *Reflection and Refraction of Progressive Seismic Waves*. Trans. by E. A. Flinn and C. H. Dix. New York: McGraw-Hill.

Červený, V.
1972 Seismic rays and ray intensities in inhomogeneous anisotropic media. *Geophysical Journal of the Royal Astronomical Society* 29:1–13.
1974 Reflection and transmission coefficients for transition layers. *Studia Geophysica et Geodaetica* 18:59–68.

Červený, V., and R. Ravindra
1971 *Theory of Seismic Head Waves*. University of Toronto Press.

Chander, R., L. E. Alsop, and J. Oliver
1968 On the synthesis of shear-coupled *PL* waves. *Bulletin of the Seismological Society of America* 58:1849–1877.

Chapman, C. H.
1971 On the computation of seismic ray travel times and amplitudes. *Bulletin of the Seismological Society of America* 61:1267–1274.
1972 Lamb's problem and comments on the paper, 'On Leaking Modes' by Usha Gupta. *Pure and Applied Geophysics* 94:233–247.
1973 The Earth flattening transformation in body wave theory. *Geophysical Journal of the Royal Astronomical Society* 35:55–70.
1974 The turning point of elastodynamic waves. *Geophysical Journal of the Royal Astronomical Society* 39:613–621.
1976a Exact and approximate generalized ray theory in vertically inhomogeneous media. *Geophysical Journal of the Royal Astronomical Society* 46:201–233.
1976b A first motion alternative to geometrical ray theory. *Geophysical Research Letters* 3:153–156.
1978a A new method for computing seismograms. *Geophysical Journal of the Royal Astronomical Society* 54:481–518.
1978b Body waves in seismology. *In* J. Miklowitz and J. D. Achenbach (editors), *Modern Problems in Elastic Wave Propagation.* New York: Wiley.

Chapman, C. H., and R. A. Phinney
1972 Diffracted seismic signals and their numerical solution. *In* B. A. Bolt (editor), *Seismology: Body Waves and Sources* (Methods in Computational Physics, Vol. 12), New York: Academic Press.

Choy, G. L.
1977 Theoretical seismograms of core phases calculated by a frequency-dependent full wave theory, and their interpretation. *Geophysical Journal of the Royal* Astronomical Society, 51:275–311.

Choy, G. L., and Paul G. Richards
1975 Pulse distortion and Hilbert transformation in multiply reflected and refracted body waves. *Bulletin of the Seismological Society of America* 65: 55–70.

Cisternas, A., O. Betancourt, and A. Leiva
1973 Body waves in a "real Earth." Part I. *Bulletin of the Seismological Society of America* 63:145–156.

Coddington, E. A., and N. Levinson
1955 *Theory of Ordinary Differential Equations.* New York: McGraw-Hill.

Cormier, V., and P. G. Richards
1977 Full-wave theory applied to a discontinuous velocity increase: The inner core boundary. *Journal of Geophysical Research* 43:3–31.

Crampin, S.
1970 The dispersion of surface waves in multi-layered anisotropic media. *Geophysical Journal of the Royal Astronomical Society* 21:387–402.
1971 The propagation of surface waves in anisotropic media. *Geophysical Journal of the Royal Astronomical Society* 25:71–87.

1975 Distinctive particle motion of surface waves as a diagnostic of anisotropic layering. *Geophysical Journal of the Royal Astronomical Society* 40:177–186.

1977 A review of the effects of anisotropic layering on the propagation of seismic waves. *Geophysical Journal of the Royal Astronomical Society* 49:9–27.

Crampin, S., and D. Bamford

1977 Inversion of *P*-wave anisotropy. *Geophysical Journal of the Royal Astronomical Society* 49:123–132.

Crampin, S., and D. W. King

1977 Evidence for anisotropy in the upper mantle beneath Eurasia from the polarization of higher mode surface waves. *Geophysical Journal of the Royal Astronomical Society* 49:59–85.

Crossley, D. J., and D. Gubbins

1975 Static deformation of the Earth's liquid core. *Geophysical Research Letters* 2:1–4.

Currie, R. G.

1974 Period and Q_w of the Chandler wobble. *Geophysical Journal of the Royal Astronomical Society* 38:179–185.

Dahlen, F. A.

1968 The normal modes of a rotating elliptical Earth. *Geophysical Journal of the Royal Astronomical Society* 16:329–467.

1969 The normal modes of a rotating, elliptical Earth. II. Near-resonance multiplet coupling. *Geophysical Journal of the Royal Astronomical Society* 18:397–436.

1972 Elastic dislocation for a self-gravitating elastic configuration with an initial static stress field. *Geophysical Journal of the Royal Astronomical Society* 28:357–383.

1973 Elastic dislocation theory for a self-gravitating elastic configuration with an initial static stress field. II. Energy release. *Geophysical Journal of the Royal Astronomical Society* 31:469–484.

1976a Seismic faulting in the presence of a large compressive stress. *Geophysical Research Letters* 3:245–248. (Correction, p. 506.)

1976b The passive influence of the oceans upon the rotation of the Earth. *Geophysical Journal of the Royal Astronomical Society* 46:363–406.

1977 The balance of energy in earthquake faulting. *Geophysical Journal of the Royal Astronomical Society* 48:239–261.

Dahlen, F. A., and M. L. Smith

1975 The influence of rotation on the free oscillations of the Earth. *Philosophical Transactions of the Royal Society of London* A279:583–629.

Davies, D., and D. P. McKenzie

1969 Seismic travel-time residuals and plates. *Geophysical Journal of the Royal Astronomical Society* 18:51–63.

Dewey, J., and P. Byerly

1969 The early history of seismometry (to 1900). *Bulletin of the Seismological Society of America* 59:183–277.

Doornbos, D. J., and E. S. Husebye

1972 Array analysis of *PKP* phases and their precursors. *Physics of the Earth and Planetary Interiors* 5:387–399.

Dorman, J., and M. Ewing
1962 Numerical inversion of seismic surface wave dispersion data and crust-mantle structure in the New York—Pennsylvania area. *Journal of Geophysical Research* 67:5227–5241.

Dunkin, J. W.
1965 Computation of modal solutions in layered, elastic media at high frequencies. *Bulletin of the Seismological Society of America* 55:335–358.

Dziewonski, A. M., A. L. Hales, and E. R. Lapwood
1975 Parametrically simple Earth models consistent with geophysical data. *Physics of the Earth and Planetary Interiors* 10:12–48.

Dziewonski, A. M., and F. Gilbert
1976 The effect of small aspherical perturbations on travel-times and a re-examination of the corrections for ellipticity. *Geophysical Journal of the Royal Astronomical Society* 44:7–17.

Eaton, J. P.
1957 Theory of the electromagnetic seismograph. *Bulletin of the Seismological Society of America* 74:37–75.

Eshelby, J. D.
1957 The determination of the elastic field of an ellipsoidal inclusion, and related problems. *Proceedings of the Royal Society of London* A241:376–396.

Ewing, M., W. Jardetzky, and F. Press
1957 *Elastic Waves in Layered Media.* New York: McGraw-Hill.

Ewing, M., F. Press, and J. L. Worzel
1952 Further study of the *T* phase. *Bulletin of the Seismological Society of America* 42:37–51.

Ewing M., and J. L. Worzel
1948 Long-range sound transmission. *Geological Society of America Memoir* 27.

Farrell, W. E.
1969 A gyroscopic seismometer: measurements during the Borrego earthquake. *Bulletin of the Seismological Society of America* 59:1239–1246.

Forsyth, D. W.
1975 The early structural evolution and anisotropy of the oceanic upper mantle. *Geophysical Journal of the Royal Astronomical Society* 43:103–162.

Frasier, C. W.
1970 Discrete time solution of plane *P-SV* waves in a plane layered medium. *Geophysics* 35:197–219.

Frasier, C. W., and D. K. Chowdury
1974 Effect of scattering on *PcP*/*P* amplitude ratios at LASA from 40° to 84° distance. *Journal of Geophysical Research* 79:5469–5477.

Frazer, L. N.
1977 Synthesis of shear-coupled *PL*. Ph.D. Thesis, Princeton University.

Friedman, B.
1951 Propagation in a non-homogeneous atmosphere. *Communications of Pure and Applied Mathematics* 4:317–350.

Fuchs, K.
1968 The reflection of spherical waves from transition zones with arbitrary depth-dependent elastic moduli and density. *Journal of Physics of the Earth*, Special Issue, 16:27–41.
1970 On the determination of velocity depth distributions of elastic waves from the dynamic characteristics of the reflected wave field. *Zeitschrift für Geophysik* 36:531–548.
1971 The method of stationary phase applied to the reflection of spherical waves from transition zones with arbitrary depth-dependent elastic moduli and density. *Zeitschrift für Geophysik* 37:89–117.

Fuchs, K., and G. Muller
1971 Computation of synthetic seismograms with the reflectivity method and comparison of observations. *Geophysical Journal of the Royal Astronomical Society* 23:417–433.

Fuchs, K., and K. Schulz
1976 Tunneling of low-frequency waves through the subcrustal lithosphere. *Journal of Geophysics* 42:175–190.

Futterman, W. I.
1962 Dispersive body waves. *Journal of Geophysical Research* 67:5279–5291.

Galitzin, B.
1914 *Vorlesungen über Seismometrie*. Leipzig: Teubner.

Gantmacher, F. R.
1959 *The Theory of Matrices* (2 vols). New York: Chelsea Publishing Co.

Gerver, M., and V. Markushevitch
1966 Determination of a seismic wave velocity from the travel time curve. *Geophysical Journal of the Royal Astronomical Society* 11:165–173.
1967 On the characteristic properties of travel time curves. *Geophysical Journal of the Royal Astronomical Society* 13:241–246.

Gilbert, F.
1967 Gravitationally perturbed elastic waves. *Bulletin of the Seismological Society of America* 57:783–794.
1971 Excitation of the normal modes of the Earth by earthquake sources. *Geophysical Journal of the Royal Astronomical Society* 22:223–226.
1976 The representation of seismic displacements in terms of travelling waves. *Geophysical Journal of the Royal Astronomical Society* 44:275–280.

Gilbert, F., and G. Backus
1966 Propagator matrices in elastic wave and vibration problems. *Geophysics* 31:326–332.

Gilbert, F., and A. M. Dziewonski
1975 An application of normal mode theory to the retrieval of structural parameters and source mechanisms from seismic spectra. *Philosophical Transactions of the Royal Society of London* A278:187–269.

Gilbert, F., and D. V. Helmberger
1972 Generalized ray theory for a layered sphere. *Geophysical Journal of the Royal Astronomical Society* 27:57–80.

Gilbert, F., and S. J. Laster

1962 Experimental investigation of *PL* modes in a single layer. *Bulletin of the Seismological Society of America* 52:59–66.

Gilbert, F., S. J. Laster, M. M. Backus, and R. Schell

1962 Observations of pulses on an interface. *Bulletin of the Seismological Society of America* 52:847–868.

Gladwin, M. T., and F. D. Stacey

1974 Anelastic degradation of acoustic pulses in rock. *Physics of the Earth and Planetary Interiors* 8:332–336.

Green, G.

1839 On the laws of reflexion and refraction of light. *Transactions of the Cambridge Philosophical Society* (Vol. 7). (Reprinted in Mathematical Papers of George Green, pp. 245–269, London, 1871.)

Gutenberg, B.

1913 Über die Konstitution des Erdinnern, erschlossen aus Erdbebenbeobachtungen. *Zeitschrift für Geophysik* 14:1217–1218.

Gutenberg, B.

1945 Amplitudes of surface waves and magnitudes of shallow earthquakes. *Bulletin of the Seismological Society of America* 35:3–12.

Gutenberg, B., and C. F. Richter

1936 Magnitude and energy of earthquakes. *Science* (New Series) 83:183–185.

1956 Earthquake magnitude, intensity, energy, and acceleration. *Bulletin of the Seismological Society of America* 46:105–145.

Haddon, R. A. W.

1972 Corrugations on the mantle-core boundary or transition layers between inner and outer cores? *EOS, Transactions of the American Geophysical Union* 53:600.

Hagiwara, T.

1958 A note on the theory of the electromagnetic seismograph. *Earthquake Research Institute Bulletin of Tokyo University* 36:139–164.

Hales, A. L., and J. L. Roberts

1974 The Zoeppritz amplitude equations: More errors. *Bulletin of the Seismological Society of America* 64:285.

Harkrider, D. G.

1964 Theoretical and observed acoustic-gravity waves from explosive sources in the atmosphere. *Journal of Geophysical Research* 69:5295–5321.

Harkrider, D. G., and D. L. Anderson

1966 Surface wave energy from point sources in plane layered Earth models. *Journal of Geophysical Research* 71:2967–2980.

Harrison, J. C.

1976a Cavity and topographic effects in tilt and strain measurements. *Journal of Geophysical Research* 81:319–328.

1976b Tilt observations in the Poorman Mine near Boulder, Colorado. *Journal of Geophysical Research* 81:329–336.

Haskell, N. A.

1953 The dispersion of surface waves in multilayered media. *Bulletin of the Seismological Society of America* 43:17–34.

1960 Crustal reflection of plane *SH* waves. *Journal of Geophysical Research* 65:4147–4150.

1962 Crustal reflection of plane *P* and *SV* waves. *Journal of Geophysical Research* 67:4751–4767.

1964 Radiation pattern of surface waves from point sources in a multi-layered medium. *Bulletin of the Seismological Society of America* 54:377–394.

Hasselman, K.

1963 A statistical analysis of the generation of microseisms. *Review of Geophysics* 1:177–210.

Haubrich, R. A.

1967 Microseisms. *In* S. K. Runcorn (editor), *International Dictionary of Geophysics*. London: Pergamon Press.

Helmberger, D. V.

1968 The crust-mantle transition in the Bering Sea. *Bulletin of the Seismological Society of America* 58:179–214.

Helmberger, D. V., and D. A. Harkrider

1978 Modeling earthquakes with generalized ray theory. *In* J. Miklowitz and J. D. Achenbach (editors), *Modern Problems in Elastic Wave Propagation*. New York: Wiley.

Helmberger, D. V., and S. D. Malone

1975 Modelling local earthquakes as shear dislocations in a layered half space. *Journal of Geophysical Research* 80:4881–4888.

Herrera, I.

1964 A perturbation method for elastic wave propagation. 1. Non-paralleled boundaries. *Journal of Geophysical Research* 69:3845–3851.

Herrin, E., and J. Richmond

1960 On the propagation of the *Lg* phase. *Bulletin of the Seismological Society of America* 50:197–210.

Hess, H.

1964 Seismic anisotropy of the uppermost mantle under oceans. *Nature* 203: 629–631.

Hildebrand, F. B.

1952 *Methods of Applied Mathematics* (2nd ed.). Englewood Cliffs, New Jersey: Prentice-Hall.

Hill, D. P.

1971a Velocity gradients and anelasticity from crustal body wave amplitudes. *Journal of Geophysical Research* 76:3309–3325.

1971b Velocity gradients in the Earth's crust from head-wave amplitudes. *American Geophysical Union Monograph*, Ed. by J. G. Heacock. 14:71–75.

Hill, D. P., F. G. Fischer, K. M. Lahr, and J. M. Coakely

1976 Earthquake sounds generated by body-wave ground motion. *Bulletin of the Seismological Society of America* 66:1159–1172.

Hobson, E. W.

1955 *The Theory of Spherical and Ellipsoidal Harmonics.* New York: Chelsea Publishing Co.

Hoop, A. T. de,

1958 Representation theorems for the displacement in an elastic solid and their applications to elastodynamic diffraction theory. D.Sc. Thesis, Technische Hogeschool, Delft.

1960 A modification of Cagniard's method for solving seismic pulse problems. *Applied Science Research* B8:349–356.

Hron, F.

1971 Criteria for selection of phases in synthetic seismograms for layered media. *Bulletin of the Seismological Society of America* 61:765–779.

1972 Numerical methods of ray generation in multilayered media. *In* B. A. Bolt (editor), *Seismology: Body Waves and Sources* (Methods in Computational Physics, Vol. 12). New York: Academic Press.

Hudson, J. A.

1962 The total internal reflection of *SH* waves. *Geophysical Journal of the Royal Astronomical Society* 6:509–531.

1969a A quantitative evaluation of seismic signals at teleseismic distances. I. Radiation from seismic sources. *Geophysical Journal of the Royal Astromical Society* 18:233–249.

1969b A quantitative evaluation of seismic signals at teleseismic distances. II. Body waves and surface waves from an extended source. *Geophysical Journal of the Royal Astronomical Society* 18:353–370.

Isacks, B., J. Oliver, and L. R. Sykes

1968 Seismology and the new global tectonics. *Journal of Geophysical Research* 73:5855–5899.

Jackson, D. D., and D. L. Anderson

1970 Physical mechanisms of seismic-wave attenuation. *Review of Geophysics and Space Physics* 8:1–63.

Jeffreys, H.

1926 On compressional waves in two superposed layers. *Proceedings of the Cambridge Philosophical Society* 23:472–481.

1931 On the cause of oscillatory movement in seismograms. *Monthly Notices of the Royal Astronomical Society, Geophysical Supplement* 2:407–416.

1958 Rock creep, tidal friction and the Moon's ellipticities. *Monthly Notices of the Royal Astronomical Society* 118:14–17.

1961 Small correction in the theory of surface waves. *Geophysical Journal of the Royal Astronomical Society* 6:115–117.

1965 *Cartesian Tensors.* Cambridge University Press.

Jeffreys, H., and B. S. Jeffreys

1972 *Methods of Mathematical Physics* (3rd ed.). Cambridge University Press.

Johnson, L. E., and F. Gilbert

1972 Inversion and inference for teleseimic ray data. *In* B. A. Bolt (editor), *Seismology: Body Waves and Sources* (Methods in Computational Physics, Vol. 12). New York: Academic Press.

Jordan, T. H., and D. L. Anderson
1974 Earth structure from free oscillations and travel times. *Geophysical Journal of the Royal Astronomical Society* 36:411–459.

Jordan, T. H., and L. N. Frazer
1975 Crustal and upper mantle structure from S_p phases. *Journal of Geophysical Research* 80:1504–1518.

Julian, B. R., and D. L. Anderson
1968 Travel times, apparent velocities and amplitudes of body waves. *Bulletin of the Seismological Society of America* 58:339–366.

Julian, B. R., and M. K. Sengupta
1973 Seismic travel time evidence for lateral inhomogeneity in the deep mantle. *Nature* 242:443–447.

Kanamori, H.
1976 Re-examination of the Earth's free oscillations excited by the Kamchatka earthquake of November 4, 1952. *Physics of the Earth and Planetary Interiors* 11:216–226.
1977 The energy release in great earthquakes. *Journal of Geophysical Research* 82:2981–2987.

Karal, F. C., and J. B. Keller
1959 Elastic wave propagation in homogeneous and inhomogeneous media. *Journal of the Acoustical Society of America* 31:694–705.

Keilis-Borok, V. I., and T. B. Yanovskaya
1962 Dependence of the spectrum of surface waves on the depth of the focus within the Earth's crust. *Bulletin of the Academy of Sciences, U.S.S.R., Geophysic Series* (English translation) 11:1532–1539.

Keith, C., and S. Crampin
1977 Seismic body waves in anisotropic media; reflection and refraction at a plane interface. *Geophysical Journal of the Royal Astronomical Society* 49:181–208.

Kennett, B. L. N., N. J. Kerry, and J. H. Woodhouse
1978 Symmetries in the reflection and transmission of elastic waves. *Geophysical Journal of the Royal Astronomical Society* 52:215–229.

Kind, R.
1976 Computation of reflection coefficients for layered media. *Journal of Geophysics* 42:191–200.
1978 The reflectivity method for a buried source. *Journal of Geophysical Research* 44:603–612.

King, D. W., R. A. W. Haddon, and J. R. Cleary
1974 Array analysis of precursors to *PKIKP* in the distance range 128° to 142°. *Geophysical Journal of the Royal Astronomical Society* 37:157–173.

Knopoff, L.
1956 Diffraction of elastic waves. *Journal of the Acoustical Society of America* 28:217–229.
1964 Q. *Review of Geophysics* 2:625–660.

Knott, C. G.
1899 Reflection and refraction of elastic waves with seismological applications. *Philosophical Magazine*, Series 5, 48:64–97.

Kostrov, B. V.
1974 Seismic moment and energy of earthquakes, and seismic flow of rock. *Izvestiya, Physics of the Solid Earth*, January, 13–21.

LaCoste, L. J. C.
1935 A simplification in the conditions for the zero-length-spring seismograph. *Bulletin of the Seismological Society of America* 25:176–179.

Lamb, H.
1904 On the propagation of tremors over the surface of an elastic solid. *Philosophical Transactions of the Royal Society of London* A203:1–42.

Landisman, M., T. Usami, Y. Satô, and R. Massé
1970 Contributions of theoretical seismograms to the study of modes, rays, and the Earth. *Review of Geophysics and Space Physics* 8:533–589.

Langer, R. E.
1951 Asymptotic solutions of a differential equation in the theory of microwave propagation. *Communications of Pure and Applied Mathematics* 3:427–438.

Langston, C. A., and D. V. Helmberger
1975 A procedure for modelling shallow dislocation sources. *Geophysical Journal of the Royal Astronomical Society* 42:117–130.

Lapwood, E. R.
1949 The disturbance to a line source in a semi-infinite elastic medium. *Philosophical Transactions of the Royal Society of London* A242:63–100.

Latham, G., M. Ewing, J. Dorman, D. Lammlein, F. Press, N. Toksöz, G. Sutton, F. Duennebier, and Y. Nakamura
1971 Moonquakes. *Science* 174:687–692.

LePichon, X.
1968 Sea-floor spreading and continental drift. *Journal of Geophysical Research* 73:3661–3697.

Lineham, D.
1940 Earthquakes in the West Indian region. *Transactions of the American Geophysical Union* 21:229–232.

Liu, H.-P., D. L. Anderson, and H. Kanamori
1976 Velocity dispersion due to anelasticity; implications for seismology and mantle composition. *Geophysical Journal of the Royal Astronomical Society* 47:41–58.

Lomnitz, C.
1956 Creep measurements in igneous rocks. *Journal of Geology* 64:473–479.
1957 Linear dissipation in solids. *Journal of Applied Physics* 28:201–205.
1970 Some observations of gravity waves in the 1960 Chile earthquake. *Bulletin of the Seismological Society of America* 60:669–670.

Longman, I. M.

1963 A Green's function for determining the deformation of the Earth under surface mass loads. 2. Computations and numerical results. *Journal of Geophysical Research* 68:485–496.

Longuet-Higgins, M. S.

1950 A theory of the origin of microseisms. *Philosophical Transactions of the Royal Society of London* A243:1–35.

Love, A. E. H.

1911 *Some Problems of Geodynamics*. Cambridge University Press.

1944 *A Treatise on the Mathematical Theory of Elasticity*. New York: Dover Publications. [First published by Cambridge University Press, 1892.]

Luh, P. C., and A. M. Dziewonski

1975 Theoretical seismograms for the Colombian earthquake of 1970 July 31. *Geophysical Journal of the Royal Astronomical Society* 43:679–695.

McHugh, S., and M. J. S. Johnston

1976 Short-period nonseismic tilt perturbations and their relation to episodic slip. *Journal of Geophysical Research* 81:6341–6346.

McKenzie, D. P.

1969 The relation between fault plane solutions for earthquakes and the directions of the principal stresses. *Bulletin of the Seismological Society of America* 59:591–601.

Major, M. W., G. H. Sutton, J. E. Oliver, and P. Metsger

1964 On elastic strain of the Earth in the period range 5 seconds to 100 hours. *Bulletin of the Seismological Society of America* 54:295–346.

Malvern, L. E.

1969 *Introduction to the Mechanics of a Continuous Medium*. Englewood Cliffs, New Jersey: Prentice-Hall.

Mantovani, E., F. Schwab, and L. Knopoff

1977 Generation of complete theoretical seismograms for *SH*-II. *Geophysical Journal of the Royal Astronomical Society* 48:531–536.

Maruyama, T.

1963 On the force equivalents of dynamic elastic dislocations with reference to the earthquake mechanism. *Bulletin of the Earthquake Research Institute, Tokyo University* 41:467–486.

Mason, W. P.

1958 *Physical Acoustics and the Properties of Solids*. London: D. Van Nostrand.

Melchior, P.

1966 *The Earth Tides*. London: Pergamon Press.

Mellman, G. R., and D. V. Helmberger

1974 High-frequency attenuation by a thin high-velocity layer. *Bulletin of the Seismological Society of America* 64:1383–1388.

Mendiguren, J.

1973 Identification of free oscillation spectral peaks for 1970 July 31, Colombian deep shock using the excitation criterion. *Geophysical Journal of the Royal Astronomical Society* 33:281–321.

Michelson, A. A., and H. G. Gale
1919 The rigidity of the Earth. *Astrophysical Journal* 50:330–345.

Minakami, T.
1974 Seismology of volcanoes in Japan. *In* L. Civetta, P. Jasparini, G. Luongo, and A. Rapolla (editors), *Physical Volcanology*. Amsterdam: Elsevier.

Molnar, P., and J. Oliver
1969 Lateral variations of attenuation in the upper mantle and discontinuities in the lithosphere. *Journal of Geophysical Research* 74:2648–2682.

Müller, G.
1973 Amplitude studies of core phases. *Journal of Geophysical Research* 78: 3469–3490.
1977 Correction. *Journal of Geophysical Research* 82:2541–2542.

Müller, G., and R. Kind
1976 Observed and computed seismogram sections for the whole Earth. *Geophysical Journal of the Royal Astronomical Society* 44:699–716.

Musgrave, M. J. P.
1970 *Crystal Acoustics*. San Francisco: Holden-Day.

Nafe, J. E.
1957 Reflection and transmission coefficients at a solid-solid interface of high velocity contrast. *Bulletin of the Seismological Society of America* 47:205–219.

Nakanishi, K., F. Schwab, and L. Knopoff
1977 Generation of complete theoretical seismograms for *SH*-I. *Geophysical Journal of the Royal Astronomical Society* 48:525–530.

Ness, N. F., J. C. Harrison, and L. B. Slichter
1961 Observations of the free oscillations of the Earth. *Journal of Geophysical Research* 66:621–629.

Nowick, A. S., and B, S. Berry
1972 *Anelastic Relaxation in Crystalline Solids*. New York: Academic Press.

Nussenzveig, H. M.
1965 High-frequency scattering by an impenetrable sphere. *Annals of Physics* (N. Y.) 34:23–95.
1972 *Causality and Dispersion Relations* (Mathematics in Science and Engineering, Vol. 95) R. Bellman, series editor. New York: Academic Press.

O'Connell, R. J., and B. Budiansky
1978 Measures of dissipation in viscoelastic media. *Geophysical Research Letters* 5:5–8.

Okal, E., and P. Mechler
1973 On the problem of the convergence of the eikonal expansion for synthetic seismograms. *Bulletin of the Seismological Society of America* 63:1315–1319.

Oliver, J.
1961 On the long-period character of shear waves. *Bulletin of the Seismological Society of America* 51:1–12.

Oliver, J., and M. Major
1960 Leaking modes and the *PL* phase. *Bulletin of the Seismological Society of America* 50:165–180.

Olver, F. W. J.
1954a The asymptotic solution of linear differential equations of the second order for large values of a parameter. *Philosophical Transactions of the Royal Society of London* A247:307–327.
1954b The asymptotic expansion of Bessel functions of large order. *Philosophical Transactions of the Royal Society of London* A247:328–368.

Orcutt, J. A., B. L. N. Kennett, and L. M. Dorman
1976 Structure of the East Pacific Rise from an ocean bottom seismometer survey. *Geophysical Journal of the Royal Astronomical Society* 45:305–320.

Page, R. A., J. A. Blume, and W. B. Joyner
1975 Earthquake shaking and damage to buildings. *Science* 189:601–608.

Pao, Y.-H., and R. R. Gajewski
1977 The generalized ray-theory and transient responses of layered elastic solids. *In* W. P. Mason (editor), *Physical Acoustics* (Vol. 13). New York: Academic Press.

Pekeris, C. L.
1948 Theory of propagation of explosive sound in shallow water. *Geological Society of America Memoirs* No. 27.

Pekeris, C. L., Z. Alterman, and H. Jarosch
1961 Rotational multiplets in the spectrum of the Earth. *Physics Review* 122: 1692–1700.

Pekeris, C. L., and Y. Accad
1972 Dynamics of the liquid core of the Earth. *Philosophical Transactions of the Royal Society of London* A273:237–260.

Peterson, J., H. M. Butler, L. G. Holcomb, and C. R. Hutt
1976 The seismic research observatory. *Bulletin of the Seismological Society of America* 66:2049–2068.

Phinney, R. A.
1961 Leaking modes in the crustal wave-guide. 1. The oceanic *PL* wave. *Journal of Geophysical Research* 66:1445–1469.

Phinney, R. A., and S. Alexander
1966 *P* wave diffraction theory and the structure of the core mantle boundary. *Journal of Geophysical Research* 71:5943–5958.

Phinney, R. A., and L. M. Cathles
1969 Diffraction of *P* by the core: A study of long-period amplitudes near the edge of the shadow. *Journal of Geophysical Research* 74:1556–1574.

Pomeroy, P. W., G. Hade, J. Savino, and R. Chander
1969 Preliminary results from high-gain wide-band long-period electromagnetic seismograph systems. *Journal of Geophysical Research* 74:3295–3298.

Poupinet, G., and C. Wright
1972 The generation and propagation of shear-coupled *PL* waves. *Bulletin of the Seismological Society of America* 62:1699–1710.

Press, F.
1965 Displacements, strains, and tilts at teleseismic distances. *Journal of Geophysical Research* 70:2395–2412.

Press, F., A. P. Crary, J. Oliver, and S. Katz
1951 Air-coupled flexural waves in floating ice. *Transactions of the American Geophysical Union* 32:166–172.

Rayleigh, Lord
1887 On waves propagated along the plane surface of an elastic solid. *Proceedings of the London Mathematical Society* 17:4–11.
1910 The problem of the whispering gallery. *Philosophical Magazine* 20:1001–1004.
1912 On the propagation of waves through a stratified medium, with special reference to the question of reflection. *Proceedings of the Royal Society of London* A86:207–226.
1945 *The Theory of Sound* (2 vols.). New York: Dover Publications.

Richards, P. G.
1971 An elasticity theorem for heterogeneous media, with an example of body wave dispersion in the Earth. *Geophysical Journal of the Royal Astronomical Society*. 22:453–472.
1973 Calculation of body waves, for caustics and tunnelling in core phases. *Geophysical Journal of the Royal Astronomical Society* 35:243–264.
1974 Weakly coupled potentials for high-frequency elastic waves in continuously stratified media. *Bulletin of the Seismological Society of America* 64:1575–1588.
1976 On the adequacy of plane-wave reflection/transmission coefficients in the analysis of seismic body waves. *Bulletin of the Seismological Society of America* 66:701–717.

Richards, P. G., and C. W. Frasier
1976 Scattering of elastic waves from depth-dependent inhomogeneities. *Geophysics* 41:441–458.

Richter, C. F.
1935 An instrumental earthquake magnitude scale. *Bulletin of the Seismological Society of America* 25:1–32.
1958 *Elementary Seismology*. San Francisco: W. H. Freeman and Company.

Rihn, W. J.
1969 The design of electromagnetic damping circuits. *Bulletin of the Seismological Society of America* 59:967–972.

Robin, L.
1957 *Fonctions Sphériques de Legendre et Fonctions Sphéroidales* (3 vols.). Paris: Gautier-Villars.

Rodgers, P. W.
1968 The response of the horizontal pendulum seismometer to Rayleigh and Love waves, tilt and free oscillations of the Earth. *Bulletin of the Seismological Society of America* 58:1384–1406.

Rodi, W. L., P. Glover, T. M. C. Li, and S. S. Alexander
1975 A fast, accurate method for computing group-velocity partial derivatives for Rayleigh and Love modes. *Bulletin of the Seismological Society of America* 65:1105–1114.

Rosenbaum, J. H.
1960 The long-time response of a layered elastic medium to explosive sound. *Journal of Geophysical Research* 65:1577–1613.

Sacks, I. S., S. Suyehiro, and D. W. Evertson
1971 Sacks-Evertson strainmeter, its installation in Japan and some preliminary results concerning strain steps. *Proceedings of the Japanese Academy* 47:707–712.

Saito, M.
1967 Excitation of free oscillations and surface waves by a point source in a vertically heterogeneous Earth. *Journal of Geophysical Research* 72:3689–3699.
1968 Synthesis of rotational and dilatational seismograms. *Journal of Physics of the Earth* 16:53–62.

Satö, Y.
1951 Study on surface waves. II. Velocity of surface waves propagated upon elastic plates. *Bulletin of the Earthquake Research Institute* 29:223–261.
1959 Numerical integration of the equation of motion for surface waves in a medium with arbitrary variation of material constants. *Bulletin of the Seismological Society of America* 49:57–77.

Satô, Y., T. Usami, and M. Ewing
1962 Basic study on the oscillation of a homogeneous elastic sphere. IV. *The Geophysical Magazine* 31:237–242.

Savage, J. C.
1966 Radiation from a realistic model of faulting. *Bulletin of the Seismological Society of America* 56:577–592.
1969a Steketee's paradox. *Bulletin of the Seismological Society of America* 59:381–384.
1969b A new method of analyzing the dispersion of oceanic Rayleigh waves. *Journal of Geophysical Research* 74:2608–2617.
1976 Anelastic degradation of acoustic pulses in rock—comments. *Physics of the Earth and Planetary Interiors* 11:284–285.

Savage, J. C., and M. E. O'Neill
1975 The relation between the Lomnitz and Futterman theories of internal friction. *Journal of Geophysical Research* 80:249–251.

Savino, J., K. McCamy, and G. Hade
1972 Structures in Earth noise beyond twenty seconds—A window for earthquakes. *Bulletin of the Seismological Society of America* 62:141–176.

Scholte, J. G. J.
1947 The range of existence of Rayleigh and Stoneley waves. *Monthly Notices of the Royal Astronomical Society, Geophysical Supplement* 5:120–126.
1956 On seismic waves in a spherical Earth. *Koninkl. Ned. Meteorol. Inst. Publ.* 65:1–55.

Schwab, F., and L. Knopoff
1970 Surface wave dispersion computations. *Bulletin of the Seismological Society of America* 60:321–344.

Seckler, B. D., and J. B. Keller
1959 Asymptotic theory of diffraction in inhomogeneous media. *Journal of the Acoustical Society of America* 31:206–216.

Shimshoni, M., and A. Ben-Menahem
1970 Computation of the divergence coefficient for seismic phases. *Geophysical Journal of the Royal Astronomical Society* 21:285–294.

Smith, M. L.
1977 Wobble and nutation of the Earth. *Geophysical Journal of the Royal Astronomical Society*, 50:103–140.

Smith, M. L., and F. A. Dahlen
1973 The azimuthal dependence of Love and Rayleigh wave propagation in a slightly anisotropic medium. *Journal of Geophysical Research* 78:3321–3333.

Smith, W. D.
1974 A non-reflecting plane boundary for wave propagation problems. *Journal of Computational Physics* 15:492–503.
1975 The application of finite element analysis to body wave propagation problems. *Geophysical Journal of the Royal Astronomical Society* 42:747–768.

Solomon, S. C.
1972 Seismic-wave attenuation and partial melting in the upper mantle of North America. *Journal of Geophysical Research* 77:1483–1502.
1973 Shear wave attenuation and melting beneath the Mid-Atlantic Ridge. *Journal of Geophysical Research* 78:6044–6059.

Sorrels, G. G.
1971 A preliminary investigation into the relationship between long-period seismic noise and local fluctuations in the atmospheric pressure field. *Geophysical Journal* 26:71–82.

Spencer, T. W.
1960 The method of generalized reflection and transmission coefficients. *Geophysics* 25:625–641.
1965 Long-time response predicted by exact elastic ray theory. *Geophysics* 30:363–368.

Spudich, P. K. P., and D. V. Helmberger
1979 Synthetic seismograms from model ocean bottoms. *Journal of Geophysical Research* 84:189–204.

Stacey, F. D., M. T. Gladwin, B. McKavanagh, A. T. Linde, and L. M. Hastie
1975 Anelastic damping of acoustic and seismic pulses. *Geophysical Survey* 2:133–157.

Stacey, F. D., J. M. W. Rynn, E. C. Little, and C. Croskell
1969 Displacement and tilt transducers of 140 db range. *Journal of Scientific Instrumentation* (*J. Phys., E*) Series 2. 2:945–949.

Stauder, W.
1968 Mechanism of the Rat Island earthquake sequence of February 4, 1965, with relation to island arcs and sea-floor spreading. *Journal of Geophysical Research* 73:3847–3858.

Stein, S., and R. J. Geller
1977 Time domain observation and synthesis of split spheroidal and torsional free oscillations of the 1960 Chilean earthquake: Preliminary results. *Bulletin of the Seismological Society of America*, 68:325–332.

Steketee, J. A.
1958 Some geophysical applications of the elasticity theory of dislocations. *Canadian Journal of Physics* 36:1168–1197.

Stephens, C., and B. L. Isacks
1977 Toward an understanding of *Sn*: Normal modes of Love waves in an oceanic structure. *Bulletin of the Seismological Society of America* 67:69–78.

Sternberg, E.
1960 On the integration of the equations of motion in the classical theory of elasticity. *Archive for Rational Mechanics* 6:34–50.

Stoneley, R.
1924 Elastic waves at the surface of separation of two solids. *Proceedings of the Royal Society of London* A106:416–428.

Strick, E.
1959 Propagation of elastic wave motion from an impulsive source along a fluid/solid interface. II. Theoretical pressure pulse. *Philosophical Transactions of the Royal Society of London* A251:465–523.
1970 A predicted pedestal effect for pulse propagation in constant-*Q* solids. *Geophysics* 35:387–403.

Su, S. S., and J. Dorman
1965 The use of leaking modes in seismogram interpretation and in studies of crust-mantle structure. *Bulletin of the Seismological Society of America* 55:989–1021.

Sykes, L. R.
1967 Mechanism of earthquakes and nature of faulting on the mid-oceanic ridges. *Journal of Geophysical Research* 72:2131–2153.

Takeuchi, H., and N. Kobayashi
1959 Surface waves propagating along a free surface of a semi-infinite elastic medium of variable density and elasticity. Part 1. *Journal of the Seismological Society of Japan* (Series 2) 12:115–121.

Takeuchi, H., and M. Saito

1972 Seismic surface waves. *In* B. A. Bolt (editor), *Seismology: Surface Waves and Earth Oscillations* (Methods in Computational Physics, Vol. 11). New York: Academic Press.

Thomson, W. T.

1950 Transmission of elastic waves through a stratified solid. *Journal of Applied Physics* 21:89–93.

Titchmarsh, E. C.

1926 Conjugate trigonometrical integrals. *Proceedings of the London Mathematical Society* (Series 2) 24:109–130.

1939 *The Theory of Functions* (2nd ed.). Oxford University Press.

Tsai, Y. B., and K. Aki

1971 Amplitude spectra of surface waves from small earthquakes and underground nuclear explosions. *Journal of Geophysical Research* 76:3440–3452.

Vaněk, J., A. Zápotek, V. Kárník, N. V. Kondorskaya, Yu. V. Rizmichenko, E. F. Savarensky, S. L. Solov'yov, and N. V. Shebalin

1962 Standardization of magnitude scales. *Izvestiya Akad. Nauk S.S.S.R., Ser. Geofiz.* 2:153–158.

Vared, M., and A. Ben-Menahem

1974 Application of synthetic seismograms to the study of low-magnitude earthquakes and crustal structure in the northern Red Sea region. *Bulletin of the Seismological Society of America* 64:1221–1237.

Vlaar, N. J.

1966 The field from an *SH*-point source in a continuously layered inhomogeneous medium. 1. The field in a layer of a finite depth. *Bulletin of the Seismological Society of America* 56:715–724.

1968 Ray theory for an anisotropic inhomogeneous elastic medium. *Bulletin of the Seismological Society of America* 58:2053–2072.

Vvedenskaya, A. V.

1956 The determination of displacement fields by means of dislocation theory. *Izvestiya Akad. Nauk. S.S.S.R., Ser. Geofiz.* pp. 227–284.

Walker, D. A., C. C. McCreely, G. H. Sutton, and F. K. Dumebier

1978 Spectral analyses of high frequency P_n and S_n phases observed at great distances in the western Pacific. *Science* 199:1333–1335.

Wesson, R. L.

1970 A time integration method for computation of the intensities of seismic rays. *Bulletin of the Seismological Society of America* 60:307–316.

White, J. E.

1965 *Seismic Waves*. New York: McGraw Hill.

Wideman, C. J., and M. W. Major

1967 Strain steps associated with earthquakes. *Bulletin of the Seismological Society of America* 57:1429–1444.

Wiechert, E.

1904 Ein astatische Pendel höher Empfindlichkeit zur mechanischen Registrierung von Erdbeben. *Beitr. Geophys.* 6:435–450.

Wiggins, R. A.

1976a Body wave amplitude calculations. II. *Geophysical Journal of the Royal Astronomical Society* 46:1–10.

1976b A fast, new computational algorithm for free oscillations and surface waves. *Geophysical Journal of the Royal Astronomical Society* 47:135–150.

Wiggins, R. A., and D. V. Helmberger

1974 Synthetic seismogram computation by expansion in generalized rays. *Geophysical Journal of the Royal Astronomical Society* 37:73–90.

Wiggins, R. A., and M. Saito

1971 Evaluation of computational algorithms for the associated Legendre polynomials by internal analysis. *Bulletin of the Seismological Society of America* 61:375–381.

Willmore, P. L.

1960 The detection of Earth movement. *In* S. K. Runcorn (editor), *Methods and Techniques in Geophysics.* New York: Interscience, pp. 230–276.

Woodhouse, J. H.

1976 On Rayleigh's principle. *Geophysical Journal of the Royal Astronomical Society* 46:11–22.

1978 Asymptotic results for elastodynamic propagator matrices in plane stratified and spherically stratified Earth models. *Geophysical Journal of the Royal Astronomical Society*, 54:263-280.

Wyss, M., and J. N. Brune

1968 Seismic moment, stress and source dimensions for earthquakes in the California-Nevada region. *Journal of Geophysical Research* 73:4681–4694.

1971 Regional variations of source properties in Southern California estimated from the ratio of short- to long-period amplitudes. *Bulletin of the Seismological Society of America* 61:1153–1167.

Young, G. B., and L. W. Braile

1976 A computer program for the application of Zoeppritz's amplitude equations and Knott's energy equations. *Bulletin of the Seismological Society of America* 66:1881–1885.

Zener, C. M.

1948 *Elasticity and Anelasticity of Metals.* The University of Chicago Press.

Zöllner, F.

1869 Veber eine neue Methode zur Messung anziehender und abstossender Kräfte. *Ber. sächs. Akad. Wis. Math.-nat. Klasse* 21:280–284.

Index

Note: Volume II begins on page 559.